1998

The
Bacteriophages
Volume 1

THE VIRUSES

Series Editors
HEINZ FRAENKEL-CONRAT, *University of California*
Berkeley, California

ROBERT R. WAGNER, *University of Virginia School of Medicine*
Charlottesville, Virginia

THE VIRUSES: Catalogue, Characterization, and Classification
Heinz Fraenkel-Conrat

THE ADENOVIRUSES
Edited by Harold S. Ginsberg

THE BACTERIOPHAGES
Volumes 1 and 2 • Edited by Richard Calendar

THE HERPESVIRUSES
Volumes 1–3 • Edited by Bernard Roizman
Volume 4 • Edited by Bernard Roizman and Carlos Lopez

THE PAPOVAVIRIDAE
Volume 1 • Edited by Norman P. Salzman
Volume 2 • Edited by Norman P. Salzman and Peter M. Howley

THE PARVOVIRUSES
Edited by Kenneth I. Berns

THE PLANT VIRUSES
Volume 1 • Edited by R. I. B. Francki
Volume 2 • Edited by M. H. V. Van Regenmortel and Heinz Fraenkel-Conrat
Volume 3 • Edited by Renate Koenig
Volume 4 • Edited by R. G. Milne

THE REOVIRIDAE
Edited by Wolfgang K. Joklik

THE RHABDOVIRUSES
Edited by Robert R. Wagner

THE TOGAVIRIDAE AND FLAVIVIRIDAE
Edited by Sondra Schlesinger and Milton J. Schlesinger

THE VIROIDS
Edited by T. O. Diener

The
Bacteriophages
Volume 1

Edited by
RICHARD CALENDAR
University of California, Berkeley
Berkeley, California

PLENUM PRESS • NEW YORK AND LONDON

Library of Congress Cataloging in Publication Data

The Bacteriophages / edited by Richard Calendar.
   p.   cm. — (The Viruses)
   Includes bibliographies and index.
   ISBN 0-306-42730-3 (v. 1)
   1. Bacteriophage. I. Calendar, Richard. II. Series
QR342.C35 1988
576′.6482 — dc19

88-9770
CIP

This limited facsimile edition has been issued
for the purpose of keeping this title available
to the scientific community.

© 1988 Plenum Press, New York
A Division of Plenum Publishing Corporation
233 Spring Street, New York, N.Y. 10013

Printed in the United States of America

Contributors

Allan Campbell, Department of Biological Sciences, Stanford University, Stanford, California 94305

Sherwood Casjens, Department of Cellular, Viral, and Molecular Biology, University of Utah Medical Center, Salt Lake City, Utah 84132

Henry Drexler, Department of Microbiology and Immunology, Wake Forest University Medical Center, Winston-Salem, North Carolina 27103

E. Peter Geiduschek, Department of Biology and Center for Molecular Genetics, University of California at San Diego, La Jolla, California 92093

Felix Gropp, Max-Planck-Institut für Biochemie, 8033 Martinsried, Federal Republic of Germany

Rasika M. Harshey, Department of Molecular Biology, Research Institute of Scripps Clinic, La Jolla, California 92037

Rudolf Hausmann, Institut für Biologie II der Universität, 78 Freiburg, Federal Republic of Germany

Roger Hendrix, Department of Biological Sciences, University of Pittsburgh, Pittsburgh, Pennsylvania 15260

George A. Kassavetis, Department of Biology and Center for Molecular Genetics, University of California at San Diego, La Jolla, California 92093

D. James McCorquodale, Department of Biochemistry, Medical College of Ohio, Toledo, Ohio 43699

Horst Neumann, Max-Planck-Institut für Biochemie, 8033 Martinsried, Federal Republic of Germany

Peter Palm, Max-Planck-Institut für Biochemie, 8033 Martinsried, Federal Republic of Germany

Wolf-Dieter Reiter, Max-Planck-Institut für Biochemie, 8033 Martinsried, Federal Republic of Germany

Michael Rettenberger, Max-Planck-Institut für Biochemie, 8033 Martinsried, Federal Republic of Germany

Margarita Salas, Centro de Biología Molecular (CSIC-UAM), Universidad Autónoma, Canto Blanco, 28049 Madrid, Spain

Nat Sternberg, E. I. DuPont de Nemours & Co., Central Research and Development Department, Experimental Station, Wilmington, Delaware 19898

Charles Stewart, Department of Biology, Rice University, Houston, Texas 77251

Jan van Duin, Department of Biochemistry, University of Leiden, Leiden, The Netherlands

Huber R. Warner, National Institute on Aging, National Institutes of Health, Bethesda, Maryland 20892

Michael B. Yarmolinsky, Laboratory of Biochemistry, National Cancer Institute, National Institutes of Health, Bethesda, Maryland 20892

Stanley A. Zahler, Section of Genetics and Development, Division of Biological Sciences, Cornell University, Ithaca, New York 14853

Wolfram Zillig, Max-Planck-Institut für Biochemie, 8033 Martinsried, Federal Republic of Germany

Preface

It has been 10 years since Plenum included a series of reviews on bacteriophages, in *Comprehensive Virology*. Chapters in that series contained physical-genetic maps but very little DNA sequence information. Now the complete DNA sequence is known for some phages, and the sequences for others will soon follow. During the past 10 years two phages have come into common use as reagents: λ phage for cloning single copies of genes, and M13 for cloning and DNA sequencing by the dideoxy termination method. Also during that period the use of alternative sigma factors by RNA polymerase has become established for SPO1 and T4. This seems to be a widely used mechanism in bacteria, since it has been implicated in sporulation, heat shock response, and regulation of nitrogen metabolism. The control of transcription by the binding of λ phage CII protein to the −35 region of the promoter is a recent finding, and it is not known how widespread this mechanism may be.

This rapid progress made me eager to solicit a new series of reviews. These contributions are of two types. Each of the first type deals with an issue that is exemplified by many kinds of phages; chapters of this type should be useful in teaching advanced courses. Chapters of the second type provide comprehensive pictures of individual phage families and should provide valuable information for use in planning experiments. During the next 10 years, at least, phages will still be attractive model systems for studies of the interactions among DNA, RNA, and proteins, since they are so easy to handle and since so much is already known about them.

These volumes are dedicated to Arthur Kornberg on the occasion of his 70th birthday. When I was a graduate student in his department, Arthur was always available, and his advice had enormous value. Recently, the success of my research on P4 phage DNA replication has depended upon generous gifts of purified proteins from Arthur's laboratory. In facing life's predictable crises, I invariably find the path to humility by asking myself "how did Arthur handle this?"

Richard Calendar

Contents

Chapter 3

Changes in RNA Polymerase

E. Peter Geiduschek and George A. Kassavetis

Chapter 4

The Single-Stranded RNA Bacteriophages

Jan van Duin

Chapter 5

Phages with Protein Attached to the DNA Ends

Margarita Salas

Chapter 6

Phage Mu

Rasika M. Harshey

Chapter 7

Bacteriophage T1

Henry Drexler

Chapter 8

The T7 Group

Rudolf Hausmann

Chapter 9

Bacteriophage P1

Michael B. Yarmolinsky and Nat Sternberg

Chapter 10

Bacteriophage T5 and Related Phages

D. James McCorquodale and Huber R. Warner

Chapter 11

Bacteriophage SPO1

Charles Stewart

Chapter 12

Viruses of Archaebacteria

*Wolfram Zillig, Wolf-Dieter Reiter, Peter Palm, Felix Gropp,
Horst Neumann, and Michael Rettenberger*

Chapter 13

Temperate Bacteriophages of *Bacillus subtilis*

Stanley A. Zahler

CHAPTER 1

Phage Evolution and Speciation

ALLAN CAMPBELL

I. THE SPECIES CONCEPT AND ITS APPLICATION TO PHAGES

One of the most important conceptual advances in evolutionary science during this century was the populational definition of the biological species (Mayr, 1969.) A species is defined not by the resemblance of individuals to some type specimen but rather by the cause of that resemblance—their genetic relatedness as members of a closed interbreeding population whose genes can be considered a common pool. Even among sexually reproducing higher eukaryotes, the species thus conceived is an ideal seldom fully realized. Attempts to apply the concept too literally have been justly criticized (Ehrlich and Raven, 1969). Nevertheless, the realization that the essence of speciation lies in reproductive isolation must qualify as one of the major insights in all of biology.

It is far from obvious that the species concept thus defined should apply either to prokaryotes or to viruses. The notion of a gene pool makes sense only if the members of that pool recombine at a significant rate under natural conditions. For the gene pool to be closed, gene flow between members of the species and other sources should ideally be zero, but should at least be small compared to intraspecific recombination. In the case of phages, the potential sources are not only other phages but also the entire panoply of DNA elements with which phages come into contact, including host chromosomes, plasmids, and transposons. Thus

ALLAN CAMPBELL • Department of Biological Sciences, Stanford University, Stanford, California 94305.

we may ask not only whether λ and P2 are distinct phage species, gene flow between which is limited at most, but also whether the rate of exchange of DNA segments between λ and the *E. coli* chromosome is significant compared to recombination between different varieties of λ.

Are the concepts of speciation and reproductive isolation at all helpful in understanding phage biology and evolution? Or do they hinder interpretation by introducing a screen of linguistic confusion along the path between observable facts and their direct interpretation? Some virologists see the populational definition of a species as clearly inapplicable to the material they study (Matthews, 1985). In the United States, this viewpoint has achieved legal status; the Environmental Protection Agency rules that "the species concept does not generally apply in virus taxonomy" (Federal Register, 1986). Other virologists imagine that homologous recombination within a viral population is so frequent that the basic units of selection are not the individual particles or clones of virus but rather the segments of their genomes that are interchangeable by recombination (Susskind and Botstein, 1978) or that the disruptive effects of phage recombination might be a significant factor in maintaining the clustering of genes whose products interact with one another or with linked DNA target sites (Dove, 1971). Still others see illegitimate recombination as such a dominant force that phage genomes can best be regarded as mosaics of genes from various nonphage sources (Hunkapiller *et al.*, 1982). Not all these viewpoints can be correct, although each may have an element of truth. The main purpose of this chapter is to make the issues explicit, not to arrive at final answers.

Application of the species concept to bacteriophages is best appreciated in the context of the applicability of the same concept to prokaryotes in general. Disparate views have been expressed on the subject, but the differences seem to be more semantic than substantive. Thus, Sonea and Panisset (1983) deny that the classification of bacteria into distinct species has the same significance as it does in eukaryotes, because bacteria exchange plasmids and phages and are therefore in genetic communication with one another, sharing a universal prokaryotic gene pool. To them, this implies that there are no distinct species and therefore that extinction of individual species, while common in eukaryotes whose gene pools are held together by sexual reproduction, is impossible in prokaryotes. At an apparently opposite extreme, Selander and Levin (1980) suggest that natural populations of *E. coli* in particular are essentially clonal, genetic exchange among them being so rare as to be inconsequential. Thus the species concept can be rejected on the one hand because genetic exchange is ubiquitous and, on the other, because it hardly occurs at all!

However, most authors, (Sonea and Panisset, 1983; Ochman and Wilson, 1986; Campbell, 1981) seem to agree on the basic facts: accessory DNA elements (plasmids, transposons, phages) do indeed get disseminated among bacteria that are very distantly related taxonomically; however, although these elements could serve as vehicles for widespread dispersal of chromosomal DNA, lateral transfer of chromosomal genes is inconsequential. Whether recombination of chromosomal DNA among closely related strains is likewise inconsequential has not yet been adequately tested; the chromosomal DNA of *E. coli* may or may not be reasonably considered as a single gene pool held together by recombination. What is clear is that *Escherichia* DNA and *Pseudomonas* DNA do not belong to the same pool; plasmids may flow between them, but extinction of one and survival of the other is plausible, because their chromosomal DNAs have each evolved far enough in their own directions that selectively advantageous recombinants are unlikely to arise between them.

If the chromosomes of *E. coli* and *P. aeruginosa* each constitute coadapted complexes that evolve separately despite the existence of pathways for gene transfer between them, so also, we might argue, do the chromosomes of *E. coli* and bacteriophage λ. At least, that is a possibility we might like to pursue.

A comprehensive evaluation of the relevant facts would include discussion of all the phages treated in other chapters of this volume. However, because my purpose here is to illustrate questions rather than to provide answers, I will confine most of my attention to the phages I know best, bacteriophage λ and its relatives; and even there I will not attempt complete coverage, only the discussion of a few examples.

II. MOLECULAR BIOLOGY AND THE ORIGINS OF PHAGE

Any discussion of viral evolution must address the origin of viruses. There are several traditional hypotheses for viral origins, including the possibility that viruses are regressed forms of cellular parasites. While such notions may still have their adherents, a rather general consensus has developed among molecular virologists during the past decade on the origin of at least the larger DNA viruses. This consensus is that a viral genome such as λ's is chimeric in origin—that the various segments of λ's genome were derived from separate source, such as disjointed chromosomal segments of one or more host strains and perhaps of plasmids or transposons that inhabit them. Thus, the replication origin might have arisen from one source (such as the bacterial replication origin), the integration genes from some transposon, the lysis genes from host genes whose products direct the turnover of cell envelope components, etc.

The general acceptance of this view of viral evolution is based on an increased appreciation of the potentialities of illegitimate recombination mechanisms (especially those effected by transposons) for creating DNA fusions. Although the existence of plausible mechanisms does not of itself demonstrate an evolutionary pathway, the origin of large DNA viruses is not really a live issue at this time.

The question that is very much alive is whether the recruitment of nonviral genes to generate new viruses is a frequent ongoing process. One extreme possibility is that the primary fusion events that contribute significantly to natural evolution are extremely rare. Bacteriophage λ occurred once, a long time ago, and all λ-related phages are derived by direct descent from the primordial λ phage. At the other extreme, new λ-like phages might be continually getting slapped together from extraneous components. Doubtless the truth lies somewhere between these two extremes. My tentative interpretation of existing evidence is much closer to the first view than to the second.

The issue, of course, is not whether phages can incorporate bacterial genes into their own genomes. The genesis of λgal phages (Morse et al., 1956) demonstrated that much. We must presume that such aberrations occur in nature as they do in the laboratory. The question is whether they go anywhere evolutionarily. If the incorporated genes are ever modified to serve viral rather than host functions, we would like to know how often that happens and how much it contributes to the genome organization of natural viruses.

III. THE GENE POOL

The most obvious components of the phage gene pool are the natural varieties, or races, of a given phage. There are, for example, numerous phages related to λ, according to the criteria of DNA homology or ability to generate viable recombinants. The genetic maps of all these phages are very similar, with analogous genes in similar positions. Heteroduplex studies indicated that any two λ-related phages generally match each other closely in some segments of their genomes and diverge completely in others (Simon et al., 1971; Fiandt et al., 1971). DNA sequencing shows that transitions from good homology to nonhomology can be as sharp as the heteroduplex results suggested (Grosschedl and Schwarz, 1979; Benedik et al., 1983; Backhaus and Petri, 1984; Franklin, 1985). Particular segmental homologies are not common to all λ-related phages but specific to each phage pair. Thus the immunity regions of phages 21 and P22 are very similar, whereas that of λ is completely different; but the tail genes of λ and 21 are homologous to one another and different from those of P22. Relations of this sort convinced Hershey (1971) that the different λ-related phages originated by recombination among yet other phage races.

Probably equally important as components of the λ gene pool are the defective remnants of λ-related phages that clutter the chromosome of the typical enterobacterial cell. The chromosome of *E. coli* K-12 has at least three such elements (Fig. 1). The *Rac* prophage (Kaiser and Murray, 1979) is bounded by functional attachment sites and contains an integrase gene, a replication origin, recombination genes, and a repressor gene that controls them. The *qsr'* prophage (Highton et al., 1985) includes

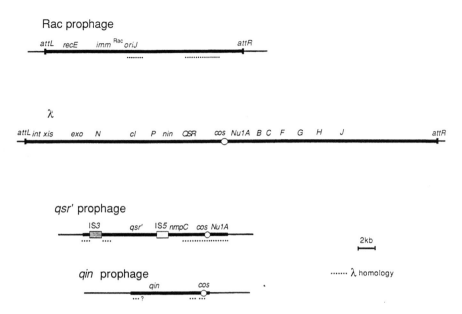

FIGURE 1. λ-related sequences in *E. coli* K-12. The defective *qsr'* prophage is at 12 min; λ prophage is at 17 min; the defective *Rac* prophage is at 29 min; and the defective *qin* prophage is at 34 min. "*qsr'*" and "*qin*" denote gene blocks analogous to λ *QSR*. From Redfield (1986).

genes analogous to the λ late gene regulator *Q*, the lysis genes *S* and *R*, and a *cos* site. In the DNA flanking these demonstrably functional elements are a segment homologous to part of the λ *redX* gene (Redfield, 1986) and at least part of the λ Nul gene. The *qin*111 prophage (Espion *et al.*, 1983) extends over about the same portion of a λ-related genome as does the *qsr'* prophage. Like the races of λ-related phages, each of these defective phages consists of some segments homologous to the corresponding segment of the λ genome interspersed with segments that are heterologous but functionally equivalent. Hybridizations with λ probes suggest that the chromosomes of many natural strains of *E. coli* and related bacteria likewise include λ-related segments (Anilionis and Riley, 1980; Riley and Anilionis, 1980). Such vestigial, inactive segments appear to represent steps in the total loss of prophage from the bacterium—"genetic debris," in the term of Strathern and Herskowitz (1975). However, if it is true that recombination plays an important role in the generation of new natural varieties of λ phage, the contribution of this vast reservoir of phage-derived sequences should not be ignored. Natural recombinants can arise under two circumstances: mixed infection of a single cell by phages of two types, or superinfection by one phage of cells harboring a prophage or a defective prophage. In nature, the former type of encounter may be much rarer than the latter.

IV. PHAGE GENES RELATED TO HOST GENES

In two cases, genes of λ or its relatives sufficiently resemble host genes to bespeak a common ancestry: (1) The repressor and *cro* genes are related to the *lexA* gene of *E. coli*, a member of a family of SOS proteins (Sauer *et al.*, 1982; Walker, 1984). (2) Gene *18* of P22 is related to the *dnaB* gene of *E. coli*. The analogous gene (*P*) of λ is not detectably related either to gene *18* or to *dnaB*; in fact, λ replication (unlike that of P22) depends on host *dnaB* function, and λ gp*P* physically interacts with host DnaB protein.

In neither case is the relationship close. In the most similar segments of gene *18* and *dnaB*, for example, there is about 50% base identity. However, this is the *dnaB* gene of *E. coli* rather than *S. typhimurium* (P22's natural host). Integration of *dnaB*-related sequences into a rather recent ancestor of P22 is thus entirely possible. Replacement of phage replication genes by host counterparts may have occurred in other phage families as well. For example, replication of phage 186 (but not of its close relative P2) is prevented by *ts* mutations in the host initiation genes *dnaA* and *dnaC* (Hooper and Egan, 1981); however, *ts* mutations of *dnaA* can give different results from null mutations (Hansen and Yarmolinsky, 1986), so the interpretation of this result is uncertain.

V. HOST FUNCTIONS REPLACEABLE BY PHAGE-DERIVED GENES

Relatedness between host and phage genes might mean that the host genes are descended from phage genes rather than the reverse. In at least two cases, laboratory models show how genes from defective prophages can assume functions of selective value to the host: (1) In *recBC* mutants of *E. coli*, selection for resistance to radiomimetic agents such as mitomycin C yields strains in which the recombination functions of the λ-related Rac prophage are derepressed (Kaiser and Murray, 1980). (2) *E. coli* mutants lacking the porin protein *ompC* grow poorly and accumulate spontaneous mutations that express a normally quiescent gene *nmpc* within the *qsr'* prophage (Highton *et al.*, 1985). This gene, whatever its ultimate origin, is present in several λ-related phages and may therefore be considered phage-derived. Its function in phage biology might be to alter the surfaces of infected or lysogenic cells, making them unable to adsorb phages of the same type. This might be advantageous either in competition with other phages of the same adsorption specificity or in the prevention of useless sequestration of liberated phages adsorbing to cell debris or to already infected cells.

It seems certain that the selective conditions imposed in these laboratory models must have their counterparts in nature. What is less clear is whether changes of this sort have had any major impact other than providing a short-term selective advantage leading to an evolutionary dead end.

VI. RECOMBINATION IN NATURAL POPULATIONS

In a purely asexual population, each local clone is reproductively isolated from its neighbors. Whatever utility the term "species" may have in the classification of such organisms, Mayr's species concept cited earlier in the chapter is clearly inapplicable. From the results of heteroduplex analysis, Hershey (1971) reasonably inferred that the various races of λ are related by recombinational events, but that fact alone does not tell us whether recombination is a frequent, significant process in the population biology of the virus (as implied by the notion of a gene pool) or recombinations between different λ races should be regarded as rare, perhaps unique, historical events.

In some cases, the DNA sequences clearly indicate that, where homologous and heterologous segments are interspersed, the homologous segments can be of quite recent common origin. These include the integrase operons of phages λ and 434 (Benedik et al., 1983), the ral genes of λ and 21 (Franklin, 1985), and the ninR genes of λ and P22 (Backhaus and Petri, 1984). The xis genes of λ and 434, for example, show about 3% DNA sequence divergence (mostly silent codon changes; Campbell et al., 1986), whereas the ral genes of λ and 21 and the 3' termini of the int genes of λ and 434 remarkably show no changes at all. Therefore, if these phage pairs arose by recombination among other λ races, at least one recombinational event must have been relatively recent. Each of these pairs also contains genes (such as the structural genes for repressor and cro) that have diverged extensively and genes such as N that are presumed to have a common ancestry, although in some cases no detectable sequence similarity remains.

One explanation for the fact that λ and 434 are so similar in their integrase operons and so different in their repressor genes would be that a recombination between int and cI occurred in the ancestry of one phage or the other, so that the left end of a recent common ancestor is juxtaposed to the right end of some other λ-related phage. Alternatively, natural selection might have favored unequal rates of evolution in different genome segments, obviating the need to invoke recombination as an explanation. None of the examples of sequence similarity cited above are likely to reflect strong selection for sequence conservation; however, intense selection for divergence in segments such as cI is plausible on the reasonable assumptions that rare immunity types have a selective advantage over common ones and that newly arisen immunity types will generally be suboptimal and capable of improvement (Campbell and Botstein,

1983). The attractiveness of that hypothesis is tempered by the fact that the minimal number of base changes required to generate a workable new repressor-operator combination is not very large (Wharton and Ptashne, 1985). Another mode of selection for divergence in λ that has been suggested in the past is selection against the ability to generate ill-adapted recombinants through recombination within heterologous functional units. However, that explanation seems extremely unlikely for a species such as λ, where encounters between heterologous types should be rare; generation of lethal or poorly adapted recombinants generates a significant selective disadvantage only in sexual species where recombination and reproduction are directly linked. At any rate, the extent of divergence between the immunity segments of the various λ-related phages far exceeds the amount required to reduce recombination drastically (Shen and Huang, 1986).

As mentioned earlier, phage 21 is very similar to P22 (and dissimilar to λ) in its repressor and *cro* genes and very similar to λ (and dissimilar to P22) in its tail genes. Such a relationship strongly indicates a recombinational origin.

The simplest explanation for all these facts collectively is that many of the λ-like phages that have been studied are related by recombinational events. Because some phages are very similar or identical in some genome segments, such recombination must sometimes have been very recent. Because there are multiple examples where such recent recombination is indicated, natural recombination among these phages must take place rather frequently.

VII. SOURCE AND DISTRIBUTION WITHIN THE GENOME OF RECOMBINABLE VARIATION

The interspersion of homologous and heterologous segments observed in pairwise comparisons of λ phages underlies the inference that natural recombination has been frequent. The heterologies serve as genetic markers subject to reassortment by recombination within the homologous segments. The conservation of overall gene order among the various phages is compatible with the notion that homologous recombination is the overriding mechanism that generates new gene combinations. However, nothing that has been said up to this point addresses the question of how the interspersion arises in the first place.

The sharp transitions between homology and heterology, inferred from heteroduplex studies and now documented by DNA sequencing, are not themselves explicable by homologous recombination. In principle, they might arise either from enormous differences in divergence rates of adjacent segments or by illegitimate recombination. If illegitimate recombination is the answer, then the possibility of heterologous replacement of segments of the λ genome from sources outside the phage gene pool cannot be discounted.

Where sequence data locate the transitions between homology and heterology on the genetic map, two features emerge. First, the transition points frequently lie strikingly close to the boundaries of functional segments of the genome; e.g., the DNAs of phages λ and 434, homologous throughout the integrase operon, diverge sharply two bases downstream of the *sib* site (the terminus of the integrase operon). At the other end of the integrase operon, four bases upstream of the *p*I promoter, the two phages differ by a 440-base insertion in 434 or deletion in λ (clearly an instance of illegitimate recombination). Likewise, the boundary between the *cro* gene and the *nutR* site is a transition point for several phage pairs, though in this case the transitions are not quite so sharp as those for the *int* operon (Campbell *et al.*, 1986).

Second, some of the best homologies lie in accessory genes (genes needed neither for lysis nor for lysogeny) such as *ral* and *nin*R. These genes present a paradox. On the one hand, they appear to be relatively unimportant; their functions, where known, are dispensable, and they are not even present in all natural λ-related phages. On the other hand, where they are present, they are highly conserved from one phage to the next.

As in the case of the *int* operon and its flanking segments in λ and 434, the transitions between the good homology observed in *ral* or *nin*R and the adjacent heterology are abrupt and best explained by illegitimate recombination events. In some ways, the situation of λ accessory genes resembles the location of accessory bacterial genes in related enterobacterial species such as *E. coli* and *S. typhimurium*. In these bacteria also, overall gene order has been generally conserved over a time span that allowed substantial sequence divergence of individual genes. Where specific DNA segments extending over several genes have been compared, the two also may differ by the presence in each of some additional genes that are absent or elsewhere in the other. These additional genes encode functions that are dispensable under optimal growth conditions. Riley (1984) reasonably interpreted this arrangement as the result of insertion of some genes brought into each species, perhaps as plasmid-borne transposons. The emergent picture of the bacterial genome (and perhaps of the λ genome as well) is one in which a conserved set of essential genes form a fixed matrix that allows intercalation of other genes serving accessory functions.

The relationship between λ and 434 differs from that between *E. coli* and *S. typhimurium* in that λ and 434 share some segments of very similar or identical sequence, whereas the genomes of *E. coli* and *S. typhimurium* seem to be divergent throughout. It is for this reason that we postulate frequent natural recombination among λ and its relatives, whereas homologous recombination between *E. coli* and *S. typhimurium*, though possible, may be inconsequential in nature. The pattern of sequence divergence of λ and its relatives occurs in some other phage families, but not in all. For example, heteroduplex studies between T3 and T7 indicate a fairly uniform rate of divergence throughout much of the genome, indicating that recombination has not been such an important factor in their pedigrees (Davis and Hyman, 1971; Hausmann, this volume).

It is entirely possible that the *ral* gene, for example, entered the λ genome by an illegitimate insertion subsequent to the divergence of λ from related phages such as 82 which have no homology to λ in the *ral* region (Simon *et al.*, 1971). It has in fact been suggested that the sequence identity of *ral* in different phages might imply that it is a rather recent addition; however, unless it added to more than one phage of the λ family in the same position on the genome, its presence in several phages requires recombination subsequent to its addition, and regardless of when addition took place, such recent recombination can by itself explain the apparent sequence conservation. The recent addition of *ral* by homologous recombination into phages that lacked it completely is further complicated by the fact that the only homology available for such a recombination lies within or very close to *ral*, the sequence whose identity in λ and other phages requires explanation. If intervening homology existed at one time, subsequent illegitimate events (such as deletions) must have eliminated it; if illegitimate events must be postulated anyway, the explanatory value of homologous recombination decreases.

Susskind and Botstein (1978) pictured the λ genome as a set of functional modules connected by linker segments, such modules having a number of allelic alternatives that differ among the various phages, the linkers remaining homologous in different phages and occupying fixed locations in the genome. The selective force maintaining the linker homology was postulated to be the potential for generating new combinations of modules through recombinational reassortment. The excellent homology observed in some accessory genes might fit this suggestion. The linker hypothesis provides no automatic explanation for the sudden transitions between good homology and nonhomology; as with other hypotheses, special explanations are required for their occurrence and/or maintenance. In general, no linker segments are ubiquitous among all λ-related phages, nor are good homologies conspicuous at frequent transition points such as the *Cro-nutR* junction; homologous modules (such as the integrase operons of λ and 434) can function as recombinational linkers in the generation of new combinations of flanking genes (with no indication of generally conserved homologies at the termini of the operon), even though in other phage pairs, such as λ and 21, the integrase operon appears as a module whose heterologous alternatives can be redistributed by recombination in segments that are homologous between these two phages. The linker hypothesis thus has its limitations, but the absence of a simple alternative explanation for all the facts prevents us from dismissing it completely. At any rate, if the linker hypothesis is even partially correct, natural recombination among λ-related phages must be both of frequent occurrence and of major importance in generating selectively advantageous gene combinations.

VIII. REPRODUCTIVE ISOLATION IN NATURE

Granting the impossibility of certain knowledge about historical events that were not directly observed, the case that recombination has played a significant role in the genesis of λ-related phages seems reasonably strong. If recombination is frequent and important, the various phages of the λ family are tied together in a manner similar to the members of a classical sexually reproducing species, drawing on a common gene pool. However, the analogy is apt only if the common pool is reproductively isolated, so that it is not shared by all phages or by all prokaryotes. The direct evidence on reproductive isolation is minimal, and investigators differ in their evaluation of the available facts.

Logically, it is clear that the λ-related phages cannot have a gene pool totally isolated from that of their hosts, the enteric bacteria. Ultimately, the phage genomes probably came from the host gene pool, and known mechanisms of gene rearrangement are adequate to ensure that phage genomes can acquire new genes from that pool as the occasion demands. The questions that can be asked are quantitative ones: How often does the occasion demand? Does the recombinational reshuffling of genes within the phage pool take place at a much higher rate than fresh acquisition of host genes? Given the fixed nature of the λ genetic map, did the heterologous alternatives at specific loci such as N, cI, or int and their cognate recognition sites arise by the accumulation of mutations in genes derived from a common ancestral phage rather than being acquired from separate sources? Have λ genes by this time achieved such a high level of specialization that almost the only new variants that are competitive in nature are those that arise by reassortment of preexisting λ genes?

My present inclination is to answer all these questions in the affirmative. The constancy of genome form among the various λ races is most simply explained on the basis of common ancestry, in which case the sequence divergence in some segments implies that the complex has existed for a long time. This is consistent with a high degree of biochemical and regulatory sophistication that suggests a long evolutionary history.

The notion that any phage species has an isolated gene pool makes sense only if such a long evolutionary history is postulated. If all the phages, prophages, and defective phages now in existence were suddenly to disappear from the face of the earth, we might expect that new phages would eventually evolve from fresh associations of host-derived genes. Recombinational isolation of the phage genes from their host components of origin could not take place instantaneously; so for perhaps a long period of time, until the phage genes had evolved under selection for their new mode of existence, considerable recombinational exchange would be expected between the phage and its cellular precursors. We must presume that such a period existed at some time for the phages we know now. Perhaps some recently evolved phages are currently at that stage. In the particular case of λ, that stage may be far in the past. However, even if that is the case, some of the genetic polymorphisms observed among the members of the λ family may have entered the species by recombination with a precursor pool early on, before the present state of reproductive isolation was reached.

In any case, I expect that the general overview of genetic exchange among prokaryotes indicated here will prove valid: the potential for unrestricted gene flow is present, but selective constraints limit its effectiveness and generate a collection of gene pools that are to some extent separate from one another. These may include the chromosomes of individual bacterial species on the one hand (although, even within a species, the extent of natural recombination among chromosomes is currently unsettled), but extrachromosomal elements such as phages or plasmids may also be regarded as having gene pools separate from those of their hosts, as well as from other phages and plasmids. The several major groups of large, double-stranded DNA coliphages (including λ, P2, Mu-1, P1, T2, T5, T7, and their respective relatives) all show considerable sequence divergence in individual genes among their members, generally with conservation of overall gene order. This suggests that all these groups represent evolutionarily successful gene complexes of relatively long duration, so it is at least plausible to imagine that each comprises a gene pool largely isolated from the others.

ACKNOWLEDGMENT. Research from the author's laboratory was supported by grant AI08573, National Institute of Allergy and Infectious Diseases.

REFERENCES

Anilionis, A., and Riley, M., 1980, Conservation and variation of nucleotide sequences within related bacterial genomes: *Escherichia coli* strains, *J. Bacteriol.* **143:**355.

Backhaus, H., and Petri, J. B., 1984, Sequence analysis of a region from the early right operon in phage P22 including the replication genes *18* and *12, Gene* **32:**289.

Benedik, M., Mascarenhas, D., and Campbell, A., 1983, The integrase promoter and T_I' terminator in bacteriophages λ and 434, *Virology* **126:**658.

Campbell, A., 1981, Evolutionary significance of accessory DNA elements in bacteria, *Annu. Rev. Microbiol.* **35:**55.

Campbell, A., and Botstein, D., 1983, Evolution of the lambdoid phages, in: *Lambda II* (R. W. Hendrix, J. W. Roberts, F. W. Stahl, and R. A. Weisberg, eds.), pp. 365–380, Cold Spring Harbor Laboratory, Cold Spring Harbor, NY.

Campbell, A., Ma, D. P., Benedik, M., and Limberger, R., 1986, Reproductive isolation in prokaryotes and their accessory DNA elements, in: *Banbury Report 24: Antibiotic*

Resistance Genes: Ecology, Transfer, and Expression, Cold Spring Harbor Laboratory, Cold Spring Harbor, NY, pp. 337–345.

Davis, R. W., and Hyman, R. W., 1971, A study in evolution: The DNA base sequence homology between coliphages T7 and T3, *J. Mol. Biol.* **62**:287.

Dove, W., 1971, Biology inference, in: *The Bacteriophage Lambda* (A. D. Hershey, ed.), pp. 297–312, Cold Spring Harbor Laboratories, Cold Spring Harbor, NY.

Ehrlich, P. R., and Raven, P. H., 1969, Differentiation of populations, *Science* **165**:1228.

Espion, D., Kaiser, K., and Dambly-Chaudiere, C., 1983, A third defective lambdoid prophage of *Escherichia coli* K-12 defined by the λ derivative, λ *qin* 111, *J. Mol. Biol.* **170**:611.

Federal Register (U.S.), 1986, **51**:23334.

Fiandt, M., Hradecna, Z., Lozeron, H. A., and Szybalski, W., 1971, Electron micrographic mapping of deletions, insertions, inversions, and homologies in the DNAs of coliphages lambda and phi 80, in: *The Bacteriophage Lambda* (A. D. Hershey, ed.), pp. 329–354, Cold Spring Harbor Laboratory, Cold Spring Harbor, NY.

Franklin, N. C., 1985, Conservation of genome form but not sequence in the transcription antitermination determinants of bacteriophages λ, φ21, and P22, *J. Mol. Biol.* **181**:75.

Grosschedl, R., and Schwarz, E., 1979, Nucleotide sequence of the Cro-cII-oop region of bacteriophage 434 DNA, *Nucleic Acids Res.* **6**:867.

Hansen, E. B., and Yarmolinsky, M. B., 1986, Host participation in plasmid maintenance: dependence upon *dnaA* of replicons derived from P1 and F, *Proc. Natl. Acad. Sci. USA* **83**:4423.

Hershey, A. D., 1971, Comparative molecular structure among related phage DNA's, *Carnegie Inst. Washington Yearb.* **1970**:3.

Highton, P. J., Chang, Y., Macotte, W. R. Jr., and Schnaitman, C. A., 1985, Evidence that the outer membrane protein *nmpC* of *Escherichia coli* K-12 lies within the defective qsr' prophage, *J. Bacteriol.* **162**:256.

Hooper, I., and Egan, J. B., 1981, Coliphage 186 infection requires host initiation functions *dnaA* and *dnaC*, *J. Virol.* **40**:599.

Hunkapiller, T. H., Huang, H., Hood, L., and Campbell, J. H., 1982, The impact of modern genetics on evolutionary theory, in: *Perspectives on Evolution* (R. Milkman, ed.), pp. 164–189, Sinauer, Sunderland, MA.

Kaiser, K., and Murray, N. E., 1979, Physical characterisation of the "Rac prophage" in *E. coli* K-12, *Mol. Gen. Genet.* **175**:159.

Kaiser, K., and Murray, N. E., 1980, On the nature of *sbcA* mutations in *E. coli* K-12. *Mol. Gen. Genet.* **179**:555.

Matthews, R. E. F., 1985, Viral taxonomy for the nonvirologist, *Annu. Rev. Microbiol.* **39**:451.

Mayr, E., 1969, *Principles of Systematic Zoology*, McGraw-Hill, New York.

Morse, M. L., Lederberg, E., and Lederberg, J., 1956, Transduction in *Escherichia coli* K-12, *Genetics* **41**:121.

Ochman, H., and Wilson, A. C., 1986, Evolutionary history of enteric bacteria, in: *Escherichia coli and Salmonella typhimurium: Molecular and Cellular Aspects* (J. Ingraham and B. Low, eds.), ASM Publications, Washington (in press).

Redfield, R., 1986, Structure of cryptic prophages. Thesis, Stanford University.

Riley, M., 1984, Arrangement and rearrangement of bacterial genomes, in: *Microorganisms as Model Systems for Studying Evolution* (R. P. Mortlock, ed.), pp. 285–316, Plenum, New York.

Riley, M., and Anilionis, A., 1980, Conservation and variation of nucleotide sequences within related bacterial genomes: Enterobacteriaceae, *J. Bacteriol.* **143**:366.

Sauer, R. T., Yocum, R., Doolittle, R., Lewis, M., and Pabo, C., 1982, Homology among DNA binding proteins suggests use of a conserved super-secondary structure, *Nature* **298**:447.

Selander, R. K., and Levin, B. R., 1980, Genetic diversity and structure in *Escherichia coli* populations, *Science* **210**:545.

Shen, P., and Huang, H. V., 1986, Homologous recombination in *Escherichia coli:* Dependence on substrate length and homology, *Genetics* **112**:441.

Simon, M. N., Davis, R. W., and Davidson, N., 1971, Heteroduplexes of DNA molecules of lambdoid phages: Physical mapping of their base sequence relationships by electron microscopy, in: *The Bacteriophage Lambda* (A. D. Hershey, ed.), pp. 313–328, Cold Spring Harbor Laboratory, Cold Spring Harbor, NY.

Sonea, S., and Panisset, M., 1983, "A New Bacteriology," Jones and Bartlett, Boston.

Strathern, A., and Herskowitz, I., 1975, Defective prophage in *Escherichia coli* K-12 strains, *Virology* **67**:136.

Susskind, M., and Botstein, D., 1978, Molecular genetics of bacteriophage P22, *Microbiol. Rev.* **42**:385.

Walker, G. C., 1984, Mutagenesis and inducible responses to deoxyribonucleic acid damage in *Escherichia coli, Microbiol. Rev.* **48**:60.

Wharton, R. P., and Ptashne, M., 1985, Changing the binding specificity of a repressor by redesigning an α-helix. *Nature* **316**:601.

CHAPTER 2

Control Mechanisms in dsDNA Bacteriophage Assembly

SHERWOOD CASJENS AND ROGER HENDRIX

I. INTRODUCTION

The introduction of the use of T-even bacteriophages as genetic and bio-chemical experimental systems by Max Delbrück in the late 1930s has led to the intense study of many aspects of bacteriophage biology. Of these, two related endeavors, the study of the structure and the assembly of the virions, have been very important models in the development of our current understanding of macromolecular assembly processes. Twenty years ago, Edgar, Kellenberger, Epstein, and collaborators nucleated these studies by showing that phage assembly follows defined pathways that can accumulate assembly intermediates when blocked and that the assembly-naive components of phage T4 thus accumulated could join properly *in vitro* (Epstein *et al.*, 1963; Edgar and Wood, 1966; Wood *et al.*, 1968). Since that time, the structure and assembly of many bacterio-phages and other viruses have been studied. The possibility of completely defining the genetic systems, and therefore the proteins involved, has made phage assembly a particularly popular and tractable area in which to study macromolecular assembly. We will not consider it the mission of this chapter to collect the details of this myriad of studies. The reader

SHERWOOD CASJENS • Department of Cellular, Viral, and Molecular Biology, University of Utah Medical Center, Salt Lake City, Utah 84132. ROGER HENDRIX • Department of Biological Sciences, University of Pittsburgh, Pittsburgh, Pennsylvania 15260.

should consult other chapters in this volume or other reviews for such details (e.g., Casjens and King, 1975; Murialdo and Becker, 1978a; Eiserling, 1979; Wood and King, 1979; King, 1980; DuBow, 1981; Mathews *et al.*, 1983; Hendrix *et al.*, 1983; Casjens, 1985c; Carrascosa, 1986). Instead we will focus on general questions currently under study and attempts to answer them in the various dsDNA phage systems. We will not discuss the problem of DNA packaging in detail and will not cover the lipid-containing dsDNA phages.

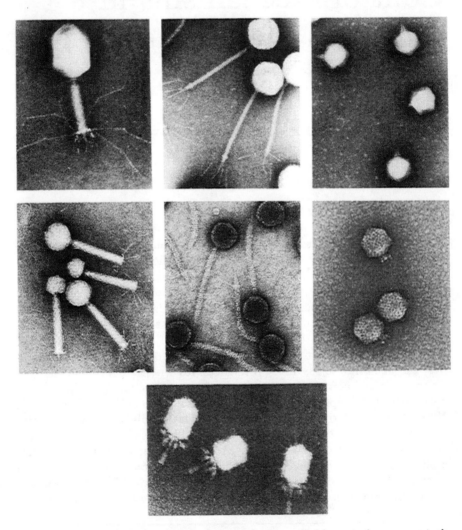

FIGURE 1. Electron micrographs of some commonly studied bacteriophages, negatively stained. Top row, left: T4; center: T5; right: T7; second row, left: P2 (large heads) and P4 (small heads); center: lambda; right: P22; bottom: ϕ29. Micrographs were kindly provided by Robley Williams (T4, T5, T7, P2/P4) and Dwight Anderson (ϕ29). Magnification is ~155,000× for all but ϕ29, which is ~310,000×.

Although a large number of tailed, dsDNA phages have been isolated from nature and observed in the electron microscope, most of the discussion in this chapter will center on the relatively few phage systems that have been studied in detail physically, genetically, and biochemically. These phages, T4, λ, P22, T7, T3, Mu, P2, P4, and φ29, include examples of each of the three major structural types observed to date (short tails— podoviridae; long contractile tails—myoviridae; and long, noncontractile tails—styloviridae; Bradley, 1967; Matthews, 1987). Figure 1 shows electron micrographs of several of these phages. Although they vary considerably in life-style and genome size (from T4's166 kbp to φ29's19 kbp), they are all built with similar overall design features. All have single, linear dsDNA chromosomes contained within a shell called a *capsid*, or *head*, which is built with coat protein molecules positioned in icosahedrally symmetric arrays. In addition, all have a single host cell adsorption apparatus called a tail attached to one corner of the coat protein shell. The tails vary considerably in form, from long and contractile (T4, P2, P4, Mu, SP01) to long and noncontractile (λ, φCbK, T5) to very short (P22, T7, T3, φ29), but all tails are thought to contain a three- or sixfold rotational symmetry axis which projects through the center of the head shell in the virion. During the initial stages of the infection process, the distal end of the tail adsorbs to the exterior of the host cell, and the phage DNA travels through the tail into the target cell. In the study of the assembly of these structures it is important to remember that these virions are not simply DNA containers but are designed to perform the act of ejection* which in most if not all cases involves movement of proteins in the structure. This requirement may account for some of the surprising complexity that is found in the structure and assembly of these particles.

II. THE STRUCTURE OF dsDNA PHAGE VIRIONS

Detailed understanding of assembly reactions necessarily requires a similar level of structural knowledge concerning the end product of the reactions, and, given their level of complexity, the dsDNA phage virions are probably the best understood biological macromolecular superstructures. Only the structures of much simpler viruses that have been studied by X-ray diffraction methods are known in greater detail. The dsDNA virions contain hundreds of molecules of a few tens of structural protein types (ranging from 6 virion protein species for φ29 to over 40 for T4). These proteins are thought to be *precisely arranged* in all cases. That is, the protein arrangement is thought to be identical in every virion of a particular species of phage (for minor exceptions to this rule see section IV.C below and Lepault and Leonard, 1985). This, along with the avail-

---

*"Ejection" is used throughout to describe the exit of the DNA from the phage particle, rather than the historically used "injection," to avoid the analogy with a syringe (after Goldberg, 1983).

ability of mutants that accumulate assembly intermediates or aberrant but related particles, has allowed many sophisticated structural analyses of these phage particles.

Negative staining electron microscopy of virions, virion parts, and assembly intermediates, occasionally in complex with specific antibodies, has elucidated many structural details of these phages and provided general locations for most proteins in the virions. Although host-encoded proteins may be involved in the assembly of virions, there is no evidence for any such protein being a structural component of the completed particle. Figure 2 shows schematic structures of the five tailed phages whose assembly has been studied in the greatest detail. Crick and

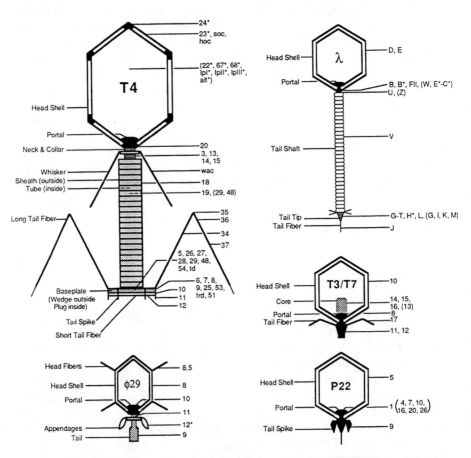

FIGURE 2. Locations of structural proteins in five bacteriophage virions. The various phage parts are indicated on the left of each schematic phage particle, and the proteins are indicated by the genes encoding them on the right. An asterisk means the protein is present as a proteolytic fragment of the primary translation product, and parentheses indicate the exact location of the enclosed protein(s) is uncertain. References for these locations are too numerous to list here; consult the text and other, more specialized reviews.

Watson (1956) recognized very early in the study of virus structure that it was impossible for a virus to encode a sufficient mass of protein to assemble a virion without using multiple copies of proteins in the virion, and they predicted that helical and cubic symmetry would be used to create repeating arrays of virion proteins. This prediction has been borne out in general, and in this type of phage we see icosahedral head shells and helical tails, both of which can contain hundreds of identical protein molecules.

In addition to the head shell building block, the "coat" protein, phage heads contain a number of proteins whose functions are known. The "portal" protein is present at a single, unique vertex (called the tail-proximal or portal vertex) and is thought to be the point of DNA entry during packaging and exit during ejection. This protein is part of a larger structure at the portal vertex structure sometimes called the "connector" or "neck". In all cases, the capsid shell is assembled from a single type of structural building block called the "coat" protein, but it may also contain a special protein at the vertices (T4) and/or "decoration" proteins on the outside of the capsid shell (λ, T4, ϕ29). Other non-coat-head proteins are "internal" proteins, "ejection" proteins, and "head completion" proteins. The tails all have a "baseplate" or "tip" structure and distal "fibers," the organelles that make initial contact with the host during adsorption. Among the phages that have long tails, those with non-contractile tails such as λ have a single protein building block for the "shaft" of the tail, whereas those with contractile tails have two—one for the inner tail "tube" and one for the outer contractile "sheath." In addition to these structural components, two types of proteins are required for the correct assembly of all dsDNA virions but are not found in the virion; these are the "scaffolding" proteins, which are internal capsid components but are removed during assembly, and the "DNA packaging" proteins, which help in the DNA packaging process.

We think of these different types of structural units as evolutionary *structure modules* which may have been *evolutionarily* mixed and matched in the different phages. Such structure modules would of course be the consequence of divergence (or convergence?) of the evolutionary *gene modules* proposed by Echols and Murialdo (1978) and Campbell and Botstein (1983). Evidence for such "mixing" can be seen in many instances—for example, closely related phages with and without decoration proteins (T4/T2—Ishii and Yanagida, 1975; Yanagida, 1977)—and thick fibers or spikes with polysaccharide cleavage activity can be found on the tips of all three major tail types (Bessler *et al.*, 1973; Reiger *et al.*, 1976; Reiger-Hug and Stirm, 1981; Wollin *et al.*, 1981).

In the different phage types, the proteins that have *analogous* (similar functional) roles during virion assembly or in virion function have not been found to have strong amino acid sequence *homology* except between very close relatives (Simon *et al.*, 1971; Fiandt *et al.*, 1971; Kim and Davidson, 1974; Studier, 1979; Hayden *et al.*, 1985; Miller and Fiess,

1985; Reide *et al.*, 1985b, 1986). For example, we have compared the coat proteins of lambda, T7, T4 and P22, as well as the portal proteins of lambda and T7, and found no strong evidence for sequence homology (our unpublished observations). Although comparison of the amino acid sequence of dsDNA phage morphogenetic proteins with protein sequence data banks has turned up a few potentially interesting amino acid sequence homologies, these have not yet been shown to be biochemically meaningful (Witkiewicz and Schweiger, 1982; Reide *et al.*, 1985c; Feiss, 1986; Kypr and Mrazek, 1986; Reide, 1986). One particularly interesting and unique case is the strong homology between a T4 tail fiber gene and a dispensible open reading frame of unknown function in the lambda late operon (Michel *et al.*, 1986). The possibility of three-dimensional structural similarity between the *analogous but not obviously homologous* (in amino acid sequence) phage structural proteins has not yet been fully explored. However, this remains a distinct possibility, since in several plant and animal viruses previously thought to be unrelated, there are startlingly conserved β-barrel tertiary structure homologies between coat proteins which were not apparent from the amino acid sequences (Harrison, 1983a; Rossmann and Erickson, 1985; Rossmann *et al.*, 1985; Hogle *et al.*, 1985; Roberts *et al.*, 1986; Fuller and Argos, 1987).

The following facts *may* indicate that such three-dimensional structural homologies occur between analogous phage structural proteins. (1) Scaffolding proteins from T4, T7, P22, and λ are all predicted from their amino acid sequence to have extremely high α-helix contents (our unpublished observations), and the P22 scaffolding protein has been shown to be highly α-helical by spectral measurements (Fish *et al.*, 1980; Thomas *et al.*, 1982). (2) Tail fiber or spike proteins from T4, T7, P22, and λ are predicted to have very high β structure (Earnshaw *et al.*, 1979; P. Arscott, personal communication; our unpublished observations). That of P22 has been shown to contain substantial β structure by spectral measurements (Thomas *et al.*, 1982; Prescott *et al.*, 1986), and in each studied case the N-terminal portion of the fiber protein is bound to the body of the virion. In the following discussion of phage structure we will emphasize the apparently conserved protein *functions* rather than the detailed properties of the individual proteins.

A. Head Structure

1. Head Shell Structure

a. Coat Protein

Figure 1 shows that the heads of these phages are generally hexagonal (sometimes elongated) in outline as is expected of an icosahedron, but *proof* of icosahedral symmetry in the arrangement of subunits depends on the identification of two-, three-, and fivefold rotational symmetry axes or their consequences in the array of coat protein subunits. This has been done for T2, T4, λ, P22, T7, ϕCbK, P2, and P4 by observing virions or related structures in the electron microscope. It is only possible to place 60 coat subunits on the surface of an icosahedron if each one is bonded to its neighboring coat subunits *identically*. Caspar and Klug (1962) suggested that by allowing small deformations, which they called *quasi-equivalences,* in the coat subunits, it would be possible to build larger structures and showed mathematically that this could only be accomplished with certain multiples of 60 subunits. For geometric reasons, they called the allowed multipliers triangulation (T) numbers (T = 1, 3, 4, 7, 9, 12, 13, 16, . . .). The allowed number of coat protein subunits per isometric virion were thus predicted to be 60, 180, 240, 420, 540 . . . , respectively. In these predicted arrangements, coat protein molecules can be viewed as forming pentamers around each icosahedral vertex and hexamers elsewhere on the surface (see Caspar, 1965, and Casjens, 1985a, for more detailed discussions of this theory). Although this theory has had rather spectacular success in predicting the overall arrangement of coat protein subunits on the surface of icosahedral virions, recent high-resolution X-ray diffraction analysis of several plant and animal viruses has shown that in fact the predicted deformations of the coat protein subunits do not seem to occur. Instead, the subunits appear very rigid, and bonding changes or *nonequivalent interactions* between coat proteins occur at various positions in the capsid (see Section IV.C). This has caused a renewed discussion of the meaning of the triangulation number and the validity of the Caspar and Klug theory (Rayment, 1984; Harrison, 1983a; Casjens, 1985a; Rossmann and Erickson, 1985; Burnett, 1985). However, in lieu of an alternate framework for discussion, we will discuss phage capsids in terms of their T number. The T numbers that have been unambiguously determined for tailed phages are given below. The T number of the elongated T4 and ϕCbK heads refers to the icosahedral "cap."

λ	7 *laevo*	Williams and Richards, 1974
P22	7 *laevo*	Casjens, 1979; Earnshaw, 1979
T7	7 *laevo*	Steven *et al.,* 1983
T4, T2	13 *laevo*	Branton and Klug, 1975; Aebi *et al.,* 1974
ϕCbK	7 *laevo*	Lake and Leonard, 1974
P4	4	Geisselsoder *et al.,* 1982
P2	9	Geisselsoder *et al.,* 1982

Possible capsomer arrangements for several of these phages are shown in Fig. 3. Although λ, P22, and T7 heads have very similar diameters and are built from identical numbers of subunits of similar size, no convincing amino acid sequence homology has been found between their coat proteins (our unpublished observations).

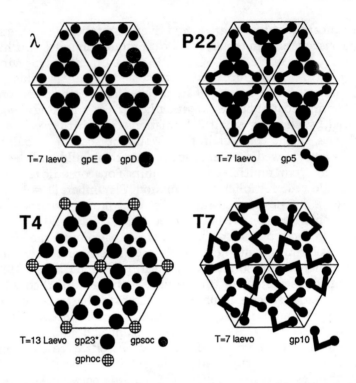

FIGURE 3. Schematic representations of capsomere structures of four bacteriophages. Six "facets" of each head shell lattice are shown (Caspar and Klug, 1962; Casjens, 1985a), and the outward protruding portions (capsomeres, clusters or mophological units) of the various capsid proteins are indicated: λ, trimer clustered decoration protein gpD and hexamer clustered coat protein gpE capsomeres are indicated (Williams and Richards, 1974; Imber et al., 1980); P22, one capsid protein, gp5, forms trimer and hexamer clustered capsomeres; the connection between these two domains has not been visualized but is shown here to emphasize the fact that there is only one polypeptide present (Casjens, 1979); T4, each individual gp23 coat protein subunit is apparently seen in electron micrographs (Steven et al., 1976; Yanagida, 1977); T7, although distinct surface structure is seen, visualization of T7 capsides and related structures has not shown obviously separated gp10 coat protein capsomeres. The subunit arrangement shown here is one of a number of possible arrangements (Steven et al., 1983; Steven and Truss, 1986).

b. Vertex Proteins

Among the phages studied, only T4 is thought to have a unique polypeptide, a cleaved form of the gene 24 protein, at its *non-tail-proximal* vertices (Muller-Salamin et al., 1977). The presence of such proteins eliminates the need for the coat protein to be able to form rings of either five or six members. Thus the T4 major coat protein need only form rings of six. It is interesting to note that mutations exist that bypass the need for the gene 24 protein. These alter the coat protein and apparently allow it to form pentamers at the vertices as well as hexamers elsewhere in the structure (McNicol *et al.*, 1977). The nucleotide sequences of genes *23* and *24* indicate that the two proteins do in fact have some regions of amino acid sequence homology (G. Yasuda, M. Parker, and D. Mooney, personal communication). This homology seems insufficient to confidently conclude that gene *24* arose from a duplication of gene *23*, although this scenario is quite possible.

c. Decoration Proteins

A second class of accessory capsid shell proteins sometimes found in head shells consists of the decoration proteins. These proteins bind to (or decorate) the outside of the coat protein shell after it is assembled. They are thought to impart greater stability or strength to the shell, but are at least under some conditions dispensable for phage growth. Phage T4 has two accessory proteins, the *soc* and *hoc* gene products, which bind to the coat protein near the local three- and sixfold axes, respectively (Ishii and Yanagida, 1977; Aebi et al., 1977). These give the phage added resistance to physical and chemical insult (Steven *et al.*, 1976; Ishii *et al.*, 1978), and gene *hoc* protein addition causes a substantial increase in the overall negative charge of the capsid surface (Yamaguchi and Yanagida, 1980). Phage T4's very close relative, T2, has no decoration proteins (Ishii and Yanagida, 1975). Phage λ has a single decoration protein, gpD†, which occupies the local threefold axes (Howatson and Kemp, 1975; Imber *et al.*, 1980). It is apparently required to give the capsid sufficient strength to be able to withstand the "internal pressure" only when a full-length chromosome is packaged inside the capsid (Sternberg and Weisberg, 1977). The function and precise positioning of the head fibers of ϕ29 are not known, but they are dispensable to the phage under laboratory conditions and so fit the definition of a decoration protein (Reilly *et al.*, 1977), and T5 may have a decoration protein (Saigo, 1978). P22, T3, and T7 appear not to have decoration proteins, but a very close relative of P22, phage L, may contain one (Hayden *et al.*, 1985). Since close relatives such as T2/T4 differ in whether or not they have decoration proteins, these proteins may be evolutionary latecomers which may or may not be advantageous, depending on the exact niche the phage occupies.

2. Internal Proteins

The heads of phages T4, T2, and probably SP01 contain proteins that are thought to be inside the head shell and perhaps bound to the DNA in

---

† gp*X* represents the primary translation product of gene *X*, and an asterisk, gp*X**, indicates a proteolytic cleavage product of that polypeptide.

the capsid, since they are released with the DNA upon various mildly disruptive treatments of virions (Black and Ahmed-Zadeh, 1971; Eiserling, 1979). Only the internal proteins of T4 have been further studied, where there are several fairly abundant (100–400 molecules per virion) proteins that are thought to be located inside the head shell. Three of these, called gpIpI*, gpIpII*, and gpIpIII*, all cleavage products of primary translation products, are nonessential, and gp68 is only partially required (Black, 1974; Black and Showe, 1983; Isobe et al., 1976a,b; Keller et al., 1984). Although these proteins are not essential in normal laboratory host strains, it is interesting to note that the gpIp proteins have DNA-binding activity (Black and Ahmed-Zadeh, 1971; Bachrach and Benchetrit, 1974), and gpIpI* appears to be ejected with the DNA during the initial stages of infection and is required in some host strains (Abremski and Black, 1979). Genes 22 and 67 encode proteins that are essential in forming the core of phage precursor particles (see Section IV.B.2). Both suffer multiple cleavages during assembly, and some of the products of these cleavages are released from the head (Kurtz and Champe, 1977; Showe, 1979; Volker et al., 1982). The precise location of the internal proteins within the head is unknown; however, the uncleaved forms of each are present in the cores of phage precursor particles that contain no DNA, indicating that at the time of assembly they are bound to other proteins of the capsid and not DNA. X-ray diffraction analysis of T4 virions suggests that the bulk of the DNA is very tightly packed, such that it would be difficult for proteins to be packed between the DNA double helices (see Section II.A.4). This may indicate that the internal proteins are located between the DNA core and the coat protein shell or possibly at the center of the DNA structure. The fact that many phages, such as λ, φ29, and P22, do not have such internal proteins implies that, although they may aid the larger phages in some as yet poorly understood way, evolution has not found them to be an absolute requirement for dsDNA phage assembly, DNA packaging or ejection.

3. The Portal Vertex

a. Portal Protein

In all tailed phages a single vertex of the icosahedral head is modified by the presence of additional proteins. These proteins form the site to which tails will bind and the "portal" through which DNA passes during packaging and ejection (Bazinet and King, 1985). The portal protein structures from phages T4, λ, φ29, T3, and P22 have been purified and analyzed by electron micrographic image enhancement techniques (Driedonks et al., 1981; Kochan et al., 1984; Carazo et al., 1985, 1986; Jimenez et al., 1986; Nakasu et al., 1985; Bazinet et. al., 1988). In all cases, they have been found to be composed of a dodecamer of a single polypeptide (mol. wt. 40–90 kD in the various phages) arranged in a ring of 12 subunits with a 3- to 4-nm hole through the center. In the virion this hole is

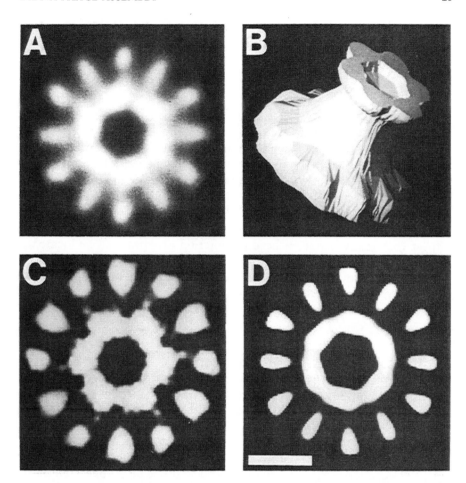

FIGURE 4. Portal protein structures. Panels A and B show a rotationally filtered electron microscopic image and a three-dimensional reconstruction (resolution 2.2 nm), respectively, of the φ29 portal protein structure. It contains 12 molecules of gp12 at the wide end and six molecules of gp11 at the narrow end; the narrow end protrudes out of the head into the tail (Carrascosa *et al.*, 1982; Carazo *et al.*, 1985; Jimenez *et al.*, 1986). Panels C and D show rotationally filtered images of the portal structures of phages P22 and T3, made up of 12 molecules of gp1 and gp8, respectively (Carazo *et al.*, 1986; Bazinet *et al.*, 1988). The magnification is similar in all panels, and the white bar in panel D is approximately 5 nm. The figure panels were kindly provided by José Carrascosa (A, B, and D) and Jonathan King (C).

aligned with the long axis of the tails so that it should form a pore in the head shell. Figure 4 shows several of these structures; it is not yet known if the true symmetry of these structures is six- or 12-fold (J. Carrascosa, personal communication). The fact that these multimers are so strikingly conserved in overall shape suggests that they play an important role in the virion. It is likely that this role includes both DNA packaging and initiation of capsid shell assembly (Murialdo and Becker, 1978; Earnshaw and Casjens, 1980; Bazinet and King, 1985; Herranz *et al.*, 1986).

b. Head Completion Proteins

Phage heads are usually quite unstable immediately after DNA is packaged, until the "head completion" proteins are added to the portal vertex. In general, these steps have not been studied in detail. In λ, it is clear that the last protein to add to the head, the gene F_{II} protein, stabilizes the head and forms at least a portion of the site to which tails will bind (Casjens *et al.*, 1972; Casjens, 1974; Tsui and Hendrix, 1980). In T4, the head completion proteins, gp13 and gp14, have also been located in the portal vertex (Coombs and Eiserling, 1977). It seems reasonable to expect that many if not all head completion proteins will be located at the portal vertex. The mechanism by which they stabilize the head is unclear, but they may prevent DNA from coming out of the portal through which it entered the head.

c. Ejection Proteins

These proteins are usually not required for (apparently) correct assembly of virions and are thought somehow to aid in the process of DNA ejection and possibly other aspects of the "establishment of infection." In general, the location of these proteins in the virion is not known directly, but since they are present in a fairly low number of molecules per virion (<50), it seems possible that they are located at the portal vertex.

The phage T4 gene 2 protein protects newly ejected DNA from the host RecBC nuclease (exonuclease V), and so it has been suggested that it is ejected with the DNA and binds the ends of the molecule in the host cell (Silverstein and Goldberg, 1976; Oliver and Goldberg, 1977; Goldberg, 1983). The location of gp2 in T4 virions is not known. In addition about 40 molecules of gp*Alt**, which ADP-ribosylates the host RNA polymerase α subunit, are also ejected by T4 and may reside at the portal vertex in the virion (Black and Showe, 1983). Three P22 proteins, the products of gene 7, 16, and 20, are classified as injection proteins, since when they are defective initiation of infection fails (Botstein *et al.*, 1973; Hoffman and Levine, 1975; Bryant and King, 1985) and in that they may be ejected with the DNA (Israel, 1977; Crowelesmith *et al.*, 1978). The location of these three P22 proteins in the virion is unknown (Hartwieg *et al.*, 1986). How ejection proteins are able to traverse the tail structure is unknown.

The observations that the T4 internal protein, gp*IpI**, is apparently ejected (Abremski and Black, 1979) and that a P22 head completion protein, gp26, may be ejected (Israel, 1977) suggest, not surprisingly, that there is overlap between the artificial categories of head protein functions we have used here for convenience of discussion. Other phages such as λ seem not to have ejection proteins that are built into the head but may have proteins with similar functions in the tail (see Sections II.B.2 and IV.B.1).

4. DNA Structure

X-ray scattering of unoriented phage particles has shown that the bulk of the DNA in phage particles is in the B form, and that DNA double helices are lying parallel and very close together with about 1.5 g water per gram of DNA in the capsid (reviewed by Earnshaw and Casjens, 1980). At this density it is difficult to imagine proteins between the DNA double helices. In addition, some phages such as P22 and λ contain no proteins that are not bound to the capsid. Therefore, current models for intraphage DNA structure assume the DNA is paracrystalline. Electron microscopy of DNA partially released from capsids has generally supported a model in which the DNA is wound as a solenoid (Tikhonenko, 1970; Richards et al., 1979; Earnshaw et al., 1978). Electron microscopy of DNA within phage heads also seems to support this model (Adrian et al., 1984). Other recent experiments are difficult to reconcile with a solenoid of DNA (Widom and Baldwin, 1983; Haas et al., 1982; Black et al., 1985; Serwer, 1986; Lepault et al., 1987). These data suggest that the solenoid model is incorrect or that a given DNA sequence is located in different places in the solenoid in different individual virions (Harrison, 1983b). The ends of the DNA molecule may occupy special positions in the capsid. In those cases where it is known, the end that is packaged last is located near or protruding through the proximal vertex into the tail (reviewed by Earnshaw and Casjens, 1980).

B. Tail Structure

Tail structure appears to be more variable than the head structure, but there are common themes. Presumably the variations reflect different strategies of DNA delivery, although adsorption and ejection are at present two of the most poorly understood aspects of the phage life cycle. In the following discussion we will emphasize the themes we currently see as common.

1. Tail Fibers

The fibers at the distal ends of the tails of most phages studied are the most important virion substructure in the process of attachment of virions to susceptible cells. In each case they are the primary site of binding of neutralizing antibodies, and they bind to specific *receptors* on the exterior of such cells. Fibers appear to be particularly variable structures, as is reflected in their shape, which ranges from the long (160 nm—Ward et al., 1970) tail fibers of phages like T4 to the much shorter (25 nm—Berget and Poteete, 1980) tail fibers of the short-tailed phages like P22 (called "spikes," "thick fibers," "appendages," or simply "tail proteins" in the various systems). T4 virions have six tail fibers, where each fiber contains 1 to 2 molecules of four different polypeptides. Other phages appear to have three (PBSI—Eiserling, 1967) or 12 tail fibers (φ29—Anderson et al., 1966). The lambdoid phages have a single tail fiber protruding "down" from the distal end of the tail, and short, very thin fibers are occasionally seen extending radially. Although all of the well-studied phages have elongated proteins that perform the initial adsorption step (Figs. 1, 2), in some phages no tail fibers have been seen in the electron microscope (e.g., SP01; see Eiserling, 1979), and the long, L-shaped fibers of T5 are dispensable in the laboratory (Saigo, 1978). It remains to be determined whether these phages have devised an alternate initial adsorption mechanism.

a. Thick Fibers or Spikes

Phage P22's six thick fibers, each a trimer of the gene 9 protein (Goldenberg and King, 1982), have an endorhamnosidase enzymic activity that cleaves the Salmonella typhimurium cell surface polysaccharide called O antigen (Iwashita and Kanegasaki, 1976a; Ericksson et al., 1979). The O antigen is the receptor for P22, and it has been suggested that P22 virions first recognize their host by binding O antigen but then must cleave their way down through the O antigen to reach the outer membrane of the cell (Israel, 1978; Bayer et al., 1979). Many phage virions whose receptors are polysaccharides on the exterior of bacteria have similar activities that cleave many different sugar-sugar bonds in the various surface polysaccharides, and some have an activity that cleaves acetyl groups from such polymers (Kwiatowski et al., 1975; Iwashita and Kanegasaki, 1976b). In all cases where this has been studied, the enzymic activity is associated with the tail fiber or spike protein (reviewed by Reiger-Hug and Strim, 1981; Wollin et al., 1981; Svenson et al., 1977; Kanegasaki and Wright, 1973).

It has not been found, however, that all thick fibers have such activities; φ29, for example, has fibers which have not been shown to have any enzymic activity, and it is unique among the phage observed to date in that it has 12 thick fibers (in this case often called appendages). Each fiber is probably a dimer of the cleaved gene 12 protein, and these fibers are responsible for attachment to cells (Tosi and Anderson, 1973; Carrascosa et al., 1981).

b. Thin Fibers

A long, bent tail fiber extends from each hexagonal vertex of the phage T4 baseplate. Its structure is fairly complex compared to others under study, in that it contains four different polypeptides (Fig. 2). The tips of the fibers are of primary importance in selection of the phage's lipopolysaccharide receptor (Beckendorf, 1973; reviewed by Wood and Crowther, 1983; see however, Granboulan, 1983). The distal half of the fiber is composed of a dimer of gene 37 protein. Its structure has been studied in some detail, and the gp37 is in an extended conformation with its C terminus distal to the baseplate (Beckendorf, 1973). A unique "cross-β" peptide backbone folding has been proposed to account for its observed structure in the electron microscope (Earnshaw et al., 1979). Secondary structure predictions of gp37 show a high degree of β structure (Oliver and Crowther, 1981; P. Arscott, personal communication). Perhaps surprisingly, the amino acid sequence of gp37 does not show any very regular repeating pattern, but it does contain six Gly-X-His-Y-His motifs, and Michel et al. (1986) have suggested these may be divalent metal binding sites.

The structure and function of the single, 23-nm-long, thin fiber on the tip of the λ tail are not well understood. Two or three molecules of the phage J gene product form the adsorption fiber (Murialdo and Siminovitch, 1972; Buchwald and Siminovitch, 1969). The C-terminal end of the protein appears to bind to a component of the host outer membrane, the lamB protein (Simon et al., 1971; Fuerst and Bingham, 1978; Thirion and Hofnung, 1971; Randall-Hazelbauer and Schwartz, 1973). One imagines it must move to allow the DNA out of the virion, but little is known about the details of ejection (Roessner and Ihler, 1984). The small lateral fibers at the base of the shaft have no known function.

Phage T3 (and probably T7) have six thin, bent, 36-nm fibers that are homotrimers of the gene 17 protein (Kato et al., 1985a,b; Serwer, 1976), and the N-terminal portion of the protein binds to the baseplate (Kato et al., 1986). Other phages with long, noncontractile (χ—Shade et al., 1967; PBP1—Lovett, 1972) and contractile tails (PBS1—Eiserling, 1967; AR9—Tikhonenko, 1970) have one, three, or more marvelous, long, curly tail fibers (see, for example, Fig. 5). At least in the case of PBP1 and χ, these are involved in adsorption to flagella. In addition, a number of phages (e.g., AR9 and PBS1) have numerous fibers of unknown function extending out from the sheath which are not present on any of the phages currently under intense study (see Fig. 5). These structures are not understood in great detail, but they do point out that not all types of structural modules have been studied or perhaps even observed.

2. Baseplates

All phages have distinct structures at the end of the tail to which the fibers are bound. These vary greatly in shape and complexity. Those whose structure is known have apparent sixfold symmetry, although phages with one, three, and 12 fibers suggest that this may not always be the case.

The T4 baseplate is a very complex structure that contains three to

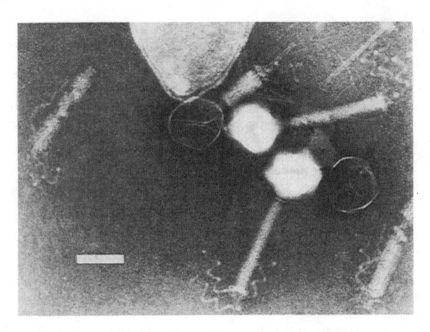

FIGURE 5. Electron micrograph of negatively stained bacteriophage PBS1. Virions of this phage have three curly tail fibers and fibers extending out from the tail sheath (especially visible in contracted tails), structural motifs not found in the phages currently under study. The micrograph was kindly provided by Fred Eiserling. The white bar represents 100 nm.

24 molecules each of 18 different proteins (King and Mykolajewycz, 1973; reviewed by Berget and King, 1983). Its structure has been analyzed by image enhancement of electron micrographs (Crowther *et al.*, 1977), and each of the proteins has been localized to the central "plug" or outer "wedge" portion of the structure. Figure 12 indicates that baseplates have an overall sixfold rotational symmetry. The detailed structural roles of most of the baseplate proteins are unknown, but the gene *12* protein is known to unfurl to form the "short tail fibers" during adsorption (Kells and Hazelkorn, 1974; Crowther, 1980), and gene *48* and/or *29* proteins may be involved in tail length determination (Section IV.B.1). It is interesting to note that several of the baseplate proteins have enzymatic activities as follows:

Gene *5* protein	Lysozyme	Kao and McClain, 1980
Gene *25* protein	Lysozyme	Szewczyk *et al.*, 1986
Gene *29* protein	Folyl hexaglutamate synthetase	Kozloff, 1983
Gene *28* protein	γ-Glutamyl carboxypeptidase	Kozloff and Zorzopulos, 1981

Gene *frd* protein   Dihydrofolate reductase   Mosher and Mathews,
                                                1979
Gene *td* protein   Thymidylate synthase   Capco and Mathews, 1973

The *frd* and *td* proteins are not required for correct assembly or virion function. The lyzozyme is thought to aid DNA entry into the cell some fashion (Nakagawa *et al.*, 1985), and the gene 28 and 29 proteins, in addition to being required structural parts of the baseplate, catalyze the metabolism of the dihydropteroyl polyglutamate found in baseplates (see Section II.C). The great complexity in the baseplate structure may well be a consequence of the fact that it undergoes a very large conformational change/rearrangement during adsorption (see Section IV.D.2).

In λ, the tail "tip" contains at least six different polypeptides present in a small number of copies each, including the gene *J* protein tail fiber (reviewed by Katsura, 1983b). This structure has not been analyzed in detail because of its small size, and its symmetry remains unknown, but the gene *J* protein is likely to move during adsorption (Roessner and Ihler, 1984). This structure also contains a protein, gp*H**, which may enter the infected cell with the DNA (Roa, 1981) and which is involved in tail length determination (Katsura and Hendrix, 1984).

Phages P22, φ29, T7, and T3 have very short tails. Those of P22, T7, and T3 have a sixfold rotational symmetry axis perpendicular to the head shell (Yamamoto and Anderson, 1961; Serwer, 1976; Matsuo-Kato *et al.*, 1981), whereas that of φ29 may be at least partly 12-fold (Anderson *et al.*, 1966; Carazo *et al.*, 1985). These have a single, central structure surrounded by a "baseplate" or "collar." The shape of this structure, although in general not well studied, seems quite variable among the short-tailed phages and is composed of several proteins (Serwer, 1976; Matsuo-Kato *et al.*, 1981; Hartwieg *et al.*, 1986). Presumably these proteins perform the combined functions of "head and tail completion" proteins (see Fig. 7) and baseplate proteins when such functions are required. This central structure is surrounded by six or 12 tail fibers or spikes. The mechanism of ejection by these phages is unclear; however, it seems probable that some protein movements must accompany adsorption and ejection.

3. The Tail Shaft

Long tails are long by virtue of their "shaft." These are found in two general types: flexible, noncontractile ones such as are present in λ and T5 (usually 100–300 nm long), and rigid, contractile ones of T4, P2, SPO1, Mu, and P1 (usually 100–450 nm long). Each of these types has been seen in phages with a variety of baseplate and tail fiber structure types.

The structures of two flexible shafts have been studied. The λ tail shaft contains about 32 rings of six gene *V* protein subunits (Buchwald *et al.*, 1970; Casjens and Hendrix, 1974a; Katsura, 1983), and that of φCbK contains about 78 rings of three subunits (Leonard *et al.*, 1973; Papadopoulos and Smith, 1982). The φCbK tail shaft is the only phage tail structure clearly shown to have threefold symmetry. In at least φCbK, these "annuli" are arranged with helical symmetry in the tail shaft. The λ and φCbK shafts have 3- to 4.5-nm holes down the center through which DNA must pass during ejection, and in the virion DNA is thought to protrude into this passage at least part way down the tail (Thomas, 1974). It is likely that the tail shaft is a passive conduit for the DNA in that there is no evidence that it actively participates in propelling the DNA from the head to the cell during ejection. Before ejection, other proteins may at least partially fill the central hole (see section IV.B.1). It is not known if there are strong selections for given tail lengths in any phages, but it is interesting to note that very close relatives can be virtually identical in appearance except for differing tail lengths (Katsura and Hendrix, 1984; Youderian, 1978; King, 1980; R. Hendrix and M. Popa, unpublished results). Schwartz (1976) has speculated that long tails allow a more rapid searching of the environment surrounding a phage for suitable receptors and thus might allow adsorption to bacteria with sparsely positioned receptors on their exterior.

The T4 tail shaft contains an internal cylinder called the tail "tube," which is composed of 144 molecules of gene *19* protein, arranged in an annular structure with most likely 24 rings of six subunits (Moody, 1971; Moody and Makowski, 1981). The central hole appears to be partially filled with another protein, possibly gene *29* and/or gene *48* proteins (Duda *et al.*, 1986). The tube is enclosed within a cylindrical "sheath," which is also composed of 144 molecules of gene *18* protein as 24 helically arranged rings of six subunits. The sheath has been the subject of a number of image enhancement studies with a resolution of about 3 nm (Fig. 6) (DeRosier and Klug, 1968; Amos and Klug, 1975; Smith *et al.*, 1976; Lepault and Leonard, 1985). It is the gene *18* protein molecules that rearrange so dramatically during contraction of the tail during ejection (see section IV.D.2 below). The structure of the sheath of one other phage with a contractile tail, *Bacillus megatherium* phage G, has been studied. It contains about 120 helically arranged rings of six subunits (Donelli *et al.*, 1972).

4. The Tail Completion Proteins

Very little is known about the head-proximal end of long tails. Proteins located here may be required to stop the growth of the shaft during assembly (in λ see Section IV.B.1) and create the site to which the head will join. In λ the gene *U* and possibly *Z* proteins are thought to be present at this end of the tail in a small number of copies, but no further details are known (Casjens and Hendrix, 1974b; Thomas *et al.*, 1978; Katsura, 1976). In T4, there are several proteins that occupy the head-proximal end of the tail and which are required for binding of the tail to

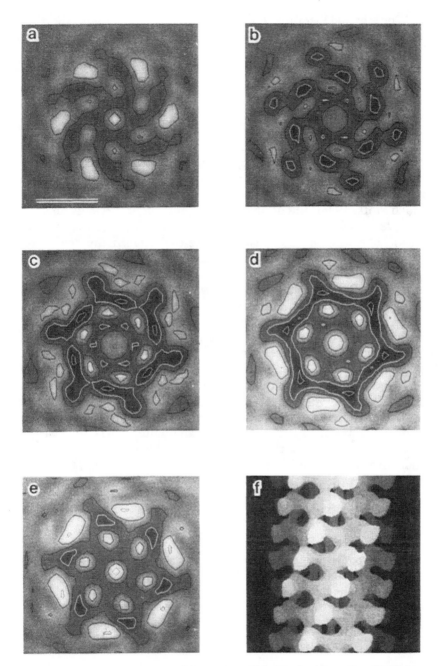

FIGURE 6. Three-dimensional reconstruction of bacteriophage T4 tails embedded in vitreous ice. Panels a through e are sections through one disk, 0.82 nm apart, progressing from the baseplate in the direction of the head. Protein is dark. Panel f shows a representation of seven gp18 sheath disks, where protein is white. The bar in panel a represents 10 nm. The figure, from Lepault and Leonard (1985), was kindly provided by Kevin Leonard and is reproduced with the permission of Academic Press.

heads (Coombs and Eiserling, 1977; King, 1968). In phages with short tails it is not useful to attempt to distinguish the head completion and tail completion proteins (see Fig. 7).

C. Small Molecules in Phage

The importance of small molecules and ions in phage virions is not well understood. One important function of these is presumed to be neutralization of the negative charge of the DNA in the head. Polyamines have been found in T4 and T5 but in insufficient amounts to neutralize even a large fraction the DNA's negative charge (Ames and Dubin, 1960; Bachrach et al., 1975), and dsDNA phages grow on host strains unable to make several polyamines (Hafner et al., 1979). The fact that divalent metal ion chelating agents cause DNA release from many phages suggests that ions such as Mg^{2+} are responsible for this neutralization, although since these heads are all at least slowly permeable to small molecules, it is difficult to determine the in vivo situation at the time of assembly (reviewed in Earnshaw and Casjens, 1980).

Even less is known about other small molecules in phage. Although a number of plant virus capsid proteins contain bound Ca^{2+}, only one phage structural metalloprotein is known. The T4 gene 12 baseplate protein contains Zn^{2+} (Kozloff and Lute, 1977; Zorzopulos and Kozloff, 1978). It has also been found that T4 baseplates contain about six molecules of dihydropteroyl polyglutamate, and it is interesting that enzymes involved in its metabolism are structural components of the baseplate (Nakamura and Kozloff, 1978; Kozloff and Lute, 1981; reviewed by Kozloff, 1983). The role of this folate compound is not known, but there are indications that (1) it is essential for baseplate function, (2) it may lie near the point of long tail fiber attachment, and (3) it may be required during the adsorption/ejection process (Dawes and Goldberg, 1973; Male and Kozloff, 1973; Kozloff et al., 1979; Kozloff, 1983). If this is true, the folate is not likely to be bound to one of the active sites of the baseplate enzymes that would be expected to bind it (gene 28 and 29 proteins), since they are components of the central hub of the baseplate (Kikuchi and King, 1975c; Kozloff and Zorzopulos, 1981). There have been reports of other virion bound ions and nucleotides, such as a GTP and Ca^{2+} bound to each tail sheath subunit of T4, but their functional importance remains obscure (Kozloff and Lute, 1960; Serysheva et al., 1984).

III. THE NATURE OF PHAGE ASSEMBLY PATHWAYS

One of the earliest lessons concerning macromolecular assembly learned from the study of phage is that such processes almost invariably follow well-defined pathways. For example, if components A, B, and C must join to form ABC, they have (ignoring 3 body collisions) three choices or "pathways" by which the heterotrimer ABC can be built:

$$A + B \rightarrow AB + C \rightarrow ABC$$
$$A + C \rightarrow AC + B \rightarrow ABC$$
$$B + C \rightarrow BC + A \rightarrow ABC$$

Usually, only one of these pathways would be chosen by the three components if this were a real assembly reaction. This strategy has obvious advantages. First, it allows *control* over each step in that the process does not continue unless the previous step is complete. In such a situation a dead-end structure cannot be assembled in which the addition of one of the components is sterically excluded from the structure (for instance, assembly of the closed head shell before the internal proteins are assembled). Edgar, Wood, and colleagues first showed clearly that phages (in their case T4) are in fact assembled by means of such pathways (Wood *et al.*, 1968). Given the structural complexity of these phages, it now seems impossible that they could be built in any other manner.

Wood *et al.* (1968) also noted that the T4 assembly pathway is branched, with "subassemblies" being built first and these structures subsequently combining to make the final virion. The "subassembly strategy" has certain advantages in increasing the accuracy and efficiency of the assembly process. This aspect of assembly has recently been discussed in detail by Berget (1985). Although T4 clearly utilizes such a strategy, it does not seem to be required to build phages, since T3, T7, and P22 do not use it nearly so extensively. Their assembly pathways are nearly linear (Studier, 1972; Roeder and Sadowski, 1977; King *et al.*, 1973).

Although there is considerable variation in the details of the assembly pathways of the various dsDNA phages, there are also remarkable overall similarities (for reviews of these pathways see Casjens and King, 1975; Murialdo and Becker, 1978; Wood and King, 1979). Figure 7 shows a "generic" pathway for these assembly processes. All dsDNA phages build the head from a precursor particle called a prohead, which contains a coat protein shell, a portal protein, and an internal scaffolding protein. About the time of DNA packaging, the scaffold is removed by either proteolysis or exit of the intact protein. DNA is then thought to enter through the portal structure, and the head is completed by the addition of several proteins. In the short-tailed phages, the tail is then built in place on the head by sequential binding of additional proteins. In the long-tailed phages, the tails are built from the distal end toward the head-proximal end, and when finished, they spontaneously join to a finished head.

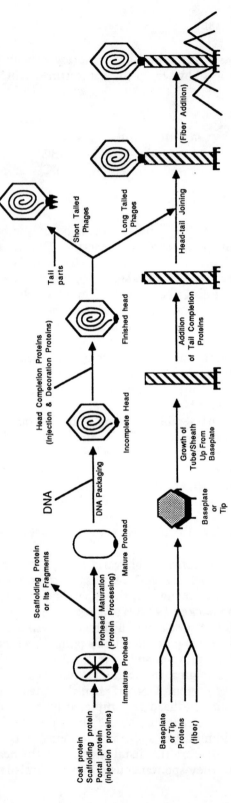

FIGURE 7. A generalized assembly pathway for the dsDNA, tailed bacteriophage virions. Parentheses indicate features not present in every pathway.

IV. CURRENT PROBLEMS IN PHAGE ASSEMBLY AND STRUCTURE

A. Assembly Pathways

1. Obligate, Sequential Assembly Steps

a. Protein-Protein Interactions

The control of phage assembly pathways lies in the proteins themselves, in that the purified proteins will only assemble in a predetermined order. This has been most elegantly studied by King, Berget, and their co-workers using the phage T4 baseplate as a model (Kikuchi and King, 1975a–c; Berget and King, 1983; Plishker and Berget, 1984). They have clearly shown that nearly all of the >20 steps in the branched-tail assembly pathway occur in an obligate order (Fig. 8). Assembly of the "wedge" portion of the baseplate is a particularly clear example of a linear, ordered pathway in which all of the steps occur with reasonable efficiency *in vitro*. But what are the details of the mechanism by which this control is exerted? Utilizing mutants in the various baseplate genes, Kikuchi and King (1975a–c) analyzed the intermediate assemblies accumulated in each mutant to define the pathway shown in Fig. 8. They also showed that if any step is blocked by a null mutation inactivating a particular protein, all the proteins that would have added later in the pathway remain unassembled. That is, the baseplate proteins are in general synthesized in a state in which they are unreactive toward one another. Reactive assembly sites to which they bind are restricted to the growing baseplate structure, so if the assembly process cannot initiate, none of the components can assemble.

An exception to this rule is the affinity of gp7 and gp10 for each other; gp7 and gp10 spontaneously join to form a complex when both are present. gp8, although it has no strong affinity for gp7 and gp10 by themselves, spontaneously binds the gp7–gp10 complex. The resulting complex is then the only substrate for gp6 addition, etc. This constitutes the wedge assembly pathway, with the gp7–gp10 joining reaction being the *nucleating* event.* Since it can be carried out with purified proteins, it is controlled by the proteins themselves. Caspar (1980), Wood (1980), King (1980), and Berget (1985) have discussed this type of control, and although the molecular details are not fully understood, it is clear that subunits already assembled somehow *cooperate* to form the binding site for the next protein in the pathway.

In some cases, especially for those proteins that form large repeating arrays in the virion, this type of control is not absolute. If the proper substrate for assembly is not present, aberrant assembly will eventually

*Note that gp11 can bind to a gp10 dimer, but since it can also add at any later stage of assembly (see below), it is not the "nucleating" event for baseplate assembly.

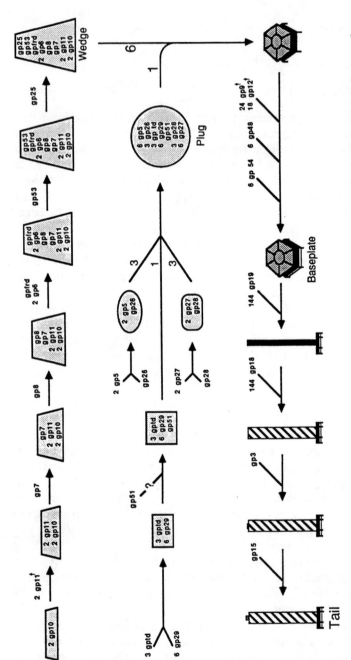

FIGURE 8. The bacteriophage T4 tail assembly pathway, the best-understood of the "protein-determined" ordered pathways. This branched pathway assembles baseplate wedges and plug independently, joins them, adds additional proteins to form the completed baseplate, assembles the shaft up from the baseplate, and caps the end with the tail completion proteins. The wedge and plug are shown as disproportionately larger than the baseplate, and the baseplate is similarly larger than the structures with the tail tube added. The steps of gp11, and gp12, and probably gp9 addition can occur at any stage after the indicated point of addition (indicated by †), except that gp12 requires gp11 to be bound before it can add. The action of gp57 which is required for proper gp12 folding is not shown in the figure. The order of the pathway steps are from Kikuchi and King (1975c), Mosher and Mathews (1979), Berger and King (1983), and Kozloff and Lute (1984). The figure is adapted from Kikuchi and King (1975c).

occur. For example, the T4 tail sheath protein gp18, at late times after an infection which lacks baseplates and/or tubes, will polymerize into a free "polysheath" (King, 1980). Incorrect assembly of phage coat proteins also often occurs if other head proteins are missing (see, e.g., Laemmli et al., 1970). Similarly, the T4 baseplate wedges will eventually assemble into aberrant hexagonal structures without the central plug (Kikuchi and King, 1975b), and T4 proximal vertex head proteins will eventually add to tails (Coombs and Eiserling, 1977). It seems reasonable to conclude (1) that when an assembly step goes awry which blocks the normal assembly of these components, their concentrations build up to some critical level and spontaneous, aberrant association occurs, and (2) that this aberrant assembly shares features with normal assembly.

The baseplate pathway also demonstrates several assemblies that are not obligately ordered. These are the addition of gp11 and gp12. Both proteins can be added to the finished baseplate structure at any stage, even after the tail is completely assembled and joined to a head. gp11 demonstrates strong affinity for gp10 in the absence of further baseplate assembly (Berget and King, 1978), and it is required for gp12 addition later in the pathway (Fig. 8). Thus, there must be a gp11 binding site on gp10 that is exposed and does not change throughout the tail assembly process, and similarly a gp12 binding site on gp11 that is exposed only by wedge-plug association. It is not clear why the phage has not ensured gp11 and gp12 addition by the control mechanism discussed above. Perhaps T4 has added these proteins late in its evolution and has not had time to build in such a mechanism, or perhaps it is advantageous not to build in an ordering mechanism if it is not required (gp12 functions in adsorption—Kells and Hazelkorn, 1974; Crowther, 1980).

The sort of cooperative control of assembly discussed above does not require multiple subunit types. This is demonstrated very well by the assembly of the T4 tail tube on the completed baseplate. Here, when the gp19 assembly site is created on the finished baseplate, 144 gp19 subunits add *sequentially* to build the tube from the baseplate outward. Again, gp19 shows no affinity for itself in the absence of a nucleating site for assembly (Meezan and Wood, 1971; King, 1971).

Since the T4 tail proteins undergo no covalent changes, it seems that this hetero- or homocooperation must occur by the following two *non-mutually exclusive* mechanisms: (1) simple assembly-dependent juxtaposition of multiple partial sites on the individual assembled proteins to create a binding site, and (2) conformational switching induced by assembly in the proteins of the growing structure which creates a new binding site. No *direct* evidence exists to prove either hypothesis; however, purified gp19 and gp18 dissociated from purified tails can spontaneously polymerize into "polytubes" and "polysheaths," respectively (Poglazov and Nicolskaya, 1969; To et al., 1969; Wagenknecht and Bloomfield, 1977; Tschopp et al., 1979; Arisaka et al., 1979), and King (1980) has argued that upon dissociation they may remain in their putative, assembly-proficient conformational state. We currently know very little about the distances over which these protein-determined cooperative effects can act, since our knowledge of the detailed arrangement of proteins in the various growing structures is quite limited. If hypothesis 1 is true, the most recently added protein must contact the subunit added in the preceding step, whereas conformational switching could (but need not) be imagined to propagate in such a way that this would not necessarily be true. There is a clear propagation of a structural change through the T4 phage head shell during expansion (see Section IV.D.3; reviewed by Black and Showe, 1983), and this conformational change does create the sites for binding of the decoration proteins gpsoc and gphoc (Steven et al., 1976; Aebi et al., 1977; Ishii and Yanagida, 1975, 1977). In T4 baseplate assembly, current structural knowledge is consistent with a model in which the wedge is built from the outside toward the center with each protein binding to the one added in the immediately preceding step (Crowther et al., 1977). One apparent exception to this is the addition of gp9 to the baseplate, since it is located near the outside of the hexagon but adds late in the pathway immediately after plug addition. The addition of gp9 is not strictly ordered, although it must add after the plug and six wedges form a hexagonal structure (Fig. 8) (Edgar and Lielausis, 1968; Berget and King, 1983). Unfortunately, there is insufficient knowledge of the baseplate structure to determine if gp9 is binding to a site at the junction of two adjacent wedges, so we cannot tell if a new gp9 binding site is created by propagation of conformational change from the inside to the outside of the wedge.

There are a few cases of phage assembly steps that appear not to be controlled by this type of protein-determined cooperative assembly mechanism but that still must occur at a *particular stage* of assembly. The best studied is in the assembly of the P22 prohead; if any one of the ejection proteins (gp7, gp16, or gp20) is missing, *in vivo* assembly proceeds, and dead-end, phagelike particles are made that are unable to eject their DNA properly (Botstein et al., 1973; Poteete et al., 1979). Unlike the T4 particles lacking gp11 and gp12, mentioned above, these particles cannot be rescued by supplying the ejection proteins, and there appears to be no protein-determined mechanism for ensuring the incorporation of these proteins into the prohead at the proper time of assembly. We do not understand how all P22 particles receive a necessary complement of ejection proteins, but differential regulation of structural gene expression might help create an excess of the ejection proteins (Casjens and Adams, 1985; see Section V.B.2).

b. Protein-Nucleic Acid Interactions

Of course, similar cooperative action could occur between nucleic acid and protein molecules to create assembly reactive sites. This has been documented for several simple ssRNA plant viruses—for example, in tobacco mosaic and southern bean mosaic virus assembly (Butler and Klug, 1971; Schuster *et al.*, 1980; Savarthi and Erickson, 1983; Driedonks *et al.*, 1976). Since DNA is packaged in an orderly manner by these phages, it is certainly not surprising that similar controls are being found to be built into the DNA recognition proteins (Feiss and Becker, 1983; Casjens, 1985b).

2. Assembly-Promoting Proteins

a. "Classical" Enzymes

There are several examples of covalent bond changes during phage assembly that are caused by what we normally consider enzymic reactions. These are hydrolytic cleavages of DNA and proteins. During assembly all of the tailed phages, except φ29 and its relatives, cleave a linear DNA chromosome from concatemeric or circular replicating DNA. The reactions by which this cleavage occurs are complex and not yet fully understood. Although the proteins that perform the cleavages are known in some cases (Wang and Kaiser, 1973; Becker and Gold, 1978; Bowden and Calendar, 1979), nothing is known concerning the details of the control of these enzymes beyond the observation that they are in fact controlled; for example, in phage where DNA is packaged from a concatemer, a second "headful" cleavage occurs only when the head shell is sufficiently full of DNA (reviewed by Feiss and Becker, 1983; Casjens, 1985b).

Proteolytic cleavage of structural proteins is also not uncommon in phage head and tail assembly (T4 head—reviewed by Black and Showe, 1983; P2 head—Lengyel *et al.*, 1973; λ head and tail—Hendrix and Casjens, 1974a,b, 1975; T5 tail—Zweig and Cummings, 1973; φ29 tail—Ramirez *et al.*, 1972; Tosi *et al.*, 1975; T2 tail fibers—Drexler *et al.*, 1986). The cleavages that are best studied are those that occur in the T4 head. In this case a specific protease zymogen, gp21, is assembled into the prohead, and later, some as yet unknown signal causes it to cleave and thereby activate itself. The activated gp21* protease cleaves at least 10 of the structural protein components of the prohead (see Fig. 2), including the removal of 65 amino acids from the N-terminal end of the coat protein, gp23 (Parker *et al.*, 1984) and the cleavage of the major scaffolding protein, gp22, into small fragments (Kurtz and Champe, 1977). Finally, gp21* destroys itself (Showe *et al.*, 1976a,b). The gp21* proteinase has specificity for the sequence

$$\text{N-terminal-} \text{- -} \text{Pho-Pho-Glu} \downarrow \text{- - -} \text{C-terminal}$$

(where Pho indicates a hydrophobic amino acid) derived from a consensus of nine known cleavage sites (reviewed by Black and Showe, 1983; Keller *et al.*, 1984). The cleavages also appear to require some aspect of the native conformation in the substrate proteins (Showe *et al.*, 1976a).

Other covalent changes in proteins may also occur during phage assembly, but in no case is the mechanism understood. In lambda, each gpC is thought to be both cleaved and fused to a gpE cleavage product (Hendrix and Casjens, 1974b; 1975), the major coat proteins of the lambdoid phage HK97 are cross-linked (R. Hendrix, unpublished observation), and a major phage P1 structural protein may have a covalent adduct (Walker and Walker, 1981).

The function of these covalent modifications of structural proteins is not known. Several possible nonmutually exclusive roles are (1) to make assembly irreversible, (2) to cause conformational changes required for further assembly, and (3) to simply remove the cleaved protein or a portion of it. It is difficult to assess the importance of possibility (1) at present, since many noncovalent assembly steps appear to be essentially irreversible under physiological conditions and since some phages have no proteolysis in their assembly pathways (e.g., P22 and T7—Botstein *et al.*, 1973; Casjens and King, 1974; Studier, 1972). The T4 coat protein also undergoes a major structural rearrangement during assembly (see Section IV.D.3), but although it may require coat protein cleavage, cleavage is not sufficient to cause the conformational change (Kistler *et al.*, 1978; Ross *et al.*, 1985). It has been suggested that possibility (3) could be important in removing several of the major scaffolding proteins of T4 and λ to create enough space inside the head shell to accommodate the DNA chromosome (Earnshaw and Casjens, 1980).

b. Assembly Templates and Jigs

The addition of tail fibers to phage T4 is the final step in virion assembly. This step requires two apparently distinct forms of catalysis. Fibers do not join with isolated tails, only with tails joined to heads. This results from the requirement for the presence of the whiskers (a neck component) for addition (Coombs and Eiserling, 1977). It is thought that the distal tips of the whiskers act as *assembly jigs* which bind the "elbow" of the fiber, thereby holding it in place or restricting its motion for joining to the baseplate (Fig. 9) (Terzaghi, 1971; Terzaghi *et al.*, 1979;

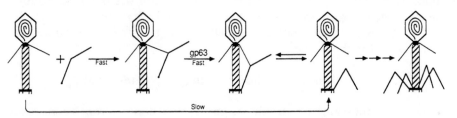

FIGURE 9. Tail fiber addition in bacteriophage T4 assembly. Whiskers cause more rapid tail fiber addition by acting as an assembly "jig" or "template." Adapted from Wood (1979).

Wood and Conley, 1979; Bloomfield, 1983). This bond must not be very strong after fiber addition is complete, however, since fibers are rarely attached to whiskers in electron micrographs of virions. Even with whiskers present, however, joining is quite slow and is accelerated at least fivefold by the presence of the gene 63 protein (Wood and Henninger, 1969; Wood et al., 1978). It has been suggested that gp63 may transiently hold one of the assembly sites in a reactive conformation or "activated" state, or it could form a template that binds the two components in proper juxtaposition with respect to each other (Wood, 1979; Berget, 1985). It is not at all clear why this assembly step should require so much help, since the joining of the two fiber halves, which one might have a priori imagined to have similar steric features, proceeds spontaneously and rapidly. One naive possibility (of many no doubt) is suggested by the fact that the tail fibers appear to swivel freely about the baseplate vertex to which they are bound (Fig. 1). Perhaps they are attached by a ball-and-socket-type joint which is held open for fiber insertion by gp63 (see Wood, 1979). It remains a biological curiosity that the gp63 protein also has RNA ligase activity (Snopek et al., 1977). No matter what the details of the reaction are, gp63 is an excellent candidate for an assembly enzyme which catalyzes a single noncovalent macromolecular assembly reaction.

c. Catalysis of Protein Folding?

Several of the fibrous protein components of the T4 virion also require catalysis for proper assembly. These are the two halves of the tail fibers and the tail spikes, built from gp34, gp37, and gp12, respectively. The assembly of the distal half-fiber from two molecules of gp37 has been most well studied. Two other T4 gene products, gp38 and gp57, are needed for the physical appearance of fibers with mature antigenic properties (Ward and Dickson, 1971; Kells and Hazelkorn, 1974; King and Laemmli, 1971). These proteins are not incorporated into the fiber and appear to be required for the parallel dimerization of the gp37 polypeptide. Since this type of dimerization necessarily results in stiff rod formation (see Section IV.C) and no thin rods are seen in the absence of dimerization, Wood (1979) has suggested that folding of the gp37 polypeptide may accompany dimerization. Thus, one could consider gp57 and gp38 to promote the concomitant folding and dimerization of gp37, but covalent modification of gp37 by gp38 remains possible (Reide et al., 1985b).

The mechanism by which this is accomplished is obscure, although Wood (1979) has speculated that they could function to transiently align critical regions of two gp37 molecules. gp57 function is particularly obscure since it is required for assembly of proximal half-fibers and spikes as well as distal half-fibers, and a host mutant has been isolated that obviates the need for gp57 (Revel et al., 1976). gp38, on the other hand, is required for assembly of only the distal half-fibers and a phage mutant that bypasses the need for gp38 maps in the C-terminal end of gp37 (Bishop and Wood, 1976). This lends tentative support to models in which there is direct contact between gp38 and a distinct portion of the gp37 polypeptide. It is not clear why assembly of these fibers requires this type of catalysis. There is no evidence for catalysis of the multimerization of the fibers or spikes in phages P22, λ, and φ29 or for the longer fibers of T7 and T3.

d. Scaffolding Proteins

Scaffolding protein is a major internal component of the P22 prohead (Earnshaw et al., 1976). It has been shown to participate in an average of six rounds of phage assembly and so clearly performs a catalytic function (King and Casjens, 1974). This catalysis is different from that performed by T4 gp38 or gp63, because it does not affect a single assembly step. Instead, the scaffolding protein adds to the growing prohead structure at an early stage, remains bound to the stable assembly intermediate (proheads), and is released at a later assembly step. This catalysis is complex with a large number of scaffolding protein molecules (200–250 in P22) participating in each event; they apparently do not form a prohead core structure in the absence of coat protein (Casjens and King, 1974; Fuller and King, 1980, 1982). If scaffolding protein is defective, head shell assembly is aberrant, and shells with incorrect curvature accumulate. A few coat protein head shells of apparently proper dimensions are made in the absence of scaffolding proteins, but they contain no proximal vertex proteins (Earnshaw and King, 1978). Thus, the scaffolding protein appears to function both in the initiation and growth of the prohead structure, perhaps as a transiently stable, coordinately assembled template (see Section IV.B.2 for a more detailed discussion of these steps). Although the scaffolding function is catalytic in the case of P22 (and probably φ29—Nelson et al., 1976), it apparently need not be. In λ and T4, the assembled scaffolding proteins are proteolytically degraded instead of being released intact.

3. Physical Chemistry of Phage Assembly and Atomic Resolution Structure

Although one of the ultimate goals of the study of phage assembly is to understand the physical principles that govern the individual reactions, only a few assembly steps have been studied with regard to their kinetics or thermodynamics, and no atomic resolution structures of dsDNA phage assembly components have been determined by X-ray crystallography. These sorts of studies have usually been severely limited by the amount of material available, but this situation is rapidly being rectified by molecular cloning techniques, and we expect progress in these areas in the near future.

a. Kinetics of Assembly

Bloomfield and co-workers have studied the detailed kinetics of head-tail joining and tail fiber attachment in T4 (reviewed by Bloomfield, 1983). Careful measurements of the rates of joining have shown that the reaction has an activation energy ($\Delta E^{\ddagger}$ = 4.1 kcal/mol) and activation entropy ($\Delta S^{\ddagger}$ = -12.6 cal/mol-degree) expected of a diffusion-controlled reaction where the activated complex is quite spatially restricted relative to the reactants. This is expected, since the T4 head-tail joining reaction occurs spontaneously, and strong restraints are expected in the relative orientation of the heads and tails. Salt and pH dependence studies indicate that even though heads and tails both have an overall negative charge, the reactive sites do not have similar charges, and a group that titrates at pH 6.8 (such as histidine) participates in the joining reaction. Also as expected, the rate of successful joining events compared to the expected rate of physical encounters is far less than 1, about 1 in 600 (Aksiyote-Benbasat and Bloomfield, 1975, 1981, 1982; Bloomfield, 1983). Bloomfield and colleagues have also found that T4 fiber addition is noncooperative (addition of one fiber does not affect the rate of addition of the next one) and that the presence of whiskers (gp*wac*) most likely accelerates the initial phage particle-fiber binding and then allows the fiber tip to contact the baseplate by rapid rotational diffusion (Bloomfield and Prager, 1979; Bloomfield, 1983). The kinetics of the addition of T4 decoration proteins have also been studied. Bindings of gp*soc* and gp*hoc* are independent of one another; gp*hoc* binding is noncooperative, and gp*soc* binding kinetics are complex, possibly indicating negative cooperativity (Ishii *et al.*, 1978).

b. Thermodynamics of Assembly

The study of the energetics of assembly of small viruses such as tobacco mosaic virus was instrumental in the discovery that solvent-ordering effects *could be* important in assembly reactions (i.e., loss of entropy due to release of bound water is a major energetic contribution to the reaction) (Lauffer and Stevens, 1968). Nonetheless, we have little knowledge concerning the thermodynamics of phage assembly steps. Studies that have been done in this area have concerned the T4 capsid. Studies with the uncleaved coat protein, gp23, in polyheads have suggested that, like tobacco mosaic virus, its assembly is "entropically driven" (Van Driel, 1977; Caldentey and Kellenberger, 1986). Ross *et al.* (1985) have studied the structural change (lattice transition) resulting in expansion of the T4 polyhead by comparing unexpanded gp23 and expanded gp23* structures. The cleavage/expansion reaction stabilized the structure considerably to thermal denaturation, owing to a substantial negative enthalpy (ΔH) change. Since the entropy change upon cleavage/expansion is negative, this expansion is enthalpically driven. This was interpreted to mean increased Van der Waals and hydrogen bonding with accompanying increases in solvent ordering due to increased surface area (see Ross and Subramanian, 1981). Binding of gpsoc caused an additional stabilization of the gp23* structure, but gphoc binding had little effect. Calorimetric measurements by Arisaka et al. (1981) have also shown that the T4 tail contraction is a strongly exothermic process ($\Delta H \approx -25$ kcal/mol gp18). These studies remind us that both head shells and contractile tails are clearly built as metastable structures, which then rearrange to more stable structures upon receiving the proper stimulus (see Section IV.D.3). Many more such thermodynamic and kinetic studies are necessary to determine the detailed mechanisms by which various macromolecular assembly and movement steps occur.

B. Size Determination

The problem of size determination in phage assembly can be considered for two different categories of structures. The first includes assemblies with subunits whose structure contains all the information needed for their correct assembly. A good example of this is the T4 tail baseplate where, at least in principle, the set of specific bonding interactions among the ~150 polypeptides that make up the baseplate are sufficient to completely specify the form (and therefore the size) of the structure. This structure is *closed* in the sense that all of the mutual bonding "appetites" of the subunits are satisfied by the bonds that are formed. In cases such as this, where specific protein-protein interactions among the components of a structure unambiguously determine the number of subunits and their arrangement in the structure, the problem of size determination is solved simply as a by-product of satisfying all the components' bonding appetites.

More complex cases of size determination are those in which the subunits that make up a structure are not intrinsically limited to assembling into that particular structure but are capable of making a variety of sizes and/or shapes of structures. A clear example of this is given by the helical tubes of the tails of phages such as λ or T4. Here, it seems likely that subunits capable of polymerizing into a helical tube would not intrinsically "know" when to stop polymerizing, since each new subunit adding to the end of the helical tube would present the same face for further polymerization as did its predecessor. This supposition is borne out by the observation that under some conditions of in vitro assembly, or in certain mutant infections, the tail tube subunits are able to polymerize to as much as 100 times the length of a normal tail (Easterbrook and Bleviss, 1969; Katsura, 1976; Tschopp and Smith, 1978).

The subunits of the phage head are also capable of incorrect assembly, as shown by the wide assortment of tubular structures and irregular "monster" forms that result from phage infections in which certain genes carry mutations. In these cases, the basic structure of the protein lattice that makes up the head is (apparently) normal; the defect lies in the absence or incorrect placement of the fivefold corners within the sixfold protein lattice, or in incorrect cylindrical folding of the lattice.

For both the head lattice and the tail tube cases, the polymerizing subunits do *not* contain all the information required for correct assembly, so the fact that they normally assemble only into the correct structures implies that something external to the subunits must be directing their assembly. In the cases where it has been identified, this external agent is another protein or set of proteins.

1. Tail Length Determination

For the dsDNA phages with long tails, tail length is determined sufficiently precisely that for most phages no variation in tail length can be measured in the population. Several different models have been proposed over the years to explain how this is accomplished. It is now likely that for all of these phages, the length of the tail shaft is determined by a length-measuring template protein, or "tape measure" protein. This mode of length determination, first suggested for T4 by King (1971), was established for λ in experiments showing that two internal in-frame deletion mutants of gene *H* (the gene that encodes the tape measure protein) produce phages with shorter tails than wild-type (Katsura and Hendrix, 1984). A comparison of several other phages of the same morphological type as λ suggests that they also have tape measure proteins homologous to λ's and that the lengths of their tails are proportional to the sizes of their tape measure proteins (Youderian, 1978; P. Youderian and J. King, personal communication; M. Popa and R. Hendrix, unpublished). Recently Katsura (1987) isolated an additional 12 in-frame deletions in λ's gene *H* which remove up to 60% of the protein sequence. These phages have tails that, though they are defective in DNA ejection, assemble normally and have lengths proportional to the sizes of their deleted gene *H* proteins. The proportionality constant relating the molecular weights of the tape measure proteins to the lengths of the tails in the series of studies just cited suggests that the tape measure protein is predominantly α-helical, and this is supported by the largely α-helical secondary structure prediction obtained from the sequence of the λ protein. The details of the λ sequence have been used to argue that the six copies of gp*H* found in each tail are wound around each other in a six-stranded coil of α helices that runs through the center of the tail tube (Katsura and Hendrix, 1984).

For phage T4 it has not yet been possible to obtain direct evidence for a protein with a tape measure function, but Duda and Eiserling (1982) and Duda *et al.* (1985) have detected material in the tail tubes that is removed by guanidine treatment, which could be a tape measure protein. Two proteins, gp48 and gp29, are possible candidates for this tape measure protein. Of these, gp29 is big enough to span the length of the tail as a largely α-helical molecule, but gp48 would need to be considerably more extended.

Although the importance of a tape measure protein in tail length determination seems clear, the mechanism by which it limits polymerization of the tube subunit is still obscure. Once the tail shaft has polymerized to the correct length, other minor "tail completion" proteins add to the end and stabilize it. In λ, one of these is gpU; in the absence of gpU, tails grow to the correct length and pause but can eventually lose the tape measure protein and continue polymerization to yield arbitrarily long tails (Katsura, 1976). For both the λ and T4 cases, the proteins implicated as tape measure proteins are assembled as part of the basal structure (baseplate) prior to tail tube polymerization (Kikuchi and King, 1975c; Katsura and Kühl, 1975). However, the hydrodynamic properties of these structures—and in the case of T4, direct electron microscopic examination as well—argue against the tape measures' being in an extended form prior to tube assembly. Thus it seems likely that the tape measure proteins are synthesized and assembled into the tail tip as predominantly globular molecules that unfold into their extended, tape-measuring conformation during the process of tube polymerization. In the case of the λ tape measure protein, there is evidence that it has an important role in ejection of the DNA and may in fact pass into the cell along with the DNA (Scandella and Arber, 1976; Elliot and Arber, 1978; Roa, 1981; Roessner and Ihler, 1984).

2. Head Size Determination

The problem of head size determination is the assembly of the major capsid protein into an icosahedral shell of the correct diameter (i.e., of the correct triangulation number) and, for phages with prolate heads, to determine the correct length of the head as well. There appear to be three main contributors to the process of size determination—the major capsid protein itself, the protein scaffold around which the capsid assembles, and the initiation process by which correct assembly begins.

a. Capsid and Scaffolding Proteins

The capsid protein carries in its own structure much but not all of the information needed for its correct assembly. In infections by λ or P22 in which the scaffolding protein has been removed by mutation, some prohead-size shells assemble (Ray and Murialdo, 1975; Earnshaw and King, 1978), arguing that the scaffold is not absolutely required for making a shell of the right dimensions. However, some of the capsid protein in these infections fails to assemble, and some of it assembles into aberrant structures. These include small-diameter, apparently closed shells (probably T=4 rather than the normal T=7) and "monster" and "polyhead" forms in which the fivefold centers in the protein lattice that make the corners of a normal icosahedral capsid are irregularly spaced or missing. These results suggest that the scaffold for these phages acts to improve the efficiency and accuracy of capsid assembly.

In all the assembled forms of capsid protein that have been observed—both normal and abnormal—the structures that are made have curved surfaces. This is taken to mean that the capsid protein has an intrinsic propensity to assemble into such curved surfaces. We think of this "intrinsic curvature" as being the primary determinant of capsid size; the other participants (scaffolding proteins and initiation proteins) appear to modify or refine the expression of this curvature in order to allow efficient and accurate assembly of the correct structure. Experiments with λ by Katsura (1983a) support this view: he isolated mutants in the major capsid protein gene, *E*, which cause efficient assembly of small (T=4) heads (Fig. 10). Like wild type, the proheads made by these mutants initially contain scaffolds, and these are smaller than the scaf-

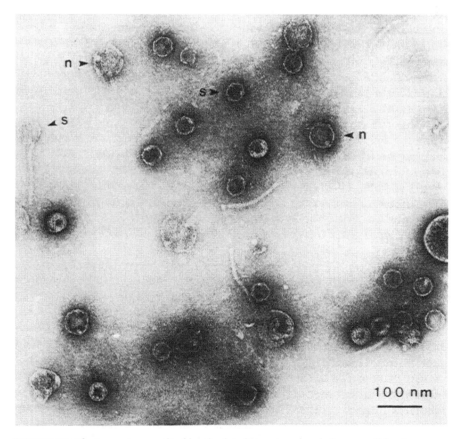

FIGURE 10. Electron micrograph of head-related structures made by a small head mutant in the λ *E* gene, taken from Katsura (1983a). Arrows indicate normal (n) and small (s) shells.

folds found in wild-type proheads, despite the fact that the scaffolding protein is itself wild-type. We can understand these results by assuming that the mutant capsid protein has a greater intrinsic curvature than the wild-type protein, and that this greater curvature imposes a smaller size on the scaffold. These experiments argue against a model of capsid assembly in which the scaffold assembles first and the capsid protein assembles around it to a size determined by the size of the scaffold.

In the phages with prolate heads, the average local curvature of the capsids varies from one part of the shell to another. As a result, we might expect the roles of the scaffolding and initiation proteins in modulating capsid curvature to be relatively more important for these phages, and available data do seem consistent with this expectation. In φ29, if the portal protein, gp10 (one of the initiation proteins), is missing, the capsid that assembles is isometric rather than prolate (Hagen et al., 1976). One way to understand this is to imagine that the portal structure holds the initial capsid subunits that polymerize around it in a less curved configuration than they would take by themselves. More capsid proteins are then required (assembling with the preferred curvature) than would be required in the absence of gp10 before the shell can close on itself. In a T4 infection with a scaffolding protein (gp22 or gp67) removed by mutation, open-ended polyheads (tubes) are assembled (Yanagida et al., 1970; Volker et al., 1982). These structures have diameters that are different from those of the normal prohead, arguing that in this case the scaffold has an important role in specifying the correct capsid diameter.

Another example showing that the size of the capsid is not determined solely by the capsid protein itself is provided by bacteriophage P2 and its satellite P4. In a P2 infection, heads of triangulation number 9 are assembled. P4 has no capsid protein gene of its own, and in cells containing P2 and P4, P4 causes the P2 capsid proteins to assemble into smaller T=4 heads (Giesselsoder et al., 1982). P4 mutations have been isolated that alter its ability to direct the assembly of the P2 capsid protein, but the mechanisms involved are unclear (Shore et al., 1978; Sauer et al., 1981).

In T4, it has been shown both in vitro and in vivo that the scaffold is capable of assembling to the correct prolate dimensions in the absence of the major capsid protein, gp23 (Van Driel and Couture, 1978; Traub and Maeder, 1984), and in fact these naked scaffolds can apparently be chased into infectious phage if gp23 is provided after they are assembled (Kuhn et al., 1987). This suggests that the scaffold must have an important role in determining head length. It could even be imagined on the basis of these results that for T4 the scaffold is the sole determinant of head length, but other experiments argue that the situation is more complex. Heads of the incorrect length ("petits" or "giants") can be caused by missense mutations in any of three genes—22 (major scaffolding protein), 23 (major capsid protein), or 24 (a capsid protein that occupies the icosahedral corners) (Black and Showe, 1983). The scaffolds within the long and short proheads are correspondingly long or short, which means that mutations in either the scaffolding protein or in the capsid proteins can alter the length of *both* the capsid and the scaffold. These results seem more consistent with the view that neither the capsid nor the scaffold is dominant in determining length but that they collaborate to do so. Other genetic studies reinforce this view. Showe and Onorato (1978) altered the ratios of gp22 to gp23 in infected cells and found that petit heads or giant heads formed. Doherty (1982) showed that mutations in gene 23 that cause assembly of giant heads could be reverted by second-site mutations in gene 22 and gene 24. These mutations were allele-specific, arguing that these capsid and scaffolding proteins interact directly with each other (presumably as components of the assembling prohead) to cooperatively determine head length.

The picture that emerges from these various studies with both isometric and prolate heads is that the scaffold and capsid *coassemble*, that both contain information about the size and shape of the final structure, and that each influences the other to assemble into the correct structure (or, in the case of mutants, the incorrect structure). The detailed natures of the physical and chemical interactions among capsid and scaffolding proteins through which this assembly takes place are almost completely unknown (see Section IV.C).

b. Initiation

Another important contributor to correct head assembly is the initiation process. Initiation of head assembly occurs at the proximal vertex of the shell. This conclusion is based largely on results of *in vitro* complementation studies, which place the proteins involved with the proximal vertex at the early part of the assembly pathway, on the phenotypes of infections in which those proteins are absent, and on demonstrated complexes between certain proteins. Phage proteins implicated in initiation by these sorts of criteria include the portal proteins, the scaffolding proteins, and various minor proteins thought to be associated with the portal vertex. Among these latter proteins are λ's gpC, which may be located in a ring around the portal structure (Murialdo and Ray, 1975); T4's gp40, a membrane protein thought to have a role in anchoring the portal protein to the membrane during the initial stages of shell assembly (Hsiao and Black, 1978); and T4's gp31, a protein required for solubilizing the major shell protein, gp23 (Laemmli *et al.*, 1969).

In addition to the phage proteins involved in initiation, two host-encoded proteins have been implicated in the process. These are the products of the *E. coli* genes groEL and groES. The GroEL and GroES proteins are abundant proteins of *E. coli*; they are among the *E. coli* heat shock class of proteins (Lemaux *et al.*, 1978; Neidhardt *et al.*, 1981; Drahos and Hendrix, 1982; Tilly *et al.*, 1983), and their synthesis is also stimulated moderately by λ infection (Kochan and Murialdo, 1982;

Drahos and Hendrix, 1982). Both proteins are found as ringlike homo-oligomers—GroEL as a 14-subunit double-ring structure with sevenfold rotational symmetry (Hohn *et al.*, 1979; Hendrix, 1979), and GroES as a ring with approximately 6–8 subunits (Chandrasekhar *et al.*, 1986). The two proteins form an ATP-dependent complex *in vitro* (Chandrasekhar *et al.*, 1986), and genetic evidence argues that they interact *in vivo* as well (Tilly and Georgopoulos, 1982). A requirement for groE function in head assembly has been demonstrated explicitly only for certain phages of *E. coli* (λ, T4, and a few λ relatives), but the GroEL protein is found to be very highly conserved among a wide variety of eubacterial species, including the hosts for phages φ29 (Carrascosa *et al.*, 1982) and P22 (R. Hendrix and S. Casjens, unpublished), so the GroE proteins' participation in phage assembly could be general. [Interestingly, groE function is also required for phage T5 assembly, but for tail rather than head assembly (Zweig and Cummings, 1973).]

The evidence that the GroE proteins have a role in initiation of head assembly is strongest for λ. Infection by λ of a cell carrying a missense mutation in one of the *groE* genes results in the production of normal tails, but the head protein is assembled into head-related monsters and proheadlike structures that cannot be matured to functional heads (Georgopoulos *et al.*, 1973). The λ portal protein, gpB, interacts directly with at least the GroEL protein, as shown by both genetic (Georgopoulos *et al.*, 1973) and biochemical (Murialdo, 1979) criteria, and there is evidence that assembly of gpB into the 12-subunit portal structure requires groE function (Tsui and Hendrix, 1980; Kochan and Murialdo, 1983). Of the steps in λ head assembly that are known, the interaction of gpB with the GroE proteins is the earliest (Murialdo and Becker, 1978b).

In a T4 infection of a *groE*⁻ cell, the major head protein, gp23, does not assemble into recognizable head-related structures as with λ, but instead accumulates in amorphous, insoluble "lumps" associated with the cell membrane (Laemmli *et al.*, 1969; Georgopoulos *et al.*, 1972). Simon (1972) observed similar lumps in a T4 infection of wild-type cells and argued on the basis of the kinetics with which they appeared that they may be a normal intermediate in T4 head assembly. This leads to the interesting suggestion that one role of the GroE proteins could be to solubilize or refold the proteins that are participating in assembly, a function similar to one proposed recently for some of the eukaryotic heat shock proteins (Pelham, 1986). At least one other phage, the lambdoid phage HK97, appears to form insoluble aggregates of the major head protein in the absence of a wild-type groE function (M. Popa and R. Hendrix, unpublished).

No head assembly initiation complex has been observed directly, and our ideas about what such a complex might look like are still rather vague. We imagine that it must include the portal protein, probably in some cases additional minor proteins associated with the proximal vertex, and the initial copies of the scaffolding protein(s) and major shell protein. The GroE proteins may also be present. As an already assembled complex, these proteins provide an attractive candidate for the site of nucleation of coat protein assembly, and this hypothesis receives indirect support from the observation that GroEL protein is sometimes found associated with immature λ proheads (Hohn et al., 1979; Kar, 1983).

Two possibilities have been discussed to rationalize why it may be advantageous to initiate head assembly from a specific point in the structure. First, it is proposed that the assembling proteins—particularly the major shell protein—are able to add to an initiation complex at a lower concentration than is needed for them to coassemble spontaneously. Thus, the initiation complex would provide a kinetically favored pathway for assembling the proteins into the correct structure, at the expense of aberrant structures that assemble without an initiation complex (and therefore require a higher subunit concentration) (Murialdo and Becker, 1978a; Black and Showe, 1983). Support for this view comes from observations of the kinetics of appearance of polyheads made in T4 mutant infections (Laemmli and Eiserling, 1968; Laemmli et al., 1970); in cases where the mutation is thought to affect initiation, the polyheads appear late, but formation of polyheads in which the initiation apparatus is intact takes place with the same kinetics as head formation in a wild-type infection. Second, if shell assembly occurs only by initiation upon a complex that will become the proximal vertex, then the "problem" of how to assure that each head has one and only one proximal vertex is solved as a direct result of the assembly scheme.

C. Nonequivalent Protein-Protein Interactions

One of the most celebrated properties of proteins is the specificity of their interactions with other molecules. For enzymes, this specificity is usually expressed in terms of the selectivity of the enzyme for a particular substrate molecule; for structural proteins, it is expressed in terms of the specificity of the bonding contacts they make with their neighbors in the structure. Phage structural proteins provide many examples of highly specific binding; however, in this section we will consider proteins that exhibit less stringent—or perhaps more accurately, *expanded*—specificity.

Two identical protein subunits that make identical bonding contacts with their neighbors are said to be "equivalent." Two equivalent subunits (along with their bonding contacts) are *exactly* superimposable on each other, and, if they are part of the same structure (a phage head or tail, for example), they are related by some symmetry operation (a rotation axis or a screw axis). A clear example of equivalent subunits is seen in the phage T4 tail sheath. Here all the individual subunits are related by a screw axis and could therefore, by translation along and rotation around the screw axis, be precisely superimposed on each other, with all their intersubunit bonds matching. On the other hand, chemically identical subunits that make different bonds to their neighbors (and which are consequently not related by a symmetry operation) are said to be "nonequivalent."

As mentioned above, Caspar and Klug (1962) introduced the term "quasiequivalent" to describe slight deviations from strict equivalence. They predicted that such quasiequivalence should be found in icosahedral virus capsids, where there is a mathematically imposed upper limit of 60 on the number of strictly equivalent (symmetry-related) subunits that can be arranged in a shell. Quasiequivalence is usually taken to mean that the subunits in question make the same or very similar bonds to their neighbors but are conformationally slightly different from each other. An easily visualized example of what is meant by quasiequivalence is seen in the long, flexible tails of phages such as λ. Here all of the subunits are strictly equivalent (symmetry-related) as long as the tail is straight. However, if the tail is allowed to bend into an arc, the perfect symmetry axis is destroyed, and the subunits can no longer all be strictly equivalent. We presume that the individual subunits accommodate the slight distortions that the bending imposes by flexing and/or by slightly distorting the intersubunit bonds but not by breaking the intersubunit bonds and forming different ones. Thus, these subunits are not equivalent but quasiequivalent while the tail remains bent. These various relationships among subunits are illustrated in Fig. 11.

1. Coat Proteins in the Head Shell

For phages with more than 60 head shell subunits per virion, it is not clear whether these subunits occupy quasiequivalent or nonequivalent positions in the capsid, and this is unlikely to be settled until high-resolution structures of these capsids are determined. Among viruses with this level of capsid complexity, only the structures of polyomavirus and adenovirus have been studied to a sufficiently high resolution by X-ray diffraction to be able to determine details of the coat protein arrangement. Both of these have the surprising property of not having hexamers in the positions that Caspar and Klug's (1962) theory predicts they should. Polyomavirus, T=7 *dextro*, has pentamers, whereas adenovirus and probably frog virus 3, T=25, and about 147, respectively, have trimers at the positions surrounded by six bonding partners (Rayment *et al.*, 1982; Burnett *et al.*, 1985; Darcy-Tripier *et al.*, 1986). Rayment (1984) and Burnett (1985) have suggested that these nonequivalent arrangements could be used in the assembly process to help dictate the structure of the larger capsids. These findings lead us to the question, is it *ever* possible to assemble a T≥7 virus with physical hexamers at these positions? Electron micrographic analyses of phages λ, P22, T7, and extralong T4 virions and calculations from the masses of the coat proteins and capsids clearly show that trimers or pentamers are not present at the hexamer positions

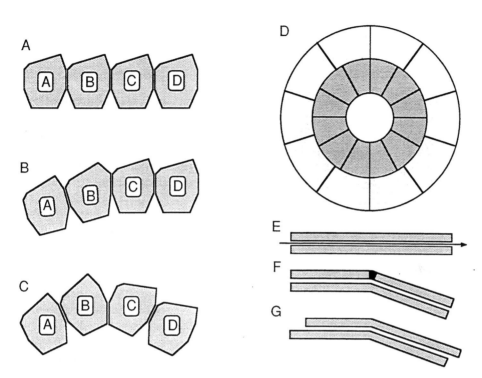

FIGURE 11. Equivalence, quasiequivalence and nonequivalence in protein superstructures. (A) Four asymmetric proteins bonded head to tail in an identical fashion. The intersubunit bonds are all identical or *"equivalent."* (B) Subunit 2 in the array from (A) above is deformed slightly to allow a bend in the array. Since bonding surfaces used by the proteins are still all the same, but the proteins are not in identical conformations, the proteins are said to be *"quasiequivalent."* (C) New bonding surfaces on subunits 2 and 3 are used to allow a bend in the array. Since the surfaces participating in the 1–2 bond and the 2–3 bond are different, these bonds are *"nonequivalent."* (D) When two multimeric protein surfaces interact, if the two different individual subunit's interacting surfaces are different sizes or the two oligomers have different symmetry, the bonds between the two oligomers (in this case 2 rings of 10 and 12 subunits, as might occur between the portal protein oligomer and the rest of the head) are necessarily of several *nonequivalent* types, since the subunits cannot stay in register. (E) Two elongated protein subunits *equivalently* bonded as might occur in T4 tail fiber subunits. The twofold rotational axis is indicated; notice that if the polypeptide has its N terminus at one end and its C terminus at the other, any portion of one polypeptide interacts with the identical region on the other. (F) The subunits in (E) bent slightly. The strict twofold symmetry of the dimer is destroyed (although 2 *local* symmetry axes remain), and one subunit must be stretched (indicated in black) to accommodate the bend. These two subunits can be said to be *quasiequivalently* bonded, since the actual contacts are nearly the same as in (E). (G) The same two proteins *nonequivalently* bonded; in this case one has been moved relative to the other so that identical regions on the two proteins do not interact with one another, and bends occur in different locations in the two subunits.

in these virions. In fact, these analyses suggest that it is very likely that hexamers are indeed present (Williams and Richards, 1974; Casjens and Hendrix, 1974a; Casjens, 1979; Steven *et al.*, 1983; Ishii and Yanagida, 1975; Aebi *et al.*, 1977).

Thus, these viruses remain candidates for having capsids containing the *quasiequivalent* coat proteins predicted by Caspar and Klug (1962). On the other hand, Burnett (1985) has pointed out that it seems easier to explain their definite icosahedral shape by postulating *equivalent* interactions within each face and different or *nonequivalent* interactions across the edges between subunits on different faces. It remains possible that preparing specimens for electron microscopy artifactually enhances the icosahedral appearance, but new methods of sample preparation have made this unlikely (Adrian *et al.*, 1984). An extension of Burnett's reasoning might suggest that the unexpanded proheads of these phages, which appear to have a nearly spherical shape, may be quasiequivalently bonded, even if the mature forms are not.

2. Symmetry Mismatches

Whereas answers to these questions for head shells must await high-resolution structural information, for various other features of phage and phage precursor structures the symmetries and shapes that can be observed at the resolution of the electron microscope allow us to infer strongly that chemically identical subunits make nonequivalent interactions—that is, that two or more chemically identical subunits in a structure interact in distinctly different ways with their neighbors. The clearest indications of nonequivalent interactions are those cases in which the phage structures show mismatches between the rotational symmetries of their parts. The first such mismatch to be noted (Moody, 1965) was the mismatch between the T4 tail, which has sixfold rotational symmetry, and the head shell, which has a fivefold axis through the corner to which the tail attaches. This symmetry mismatch is found in all the tailed phages for which information is available and is thought to occur generally. The actual transition between these "clashing" symmetries is now known to be located within the head, at the interface between the shell lattice (5-fold) and the portal structure (6- or 12-fold) to which the tail attaches (see Fig. 2).

The portal structure, in all five cases that have been characterized, is made of 12 protein subunits arranged with apparent 12-fold rotational symmetry (Fig. 4). The 12-fold symmetry axis coincides with one of the fivefold axes of the head shell; this means that each of the 12 portal subunits is in a different geometric relationship to the apposed subunits of the head shell. A structural feature that is as well conserved among phages as is this symmetry mismatch would seem likely to be of some functional importance, but direct evidence for such a function does not exist. As one approach to understanding the function of the mismatch, it has been noted (Hendrix, 1978) that the existence of the symmetry mismatch means that the portal structure may be able to rotate with respect to the shell. This results from the fact that the bonding energy with which the shell and portal interact is the sum of interactions between individual shell and portal subunits. If the portal rotates around its axis (and the shell remains stationary), the strengths of these individual interactions will change out of phase with each other. As a consequence, the total bonding energy between shell and portal structure should be relatively independent of angle, and therefore there should be no strongly preferred orientation that would block rotation. As one possible consequence of this, we have postulated that the hypothesized ability of the portal structure to rotate may be a central feature of the mechanism of DNA packaging in these phages (Hendrix, 1978). In a more general context, we imagine that mismatched rotational symmetries could provide a useful design for constructing rotating joints in other biological structures as well.

A related symmetry mismatch occurs between the head shell and the scaffold. The scaffold has a sixfold symmetry axis aligned with the same fivefold axis of the shell that clashes with the portal structure (Paulson and Laemmli, 1977; Engel et al., 1982). (The 6-fold symmetry of the scaffold has thus far been shown only for T4, but it could be a general feature.) This mismatch is especially interesting in light of the role of the scaffold discussed above. We indicated that the scaffold and the shell probably polymerize simultaneously and interact in some way that promotes correct assembly of both. As with the portal-shell mismatch, the mismatch in this case means that individual shell-scaffold subunit interactions are nonequivalent. This situation places constraints on how we can imagine the assembly process taking place; for example, it argues very strongly against a model in which a shell and a scaffold subunit form a heterodimer which then assembles into the growing prohead (and there is in fact no evidence for such heterodimers for any of these phages). We suggest that it may be the global properties of the scaffold surface (e.g., charge, hydrophobicity, etc.), rather than specific shell-scaffold subunit interactions, that help direct the course of shell polymerization.

A third mismatch in rotational symmetries *may* occur in the early stages of head assembly, between the host *GroEL* protein and the portal structure. *GroEL* protein appears to be required for assembly of gpB, the λ portal protein, into the 12-fold symmetrical portal structure, and the two proteins have been shown to form a transient complex (Murialdo, 1979). The *GroEL* protein oligomer has 14 identical subunits arranged with sevenfold rotational symmetry (Hendrix, 1979; Hohn et al., 1979). What bearing the nonequivalent gp*GroEL*-gpB interactions suggested by this symmetry difference may have on the mechanism of gpB assembly may become clearer when more is learned about the structure of the gp*GroEL*-gpB complex.

A somewhat different type of relationship among subunits that might also be classified as a symmetry mismatch takes place between the T4 gp23* (major shell subunit) and gphoc proteins. A single subunit of gphoc attaches to the shell at the center of each gp23* hexamer (Aebi et al., 1974), giving a sixfold versus onefold "mismatch." Pending more detailed structural information, it seems reasonable to postulate that there are gphoc-binding sites on each of the gp23* subunits in the hexamer but that the first gphoc to bind to one of them occludes the other five sites, thus limiting to one the number of gphoc subunits able to bind to the hexamer. An interesting consequence of this view is that it suggests that the specific site among the six that is used by gphoc may be chosen randomly and therefore that an individual T4 virion may differ from its siblings with respect to the constellation of binding sites used by the ~160 gphoc molecules on its surface. Although the details are less well understood, a similar description may apply to the attachment of the head fibers of φ29 (Reilly et al., 1977).

3. Asymmetric Oligomers

Another type of nonequivalent bonding is possible for certain of the phage tail fibers; this involves a small number of identical protein subunits bonding nonequivalently to each other. The prototype for this type of structural arrangement is the yeast enzyme hexokinase, for which a high-resolution structure is available (Steitz et al., 1977). Hexokinase forms a dimer of two chemically identical polypeptide chains that are bound to each other in such a way that the dimer lacks a dyad symmetry axis—that is, the subunits are nonequivalent. The effect of this asymmetric binding on the biochemical activities of the subunits is substantial: the catalytic sites on the two subunits appear to bind the substrate differently, and there is a single, asymmetric allosteric effector binding site made from parts of both subunits.

Among phage tail fibers, there are examples of apparently equivalently bonded as well as nonequivalently bonded structures. What allows us to distinguish these types at the electron microscopic level of resolution is that equivalently bonded subunits *must* be related by a symmetry axis. For extended structures like tail fibers, it is often possible to tell by inspection whether the shape of the structure is consistent with the presence of such a symmetry axis. Phages T3 and T7 have six tail fibers (Serwer, 1976; Kato et al., 1985b), each consisting of three identical subunits; each tail fiber is bent at a point near the middle so that the tip points away from the head. For the three subunits to be equivalent, they would have to be symmetrically arrayed around a threefold axis; such a structure *could not* be bent (Kocher, 1979), so we are led to the conclusion that the three subunits are not strictly equivalent. We believe that the apparently consistent direction of the bend and its relatively constant magnitude argue that the trimers are *nonequivalently* bonded in the sense described in Fig. 11G; however, at the current level of resolution, the possibility also remains that they are *quasiequivalent*, as shown in Fig. 11F. In either case, it appears that for these fibers, as with hexokinase, nonequivalent (or quasiequivalent) bonding of identical subunits allows construction of a structure that is more complex (and, we presume, better suited to its function) than would be possible with strictly equivalent bonding.

The tail fibers of T4 are also bent, but in this case they are constructed of not one but four separate types of protein subunits (King and Laemmli, 1971; Ward and Dickson, 1971). The bend probably occurs at the position occupied by the single copy of gp35. The best-studied part of this tail fiber, the distal half fiber, contains two copies each of gp36 and gp37. This part of the fiber appears to be perfectly straight, consistent with the idea that the two members of each pair of subunits are equivalent as well as chemically identical; the dyad axis that is demanded by this view of the fiber's structure would be coincident with the half-fiber axis.

D. Structural Movements and Molecular Machines

Unquestionably, a major function of the virion proteins is to act as a shipping container for the phage genome during its journey from one host cell to the next. However, to think of the capsid as no more than an inert box is to ignore some of its most interesting and informative aspects. Phages have provided several clear examples in which protein assemblages, either in the mature virion or in virion assembly intermediates, undergo elaborate structural rearrangements as essential parts of the assembly process or of virion function. In some of these cases at least, it seems legitimate to refer to the rearranging structures as "molecular machines." As with other aspects of phage structure and function, it is likely that as we learn more about the mechanisms by which these machines operate, we will come to understand general principles about what proteins and the macromolecular structures they form are capable of doing.

1. DNA Packaging Apparatus

One of the most intriguing of the phage's machines is the one that packages DNA. It now seems most likely that transport of the DNA into the head shell is mediated by a complex of proteins at the portal vertex, including at least the portal structure and terminase enzymes, and probably other proteins as well; in addition, ATP is required for packaging, presumably to provide energy for driving the DNA into the head against a thermodynamic gradient. Whatever the chemical and physical mechanisms by which this machine carries out its job, the things it must do include locating the correct place on the replicated DNA to begin packaging, cutting the DNA (except φ29), transporting the DNA into the shell in such a way that it becomes compacted several hundredfold, sensing how full the head is becoming, and determining the correct place on the DNA to make a second cut at the termination of packaging. Various suggestions have been made about how these several integrated functions are accomplished (for which see Casjens, 1985b), and some of the properties of a number of the participating proteins have been studied, but the basic mechanisms of DNA packaging are still elusive.

2. Baseplate Rearrangements and Tail Contraction

The contractile tail of T4 is a particularly flamboyant and well-characterized example of a multiprotein machine that carries out an important function in the phage life cycle. DNA ejection by T4 (and other phages with contractile tails) is accompanied by an extensive rearrangement of subunits in the baseplate, followed by a contraction of the tail sheath, which causes the tail core to be driven into the surface of the cell. The transition the baseplate undergoes is seen most clearly in isolated baseplates; they have a characteristic hexagonal shape that can change spontaneously to a star shape (Simon and Anderson, 1967b) as the result of coordinated movements of the component protein subunits. This transition is thought to be the same as the transition that occurs in the intact tail during contraction. The structures of the two baseplate forms have been analyzed by Crowther *et al.* (1977), using rotational filtration of individual baseplate images (Fig. 12). In addition to defining the mor-

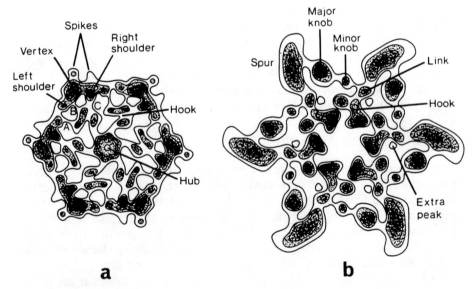

FIGURE 12. Schematic drawings of the anatomy of the hexagon (left) and star (right) forms of the T4 tail baseplate. From Crowther *et al.* (1977).

phology of these structures, they used mutant baseplates to assign three polypeptides—gp9, gp11, and gp12—to specific morphological features. The transition from hexagon to star entails extensive morphological rearrangements. This includes an apparent inward rotation of gp9, the protein to which the long tail fibers are thought to attach; a splaying out of gp12 (short tail fibers) and gp11 (tail spikes); and considerable rearrangement of the other, less specifically identified features. Among the latter changes is the appearance of a hole in the center of the baseplate at the location occupied by the "hub" prior to rearrangement.

Crowther et al. (1977) combined their morphological data with earlier results (Flatgaard, 1969; Yamamoto and Uchida, 1975; Arscott and Goldberg, 1976; Crawford and Goldberg, 1977) suggesting roles for the long and short tail fibers in triggering tail contraction to present a speculative mechanism by which the tail "decides" on the correct time to eject the DNA. In this model, the long tail fibers make contact with their cellular receptors and transmit the information that they have done so to the baseplate through their attachment point, gp9. This activates the baseplate so that it is susceptible to being triggered to enter the hexagon to star transition. The actual transition is triggered by the short tail fibers (gp12), but not until they contact their receptors, thereby completing the cellular contacts needed for successful ejection.

Following the triggering of the baseplate transition, the tail sheath contracts. The bottom annulus of the sheath is in intimate contact with the top surface of the baseplate, and sheath contraction is presumed to be set off by the baseplate rearrangement. Moody (1973) trapped tails partway through contraction by treating them with alkaline formaldehyde. The contracted part of the sheath in these structures was always adjacent to the baseplate, arguing that sheath contraction starts at the baseplate and moves progressively along the sheath toward the head. Contraction results in a dramatic change in the helical parameters of the sheath, resulting in a length decrease of about 2.5-fold, but no apparent change in the tail tube. During contraction, the tube and sheath remain attached at their head ends, and the sheath remains attached to the baseplate (which is attached to the cell), with the result that the tube is driven through the hole that has opened in the middle of the baseplate and through the surface layers of the cell (Simon and Anderson, 1967a).

3. Lattice Transformations

Sheath contraction in a T4-style tail, as described above, is an example of a *lattice transformation*. In this sort of molecular rearrangement, the protein subunits that make up the lattice all change their conformation in such a way that there is a change in the lattice constants that define the spacings and angles in the subunit array, with a consequent change in the overall size and shape of the structure.

The other clear example of a lattice transformation during phage assembly is the expansion of the head shell that occurs at about the time of DNA packaging. In this case, the lattice constants of the (primarily) hexagonal lattice of head protein subunits increases by ~20% (Aebi et al., 1974; Wurtz et al., 1976), which results in an increase of 10–15% in shell diameter and a roughly twofold increase in internal volume of the shell. Since the mass of the shell remains constant, the expansion results in a thinner shell wall (Kistler et al., 1978; Earnshaw et al., 1979). Expansion has been studied extensively by investigating the properties of "polyheads" of T4 and λ, which are much more amenable to image analysis than are the normal head structures; polyhead lattices undergo expansion in a way that is assumed to be analogous to normal prohead expansion. The change in the subunits that is associated with expansion is apparently rather dramatic. Besides the change in lattice dimensions, the surface texture of the polyhead changes (Yanagida et al., 1970; Wurtz et al., 1976), and at least one antigenic determinant that is unavailable prior to expansion appears on the outside surface (Kistler et al., 1978). Similarly, the binding sites for the decoration proteins of the T4 and λ heads become available only with expansion (Imber et al., 1980; Ishii and Yanagida, 1975). Expansion is apparently energetically favorable, since it will happen spontaneously in proheads that are mildly mistreated [e.g., 4 M urea for λ (Hohn et al., 1976), or 0.8% sodium dodecylsulfate for P22 (Casjens and King, 1974)], and the reverse reaction, contraction of expanded shells, has not been observed. As mentioned above, calorimetric measurements of the expansion of T4 polyheads show that this reaction is driven by a substantial change in enthalpy (Ross et al., 1985). In the case of T4 it has been shown that expansion makes the shell more resistant to disruption by sodium dodecylsulfate (Steven et al., 1976). It is not known how normal expansion is triggered in vivo; however, DNA packaging may have a role: λ proheads can package a 5.5 kbp DNA restriction fragment without expanding, but if they instead package a 21-kbp fragment, they expand (B. Hohn, cited in Hohn, 1983). There is also indirect evidence that the λ protein gpC, which is thought to be located at the portal vertex (Murialdo and Ray, 1975), may have some role in triggering expansion (Earnshaw et al., 1979). T4 polyheads that are partially expanded have been seen in the electron microscope (Steven et al., 1976). They show a rather sharp line between the region in which the lattice is expanded and the region where it is not, suggesting that the polyhead has been caught as a wave of lattice transformation was sweeping across it. We imagine that a similar wave of structural change must sweep across the prohead during expansion, possibly starting at the portal vertex, and possibly triggered by events at that vertex associated with DNA packaging.

What function the shell expansion serves is not known, although its universality among these phages suggests that it has some important role. One suggestion (Hohn et al., 1974; Serwer, 1980; Gope and Serwer, 1983) has been that the expansion drives DNA packaging by "sucking" the DNA into the head, but more recent evidence argues against this view (see Casjens, 1985b). Another view, which we favor, is that expansion is an important feature of the assembly of the shell itself. We can imagine that the original conformation of the shell subunit is optimized for efficient assembly but not for maximal stability of the shell that is assembled. The lattice transformation would be the result of a conformational change designed to "tighten up" the structure, and the expansion a secondary by-product of that change. This view seems especially interesting if, as suggested above, the transformation is a change from a structure made of quasiequivalently bonded subunits into a structure in which the subunits are nonequivalently bonded. In this case, the unexpanded form of the structure could provide a route to assembly of the nonequivalently bonded (expanded) form, which it might not be possible to achieve by direct assembly of the subunits.

V. GENES AND MORPHOGENESIS

A. Control of Gene Expression

1. Turning Morphogenetic Genes On

It has been abundantly clear from early in the study of phage assembly that the exact time of expression of the genes that encode the virion structural proteins is *not* generally a critical factor in controlling the order of steps in phage assembly pathways. In the dsDNA phages studied, nearly all of the morphogenetic proteins are coordinately synthesized and made only at late times after infection. Although this coordinate regulation is accomplished by different mechanisms in different phages, in all cases the time of turn-on of the morphogenetic genes is transcriptionally controlled (for the single known exception to this rule, the T4 decoration protein gene, *soc*, see MacDonald *et al.*, 1984). In some cases, for example λ and P22, all of the structural genes lie within a single operon; in other cases, such as T4 and P2, there are multiple late operons, each encoding several morphogenetic proteins.

Large lytic phages, such as T4 and SP01, synthesize proteins early in infection which modify the host RNA polymerase's promoter specificity, thereby causing a switch to late promoters partway through the lytic cycle (reviewed by Geiduschek *et al.*, 1983; Elliot *et al.*, 1984; Losick and Pero, 1981). T4 has a number of late promoters that control the morphogenetic genes, most of which appear to lie in relatively small operons (reviewed by Christensen and Young, 1983). Phage P2 has four late operons which are turned on by an early gene product (Christie and Calendar, 1985). The medium-size lytic phages, T3 and T7, encode as an early protein an entirely new RNA polymerase which expresses the late genes from several promoters not utilized by the host polymerase (Oakley and Coleman, 1977; Dunn and Studier, 1983). The small lytic phage ϕ29 has a single promoter for morphogenetic gene expression that is turned on at late times by a phage-encoded early protein (Mellado *et al.*, 1986). Late gene expression in λ and P22 is controlled in a more complex manner (reviewed by Friedman and Gottesman, 1983; Wulff and Rosenberg, 1983). This is likely due to the fact that these phages can lysogenize their host and must not express the late genes when a decision to become lysogenic has been made. The host RNA polymerase constitutively recognizes the late operon promoter, P_R', but normally terminates at t_R' before it reaches the genes of the late operon. The efficiency of termination at this terminator is controlled in an unknown way by the Q gene protein, whose structural gene is expressed from the early promoter P_R (Grayhack *et al.*, 1985). The synthesis of the Q protein is in turn dually controlled by the *cII* gene protein. The *cII* protein is thought to be central in the lytic/lysogenic growth decision, with high levels of gp*cII* promoting the establishment of lysogeny (Herskowitz and Hagen, 1980). It (1) stimulates the rate of prophage repressor synthesis which in turn shuts off P_R, the promoter from which the Q gene is expressed, and (2) directly stimulates the rate of expression of the promoter P_{aQ}. P_{aQ} initiates an RNA which is antisense to the 5' portion of the gene Q message and which acts as a translational repressor of the Q gene (Hoopes and Mc-Clure, 1985; Ho and Rosenberg, 1985; Stephenson, 1985). Thus, when gp*cII* is made at high levels, the phage makes more prophage repressor, which in turn shuts off P_R and Q gene mRNA synthesis. High gp*cII* also stimulates P_{aQ}, whose RNA transcript depresses translation of any Q message that has been made, resulting in little if any Q gene expression and no late gene expression. Conversely, when gp*cII* is low, P_R is expressed well and P_{aQ} is not expressed, which results in effective Q gene expression, antitermination at t_R', and strong late operon expression.

2. Relative Rates of Expression of Individual Morphogenetic Genes

It has been known for some time that the relative rates of synthesis of the various virion structural proteins usually reflect the amounts of these proteins needed to build virions (e.g., the coat proteins are made in much greater amounts than tail fiber proteins). There are several instances in phages T4 and λ where deviation from the normal protein ratios appears to cause assembly defects (Floor, 1970; Georgopoulos *et al.*, 1973; Showe and Onorato, 1978; Eiserling *et al.*, 1984), but in general it is not known if this careful control of the expression of the morphogenetic genes is done for reasons of economy (it is wasteful to make excess "minor" proteins) or because it would be detrimental to the assembly process *per se* to have the proteins present in incorrect ratios. It should be mentioned that there are also a few cases of structural proteins being made in apparently large excess. One clear example is the λ gene U protein, which is made in about 50-fold excess over the amount actually incorporated into the particle. The assembly step performed by gpU is

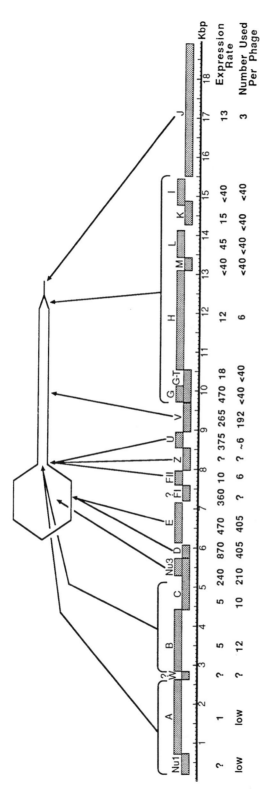

FIGURE 13. The bacteriophage λ morphogenetic genes. The physical size and location of all the λ morphogenetic genes are shown (with a scale in kbp) in the center (Sanger *et al.*, 1982). Normal transcription of the region initiates about 4 kbp to the left of position 0 and proceeds from left to right. The λ virion is shown schematically above, and the arrows indicate the location where the various proteins reside or must act (for references see Georgopoulos *et al.*, 1983; Feiss and Becker, 1983; Katsura, 1983). Below we show the relative translation rates of the various genes and our current best estimates of the number of molecules of each gene product required to construct one virion (Murialdo and Siminovitch, 1972; Casjens, 1974; Casjens and Hendrix, 1974a; Katsura and Tsugita, 1977; Georgopoulos *et al.*, 1983; Katsura, 1983).

one of those which are not absolutely controlled (see Sections IV.A.1.a and B.1), in that if gpU is absent, defective overlength tails are eventually built. Thus, it could be important to have an excess of gpU to make sure this step is not accidentally skipped (see also Katsura, 1983b).

This aspect of morphogenetic gene expression is particularly well studied in λ (see Fig. 13). Since there is a single late promoter and no terminators within the operon, the rate of synthesis of the messengers for the various morphogenetic genes is identical (Murialdo and Siminovitch, 1972; Ray and Pearson, 1974, 1975, 1976). The differences seen in the individual rates of synthesis of the λ late proteins (up to 800-fold) must be determined at a translational level, since the functional decay rates of the messages from the various cistrons do not differ more than about two- or threefold. The differences in protein synthetic rates must be caused by different rates of either ribosome loading or ribosome movement (the latter could perhaps be dictated by differential codon usage or mRNA secondary structure) (Ray and Pearson, 1975; Sanger et al., 1982). Similar conclusions have been drawn for P22 (Weinstock et al., 1980; Casjens and Adams, 1985). Recent gene fusion experiments with the λ A, D, E and F_{II} genes, which show that the information required for proper translational expression of these genes lies quite close to the start codon, strongly support the differential ribosome loading hypothesis for these four genes (L. Sampson, personal communication). In phages such as T4, T7, and P2, where late genes are transcribed from multiple promoters, we presume any transcriptional rate differences will be superimposed upon variations in translation efficiency to give the observed expression rates.

Other aspects of translation are also known to be important in the control of expression of virion structural genes, but few cases have been analyzed in sufficient detail to comment on here. One such aspect is *translational coupling,* in which the translation of a downstream gene depends on successful translation of the immediately upstream gene. For example, in λ, efficient translation of the H gene requires efficient translation of the immediately upstream V and G genes (reviewed by Georgopoulos et al., 1983). This type of mechanism could be useful in maintaining constant relative synthesis rates for two structural proteins.

3. One "Gene," Two Proteins

a. Translational Frame Shifting

There are now a number of situations known where ribosomes must perform a frame shift or read through a termination codon in order to translate a gene successfully (Kastelein et al., 1982; Dunn and Studier, 1983; Craigen et al., 1985; Fox and Weiss-Brummer, 1980; Clare and Farabaugh, 1985; Jacks and Varmus, 1985). Two examples of *programmed frame shifting* (frame shifting at a particular location that is programmed into the system by nucleotide sequence information) are known in

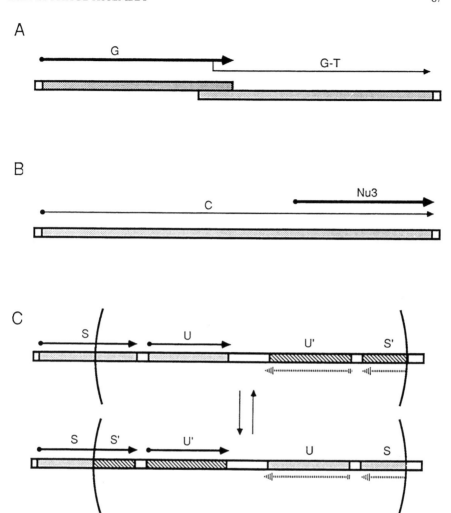

FIGURE 14. One "gene" can synthesize two different polypeptides. (A) Programmed translational frameshifting. (B) Nested genes. (C) DNA segment inversion. In all cases, transcription passes through the indicated regions from left to right, and the horizontal black arrows represent active translational reading frames (solid circles indicate translation start sites and the point of the arrows indicates the termination sites). The vertical portion of the narrow arrow in panel (A) indicates a low-frequency ribosomal frameshift. In panel (C) the stippled arrows represent inactive reading frames, and the vertical arcs indicate the ends of the invertible DNA segment. The various gene sizes are not drawn accurately to scale.

dsDNA phage morphogenetic genes. The phage T7 coat gene is one such case. A small proportion (5–10%) of the ribosomes reading its coat protein gene, *10*, undergo a −1 frame shift somewhere within the last 10 codons and continue to read 53 codons beyond the normal terminator before encountering a termination codon in the new frame (Dunn and

Studier, 1983). Thus, two types of coat protein are made which differ in amino acid sequence at their C termini. It is not known whether the resulting "frame-shifted" protein is required for proper T7 assembly, but it is present in small numbers in virions (10–50 molecules per virion) (Studier, 1972). A frame shift is also thought to occur in the λ tail gene G. In this case, a considerably longer stretch of about 175 amino acids is added to the C terminus of the gene G protein by an infrequent frame shift late in the G gene to make the G-T protein (Fig. 14A) (M. Levin, U. Ripepi, and R. Hendrix, unpublished observations).

b. Nested Genes

A second type of genetic arrangement which gives rise to two proteins with common sequence elements is exemplified by the nesting of the λ Nu3 gene within the C gene (Fig. 14B). The Nu3 translation start lies in-frame, within gene C, and Nu3 translation terminates at the gene C termination codon, so gpNu3 has the same amino acid sequence as the carboxy-terminal portion of gpC (Shaw and Murialdo, 1980). Both proteins are thought to be required for λ head assembly and are prohead components. Both frame shifting and gene nesting result in the simultaneous synthesis of two different proteins which have considerable amino acid sequence in common. Although in no case is there a detailed understanding of why two such proteins are made, it would allow two "different" proteins to interact in very similar ways with other proteins. Alternatively, consider the following (speculative) scenario. If the smaller protein such as gpNu3 could form a homomultimeric structure, the unique portion of the other protein, here the N-terminal end of gpC, could bind to a specific site (the portal?), and the gpNu3-like end of gpC could act as the "decoy on the duck pond" to recruit gpNu3 molecules for assembly into a prohead scaffold in the correct location relative to the site to which the gpC N terminus binds. This case seems particularly interesting because of the apparent nonequivalent bonding between the coat and scaffold subunits (see Section IV.C). Thus, the existence of such complex "genes" has potentially profound implications for assembly.

c. DNA Rearrangements

A third type of synthesis of two different proteins from one gene occurs in phages Mu, D108, P1, and P7. In Mu, the only case studied in detail, the two proteins are alternate, mutually exclusive gene products. These are the S and U genes (and their alternates S' and U') which are involved in tail fiber biosynthesis and/or assembly and are responsible for determining Mu's host range (Toussaint et al., 1978; Grundy and Howe, 1984). The Mu chromosome includes a DNA segment which undergoes fairly slow stochastic inversion when it is in the prophage state in E. coli and is promoted by the phage gin gene product (Symonds and Coelho,

1978). In one inversion orientation, transcription from the late promoter expresses the S and U genes; in the other orientation, a new S' C terminus is fused to the N-terminal portion of the S gene, and the U' gene replaces the U gene (Fig. 14C). It has been suggested but not yet fully proved that the N-terminal portion of the S gene encodes the portion of the fiber that binds to baseplates, and the S' and U' reading frames encode the outer parts of the fiber, which function during adsorption (Grundy and Howe, 1984). Phages grown in the two orientations do have differences in their structural proteins (Giphart-Gassler et al., 1982), and phages with S and U proteins adsorb to E. coli whereas phages with S' and U' proteins adsorb to Citrobacter freundiii, Serratia marcescens, and other bacterial strains (Bukhari and Ambrosio, 1978; Kamp and Sandulache, 1983; Van de Putte et al., 1980). Phages D108, P1, and P7 also have DNA inversions that affect host range and are presumed to be similar to that in Mu (Gill et al., 1981; Iida, 1984; Iida et al., 1985). Thus, these phages occasionally segregate genomes which express alternate tail fiber genes and which therefore give rise to virions with an alternate host range.

d. Differential Covalent Modification

Phage λ has two examples of differential protein cleavage during assembly. About three-quarters of gpB is cleaved during head assembly, and gpC appears to undergo two alternate cleavages (Hendrix and Casjens, 1974a, 1975). It is not known if the two forms of either of these differentially processed proteins have different functions in the virion, although it has been suggested that the two forms of modified gpC might occupy structurally similar but distinct positions around the portal vertex (Murialdo and Ray, 1975).

4. Assembly-Controlled Gene Expression

In light of the complexities and redundancies built into the control of dsDNA phage early gene expression, it might be considered surprising that very few examples of feedback regulation by the assembly process on late gene expression have been found. Presumably this reflects the fact that once a phage is committed to grow lytically, the cell is dying. Thus, it would do little good, for example, to turn off late gene expression if assembly was not proceeding well. There is only one well-studied case of assembly affecting the synthesis of a morphogenetic protein. Expression of the phage P22 scaffolding protein gene is turned down about sixfold by unassembled scaffolding protein, whereas scaffolding protein assembled in proheads has no such effect (King et al., 1978; Casjens et al., 1985). Such control seems appropriate, since the scaffolding protein acts catalytically (see Section IV.A.2) but is still required in quite large amounts. The effect of this regulation is that just enough scaffolding protein molecules are made to assemble into proheads with the available coat protein, since any excess unassembled scaffolding protein depresses additional synthesis. This modulation of expression appears to be posttranscriptional, but the details of the regulatory mechanism are not known (Casjens and Adams, 1985; Wyckoff and Casjens, 1985). Other phages such as φ29 (and perhaps T3 and T7) have scaffolding proteins that recycle, but they are not known to be similarly regulated (Nelson et al., 1976). In the dsDNA phages, there is one other case of gene expression being affected by assembly—the overproduction of the P22 tail spike protein when DNA packaging fails (Adams et al., 1985). The mechanism of this regulation and the reason for it are not known.

B. Gene Position

1. Assembly and Gene Position

Epstein et al. (1963) and Weigle (1966) first drew attention to the fact that the T4 morphogenetic genes are nonrandomly distributed and that the phage λ head and tail genes map in two nonoverlapping, adjacent clusters. Such linkage is thought to have evolutionary advantages beyond allowing easy coordinate regulation of the genes involved, in that gene modules can be evolutionarily exchanged among phages while minimizing the formation of inactive hybrids (Fisher, 1930; Stahl and Murray, 1966; Dove, 1971; Casjens and Hendrix, 1974b; Echols and Murialdo, 1978; Campbell and Botstein, 1983). Stahl and Murray (1966) stated it particularly clearly: Evolution imposes a "selective value upon linkage arrangements which minimize recombination between genes whose products most intensely interact." Thus, the genes whose protein products interact directly during assembly and in the virion might be expected to map nearest one another. This is strikingly true in many of the phage morphogenetic genes. Again, λ is an excellent example of such a gene arrangement, although other phages show strong elements of such positioning as well (King and Laemmli, 1971; King et al., 1973; Studier, 1972). In addition, genes encoding proteins that are involved in recognition of the DNA to be packaged usually map near the DNA site to which they bind (Thomas, 1964; Backhaus, 1985). We speculate below on the importance of gene positioning to assembly.

Figure 13 shows schematically that λ genes whose protein products interact in the virion or during virion assembly are indeed often adjacent. The tail genes whose functions are known map essentially linearly relative to the tail structure. The arrangement of the head genes is not as simply interpreted as that of the tail genes. The F_{II} gene, whose product forms the site to which tails bind (Casjens, 1974; Tsui and Hendrix, 1980), is adjacent to the tail gene cluster; the D and E genes are adjacent, and their products clearly interact (Casjens and Hendrix, 1974a; Williams and Richards, 1974); and the Nu1 and A gene products bind to each other and the *cos* site on the DNA, and their genes map adjacent to that site (reviewed by Feiss and Becker, 1983). The functions of the *W* and F_I gene products are not yet well understood. The gene *Nu3* scaffolding protein probably interacts with the coat protein (gp*E*), and perhaps gp*B* and/or gp*C* (Murialdo and Ray, 1975), but presumably its gene position must be fixed partly by the gene *C* function. The portal protein is encoded by the *B* gene (Kochan *et al.*, 1984), and although it is located near the tail joining site (Tsui and Hendrix, 1980), it maps near the left end of the head gene cluster. This could be the result of the fact that it may interact with the gp*Nu1*-gp*A* terminase during the initiation of DNA packaging. The location of the (modified) *C* gene protein in the virion is not known, but it has been suggested that it is also at the portal vertex (Murialdo and Ray, 1975). We should note that in some other phages the morphogenetic genes seem (at least at our present stage of understanding) to be somewhat more loosely associated. The T4 baseplate genes, for example, map only approximately in the order of the assembly of their proteins, but phage parts are three-dimensional and chromosomes are one-dimensional structures.

This type of analysis can in some instances be extended to portions of genes as well as whole genes. In general, very little information is available on what parts of the various proteins interact with their neighbors in the virion, but in those few cases where such information is available, the rule under discussion here does seem to apply. The λ *A* gene is best studied in this regard. Its N-terminal portion has been shown to interact with the *Nu1* protein, and its C-terminal portion to interact with proheads, most likely gene *B* or *C* protein, during DNA packaging (Frackman *et al.*, 1985; Feiss, 1986). Thus, both ends of the *A* gene lie close to the genes with whose products those ends of the gene *A* protein must interact. Also, the finding that host range mutants, presumably changes in the distal portion of the tail fiber, alter the C-terminal end of gene *J* supports this notion (Fuerst and Bingham, 1978). The N-terminal portion of the protein, which should therefore interact with the tail "tip" proteins, is in fact proximal to the tail tip genes. The position of the T3 tail fiber gene, *17*, is also positioned so that the end of the protein that interacts with the baseplate is encoded by the portion of the gene that maps nearer to those genes (Kato *et al.*, 1986). The T4 tail fiber gene group is also a striking example of such an arrangement (Revel, 1981). The 3' end of the distal fiber gene, *37*, maps adjacent to gene *38*:

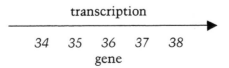

which is likely to interact with the C-terminal region of gp37 to promote its proper folding (Beckendorf *et al.*, 1973; Bishop and Wood, 1976); the 5' end of gene *37* is next to genes *35* and *36*, which are present at the fiber elbow and must interact with the N-terminal region of gp37 (Yanagida and Ahmed-Zadeh, 1970; Oliver and Crowther, 1981; Wood and Crowther, 1983; Reide *et al.*, 1985a). These genes in turn map adjacent to the 3' end of gene *34* which is thought to be arranged with its C-terminal end at the fiber "elbow" and N terminus at the baseplate (Wood and King, 1979). One observation that appears not to fit this type of arrangement is that the C terminus of T4 gp*11* is required to assemble with gp*10*, whereas the N-terminal encoding end of gene *11* is adjacent to gene *10* (Plishker and Berget, 1984).

It is not yet clear if the gene positions are meaningful enough to have real predictive value in learning the details of the phage assembly processes. For example, should the C-terminal region of the λF_{II} protein be expected to interact with tails, and is it significant that in all cases known (ϕ29, λ, T3, T7, P22, and T4), the scaffolding protein gene(s) maps immediately 5' to the coat protein gene? We feel that, although no direct conclusions should be drawn from them, the sorts of findings discussed above are sufficiently suggestive to at times warrant the use of this information to give direction to more detailed genetic and physical analyses.

2. Expression Time and Assembly

As has been stated above, there is no evidence that the time of expression of any morphogenetic gene is critical in any phage assembly process; however, there are clear examples of morphogenetic proteins whose synthesis does not parallel typical late synthesis. In phage T4, where transcriptional control of phage gene expression is very complex, several morphogenetic genes (*IpI, IpII, IpIII, 31, 40,* and *63*) are known to be expressed before typical late morphogenetic genes are turned on. Curiously, the *Ip* proteins and gp63 are made throughout infection whereas gp31 is made only at early times (Black, 1974; Castillo *et al.*, 1977; Wood *et al.*, 1978). Why these proteins are made before the bulk of the virion structural proteins is not clear, but gp31 and gp40 function at the earliest stages of prohead assembly, and it *may* be advantageous for the phage to accumulate these proteins before coat protein is made in large amounts. This could be especially important with T4, since its coat protein seems quite prone to assemble aberrantly if its concentration is too high relative to the other head proteins (Showe and Onorato, 1978; Eiserling *et al.*, 1984). Several other T4 genes involved in virion assembly—*49, 57, td,* and *frd*—are known to be expressed early or map in positions expected for early genes. Most of these and gp63 appear to have other functions in addition to their role in assembly. For example gp49 may function in recombination, gp63 has RNA ligase activity in addition to its tail fiber joining activity, gp*frd* is a dihydroforlate reductase, and gp*td* is a thymidine synthetase (Kemper *et al.*, 1981; Herrmann, 1982; Snopek *et al.*, 1977; Purobit *et al.*, 1981; Trimble *et al.*, 1972; Chu *et al.*, 1984). In P1, another very complex phage, at least one of the DNA packaging genes, 9, is expressed early (Sternberg and Coulby, 1986). It is not known if early expression is actually required for proper assembly in any of these cases.

Prokaryotic RNA polymerases transcribe at about 50 nucleotides per second *in vivo* (Rose *et al.*, 1970; Ray and Pearson, 1975; Casjens and Adams, 1985). Thus, in the phages with very long late operons, the time of appearance of proteins encoded by genes early in the operon is significantly earlier than those encoded by the more distal genes. P22 and λ, for example, have late operons about 20 and 23 kbp long, respectively, so it takes RNA polymerase 7–8 min to traverse these operons (Ray and Pearson, 1975; Casjens and Adams, 1985). This is a significant time period relative to the total time of late protein synthesis, about 30–50 min, and the delayed appearance of proteins encoded by the distal genes has been observed directly in λ-infected cells (Murialdo and Siminovitch, 1972). Could this be responsible for determining the overall order of the morphogenetic genes (i.e., why not invert the order of the late genes relative to transcription)? At this time we cannot answer this question, but given the lengths to which λ has gone to control all aspects and minute details of its life cycle, we can imagine reasons why it might be true. For example, it could be advantageous to accumulate the tail shaft subunit, gpV, before initiating tail assembly by making the gene *J* protein, thereby minimizing the number of partially assembled tail shafts in infected cells.

As mentioned above, there are phage assembly steps that appear to be "uncontrolled." That is, if certain structural proteins are not present, the assembly process continues, and nonrescuable, aberrant particles are made. An apparent example of such an uncontrolled step is the initiation of P22 proheads. The portal protein, gp1, and three ejection proteins— gp7, gp16, and gp20—must be present, or aberrant particles are made that cannot be rescued by the later addition of these proteins (Botstein *et al.*, 1973; Poteete *et al.*, 1979). Although we do not fully understand these assembly steps, at present there seems to be no protein-dependent cooperative mechanism (Section IV.A.1.a) for ensuring that these proteins participate in the assembly process (King *et al.*, 1973; Poteete and King, 1977). Measurements of the rates of expression of the P22 late genes showed that, even though its gene is located in about the middle of the late operon, the rate of synthesis of coat protein does not precisely parallel the synthesis of the other late proteins. Changes in the stability of the coat cistron message seem to dictate a three- to fourfold lower rate of coat protein synthesis early in the late period as compared to later times (Casjens and Adams, 1985). Thus, a three- to fourfold excess of gp1, gp7, gp16, and gp20 early in the late period of infection could help ensure that these proteins become incorporated into the prohead. However, the promoter distal position of genes 7, 16, and 20 does not help to alleviate this situation, and it points out the extremely speculative nature of the foregoing discussion.

VI. PROSPECTS

In the above discussion we have tried to point out our current view of *themes* and *variations* in our current knowledge of dsDNA bacteriophage assembly. Some of the common themes such as (1) the use of repeating subunits to build large structures, (2) assembly via specific (often branched) pathways whose steps are kinetically controlled by protein-protein interactions among the assembling components, and (3) the use of "accessory" proteins not present in the final structure to direct proper assembly of subunits that are potentially able to form many different structures have been known for several years and have contributed substantially to our ideas about macromolecular assembly processes in general. Other general ideas concerning (1) nonequivalent protein-protein interactions in virions, (2) protein movements in supermolecular structures, (3) protein folding, and possibly (4) DNA packaging seem to be on the verge of "crystallization" into such themes. No doubt additional potential generalities are lurking just out of our current vision. All of the above "themes" are *strategic* rather than *mechanistic* in scope. In no case do we know the details of the underlying *molecular mechanisms* by which these processes occur, so we are unable to attribute the strategic commonalities seen to the existence of only a limited number of available strategies or to a limited number of available molecular mechanisms. More detailed structural information and assembly mechanism studies should help determine whether the analogies between the dsDNA phages discussed above are the result of convergent or divergent evolution. Standard genetic and biochemical approaches have been very successful in allowing analysis of the biological/biochemical/genetic strategies of many aspects of bacteriophage assembly, and we now have a relatively clear picture of the overall way in which a number of dsDNA phages are built. However, to date these methods have been much less useful in the analysis of underlying physicochemical mechanisms. This is due primarily to the fact that one must almost certainly know the structure of the components to atomic resolution in order to understand the detailed mechanisms, and no phage morphogenetic protein structures are known with such precision. Attempts to crystallize these proteins have been thwarted by the facts that the major components are often difficult to keep in solution as well-defined, unassembled, nondenatured entities (since they tend to aggregate into phage-related structures at high concentrations); the minor components have been difficult to obtain in large amounts; and these proteins in general do not have traditional active sites or bind small molecules and so have not been particularly attractive to crystallographers. Application of the methods of genetic engineering is clearly solving the second problem, continued biochemical study should solve the first, and increased knowledge about the various assembly processes appears to be solving the third. An overall understanding of the physical and energetic aspects of macromolecular assembly processes will necessarily be obtained from the analysis of many individual systems in minute detail. Eventually, themes and a "library" of possible variations on those themes will become evident. Therefore, we are optimistic that in addition to developing more information about assembly strategies, the study of phage assembly will soon contribute fundamentally to our ideas concerning the actual molecular mechanisms involved.

ACKNOWLEDGMENTS. We thank all of our colleagues in the field of virus structure and assembly for many open, fruitful discussions. We also thank Dwight Anderson, José Carrascosa, Gus Doermann, Fred Eiserling, Hisao Fugisawa, Isao Katsura, Jon King, Kevin Leonard, Nat Sternberg, Robley Williams, and Glenn Yasuda for supplying figures and unpublished results. Our research in this area is supported by NIH research grants GM36529 and GM21975 (S.C.) and AI12227 and GM36523 (R.H.).

REFERENCES

Abremski, K., and Black, L., 1979, The function of bacteriophage T4 internal protein I in a restrictive strain of *Escherichia coli*, *Virology* **97**:439.

Adams, M., Brown, H., and Casjens, S., 1985, Bacteriophage P22 tail gene expression, *J. Virol.* **53**:180.

Adrian, M., Dubochet, J., Lepault, J., and McDowall, A., 1984, Cryoelectron microscopy of viruses, *Nature* **308**:32.

Aebi, U., Bijlinga, R., Van der Broek, J., Van der Broek, R., Eiserling, F., Kellenberger, C., Kellenberger, E., Mesyanzihnov, V., Muller, L., Showe, M., Smith, R., and Steven, A., 1974, The transformation of tau-particles into T4 heads. II. Transformations of the surface lattice and related observations on form determination, *J. Supramol. Str.* **2**:253.

Aebi, U., Bijlinga, R., ten Heggler, B., Kistler, J., Steven, A., and Smith, P., 1976, Comparison of the structural and chemical composition of giant T-even phage heads, *J. Supramol. Str.* **5**:475.

Aebi, U., Van Driel, R., Bijlenga, R., ten Heggler, B., Van der Broek, R., Steven, A., and Smith, P., 1977, Capsid fine structure of T-even bacteriophages. Binding and localization of two dispensible capsid proteins into P23 surface lattice, *J. Mol. Biol.* **110**:687.

Aksiyote-Benbasat, J., and Bloomfield, V., 1975, Joining of bacteriophage T4D heads and tails: A kinetic study based on inelastic light scattering, *J. Mol. Biol.* **95**:335.

Aksiyote-Benbasat, J., and Bloomfield, V., 1981, Kinetics of head-tail joining in bacteriophage T4D studied by quasi-elastic light scattering: Effects of temperature, pH, and ionic strength, *Biochemistry* **20**:5018.

Aksiyote-Benbasat, J., and Bloomfield, V., 1982, Hydrodynamics, size, and shape of bacteriophage T4D tails and baseplates, *Biopolymers* **21**:797.

Ames, B., and Dubin, D., 1960, The role of polyamines in the neutralization of bacteriophage deoxyribonucleic acid, *J. Biol. Chem.* **235**:769.

Amos, L., and Klug, A., 1975, Three dimensional image reconstructions of the contractile tail of the T4 bacteriophage, *J. Mol. Biol.* **99**:51.

Anderson, D., Hickman, D., and Reilly, B., 1966, Structure of *Bacillus subtilis* bacteriophage φ29 and the length of the φ29 deoxyribonucleic acid, *J. Bacteriol.* **91**:2081.

Arisaka, F., Tscopp, J., Van Driel, R., and Engel, J., 1979, Reassembly of the bacteriophage T4 tail from the core baseplate and the monomeric sheath protein P18: A co-operative process, *J. Mol. Biol.* **132**:369.

Arisaka, F., Engel, J., and Klump, H., 1981, Contraction and dissociation of the bacteriophage tail sheath induced by heat and urea, in: *Bacteriophage Assembly* (M. DuBow, ed.), pp. 365–379, Alan R. Liss, New York.

Arscott, P. G., and Goldberg, E. B., 1976, Cooperative action of the T4 tail fibers and baseplate in triggering conformational changes and in determining host range, *Virology* **69**:15.

Bachrach, U., and Benchetrit, L., 1974, Studies on phage internal proteins. III. Specific binding of T4 internal proteins to T4 DNA, *Virology* **59**:443.

Bachrach, U., Fischer, R., and Klein, I., 1975, Occurrence of polyamines in coliphages T5, φX174 and in phage-infected bacteria, *J. Gen. Virol.* **26**:287.

Backhaus, H., 1985, DNA packaging initiation of *Salmonella* bacteriophage P22: Determination of cut sites within the DNA sequence coding for gene *3*, *J. Virol.* **55**:458.

Bayer, M. E., Thurow, H., and Bayer, M. H., 1979, Penetration of the polysaccharide capsule of *Escherichia coli* (Bi 161/42) by bacteriophage K29, *Virology* **94**:95.

Bazinet, C., and King, J., 1985, The DNA translocating vertex of dsDNA bacteriophage, *Annu. Rev. Microbiol.* **39**:109.

Bazinet, C., Benbasat, J., King, J., Carazo, J., and Carrascosa, J., 1988, Purification and organization of the gene *1* protoprotein required for phage P22 DNA packaging, *Biochemistry* (in press).

Beckendorf, S., 1973, Structure of the distal half of the T4 tail fiber, *J. Mol. Biol.* **73**:37.

Beckendorf, S., Kim, J., and Lielausis, I., 1973, Structure of bacteriophage T4 genes *37* and *38*. *J. Mol. Biol.* **73**:17.

Becker, A., and Gold, M., 1978, Enzymatic breakage of the cohesive end site of phage λ DNA: terminase (ter) reaction, *Proc. Natl. Acad. Sci. USA* **75**:4199.

Berget, P., 1985, Pathways in viral morphogenesis, in: *Virus Structure and Assembly* (S. Casjens, ed.), pp. 149–168, Jones and Bartlett, Boston.

Berget, P., and King, J., 1978, The isolation and characterization of precursors in T4 baseplate assembly. The complex of gene *10* and gene *11* products, *J. Mol. Biol.* **124**:469.

Berget, P., and King, J., 1983, in: *Bacteriophage T4* (C. Mathews, E. Kutter, G. Mosig, and P. Berget, eds.), pp. 246–258, ASM Publications, Washington.

Berget, P., and Poteete, A., 1980, Structure and function of the bacteriophage P22 tail protein, *J. Virol.* **34**:234.

Bessler, W., Freund-Molbert, E., Knufermann, H., Rudolph, C., Thurow, H., and Stirm, S., 1973, A bacteriophage-induced depolymerase active on *Klebsiella* K11 capsular polysaccharide, *Virology*, **56**:134.

Bishop, R., and Wood, W., 1976, Genetic analysis of T4 tail fiber assembly. I. A gene *37* mutation that allows bypass of gene *38* function, *Virology* **72**:244.

Black, L., 1974, Bacteriophage T4 internal protein mutants: Isolation and properties, *Virology* **60**:166.

Black, L., and Ahmed-Zadeh, C., 1971, Internal proteins of bacteriophage T4: Their characterization and relation to head structure and assembly, *J. Mol. Biol.* **57**:71.

Black, L., and Showe, M., 1983, Morphogenesis of the T4 head. in: *Bacteriophage T4* (C. Mathews, E. Kutter, G. Mosig, and P. Berget, eds.), pp. 219–245, ASM Publications, Washington.

Black, L., Newcomb, W., Boring, J., and Brown, J., 1985, Ion etching of bacteriophage T4: Support for a spiral-fold model of packaged DNA, *Proc. Natl. Acad. Sci. USA* **82**:7960.

Bloomfield, V., 1983, Physical studies of morphogenetic reactions, in: *Bacteriophage T4* (C. Mathews, E. Kutter, G. Mosig, and P. Berget, eds.), pp. 270–276, ASM Publications, Washington.

Bloomfield, V., and Prager, S., 1979, Diffusion-controlled reactions on spherical surfaces: Application to bacteriophage tail fiber attachment, *Biophys. J.* **27**:227.

Botstein, D., Waddell, C., and King, J., 1973, Mechanism of head assembly and DNA encapsulation in *Salmonella* phage P22. I. Genes, proteins, structures, and DNA maturation, *J. Mol. Biol.* **80**:669.

Bowden, D., and Calendar, R., 1979, Maturation of bacteriophage P2 DNA *in vitro*: A complex, site-specific system of DNA cleavage, *J. Mol. Biol.* **129**:1.

Bradley, D., 1967, Ultrastructure of bacteriophages and bacteriocins, *Bacteriol. Rev.* **3**:230.

Branton, D., and Klug, A., 1975, Capsid geometry of bacteriophage T2: Freeze-etching study, *J. Mol. Biol.* **92**:559.

Bryant, J., and King, J., 1985, Injection proteins as targets of acridine-sensitized photoinactivation of bacteriophage P22, *J. Mol. Biol.* **180**:837.

Buchwald, M., and Siminovitch, L., 1969, Production of serum blocking material by mutants of the left arm of the λ chromosome, *Virology* **38**:1.

Buchwald, M., Steed-Glaister, P., and Siminovitch, L., 1970, The morphogenesis of bacteriophage lambda. I. Purification and characterization of heads and tails, *Virology* **42**:375.

Bukhari, A., and Ambrosio, L., 1978, The invertible segment of bacteriophage Mu DNA determines the adsorption properties of Mu particle, *Nature* **271**:575.

Burnett, R., 1985, The structure of the adenovirus capsid. II. The packing symmetry of hexon and its implications for viral architecture, *J. Mol. Biol.* **185**:125.

Burnett, R., Grutter, M., and White, J., 1985, The structure of the adenovirus capsid. I. An envelope model of hexon at 6 Å resolution, *J. Mol. Biol.* **185**:105.

Butler, P., and Klug, A., 1971, Assembly of the particle of tobacco mosaic virus from RNA and disks of protein, *Nature New Biol.* **229**:48.

Caldentey, J., and Kellenberger, E., 1986, Assembly and disassembly of bacteriophage T4 polyheads, *J. Mol. Biol.* **188**:39.

Campbell, A., and Botstein, D., 1983, Evolution of the lambdoid phages, in: *Lambda II* (R. Hendrix, J. Roberts, F. Stahl, and R. Weisberg, eds.), pp. 365–380, Cold Spring Harbor Laboratory, Cold Spring Harbor, NY.

Capco, G., and Mathews, C., 1973, Bacteriophage-coded thymidine synthetase: Evidence that the T4 enzyme is a capsid protein, *Arch. Biochem. Biophys.* **158**:736.

Carazo, J., Santisteban, A., and Carrascosa, J., 1985, Three-dimensional reconstruction of the bacteriophage φ29 neck particles at 2.2 nm resolution, *J. Mol. Biol.* **183**:79.

Carazo, J., Fujisawa, H., Nakasu, S., and Carrascosa, J., 1986, Bacteriophage T3 gene 8 product oligomer structure, *J. Ultrastr. Res.* **94**:105.

Carrascosa, J., 1986, Bacteriophage morphogenesis, in: *Electron Microscopy of Proteins*, Vol. 5: *Viral Structure* (J. Harris and R. Horne, eds.), pp. 37–70, Academic Press, London.

Carrascosa, J., Mendez, E., Corral, J., Rubio, V., Ramirez, G., Vinuela, E., and Salas, M., 1981, Structural organization of *Bacillus subtilis* phage φ29. A model, *Virology* **111**:401.

Carrascosa, J., Vinuela, E., Garcia, N., and Santisteban, A., 1982, Structure of the head-tail connector of φ29, *J. Mol. Biol.* **154**:311.

Casjens, S., 1974, Bacteriophage lambda F_{II} protein: Role in head assembly, *J. Mol. Biol.* **90**:1.

Casjens, S., 1979, Molecular organization of the bacteriophage P22 coat protein shell, *J. Mol. Biol.* **131**:1.

Casjens, S., 1985a, An introduction to virus structure and assembly, in: *Virus Structure and Assembly* (S. Casjens, ed.), pp. 1–28, Jones and Bartlett, Boston.

Casjens, S., 1985b, Nucleic acid packaging by viruses, in: *Virus Structure and Assembly* (S. Casjens, ed.), pp. 75–147, Jones and Bartlett, Boston.

Casjens, S. (ed.), 1985c, *Virus Structure and Assembly*, Jones and Bartlett, Boston.

Casjens, S., and Adams, M., 1985, Posttranscriptional modulation of bacteriophage P22 scaffolding protein gene expression, *J. Virol.* **53**:185.

Casjens, S., and Hendrix, R., 1974a, Locations and amounts of the major structural proteins in bacteriophage lambda, *J. Mol. Biol.* **88**:535.

Casjens, S., and Hendrix, R., 1974b, Comments on the arrangement of the morphogenetic genes of bacteriophage lambda, *J. Mol. Biol.* **90**:20.

Casjens, S., and King, J., 1974, P22 morphogenesis: Catalytic scaffolding protein in capsid assembly, *J. Supramol. Str.* **2**:202.

Casjens, S., and King, J., 1975, Virus assembly, *Annu. Rev. Biochem.* **44**:55.

Casjens, S., Hohn, T., and Kaiser, A. D., 1972, Head assembly steps controlled by genes *F* and *W* in bacteriophage lambda, *J. Mol. Biol.* **64**:551.

Casjens, S., Adams, M., Hall, C., and King, J., 1985, Assembly-controlled autotogenous modulation of bacteriophage P22 scaffolding protein gene expression, *J. Virol.* **53**:174.

Caspar, D., 1965, Design principles in virus construction, in: *Viral and Rickettsial Diseases of Man*, 4th Ed. (P. Horsfall and I. Tamm, eds.), pp. 51–93, Lippincott, Philadelphia.

Caspar, D., 1980, Movement and self-control in protein assemblies. Quasi-equivalence revisited, *Biophys. J.* **32**:103.

Caspar, D., and Klug, A., 1962, Physical principles in the construction of regular viruses, *Cold Spr. Harbor Symp. Quant. Biol.* **27**:1.

Castillo, C., Hsiao, C., Coon, P., and Black, L., 1977, Identification and properties of bacteriophage T4 capsid-forming gene products, *J. Mol. Biol.* **110**:585.

Chandrasekhar, G. N., Tilly, K., Woolford, C., Hendrix, R., and Georgopoulos, C., 1986, Purification and properties of the groES morphogenetic protein of *Escherichia coli*, *J. Biol. Chem.* **261**:12414.

Christie, G., and Calendar, R., 1985, Bacteriophage P2 late promoters. II. Comparison of four late promoter sequences, *J. Mol. Biol.* **181**:373.

Christensen, A., and Young, E., 1983, Characterization of T4 transcripts, in: *Bacteriophage T4* (C. Mathews, E. Kutter, G. Mosig, and P. Berget, eds.), pp. 184–188, ASM Publications, Washington.

Chu, F., Maley, G., Maley, F., and Belfort, M., 1984, Intervening sequence in the thymidylate synthetase gene of bacteriophage T4, *Proc. Natl. Acad. Sci. USA* **81**:3049.

Clare, J., and Farabaugh, P., 1985, Nucleotide sequence of a yeast Ty element: Evidence for an unusual mechanism of gene expression, *Proc. Natl. Acad. Sci. USA* **82**:2829.

Coombs, D., and Eiserling, F., 1977, Studies on the structure, protein composition, and assembly of the neck of bacteriophage T4, *J. Mol. Biol.* **116**:375.

Craigen, W., Cook, R., Tate, W., and Caskey, T., 1985, Bacterial chain release factors: Conserved primary structure and possible frameshift regulation of release factor two, *Proc. Natl. Acad. Sci. USA* **82**:3616.

Crawford, J. T., and Goldberg, E. B., 1977, The effect of baseplate mutations on the requirement for tail fiber binding for irreversible adsorption of bacteriophage T4, *J. Mol. Biol.* **111**:305.

Crick, F., and Watson, J., 1956, Structure of small viruses, *Nature* **177**:473.

Crowelesmith, I., Schindler, M., and Osborn, M., 1978, Bacteriophage P22 is not a likely probe for zones of adhesion between the inner and outer membranes of *Salmonella typhimurium*, *J. Bacteriol.* **135**:259.

Crowther, R. A., 1980, Mutants of bacteriophage T4 that produce infective fiberless particles, *J. Mol. Biol.* **137**:159.

Crowther, R. A., Lenk, E., Kikuchi, Y., and King, J., 1977, Molecular reorganization in the hexagon to star transition of the baseplate of bacteriophage T4, *J. Mol. Biol.* **116**:489.

Darcy-Tripier, F., Nermut, M., Brown, E., Nonnemacher, H., and Braunwald, J., 1986, Ultrastructural and biochemical evidence of the trimeric nature of frog virus 3 (FV3) six-coordinated capsomers, *Virology* **149**:44.

Dawes, J., and Goldberg, E., 1973, Functions of baseplate components in bacteriophage T4 infection. I. Dihydrofolate reductase and dihydropterylhexaglutamate, *Virology* **55**:380.

DeRosier, D., and Klug, A., 1968, Reconstruction of three-dimensional structures from electron micrographs, *Nature* **217**:130.

Doherty, D. H., 1982, Genetic studies on capsid-length determination in bacteriophage T4. II. Genetic evidence that specific protein-protein interactions are involved, *J. Virol.* **43**:655.

Donelli, G., Guglielmi, F., and Paoletti, L., 1972, Structure and physiochemical properties of bacteriophage G. 1. Arrangement of protein subunits and contraction process of tail sheath, *J. Mol. Biol.* **71**:113.

Dove, W., 1971, Biological inferences, in: *The Bacteriophage Lambda* (A. Hershey, ed.), pp. 297–312, Cold Spring Harbor Laboratory, Cold Spring Harbor, NY.

Drahos, D. J., and Hendrix, R. W., 1982, Effect of bacteriophage λ infection on synthesis of groE protein and other *Escherichia coli* proteins, *J. Bacteriol.* **149**:1050.

Drexler, K., Riede, I., and Henning, U., 1986, Morphogenesis of the long tail fibers of bacteriophage T2 involved proteolytic processing of the polypeptide (gene product 37) constituting the distal part of the fiber, *J. Mol. Biol.* **191**:267.

Driedonks, R., Krigsman, P., and Mellema, J., 1976, A study of the states of aggregation of alfalfa mosaic virus, *Phil. Trans. R. Soc. Lond.* B **276**:131.

Driedonks, R., Engel, A., ten Heggler, B., and Van Driel, R., 1981, Gene *20* product of bacteriophage T4. Its purification and structure, *J. Mol. Biol.* **152**:641.

DuBow, M. (ed.), 1981, *Bacteriophage Assembly: Progress in Clinical and Biological Research*, Vol. 64, Alan R. Liss, New York.

Duda, R., and Eiserling, F., 1982, Evidence for an internal component of the bacteriophage T4D tail core: A possible length-determining template, *J. Virol.* **43**:714.

Duda, R., Wall, J., Hainfeld, J., Sweet, R., and Eiserling, F., 1985, Mass distribution of a probable tail-length-determining protein in bacteriophage T4, *Proc. Natl. Acad. Sci. USA* **82**:5550.

Duda, R., Gingery, M., and Eiserling, F., 1986, Potential length determiner and DNA injection protein is extruded from bacteriophage T4 tail tubes *in vitro*, *Virology* **151**:296.

Dunn, R., and Studier, W., 1983, Complete nucleotide sequence of bacteriophage T7 DNA and the locations of T7 genetic elements, *J. Mol. Biol.* **166**:477.

Earnshaw, W., 1979, Modelling of the small-angle X-ray diffraction arising from the surface lattices of phages lambda and P22, *J. Mol. Biol.* **131**:14.

Earnshaw, W., and Casjens, S., 1980, DNA packaging by the double-stranded DNA bacteriophages, *Cell* **21**:319.

Earnshaw, W., and King, J., 1978, Structure of phage P22 protein aggregates formed in the absence of the scaffolding protein, *J. Mol. Biol.* **126**:721.

Earnshaw, W., Casjens, S., and Harrison, S., 1976, Assembly of the head of bacteriophage P22: X-ray diffraction from heads, proheads and related structures, *J. Mol. Biol.* **104**:387.

Earnshaw, W., King, J., Harrison, S., and Eiserling, F., 1978, The structural organization of DNA packaged within heads of T4 wild-type, isometric and giant bacteriophages, *Cell* **14**:559.

Earnshaw, W., Goldberg, E., and Crowther, R. A., 1979, The distal half of the tail fiber of bacteriophage T4, *J. Mol. Biol.* **132**:101.

Earnshaw, W., Hendrix, R., and King, J., 1979, Structural studies of bacteriophage λ heads and proheads by small angle X-ray scattering, *J. Mol. Biol.* **134**:575.

Easterbrook, K. B., and Bleviss, M., 1969, In vitro polymerization of bacteriophage λ tails, *Virology* **39**:331.

Echols, H., and Murialdo, H., 1978, Genetic map of bacteriophage lambda, *Microbiol. Rev.* **42**:577.

Edgar, R., and Lielausis, I., 1968, Some steps in the assembly of bacteriophage T4, *J. Mol. Biol.* **32**:263.

Edgar, R., and Wood, W., 1966, Morphogenesis of bacteriophage T4 in extracts of mutant-infected cells, *Proc. Natl. Acad. Sci. USA* **55**:498.

Eiserling, F., 1967, The structure of *Bacillus subtilis* bacteriophage PBS1, *J. Ultrastr. Res.* **17**:342.

Eiserling, F., 1979, Bacteriophage structure, in: *Comprehensive Virology*, Vol. 13 (H. Fraenkel-Conrat and R. Wagner, eds.), pp. 534–580, Plenum Press, New York.

Eiserling, F., Corso, J., Feng, S., and Epstein, R., 1984, Intracellular morphogenesis of bacteriophage T4. II. Head morphogenesis, *Virology* **137**:95.

Elliot, J., and Arber, W., 1978, *E. coli* K-12 *pel* mutants, which block phage λ DNA injection, coincide with *ptsM*, which determines a component of a sugar transport system, *Mol. Gen. Genet.* **161**:1.

Elliot, T., Kassavetis, G., and Geiduschek, E. P., 1984, The complex pattern of transcription in the segment of the bacteriophage T4 genome containing three of the head protein genes, *Virology* **139**:260.

Engel, A., Van Driel, R., and Driedonks, R., 1982, A proposed structure of the prolate phage T4 prehead core, *J. Ultrastruct. Res.* **80**:12.

Epstein, R., Bolle, A., Steinberg, C., Kellenberger, E., Boy de la Tour, E., Chevalley, R., Edgar, R., Sussman, M., Denhardt, C., and Lielausis, I., 1963, Physiological studies of condi-

tional lethal mutants of bacteriophage T4D, *Cold Spr. Harbor Symp. Quant. Biol.*
    28:375.
Eriksson, U., Svenson, S., Lonngren, J., and Lindberg, A., 1979, *Salmonella* phage glycanas-
    es: Substrate specificity of the phage P22 endo-rhamnosidase, *J. Gen. Virol.* **43**:503.
Feiss, M., 1986, Terminase and the recognition, cutting and packaging of λ chromosomes,
    Trends Genet. **2**:100.
Feiss, M., and Becker, A., 1983, DNA packaging and cutting, in: *"Lambda II"* (R. Hendrix, J.
    Roberts, F. Stahl, and R. Weisberg, eds.), pp. 305–330, Cold Spring Harbor Laboratory,
    Cold Spring Harbor, NY.
Fiandt, M., Hradecna, Z., Lozeron, H., and Szybalski, W., 1971, Electron micrographic
    mapping of deletions, insertions, inversions, and homologies in the DNA's of col-
    iphages lambda and phi80, in: *The Bacteriophage Lambda* (A. Hershey, ed.), pp. 329–
    354, Cold Spring Harbor Laboratory, Cold Spring Harbor, NY.
Fish, S., Hartman, K., Fuller, M., King, J., and Thomas, G., 1980, Investigation of secondary
    structures and macromolecular interactions in bacteriophage P22 by laser raman spec-
    troscopy, *Biophys. J.* **32**:234.
Fisher, R., 1930, *The Genetical Theory of Natural Selection*, Clarendon Press, Oxford,
    U.K.
Flatgaard, J. E., 1969, PhD thesis, California Institute of Technology.
Floor, E., 1970, Interaction of morphogenetic genes of bacteriophage T4, *J. Mol. Biol.* **47**:293.
Fox, T., and Weiss-Brummer, B., 1980, Leaky +1 and −1 frameshift mutations at the same
    site in a yeast mitochondrial gene, *Nature* **288**:60.
Frackman, S., Siegele, D., and Feiss, M., 1985, The terminase of bacteriophage λ. Functional
    domains of *cos*B binding and multimer assembly, *J. Mol. Biol.* **183**:225.
Friedman, D., and Gottesman, M., 1983, Lytic mode of lambda development, in: *Lambda II*
    (R. Hendrix, J. Roberts, F. Stahl, and R. Weisberg, eds.), pp. 21–51, Cold Spring Harbor
    Laboratory, Cold Spring Harbor, NY.
Fuerst, C., and Bingham, H., 1978, Genetic and physiological characterization of the *J* gene
    of bacteriophage lambda, *Virology* **87**:437.
Fuller, M., and King, J., 1980, Regulation of coat protein polymerization by the scaffolding
    protein of bacteriophage P22, *Biophys. J.* **23**:381.
Fuller, M., and King, J., 1982, Assembly *in vitro* of bacteriophage P22 procapsids from
    purified coat and scaffolding subunits, *J. Mol. Biol.* **156**:633.
Fuller, S., and Argos, P., 1987, Is sinbis a simple picornavirus with an envelope? *EMBO J.*
    6:1099.
Geiduschek, E. P., Elliott, T., and Kassavetis, G., 1983, Regulation of late gene expression,
    in: *Bacteriophage T4* (C. Mathews, E. Kutter, G. Mosig, and P. Berget, eds.), pp. 189–
    192, ASM Publications, Washington.
Geisselsoder, J., Sedivy, J., Walsh, R., and Goldstein, R., 1982, Capsid structure of satellite
    phage P4 and its P2 helper, *J. Ultrastr. Res.* **79**:165.
Georgopoulos, C., Hendrix, R., Kaiser, A. D., and Wood, W. B., 1972, Role of the host cell in
    bacteriophage morphogenesis: Effect of a bacterial mutation on T4 head assembly,
    Nature New Biol. **239**:38.
Georgopoulos, C., Hendrix, R., Casjens, S., and Kaiser, A., 1973, Host participation in
    bacteriophage lambda head assembly, *J. Mol. Biol.* **76**:45.
Georgopoulos, C., Tilly, K., and Casjens, S., 1983, Lambdoid phage head assembly, in:
    Lambda II (R. Hendrix, J. Roberts, F. Stahl, and R. Weisberg, eds.), pp. 279–304, Cold
    Spring Harbor Laboratory, Cold Spring Harbor, NY.
Gill, G., Hull, R., and Curtiss, R., 1981, Mutator bacteriophage D108 and its DNA: An
    electron microscopic characterization, *J. Virol.* **37**:420.
Giphart-Gassler, M., Plasternak, R., and Van de Putte, P., 1982, G inversion in bacterio-
    phage Mu: A novel way of gene splicing, *Nature* **297**:339.
Goldberg, E., 1983, Recognition, attachment and injection, in: *Bacteriophage T4* (C.
    Mathews, E. Kutter, G. Mosig, and P. Berget, eds.), pp. 32–39, ASM Publications,
    Washington.
Goldenberg, D., and King, J., 1982, Trimeric intermediate in the *in vivo* folding and subunit

assembly of the tail spike endorhamnosidase of the tail spike of bacteriophage P22, *Proc. Natl. Acad. Sci. USA* **79**:3403.

Gope, R., and Serwer, P., 1983, Bacteriophage P22 *in vitro* packaging monitored by agarose gel electrophoresis: Rate of DNA entry into capsids, *J. Virol.* **47**:96.

Granboulan, P., 1983, The tail fiber of bacteriophage T4 is sensitive to proteases at elevated temperatures, *J. Gen. Microbiol.* **129**:2217.

Grayhack, E., Yang, X., Lan, L., and Roberts, J., 1985, Phage lambda gene Q antiterminator recognizes RNA polymerase near the promoter and accelerates it through a pause site, *Cell* **42**:259.

Grundy, F., and Howe, M., 1984, Involvement of the invertible G segment in bacteriophage Mu tail fiber biosynthesis, *Virology* **134**:296.

Haas, R., Murphy, R., and Cantor, C., 1982, Testing models of the arrangement of DNA inside λ by crosslinking the packaged DNA, *J. Mol. Biol.* **159**:71.

Hafner, E., Tabor, C., and Tabor, H., 1979, Mutants of *Escherishia coli* that do not contain 1,4-diaminobutane (putrescine) or spermidine, *J. Biol. Chem.* **154**:12419.

Hagen, E. W., Reilly, B. E., Tosi, M. E., and Anderson, D. L., 1976, Analysis of gene function of bacteriophage φ29 of *Bacillus subtilis:* Identification of cistrons essential for viral assembly, *J. Virol.* **19**:501.

Harrison, S., 1983a, Virus structure: High-resolution perspectives, *Adv. Virus Res.* **28**:175.

Harrison, S., 1983b, Packaging of DNA into bacteriophage heads: A model, *J. Mol. Biol.* **171**:577.

Hartwieg, E., Bazinet, C., and King, J., 1986, DNA injection apparatus of phage P22, *Biophys. J.* **49**:24.

Hayden, M., Adams., M., and Casjens, S., 1985, Bacteriophage L: Chromosome physical map and structural proteins, *Virology* **147**:431.

Hendrix, R., 1978, Symmetry mismatch and DNA packaging in large DNA bacteriophages, *Proc. Natl. Acad. Sci. USA* **75**:4779.

Hendrix, R., 1979, Purification and properties of groE, a host protein involved in bacteriophage assembly, *J. Mol. Biol.* **129**:375.

Hendrix, R., and Casjens, S., 1974a, Protein cleavage in bacteriophage λ tail assembly, *Virology* **61**:156.

Hendrix, R., and Casjens, S., 1974b, Protein fusion: A novel reaction in bacteriophage λ head assembly, *Proc. Natl. Acad. Sci. USA* **71**:1451.

Hendrix, R., and Casjens, S., 1975, Protein processing and its genetic control in petit λ assembly, *J. Mol. Biol.* **91**:187.

Hendrix, R., Roberts, J., Stahl, F., and Weisberg, R. (eds.), 1983, *Lambda II*, Cold Spring Harbor Laboratory, Cold Spring Harbor, NY.

Herranz, L., Salas, M., and Carrascosa, J., 1986, Interaction of bacteriophage φ29 connector protein with the viral DNA, *Virology* **155**:289.

Herrmann, R., 1982, Nucleotide sequence of bacteriophage T4 gene 57 and deduced amino acid sequence, *Nucleic Acids Res.* **10**:1105.

Herskowitz, I., and Hagen, D., 1980, The lysis-lysogeny decision of phage lambda: Explicit programming and responsiveness, *Annu. Rev. Genet.* **14**:399.

Ho, Y., and Rosenberg, M., 1985, Characterization of a third, cII-dependent, coordinately activated promoter on phage lambda involved in lysogeneic development, *J. Biol. Chem.* **260**:11838.

Hoffman, B., and Levine, M., 1975, Bacteriophage P22 virion protein which performs an essential early function. II. Characterization of the gene 16 function, *J. Virol.* **16**:1547.

Hogle, J., Chow, M., and Filman, D., 1985, Three-dimensional structure of poliovirus at 2.9Å resolution, *Science* **229**:1358.

Hohn, B., 1983, DNA sequences necessary for packaging of bacteriophage λ DNA, *Proc. Natl. Acad. Sci. USA* **80**:7456.

Hohn, B., Wurtz, M., Klein, B., Lustig, A., and Hohn, T., 1974, Phage lambda DNA packaging in vitro, *J. Supramol. Struct.* **2**:302.

Hohn, T., Wurtz, M., and Hohn, B., 1976, Capsid transformation during packaging of bacteriophage λ DNA, *Phil. Trans. R. Soc. Lond. B* **276**:51.

Hohn, T., Hohn, B., Engel, A., Wurtz, M., and Smith, P. R., 1979, Isolation and characterization of the host protein groE involved in bacteriophage lambda assembly, *J. Mol. Biol.* **129**:359.

Hoopes, B., and McClure, W., 1985, A cII-dependent promoter is located within the Q gene of bacteriophage lambda, *Proc. Natl. Acad. Sci. USA* **82**:3134.

Howatson, A., and Kemp, C., 1975, The structure of tubular head forms of bacteriophage λ: Relation to the capsid structure of petit λ and normal λ heads, *Virology* **67**:80.

Hsiao, C. L., and Black, L. W., 1978, Head morphogenesis of bacteriophage T4. II. The role of gene 40 in initiating prehead assembly, *Virology* **91**:15.

Iida, S., 1984, Bacteriophage P1 carries two related sets of genes determining its host range in the invertible C segment of its genome, *Virology* **134**:421.

Iida, S., Hiestand-Nauer, R., Meyer, J., and Arber, W., 1985, Crossover sites *cix* for inversion of invertible DNA segment C on the bacteriophage P7 genome, *Virology* **143**:347.

Imber, R., Tsugita, A., Wurtz, M., and Hohn, T., 1980, The outer surface protein of bacteriophage lambda, *J. Mol. Biol.* **139**:277.

Ishii, T., and Yanagida, M., 1975, Molecular organization of the shell of T-even bacteriophage heads, *J. Mol. Biol.* **97**:655.

Ishii, T., and Yanagida, M., 1977, The two dispensable structural proteins (soc and hoc) of the T4 phage capsid: Their properties, isolation and characterization of defective mutants, and their binding to defective heads *in vitro*, *J. Mol. Biol.* **109**:487.

Ishii, T., Yamaguchi, Y., and Yanagida, M., 1978, Binding of the structural protein soc to the head shell of bacteriophage T4, *J. Mol. Biol.* **120**:533.

Isobe, T., Black, L., and Tsugita, A., 1976a, Primary structure of bacteriophage T4 internal protein II and characterization of the cleavage upon phage maturation, *J. Mol. Biol.* **102**:349.

Isobe, T., Black, L., and Tsugita, A., 1976b, Protein cleavage during virus assembly: A novel specificity of assembly dependent cleavage in bacteriophage T4, *Proc. Natl. Acad. Sci. USA* **73**:4209.

Israel, V., 1977, E proteins of bacteriophage P22 I. Identification and ejection from wild-type and defective particles, *J. Virol.* **23**:91.

Israel, V., 1978, A model for the adsorption of phage P22 to *Salmonella typhimurium*, *J. Gen. Virol.* **40**:669.

Iwashita, S., and Kanegasaki, S., 1976a, Enzymic and molecular properties of baseplate parts of bacteriophage P22, *Eur. J. Biochem.* **65**:87.

Iwashita, S., and Kanegasaki, S., 1976b, Deacetylation reaction catalyzed by *Salmonella* phage c341 and its baseplate parts, *J. Biol. Chem.* **251**:5361.

Jacks, T., and Varmus, H., 1985, Expression of the Rous sarcoma virus *pol* gene by ribosomal frameshifting, *Science* **230**:1237.

Jimenez, J., Santisteban, A., Carazo, J., and Carrascosa, J., 1986, Computer graphic display method for visualizing three-dimensional structures, *Science* **232**:113.

Kamp, D., and Sandulache, R., 1983, Recognition of cell surface receptors is controlled by invertible DNA of Mu, *FEMS Microbiol. Lett.* **16**:131.

Kanegasaki, S., and Wright, A., 1973, Studies on the mechanism of phage adsorption: Interaction between phage ε[15] and its cellular receptor, *Virology* **52**:160.

Kao, S., and McClain, W., 1980, Baseplate protein of bacteriophage T4 with both structural and lytic functions, *J. Virol.* **34**:95.

Kar, S., 1983, Structural proteins of bacteriophage lambda: Purification, characterization and localization, PhD thesis, University of Pittsburgh.

Kastelein, R., Remaut, E., Feirs, W., and Van Duin, J., 1982, Lysis gene expression of RNA phage MS2 depends on a frameshift during translation of the overlapping coat protein gene, *Nature* **295**:35.

Kato, H., Fujisawa, H., and Minagawa, T., 1985a, Genetic analysis of subunit assembly of the tail fiber of bacteriophage T3, *Virology* **146**:12.

Kato, H., Fujisawa, H., and Minagawa, T., 1985b, Purification and characterization of gene 17 product of bacteriophage T3, *Virology* **146**:22.

Kato, H., Fujisawa, H., and Minagawa, T., 1986, Subunit arrangement of the tail fiber of bacteriophage T3, *Virology* **153**:80.

Katsura, I., 1976, Morphogenesis of bacteriophage lambda tail: Polymorphism in the assembly of the major tail protein, *J. Mol. Biol.* **107**:307.

Katsura, I., 1983a, Structure and inherent properties of bacteriophage lambda head shell. IV. Small head mutants, *J. Mol. Biol.* **171**:297.

Katsura, I., 1983b, Tail assembly and injection, in: *Lambda II* (R. Hendrix, J. Roberts, F. Stahl, and R. Weisberg, eds.), pp. 331–346, Cold Spring Harbor Laboratory, Cold Spring Harbor, NY.

Katsura, I., 1987, Determination of bacteriophage λ tail length by a protein ruler, *Nature* **327**:73.

Katsura, I., and Hendrix, R., 1984, Length determination in bacteriophage lambda tails, *Cell* **39**:691.

Katsura, I., and Kühl, P. W., 1975, Morphogenesis of the tail of bacteriophage λ. III. Morphogenetic pathway, *J. Mol. Biol.* **91**:257.

Katsura, I., and Tsugitsa, A., 1977, Purification and characterization of the major protein and the termination protein of the bacteriophage lambda tail, *Virology* **76**:129.

Keller, B., Senstag, C., Kellenberger, E., and Bickel, T., 1984, Gene *68*, a new bacteriophage T4 gene which codes for the 17 Kd prohead core protein is involved in head size determination, *J. Mol. Biol.* **179**:415.

Kells, S., and Hazelkorn, R., 1974, Bacteriophage T4 short tail fibers are the product of gene *12, J. Mol. Biol.* **83**:473.

Kemper, B., Garabett, M., and Courage, U., 1981, Studies on T4 head maturation. II. Substrate specificity of gene-49-controlled endonuclease, *Eur. J. Biochem.* **115**:133.

Kikuchi, Y., and King, J., 1975a, Genetic control of bacteriophage T4 baseplate morphogenesis. I. Sequential assembly of the major precursor, *in vivo* and *in vitro, J. Mol. Biol.* **99**:645.

Kikuchi, Y., and King, J., 1975b, Genetic control of bacteriophage T4 baseplate morphogenesis. II. Mutants unable to form the central part of the baseplate, *J. Mol. Biol.* **99**:673.

Kikuchi, Y., and King, J., 1975c, Genetic control of bacteriophage T4 baseplate morphogenesis. III. Formation of the central plug and overall assembly pathway, *J. Mol. Biol.* **99**:695.

Kim, J., and Davidson, N., 1974, Electron microscope heteroduplex study of sequence relations of T2, T4, and T6 bacteriophage DNA's, *Virology* **57**:93.

King, J., 1968, Assembly of the tail of bacteriophage T4, *J. Mol. Biol.* **32**:231.

King, J., 1971, Bacteriophage T4 tail assembly: Four steps in core formation, *J. Mol. Biol.* **58**:693.

King, J., 1980, Regulation of structural protein interactions as revealed in phage morphogenesis, in: *Biological Regulation and Development* (R. Goldberger, ed.), pp. 101–132, Plenum Press, New York.

King, J., and Casjens, S., 1974, Catalytic head assembling protein in virus morphogenesis, *Nature* **251**:112.

King, J., and Laemmli, U., 1971, Polypeptides of the tail fibers of bacteriophage T4, *J. Mol. Biol.* **62**:465.

King, J., and Mykolajewycz, N., 1973, Bacteriophage T4 tail assembly: Proteins of the sheath, core and baseplate, *J. Mol. Biol.* **75**:39.

King, J., Lenk, E., and Botstein, D., 1973, Mechanism of head assembly and DNA encapsidation in *Salmonella* phage P22. II. Morphogenetic pathway, *J. Mol. Biol.* **80**:697.

King, J., Hall, C., and Casjens, S., 1978, Control of the synthesis of phage P22 scaffolding protein is coupled to capsid assembly, *Cell* **15**:551.

Kistler, J., Aebi, U., Onorato, L., ten Heggler, B., and Showe, M., 1978, Structural changes during transformation of T4 polyheads. I. Characterization of the initial and final states by Fab-fragment labelling of freeze-dried and shadowed preparations, *J. Mol. Biol.* **126**:571.

Kochan, J., and Murialdo, H., 1982, Stimulation of groE synthesis in *Escherichia coli* by bacteriophage lambda infection, *J. Bacteriol.* **149**:1166.

Kochan, J., and Murialdo, H., 1983, Early intermediates in bacteriophage lambda prohead assembly. II. Identification of biologically active intermediates, *Virology* **131**:100.

Kochan, J., Carrascosa, J., and Murialdo, H., 1984, Bacteriophage lambda preconnectors: Purification and structure, *J. Mol. Biol.* **174**:433.

Kocher, F., 1979, Two theorems in solid geometry, *J. Mol. Biol.* **127**:39.

Kozloff, L., 1983, The T4 particle: Low-molecular weight compounds and associated enzymes, in: *Bacteriophage T4* (C. Mathews, E. Kutter, G. Mosig, and P. Berget, eds.), pp. 25–31, ASM Publications, Washington.

Kozloff, L., and Lute, M., 1960, Calcium content of bacteriophage T2, *Biochim. Biophys. Acta* **37**:420.

Kozloff, L., and Lute, M., 1977, Zinc, an essential component of the baseplates of T-even bacteriophages, *J. Biol. Chem.* **252**:7715.

Kozloff, L., and Lute, M., 1981, Dual functions of bacteriophage T4D gene 28 product. II. Folate and polyglutamate cleavage activity of uninfected and infected *Escherichia coli* cells and bacteriophage particles, *J. Virol.* **40**:645.

Kozloff, L., and Lute, M., 1984, Identification of bacteriophage T4D gene products 26 and 51 as baseplate hub structural components, *J. Virol.* **52**:344.

Kozloff, L., and Zorzopulos, J., 1981, Dual functions of T4D gene 28 product: Structural components of the viral baseplate central plug and cleavage enzyme for folyl polyglutamates. I. Identification of T4D gene 28 product in the tail plug, *J. Virol.* **40**:635.

Kozloff, L., Crosby, L., and Lute, M., 1979, Structural role of the polyglutamate portion of the folate found in T4D bacteriophage baseplate, *J. Virol.* **32**:497.

Kuhn, A., Keller, B., Maeder, M., and Traub, F., 1987, Prohead core of bacteriophage T4 can act as an intermediate in the T4 head assembly pathway, *J. Virol.* **61**:113.

Kurtz, M., and Champe, S., 1977, Precursors of the T4 internal peptides, *J. Virol.* **22**:412.

Kwiatowski, B., Beilharz, H., and Stirm, S., 1975, Disruption of Vi bacteriophage III and localization of its deacetylase activity, *J. Gen. Virol.* **29**:267.

Kypr, J., and Mrazek, J., 1986, Lambda phage protein Nu1 contains the conserved DNA binding fold of repressors, *J. Mol. Biol.* **191**:139.

Laemmli, U., and Eiserling, F., 1968, Studies on the morphopoiesis of the head of phage T-even. IV. The formation of polyheads, *J. Mol. Biol.* **80**:575.

Laemmli, U., Beguin, F., and Gujer-Kellenberger, B., 1969, A factor preventing the major head protein of bacteriophage T4 from random aggregation, *J. Mol. Biol.* **47**:69.

Laemmli, U., Molbert, E., Showe, M., and Kellenberger, E., 1970, Form determining function of the genes required for the assembly of the head of bacteriophage T4, *J. Mol. Biol.* **49**:99.

Lake, J., and Leonard, K., 1974, Structure and protein distribution for the capsid of *Caulobacter crescentus* bacteriophage φCbK, *J. Mol. Biol.* **86**:499.

Lauffer, M., and Stevens, C., 1968, Structure of the tobacco mosaic virus particle and polymerization of tobacco mosaic virus protein, *Adv. Virus Res.* **13**:1.

Lemaux, P. G., Herendeen, S. L., Bloch, P. L., and Neidhardt, F. C., 1978, Transient rates of synthesis of individual polypeptides in *E. coli* following temperature shifts, *Cell* **13**:427.

Lengyel, J., Goldstein, R., Marsh, M., Sunshine, M., and Calendar, R., 1973, Bacteriophage P2 morphogenesis: Cleavage of the major capsid protein, *Virology* **53**:1.

Leonard, K., Kleinschmidt, A., and Lake, J., 1973, *Caulobacter crescentus* bacteriophage φCbK: Structure and *in vitro* self assembly of the tail, *J. Mol. Biol.* **81**:349.

Lepault, J., and Leonard, K., 1985, Three-dimensional structure of unstained frozen-hydrated extended tails of bacteriophage T4, *J. Mol. Biol.* **182**:431.

Lepault, J., Dubocet, J., Baschong, W., and Kellenberger, E., 1987, Organization of double-stranded DNA in bacteriophages: A study by cryoelectron microscopy of vitrified samples, *EMBO J.* **6**:1507.

Losick, R., and Pero, J., 1981, Cascades of sigma factors, *Cell* **25**:582.

Lovett, P., 1972, PBP1: A flagella specific bacteriophage mediating transduction in *Bacillus pumilus*, *Virology* **47**:743.

MacDonald, P., Kutter, E., and Mosig, G., 1984, Regulation of a bacteriophage T4 late gene, *soc*, which maps in the early region, *Genetics* **106**:17.

Male, C., and Kozloff, L., 1973, Function of T4D structural dihydrofolate reductase in bacteriophage infection, *J. Virol.* **11**:840.

Mathews, C., Kutter, E., Mosig, G., and Berget, P. (eds.), 1983, *Bacteriophage T4*, ASM Publications, Washington.

Matsuo-Kato, H., Fujisawa, H., and Minagawa, T., 1981, Structure and assembly of bacteriophage T3 tails, *Virology* **109**:157.

Matthews, R., (Ed.) 1987, A critical appraisal of viral taxonomy, CRC Press, Boca Raton, Florida.

McNicol, L., Simon, L., and Black, L., 1977, A mutation which bypasses the requirement for p24 in bacteriophage T4 capsid morphogenesis, *J. Mol. Biol.* **116**:261.

Meezan, E., and Wood, W., 1971, The sequence of gene product interaction in bacteriophage T4 tail core assembly, *J. Mol. Biol.* **58**:685.

Mellado, R., Barthelemy, I., and Salas, M., 1986, *In vivo* transcription of bacteriophage ϕ29 DNA early and late promoter sequences, *J. Mol. Biol.* **191**:191.

Michel, C., Jacq, B., Arques, D., and Bickle, T., 1986, A remarkable amino acid sequence homology between a phage T4 tail fibre protein and ORF314 of phage λ located in the tail operon, *Gene* **44**:147.

Miller, G., and Fiess, M., 1985, Sequence of the left end of phage 21 DNA, *J. Mol. Biol.* **183**:246.

Moody, M., 1965, The shape of the T-even bacteriophage head, *Virology* **26**:567.

Moody, M., 1971, Application of optical diffraction to helical structures in the bacteriophage tail, *Phil. Trans. R. Soc. Lond. B* **261**:181.

Moody, M., 1973, Sheath of bacteriophage T4. III. Contraction mechanism deduced from partially contracted sheaths, *J. Mol. Biol.* **80**:613.

Moody, M., and Makowski, L., 1981, X-ray diffraction study of tail tubes from bacteriophage T2L, *J. Mol. Biol.* **150**:217.

Mosher, R., and Mathews, C., 1979, Bacteriophage T4-coded dihydrofolate reductase: Synthesis, turnover, and location of the virion protein, *J. Virol.* **31**:94.

Muller-Salamin, L., Onorato, L., and Showe, M., 1977, Localization of the minor head components of bacteriophage T4, *J. Virol.* **24**:121.

Murialdo, H., 1979, Early intermediates in bacteriophage lambda prohead assembly, *Virology* **96**:341.

Murialdo, H., and Becker, A., 1978a, Head morphogenesis of complex double-stranded desoxyribonucleic acid bacteriophages, *Microbiol. Rev.* **42**:529.

Murialdo, H., and Becker, A., 1978b, A genetic analysis of bacteriophage lambda prohead assembly in vitro, *J. Mol. Biol.* **125**:57.

Murialdo, H., and Ray, P., 1975, Model for the arrangement of minor structural proteins in the head of bacteriophage lambda, *Nature* **257**:815.

Murialdo, H., and Siminovitch, L., 1972, The morphogenesis of bacteriophage lambda. IV. Identification of gene products and control of the expression of the morphogenetic information, *Virology* **48**:785.

Nakagawa, H., Arisaka, F., and Ishii, S., 1985, Isolation and characterization of bacteriophage T4 tail-associated lysozyme, *J. Virol.* **54**:460.

Nakamura, K., and Kozloff, L., 1978, Folate polyglutamates in T4D bacteriophage and T4D-infected *Escherichia coli*, *Biochim. Biophys. Acta* **540**:313.

Nakasu, S., Fujisawa, H., and Minagawa, T., 1985, Purification and characterization of gene 8 product of bacteriophage T3, *Virology* **143**:422.

Neidhardt, F. C., Phillips, T. A., VanBogelen, R. A., Smith, M. W., Georgalis, Y., and Subramanian, A. R., 1981, Identity of the B56.5 protein, the A-protein, and the groE gene product of *Escherichia coli*, *J. Bacteriol.* **145**:513.

Nelson, R., Reilly, B., and Anderson, D., 1976, Morphogenesis of bacteriophage φ29 of *Bacillus subtilis:* Preliminary isolation and characterization of intermediate particles of the assembly pathway, *J. Virol.* **19:**518.

Oakley, J., and Coleman, J., 1977, Structure of a promoter for T7 RNA polymerase, *Proc. Natl. Acad. Sci. USA* **74:**4266.

Oliver, D., and Crowther, R. A., 1981, DNA sequence of the tail fiber genes 36 and 37 of bacteriophage T4, *J. Mol. Biol.* **153:**545.

Oliver, D., and Goldberg, E., 1977, Protection of parental T4 DNA from a restriction exonuclease by the product of gene 2, *J. Mol. Biol.* **116:**877.

Papadopoulos, S., and Smith, P. R., 1982, The structure of the tail of φCbK, *J. Ultrastr. Res.* **80:**62.

Parker, M., Christensen, A., Boosman, A., Stackard, J., Young, E., and Doermann, G., 1984, Nucleotide sequence of bacteriophage T4 gene 23 and the amino acid sequence of its product, *J. Mol. Biol.* **180:**399.

Paulson, J. R., and Laemmli, U. K., 1977, Morphogenetic core of the bacteriophage T4 head. Structure of the core in polyheads, *J. Mol. Biol.* **111:**459.

Pelham, HRB., 1986, Speculation on the functions of the major heat shock and glucose-regulated proteins, *Cell* **46:**959.

Plishker, M., and Berget, P., 1984, Isolation and characterization of precursors in bacteriophage T4 baseplate assembly pathway. III. The carboxyl termini of proteins P11 are required for assembly activity, *J. Mol. Biol.* **178:**699.

Poglazov, B., and Nicoskaya, T., 1969, Self-assembly of the protein of bacteriophage T2 tail cores, *J. Mol. Biol.* **43:**231.

Poteete, A., and King, J., 1977, Functions of two new genes in *Salmonella* phage P22 assembly, *Virology* **76:**725.

Poteete, A., Jarvick, V., and Botstein, D., 1979, Encapsidation of P22 DNA *in vitro, Virology* **95:**550.

Prescott, B., Yu, M.-H., King, J., and Thomas, G. Jr., 1986, Thermostability of the secondary structure of wild type and mutant forms of phage P22 tail spike protein, *Biophys. J.* **49:**438a.

Purobit, S., Bestwick, R., Lasser, G., Rogers, C., and Mathews, C., 1981, T4 phage-coded dihydrofolate reductase. Subunit composition and cloning of its structural gene, *J. Biol. Chem.* **256:**9121.

Ramirez, G., Mendez, E., Salas, M., and Vinuela, E., 1972, Head-neck connecting protein in phage φ29, *Virology* **48:**263.

Randall-Hazelbauer, L., and Schwartz, M., 1973, Isolation of the bacteriophage lambda receptor from *Escherichia coli, J. Bacteriol.* **116:**1436.

Ray, P., and Murialdo, H., 1975, The role of gene Nu3 in bacteriophage lambda head morphogenesis, *Virology* **64:**247.

Ray, P., and Pearson, M., 1974, Evidence for the posttranscriptional control of the morphogenetic genes of lambda, *J. Mol. Biol.* **85:**163.

Ray, P., and Pearson, M., 1975, Functional inactivation of bacteriophage lambda morphogenetic gene mRNA, *Nature* **253:**647.

Ray, P., and Pearson, M., 1976, Synthesis of morphogenetic proteins by mutants of bacteriophage lambda carrying tandem duplications, *Virology* **73:**381.

Rayment, I., 1984, Animal virus structure, in: *Biological Macromolecules and Assemblies,* Vol. 1: *Virus Structure* (F. Jurnak and A. McPherson, eds.), pp. 255–298, John Wiley and Sons, New York.

Rayment, I., Baker, T., and Caspar, D., 1982, Polyoma virus capsid structure at 22.5Å resolution, *Nature* **295:**110.

Reide, I., 1986, T-even type phages can change their host range by recombination with gene 34 (tail fibre) or gene 23 (head), *Mol. Gen. Genet.* **205:**160.

Reide, I., Degen, M., and Henning, U., 1985a, The receptor specificity of bacteriophage can be determined by a tail fiber modifying protein, *EMBO J.* **4:**2343.

Reide, I., Drexler, K., and Eschbach, M., 1985b, Nucleotide sequence of the tail fiber gene 36 of bacteriophage T2 and of 36 genes of T-even type *Escherichia coli* bacteriophages K3 and Ox2, *Nucleic Acids Res.* **13**:605.

Reide, I., Drexler, K., and Eschbach, M., 1985c, Presence of DNA, encoding parts of bacteriophage tail fiber genes, in chromosome of *Escherichia coli* K12, *J. Bacteriol.* **163**:832.

Reide, I., Drexler, K., Eschbach, M., and Henning, U., 1986, DNA sequence of the tail fiber genes 37, encoding the receptor recognizing part of the fiber, of bacteriophages T2 and K3, *J. Mol. Biol.* **191**:255.

Reiger, D., Freund-Molbert, E., and Stirm, S., 1976, *Escherichia coli* capsule bacteriophages. VIII. Fragments of bacteriophage 28-1, *J. Virol.* **17**:859.

Reiger-Hug, D., and Stirm, S., 1981, Comparative study of host capsule depolymerases associated with *Klebsiella* bacteriophages, *Virology* **113**:363.

Reilly, B., Nelson, R., and Anderson, D., 1977, Morphogenesis of ϕ29 of *Bacillus subtilis:* Mapping and functional analysis of the head fiber gene, *J. Virol.* **24**:363.

Revel, H., 1981, Organization of the bacteriophage T4 tail fiber gene cluster 34-38, in: *Bacteriophage Assembly* (M. DuBow, ed.), pp. 353–364, Alan R. Liss, New York.

Revel, H., Herrmann, R., and Bishop, J., 1976, Genetic analysis of T4 tail fiber assembly. II. Bacterial host mutants that allow bypass of T4 gene 57 function, *Virology* **72**:255.

Richards, K., Williams, R., and Calendar, R., 1979, Mode of DNA packing within bacteriophage heads, *J. Mol. Biol.* **78**:255.

Roa, M., 1981, Receptor-triggered ejection of DNA and protein in phage lambda, *FEMS Microbiol. Lett.* **11**:257.

Roberts, M., White, J., Grutter, M., and Burnett, R., 1986, Three-dimensional structure of the adenovirus major coat protein hexon, *Science* **232**:1148.

Roeder, G., and Sadowski, P., 1977, Bacteriophage T7 morphogenesis: Phage-related particles in cells infected with wild-type and mutant T7 phage, *Virology* **76**:263.

Roessner, C., and Ihler, G., 1984, Proteinase sensitivity of bacteriophage lambda tail protein gpJ and gpH* in complexes with lambda receptor, *J. Bacteriol.* **157**:165.

Rose, J., Mosteller, R., and Yanofsky, C., 1970, Tryptophan messenger ribonucleic acid elongation rates and steady-state levels of tryptophan operon enzymes under various growth conditions, *J. Mol. Biol.* **51**:540.

Ross, P., and Subramanian, S., 1981, Thermodynamics of protein association reactions: Forces contributing to stability, *Biochemistry* **20**:3096.

Ross, P., Black, L., Bisher, M., and Steven, A., 1985, Assembly-dependent conformational changes in a viral capsid protein. Calorimetric comparison of successive conformational states of the gp23 surface lattice of bacteriophage T4, *J. Mol. Biol.* **183**:353.

Rossmann, M., and Erickson, J., 1985, Structure and assembly of icosahedral shells, in: *Virus Structure and Assembly* (S. Casjens, ed.), pp. 27–73, Jones and Bartlett, Boston.

Rossmann, M., Arnold, E., Erickson, J., Frankenberger, E., Griffith, J., Hecht, H., Johnson, J., Kramer, G., Luo, M., Mosser, A., Reuckert, R., Sherry, B., and Vriend, G., 1985, Structure of a human common cold virus and functional relationship to other picornaviruses, *Nature* **317**:145.

Saigo, K., 1978, Isolation of high density mutants and identification of nonessential structural proteins in bacteriophage T5: Dispensibility of L shaped fibers and a secondary major head protein, *Virology* **85**:422.

Sanger, F., Coulson, G., Hong, G., Hill, D., and Petersen, G., 1982, Nucleotide sequence of bacteriophage λ DNA, *J. Mol. Biol.* **162**:729.

Sauer, B., Ow, D., Ling, L., and Calendar, R., 1981, Mutants of bacteriophage P4 that are defective in the suppression of transcriptional polarity, *J. Mol. Biol.* **145**:29.

Savarthi, H., and Erickson, J., 1983, The self-assembly of the cowpea strain of southern bean mosaic virus: Formation of T=1 and T=3 nucleoprotein complexes, *Virology* **126**:328.

Scandella, D., and Arber, W., 1976, Phage λ DNA injection into *Escherichia coli* pel⁻ mutants is restored by mutations in phage genes *V* or *H*, *Virology* **69**:206.

Schuster, T., Scheele, R., Adams, M., Shire, S., Steckert, J., and Potschka, M., 1980, Studies of the mechanism of assembly of tobacco mosaic virus, *Biophys. J.* **32**:313.

Schwartz, M., 1976, The adsorption of coliphage lambda to its host: Effect of variations in the surface density of receptor and in phage-receptor affinity, *J. Mol. Biol.* **103**:521.

Serwer, P., 1976, The internal proteins of bacteriophage T7, *J. Mol. Biol.* **107**:271.

Serwer, P., 1980, A metrizamide-impermeable capsid of the DNA packaging pathway of bacteriophage T7, *J. Mol. Biol.* **138**:65.

Serwer, P., 1986, Arrangement of double-stranded DNA packaged in bacteriophage capsids. An alternative model, *J. Mol. Biol.* **190**:509.

Serysheva, I., Tourkin, A., Venyaminov, A., and Poglazov, B., 1984, On the presence of guanosine phosphate in the tail of bacteriophage T4, *J. Mol. Biol.* **179**:565.

Shade, S., Adler, J., and Ris, H., 1967, How bacteriophage χ attacks motile bacteria, *J. Virol.* **1**:599.

Shaw, J., and Murialdo, H., 1980, Morphogenetic genes *C* and *Nu3* overlap in bacteriophage lambda, *Nature* **283**:30.

Shore, D., Deho, D., Psipis, J., and Goldstein, R., 1978, Determination of capsid size by satellite bacteriophage P4, *Proc. Natl. Acad. Sci. USA* **75**:400.

Showe, M., 1979, Limited proteolysis during the maturation of bacteriophage T4, in: *Limited Proteolysis in Microorganisms: Biological Function, Use in Protein Structural and Functional Studies*, pp. 151–155, DHEW publication (NIH) 78-1591, U.S. Government Printing Office, Washington.

Showe, M., and Onorato, L., 1978, A kinetic model for form-determination of the head of bacteriophage T4, *Proc. Natl. Acad. Sci. USA* **75**:4165.

Showe, M., Isobe, E., and Onorato, L., 1976a, Bacteriophage T4 prehead proteinase. I. Purification and properties of a bacteriophage enzyme which cleaves the capsid precursor proteins, *J. Mol. Biol.* **107**:35.

Showe, M., Isobe, E., and Onorato, L., 1976b, Bacteriophage T4 prehead proteinase. II. Its cleavage from the product of the *21* and regulation in phage infected cells, *J. Mol. Biol.* **107**:55.

Silverstein, J., and Goldberg, E., 1976, T4 DNA injection. II. Protection of entering DNA from host exonuclease V, *Virology* **72**:212.

Simon, L. D., 1972, Infection of *Escherichia coli* by T2 and T4 bacteriophages as seen in the electron microscope: T4 head morphogenesis, *Proc. Natl. Acad. Sci. USA* **69**:907.

Simon, L. D., and Anderson, T., 1967a, Infection of *Escherichia coli* by T2 and T4 bacteriophages as seen in the electron microscope. I. Attachment and penetration, *Virology* **32**:279.

Simon, L. D., and Anderson, T., 1967b, Infection of *Escherichia coli* by T2 and T4 bacteriophages as seen in the electron microscope. II. Structure and function of the baseplate, *Virology* **32**:298.

Simon, M., Davis, R., and Davidson, N., 1971, Heteroduplexes of DNA molecules of lambdoid phages: Physical mapping of their base sequence relationships by electron microscopy, in: *The Bacteriophage Lambda* (A. Hershey, ed.), pp. 313–328, Cold Spring Harbor Laboratory, Cold Spring Harbor, NY.

Smith, P. R., Aebi, U., Josephs, R., and Kessel, M., 1976, Studies of the structure of the T4 bacteriophage tail sheath, *J. Mol. Biol.* **106**:275.

Snopek, T., Wood, W., Conley, M., Chen, P., and Cozzarelli, N., 1977, Bacteriophage T4 RNA ligase is gene *63* product, the protein that promotes tail fiber attachment to the baseplate, *Proc. Natl. Acad. Sci. USA* **74**:3355.

Stahl, F., and Murray, N., 1966, The evolution of gene clusters and genetic circularity in microorganisms, *Genetics* **58**:569.

Steitz, T. A., Anderson, W. F., Fletterick, R. J., and Anderson, C. M., 1977, High resolution crystal structures of yeast hexokinase complexes with substrates, activators, and inhibitors, *J. Biol. Chem.* **252**:4494.

Stephenson, F., 1985, A cII-responsive promoter in the *Q* gene of bacteriophage lambda, *Gene* **35**:313.

Sternberg, N., and Coulby, J., 1987, Recognition and cleavage of the bacteriophage P1 packaging site (pac) I. Differential processing of the cleaved ends *in vivo, J. Mol. Biol.* **194**:453.

Sternberg, N., and Weisberg, R., 1977, Packaging of coliphage λ DNA. II. The role of the *D* gene protein, *J. Mol. Biol.* **117**:733.

Steven, A., and Truss, B., 1986, The structure of bacteriophage T7, in: *Electron Microscopy of Proteins*, Vol. 5, *Viral Structure* (J. Harris and R. Horne, eds.), pp. 1–36, Academic Press, London.

Steven, A., Couture, E., Aebi, U., and Showe, M., 1976, Structure of T4 polyheads. II. A pathway of polyhead transformations as a model for T4 capsid maturation, *J. Mol. Biol.* **106**:187.

Steven, A., Serwer, P., Bisher, M., and Trus, B., 1983, Molecular architecture of bacteriophage T7 capsid, *Virology* **124**:109.

Studier, W., 1972, Bacteriophage T7. Genetic and biochemical analysis of this simple phage gives information about basic genetic processes, *Science* **176**:367.

Studier, W., 1979, Relationships among different strains of T7 and among T7-related bacteriophages, *Virology* **95**:70.

Svenson, S., Lonngren, J., Carlin, N., and Lindberg, A., 1979, *Salmonella* bacteriophage glycanases: Endorhamnosidases of *Salmonella typhimurium* bacteriophages, *J. Virol.* **32**:583.

Symonds, N., and Coelho, A., 1978, Role of the G segment in the growth of phage Mu, *Nature* **271**:573.

Szewczyk, B., Bienkowska-Szweczyk, C., and Kozloff, L., 1986, Identification of T4 gene *25* product, a component of the tail baseplate, as a 15K lysozyme, *Mol. Gen. Genet.* **202**:363.

Terzaghi, B., Terzaghi, E., and Coombs, D., 1979, Mutational alteration of the T4D tail fiber attachment process, *J. Mol. Biol.* **127**:1.

Terzaghi, E., 1971, Alternative pathways of tail fiber assembly in bacteriophage T4, *J. Mol. Biol.* **59**:319.

Thirion, J., and Hofnung, M., 1971, On some aspects of phage λ resistance to *E. coli* K12, *Genetics* **71**:702.

Thomas, G. Jr., Li, Y., Fuller, M., and King, J., 1982, Structural studies of P22 phage, precursor particles, and proteins by laser Raman spectroscopy, *Biochemistry* **21**:3866.

Thomas, J., 1974, Chemical linkage of the tail to the right-hand end of bacteriophage lambda DNA, *J. Mol. Biol.* **87**:1.

Thomas, J., Sternberg, N., and Weisberg, R., 1978, Altered arrangement of the DNA in injection-defective lambda bacteriophage, *J. Mol. Biol.* **123**:149.

Thomas, R., 1964, On the genetic segment controlling immunity in temperate bacteriophages, *J. Mol. Biol.* **8**:247.

Tikhonenko, A., 1970, *Ultrastructure of Viruses*, Plenum Press, New York.

Tilly, K., and Georgopoulos, C., 1982, Evidence that the two *Escherichia coli* groE morphogenetic gene products interact *in vivo, J. Bacteriol.* **149**:1082.

Tilly, K., VanBogelen, R. A., Georgopoulos, C., and Neidhardt, F. C., 1983, Identification of the heat-inducible protein C15.4 as the groES gene product in *Escherichia coli, J. Bacteriol.* **154**:1505.

To, C., Kellenberger, E., and Eisenstark, A., 1969, Disassembly of T-even bacteriophage into structural parts and subunits, *J. Mol. Biol.* **46**:493.

Tosi, M., and Anderson, D., 1973, Antigenic properties of bacteriophage φ29. Structural proteins, *J. Virol.* **12**:1548.

Tosi, M., Reilly, B., and Anderson, D., 1975, Morphogenesis of bacteriophage φ29 of *Bacillus subtilis:* Cleavage and assembly of the neck appendage protein, *J. Virol.* **16**:1282.

Toussaint, A., Lefebvre, N., Scott, J., Cowan, J., DeBruijn, F., and Bukhari, A., 1978, Relationships between temperate phages Mu and P1, *Virology* **89**:146.

Traub, F., and Maeder, M., 1984, Formation of the prohead core of bacteriophage T4 *in vivo, J. Virol.* **49**:892.

Trimble, R., Galivan, J., and Maley, F., 1972, The temporal expression of T2r+ bacterio-phage genes *in vivo* and *in vitro, Proc. Natl. Acad. Sci. USA* **69**:1659.

Tschopp, J., and Smith, P. R., 1978, Extra long T4 tails produced in *in vitro* conditions, *J. Mol. Biol.* **114**:281.

Tschopp, J., Arisaka, F., Van Driel, R., and Engel, J., 1979, Purification, characterization and reassembly of the bacteriophage T4D tail sheath protein, *J. Mol. Biol.* **114**:281.

Tsui, L., and Hendrix, R., 1980, Head-tail connector of bacteriophage lambda, *J. Mol. Biol.* **142**:419.

Van de Putte, P., Kramer, S., and Giphart-Gassler, M., 1980, Invertible DNA determined host specificity of bacteriophage Mu, *Nature* **286**:218.

Van Driel, R., and Couture, E., 1978, Assembly of the scaffolding core of bacteriophage T4 preheads, *J. Mol. Biol.* **123**:713.

Volker, T., Gafner, J., Showe, M., and Bickle, T., 1982, Gene 67, a new essential bacterio-phage T4 head gene codes for a pre-head core component, PIP. I. Genetic mapping and DNA sequence, *J. Mol. Biol.* **161**:491.

Wagenknecht, T., and Bloomfield, V., 1977, *In vitro* polymerization of bacteriophage T4D core subunits, *J. Mol. Biol.* **116**:347.

Walker, J., and Walker, D., 1981, Structural proteins of coliphage P1, in: *Bacteriophage Assembly* (M. DuBow, ed.), pp. 67–77, Alan R. Liss, New York.

Ward, S., and Dickson, R., 1971, Assembly of bacteriophage T4 tail fibers. III. Genetic control of the major tail fiber polypeptides. *J. Mol. Biol.* **62**:479.

Ward, S., Luftig, R., Wilson, J., Eddleman, H., Lyle, H., and Wood, W., 1970, Assembly of bacteriophage T4 tail fibers. II. Isolation and characterization of tail fiber precursors, *J. Mol. Biol.* **54**:15.

Wang, J., and Kaiser, A. D., 1973, Evidence that the cohesive ends of mature λ DNA are generated by the A gene product, *Nature New Biol.* **241**:16.

Weigle, J., 1966, Assembly of phage lambda *in vitro, Proc. Natl. Acad. Sci. USA* **55**:1462.

Weinstock, G., Riggs, P., and Botstein, D., 1980, Genetics of bacteriophage P22. III. The late operon, *Virology* **106**:82.

Williams, R., and Richards, K., 1974, Capsid structure of bacteriophage lambda, *J. Mol. Biol.* **88**:547.

Widom, J., and Baldwin, R., 1983, Tests of spool models for DNA packaging in phage lambda, *J. Mol. Biol.* **171**:419.

Witkiewicz, H., and Schweiger, M., 1982, The head protein D of bacterial virus λ is related to eukaryotic chromosomal proteins, *EMBO J.* **1**:1559.

Wollin, R., Ericksson, U., and Lindberg, A., 1981, *Salmonella* bacteriophage glycanases: Endorhamnosidase activity of bacteriophages P27, 9NA and KB1, *J. Virol.* **38**:1025.

Wood, W., 1979, Bacteriophage T4 assembly and the morphogenesis of subcellular struc-ture, *Harvey Lect.* **73**:203.

Wood, W., 1980, Bacteriophage T4 morphogenesis as a model for assembly of subcellular structure, *Q. Rev. Biol.* **55**:353.

Wood, W., and Conley, M., 1979, Attachment of tail fibers in bacteriophage T4 assembly: Role of the phage whiskers, *J. Mol. Biol.* **127**:15.

Wood, W., and Crowther, R. A., 1983, Long tail fibers: Genes, proteins, assembly, and structure, in: *Bacteriophage T4* (C. Mathews, E. Kutter, G. Mosig, and P. Berget, eds.), pp. 259–269, ASM Publications, Washington.

Wood, W., and Henninger, 1969, Attachment of tail fibers in bacteriophage T4 assembly: Some properties of the reaction *in vitro, J. Mol. Biol.* **39**:608.

Wood, W., and King, J., 1979, Genetic control of complex bacteriophage assembly, in: *Comprehensive Virology*, Vol. 13 (H. Fraenkel-Conrat and R. Wagner, eds.), pp. 581–633, Plenum Press, New York.

Wood, W., Edgar, R., King, J., Henninger, M., and Lielausis, I., 1968, Bacteriophage assembly, *Fed. Proc.* **27**:1160.

Wood, W., Conley, M., Lyle, H., and Dickson, R., 1978, Attachment of tail fibers in bacteriophage T4 assembly. Purification, properties and site of action of the accessory protein coded by gene 63, *J. Biol. Chem.* **253**:2437.

Wulff, D., and Rosenberg, M., 1983, Establishment of repressor synthesis, in: *Lambda II* (R. Hendrix, J. Roberts, F. Stahl, and R. Weisberg, eds.), pp. 52–73, Cold Spring Harbor Laboratory, Cold Spring Harbor, NY.

Wurtz, M., Kistler, J., and Hohn, T., 1976, Surface structure of *in vitro* assembled bacteriophage lambda polyheads, *J. Mol. Biol.* **101**:39.

Wyckoff, E., and Casjens, S., 1985, Autoregulation of the bacteriophage P22 scaffolding protein gene, *J. Virol.* **53**:192.

Yamaguchi, Y., and Yanagida, M., 1980, Head shell protein hoc alters the surface charge of bacteriophage T4, *J. Mol. Biol.* **141**:175.

Yamamoto, N., and Anderson, T., 1961, Genomic masking and recombination between serologically unrelated phages P22 and P221, *Virology* **14**:430.

Yamamoto, M., and Uchida, H., 1975, Organization and function of the tail of bacteriophage T4. II. Structural control of the tail contraction, *J. Mol. Biol.* **92**:207.

Yanagida, M., 1977, Molecular organization of the shell of T-even bacteriophage head. II. Arrangement of subunits in the head shells of giant phages, *J. Mol. Biol.* **109**:515.

Yanagida, M., and Ahmed-Zadeh, C., 1970, Determination of gene product positions in bacteriophage T4 by specific antibody association, *J. Mol. Biol.* **51**:411.

Yanagida, M., Boy de la Tour, E., Alff-Steinberger, C., and Kellenberger, E., 1970, Studies on the morphopoeisis of the head of bacteriophage T-even. VIII. Multilayered polyheads, *J. Mol. Biol.* **50**:35.

Youderian, P., 1978, Genetic analysis of the length of the tails of lambdoid bacteriophages, Ph.D. thesis, Massachusetts Institute of Technology.

Zorzopulos, J., and Kozloff, L., 1978, Identification of T4D bacteriophage gene product *12* as the baseplate zinc metalloprotein, *J. Biol. Chem.* **253**:5593.

Zweig, M., and Cummings, D., 1973, Cleavage of head and tail proteins during bacteriophage T5 assembly: Selective host involvement in the cleavage of a tail protein, *J. Mol. Biol.* **80**:505.

CHAPTER 3

Changes in RNA Polymerase

E. PETER GEIDUSCHEK AND GEORGE A. KASSAVETIS

I. INTRODUCTION

The bacteriophages generate developmental time sequences of gene expression by a variety of mechanisms, primarily but not exclusively involving transcription. There are many ways to modulate transcription in prokaryotes; phages are known to use many of these and may use them all. The viral infections we discuss are terminal events for their bacterial host cells, comparable, in that sense, with noninfectious terminal differentiations such as heterocyst and spore formation. From a regulatory point of view, the principal difference is that the commitment to bacterial differentiation involves more or less complex interplays of metabolic signals that the initiating event of infection—virus attachment and the injection of the foreign DNA—in no way resembles. However, similar alterations of transcription machinery are called into action during bacterial differentiation and during phage infection.

We interpret the topic of this chapter—RNA polymerase modification in virus development—as including noncovalent along with covalent alterations of the host's enzyme during viral development, but excluding virus-coded RNA polymerases that are newly synthesized during phage infection. Since many of the chapters in this book deal categorically with individual groups of phages, our emphasis is, where possible, comparative and mechanistic.

E. PETER GEIDUSCHEK AND GEORGE A. KASSAVETIS • Department of Biology and Center for Molecular Genetics, University of California, San Diego, La Jolla, California 92093.

II. COVALENT MODIFICATIONS OF RNA POLYMERASES

Knowledge about covalent polymerase modification comes from studies of bacteriophages T4 and T7. The *E. coli* RNA polymerase is ADP ribosylated in its α subunits after phage T4 infection. Two distinctive phage coded ADPR transferases are responsible for this modification. One of the transferases is an encapsidated 70-kD protein, the product of gene *alt*, and is injected at infection. This enzyme has a relatively broad range of substrates, transfers ADPR to Arg 265 of *one* α subunit, but is also capable of transferring ADPR to other sites in RNA polymerase and to other substrates. The *alt* protein is evidently not designed to act in the concurrent infectious cycle: it is synthesized as an inactive precursor and activated only upon virion assembly (quoted by Rabussay, 1983; Horvitz, 1974). The second enzyme, a 27-kD protein encoded by gene *mod*, is strictly specific with regard to substrate and transfers ADPR to Arg 265 of both α subunits. In a normal infection, both α subunits are fully modified within 4–5 min (at 30°C). Evidence that the half-modification of α (by the *alt* enzyme) is reversible *in vivo* after a time interval suggests the possible involvement of a phage-specific cleaving enzyme, but this has not been characterized. Either the fully modified enzyme is not sensitive to this activity or else the balance between cleavage and retransfer of ADPR greatly favors full enzyme modification.

E. coli RNA polymerase is phosphorylated after phage T7 infection by the T7 gene 0.7 protein, a threonine phosphokinase that has many substrates in the *E. coli* cell. In RNA polymerase, β′ is its principal target, with several sites of phosphorylation; β is much less, and σ only barely, phosphorylated. *In vitro*, holoenzyme is a better substrate than core (Rothman-Denes *et al.*, 1973: Rahmsdorf *et al.*, 1974; Zillig *et al.*, 1975). The time span of action of this kinase, which is a T7 early protein, is limited by its duration of synthesis and by the (auto)phosphorylation that ultimately inactivates it. The extent of phosphorylation of RNA polymerase, among other substrates, during T7 infection ultimately decreases, presumably owing to the action of a phosphatase (Zillig *et al.*, 1975). However, by that time of the infection, another phage-specific barrier to the activity of *E. coli* RNA polymerase has been erected through the synthesis of gp2 (discussed below, in Section III).

The phage T4 and T7 RNA polymerase modifications must be seen in their different biological contexts: Phage T7 codes for its own RNA polymerase, shuts off the activities of, and then totally degrades, the *E. coli* genome, and eventually dispenses with the function of the cellular RNA polymerase. In contrast, bacteriophage T4 employs the core subunits of *E. coli* RNA polymerase throughout infection for all of its own transcription. Neither covalent modification of RNA polymerase is absolutely essential to the execution of the phage-specific programs of gene expression: $alt^- mod^-$ T4 phage and gene 0.7^- T7 phage grow well on various laboratory strains of *E. coli*. In the T4 case, ADP ribosylation does change properties of RNA polymerase that are discernible *in vitro* and have plausible connections to the transcription-regulatory events of phage T4 development.

1. ADP ribosylation inhibits transcription from certain *E. coli* and T4 early promoters (Mailhammer *et al.*, 1975, Goldfarb, 1981; Goldfarb and Palm, 1981) and thus may be a factor in host transcription shutoff and in the shutoff of some T4 early promoters, although it is clearly not the sole agent of shutoff (see below), since $alt^- mod^-$ phages are host shutoff-competent (Goff and Setzer, 1980).
2. It has been suggested that ADP ribosylation changes chain termination and pausing by RNA polymerase (Schäfer and Zillig, 1973; Rabussay, 1983; Goldfarb and Malik, 1984). Experiments involving other possible components of RNA chain termination, such as *nus* A protein, or the recently described factor, τ (Briat and Chamberlin, 1984), have not yet been done. It has been surmised, on the basis of extensive physiological and genetic experiments, that the switch from the first phase of transcription (traditionally called "pre-early") to the second, middle phase (traditionally called "early") includes an antitermination mechanism (reviewed by Brody *et al.*, 1983). Thus, the above-cited experiments also connect ADP ribosylation with this second phase-specific regulatory event (although the specific details are difficult to fit into a consistent picture). There exists an as yet ill-defined T4 function, *com-Cα*, which is distinct from *mod* and *alt* and is implicated in phage-host interactions affecting transcriptional termination (Caruso *et al.*, 1979; Pulitzer *et al.*, 1979). Thus, ADP ribosylation of the RNA polymerase α subunits is not likely to be the sole means by which phage T4 modify the host's apparatus for RNA chain termination.
3. ADP ribosylation of RNA polymerase changes interactions with other RNA polymerase-binding proteins. It decreases affinity for σ (Rabussay *et al.*, 1972), enhances the anti-σ action of a 10-kD phage-coded RNA polymerase binding protein (Stevens, 1976, 1977), and must thus be capable of affecting promoter strength (Khesin *et al.*, 1972, 1976).

III. PROTEINS BINDING TO THE BACTERIAL RNA POLYMERASE CORE

Many larger lytic phages code for RNA polymerase-binding proteins. The list includes *E. coli* phages T7 and T3, T5, T4 and T2 and *B. subtilis* phages SP01 and SP82. It is useful to divide these proteins, which we next describe, into two groups: (1) the σ-like initiation factors, which positively regulate gene expression by specifically altering promoter recognition; quite a lot is known about the initiation factors; and (2) all the other proteins, about which relatively little is known.

A. Phage-Coded Initiation Proteins

The proteins that we now consider are part of a highly diverged family that includes the bacterial σ proteins (Gribskov and Burgess, 1986). These phage-coded, σ-like proteins retain the promoter and core interaction functions of the much larger major σ^{70} of E. coli and σ^{43} of B. subtilis but lack domains that surely have some other significance for transcription (Fig. 1).

1. T4 Gene 55 Protein

This protein, which is required for transcribing the T4 late genes, binds to E. coli RNA polymerase core (Ratner, 1974a). One segment of gp55, located near the middle of its 185 amino acid polypeptide chain (Gram and Rüger, 1985), is homologous to other σ-like proteins (Gribskov and Burgess, 1986) (Fig. 1). When either the ADP-ribosylated or the unmodified form of E. coli RNA polymerase core binds gp55 in place of E. coli σ^{70}, T4 late promoters, with their characteristic sequence RXCXX-TATAAATAX$_{1-3}$ AYT$\boxed{\text{R}}$A/T (the box marks the preferred starting site

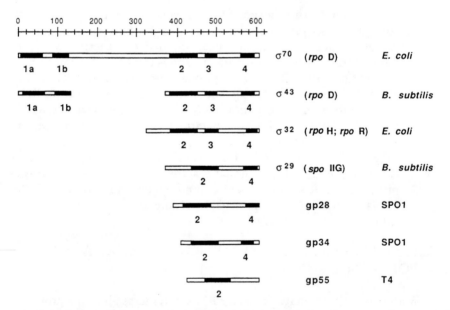

FIGURE 1. The locations of conserved amino acid sequence in the primary structures of transcription-initiation proteins (from Gribskov and Burgess, 1986). σ^{70} and σ^{43} are the principal initiation factors of E. coli and B. subtilis, respectively. σ^{32} is the E. coli htpR (rpoH) protein, regulating heat shock-specific transcription. σ^{29} is the B. subtilis spoIIG protein (Trempy et al., 1985; Stragier et al., 1984). The scale at the top of the figure shows the number of amino acid residues in these polypeptide chains. The conserved segments are numbered, and their lengths are indicated by the solid bars.

TABLE I. Promoter Sequences

σ Family	$-35^{a,c}$	Spacing[b]	$-10^{a,c}$
E. coli σ^{70d}	TTGACA	17 ± 1	TATAAT
B. subtilis σ^{43}			
SPO1 gp28	AGGAGA[d]	18 ± 1	hhhXhhh[d,e]
SPO1 late	CGhhAGA	18 ± 1	GAhAhh[e]
(gp34 or 33/34)			
T4 gp55	None	—	TATAAATA[d]
Within–promoter			
positive control			
λ cII	TTGCXXXXXXTTGC[f]	13	$(\sigma^{70}\text{—}10)$
T4 motA	$\frac{\text{AAA}}{\text{TTT}}$TGCTT	14 ± 1	$(\sigma^{70}\text{—}10)$

[a] All sequences in the nontranscribed strand.

[b] Number of base pairs between the 3' end of the upstream consensus sequence and the 5' end of the downstream consensus sequence (as written).

[c] Consensual uniformity is not constant across these sequences. (See Hawley and McClure, 1983, for extensive data on E. coli σ^{70}.)

[d] These are the core consensus sequences. The conservation of sequence extends as follows:

SPO1 gp28 (-35): $\frac{\text{A}}{\text{h}}$XAGGAGAXRAXhh

SPO1 gp28 (-10): hhhXhhhhXXXYRA

T4 gp55 (-10):   RXCXXTATAAATAX$_{1-3}$AYT$\boxed{\text{R}}\frac{\text{A}}{\text{T}}$(preferred transcriptional start site boxed in).

[e] Preferential transcription of hmUra (h)-containing DNA.

[f] The cII consensus site flanks the -35 promoter site, denoted by X$_6$.

Taken from the following references: E. coli σ^{70} and B. subtilis σ^{43}: Hawley and McClure (1983), Doi and Wang (1986); SPO1 gp28: Losick and Pero (1981), V. Scarlato (unpublished); SPO1 late: Costanzo (1984); T4 gp55: Christensen and Young (1982, 1983), Kassavetis et al., (1986); λ cII: Ho et al., (1986); T4 mot A: Guild (1986).

in vitro; Table I), are recognized instead of E. coli promoters (Kassavetis and Geiduschek, 1984; Malik et al., 1985). The transcription activity of RNA polymerase core associated with gp55 is greatly enhanced in vitro in underwound DNA (Kassavetis and Geiduschek, 1984). No detailed studies of kinetics or binding in T4 late promoter-RNA polymerase interactions, comparable with the work on E. coli σ^{70} promoters (reviewed by McClure, 1985), have been done. Like E. coli σ^{70}, T4 gp55 is released from RNA polymerase core during RNA chain elongation (Kassavetis et al., 1987) and, like σ^{70}, gp55 is substoichiometric to RNA polymerase core in the cell: it is made from 5 until after 18 min of T4 infection at 30°C (Stevens, 1972; Horvitz, 1973; Ratner, 1974b), and an analysis with specific antibodies to gp55 provides the estimate that there are only ~ 550 molecules of gp55 per cell at 12 min after phage T4 infection at 37°C (equivalent to ~ 18 min at 30°C; Williams, unpublished). Given these properties, it is evident that gp55 must compete for RNA polymerase core with other initiation subunits during each round of transcription. The host σ^{70} subunit is not inactivated or destroyed after T4 infection, yet in vitro experiments suggest that gp55 is intrinsically at a disadvantage in competing with σ^{70}; gp55 is displaced by, rather than displacing, σ^{70} when these two proteins

compete for *E. coli* unmodified or ADP ribosylated RNA polymerase core
(Kassavetis *et al.*, 1987). Presumably, therefore, other proteins serve
eventually to block the participation of σ^{70} in transcription. T4 gp33 is
discussed in this context later on in the chapter. The function of gp33,
which is also a RNA polymerase-binding protein, is required for T4 late
transcription *in vivo* but not for T4 late promoter recognition by pure RNA
polymerase *in vitro*.

The *in vitro* analysis of RNA chain initiation at T4 late promoters
shows that the distance between the recognition sequence and the transcriptional start site is quite precisely fixed; for example, at the gene 23
promoter, chains start almost exclusively at G, 13 bp downstream of the *T*
in *TATAAATACTCCTGA* (Kassavetis and Geiduschek, 1984; Kassavetis
et al., 1986). At two other gp55-dependent promoters, initiation *in vitro*
was found to occur either at this 13-bp spacing or at both a 12- and a 13-bp
spacing (Kassavetis *et al.*, 1986). Analysis of transcriptional start sites *in
vivo* is complicated by subsequent RNA processing and by various problems of analysis, particularly of the much used S1 nuclease analysis for
mapping RNA 5' ends. Thus, the assignment of heterogeneous *in vivo* start
sites to various promoters (Christensen and Young, 1983) including the
gene 23 promoter does not *a priori* signify a conflict. However, RNA 5'
ends at two late promoters for the lysozyme gene have been mapped by
primer extension with reverse transcriptase in conjunction with an examination of mRNA turnover (McPheeters *et al.*, 1986). One possibility is
that the observed spread of RNA 5' ends at these two promoters may,
indeed, reflect heterogeneous transcriptional initiation. The reason for
considering this detail significant or interesting is that the transcription of
T4 late genes *in vivo* is normally coupled to concurrent replication, and the
nature of the DNA template on which transcription occurs *in vivo* is not
known. *In vitro* transcription is currently done with (underwound) double-
stranded DNA. If there really is a difference in detail between *in vivo* and
in vitro initiation, it is likely that it reflects significant differences between the conformations and structures of promoters.

2. SP01 Gene 28 and 34 Proteins

The phage SP01 genome codes for at least six RNA polymerase-
binding proteins (Fox and Pero, 1974; Fox *et al.*. 1976; Pero *et al.*, 1975).
Three of these proteins execute successive steps in the viral transcription
program. The function of gene 28, which codes for a 220 amino acid *B.
subtilis* RNA polymerase core-binding protein, is required for the middle
phase of viral transcription. The functions of genes 33 and 34 are required
for the late phase of viral transcription, and gene 34 codes for a 197 amino
acid RNA polymerase core-binding protein (Costanzo *et al.*, 1984; prior
work reviewed by Geiduschek and Ito, 1982).

B. subtilis RNA polymerase core with gp28 attached (E.gp28) recognizes phage SP01 middle promoters, whose characteristic sequence is (A/h)AGGAGAXRAXhh (12 ± 1 bp) hhhXhhhh (3 bp) YRA (Talkington and Pero, 1978, 1979; Losick and Pero, 1981). Gp28 has some homology to segments of the other σ-like proteins (Fig. 1). Since gp28 positively controls viral middle gene expression, and since the utilization of viral early promoters declines at about the time that viral middle transcription starts, gp28 might be thought capable of directly displacing the major initiation subunit, σ^{43}, from B. subtilis RNA polymerase holoenzyme. The available evidence goes counter to that idea. Gp28, like T4 gp55, is ineffectual as an anti-σ protein, at least at moderate ionic strengths, in vitro. Instead, gp28 remains bound to core in ternary transcription complexes, rather than dissociating from core as do E. coli σ^{70}, B. subtilis σ^{43}, and T4 gp55. Thus, when a molecule of gp28 attaches to a unit of RNA polymerase core, it is able to bias or perhaps even to sequester the activity of that molecule for middle gene transcription (Chelm et al., 1982).

As it confers SP01 middle promoter recognition on B. subtilis RNA polymerase core (Duffy and Geiduschek, 1976, 1977), gp28 also confers preferential binding to hmUra-containing DNA. This preference for hmUra over T in transcriptional initiation at middle promoters (Lee et al., 1980) is not absolute (Romeo et al., 1986): specific and selective initiation at T-containing promoters can be detected in underwound DNA with sufficiently sensitive methods, although middle promoter strength is much reduced in T-containing DNA. This nucleotide modification specificity has been explored further by analyzing the properties of hybrid middle promoters in which either strand is T-containing and the complementary strand contains hmUra. Since the core consensus sequence of middle promoters is (in the nontranscribed strand) AGGAGA (18 ± 1 bp) hhhXhhh, it consists entirely of purines at −35 and at the transcription start, and entirely of pyrimidines at −10. T-for-hmUra substitution in the nontranscribed strand therefore affects the −10 sequence selectively, while the same substitution in the transcribed strand affects the −35 core recognition and start sequences selectively.

It has been found that both hybrid promoters are weaker than the normal promoter containing only hmUra. The promoter that is T-substituted at −10 (i.e., in the nontranscribed strand) is weaker than the promoter that is T-substituted at −35 (i.e., in the transcribed strand). In general, the transcription and polymerase-binding properties of these hybrid promoters and the normal all-hmUra promoter correlate, in the sense that the normal promoter binds E.gp28 most tightly and displays the most electrolyte-resistant transcription whereas the non-coding-strand-T-substituted promoter binds RNA polymerase most weakly, apparently does not form stable open promoter complexes at any temperature, and is least well transcribed (Choy et al., 1986). The relative weakness of both hybrid promoters indicates that the preferential hmUra interactions of enzyme containing gp28 are dispersed along the promoter rather than restricted to only one cluster of pyrimidines and that they are distributed on both strands. That is remarkable, considering the small size of gp28, but consistent with the proposal (Losick and Pero, 1981) that gp28 interacts directly with both promoter regions. Preferential binding by a protein to hmUra-containing B helix DNA could have its source either in the substitution of $-CH_2OH$ for $-CH_3$ in the major groove of B DNA or, if the hmUra-for-T substitution affects the detailed structure of DNA, in the changed width and shape of the major and minor helical grooves. HmUra-containing phage DNA is B-type DNA, but no detailed structural comparison of oligonucleotides containing T and hmUra has yet been attempted.

In comparing SP01 sp28 with T4 gp55, it is interesting to note that gp28, which recognizes a bipartite $(-35/-10)$ sequence has two regions of homology with the family of initiation factor proteins, whereas the slightly smaller gp55 has only one region of homology and recognizes a nonsegmented promoter (-10 only). What role might the conserved segments of these polypeptide chains play? At least one conserved segment might represent an attachment site to core enzyme. If gp28 has two RNA polymerase core attachment sites, then perhaps each one is associated with a segment (or domain) of the protein that interacts, separately, with the -35 or -10 promoter sequence.

RNA polymerase core with gp33 and 34 attached recognizes phage SP01 late promoters, whose consensus is CGhhAGA (18 ± 1 bp) GAhAhh (Table I). Thus, at least one of these two phage-coded proteins is an analogue of gp28. Genes 33 and 34 have been sequenced: gp34 (197 amino acids) is homologous with other transcription-initiating proteins (Table I), but gp33 (101 amino acids) is not (Costanzo et al., 1984; Gribskov and Burgess, 1986). It is thus possible that only SP01 gp 34, like T4 gp55, is required for late promoter recognition and that SP01 gp33, like T4 gp33, has some other role in determining the ability to transcribe late genes.

B. Other RNA Polymerase Core-Binding Proteins

We turn next to a heterogeneous and much less studied group of proteins. For the sake of the general interest of the discussion, comparative and general considerations will be introduced where possible, at the risk of imposing order without knowing the proper rules.

1. A T4-Coded Anti-σ Protein

If positive transcriptional regulation, permitting the recognition of different promoters in temporal sequence, can be generated by batteries of σ-like RNA polymerase-binding proteins, then repression sequences might be similarly generated by proteins that block σ function (Bogdanova et al., 1970; Khesin et al., 1972). Such anti-σ substances might include inhibitors whose only function is to block the access of specific σ-like initiation factors to their binding sites on RNA polymerase core and initiation factors that compete directly for a common σ-binding sites on RNA polymerase core. Thus, *a priori*, the phage SP01 gp28 and T4 gp55 might serve as anti–*B. subtilis* σ^{43} and anti–*E. coli* σ^{70}, respectively. As already mentioned, the present evidence goes against this attractive notion: SP01 gp28 is ineffectual at displacing σ^{43} from *B. subtilis* holoenzyme, tending instead to be displaced by σ^{43} at low ionic strength, and T4 gp55 is also unable to alone displace *E. coli* σ^{70} from RNA polymerase core. T4 gp55, like *E. coli* σ^{70}, recycles, although SP01 gp28 remains with transcription complexes during RNA chain elongation, which should bias the attached molecule of RNA polymerase core to repeated cycles of gp28-specific transcriptional activity. It is not yet known whether minor σ-like proteins of uninfected bacteria work in the same way as SP01 gp28.

How, then, is the activity of the major σ of *E. coli* and *B. subtilis*, which can be recovered functionally intact and chemically either entirely or mostly unmodified from T4- or SP01-infected bacteria, blocked *in vivo* at late stages of viral development? It is in this context that we consider one of the phage T4-specific RNA polymerase-binding proteins (Stevens, 1970, 1972). This small protein (molecular weight ~ 10,000 according to its electrophoretic mobility in denaturing gels) copurifies with RNA polymerase, is detached with σ from core enzyme on phosphocellulose, but can be dissociated from σ in urea (Stevens and Rhoton, 1975; Stevens, 1976). This protein inhibits σ^{70}-mediated initiation with unmodified or ADP ribosylated RNA polymerase at relatively high ionic strength and, unlike σ or RNA polymerase holoenzyme, is subject to inhibition by a neutral detergent (Stevens, 1974, 1976, 1977). The T4 gene coding for this protein has not yet been identified. As already discussed in connection with RNA polymerase core modification after T4 infection, the antagonism to *E. coli* σ^{70} is probably multifactorial. We suggest another candidate for participation in this activity in the next section.

2. T4 and SP01 Gene 33 Proteins

These two proteins are considered together on the basis of relatively superficial resemblances. The previously mentioned 12.8-kD T4 gp33 (112 amino acids; Hahn et al, 1986), which binds relatively loosely to RNA polymerase core, is required for late transcription *in vivo* yet not required for T4 late promoter recognition *in vitro*. The function of gp33 has not yet been analyzed *in vitro* with pure preparations of the protein. It is evidently not absolutely required for T4 late promoter recognition (Kassavetis and Geiduschek, 1984) or for gp55 binding to RNA polymerase core (Kassavetis *et al.*, 1987). One function of gp33 is suggested by the following observations: (1) At the outset of T4 DNA replication, the RNA primer is generated at replication origins by RNA polymerase. Eventually, DNA relication becomes strongly dependent on recombination, but that dependence is relieved if the function of gene 33 or gene 55 fails. A simple molecular interpretation, with regard to gp33, is that it might participate in displacing *E. coli* σ, thereby inactivating the initiation of RNA primer synthesis in the vicinity of replicative origins. (2) The fact that a gene 33 mutant is also defective in shutting off certain early transcription units (Bolund, 1973; Sköld, 1970) is also consistent with this interpretation. Neither observation requires that gp33, *by itself*, be capable of acting as an anti-σ protein; with the cloning of gene 33 (Hahn *et al.*, 1986), experiments to test directly for that function are, in principle, relatively accessible.

The rationale for connecting SP01 gp33 with T4 gp33 is weak. The SP01 protein is also small (101 amino acids; Costanzo *et al.*, 1984), is also an RNA polymerase-binding protein, and is also required for late transcription *in vivo*, yet its amino acid sequence does not place it in the σ family (Gribskov and Burgess, 1986). While *B. subtilis* RNA polymerase containing gp33 and 34 together transcribes late genes *in vitro* (Tjian and Pero, 1976), experiments to examine the ability of SP01 sp34 alone to act in the same way as T4 gp55 have not been done.

3. Other Genetically Identified RNA Polymerase-Binding Proteins

We turn next to the products of adjacent T4 genes coding for two other RNA polymerase-binding proteins. One of these proteins sticks very tightly to RNA polymerase core. The gene coding for this protein has recently been mapped (Williams *et al.*, 1987) to a previously sequenced open reading frame next to gene 45 (Hsu *et al.*, 1987). The molecular weight of this 100–amino acid protein is 11,400; its name, 15K, derives from its gel-electrophoretic apparent molecular weight, and it has been renamed the *rpb*A (for *R*NA *p*olymerase *b*inding) protein. The *rpb*A protein changes RNA polymerase-promoter interactions, whether attached to unmodified or ADP ribosylated RNA polymerase core. It increases the thermal transition temperature for forming open complexes at T4 early promoters (tested in total T4 DNA with *E. coli* σ70-containing enzyme) by, on the average, approximately 15°C (Malik and Goldfarb, 1984). Thus, at certain temperatures within that transition range, early promoter utilization is depressed by the *rpb*A protein. However, it remains to be shown whether this effect on open complex formation is selective, against early gene expression only, or general. The *rpb*A protein may also be responsible for the lower affinity of modified and unmodified RNA polymerase from T4-infected cells for *E. coli* ribosomal promoters (Baralle and Travers, 1976). Apparently, the *rpb*A protein does not (alone) allow gp55 to displace *E. coli* σ70 from RNA polymerase core (Kassavetis *et al.*, 1987).

The *rpb*A protein remains associated with elongating ternary transcription complexes and thus is capable of phage-specifically modifying interactions of RNA polymerase with other proteins at every stage of transcription. ADP ribosylated RNA polymerase carrying the *rpb*A protein reads through a ρ-independent terminator in the tRNA region of the T4 genome (Broida and Abelson, 1985; Goldfarb and Malik, 1984).

Although it is known that the T4 gene 45 protein (222 amino acids, molecular weight 24,700) is absolutely required for T4 late transcription, the mechanism of its action is not understood. Gp45 is also an essential component of the T4 replisome (reviewed by Nossal and Alberts, 1983) constituting a DNA-dependent ATPase with gp44 and 62. Since T4 late transcription is normally strongly coupled to concurrent DNA replication, gp45 might be involved in that coupling, but it is, in fact, also absolutely required for replication-independent late transcription (Wu et al., 1975; Jacobs and Geiduschek, 1981).

Three lines of evidence suggest that gp45 interacts with RNA polymerase: (1) Suppressor genetic analysis (Coppo et al., 1975a,b) identifies an interaction between T4 genes 45 and 55 on the one hand and the E. coli gene coding for RNA polymerase subunit β on the other. The difficulty with this compelling evidence for a functional connection is that it does not a priori distinguish between physical and physiological levels of connection (although it is common to ignore the physiological alternative in interpreting genetic suppression). (2) Gp45 has been shown to bind to immobilized T4-modified RNA polymerase but not to unmodified RNA polymerase core or holoenzyme (Ratner, 1974b). This direct experiment, which suggests binding that is dependent on the ADP ribosylation of α subunits, is also not unequivocal. Since it does not establish stoichiometry, it fails to distinguish between binding to RNA polymerase core and binding to RNA polymerase core-binding proteins or even to minor contaminants of the preparation. (3) A protein comigrating on polyacrylamide gels with purified gp45 (but referred to as having an equivalent molecular weight of 29,000 instead of its real molecular weight of 24,700) copurifies with RNA polymerase from cells infected with particular T4 mutants (in genes reg A, 42, and 30; Malik and Goldfarb, 1984). These cells are replication- and late transcription-defective and DNA ligation-defective and overproduce many early proteins. The attachment of gp45 to RNA polymerase from wild-type phage-infected cells and from cells infected with various other mutants does not survive the same preparation method. Gp45 has not been identified as a component of highly purified RNA polymerase made by other methods (Rabussay, 1983). Thus, copurification of gp45 with RNA polymerase may depend on a complex balance of mutually exclusive and mutually dependent interactions of RNA polymerase-binding proteins.

4. A Virus-Coded Inhibitor of Bacterial RNA Polymerase

Bacteriophage T7 codes for a 64 amino acid RNA polymerase-binding protein that inactivates the enzyme (Hesselbach and Nakada, 1975, 1977a,b; De Wyngaert and Hinkle, 1979; Dunn and Studier, 1981). Since this protein is the product of gene 2, a middle gene (group II in the standard nomenclature, see Chapter 12), it acts after the already discussed T7 gene 0.7 protein kinase, which inactivates *E. coli* RNA polymerase by phosphorylation, primarily of subunit β'. The small (62 amino acids) gp2 forms salt-dissociable complexes with holoenzyme and also binds somewhat less tightly to core enzyme. Upon binding, gp2 almost completely blocks the transcriptional activity of holoenzyme by preventing the formation of open promoter complexes. The RNA polymerase activity of core enzyme is not inactivated by gp2 (De Wyngaert and Hinkle, 1979). Its interactions with *E. coli* RNA polymerase-T7 early promoter complexes under various conditions can, with hindsight, be interpreted as suggesting that gp2 inactivates preformed closed but not open promoter complexes. It seems, therefore, that gp2 might be an interesting probe of the structure of promoter complexes. Since it inactivates the σ^{70}-bearing holoenzyme but not core enzyme, a part of the binding specificity of holoenzyme for T7 gp2 might reside in σ^{70}. If that is the case, then T7 gp2 might be capable of discriminating between enzymes carrying different transcription-initiation factors—e.g., the *E. coli rpo* D, H, and N proteins (Fig. 1). From a general point of view, one source of interest in this question relates to the possible existence of anti-σ proteins acting at the level of RNA polymerase to negatively regulate the activity of minor σ proteins. However, the primary function of gp2—that is, the function that makes it indispensable—involves virion assembly rather than gene regulation: it appears to be required to strip bound *E. coli* RNA polymerase off viral DNA as a prerequisite to packaging in the phage head (LeClerc and Richardson, 1979).

Be that as it may, at least a part of the recognition for gp2 has been shown to reside in the β' subunit of RNA polymerase core, coded for by gene *rpo* C. Certain *E. coli rpo* C mutants are nonpermissive (*tsn* B) for T7 phage (Buchstein and Hinkle, 1982). The altered RNA polymerase of these mutants is resistant to inhibition by gp2 (Shanblatt and Nakada, 1982; De Wyngaert and Hinkle, 1979), and a T7 mutant that overcomes the *tsn* B block has 3 amino acid changes in gp2 (Schmitt *et al.*, 1987).

Phage T3 also codes for an inhibitor of RNA polymerase (Mahadik *et al.*, 1972, 1974) with similar properties. The T3 gene 2 has recently been sequenced: it should code for a 52 amino acid protein that is highly conserved, relative to T7 pg2, along a 40 amino acid stretch (Schmitt *et al.*, 1987), but it has not yet been shown that this protein is the active RNA polymerase inhibitor.

IV. DNA-BINDING PROTEINS THAT INTERACT WITH, AND ENHANCE THE ACTIVITY OF, RNA POLYMERASE

A discussion of phage-specific modifications of the activities of cellular RNA polymerase must include consideration of the effects that can be generated by DNA-binding proteins. The prototypes are the phage λ cII and cI proteins. We focus here on the cII protein.

The cII protein is a positive regulator of transcriptional initiation at three phage λ promoters by virtue of its ability to bind to DNA, independently of RNA polymerase, 27–40 bp upstream of the transcriptional start sites (Wulff and Rosenberg, 1983; Ho et al., 1983; Shih and Gussin, 1984a,b; Ho and Rosenberg, 1985; Hoopes and McClure, 1985; Stephenson, 1985). These three λ promoters conform relatively poorly to the E. coli σ^{70} consensus at -10 and not at all at -35. They are extremely weak in the absence of cII protein but greatly activated by cII binding: the second-order rate constant for open promoter formation (i.e., the product, $K_B k_2$, of the equilibrium constant for RNA polymerase binding and the first-order rate constant for promoter opening, respectively) is increased 600-fold under standard conditions at the P_{RE} and 10^4-fold at the P_I promoter (Shih and Gussin, 1984b; McClure, 1985). In each of these three promoters, the noncanonical -35 sequence is flanked by a characteristic sequence: $TTGCX_6TTGC$. That the cII protein, a tetramer of 11-kD subunits (Ho et al., 1982) binds specifically to the directly repeated TTGC sequence in the absence of RNA polymerase has been shown by mutagenesis and by DNA footprinting (Ho et al., 1983). Stereochemical considerations and model building lead to the conclusion that cII protein and RNA polymerase can bind to opposite sides of the same segment of DNA helix (Ho et al., 1983). Since each protein increases the affinity of the other in binding to DNA, direct protein-protein contacts between cII and RNA polymerase are envisaged (Ho et al., 1983, 1986).

CII protein works in conjunction with, rather than in substitution for, σ. Although cII binding sites at these promoters functionally substitute for RNA polymerase holoenzyme -35 binding sites, RNA polymerase still engages the -35 sequence. This is made clear by the footprinting analysis of RNA polymerase-DNA contacts and by finding that certain mutations between the TTGC direct repeats, in the -35 site, lower the activity of the cII-activated P_{RE} promoter (Shih and Gussin, 1983).

Protein alignment on a helical DNA lattice is extremely sensitive to the spacing of binding sites. Each base pair added or removed between two binding sites changes relative orientations by $1/10$ of a turn, which translates into large changes in the locations of functional groups on the distal surfaces of proteins binding to the DNA. If each of the components from which a functional cII-activated promoter complex must be assembled is relatively rigid, it is easy to understand why the cII binding site is precisely fixed relative to the -35 site of the promoter. In that respect, the cII and ntrC transcriptional activators are precisely antithetical: the ntrC (also designated as NRI or glnG) protein contributes to positively regulated transcription at the P_2 promoter of the E. coli gln ALG operon by binding to sites that can be moved around, relative to the promoter, and inverted. These binding sites for the ntrC protein resemble enhancers in eukaryotic cells (Hunt and Magasanik, 1985; Reitzer and Magasanik, 1986).

A mutational analysis of the cII protein (Wulff and Rosenberg, 1983;

Wulff et al., 1984) and examination of its amino acid sequence (Schwartz et al., 1978) help to define functional domains. Mutations in two relatively hydrophobic segments comprising almost half the amino acids affect the ability to assemble the cII tetramer. Between these domains lies a 19–amino acid sequence that shows homology to the helix-turn-helix motif of several other site-specific DNA-binding proteins. Several mutations within this segment affect DNA binding (Ho et al., 1986). Mutations that affect promoter activation without changing DNA binding can pinpoint the part of the activator protein surface that touches RNA polymerase. Such mutations have been identified in the λ cI gene and provide the basis for a specific model of how the cI protein acts as an activator of RNA polymerase at the λ P_{RM} promoter (Guarente et al., 1982; Hochschild et al., 1983; Ptashne, 1986). No comparable mutations have been reported for the cII gene.

The subtlety and precise adjustment of protein contacts required for cII function are suggested by the following observation (Wulff and Mahoney, 1985). The lambdoid phages, 21 and P22, also code for cII-like proteins. (Fortune would have it that the P22 homologue of λ cII is called cI!) Each of these related proteins optimally activates its homologous P_{RE} promoter and cross-activates the other two P_{RE} promoters weakly. Nevertheless, the phage 21 and P_{RE} promoters have identical TTGC repeats, while the single bp change of the P_{22} P_{RE} promoter does not account for the observed difference of its properties. Thus, species specificity of interaction must be generated by additional cII-DNA interactions outside the TTGC repeat or, conceivably, by shape-determining differences of DNA fine structure among the three P_{RE} promoters (Dickerson, 1983; Koo et al., 1986) which might subtly change contacts between RNA polymerase and the activator proteins. Indeed, if, as recently reported (Ho et al., 1986), the heterologous, weakly activating P22 cI protein binds more strongly than the homologous, strongly activating λ cII protein to the λ P_{RE} promoter, then a role for protein-protein contacts in determining species specificity of cII action is highly plausible.

The bacteriophage T4 motA gene codes for a DNA-binding protein that positively regulates initiation of transcription. The function of the T4 motA gene is not essential on many E. coli strains. Accordingly, mutants in this gene, also called far or sib, have been selected on the basis of relatively complex phenotypic characteristics, such as mitigation of the toxicity of metabolic analogues or the lethality of other mutations (Mattson et al., 1974, 1978; Chace and Hall, 1975; Homyk et al., 1976). Certain E. coli strains are nonpermissive for motA mutants by virtue of mutations in the RNA polymerase β subunit (rpoB) (Pulitzer et al., 1979; Hall and Snyder, 1981).

The motA gene codes for a 25.5-kD DNA-binding protein (Uzan et al., 1983, 1985). There is clear evidence that this protein stimulates transcription at T4 middle promoters. Fifteen of these promoters have now been analyzed (Guild, 1986). At each promoter, the consensus sequence (A/T)$_3$TGCTT (in the nontranscribed strand) appears 14 ± 1 bp upstream of an E. coli −10 consensus sequence, and there is no satisfactory fit to the *E. coli* −35 consensus. Initiation of transcription at these promoters is almost completely abolished *in vivo* in the absence of *mot*A gene function. Decreasing the spacing between the "mot box," and the −10 sequence of one of these promoters (rIIB$_2$) from 13 to 12 bp destroys promoter activity. *In vitro* experiments on *mot*-dependent transcription at T4 middle promoters have thus far utilized DNA-protein complexes from *mot*-active and *mot*-defective phage infected bacteria (De Franciscis *et al.*, 1982; Uzan *et al.*, 1985). On the basis of these experiments and the DNA-binding properties of the *mot*A protein, it has been suggested that gp *mot*A, like the cII protein, participates in transcription initiation in addition to, rather than substitution for, *E. coli* σ (Brody *et al.*, 1983). The relatively strict requirement for positioning the putative *mot*A binding sites strengthens that supposition. Nevertheless, *in vitro* experiments with purified *mot*A protein and RNA polymerase to generate transcripts initiated at single T4 middle promoters remain to be done.

The *mot*A protein functions in cytosine-containing and glucosylated hydroxymethylcytosine-containing T4 DNA, although the recognition site includes two, and can perhaps tolerate three GC base pairs. As Guild (1986) has pointed out, that adds further interest to the question of how *mot*A protein recognizes its DNA binding site: the glucosylated hydroxymethyl substituent in normal T4 DNA is bulky and projects into the major groove of the DNA B helix. Minor groove recognition of the entire mot box seems unlikely because of the almost perfect conservation of sequence, rather than merely AT/GC distinction, at five nucleotides (TGCTT). Moreover, if the *mot*A protein binds to the entire mot box consensus sequence in helical DNA, it has to wrap around the DNA rather than being confined to one face of the helix, and it must overlap the −35 RNA polymerase-binding site. Thus, considerable interest attaches to the elucidation of the mechanism of DNA binding and action of *mot*A protein from a structural point of view, because, although it shares some functional properties with cII protein, it is likely different in significant, and therefore novel, ways.

Why does a defective T4 *mot*A gene not generate drastic consequences for phage development? A part of the answer to that question lies with the complex and overlapping layout of T4 early and middle transcription units. Most T4 genes that are transcribed from a middle, *mot*A-dependent promoter can also be accessed from early promoters in at least some measure, so long as two other T4 regulatory functions, *mot*B and *mot*C, are effective (Pulitzer *et al.*, 1985). The mode of *mot*B and C action is not known.

V. OTHER RNA POLYMERASE-BINDING PROTEINS AND OTHER PHAGES

The common property of these RNA polymerase-binding proteins is that too little is known about them to permit any functional classification. Interest attaches to these proteins because of their association with problems in the regulation of gene expression whose solution is likely to provide new insights and may even provide new paradigms.

As described in Chapter 10, several phage-specific proteins associate more or less tightly with *E. coli* RNA polymerase during phage T5 infection. Two of these proteins are pre-early (see Chapter 10 for a definition of terms and for a description of the temporal program of phage T5 transcription); one has a molecular weight of 11 kD (Szabo *et al.*, 1975), and the other, the product of the A1 gene, is a 60-kD protein (McCorquodale *et al.*, 1981). An early protein (molecular weight 15,000) binds tightly to RNA polymerase core or holoenzyme (Szabo and Moyer, 1975). A second early protein ($\sim$ 90 kD), which is the product of the late transcription-regulatory C2 gene, binds loosely to RNA polymerase. The major problem of T5 gene expression—how the late genes, located in one cluster and covering $\sim$ 25% of the genome, are regulated—is now sharply defined. The late promoters are *E. coli* σ^{70} promoters and include two of the strongest known *E. coli* promoters. These two late promoters have good to perfect -10 or -35 consensus sequences, and one of them has a downstream segment lending additional promoter strength *in vivo* and *in vitro* (Gentz and Bujard, 1985; Bujard *et al.*, 1985). They are active on plasmids in uninfected cells and thus require no positive regulatory element in that structural context (Ksenzenko *et al.*, 1982; Brunel *et al.*, 1983; Gentz and Bujard, 1985). The two puzzles about these promoters are their inactivity at *any* time after they enter the host cell's cytoplasm, and their eventual reactivation, which requires the function of the RNA polymerase-interacting C2 protein and, as also stated in Chapter 10, some uncharacterized but special state of the viral chromatin holding the late transcription units.

The late genes of the small *B. subtilis* phage 29 genome (11 kbp) are part of a single transcription unit, under the positive control of phage 29 gene 4. The 15-kD gene 4 protein affects transcription by *B. subtilis* RNA polymerase *in vitro* at specific sites (Mellado *et al.*, 1985), but apparently this activation is not restricted to the single late promoter that has been identified for *in vivo*–synthesized late RNA (Barthelemy *et al.*, 1986; Mellado *et al.*, 1986). The sequence of the single late promoter shows a perfect consensus to *B. subtilis* σ^{43} promoters at -10 but none at -35. Moreover, the gene 4 protein's amino acid sequence is not detectably homologous with members of the σ family (Fig. 1). The elucidation of the mechanism of action of phage 29 gp4 appears now to be accessible, but the nature of its interactions (i.e., whether it binds independently to DNA or to RNA polymerase) and its mode of action are still unknown.

As described in Bertani and Six (1988), the *E. coli* phage, P2, and its satellite phage, P4, code for proteins that positively regulate late transcription. The two proteins are cross-specific *in vivo* and are responsible for transactivation of late functions between the two phages. A suppressor-genetic argument strongly implies that each protein interacts with RNA polymerase core, specifically with the α subunit (but see Section III.B.3 for comments on this kind of argument). The 19-kD P4 δ

protein is strongly homologous with the 9-kD P2 *ogr* protein. The P2-related phage, 186, codes for a functionally related protein. Neither of these proteins shows any homology with the σ family of proteins, and both are smaller than the proteins listed in Fig. 1; it is argued in Bertani and Six (1988) that neither protein substitutes for σ^{70}.

VI. CONCLUDING COMMENTS

In this chapter we have discussed, or referred to, many modifications of RNA polymerase during phage infection. Although there are numerous gaps in our understanding of mechanisms, it is clear that the diversity of modes of action that are represented by our chosen examples must be very great. In closing, it is therefore worth shifting attention, briefly, to examples of mechanisms or targets of action that have not yet been encountered but that would belong in this chapter if they existed.

No phage transcription unit is known to depend on a minor vegetative σ. We can think of no structural or regulatory principle excluding such a possibility, although it may not be realized among those phages that are the narrow focus of current research. The *Streptomyces* phages, whose bacterial hosts have multiple polymerases (Westpheling *et al.*, 1985), might be the place to look for examples.

No phage-induced activation, or induction, of a specialized bacterial-positive regulator of transcription for phage-specific transcription has been identified. Some of these positive regulators (*rpo*H, heat shock; *rpo*N/*ntr*A, nitrogen assimilation) are σ-like proteins. The regulatory systems that are cocontrolled by these transcription factors respond, in the uninfected cell, to environmental stimuli that are probably not capable of being literally duplicated during phage infection. One can, however, imagine a kind of molecular mimicry in which a phage-coded protein is capable of triggering activation by interacting with the environmental sensor of such a regulatory system. The viral protein thereby serves as the indirect positive regulator of a group of viral genes. The complications and restrictions associated with such a mechanism are readily imagined. They include the metabolic consequences of the activation, which may adversely affect the yield of the infection, and in the case of sporulation, the coinduction of the synthesis of bacterial proteins interfering with viral development.

REFERENCES

Baralle, F. E., and Travers, A. A., 1976, Phage T4 infection restricts rRNA synthesis by *E. coli* RNA polymerase, *Mol. Gen. Genet.* **147**:291.

Barthelemy, I., Salas, M., and Mellado, R. P., 1986, *In vivo* transcription of bacteriophage φ29 DNA: Transcription initiation sites, *J. Virol.* **60**:874.

Bertani, L. E., and Six, E. W., 1988, the P2-like phages and their parasite, P4, in *The Bacteriophages*, vol. 2 (R. Calendar, ed.), pp. 73–143, Plenum Press, New York.

Bogdanova, E. S., Zograff, Y. N., Bass, I. A., and Shemyakin, M. F., 1970, Free subunits of RNA polymerase in normal and phage-infected cells of *E. coli*, *Mol. Biol.* **4**:435 (in Russian).

Bolund, C., 1973, Influence of gene 55 on the regulation of synthesis of some early enzymes in bacteriophage T4-infected *E. coli*, *J. Virol.* **12**:49.

Briat, J.-F., and Chamberlin, M. J., 1984, Identification and characterization of a new transcriptional termination factor from *Escherichia coli*, *Proc. Natl. Acad. Sci. USA* **81**:7373.

Brody, E., Rabussay, D., and Hall, D. H., 1983, Regulation of transcription of prereplicative genes, in: *Bacteriophage T4* (C. K. Mathews, E. M. Kutter, G. Mosig, and P. B. Berget, eds.), p. 174, American Society for Microbiology, Washington.

Broida, J., and Abelson, J., 1985, Sequence organization and control of transcription in the bacteriophage T4 tRNA region, *J. Mol. Biol.* **185**:545.

Brunel, F., Thi, V. H., Pilaete, M. F., and Davison, J., 1983, Transcription regulatory elements in the late region of bacteriophage T5 DNA, *Nucleic Acids Res.* **11**:7649.

Buchstein, S. R., and Hinkle, D. C., 1982, Genetic analysis of two bacterial RNA polymerase mutants that inhibit the growth of bacteriophage T7, *Mol. Gen. Genet.* **188**:211.

Bujard, H., Deuschle, U., Kammerer, W., Gentz, R., Bannworth, W. and Stüber, D., 1985, in: *Sequence Specificity in Transcription and Translation* (R. Calendar and L. Gold, eds.), p. 21, A. R. Liss, New York.

Caruso, M., Coppo, A., Manzi, A., and Pulitzer, J., 1979, Host-virus interactions in the control of T4 prereplicative transcription. I. tabC(rho) mutants, *J. Mol. Biol.* **135**:950.

Chace, K. V., and Hall, D. H., 1975, Characterization of new regulatory mutants of bacteriophage T4. II. New class of mutants, *J. Virol.* **15**:929.

Chelm, B. K., Duffy, J. J., and Geiduschek, E. P., 1982, Interaction of *B. subtilis* RNA polymerase core with two specificity-determining subunits, *J. Biol. Chem.* **257**: 6501.

Choy, H. A., Romeo, J. M., and Geiduschek, E. P., 1986, Activity of a phage-modified RNA polymerase at hybrid promoters: Effects of substituting thymine for hydroxymethyl uracil in a phage SP01 middle promoter, *J. Mol. Biol.* **191**:59.

Christensen, A. C., and Young, E. T., 1982, T4 late transcripts are initiated near a conserved DNA sequence, *Nature (Lond.)* **299**:369.

Christensen, A. C., and Young, E. T., 1983, Characterization of T4 transcripts, in: *Bacteriophage T4* (C. K. Mathews, E. M. Kutter, G. Mosig, and P. B. Berget, eds.), p. 184, American Society for Microbiology, Washington.

Coppo, A., Manzi, A., Pulitzer, J. F., and Takahashi, H., 1975a, Host mutant (*tabD*)-induced inhibition of bacteriophage T4 late transcription. I. Isolation and phenotypic characterization of the mutants, *J. Mol. Biol.* **96**:579.

Coppo, A., Manzi, A., and Pulitzer, J. F., 1975b, Host mutant (*tabD*)-induced inhibition of bacteriophage T4 late transcription. II. Genetic characterization of mutants, *J. Mol. Biol.* **96**:601.

Costanzo, M., 1984, Ph.D. thesis, Harvard University.

Costanzo, M., Hannett, N., Brzustowicz, L., and Pero, J., 1983, Bacteriophage SP01 gene 27: Location and nucleotide sequence, *J. Virol.* **48**:555.

Costanzo, M., Brzustowicz, L., Hannett, N., and Pero, J., 1984, Bacteriophage SP01 genes 33 and 34, *J. Mol. Biol.* **180**:533.

De Franciscis, V., and Brody, E., 1982, *In vitro* system for middle T4 RNA. I. Studies with *Escherichia coli* RNA polymerase, *J. Biol. Chem.* **257**:4087.

De Franciscis, V., Favre, R., Uzan, M., Leautey, J., and Brody, E., 1982, *In vitro* system for T4 RNA. II. Studies with T4-modified RNA polymerase, *J. Biol. Chem.* **257**:4097.

De Wyngaert, M. A., and Hinkle, D. C., 1979, Bacterial mutants affecting phage T7 DNA replication produce RNA polymerase resistant to inhibition by the T7 gene 2 protein, *J. Biol. Chem.* **254**:11247.

Dickerson, R. E., 1983, Base sequence and helix structure variation in B and A DNA, *J. Mol. Biol.* **166**:419.

Doi, R. H., and Wang, L.-F., 1986, Multiple procaryotic ribonucleic acid polymerase sigma factors, *Microbiol. Rev.* **50**:227.

Duffy, J. H., and Geiduschek, E. P., 1976, The virus-specified subunits of a modified *B. subtilis* RNA polymerase are determinants of DNA binding and RNA chain initiation, *Cell* **8**:595.

Duffy, J. H., and Geiduschek, E. P., 1977, Purification of a positive regulatory subunit from phage SP01-modified RNA polymerase, *Nature* **270**:28.

Dunn, J. J., and Studier, F. W., 1981, Nucleotide sequence from the genetic left end of bacteriophage T7 DNA to the beginning of gene 4, *J. Mol. Biol.* **148**:303.

Fox, T. D., and Pero, J., 1974, New phage-SP01-induced polypeptides associated with *Bacillus subtilis* RNA polymerase, *Proc. Natl. Acad. Sci. USA* **71**:2761.

Fox, T. D., Losick, R., and Pero, J., 1976, Regulatory gene 28 of bacteriophage SP01 codes for a phage-induced subunit of RNA polymerase, *J. Mol. Biol.* **101**:427.

Fujita, D. J., Ohlsson-Wilhelm, B. M., and Geiduschek, E. P., 1971, Transcription during bacteriophage SP01 development: Mutations affecting the program of viral transcription, *J. Mol. Biol.* **57**:301.

Geiduschek, E. P., and Ito, J., 1982, Regulatory mechanisms in the development of lytic bacteriophages in *Bacillus subtilis*, in: *The Molecular Biology of Bacilli 1* (D. Dubnau, ed.), Chap. 7, Academic Press, New York.

Gentz, R., and Bujard, H., 1985, Promoters recognized by *Escherichia coli* RNA polymerase selected by function: Highly efficient promoters from bacteriophage T5, *J. Bacteriol.* **164**:70.

Goff, C. G., 1979, Bacteriophage T4 *alt* gene maps between genes 30 and 54, *J. Virol.* **29**:1232.

Goff, C. G., and Setzer, J., 1980, ADP ribosylation of *Escherichia coli* RNA polymerase is nonessential for bacteriophage T4 development, *J. Virol.* **33**:547.

Goldfarb, A., 1981, Changes in the promoter range of RNA polymerase resulting from bacteriophage T4-induced modification of core enzyme, *Proc. Natl. Acad. Sci. USA* **78**:3454.

Goldfarb, A., and Malik, S., 1984, Changed promoter specificity and antitermination properties displayed *in vitro* by bacteriophage T4-modified RNA polymerase, *J. Mol. Biol.* **177**:87.

Goldfarb, A., and Palm, P., 1981, Control of promoter utilization by bacteriophage T4-induced modification of RNA polymerase alpha-subunit, *Nucleic Acids Res.* **9**:4863.

Gram, H., and Rüger, W., 1985, Genes 55, αgt, 47 and 46 of bacteriophage T4: The genomic organization as deduced by sequence analysis, *EMBO J.* **4**:257.

Gribskov, M., and Burgess, R. R., 1986, Sigma factors from *E. coli*, *B. subtilis*, phage SP01, and phage T4 are homologous proteins, *Nucleic Acids Res.* **14**:6745.

Guarente, L., Nye, J. S., Hochschild, A., and Ptashne, M., 1982, A mutant λ repressor with a specific defect in its positive control function, *Proc. Natl. Acad. Sci. USA* **79**:2236.

Guild, N., 1986, Ph.D. thesis, University of Colorado.

Gussin, G. N., 1984a, Kinetic analysis of mutations affecting the cII activation site at the P_{RE} promoter of bacteriophage lambda, *Proc. Natl. Acad. Sci. USA* **81**:6432.

Gussin, G. N., 1984b, Role of cII protein in stimulating transcription initiation at the lambda P_{RE} promoter. Enhanced formation and stabilization of open complexes, *J. Mol. Biol.* **172**:489.

Hahn, S., Kruse, U., and Rüger, W., 1986, The region of phage T4 genes 34, 33 and 59: Primary structures and organization on the genome, *Nucleic Acids Res.* **14**:9311.

Hall, D. H., and Snyder, R. D., 1981, Suppressors of mutations in the rII gene of bacteriophage T4 affect promoter utilization, *Genetics* **97**:1.

Hawley, D. K., and McClure, W. R., 1983, Compilation and analysis of *Escherichia coli* promoter DNA sequences, *Nucleic Acids Res.*, **11**:2237.

Hesselbach, B. A., and Nakada, D., 1975, Inactive complex formation between *E. coli* RNA polymerase and an inhibitor protein purified from T7 phage infected cells, *Nature (Lond.)* **258**:354.

Hesselbach, B. A., and Nakada, D., 1977a, "Host shutoff" function of bacteriophage T7: Involvement of T7 gene 2 and gene 0.7 in the inactivation of *Escherichia coli* RNA polymerase, *J. Virol.* **24:**736.

Hesselbach, B. A., and Nakada, D., 1977b, I protein: Bacteriophage T7-coded inhibitor of *Escherichia coli* RNA polymerase, *J. Virol.* **24:**746.

Ho., Y.-S., and Rosenberg, M., 1985, Characterization of a third, cII-dependent, coordinately activated promoter on phage lambda involved in lysogenic development. *J. Biol. Chem.* **260:**11838.

Ho, Y.-S., Lewis, M., and Rosenberg, M., 1982, Purification and properties of a transcriptional activator: The cII protein of phage λ, *J. Biol. Chem.* **257:**9128.

Ho, Y.-S., Wulff, D. L., and Rosenberg, M., 1983, Bacteriophage lambda protein cII binds promoters on the opposite fact of the DNA helix from RNA polymerase. *Nature (Lond.)* **304:**703.

Ho, Y.-S., Wulff, D.L., and Rosenberg, M., 1986, Protein–nucleic acid interactions involved in transcription activation by the phage lambda regulatory protein cII, in: *Regulation of Gene Expression* (I. Booth and C. Higgins (eds.), *Symposium of the Society for General Microbiology*, p. 79, Cambridge University Press, London.

Hochschild, A., Irwin, N., and Ptashne, M., 1983, Repressor structure and the mechanism of positive control, *Cell* **32:**319.

Homyk, T., Rodriguez, A., and Weil, J., 1976, Characterization of T4 mutants that partially suppress the inability of T4 rII to grow in lambda lysogens, *Genetics* **83:**477.

Hoopes, B. C., and McClure, W. R., 1985, A cII-dependent promoter is located within the Q gene of bacteriophage lambda. *Proc. Natl. Acad. Sci. USA* **82:**3134.

Horvitz, H. R., 1973, Polypeptide bound to the host RNA polymerase is specified by T4 control gene 33, *Nature New Biol. (Lond.)* **244:**137.

Horvitz, R. H., 1974, Bacteriophage T4 mutants deficient in alteration and modification of the *E. coli* RNA polymerase, *J. Mol. Biol.* **90:**739.

Hsu, T., Wei, R., Dawson, M., and Karam, J., 1987, Identification of two new bacteriophage T4 genes that may have roles in transcription and DNa replication, *J. Virol.* **61:**366.

Hunt, T. F., and Magasanik, B., 1985, Transcription of *glnA* by purified *Escherichia coli* components: Core RNA polymerase and the products of *glnF, glnG* and *glnL, Proc. Natl. Acad. Sci. USA* **82:**8453.

Jacobs, K. A., and Geiduschek, E. P., 1981, Regulation of expression of cloned bacteriophage T4 late gene 23, *J. Virol.* **39:**46.

Jacobs, K. A., Albright, L. M., Shibata, D. K., and Geiduschek, E. P., 1981, Genetic complementation by cloned bacteriophage T4 late genes, *J. Virol.* **39:**31.

Kassavetis, G. A., and Geiduschek, E. P., 1984, Defining a bacteriophage T4 late promoter: Bacteriophage T4 gene 55 protein suffices for directing late promoter recognition, *Proc. Natl. Acad. Sci. USA* **81:**5101.

Kassavetis, G. A., Zentner, P. G., and Geiduschek, E. P., 1986, Transcription at bacteriophage T4 variant late promoters, *J. Biol. Chem.* **261:**14256.

Kassavetis, G. A., Williams, K. P., and Geiduschek, E. P., 1987, Interactions of bacteriophage T4 gene 55 product with *Escherichia coli* RneA polymerase, *J. Biol. Chem.* **262:**12365.

Khesin, R. B., Bogdanova, E. S., Goldfarb, A. D., and Zograff, Y. N., 1972, Competition for the DNA template between RNA polymerase molecules from normal and phage-infected *E. coli, Mol. Gen. Genet.* **119:**299.

Khesin, R. B., Nikiforov, V. G., Zograff, Y. N., Danilevskaya, O. N., Kalayaeva, E. S., Lipkin, V. M., Modyanov, N. N., Dmitriev, A. D., Velkov, V. V., and Gintsburg, A. L., 1976, Influence of mutations and phage infection on *E. coli* RNA polymerase, in: *RNA Polymerase* (R. Losick and M. Chamberlin, eds.), p. 629, Cold Spring Harbor Laboratory, Cold Spring Harbor, NY.

Koo, H.-S., Wu, H.-M., and Crothers, D. M., 1986, DNA bending at adenine thymine tracts, *Nature* **320:**501.

Ksenzenko, V. N., Kamynina, T. P., Pustoshilova, N. M., Kryukov, V. M., and Bayev, A. S., 1982, Cloning of bacteriophage T5 promoters *Mol. Gen. Genet.* **185:**520.

LeClerc, J. E., and Richardson, C. C., 1979, Bacteriophage T7 DNA replication *in vitro*. XVI. Gene 2 protein of bacteriophage T7: Purification and requirement for packaging T7 DNA *in vitro*, *Proc. Natl. Acad. Sci. USA* **76**:4852.

Lee, G., Hannett, N. M., Korman, A., and Pero, J., 1980, Transcription of cloned DNA from *Bacillus subtilis* phage SP01, *J. Mol. Biol.* **139**:407.

Lee, G., and Pero, J., 1981, Conserved nucleotide sequences in temporally controlled bacteriophage promoters, *J. Mol. Biol.* **152**:247.

Losick, R., and Pero, J., 1981, Cascades of sigma factors, *Cell* **25**:582.

Mahadik, S. P., Dharmgrongartama, B., and Srinivasan, P. R., 1972, An inhibitory protein of *Escherichia coli* RNA polymerase in bacteriophage T 3-infected cells, *Proc. Nat. Acad. Sci. USA* **69**:162.

Mahadik, S. P., Dharmgrongartama, B., and Srinivasan, P. R., 1974, Regulation of host RNA synthesis in bacteriophage T 3-infected cells. Properties of an inhibitory protein of *E. coli* ribonucleic acid polymerase, *J. Biol. Chem.* **249**:1787.

Mailhammer, R., Yang, H.-L., Reiness, G., and Zubay, G., 1975, Effects of bacteriophage T4-induced modification of *Escherichia coli* RNA polymerase on gene expression *in vitro*, *Proc. Natl. Acad. Sci. USA* **72**:4928.

Malik, S., and Goldfarb, A., 1984, The effect of a bacteriophage T4-induced polypeptide on host RNA polymerase interaction with promoters, *J. Biol. Chem.* **259**:13292.

Malik, S., Dimitrov, M., and Goldfarb, A., 1985, Initiation of transcription by bacteriophage T4-modified RNA polymerase independently of host sigma factor, *J. Mol. Biol.* **185**:83.

Mattson, T., Richardson, J., and Goodin, D., 1974, Mutant of bacteriophage T4D affecting expression of many early genes, *Nature (Lond.)* **250**:48.

Mattson, T., Van Houwe, G., and Epstein, R. H., 1978, Isolation and characterization of conditional lethal mutations in the *mot* gene of bacteriophage T4, *J. Mol. Biol.* **126**:551.

McClure, W. R., 1985, Mechanism and control of transcription initiation in prokaryotes, *Annu. Rev. Biochem.* **54**:171.

McCorquodale, D. J., Chen, C. W., Joseph, M. K., and Woychik, R., 1981, Modification of RNA polymerase from *Escherichia coli* by pre-early gene products of bacteriophage T5, *J. Virol.* **40**:958.

McPheeters, D. S., Christensen, A., Young, E. T., Stormo G., and Gold, L., 1986, Translational regulation of expression of the bacteriophage T4 lysozyme gene, *Nucleic Acids Res.* **14**:5813.

Mellado, R. P., Barthelemy, I., and Salas, M., 1986, *In vitro* transcription of bacteriophage φ29 DNA. Correlation between *in vitro* and *in vivo* promoters. *Nucleic Acids Res.* **14**:4731–4741.

Mellado, R. P., Carrascosa, J. L., and Salas, M., 1985, Control of the late transcription of B. subtilis phage φ29, in: *Sequence Specificity in Transcription and Translation* (R. Calendar and L. Gold, eds.), p. 219, A. R. Liss, New York.

Nossal, N. G., and Alberts, B. M., 1983, Mechanism of DNA replication catalyzed by purified T4 replication proteins, in: *Bacteriophage T4* (C. K. Mathews, E. M. Kutter, G. Mosig, and P. B. Berget, eds.), p. 71, American Society for Microbiology, Washington.

Pero, J., Tjian, R., Nelson, J., and Losick, R., 1975, *In vitro* transcription of a late class of phage SP0 1 genes, *Nature (Lond.)* **257**:248.

Ptashne, M., 1986, *A Genetic Switch*, Cell Press and Blackwell Scientific Publications, London.

Pulitzer, J. F., Coppo, A., and Caruso, M., 1979, Host-virus interactions in the control of T4 prereplicative transcription. II. Interaction between *tab(rho)* mutants and T4 *mot* mutants, *J. Mol. Biol.* **135**:979.

Pulitzer, J. F., Colombo, M., and Ciaramella, M., 1985, New control elements of bacteriophage T4 pre-relicative transcription, *J. Mol. Biol.* **182**:249.

Rabussay, D., 1983, Regulation of gene expression, in: *Bacteriophage T4* (C. K. Mathews, E. M. Kutter, G. Mosig, and P. B. Berget, eds.), p. 167, American Society for Microbiology, Washington.

Rabussay, D., Mailhammer, R., and Zillig, W., 1972, Regulation of transcription by T4 phage-induced chemical alteration and modification of transcriptase (EC 2.7.7.6), in: *Metabolic Interconversion of Enzymes* (O. Wieland, E. Helmreich, and H. Holzer, eds.), p. 213, Springer-Verlag, New York.

Rahmsdorf, H. J., Pai, S. H., Ponta, H., Herrlich, P., Roskoski, R. Jr., Schweiger, M., and Studier, F. W., 1974, Protein kinase induction in *Escherichia coli* by bacteriophage T7, *Proc. Natl. Acad. Sci. USA* **71**:586.

Ratner, D., 1974a, Bacteriophage T4 transcriptional control gene 55 codes for a protein bound to *Escherichia coli* RNA polymerase, *J. Mol. Biol.* **89**:803.

Ratner, D., 1974b, The interactions of bacterial and phage proteins with immobilized *E. coli* RNA polymerase, *J. Mol. Biol.* **88**:373.

Reitzer, L. J., and Magasanik, B., 1986, Transcription of *glnA* in *E. coli* is stimulated by activator bound to sites far from the promoter, *Cell* **45**:785.

Romeo, J. M., Greene, J. R., Richards, S. H., and Geiduschek, E. P., 1986, The phage SP01-specific RNA polymerase, E. gp28, recognizes its cognate promoters in thymine-containing DNA, *Virology* **153**:46.

Rothman-Denes, L., Muthukrishnan, S., Haselkorn, R., and Studier, F. W., 1973, A T7 gene function required for shut-off of host and early T7 transcription, in: *Virus Research* (C. F. Fox and W. S. Robinson, eds.), p. 227, Academic Press, New York.

Schäfer, R., and Zillig, W., 1973, The effects of ionic strength on termination of transcription of DNAs from bacteriophages T4, T5 and T7 by DNA-dependent RNA polymerase from *Escherichia coli* and the nature of termination by factor rho, *Eur. J. Biochem.* **33**:215.

Schmitt, M. P., Beck, P. J., Kearney, C. A., Spence, J. L., DiGiovanni, D., Condreay, J. P., and Molineux, I. J., 1987, Evolution of a conditionally essential region, including the primary origin of DNA replication, of bacteriophage T3, *J. Mol. Biol.* **193**:479.

Schwartz, E., Scherer, G., Hobom, G., and Kössel, H., 1978, Nucleotide sequence of *cro*, *cII* and part of the *O* gene in phage λ DNA, *Nature* **272**:410.

Shanblatt, S. H., and Nakada, D., 1982, *Escherichia coli* mutant which restricts T7 bacteriophage has an altered RNA polymerase, *J. Virol.* **42**:1123.

Shih, M.-C., and Gussin, G. N., 1983, Differential effects of mutations on discrete steps in transcription initiation at the λP_{RE} promoter, *Cell* **34**:941.

Shih, M. C., and Gussin, G. N., 1984a, Kinetic analysis of mutations affecting the cII activation site at the P_{RE} promoter of bacteriophage lambda, *Proc. Natl. Acad. Sci. USA* **81**:6432.

Shih, M. C., and Gussin, G. N., 1984b, Role of cII protein in stimulating transcription initiation at the lambda P_{RE} promoter. Enhanced formation and stabilization of open complexes, *J. Mol. Biol.* **172**:489.

Sköld, O., 1970, Regulation of early RNA synthesis in bacteriophage T̄4-infected *Escherichia coli* cells, *J. Mol. Biol.* **53**:339.

Stephenson, F. H., 1985, A cII-responsive promoter within the Q gene of bacteriophage lambda, *Gene* **35**:313.

Stevens, A., 1970, An isotopic study of DNA-dependent RNA polymerase of *E. coli* following T4 infection, *Biochem. Biophys. Res. Commun.* **41**:367.

Stevens, A., 1972, New small polypeptides associated with DNA-dependent RNA polymerase of *Escherichia coli* after infection with bacteriophage T4, *Proc. Natl. Acad. Sci. USA* **69**:603.

Stevens, A., 1974, Deoxyribonucleic acid dependent ribonucleic acid polymerases from two T4 phage-infected systems, *Biochemistry* **13**:493.

Stevens, A., 1976, A salt-promoted inhibitor of RNA polymerase isolated from T4-infected E. coli, in: "RNA Polymerase (R. Losick and M. Chamberlin, eds.), p. 617, Cold Spring Harbor Laboratories, Cold Spring Harbor, NY.

Stevens, A., 1977, Inhibition of DNA-enzyme binding by an RNA polymerase inhibitor from T4 phage-infected *Escherichia coli*, *Biochim. Biophys. Acta* **475**:193.

Stevens, A., and Rhoton, J. C., 1975, Characterization of an inhibitor causing potassium chloride sensitivity of an RNA polymerase from T4 phage-infected *Escherichia coli*, *Biochemistry* **14**:5074.

Stragier, P., Bouvier, J., Bonamy, C., and Szulmajster, J., 1984, A developmental gene product of *Bacillus subtilis* homologous to the sigma factor of *Escherichia coli, Nature (Lond.)* 312:376.

Szabo, C., and Moyer, R. W., 1975, Purification and properties of a bacteriophage T5-modified form of *Escherichia coli* RNA polymerase, *J. Virol.* 15:1042.

Szabo, C., Dharmgrongartama, B., and Moyer, R. W., 1975, The regulation of transcription in bacteriophage T5-infected *Escherichia coli, Biochemistry* 14:989.

Talkington, C., and Pero, J., 1978, Promoter recognition by phage SP01-modified RNA polymerase, *Proc. Natl. Acad. Sci. USA* 75:1185.

Talkington, C., and Pero, J., 1979, Distinctive nucleotide sequences of promoters recognized by RNA polymerase containing a phage-coded "sigma-like" protein, *Proc. Natl. Acad. Sci. USA* 76:5465.

Tjian, R., and Pero, J., 1976, Bacteriophage SP01 regulatory proteins directing late gene transcription *in vitro, Nature* 262:753.

Trempy, J. E., Bonamy, C., Szulmajster, J., and Haldenwang, W. G., 1985, *Bacillus subtilis* σ factor σ^{29} is the product of the sporulation-essential gene *spoIIG, Proc. Natl. Acad. Sci. USA* 82:4189.

Uzan, M., Leautey, J., d'Aubenton-Carafa, Y., and Brody, E., 1983, Identification and biosynthesis of the bacteriophage T4 mot regulatory protein, *EMBO J.,* 2:1207.

Uzan, M., d'Aubenton-Carafa, Y., Favre, R., De Franciscis, V., and Brody, E., 1985, The T4 mot protein functions as part of a pre-replicative DNA-protein complex, *J. Biol. Chem.* 260:633.

Westpheling, J., Ranes, M., and Losick, R., 1985, RNA polymerase heterogeneity in *Streptomyces coelicolor, Nature* 313:22.

Williams, K. P., Kassavetis, G. A., Esch, F. S., and Geiduschek, E. P., 1987, Identification of the gene encoding an RNA polymerase-binding protein of bacteriophage T4, *J. Virol.* 61:597.

Wu, R., Geiduschek, E. P., and Cascino, A., 1975, The role of replication proteins in the regulation of bacteriophage T4 transcription. II. Gene 45 and late transcription uncoupled from replication, *J. Mol. Biol.* 96:539.

Wu, R., and Geiduschek, E. P., 1975, The role of replication proteins in the regulation of bacteriophage T4 transcription. I. Gene 45 and HMC containing DNA, *J. Mol. Biol.* 96:513.

Wulff, D. L., and Mahoney, M. E., 1985, Cross-specificities of functionally identical DNA-binding proteins from different lambdoid bacteriophages, in: *Sequence Specificity in Transcription and Translation* (R. Calendar and L. Gold, eds.), pp. 219–227, A. R. Liss, New York.

Wulff, D. L., Mahoney, M., Shatzman, A., and Rosenberg, M., 1984, Mutational analysis of a regulatory region in bacteriophage λ that has overlapping signals for the initiation of transcription and translation, *Proc. Nat. Acad. Sci. USA* 81:555.

Wulff, D., and Rosenberg, M., 1983, The lysogenic program of bacteriophage lambda; establishment of repression, in: *Bacteriophage Lambda II* (R. Hendrix, J. Roberts, F. Stahl, and R. Weissberg, eds.), p. 53, Cold Spring Harbor Laboratory, Cold Spring Harbor, NY.

Zillig, W., Fujiki, H., Blum, W., Janekovíc, D., Schweiger, M., Rahmsdorf, H.-J., Ponta, H., and Hirsch-Kauffmann, M., 1975, *In vivo* and *in vitro* phosphorylation of DNA-dependent RNA polymerase of *Escherichia coli* by bacteriophage-T7-induced protein kinase, *Proc. Natl. Acad. Sci. USA* 72:2506.

CHAPTER 4

Single-Stranded RNA Bacteriophages

JAN VAN DUIN

I. INTRODUCTION

Since their discovery in 1961 by Loeb and Zinder, the RNA phages have served as a model system to explore a variety of problems in molecular biology. As a source of homogeneous and readily obtainable messenger RNA, they have been particularly helpful in solving questions on initiation of translation, and they have provided good insight into regulation of gene expression at the level of translation. The concepts of translational polarity and translational control by repressor proteins resulted from early studies on bacteriophage RNA.

A second, much less explored area in which RNA phages can contribute to our knowledge is the origin and evolution of self-replicating RNA molecules. In particular, recent findings on the enzymatic potential of RNA (Zaug and Cech, 1986), the clear demonstration that RNA recombination takes place in plant viral RNA (Bujarski and Kaesberg, 1986), and the claim that the RNA-dependent RNA polymerases involved in coliphage amplification are capable of noninstructed RNA synthesis (Sumper and Luce, 1975) should stimulate efforts to better understand the genesis of these informational polynucleotides. The mature RNA phages together with their abridged variants and the other small replicating RNA molecules associated with phage infection provide ideal systems to pursue these questions. The aforementioned polynucleotides have been cloned and sequenced, and recombinant DNA technology now offers the

JAN VAN DUIN • Department of Biochemistry, University of Leiden, Leiden, The Netherlands.

possibility to release RNA molecules of the experimenter's design into the live environment of a (bacterial) cell to test their "fitness" and "adaptation" (Taniguchi *et al.*, 1978).

Another promising development is the increasing number of phage RNA sequences that become available. Comparative sequence analysis, which has been so helpful in establishing the higher-order structure of ribosomal RNA, should be even more powerful in this area. There are several reasons for this optimistic view. (1) We know more about the function of phage RNA than about that of ribosomal RNA, because most of the viral sequence is earmarked to code for the phage proteins and is consequently constrained for this purpose. (2) Many features necessary for replication must also show up in copy RNA. (3) There are only a few specific protein-phage RNA interactions. (4) The various phage RNAs can function in the same organism. Therefore, differences in primary and secondary structure between phage RNAs need only be interpreted in terms of RNA functioning and not in terms of differential host requirements. A disadvantage of phage RNA could be that its diversification in length and sequence may be much less than that of ribosomal RNA.

In this chapter I have not tried to present a comprehensive account on all aspects of the life cycle of RNA phages. Several aspects have not received much attention over the past few years, and for this reason they are not extensively covered here. For information on these issues the reader is referred to excellent books such as *RNA Phages* edited by N. D. Zinder (1975) or Volume 13 of *Comprehensive Virology* (Fiers, 1979). In addition, there is a good review on RNA replication by Blumenthal and Carmichael (1979). I have chosen to concentrate on several recent activities that have produced interesting results and on areas that have not been chronicled extensively in the past.

II. CLASSIFICATON OF RNA PHAGES

A. RNA Coliphages

The RNA coliphages are divided into four groups. The best-known representatives of each group are MS2 (group I), GA (group II). Qβ (group III) and SP (group IV). These four phages have now been sequenced (Fig. 1). RNA phage classification rests on several criteria, such as serological cross-reactivity, density and molecular weight of the virus particle, and sedimentation velocity of the viral RNA. The serological type is determined by measuring the degree of inactivation of a phage by serum raised against another phage. An example is given in Table I, showing that in general serum raised against a member of one group does not inactivate members of other groups. It is, however, possible using the same technique to identify subgroups per group. This kind of ordering is shown in Table II and establishes the order of relatedness within group I as MS2>M12>R17>f2>β>fr>f4.

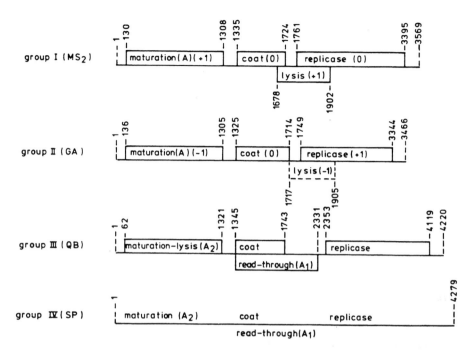

FIGURE 1. Genetic map of group I–IV RNA coliphages. The MS2 sequence was determined by Fiers *et al.* (1976), GA is from Inokuchi *et al.* (1986), and Qβ from Mekler (1981). SP is highly homologous to Qβ, was lately sequenced and found to be 4279 nucleotides long (Hirashima, personal communication). For group I and II phages the relative reading frames of the genes are indicated by −1, 0, and +1, respectively. The lysis gene in GA has been tentatively identified.

Knowledge of serological typing is very valuable in setting up the phylogenetic tree of RNA phages and to begin to understand aspects of their evolution and diversification. It also provides the criteria for deciding which phage RNA should be (partly) sequenced to answer specific questions on higher-order structure of RNA segments or on regulatory mechanisms. A recent extensive survey has, for instance, resulted in the discovery of several RNA phages that may represent intermediates be-

TABLE I. Serological Relationships among Four Groups of RNA Phages[a]

Phage	Antiserum			
	MS2	GA	Qβ	SP
Group I (MS2)	1×10^{-4b}	0.7	1.0	1.3
Group II (GA)	0.6	1×10^{-4}	1.3	1.3
Group III (Qβ)	1.1	1.0	1×10^{-5}	0.9
Group IV (SP)	1.0	1.2	1.2	8×10^{-4}

[a] From Furuse *et al.* (1978).
[b] The degree of inactivation by antiserum was obtained from [PFU on serum plate (K=1/plate) per PFU on serum-free plate].

TABLE II. Relative Rates of Neutralization of the RNA Coliphages[a]

Phage	K' values with antiserum to phage				
	MS2	R17	f2	fr	Qβ
MS2	1.00	0.68	0.58	0.21	<0.01
M12	0.84	0.84	0.70	0.38	<0.01
R17	0.68	1.00	0.86	0.54	0.03
f2	0.52	0.50	1.00	0.72	0.08
β	0.41	0.40	0.40	0.85	0.22
fr	0.20	0.24	0.22	1.00	0.50
f4	0.10	0.12	0.10	0.20	0.70
Qβ	0.02	0.03	0.03	0.11	1.00

[a] From Krueger (1969).

tween the four groups. JP34, classified as a group II member by the sedimentation value of its RNA, is inactivated by anti-MS2 as well as by anti-GA serum. Likewise MX, group IV by RNA size, is neutralized by anti-SP and by anti-Qβ serum. In addition, phage ID2, group IV by RNA size, was inactivated significantly by group I, II, and III phage antisera. As will be discussed later, the existence of a group IV-type phage with serological relations to the other three groups has been used as an argument that the shorter phages descend from the longer by deletions (Furuse, 1986).

Distinction of the phages by the size of their RNAs has revealed the same four groups. Groups I and II have short RNA with S values of 24S and 23S, respectively, whereas groups III and IV have longer RNAs with S values of 27S and 28S. These early characterizations have in recent times been fully confirmed by the primary sequence. The genetic map of these phages (Fig. 1) shows indeed that groups III and IV contain about 600 nucleotides more than groups I and II. In agreement with the physicochemical data, GA is shorter than MS2, and SP is larger than Qβ. The extra information present in Qβ and SP is dedicated to the synthesis of a carboxy-extended coat protein, named the read-through protein. This is a structural constituent of the virus particle. The coat protein gene in Qβ is terminated by a leaky UGA stop codon. Statistically, 5% of the ribosomes insert tryptophan at this codon, leading to a small amount of carboxy-extended coat protein (Hofstetter et al., 1974; Weiner and Weber, 1971). It is not known how uniform the genome sizes are—i.e., whether subgroups may have genomes that differ from the archetypes shown by a few nucleotides.

Based on strong similarities it is convenient to divide the single-stranded RNA phages into two major groups, which coincide with the absence or presence of extra information for the read-through protein. Accordingly, groups I and II are collectively called group A, and group B refers to the remaining two groups (Furuse, 1986).

TABLE III. Properties of ssRNA Bacteriophages[a]

Host	Phage	Morphology	$S_{20,w}$ (S)	Diameter (nm)	Host receptor site	Proteins	
						Coat (daltons)	A protein (daltons)
Escherichia coli	MS2-R17-f2	Icosahedral	79–80	26.0–26.6	F pili	13,731	43,988[b]
Escherichia coli	Qβ	Icosahedral	83–84	26.0	F pili	14,125	41,000[c]
Pseudomonas aeruginosa	7S[d,e]	Icosahedral	88	25.0	Polar pili	+[h,i]	+[h,i]
	PP7[f,g]	Icosahedral		25.0	Polar pili		
Caulobacter crescentus	Cb5[j]	Icosahedral	70–71	23.0	Polar pili	12,000	40,000
	Cb12r[k,l]	Icosahedral		22.0–23.0	Polar pili		
Caulobacter bacteroides	Cb8r[k,l]	Icosahedral	—	22.0–23.0	Polar pili		
Caulobacter fusiformis	Cb23r[k,l]	Icosahedral	—	21.0–23.0	Polar pili		
RP plasmid[m]	PRR1[n]	Icosahedral	80	25.0	RP pili	+[h,o]	+[h,o]

[a] Adapted from Shapiro and Bendis (1975).
[b] Fiers et al. (1975).
[c] Weber and Konigsberg (1975).
[d] Feary et al. (1963).
[e] Feary et al. (1964).
[f] Bradley (1966).
[g] Lin and Schmidt (1972).
[h] +, Present and similar in size to the group I coliphages (MS2-R17-f2).
[i] Davis and Benike (1974).
[j] Bendis and Shapiro (1970).
[k] Schmidt and Stanier (1965).
[l] Schmidt (1966).
[m] This plasmid has a wide host range (see text).
[n] Olson and Thomas (1973).
[o] Dhaese et al. (1977).

B. RNA Phages of Other Genera

RNA phages have also been found in several other gram-negative bacteria such as *Pseudomonas* and *Caulobacter*. As judged by several criteria, these phages must be very similar to the coliphages. They also infect via pili and have the same morphology, diameter, and molecular weight range, and it would not be surprising to find that they also fall into four groups. Some parameters of these phages are presented in Table III, which is taken from Fiers (1979). As shown, these noncoliphages also contain a major coat protein with the standard size, together with a minor structural component that corresponds to the maturation (A) protein. In view of the established role of the A protein in absorption to the pili (see below), the presence of this sort of protein is expected.

The dependence of RNA phages on pili as an entry to the bacterial cytoplasm is also reflected in the existence of phage PRR1 that will infect many genera such as *Pseudomonas*, *E. coli* (Hfr and F^+), *Salmonella typhimurium*, and *Vibrio cholerae*. provided the host expresses the pili encoded in the drug resistance factor RP (P compatibility group—e.g., RP1, RP4, or R1822). In this connection, we should mention that some members of the Enterobacteriaceae have been artificially converted to coliphage sensitivity. If the F factor of *E. coli* is introduced in *Shigella*, *Proteus*, or *Salmonella*, these genera can be infected by the *E. coli* phages. Recently Havelaar *et al.* (1984) exploited this phenomenon to screen sewage samples for the occurrence of RNA phages. They introduced the F factor into *Salmonella*. As *Salmonella* is not readily infected by DNA phages, the majority of plaques obtained represent the RNA phages.

The RNA bacteriophage φ6 will be dealt with elsewhere in this work.

III. ECOLOGY OF COLIPHAGES

Furuse (1986) has examined and reviewed the geographical distribution of the single-stranded RNA phages as well as their present-day habitat. They are most frequently encountered in sewage and feces of mammals, and their titers in sewage samples may be as high as 10^7 PFU/ml. RNA phages may constitute to up to 90% of total coliphages present in these samples. but the number can vary substantially. It is about 50% for Japan, Korea, and the Philippines but as low as 5% in, for example, India, Mexico, or Brazil. An explanation for the difference is not available.

Group IV phages have the broadest habitat and are found in human and animal sewage, whereas group I occurs predominantly in animals. Groups II and III are associated with humans almost exclusively. The geographical distribution of groups II and III shows a strong bias. In northern Japan, there is a relative abundance of group II over group III (6:1) per sampling site. Moving southward, this ratio drops dramatically until in

Southeast Asia group II becomes rare (Furuse *et al.*, 1978). It is suggested that the north-south gradient is related to differences in climate. Group III (and also I and IV) propagates well at 40°C (but not at 20°C), whereas for group II the situation is reversed (Furuse, 1982).

Although the natural host for the coliphages is not known with certainty, it has now been shown that RNA phages survive passage through the gastrointestinal tract of gnotobiotic mice and propagate stably in the intestines if *E. coli* is present as host. Thus *E. coli* can sustain the life cycle of the phage under "natural" circumstances (Furuse, 1986).

IV. THE INFECTION PROCESS

The single-stranded RNA phages, but also some single-stranded male-specific DNA phages, gain access to the interior of the cell by attachment to and transport via long, filamentous structures called pili. These can be of various origins and serotypes, as shown in Table III. In *E. coli*, the sex (or F) pili are used as vehicle. These pili are designed to promote the import and export of single-stranded DNA from one cell to another and they provide the phage with an effective way to get its RNA across the bacterial envelope. Transfer of RNA through pili is not essential for infection, however. Cells that lack pili can be infected if they are converted to spheroplasts.

The attachment of the phage to the sides of the pili proceeds via the maturation (A) protein. The coat protein plays no role in this process, since binary complexes consisting of one copy of the A protein and one copy of (group I) RNA can infect a piliated cell (Leipold and Hofschneider, 1975; Shiba and Miyake, 1975). In group III phages, such minimal infection set requires the additional presence of the read-through (A1) protein (Shiba, 1975).

For group I phages, it was shown by Krahn *et al.* (1972) that contact of the phage with the pilus results in cleavage of the A protein into a 15-kD and a 24-kD fragment. Paranchych (1975) speculated that this cleavage triggers the ordered ejection of the tightly packed RNA from the virus shell. There is some recent support for this idea. Shiba and Suzuki (1981) identified the binding sites of the A protein on MS2 RNA. It was found that the protein has two binding sites, one for a region centered around position 3515 and the other for the 5' end (position 393 and neighboring nucleotides). We suggest that cleavage of the A protein occurs between these two RNA binding domains, thus potentially liberating the two ends. We also know that nonsense mutations in the A protein gene yield phage particles that lack the A protein. Such virus particles, if prepared under the usual (not RNase-free) conditions, contain RNA that misses the 5' terminal part. Thus, it can be assumed that A protein prevents this part of the RNA from protruding from the viral shell. It is not inconsistent with the facts to propose that cleavage of the A protein releases the

5' end of the RNA which may begin moving along the pilus toward the cell. This stage of infection presumably corresponds to the RNase-sensitive step. The two A protein fragments have been shown to remain associated with the RNA during penetration of the cell envelope. However, it is unlikely that these protein fragments play any further role, since the naked RNA is fully able to generate infectious progeny in the spheroplast infection assay.

V. VIRION STRUCTURE

In addition to one molecule of plus strand RNA, each virion contains 180 copies of the coat protein and one copy of the maturation, or A, protein (Weber and Konigsberg, 1975). The presence of about 1000 molecules of spermidine per virion has been reported (Fukuma and Cohen, 1975). Presumably, the polyamine serves to neutralize the phosphate charges on the RNA. The group III phages contain also between 3 and 14 copies of the read-through protein in their capsid (Weiner and Weber, 1971), and this presumably also holds for group IV. The diameter of the phage is 26 nm, and the protein shell is 2.3 nm wide (Zipper *et al.*, 1971; Crowther *et al.*, 1975). The icosahedral shell has a T-3 surface lattice.

VI. REPLICATION OF PHAGE RNA

A. Mature Phage RNA

Most of what is known about phage RNA replication derives from studies with Qβ replicase, which is relatively easily purified. The isolation of the active enzyme from group I phages has proved very difficult (Federoff, 1975). However, the purification of the replicase from phage GA has been reported (Yonesaki and Haruna, 1981). It has the same subunit composition as Qβ replicase, so we may assume that replication in the four groups proceeds along the same lines.

The fully geared replicase complex contains five different proteins called subunits I to IV, or subunits α, β, γ, and δ; the fifth subunit is called host factor (HF). Subunit I was identified as ribosomal protein S1 (Kamen *et al.*, 1972; Inouye *et al.*, 1974, Wahba *et al.*, 1974), and subunits III and IV are the translation elongation factors EF-Tu and EF-Ts (Blumenthal *et al.*, 1972). Thus, three proteins that normally function in the synthesis of proteins are recruited by the phage to assist in RNA synthesis. Subunit II, also called replicase, or sometimes RNA synthetase, is encoded in the viral genome, and all these four subunits occur in the enzyme complex in one copy. The fifth component, the host factor, is present as a hexamer (Franze de Fernandez *et al.*, 1968). It is a small, heat-stable protein (12 kD) whose function in the uninfected cell is not known. *In vitro* experiments with HF neutralizing antibodies suggest that the protein is not involved in protein synthesis (M. S. Dubov and A. Wahba, personal communication, quoted in Blumenthal and Carmichael, 1979). There is a claim that the host factor for GA replicase is different from that of Qβ replicase (Yonesaki and Aoyama, 1981).

Initially, when only Qβ RNA replicase was available for testing, it was thought that the RNA-dependent RNA polymerases were highly specific. Now that the GA and SP replicases have also been purified, it has become clear that template specificity is not nearly so tight. For instance, GA replicase works very well with group I RNA and shows significant activity with RNA templates from group IV (SP, TW19, and TW28), but Qβ RNA is not amplified. Similarly, Qβ replicase does not accept group A RNA as template, but SP RNA is replicated with limited efficiency (25%). SP replicase amplifies Qβ RNA (20%) but not the group A RNAs. Evidently, the specificity of replication of phage RNA can be used as a criterion for RNA classification and as a measure of evolutionary distance (Yonesaki et al., 1982).

The requirements for copying the positive and the negative strand are different. To copy the (−) strand, only subunits II, III, and IV are required, whereas (+) strand replication needs also subunit I and the host factor (Kamen, 1975). There is evidence (and it would make sense) that the requirement for S1 (subunit I) is related to the fact that the plus strand serves as template not only for replication but also for translation. Specially designed experiments in Weissmann's laboratory showed that Qβ replicase, on its way copying the plus strand, cannot dislodge an oncoming elongating ribosome heading the opposite direction. Thus, some mechanism must exist to prevent a situation in which phage RNA is inactivated by an encounter between replicase and ribosome. The experiments have suggested that the "dead-end" situation is most probably avoided by competition between the ribosome and the replicase for common sites on plus strand RNA (Kolakofsky and Weissmann, 1971). RNase protection studies have shown that in the absence of the initiator nucleotide GTP, the replicase binds to two internal RNA sequences. One site, the M site, consists of three closely spaced regions, spanning the Qβ RNA sequence 2546–2873 (M5:2546–2614, M26:2638–2812, and M11:2845–2873) (Meyer, 1978; Meyer et al., 1981). This site is bound in the presence of Mg^{2+} and is essential for replication. The other site, the S site, binds the replicase in the absence of Mg^{2+} and spans the region 1248–1347, which overlaps the ribosome-binding site of the Qβ coat cistron (Weber et al., 1972). Occupation of the S site would therefore exclude the ribosome from the start of the coat gene. As we will discuss below, the only available ribosomal binding site on unperturbed phage RNA is that of the coat gene. Consequently, the binding of the replicase to the S site is sufficient to clear the messenger RNA from ribosomes (Kolakofsky and Weissmann, 1971).

Similarly, ribosome binding to the coat gene start is expected to interfere with replicase attachment. There is, however, a small problem here. *In vitro* studies have shown that the S site is not strictly required for replicase binding and replication. A contribution to solving this dilemma was recently made by Boni *et al.* (1986). The authors analyzed the binding domain of a 30S ribosome on MS2 RNA and Qβ RNA in the absence of initiator tRNA. Surprisingly, parts of the Qβ RNA M site were found cross-linked to S1 by UV irradiation. For MS2 the probable equivalent of the M site was found, consisting of a pyrimidine rich region in the proximal portion of the replicase gene at positions 2030–2056 (Boni *et al.*, 1986; Boni, personal communication). These results strongly indicate that initiating ribosomes and replicase enzymes bind to the same two regions on phage RNA.

Protection of the M and S sites against ribonuclease is also offered by the enzyme complex containing only subunits I and II (Weber *et al.*, 1972). It has further been shown that ribosomal protein S1 by itself, binds to two sequences in Qβ RNA. One is part of the S site (nucleotides 1291–1311); the other, whose significance is questionable, is located near the 3' end (Senear and Steitz, 1976; Goelz and Steitz, 1977).

The exact function of ribosomal protein S1 in mRNA translation has not been fully established, but some characteristics that could help to understand its role in replication are worth mentioning. Protein S1 is only required at the initial stage of translation—i.e., messenger binding to the ribosome. It is not required for the translation of artificial messengers such as poly(U), poly(A,G,U), etc. or for the translation of phage RNA whose higher-order structure has been destroyed by formaldehyde. Neither is the protein needed for elongation of translation (Van Dieyen *et al.*, 1975, 1976, 1978). There is thus a remarkable symmetry between the functioning of S1 in translation and replication; the protein is only required to bind the replicase or the ribosome to the native template. Conceivably, it does so by melting an ordered structure in which the M and S sites participate. Protein S1 thus permits the ribosome to bind to the start of the Qβ coat cistron. By adopting S1 as a subunit, the replicase acquires the property to compete with ribosomes for phage RNA. From a teleological viewpoint, it is then clear that phage RNA chains that cannot serve as template for the ribosome, such as the Qβ RNA (−) strand and the "6S" variants do not require replicase equiped with S1 for their replication. This sort of reasoning suggests that the phage has made an extra investment in its RNA structure to make replication dependent on S1, but of course this does not tell us what exactly the role of S1 in the enzyme is (see Subramanian, 1983, for a recent review on protein S1).

It has been established that the elongation of Qβ (+) strand synthesis does not require subunit I (Landers *et al.*, 1974). Such a behavior is reminiscent of protein synthesis, where as mentioned, protein S1 only functions in the selection of and binding to the start site.

As long as the role of the host factor in the uninfected cell has not been clarified, its contribution to replication must remain obscure.

The function of EF-Tu and EF-Ts in viral RNA replication has been extensively studied but not resolved (Blumenthal and Carmichael, 1979). It has been shown that the workings of the elongation factors in protein synthesis and in replication are quite different; treatments that inactivate their role in protein synthesis do not affect replication. For instance, in Qβ replicase the elongation factors will function in the cross-linked state, but in polypeptide synthesis the covalently linked EF-Tu·Ts complex is inactive (Blumenthal, 1977; Brown and Blumenthal, 1976). It is felt, however, that the common functional basis must lie in the structure of the viral 3' end. This shows resemblance to tRNA, which is the natural reaction partner of the elongation factors. Nevertheless, attempts to demonstrate an interaction between EF-Tu or Ts and Qβ RNA have failed. Another role for EF-Tu in replication might be that of GTP carrier. With no exception, all newly synthesized RNAs start with a row of G's, and EF-Tu has a GTP binding site. Such a role would be consistent with the fact that the EF-Tu·Ts complex is only required to get the polymerization reaction going and is not needed to elongate the RNA chains (Landers et al., 1974).

As it is difficult to imagine that the phage has "discovered" a property of the EF-Tu·Ts complex that the cell does not exploit, we must assume that the binary complex has functions that have not yet been identified. In this respect we may recall that quite some time ago the EF-Tu·Ts complex (called the ψ factor) was proposed to modulate cellular RNA synthesis (Travers et al., 1970) and that phage infections have been reported to inhibit the synthesis of ribosomal RNA (Watanabe et al., 1968). On the other hand, there are now many examples of proteins that contain binding sites that function in the regulation of their own synthesis (Nomura et al., 1982). Since the synthesis of EF-Tu is autoregulated, it may be that the phage takes advantage of such properties.

B. 6S RNA and Qβ RNA Variants

Infection of *E. coli* by Qβ leads, apart from phage multiplication, to the accumulation of what has been termed "6S" RNA. This is a non-homogeneous collection of RNA molecules that vary in size from about 100 to 200 nucleotides and serve together with their (−) strands as template for Qβ replicase. They do not code for any protein and do not contribute to the infection process. Three "6S" RNA representatives have been fully sequenced: midivariant-1 (MDV-1) (Mills et al., 1973); microvariant (MCV) (Mills et al., 1975); and nanovariants I, II, and III (NNV) (Schaffner et al., 1977). Their proposed secondary structures are shown in Fig. 2. MDV-1 has a length of 220 nucleotides, and MCV and NNV contain 114 and 91 nucleotides, respectively. The nanovariants (NNV), which were called WSI, WSII, and WSIII by the authors, are the shortest self-replicating molecules with known primary structure; WSII and WSIII are base substitution mutants of WSI.

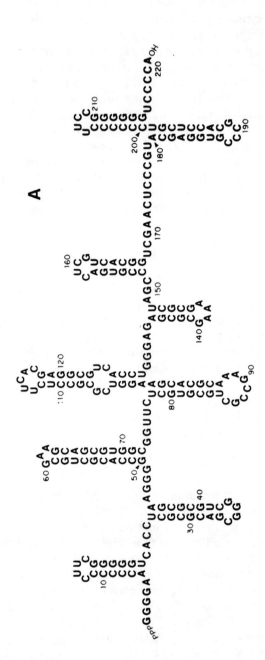

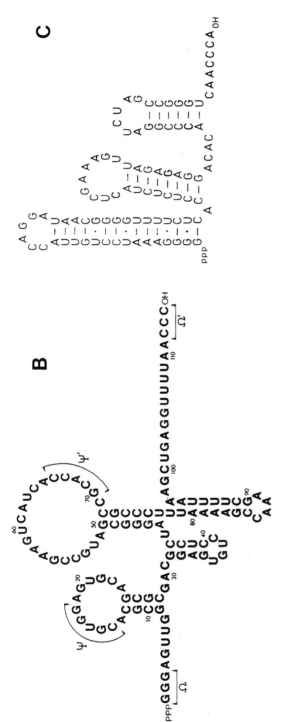

FIGURE 2. Secondary structure models for three 6S RNA species. (A) MDV-1 RNA. From Mills *et al.* (1975). (B) WSI RNA. From Schaffner *et al.* (1977). (C) Microvariant RNA. From Mills *et al.* (1978). We have lately run the 6S RNA sequences through a computer program that allows the formation of pseudoknots (Pley *et al.*, 1985; Abrahams, unpublished results). Only slightly different secondary structures were obtained, and no common structural features emerged. Ψ and Ψ′ and Ω and Ω′ identify regions of additional complementarity in MCV-RNA.

From an evolutionary point of view, it is interesting that these molecules quickly respond to selective pressure. Mutants of MDV-1 could be obtained that had adapted to replication under adverse conditions, such as the presence of ethidium bromide or limiting amounts of one of the building blocks (Kramer et al., 1974). Another member of the family of small self-replicating RNA molecules was itself the product of a man-made selection procedure. Mills et al. (1967) set up a serial transfer experiment designed to enrich for RNA chains that would replicate in progressively shorter times. A sample of an incubation mixture that contained Qβ RNA and all the ingredients for Qβ RNA replication was transferred to a fresh tube containing all ingredients except template RNA. As in each fresh tube enzyme is in excess over template, the fastest-replicating molecules will outgrow other self-replicating species. By the 74th serial transfer, the lower size limit of 550 nucleotides attainable by this device was reached. That the abridged variant (V-1 RNA) was in fact derived from Qβ RNA was established by hybridization experiments. V-1 RNA representing only 12% of the original molecule was also sensitive to selection pressure. Each new set of replication conditions would yield another RNA molecule, but none of these have been sequenced (Mills et al., 1973). In fact, there seems no basic difference between V-1 RNA and the RNA present in the defective interfering (DI) virus particles that accompany, for instance, *influenza* infection (Jennings et al., 1983). Also here, the abridged molecules once created by replication errors can survive, since they are templates for the replicase, and their multiplication does not endanger the survival of the virus population as a whole.

The rapid yield to environmental pressure by 6S RNA and the Qβ RNA variants reflects the inaccuracy of the Qβ replicase, which has been estimated as between 10^{-3} and 10^{-4} per nucleotide per replication (Domingo et al., 1978). The presumed absence of a 3′–5′ exonuclease editing activity in Qβ replicase would be consistent with this relatively low copying fidelity. At the same time the frequency with which deletions occur must also be unusually high. Deletions may arise by a copy choice mechanism, where regions of homology on the same or on a different template serve as the starting point to continue the replication process. Alternatively, instead of homology, hairpin loops or the relative spatial orientation of RNA segments may facilitate the "jumping" of the replicase (Jennings et al., 1983).

A general property of all variants and also of mature phage RNA is the relatively high degree of secondary structure. This feature is required to prevent the annealing of (+) and (−) strands, which is known to halt the net synthesis of template molecules (Kamen, 1975).

Because of its small size, 6S RNA has been used as a model system to study the recognition between template and replicase. Schaffner et al. (1977) and Nishihara et al. (1983) have identified regions on WS RNA and MDV-1 RNA that bind the enzyme. Similar to what was found for full-length Qβ RNA, internal regions, but not the 3′ end, were protected

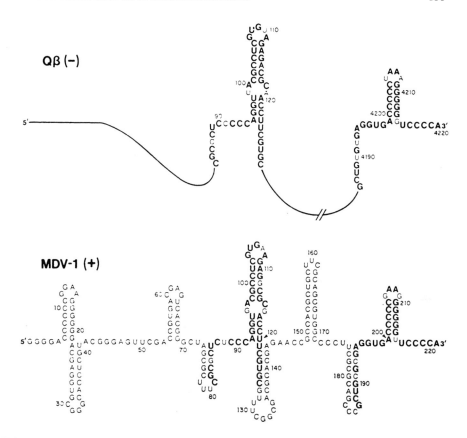

FIGURE 3. Comparison of MDV-1 RNA and Qβ RNA (−) strand. Homologous regions are in boldface type. From Nishihara et al. (1983).

against RNase in the absence of GTP. The enzyme-binding site in WSI RNA extends from residue 20 to residue 50. This is much smaller than in Qβ RNA, but it may be not so surprising, since the variants are replicated in the absence of S1 (and host factor), whereas Qβ RNA requires the presence of these proteins for formation of the enzyme-template complex. It may be more appropriate to compare the RNase protection patterns of the variants to those obtained with Qβ RNA (−) strand.

The availability of some 10 sequences of amplifiable RNA (Qβ, MDV, MCV, WS, and their complements and mutants) had raised hope that the interaction between enzyme and template might be understood or at least described in terms of a consensus sequence or consensus structure. For instance, very striking homology of some 40 nucleotides was observed between two regions in Qβ RNA (−) strand and two regions in MDV-1 (+) strand (Nishihara et al., 1983) (Fig. 3). Although significant homologies have also been noticed between some other pairs of templates, no primary sequences of significant length are common to all, except for a set of three C's at the 3' end (Schaffner et al., 1977). Thus the primary sequence does not seem to hold any secrets, and the observed homologies may reflect the genesis of the molecules.

It is interesting that the central helix in MDV-1 RNA and its counterpart in the Qβ RNA (−) strand show a coordinated base substitution pattern (Fig. 3). The C·A mismatch in Qβ (position 103) has become a C-G match in MDV-1 (position 100). On the other hand, the U-A match in Qβ (position 99) has turned into a G·G mismatch in MDV-1 (position 96). This suggests that maintenance of the stability of the helix has priority over the primary sequence. There are many examples of this type of energy conservation in RNA phage RNA. A few will be discussed later.

The prospects for finding clues on template recognition in the secondary structure are not much better. Enzymatic digestions and chemical modifications have provided experimental evidence for the proposed secondary structures, but the available models as depicted in Figs. 2 and 3 do not show obvious commonalities. It was therefore suggested that "there may be a variety of ways in which an RNA can attach to the enzyme and allow initiation and that the various families of '6S' RNA have evolved by taking advantage of these possibilities" (Schaffner et al., 1977). One could indeed take the view that the specificity does not reside with the replicase but with the RNA. Specificity would then be negatively defined as the ability of cellular RNAs such as ribosomal, messenger, and tRNA to avoid copying.

The work of Schaffner et al. (1977) has shown that the replicase binds to WS RNA sequences that, in the unperturbed RNA, are almost certainly part of hairpin structures. As a first step to understand template binding, one would have to know if the bound sequence maintains its original structure, as is generally the case when ribosomal proteins bind to ribosomal RNA, and also when the coat protein of the RNA phages binds to the hairpin at the start of the replicase cistron (see below). If this is so, conservation of secondary structure elements should be expected as was found for hairpins in ribosomal RNA (El-Baradi et al., 1985) and for the replicase start region (see below and Fig. 5).

The other possibility is that the replicase has a high affinity for single-stranded regions and binds by melting weak secondary structures. As an example for such an interaction, we may mention here ribosomal protein S1, which indeed melts ordered structures upon binding (Bear et al., 1976; Kolb et al., 1977). The fact that poly C is readily copied and poly U, which is a random coil polynucleotide, acts as a competitive inhibitor of Qβ replicase is consistent with this mode of binding. If the replicase would indeed prefer binding to single-stranded regions regardless of the sequence, template specificity would be reduced to a compromise between a secondary structure that is sufficiently labile to allow replicase attachment and sufficiently stable to resist ribonuclease and prevent annealing between copy and template.

As pointed out by Meyer *et al.* (1981), another major condition for successful RNA amplification must be met. The 3' end should (1) contain a row of C residues, (2) be single-stranded, and (3) be properly placed in space or flexible enough to reach the active site on the enzyme bound to the internal region.

Though our view on enzyme-template binding is undoubtedly an oversimplification, it contains some elements that can be experimentally tested. One of the predictions is that "salt pressure" will force a 6S RNA molecule to evolve hairpins that have a lower stability. One other is that at low salt concentrations, where helices are less stable, the enzyme-template specificity will be more relaxed. Such a decreased discrimination has indeed been found for Qβ and SP replicases (Fukami and Haruna, 1979). That specificity may reside with the RNA template rather than with the enzyme system may also be illustrated by a comparison with initiation of translation. The ribosome will not translate any ribosomal RNA or tRNA, and even on the 4000-nucleotides-long RNA from the RNA phages, there is only one start site available. Yet, the ribosome will readily initiate protein synthesis on random poly(A,G,U) copolymers. The suggestion from such data is that much effort is invested in avoiding being a substrate.

VII. ORIGIN AND EVOLUTION OF THE 6S RNA FAMILIES

As mentioned, 6S RNA is found in large amounts of Qβ-infected cells, and two possibilities for its origin can be envisaged. One is that this RNA is (co)packaged into otherwise normal phage particles and propagated as a molecular parasite. This view places its origin far back in time. The other possibility is that these self-replicating molecules are generated anew every time and progress from creation to extinction within the span of one infectious cycle. The last possibility now seems to be correct since Qβ replicase introduced into the noninfected cell via a plasmid carrying its cDNA also gives rise to 6S RNA (Biebricher *et al.*, 1986). This, however, does not yet tell us what the intracellular origin of this molecule is.

6S RNA can also be generated by the *in vitro* incubation of Qβ replicase with ribonucleoside triphosphates in the absence of added primer (Banerjee *et al.*, 1969). It was soon realized that the enzyme contained sufficient endogenous template to account for the synthesis of the variant RNAs. However, even if all detectable primer RNA had been removed, the enzyme appeared able to make RNA after a long lag period, and the crucial question arose whether undetectable traces of template were being amplified or the replicase was able to produce RNA without instruction. Sumper and Luce (1975) and Biebricher *et al.* (1981a,b) have taken great pain to demonstrate the noninstructed synthesis of RNA by Qβ replicase. They present several strong arguments: (1) Template-instructed RNA synthesis is proportional to the nucleoside triphosphate concentration with the power <1, but *de novo* synthesis responds to the substrate concentration to the third or fourth power, hinting at a "nucleation process" involving 3 to 4 triphosphate molecules (*de novo* synthesis defined for the moment as the condensation of nucleoside triphosphates by replicase rigorously freed of endogenous template). (2) There is a differential dependence on enzyme concentration; *de novo* synthesis to the power of 2 or 3, while instructed synthesis is linearly proportional to replicase concentration. This suggests cooperativity between several enzyme molecules in template-free RNA synthesis (3) Different RNA species emerge when different enzyme concentrations are used, even though the absolute amount of enzyme remains the same in these experiments. (4) The nature of the RNA synthesized in *de novo* replication is sensitive to the conditions prevailing during the lag phase; however, the addition of 5 to 10 molecules of 6S RNA during this lag phase always leads to a uniform product—i.e., the added self-replicating species. The added template apparently easily overgrows any new creation. (5) *De novo* RNA synthesis cannot be obtained below a critical NTP concentration, but template-dependent replication proceeds unabated at this concentration.

Biebricher *et al.* (1981a) propose that during the lag phase there is semirandom polymerization of nucleotides. Once a few polymers are formed and released, natural selection will, assisted by the rather low copying fidelity of the enzyme, mold and model these RNAs, resulting in the macroscopic appearance of several "fit" candidates. At this stage, the products have been separated and characterized by length and by ribonuclease T1 fingerprints. If this mixture is subjected to internal competition by supplying limiting amounts of replicase, one quasispecies is obtained. It is interesting that the evolution from the first detectable specimens to stable quasispecies is usually accompanied by an increase in chain length. The RNAs obtained this way were termed minivariants (MNV) and have a length of about 120–150 nucleotides. None of these have so far been sequenced. The properties of the MNV RNAs can also be changed by environmental pressure. If salt is the strain, MNV mutates to a longer molecule with the same mobility as MDV-1 and possibly identical to it.

De novo synthesis was recently challenged again by Hill and Blumenthal (1983). They report a new procedure to free Qβ replicase of endogenous template, which presumably leads to a preparation incapable of self-instructed RNA synthesis. It has been questioned, however, whether the polymerization conditions used by Hill and Blumenthal permit uninstructed RNA synthesis (Biebricher *et al.*, 1986).

If uninstructed RNA synthesis by Qβ RNA replicase exists, it suggests a simple explanation for the origin of the 6S RNA family; during infection, this RNA is invented by the replicase by trial and error. This process may or may not exploit Qβ RNA fragments present in the cell. The creation of such "selfish RNA" molecules can apparently not be avoided.

Although probably useless to the phage, the 6S RNAs and the Qβ variants will be of great advantage to study evolutionary problems that involve RNA as genetic material.

VIII. GENE EXPRESSION

The genetic map of representatives of the four phage groups is given in Fig. 1. The appearance of the phage-coded proteins is carefully controlled in timing and amount. As shown in Fig. 4, the replicase is an early product, and the amount of coat protein exceeds by far that of the other products. Since no DNA intermediates occur in the life cycle of the phage, control is predominantly exerted at the level of translation, sometimes assisted by replication intermediates.

Upon entry in the cell, the RNA first serves as messenger, because multiplication requires the product of the replicase gene. After subunit II is made, the holoenzyme with the accessory host factors can be assembled and amplification can start.

A. The A Protein

The A, or maturation, protein is needed in small amounts, one copy per virion, and its translation is accordingly kept at a low level. Some-

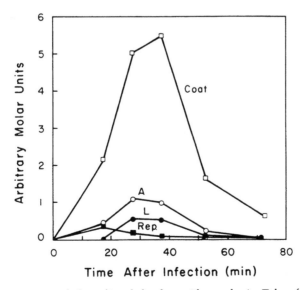

FIGURE 4. Time course of phage f2-coded polypeptide synthesis. Taken from Beremand and Blumenthal (1979).

what more is produced than is actually needed. It has been suggested that regulation of the A protein in group I phages is achieved by a mechanism whereby only nascent RNA chains can serve as messenger for this product. This is supposedly allowed for by base-pairing between the ribosomal binding site of the A protein gene and an internal A protein gene coding sequence (Fiers et al., 1975). As long as the internal complementary sequence is not yet produced, the A protein gene start region will be exposed. The data leading to this concept have been mainly obtained by comparison of in vitro translation products made on replicating intermediates and on full-size RNA. The intermediates made much more A protein than the full-size RNA (Robertson and Lodish, 1970; Lodish, 1975). It may also be that the initiation codon of the A protein assists in lowering expression. This codon is GUG in MS2, and it has been shown that GUG results in lower expression than AUG (Wulff et al., 1984; Munson et al., 1984; Looman and Van Knippenberg, 1986). That other group I phages like R17 and M12 have AUG as start codon for the A protein indicates that low expression can also be achieved with the optimal initiation codon.

The RNA of several phages has been cloned, and the proposed masking of the A protein gene start region by secondary structure is now open to experimental testing.

B. The Replicase (Subunit II)

Replicase is needed in small amounts and early in infection. Its synthesis was found to be fully turned down 20 min after infection. This inhibition is effectuated by the coat protein which binds to the start region of the replicase gene (Capecchi and Webster, 1975; Weber, 1976). For MS2 RNA, it has been found that this region can adopt an irregular helical structure that, once formed, will be stabilized by the binding of one copy of the coat protein (Gralla et al., 1974). This control mechanism has also been established for Qβ, and it is likely that it also operates for the other two groups. One might thus expect a higher degree of sequence or secondary structure conservation for this area. The corresponding structures for MS2, fr, GA, and Qβ are displayed in Fig. 5. In spite of the fact that MS2, GA, and fr have substantial overall sequence differences (50% between GA and MS2), there is strong primary and secondary structure homology here.

There is firm experimental evidence for the displayed MS2 structure (Gralla et al., 1974; Hilbers et al., 1974), and the sequences of GA and fr are so close that the proposed structures are convincing. The more distant relationship of Qβ RNA is reflected in the fact that the primary structure in this region diverges substantially. Yet, the secondary structure proposed for this area by Weber (1976) has similarities that increase its probability. For instance, it has the bulged-out A that is essential for coat

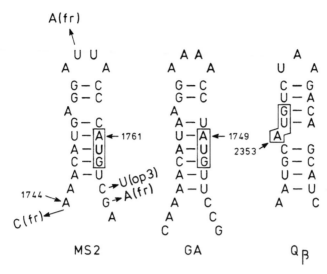

FIGURE 5. Secondary RNA structure at the start region of the replicase cistron of MS2 (Steitz, 1969), fr (Berzin *et al.*, 1982), GA (Inokuchi *et al.*, 1986), and Qβ (Weber, 1976). The sequence of MS2, f2, and R17 is identical in the region.

protein binding to the MS2 hairpin. Also, the structure sequesters the AUG start codon, precluding its functioning in ribosome binding when the helix becomes stabilized by Qβ coat protein.

Comparison between MS2 and GA reveals several interesting points. The G-C pair in MS2 (1751 <–> 1760) is replaced by the weaker A-U pair, but an additional A-U pair at the bottom of the stem appears, probably to compensate for the loss of stability. This is another example showing the tendency to maintain narrow stability limits on individual hairpins. Here, a connection with regulation of translation is indicated by the phenotype of the *op3* mutant of phage f2 (=MS2) where C1765 has changed into a U. In this mutant, the replicase hairpin is extended with one A-U pair (Fig. 5) (Atkins *et al.*, 1979). The translation of the replicase gene in the *op3* mutant is decreased, which was attributed to the extra base pair. The binding strength of the coat protein to the hairpin in the *op3* mutant is not affected by the extra base pair (Carey *et al.*, 1984). We note that the G-C (<–>) A-U covariance between MS2 and GA constitutes independent evidence for the existence of the hairpin.

Uhlenbeck *et al.* (1983) have carefully studied the elements of the R17 hairpin structure that contribute to coat protein binding by preparing sequence and length variants of the RNA region. The basis of all experiments is an oligomer of 21 nucleotides prepared by organic synthesis. This oligomer displays the same binding characteristics as full-length R17 RNA and comprises the sequence from residue 1744 to residue 1764 (Fig. 5). Taking two nucleotides off the 3' side (1764 and 1763) does not affect the binding constant ($K_a = 3 \times 10^8$ mol/liter). Removing more than two nucleotides from the 3' side progressively reduces the binding constant. It is possible that this drop in affinity is caused in an indirect way by weakening the hairpin beyond the allowable limit. Taking away A1744 reduces the binding constant by 3 orders of magnitude, whereas extending the helix by adding one or two U's at position 1765 and 1766 has no effect. Thus A1744 is crucial regardless of its base-paired state, suggesting that the adenine functional groups are not involved in coat protein binding. That helix extension does not affect the binding constant is consistent with the fact that the hairpin in GA and in the *op3* mutant is actually 1 base pair longer.

Substitutions that affect the two G-C base pairs in the upper part of the helix destroy coat protein binding. Presumably, the protein contacts functional groups on the nucleic acid that are held in position by the secondary structure. Alternatively, the protein interacts directly with the G-C base pairs. Of particular interest is the bulged-out A residue. Such irregularities have been shown to occur also in ribosomal RNA helices that function as binding sites for proteins. Removal of A1751 in the replicase stem structure or replacing it by a C residue abolishes binding of the coat protein (Uhlenbeck *et al.*, 1983). This result has a perfect parallel. Removing the bulged-out A66 from *E. coli* 5S RNA led to drastic loss of binding of ribosomal protein L18 (Christiansen *et al.*, 1985).

Substitutions in the hairpin loop gave the following results: (1) Changing U1755 to any other base has no effect on the binding, which fits nicely the fact that fr phage has an A at this position. (2) Changing A1754 or A1757 to any other base decreases K_a 100- to 1000-fold. (3) Substituting U1756 for a C increases K_a approximately 50 times, but a change to purine again lowers K_a 10 to 100 times. Apparently, the high specificity of the interaction originates from precise contacts at a large number of positions spread over the entire region, most but not all of which are sequence-specific (Uhlenbeck *et al.*, 1983; Lowary and Uhlenbeck, 1987).

Romaniuk and Uhlenbeck (1985) have also presented evidence that a transient covalent bond may be formed between a cysteine residue in the R17 coat protein and U1756. Substitution of this U residue for a C (see above) does not change the association constant much, in agreement with the expectation that a similar cysteine-cytidine adduct can be formed. Interestingly, GA RNA has an adenine at the corresponding position, but the GA coat protein does not contain cysteine residues. It is not known how GA compensates for the loss of this contact point (Inokuchi *et al.*, 1986).

The MS2 replicase start region shown in Fig. 5 can assume an alternative structure involving a long-range interaction with a part of the coat gene. as shown in Fig. 6. This structure was suggested by Min Jou *et al.* (1972), and it explains two sets of data. First, the regions shown in Fig. 6 (together with several other fragments) comigrate when a limited RNase

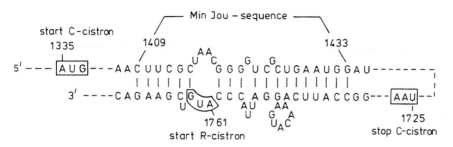

FIGURE 6. Long-distance interaction between the MS2 replicase start region and a portion of the coat gene coding sequence. The interaction was proposed by Min Jou *et al.* (1972). The illustration is from Berkhout and Van Duin (1985).

T1 digest is separated on nondenaturing polyacrylamide gels. Second, it had been found that the early coat amber mutant (amino acid 6) produced very little replicase, whereas late coat ambers (amino acid 50 and further) showed normal rates of replicase synthesis and eventually overproduced replicase owing to the absence of a functional coat protein repressor (Robertson, 1975). The idea is therefore that ribosome movement over the coat gene through the base-paired region is required to activate the replicase start.

Using MS2 cDNA subgenomic fragments and a suitable expression vector, the translational coupling between coat and replicase genes has been fully confirmed (Berkhout and Van Duin, 1985). cDNA constructs that lack the start signals of the coat gene synthesize very little replicase. Recombinant DNA procedures have also allowed to confirm the base-pairing scheme proposed by Min Jou *et al.* (1972) (Fig. 6). MS2 cDNA clones starting before or at position 1420 synthesize very little replicase, but clones that begin at position 1432 or further downstream produce 10 times more. Thus, the nucleotides between position 1420 and 1432 are responsible for the polarity phenomenon. Their removal results in uncoupled replicase synthesis.

In clones where coat protein is synthesized, the replicase protein is virtually undetectable. Control of replicase synthesis by the long-distance interaction is thus less tight than by coat protein repression. In the fully unrepressed state (clones starting at position 1432), replicase is synthesized at half the rate of the coat protein, showing that although very little of this protein is required and actually present, the ribosomal binding site is in itself very strong. It may be noted that the two control mechanisms serve different purposes. Coat protein binding to the replicase start is used to fully block further production of this enzyme. The long-distance base-pairing, on the other hand, does not really restrict the replicase concentration but merely establishes the translational coupling to the coat gene. As discussed in section VI.A, this translational coupling has most likely evolved to avoid uncontrolled access of ribosomes to the phage RNA cistrons. This design safeguards the switch from template for translation to template for replication.

Induction of the replicase gene for a few minutes leads to very rapid cell death (Remaut et al., 1982; our unpublished results). The replicase gene of MS2 (and also of GA) contains several internal open reading frames, and it seemed possible that the encoded polypeptides were expressed and responsible for killing the host. Deletion mapping, however, excluded the role of these open-reading frames (our unpublished results). Apparently, the replicase itself is a very poisonous protein, which may be related to its membrane association. In view of the small amounts of replicase needed to kill the cell, it is unlikely that depletion of the host factors S1, EF-Tu, EF-Ts, and HF is responsible. The biological role of killing, if any, is unknown.

C. The Coat Protein

The coat protein is abundantly expressed during phage infection. Also, upon induction of the appropriate cDNA clones up to about 40% of newly synthesized product is viral coat protein (Remaut et al., 1982; Kastelein et al., 1983b). Except for the earlier-described competition between ribosome and replicase for the ribosomal binding site of the coat gene, the synthesis of this protein seems not to be negatively controlled in any way. A positive control mechanism has been proposed whereby the intact coat protein promotes the formation of a replicating intermediate lacking most of the replicase sequences. This intermediate was thought to synthesize higher amounts of coat protein than the full-size template (Robertson, 1975). However, experiments with MS2 cDNA clones lacking the complete or partial replicase sequences have not confirmed this model; the 3' cutoff point of the MS2 cDNA did not affect the rate of coat protein synthesis (Kastelein et al., 1983b). Also in these studies the presence of the native coat protein did not affect the frequency of translational starts at the coat gene (Berkhout and Van Duin, unpublished). Remarkably, it was found in the same study that the 5' cutoff point had a pronounced influence on coat gene expression. For instance, if the rate is set at 100% when the 5' boundary is located at position 103 on the MS2 map, it is 50% for clones starting at 869, 25% for 1221, and less than 10% if the MS2 subgenomic fragment starts with nucleotide 1308 (see also Remaut et al., 1982). We note that even in the shortest clone, the complete ribosome binding site is encoded. This behavior is exceptional. If this sort of experiment is carried out for the replicase or the lysis genes only, a negative contribution to translation is found by upstream leader sequences (Berkhout and Van Duin, 1985; Kastelein et al., 1983a). It is possible that the upstream sequences that enhance the expression of the coat protein assist in exposing the start region of the coat gene at the surface of the MS2 RNA molecule.

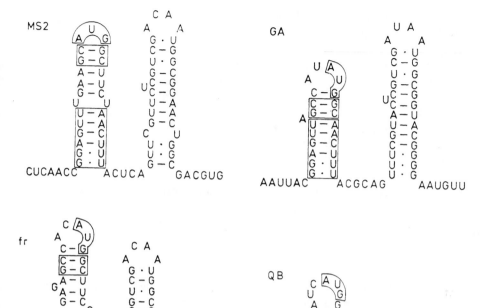

FIGURE 7. Proposed secondary structures for the coat cistron start region of four RNA phages. For MS2 and Qβ the initiator hairpin structure was initially proposed by Steitz (1969) and Weber *et al.* (1972). For GA and fr, the structure is based on sequence comparison. The second stem loop structure shown for MS2, GA, and fr is also proposed on the basis of comparative analysis and differs somewhat from the suggested structure of Min Jou *et al.* (1972).

In Fig. 7 the proposed structure of the coat gene start region is given for the phages MS2, fr, GA, and Qβ. For MS2, the structure is consistent with the RNase T1 sensitivity pattern determined by Min Jou *et al.* (1972). We also feel that phylogenetic comparison between MS2 and GA supports the secondary structure shown. There is less certainty for Qβ. Assuming that the structures are correct, we would like to point out a similarity that may be related to the high expression of this gene. The double helical structure is quite irregular and has a low thermodynamic stability. This property is also shared by the initiator hairpin of the replicase gene where the link between stability and efficiency of translation was supported by the phenotype of the *op3* mutant. Supposedly, such labile secondary structures are a compromise between protection against ribonuclease and efficiency of ribosome binding.

Comparison of the MS2 and GA hairpins reveals a quite interesting choice between the evolutionary conservation of protein versus RNA structure. The lower part of the helix structure is fully conserved, but the reading frame over this conserved part is not. This suggests that the secondary structure here is more important to the phage than the N-terminal amino acids of the coat protein.

Figure 7 also shows what the origin of the changed reading frame is. In GA, four nucleotides, UUCU, have been deleted shortly after the AUG start codon. Restoration of the frame occurs by the insertion of a U residue some 15 nucleotides further downstream. This still leaves the GA coat protein with one amino acid less than MS2. Though highly speculative, it seems that GA compensates for this deficit in a very unexpected way; there is a triplet insertion in an otherwise homologous region at the very end of coat gene leading to the addition of an alanine residue at the carboxy terminus (Figs. 12, 15). It is intriguing to try to reconstruct intermediate life forms between MS2 and GA. As it seems impossible that any viable intermediate carried a frame shift mutation, one may consider that the insertion and deletion have occurred simultaneously. The secondary structure of the coat gene initiator hairpin renders this possibility unlikely though. We see in Fig. 7 that the UUCU deletion in phage GA has required additional compensation in the opposite strand of the helix in the form of a three-nucleotide deletion (UGA). Given our present understanding of how these RNAs function as template in protein synthesis and the fact that RNA recombination has not been found in these RNA phages, it is not easy to perceive how a phage can pass through these metastable intermediates in the face of selection pressure. Possibly, a metastable intermediate may escape such pressure if it is present as a minor fraction of the total phage RNA population in the infected cell. It is conceivable that at high multiplicity of infection such fellow travelers enjoy a free ride based on the fitness of the wild-type coinfectors. Indeed, the temporary coexistence of slightly different sequences in an RNA population has been demonstrated (Domingo et al., 1978). Eventually, additional mutations in these aberrant molecules may produce an RNA variant that is viable on its own and may form a separate subgroup or outgrow the original population.

Phage fr seems to take an intermediate position between MS2 and GA. Its relatively distant relationship to MS2 within group I was already clear from the immunological data presented in Table II. The resemblance between GA and fr is attested to by the triplet insertion in front of the AUG codon that results in the typical asymmetric placing of the start codon. Homologies between MS2 and fr, on the other hand, are apparent from features in the second hairpin, where GA has a different wobble base at the hairpin loop and shows the inversion of an U-A <-> A-U wobble base pair (see Fig. 12 for a full sequence comparison). It is surprising that the initiator hairpins diverge much more than, for example, the second hairpins shown in Fig. 7. One would expect that ribosome recognition would place extra constraints on such a sequence. An important point seems that the structure is sufficiently labile to pass on to the random coil.

As mentioned above, the tolerated local frame shift in GA indicates that the identity of the first few amino acids of the coat protein is subsidiary to the desired RNA structure. This observation may have general validity, since a survey of codon usage in the first few amino acids of a protein has shown that codon bias here diverges from that used in the rest of the gene (Looman, 1986). The base changes in the 5' terminal part of the lysis gene of the group I RNA phages f2, M12, fr, and R17 also show the overriding weight of structure conservation over amino acid sequence (Berkhout et al., 1985b, and below).

Jacobson et al. (1985) have attempted to identify secondary and tertiary structure features in MS2 RNA using electron microscopy. Spreading of the RNA in the presence of spermidine yields distinct reproducible images which display long-distance as well as local interactions. One feature is of particular relevance for the present discussion, because it proposes an alternative to the structure presented in Fig. 7. The region spanning the nucleotides 1293–1345 and containing the start of the coat gene is thought to base-pair with the replicase region 2131–2178 forming an irregular helix with a calculated stability of about 50 kcal. In fact, this helix is part of a more extensive interaction that embraces the region 1293–1379 and 1928–2178.

As the relationships between the structure of RNA in solution and its appearance under the electron microscope have only been poorly examined, it is difficult to evaluate the results presented by Jacobson et al. (1985). In its rigid form, the suggested interaction seems to pose several problems. First, the structure seems too stable to serve as an efficient ribosomal binding site. Second, the interaction appears to lead to a reversed translational polarity; i.e., translation of the replicase gene is now expected to liberate the coat gene start region. Finally, deleting the complementary replicase region should affect coat gene expression. The experiments of Kastelein et al. (1983b) have shown, however, that deleting parts or all of the replicase gene hardly changes the rate of coat protein synthesis. Recall that deletion of the "Min Jou sequence" (Fig. 6) did increase replicase synthesis 10-fold. It may very well be, however, that the structure proposed by Jacobson et al. (1985) precedes the one shown in Fig. 7 at initial stages of translation or replication. In fact, evidence for the long-range interaction comes from the experiments of Boni et al. (1986).

As already mentioned, the region 2030–2056 was found UV-cross-linked to protein S1 in the binary complex consisting of 30S ribosomes and MS2 RNA. As it was shown (Van Duin et al., 1980) that this complex is a precursor to coat gene translation, the data indicate that at least some of the regions proposed by Jacobson et al. (1985) are indeed close together in space and may form a functional domain recognized by S1 either as part of the replicase or the ribosome. In this connection I should mention an experiment carried out in our laboratory but not understood at the time. We cleaved MS2 RNA in vitro in the intercistronic region between coat and replicase genes with RNase H and a deoxyoctanucleotide covering the region 1745–1752. In agreement with the *in vivo* results, we also found *in vitro* that the half-molecules sustained synthesis of the coat protein at the same rate as the intact RNA (Kastelein *et al.*, 1983b).

There was one difference, however, between half- and full-size MS2 RNA. We have shown that in the 30S ribosome·MS2 RNA precursor complex, protein S1 is protected from inactivation by antibodies against S1. In the complex containing the split MS2 RNA and 30S ribosomes, protein S1 is not protected against the antibodies. (The assay is whether or not the complex can synthesize the coat protein in the presence of anti-S1 upon supplementation with missing components for protein synthesis.) This result together with those of Boni *et al.* (1986) suggests that protein S1 is bound to and protected by the MS2 RNA M site. There seems a peculiar mirror symmetry in the interaction of phage RNA with the ribosome and the replicase. Both S and M sites are bound, but S site binding is dispensable for replicase function, and the M site binding is for ribosome function. One may speculate that protein S1 unfolds the domain formed by the interaction of M and S sites leading to the structure proposed in Fig. 7.

D. The Lysis Protein of Group A Phages

The group I phages have an overlapping gene encoding a small hydrophobic protein that is responsible for cell lysis at the end of the infection cycle (Beremand and Blumenthal. 1979; Model *et al.*, 1979; Atkins *et al.*, 1979). Just as was found for the replicase, the lysis gene is under translational control of the coat gene. If coat gene translation is precluded by early nonsense mutations or by the absence of the start signals on the appropriate MS2 DNA subgenomic segments, the lysis protein is not produced. This finding again stresses the phage's vital interest in avoiding the free admission of ribosomes to its genes. The mechanism by which coat gene translation brings about lysis gene expression has been studied by Kastelein *et al.* (1982) using cloned MS2 cDNA. It was found that frame shift deletions in the coat gene, leading to premature termination at either a −1 or a +1 phase stop codon present at positions 1652 and 1672, resulted in cell lysis. Based on a variety of supporting data, a model was put forward in which reading-frame errors were supposed to divert a fraction of the coat gene translating ribosomes to the above-mentioned out-of-phase stop codons (see Fig. 8 for their location in the MS2 sequence). Lately, we have critically tested this frame shift model. Site-directed mutagenesis was employed to inactivate the two out-of-phase stop codons while maintaing the in-phase encoded coat protein sequence:

$$
\begin{array}{cc}
1652 & 1673 \\
\downarrow & \downarrow
\end{array}
$$

GUU·AAG → GUC·AAG and CUA·AAA → CUC·AAA

According to the model the substitutions depicted above should abolish the expression of the lysis gene. However, the two single mutants as well as the double mutant showed an unchanged lysis (+) phenotype. The conclusion is therefore that although termination at the indicated out-of-frame stops is sufficient to activate the lysis gene, it is not a necessary condition (Berkhout *et al.*, 1987; Berkhout, 1986).

Our further analysis of this problem has shown that termination of coat gene translation at the natural position is the trigger to lysis gene expression. This conclusion rests mainly on the following observations: (1) Removal of the two tandem stop codons by site-directed mutagenesis, which leads to termination 8 codons further downstream, abolishes cell lysis. (2) If ribosomes translating the coat cistron are diverted to another phase and thus pass over the coat gene terminator in the wrong frame, the lysis gene is not activated. (3) If in the construct just mentioned a stop codon is introduced in the reading frame used and placed at the position of the coat gene terminator, the lysis (+) phenotype is reestablished (Berkhout *et al.*, 1987). The result mentioned under (2) also shows that the mere movement of ribosomes over the lysis gene start is not sufficient to activate the gene.

The RNA secondary structure that keeps the lysis cistron closed in the absence of coat gene translation has been determined with structure-

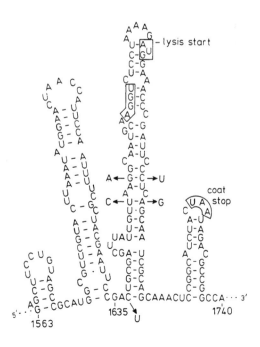

FIGURE 8. Secondary structure of MS2 RNA around the start of the lysis gene. Substitutions in M12 are only shown in the lysis hairpin.

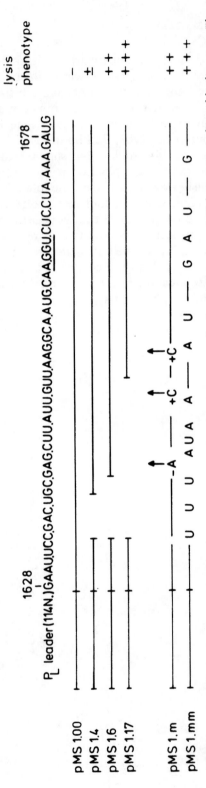

FIGURE 9. Partial nucleotide sequences of pMS1.00 and derived deletion, insertion, and substitution mutants. Deletions are indicated by line gaps. The Shine and Dalgarno region and the initiation codon of the L gene are underlined. Dots indicate the reading frame of the coat protein gene. MS2 DNA is under transcriptional control of the P_L promoter of phage λ.

specific enzymes and chemicals (Schmidt *et al.*, 1987). The secondary structure that fits the data obtained is presented in Fig. 8. The stem structure containing the lysis gene start is supported by phylogenetic comparison with the close relative M12, which has five base substitutions that preserve the base-pairing scheme. There is substantial evidence that this secondary structure prevents ribosomal binding in the absence of coat gene translation. For instance, clone pMS1.00 containing the MS2 DNA sequence 1628–2057 does not express the lysis gene; the start signals for coat translation are absent, and all the sequences to form the lysis hairpin are present (Fig. 9). Small deletions like those present in pMS1.6 or pMS1.17 that reach a few base pairs into the lysis helix lead to uncoupled synthesis of the lysis protein (uncoupled defined as not dependent on coat gene translation). Similarly, two insertions and one deletion in the left strand of the lysis helix (pMS1.m) that are expected to destabilize the helix also lead to a lysis (+) phenotype. Finally, random substitutions at coat gene wobble positions result in uncoupled translation of the lysis gene (pMS1.mm).

   More subtle changes have also been made. These are shown in Fig. 10. Several wobble positions of coat protein codons occurring in the lysis gene leader region have been exchanged for nucleotides that give synonymous codons but are expected to destabilize the hairpin (pMS1.d1, pMS1.d2, and pMS1.q1). Every destabilization leads to uncoupled expression of the L gene, whereas substitutions that increase or maintain

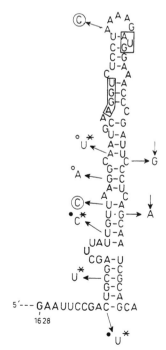

FIGURE 10. Identification of base substitutions in the lysis hairpin. The double mutants are pMS1.d1 (●), pMS1.d2 (○), and pMS1.d4 (↓). Circled C residues indicate pMS1.d3. The quadruple mutant pMS1.q1 is identified by asterisks (*).

the stability of the stem do not (pMS1.d4 and pMS1.d3) (Schmidt et al., 1987).

The two substitutions present in pMS1.d4 have an interesting aspect. They stabilize the hairpin, and, as mentioned, the closed state of the lysis gene is maintained by these mutations. However, when the two stabilizing mutations are present in clones where coat gene translation can take place, then termination of translation at the end of the coat gene fails to activate the lysis start. Therefore, for the coat lysis coupling to work, the stability of the lysis helix must be kept within narrow limits. The importance of finely tuned helix stability is also apparent in the corresponding hairpin strcture in M12 (Fig. 8). Here the upper base pair change weakens and the lower strengthens the helix. The remaining question is how termination at the coat gene terminator activates the lysis start. The coupling of initiation to termination of translation is not unusual in prokaryotes or eukaryotes and has been described for several systems (Yates and Nomura, 1981; Schumperli et al., 1982: Das and Yanofsky,

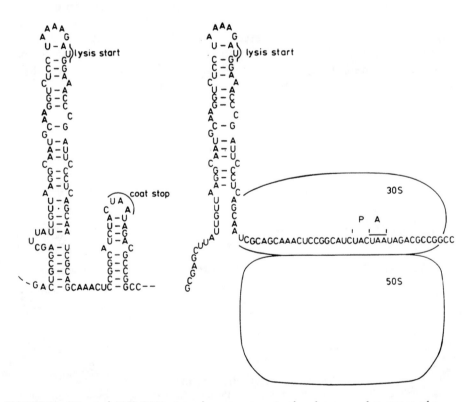

FIGURE 11. Proposed MS2 RNA secondary structure in the absence and presence of a ribosome terminating at the coat gene stopcodon.

1984; Kozak, 1984; Cone and Steege, 1985a,b). The unique aspect in the RNA phage is that the activated site lies far upstream of the termination point.

There are two possibilities for the mechanics of the termination-initiation cycle: the participation of two ribosomes, and the essence of the activation mechanism is shown in Fig. 11. A ribosome protects about 30–35 nucleotides of the bound messenger against nuclease (Steitz, 1979). Some 20 residues of these are located upstream of the ribosomal P-site. Though these numbers have been mostly obtained from initiating ribosomes, they are probably also correct for a terminating ribosome. As a consequence, approximately six base pairs of the lysis helix will be melted by the termination event. The phenotypes of our deletion mutants such as pMS1.6 show that such a destabilization is quite sufficient to permit binding of a free ribosome to the lysis start. The proposal is also consistent with the observation that in pMS1.d4 the lysis gene cannot be activated by the terminating ribosome; the helix-stabilizing substitutions are outside the reach of the ribosome. This model requires that the terminating ribosome is somehow capable of blocking the advance of its 5′ elongating neighbor, since unabated translation over a potentially active ribosomal binding site was shown to inhibit its function (Berkhout et al., 1985a).

A second way to describe the activation of the lysis start assumes the "phaseless walk" of the ribosome terminated at the end of the coat gene. It is known that the release of the peptide and the corresponding tRNA at a stop codon does not spontaneously free the ribosome from the messenger. Instead, an active process requiring GTP, the G-factor, and the ribosome-releasing factor (RRF) is needed (Ryoji et al., 1981). Thus, limited lateral movements following peptide chain termination seem an acceptable way to explain translational restarts at positions close to the termination site—for example, in the rIIB gene of phage T4 and at the boundaries of the E/D and B/A genes of the tryptophan operon (Napoli et al., 1981; Das and Yanofsky, 1984).

In spite of the relatively great distance from the coat gene terminator to the lysis gene initiator codon and the upstream position of the latter, we feel that the possibility that the lysis gene is translated by the same ribosome that terminates needs consideration, because there is evolutionary support for it. In the sequence of the group II phage GA, a lysis gene has been tentatively identified, as indicated in Fig. 1. The interesting feature is that the termination codon of the GA coat gene overlaps the start of the lysis gene (Fig. 12). It can be safely assumed that in GA the translational coupling between coat and lysis genes exists also. The overlap in GA, however, excludes the simultaneous involvement of two ribosomes. Thus, if the activation mechanism is preserved during evolution, the GA sequence argues for the involvement of a single ribosome in MS2.

MS2 - fr - GA

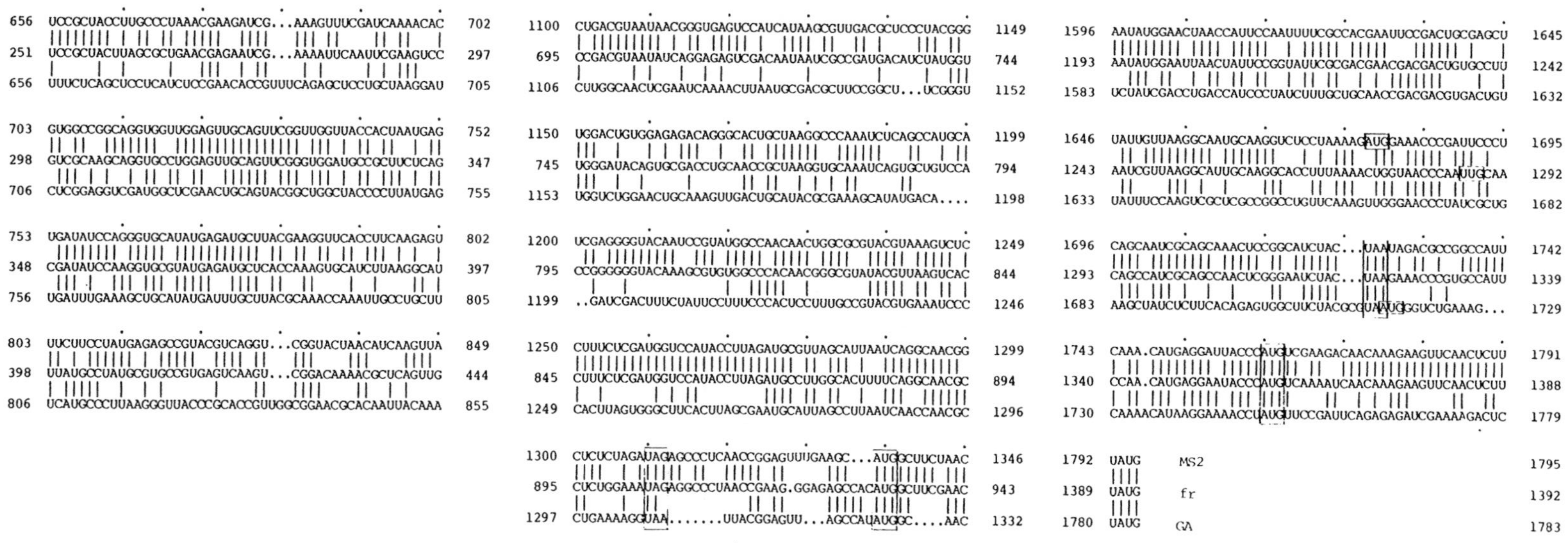

FIGURE 12. Partial sequence comparison of MS2, fr, and GA RNA. MS2 is from Fiers *et al.* (1976), GA is from Inokuchi *et al.* (1986), and fr was partially sequenced by Berzin *et al.* (1986). Stop and start codons are boxed. The fr₁ sequence is arbitrarily started at 1. The deduced fr coat protein sequence differs from the published sequence at four positions (116–119) (Wittmann- Liebold and Wittmann, 1967).

```
                  -10              -20              -30              -40              -50
Qβ   - - - - - - - - - - - - - - - - - - - - - - - - - - - - - - - - - - - - - - - - - - - - - - - - - - - -
MS2  M R A F S T L D R E N E T F V P S V R V Y A D G E T E D N S F S L K Y R S N W T P G R F N S T G A K T K
GA   M F P K S N I D R N Y K V K L I S Y D K K G K L V S D D S F E Q V E N Y L F Q N R S T T Y K P G Y I R R


                  -60              -70              -80              -90              -100
Qβ   - - - - - - - - - - M P K L P R G L R F G A D N E I L N D F Q E L W F P D L⌐F I E S S D T H P W Y T L⌐
MS2  Q W H Y P S P Y S R G A L S V T S I D Q G A Y K R S G S S S W G R P Y E E K A G⌐F G F S L D A R S C Y S L⌐
GA   D F R R P T N F W N G Y R C F N Q P - V G T F T R K L S D G G R Q V A D Y G I V N P N K F T A N S Q H L


                  -110             -120             -130             -140             -150
Qβ   K G R V L N A H L D D R L P N V G G R Q V R R T P H R V T V P I A S S G L R P V T T V Q Y D P A A L S F
MS2  F P V S Q N L T Y I E V P Q⌐N V A N R A S T E V L Q K V T Q G N F N L G V A L A E A R S T A S Q L A T Q
GA   G D N M V I Y P G P F S - I⌐N I D Q R A S V E V L N K L S Q S N L N I G V A I A E A K M T A S L L A K Q


                  -160             -170             -180             -190             -200
Qβ   L L N A R V D W D F G N G D S A N L V I N D F L F R T F A P K E - F D F S N S L V P R Y T Q A F S A F N
MS2  T I A L V K A Y T A A R R G N W R Q A L R Y L A L N E D R K F R - S K H V A G R W L E L Q F G W L P L M
GA   S I A L I R A Y T A A K R G N W R E V L S Q L L I S E H R F R A P A K D L G G R W L E L Q Y G W L P L M


                  -210             -220             -230             -240             -250
Qβ   A K Y - G T M I G E G L E T I K Y L G L L L R R L R E G Y R A V K R G D L R A L R R V I Q S Y H - N G K
MS2  S D I Q G A Y E M L T K V H L Q E F L P M R A V R Q V G⌐T N I K L D G R L S Y P A A N F Q T T C - N I S
GA   S D L K A A Y D L L T Q T K L P A F M P L R V T R T V G G T H N Y K V R N V E S A G D T W S Y R H R L S


                  -260             -270             -280             -290             -300
Qβ   W K P A T A G N L W L E F R Y G L M P L F Y D I R D V M L D W Q N R H D K I Q R L L R F S V G H G E D Y
MS2  R R I V⌐I W F Y I N D A R L A W L S S L G I L N P L G I V W E K V P F S F V V D W L - L P V G N M L E⌐G
GA   V N Y R⌐I W Y F I S D P R L A W A S S L G L L N P L E I Y W E K T P W S F V V D W F - L P V G N L I E⌐A


                  -310             -320             -330             -340             -350             -360
Qβ   V V E F D N L Y P A V A Y F K L K G E I - T L E R R H R⌐H G I S Y A N R E G Y A⌐V F D N G S L R P V S D
MS2  L T A P V G C S Y M S G T V T D V I T G E S I I S V D A⌐P Y G W T V E R Q G T A⌐- K A Q I S A M H R G V
GA   M S N P L G L D I I S G T K T W Q L - - E S K L N A T L P A S G W S G T A K L T - - A Y A K A Y D R S T


Qβ   W K E L A T A F I N P H E V A W E L T P Y S F V V D W F L N V G D I L A Q Q G Q L Y H N I D I V D G F D
MS2  Q S V W P T T G - - - - - - - - - - - - - - - - - - - - - - - - - - - - - - - - - - - - - - - - - - - -
GA   F Y S F P T P L - - - - - - - - - - - - - - - - - - - - - - - - - - - - - - - - - - - - - - - - - - - -


                                                                                       -370
Qβ   R R D I R L K S F T I K G E R N G R P V N V S A S L S A V D L F Y S R L H T S N L P F A T L D L D⌐T T F
MS2  - - - - - - - - - - - - - - - - - - - - - - - - - - - - - - - - - - - - - - - - - - - - - - A Y V K⌐S P F
GA   - - - - - - - - - - - - - - - - - - - - - - - - - - - - - - - - - - - - - - - - - - - - - - P Y V K⌐S P L⌐


                  -380             -390
Qβ   S S F K H V L D S I F L L T Q R V K R
MS2  - S M V H T L D A L A L I R Q R L S R
GA   - S G L H L A N A L A L I N Q R L K R
```

E. The Lysis Protein of Group B Phages

The strong similarity in gene arrangement and control of gene expression in the single-stranded RNA phages is fully lost in the way the phages organize their escape from the wasted host. Phenotypic analysis of subgenomic Qβ DNA fragments has shown that the overproduction of the maturation protein causes cell lysis (Karnik and Billeter, 1983; Winter and Gold, 1983). This finding is consistent with the fact that Qβ mutants carrying an amber mutation in the maturation protein do not lyse (Horiuchi and Matsuhashi, 1970). The implications of this different lysis mechanism for the evolutionary relationship between the two major phage groups will be discussed later.

IX. SEQUENCE COMPARISON BETWEEN GROUP I, II, AND III PHAGES

One of the valuable aspects of nucleic acid sequence comparison is that it allows us to distinguish between coincidence and necessity. It reveals which features in the primary and secondary structure of the RNA are essential to the survival of the phage and which are accidental.

As already mentioned, the special status of the GUG start codon of the MS2 maturation protein fell when related phages showed AUG at the same position. When comparing GA to MS2, some general remarks to this effect can be made. For instance, the double stop codon at the end of the MS2 coat gene is not a conserved feature. Besides in GA it also fails in fr (Borisova et al., 1979), and hence it may not be a crucial property. Similarly, the fact that coat and replicase genes have the same frame in MS2 is not important, as it is not conserved in GA. On the other hand, the relative frames of the lysis and replicase genes are conserved, in keeping with the finding that the overlap with the replicase gene provides the essential amino acid sequence for the lysis function (Berkhout et al., 1985b). We present below the amino acid sequence comparison for the group I, II, and III phage proteins. The homologies may identify the parts of the protein that serve a common function in the three phages.

A. The Maturation Protein

The homology between the phage proteins is the least for the maturation protein (Fig. 13). Between the Qβ A2 protein and the corresponding proteins from the group A phages, there is only homology in the carboxy-

---

FIGURE 13. Amino acid sequences of the maturation protein of Qβ, MS2, and GA. The numbering is that of the MS2-coded protein. The relative positioning of the MS2 and Qβ residues is chosen to show the homology proposed by Mekler (1981) and indicated here by boxes in broken lines. The sequence in the maturation gene where the MS2 A protein binds is underlined.

terminal 20 or so amino acids. This could suggest that this part of the protein has an identical function in the three phages—for instance, the attachment of the phage particle to the F pili. The alternative is that this part of the RNA is subject to structural constraints that relate either to the translation termination process or to the initiation events at the coat gene start site which lies only about 15 nucleotides downstream. It should be remembered that the region of protein homology (Qβ positions 1256–1321) lies within the S site (1245–1342) that is recognized by Qβ replicase. Also the S1-binding site on Qβ RNA is contained within the same region (1291–1312). A connection between the amino acid sequence conservation and a translational or replicational requirement is therefore conceivable. Of course the two explanations suggested here are not mutually exclusive.

In the initial comparison between Qβ and MS2 maturation protein, weak homology was proposed for two additional regions, as indicated in Fig. 13 (Mekler, 1981). However, these two homologies do not extend to the GA sequence and are therefore suspected to be fortuitous. The absence of similarities between Qβ and group A maturation proteins may reflect the development of "lysis" properties in the Qβ A2 protein.

The homology between GA and MS2 is 38%. It is particularly strong between amino acids 120 and 310 but almost nonexisting for the first 120 residues (Inokuchi et al., 1986). The A protein of MS2 binds to the RNA region that encodes amino acids 84–98 (Shiba and Suzuki, 1981), which is

REPLICASE PROTEIN

```
                    10          20          30          40          50
QB   M S K T A S S R N S L S A Q L R R A A N T R I E V E G N L A L S I A N D L L L A Y G Q S P F N S E A E
MS2  M S K T T K K F N S L C I D L P R D L S L E I Y Q S - - - I A S V A - - - - - - - - - - T G S G D P H
GA   M F R F R E I E K T L C M D R T R D C A V R F H V Y - - - L Q S L D - - - - - - - - - - L G S S D P L

                    60          70          80          90          100
QB   C I S F S P R F D G T P D D F R I N Y L K A E I M S K Y D D F S L G I D T E A V A W E K F L A A E A E
MS2  S D D F T A I A Y L R D E L L T K H P T L G S G N D E A T R R T L A I A K L R E A N D R C G Q I N R E
GA   S P D F D G L A Y L R D E C L T K H P S L G D S N S D A R R K E L A Y A K L M D S D Q R C K I Q N S N

                   110         120         130         140         150
QB   C A L T N A R L Y R P D Y S E D F N F S L G E S C I H M A R R K I A K L I G D V P S V E G M L R H C R
MS2  G F L H D K S L S W D P D V L Q T S I R S L I G N L L S G Y R S S L F G Q C T F S N G A P M G H K L Q
GA   G Y D Y S H I E S G V L S G I L K T A Q A L V A N L L T G F E S H F L N D C S F S N G A S Q G F K L R
```

FIGURE 14. Amino acid sequences of the replicase protein (subunit β) of Qβ, MS2 and GA. The numbering is that of the Qβ-coded protein. Fortuitous amino acid sequence homologies are indicated by a broken-line box. Regions of strong homologies are boxed by solid lines. We report here a correction on the MS2 RNA sequence around position 2000. Old sequence: UGGUGAUCGC. New sequence: UGAUCGGUGC. There is now a better homology with GA. The revised tripeptide sequence in MS2 (DRC, amino acids 81, 82, 83) is underlined.

```
                  160        170         180              190            200
QB   F S G G A T T T N N R S Y G H P S F K F A L P Q A C T P R A L K Y V L A L R A S T H F D T R I S D I S
MS2  D A A P Y K K F A E Q A T V T P R A L R A A L L V R D Q C A P W I R H A V R Y N E S Y E - - F R - L V
GA   D A A P F K K I A G Q A T V T A P A Y D I A V A A V K T C A P W Y A Y M Q E T Y G D E T K W F R - R V


                  210        220         230              240            250
QB   P F N K A V T V P K N S K T D R C I A I E P G W N M F F Q L G I G G I L R D R L R C W G I D L N D Q T
MS2  V G N G V F T V P K N N K I D R A A C K E P D M N M Y L Q K G V G A F I R R R L K S V G I D L N D Q S
GA   Y G N G L F S V P K N N K I D R A A C K E P D M N M Y L Q K G A G S F I R K R L R S V G I D L N D Q T


                  260        270         280              290            300
QB   I N Q R R A H E G S V T N N L A T V D L S A A S D S I S L A L C E L L L P L G W F E V L M D L R - S P
MS2  I N Q R L A Q Q G S V D G S L A T I D L S S A S D S I S D R L V W S F L P P E L Y S Y L D R I R - S H
GA   R N Q E L A R L G S I D G S L A T I D L S S A S D S I S D R L V W D L L P P H V Y S Y L A R I R T S F


                  310        320         330              340            350
QB   K G R L P D G S V V T Y E K I S S M G N G Y T F E L E S L I F A S L A R S V C E I L D L D S S E V T V
MS2  Y G - I V D G E T I R W E L F S T M G N G F T F E L E S M I F W A I V K A T Q I H F G N A G T I G I -
GA   T M - I - D G R L H K W G L F S T M G N G F T F E L E S M I F W A L S K S I M L S M G V T G S L G I -


                  360        370         380              390            400
QB   Y G D D I I L P S C A V P A L R E V F K Y V G F T T N T K T F S E G P F R E S C G K H Y Y S G V D V
MS2  Y G D D I I C P S E I A P R V L E A L A Y Y G F K P N L R K T F V S G L F R E S C G A H F Y R G V D V
GA   Y G D D I I V P V E C R P T L L K V L S A V N F L P N E E K T F T T G Y F R E S C G A H F F K D A D M


                  410        420         430              440            450
QB   T P F Y I R H R I V S P A D L I L V L N N L Y R W A T I D G V W D P R A H S V Y L K Y R K L P P K Q L
MS2  K P F Y I K K P V D N L F A L M L I L N R L R G W G V V G G M S D P R L Y K V W V R L S S Q V P S M F
GA   K P F Y C K R P M E T L P D V M L L C N R I R G W Q T V G G M S D P R L F P I W K E F A D M I P P K F


                  460        470         480              490            500
QB   Q R N T I P D G Y G D G A L V G S V L I N P F A K N R G W I R Y V P V I T D H T R D R E R A E L G S Y
MS2  F G G T D L A A D Y Y V V S P P T A V S V Y T - K T P Y G R L L A D T R T S G F R L A R I A R E R K F
GA   K G G C N L D R D T Y L V S P D K P G V S L V - R I A K V R - - - - - - S G F N H A - - - - - - - -


                  510        520         530              540            550
QB   L Y D L F S R C L S E S N D G L P - L R G P S G C D S A D L F A I D Q L I C R S N P T K I S R S T G K
MS2  F S E K H D S G R Y I A W F H T G - G E I T D S M K S A G V R V I R T S E W L T P V P T F P - - - - -
GA   F P Y G H E N G R Y V H W L H M G S G E V L E T I S S A R Y R C K P N S E W R T Q I P L F P - - - - -


                  560        570         580
QB   F D I Q Y I A C - S S - R V L A P Y G V F Q G T K V A S L H E A
MS2  - - - Q E C G P A S S P R - - - - - - - - - - - - - - - - - -
GA   - - - Q E L E A C V L S - - - - - - - - - - - - - - - - - - -
```

FIGURE 14. (Continued)

indicated by a solid line in Fig. 13. This sequence has not been conserved between GA and MS2, although it cannot be excluded that the secondary structure is conserved.

B. The Replicase

In contrast to the maturation protein, the amino acid sequences of the replicases show strong resemblances. As expected, these are more pronounced between GA and MS2 (50%) than between Qβ and group A. The full sequences are presented in Fig. 14, and the more extensive homologies are boxed. The least correspondence is found at the N and C termini.

The first phage RNA sequences to be known were MS2 and Qβ, and at that time it seemed reasonable to attribute significance to the indicated homology at the very N terminus of Qβ and MS2 replicase (Mekler, 1981). However, in the light of new data on GA and the partial fr sequence, the homology does not stand up, and the resemblances between Qβ and MS2 must be considered accidental. That the replicases show stronger homology among themselves than the A proteins probably reflects the fact that they interact with the same host proteins and have the same enzymatic activity.

C. The Coat Protein

GA and MS2 have in 60% of the corresponding positions the same amino acids. Between MS2 and fr the homology is 80%. The homology with Qβ is restricted to a region around positions 30 and 60, as shown in Fig. 15, where we present the coat protein sequences for the three phages. It is not unexpected to find that much of the divergence between GA and MS2 occurs at the N terminus. We have discussed this point before in section VIII.C. Figure 15 also provides some interesting differences between the group A coat protein genes. Apart from what has already been mentioned, we see for instance at position 23 of GA the insertion of valine, which seems compensated for by the deletion of phenylalanine 3 positions further downstream. The supposed insertion-deletion events can be traced back in the nucleic acid sequence in Fig. 12. They are not explicitly indicated here.

D. The Lysis Protein

The complete amino acid sequence of the lysis protein is known for two group A phages, GA and MS2. Partial sequences are available for group I phages fr, R17, f2, and M12. In Fig. 16, MS2, fr, and GA are

COAT PROTEIN

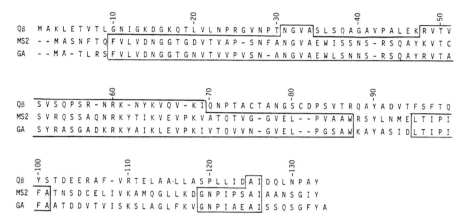

FIGURE 15. Amino acid sequences of the coat protein of Qβ, MS2, and GA. The numbering is that of the Qβ coat protein. Regions of strong homologies are boxed.

compared. Several features are noteworthy. Significant conservation of residues is found in the middle part of the molecule (amino acid 34–52 of the MS2 sequence). The homologies can be further extended if analogous amino acids are included such as phenylalanine, valine, and alanine. There is also conservation of a histidine at position 24 and a cysteine residue at position 29. We consider these as fortuitous in the sense that they result from constraints placed on the RNA sequence. Histidine and cysteine are encoded in the complementary parts of the replicase hairpin (Fig. 5).

Berkhout *et al.* (1985b) have deleted increasing 5′ parts of the MS2 lysis protein gene by recombinant DNA procedures and have found that the first odd 40 amino acids are dispensable for function. This result is consistent with the high variability found between the N-terminal regions of MS2 and fr. It also is in agreement with the complete absence of

```
                 -10           -20           -30           -40           -50
MS2  M E T R F P Q Q S Q Q T P A S T N R R R P F K H E D Y P C R R Q Q R S S T L Y V L I F L A I F L S K F T N
GA   - - - - - - - - - - - - - - - - - M G L K A K H K E N L C S D S E R S K R L Y V W I A L A I V L S D F T S
fr        ? Q Q P S Q P T R E S T K K P V P F Q H E E Y P C Q N Q Q R S S T L Y

          -60           -70
MS2  Q L L L S L L E A V I R T V T T L Q Q L L T
GA   I F S H W I W G L L I L Y L Q T L M D L P T F V M N V
```

FIGURE 16. Amino acid sequences of the lysis protein of MS2, GA, and fr (partial). The numbering is that of MS2. The fr sequence does not contain an AUG or GUG codon in the lysis reading frame. At the position of the question mark there is a UUG codon that may function as a start signal.

```
Replicase  M  S  K  T  T  K  K  F  N  S  L  C  I  D  L  P  R  D  L  S  L  E  I  Y
MS2        AUGUCGAAGACAACAAACAAGUUCAACUCUUUAUGUAUUGAUCUUCCUCGCGAUCUUUCUCUCGAAAUUUACC----

           C  R  R  Q  Q  R  S  S  T  L  Y  V  L  I  F  L  A  I  F  L  S  K  F  T
Lysis      C  S  D  S  E  R  S  K  R  L  Y  V  W  I  A  L  A  I  V  L  S  D  F  T

GA         AUGUUCCGAUUCAGAGAGAUCGAAAAGACUCUAUGUAUCGAUCGCACUCGCGAUUGUGCUGUCCGAUUUCACG----
Replicase  M  F  R  F  R  E  I  E  K  T  L  C  M  D  R  T  R  D  C  A  V  R  F  H
```

FIGURE 17. Amino acid choice in a part of the lysis-replicase overlap region of the phages MS2 and GA. The nucleotide sequence starts at the replicase initiation codon. Homologous amino acids are boxed.

the first 17 amino acids in the GA lysis protein and the appearance of significant homology beyond amino acid 33.

It is instructive to have a closer look at the major conserved lysis protein sequence (amino acid 34–51) (Fig. 17). The area shows several regions where the lysis protein sequence is conserved at the expense of that of the replicase. This may mean that the optimal replicase sequence has not been realized but that a compromise with the selective advantage of the lysis function has been reached.

Another interesting result emerging from sequence comparison is that the sequence leu leu leu ser leu leu that is common to ϕX_{174}, G4, and MS2 is not conserved in GA. Therefore, this structure element may be less crucial in cell lysis than initially thought (Inokuchi et al., 1986).

E. The Noncoding Regions

The homology between Qβ and group A in the 5′ and 3′ extra cistronic regions is virtually nonexistent except for the obligatory row of C's at the 3′ terminus. The lack of homology between group A and group B phages in these parts of the RNA is not unexpected, because there are not even sequence similarities within the 6S RNA family. The inter-cistronic regions of Qβ and group A also have a low degree of similarity in primary sequence except, of course, for the signals required for initiation of translation.

For GA and MS2 the situation is different. The 3′ extracistronic region has 71% of its sequence conserved, but this value is much lower for the 5′ noncoding part, where it is only 46%. There is no satisfactory explanation for this difference.

F. The Savings between MS2 and GA

GA is 103 nucleotides shorter than MS2. Assuming for the moment that the difference is the result of deletions motivated by selective advantage, one may wonder where such savings are possible. The GA replicase has 39 bases less than the MS2 counterpart, but the shortening is not the result of a single deletion. In the 3' end of the GA replicase gene, two regions of seven and eight amino acids are missing around positions 470 and 480, respectively (GA replicase amino acid numbers). On the other hand, GA shows an insertion of two amino acids with respect to MS2 at position 175 of the GA replicase (see Fig. 14). This situation thwarts any attempt to reconstruct the historical course of events that led to the appearance of the two phage groups.

The second major difference in length originates from the 3' untranslated region, which is 54 nucleotides shorter in GA than in MS2. Here GA lacks sequences corresponding to regions between nucleotide numbers 70 and 90 and between 100 and 140 from the 3' end of MS2 RNA. The extra sequences in MS2 do not show clear repeats with neighboring regions, which decreases the probability that the differences are the result of duplications.

X. PHYLOGENY OF RNA PHAGES

An interesting but necessarily most difficult question is that of the origin and kinships of the RNA bacteriophages. It is generally assumed that the four groups derive from a common ancestor. The basis for this assumption is the identical genetic organization, the strong resemblances of the replicases, the use of the same host proteins as auxiliaries in the copying reaction, and the similarities of several control mechanisms, such as the translational coupling between coat and replicase genes and the repression of replicase synthesis by the coat protein. The properties mentioned are more easily explained by divergent than by convergent evolution.

We will discuss two views at the relationships: the longer phages derive from the shorter by insertions, or the shorter derive from the longer ones by deletions. The data obtained on 6S RNA replication indicate that the replicase is capable of insertions as well as deletions. Thus the potential of the polymerizing enzyme does not help to decide on the direction of evolution.

Furuse (1986) has proposed that group IV stands closest to the forebear, implying that "recent" evolution was accompanied by deletions. Several arguments exist. Group IV contains the most diverse subgroups and has the broadest habitat. One of its members, ID2, shows serological cross-reactivity with all other groups. Some amber mutants of SP phage produce viable progeny in intergroup complementation with MS2 and Qβ. Also within group IV there is more variability in the molecular weight of the coat protein and in several other physical parameters.

The tentative reconstruction would be that group IV has spawned group III by small deletions and group A by deleting the read-through part of the coat protein in combination with several other small losses. The poor serological relationship between Qβ and group A makes it unlikely that group A is directly derived from group III. If indeed during younger times the selection pressure has favored shorter genomes, then of course group II derives from group I. Also, the strong conservation between GA and MS2 seems to exclude that these two groups arose independently from group IV (Furuse, 1986).

Mekler (1981) came to a different conclusion. He observed that several coat gene sequences in Qβ are repeated in the same order in the coat gene read-through section, which suggests that a gene duplication has occurred. The appearance of a conditional stop codon between the two coat gene copies would then contribute to the genetic potential.

The position of the lysis gene in group A has interesting aspects against this background. In the gene contraction view, this gene must have evolved after group A emerged from group IV. This is consistent with the observation that codon usage in the lysis gene qualifies it as a late addition (Normark et al., 1983; Shepherd. 1981). Also in the gene expansion view the development of the lysis function must be a late event. If not, one would have to explain what possible advantage could be gained by inactivating the lysis gene through an insertion, whose usefulness still had to be crafted.

It may be surprising that such a simple protein arose only late in evolution. One speculation on this point is that its synthesis and development could only start when the protein synthetic machinery of the bacterial cell had acquired the properties necessary for termination-dependent initiation. This new property would also stimulate the contraction of the bacterial genome, in particular the intercistronic regions in polycistronic messenger RNA.

XI. CONCLUDING REMARKS

The phage RNAs are about the smallest "live" molecules carrying the information necessary for their amplification and transmission. The complete nucleotide sequence of four different phages has now been determined, and, given the available technology, more sequences will follow soon. These data will enable us to make meaningful suggestions on the vital secondary and tertiary structure of these RNAs. In addition, there is a chance that we will begin to understand the lines along which these phages have diversified.

Also our ability to generate live phages from their cloned cDNA (Taniguchi et al., 1978) opens the way to infect a bacterium with the sequence of our choice and let it evolve under the selective pressure we want. These tools may also be used to reexamine the occurrence of RNA recombination.

The construction of hybrid phages that derive their 3' and 5' segments from different but closely related group members may aid our understanding of the physical communication between such distant segments and may help identifying the distant contact points.

There is no hope that we can improve on the viability of these highly elaborate structures, in which the nucleotide choice at every position has been carefully decided on. It is also unlikely that we will understand in the near future why the gene order is as it is. In spite of such limitations, it will remain a pleasure to accept the challenge to understand the design and workings of these sophisticated forms of life.

ACKNOWLEDGMENTS. I thank the many colleagues who made their unpublished results available for use in this chapter. Maarten de Smit is acknowledged for preparing Fig. 12.

REFERENCES

Atkins, J. F., Steitz, J. A., Anderson, C. W., and Model, P., 1979, Binding of mammalian ribosomes to MS2 phage RNA reveals an overlapping gene encoding a lysis function, *Cell* **18**:247.

Banerjee, A. K., Rensing, U., and August, J. T., 1969, Replication of RNA viruses. X. Replication of a natural 6S RNA by the Qβ RNA polymerase, *J. Mol. Biol.* **45**:181.

Bear, D. G., Ng, R., Van der Veer, D., Johnson, N. P., Thomas, G., Schleich, T., and Noller, H. F., 1976, Alteration of polynucleotide secondary structure by ribosomal protein S1, *Proc. Natl. Acad. Sci. USA* **73**:1824.

Bendis, I., and Shapiro, L., 1970, Properties of *Caulobacter* RNA bacteriophage φCb5, *J. Virol.* **6**:847.

Beremand, M. W., and Blumenthal, T., 1979, Overlapping genes in RNA phage: A new protein implicated in cell lysis, *Cell* **18**:257.

Berkhout, B., 1986, Translational control mechanisms in RNA bacteriophage MS2, Ph.D. thesis, University of Leiden.

Berkhout, B., and Van Duin, J., 1985, Mechanism of translational coupling between coat protein and replicase genes of RNA bacteriophage MS2, *Nucleic Acids Res.* **13**:6955.

Berkhout, B., Kastelein, R. A., and Van Duin, J., 1985a, Translational interference at overlapping reading frames in prokaryotic messenger RNA, *Gene* **37**:171.

Berkhout, B., De Smith, M. H., Spanjaard, R. A., Blom, T., and Van Duin, J., 1985b, The amino terminal half of the MS2-coded lysis protein is dispensable for function: Implications for our understanding of coding region overlaps, *EMBO J.* **4**:3315.

Berkhout, B., Schmidt, B. F., Van Strien, A., Van Boom, J. H., Van Westrenen, J., and Van Duin, J., 1987, Lysis gene of bacteriophage mS2 is activated by translation termination at the overlapping coat gene, *J. Mol. Biol.* **195**:517.

Berzin, V., Cielens, I., Jansone, I., and Gren, E. J., 1982, The regulatory region of phage fr replicase cistron. III. Initiation activity of specific fr RNA fragments, *Nucleic Acids Res.* **10**:7763.

Berzin, V., Awots, A., Jansone, I., and Zimanis, A., 1986, Primary structure of a cDNA fragment of phage fr, *Bioorg. Khim.* **12**:149 (in Russian).

Biebricher, C. K., Eigen, M., and Luce, R., 1981a, Product analysis of RNA generated *de novo* by Qβ replicase, *J. Mol. Biol.* **148**:369.

Biebricher, C. K., Eigen, M., and Luce, R., 1981b, Kinetic analysis of template-instructed and *de novo* RNA synthesis by Qβ replicase, *J. Mol. Biol.* **148**:391.

Biebricher, C. K., Eigen, M., and Luce, R., 1986, Template-free RNA synthesis by Qβ replicase, *Nature* **321**:89.

Blumenthal, T., 1977, Interaction of Qβ replicase with guanine nucleotides. Different modes of inhibition and inactivation, *Biochim. Biophys. Acta* **478**:201.

Blumenthal, T., and Carmichael, G. G., 1979, RNA replication: Function and structure of Qβ replicase, *Annu. Rev. Biochem.* **48**:525.

Blumenthal, T., Landers, T. A., and Weber, K., 1972, Bacteriophage Qβ replicase contains
    the protein biosynthesis elongation factors EF-Tu and EF-Ts, *Proc. Natl. Acad. Sci.
    USA* **69**:1313.
Boni, I. V., Isaeva, D. M., and Budowsky, E. I., 1986, Ribosomal protein S1 binds to the
    internal region of the replicase gene within the complex of *E. coli* 30S ribosomal sub-
    unit with MS2 phage RNA, *Bioorg. Khim.* **12**:293 (in Russian).
Borisova, G. P., Volkova, T. M., Berzin, V., Rosenthal, G., and Gren, E. J., 1979, The reg-
    ulatory region of MS2 phage RNA replicase cistron. IV. Functional activity of specific
    MS2 RNA fragments in formation of the 70S initiation complex of protein biosynthesis,
    Nucleic Acids Res. **6**:1761.
Bradley, D. E., 1966, The structure and infective process of a *Pseudomonas aeruginosa*
    bacteriophage containing ribonucleic acid, *J. Gen. Microbiol.* **45**:83.
Brown, S., and Blumenthal, T., 1976, Reconstitution of Qβ RNA replicase from a covalently
    bonded elongation factor Tu·Ts complex, *Proc. Natl. Acad Sci. USA* **73**:1131.
Bujarski, J. J., and Kaesberg, P., 1986, Genetic recombination between RNA components of a
    multipartite plant virus, *Nature* **321**:528.
Capecchi, M. R., and Webster, R. E., 1975, Bacteriophage RNA as template for *in vitro*
    protein synthesis, in: *RNA Phages* (N. D. Zinder, ed.), p. 279, Cold Spring Harbor
    Laboratory, Cold Spring Harbor, NY.
Carey, J., Cameron, V., Krug, M., De Haseth, P. L., and Uhlenbeck, O. C., 1984, Failure of
    translational repression in phage f2 *op3* mutant is not due to an altered coat protein-
    RNA interaction, *J. Biol. Chem.* **259**:20.
Christiansen, J., Douthwaite, S. R., Christensen, A., and Garret, R. A. 1985, Does unpaired
    adenosine-66 from helix II of *E. coli* 5S bind to protein L18?, *EMBO J.* **4**:1019.
Cone, K. C., and Steege, D. A., 1985a, Messenger RNA conformation and ribosome selection
    of translation reinitiation sites in the *lac* repressor mRNA, *J. Mol. Biol.* **186**:725.
Cone, K. C., and Steege, D. A., 1985b, Functional analysis of *lac* repressor restart sites in
    translational initiation and reinitiation, *J. Mol. Biol.* **186**:733.
Crowther, R. A., Amost, L. A., and Finch, J. T., 1975, Three dimensional image reconstruc-
    tions of bacteriophages R17 and f2, *J. Mol. Biol.* **98**:631.
Das, A., and Yanofsky, 1984, A ribosome-binding sequence is necessary for efficient ex-
    pression of the distal gene of a translationally coupled gene pair, *Nucleic Acids Res.*
    12:4757.
Davis, J. W., and Benike, C., 1974, Translation of virus mRNA: Synthesis of bacteriophage
    PP7 proteins in cell free extracts from *Pseudomonas aeruginosa, Virology* **61**:450.
Dhaese, P., Vanderkerckhove, J., Vingerhoed. J. P., and Van Montagu, M., 1977, Studies on
    PRR1, an RNA bacteriophage with broad host range, *Arch. Int. Physiol. Biochim.*
    85:168.
Domingo, E., Sabo, D., Taniguchi, T., and Weissmann, C , 1978, Nucleotide sequence
    heterogeneity of an RNA phage population, *Cell* **13**:735.
El-Baradi, T. T. A. L., Raué, H. A., De Regt, V. C. H. F., Verbree, E. C., and Planta, R. J., 1985,
    Yeast ribosomal protein L25 binds to an evolutionary conserved site an yeast 26S and *E.
    coli* 23S rRNA, *EMBO J.* **4**:2101.
Feary, T. W., Fisher, E. Jr., and Fisher, T. N., 1963, A small RNA containing *Pseduomonas
    aerguinosa* bacteriophage, *Biochem. Biophys. Res. Commun.* **10**:359.
Feary, T. W., Fisher, E. Jr., and Fisher, T. N., 1964, Isolation and preliminary characteristics
    of three bacteriophages associated with a lysogenic strain of *Pseudomonas aeruginosa,
    J. Bacteriol.* **87**:196.
Federoff, N., 1975, Replicase of the phage f2, in: *RNA Phages* (N. D. Zinder, ed.), p. 235, Cold
    Spring Harbor Laboratory, Cold Spring Harbor, NY.
Fiers, W., 1979, RNA Bacteriophages, in: *Comprehensive Virology*, Vol. 13 (H. Fraenkel-
    Conrat and R. R. Wagner, eds.), p. 69, Plenum Press, New York.
Fiers, W., Contreras, R., Duerinck, F., Haegeman, G., Merregaert, J., Min Jou, W., Raey-
    maekers, A., Volckaert, G., Ysebaert, M., Van der Kerckhove, J., Nolf, F., and Van
    Montagu, M., 1975, A-protein gene of bacteriophage MS2, *Nature (Lond.)* **256**:273.
Fiers, W., Contreras, R., Duerinck, F., Haegeman, G., Iserentant, D., Merregaert, J., Min Jou,

W., Molemans, F., Raeymakers, A., Vandenberghe, A., Volckaert, G., and Isebaert, M., 1976, Complete nucleotide sequence of bacteriophage MS2-RNA: Primary and secondary structure of the replicase gene, *Nature (Lond.)* **260**:500.

Franze de Fernandez, M. T., Eoyang, L., and August, J. T., 1968, Factor fraction required for the synthesis of bacteriophage Qβ RNA, *Nature* **219**:588.

Fukami, Y., and Haruna, I., 1979, Template specifitiy of Qβ and SP phage RNA replicases as studied by replication of small variant RNAs, *Mol. Gen. Genet.* **169**:173.

Fukuma, I., and Cohen, S. S., 1975, Polyamines in bacteriophage R17 and its RNA, *J. Virol.* **16**:222.

Furuse, K., 1982, Phylogenetic studies on RNA coliphages, *J. Keio Med. Soc.* **59**:265.

Furuse, K., 1987, Distribution of coliphages in the environment: General considerations, in: *Phage Ecology* (S. M. Goyal, ed.), p. 87. John Wiley and Sons, New York.

Furuse, K., Sakurai, T., Hirashima, A., Katsuki, M., Ando, A., and Watanabe, I., 1978, Distribution of RNA coliphages in south and east Asia, *Appl. Environ. Microbiol.* **35**:995.

Garrett, R. A., Vester, B., Leffers, H., Sorensen, P. M., Kjems, J., Olesen, S. O., Christensen, A., Christiansen, J., and Douthwaite, S., 1984, Mechanisms of protein-RNA recognition and assembly in ribosomes, in: *Gene Expression, Alfred Benson Symposium 19* (B. F. C. Clark and H. U. Petersen, eds.), p. 331, Munksgaard, Copenhagen.

Goelz, S., and Steitz, J. A., 1977, *Escherichia coli* ribosomal subunit S1 recognizes two sites in bacteriophage Qβ RNA, *J. Biol. Chem.* **252**:5177.

Gralla, J., Steitz, J. A., and Crothers, D. M., 1974, Direct physical evidence for secondary structure in an isolated fragment of R17 bacteriophage mRNA, *Nature* **240**:204.

Havelaar, A. H., Hogeboom, W. M., and Pot, R., 1984, F specific RNA bacteriophages in sewage: Methodology and occurrence, *Wat. Sci. Tech.* **17**:645.

Hilbers, C. W., Shulman, R. G., Yamane, T., and Steitz, J. A., 1974, High resolution proton NMR study of an isolated fragments of R17 bacteriophage mRNA, *Nature* **248**:225.

Hill, D., and Blumenthal, T., 1983, Does Qβ replicase synthesize RNA in the absence of template? *Nature* **301**:350.

Hindley, J., and Saples, D. H., 1969, Sequence of a ribosome binding site in bacteriophage Qβ RNA, *Nature* **224**:964.

Hofstetter, H., Monstein, H. J., and Weissmann, C., 1974, The read-through protein A1 is essential for the formation of viable Qβ particles, *Biochim. Biophys. Acta* **374**:238.

Horiuchi, K., and Matsuhashi, S., 1970, Three cistrons in bacteriophage Qβ, *Virology* **42**:49.

Inokuchi, Y., Takahashi, R., Hirose, T., Inayama, S., Jacobson, B., and Hirashima, A., 1986, The complete nucleotide sequence of group II RNA coliphage GA, *J. Biochem.* **99**:1169.

Inouye, H., Pollack, Y., and Petre, J., 1974, Physical and functional homology between ribosomal protein S1 and interference factor i, *Eur. J. Biochem.* **45**:109.

Jacobson, A. B., Kumar, H., and Zuker, M., 1985, Effect of sperimidine on the conformation of bacteriophage MS2 RNA. Electron microscopy and computer modeling, *J. Mol. Biol.* **181**:517.

Jennings, P. A., Firsch, J. T., Winter, G., and Robertson, J. S., 1983, Does the higher order structure of the influenza virus ribonucleoprotein guide sequence arrangements in influenza viral RNA? *Cell* **34**:619.

Kamen, R., Kondo, M., Romer, W., and Weissmann, C., 1972, Reconstitution of Qβ replicase lacking subunit α with protein-synthesis-interference factor i, *Eur. J. Biochem.* **31**:44.

Kamen, R. I., 1975, Structure and function of the Qβ replicase, in: *RNA Phages* (N. D. Zinder, ed.), p. 203, Cold Spring Harbor Laboratory, Cold Spring Harbor, NY.

Karnik, S., and Billeter, M., 1983, The lysis function of RNA bacteriophage Qβ is mediated by the maturation (A2) protein, *EMBO J.* **2**:1521.

Kastelein, R. A., Remaut, E., Fiers, W., and Van Duin, J., 1982, Lysis gene expression of RNA phage MS2 depends on a frameshift during translation of the overlapping coat protein gene, *Nature* **295**:35.

Kastelein, R. A., Berkhout, B., and Van Duin, J., 1983a, Opening the closed ribosome-binding site of the lysis cistron of bacteriophage MS2, *Nature* **305**:741.

Kastelein, R. A., Berkhout, B., Overbeek, G. P., and Van Duin, J., 1983b, Effect of the sequences upstream from the ribosome-binding site on the yield of protein from the cloned gene for phage MS2 coat protein, *Gene* **23**:245.

Kolakofsky, D., and Weissmann, C., 1971, Possible mechanism for transition of viral RNA from polysome to replication complex, *Nature New Biol.* **231**:42.

Kolb, A., Hermoso, J. M., Thomas, J. O., and Szer, W., 1977, Nucleic acid unwinding properties of ribosomal protein S1 and the role of S1 in mRNA binding to ribosomes, *Proc. Natl. Acad. Sci. USA* **74**:2397.

Kozak, M., 1984, Selection of initiation sites by eukaryotic ribosomes: Effect of inserting AUG triplets upstream from the coding sequence for preproinsulin, *Nucleic Acids Res.* **12**:3873.

Krahn, P. M., O'Callaghan, R. J., and Paranchych, W., 1972, Stages in phage R17 infection. VI. Injection of A protein and RNA into the host cell, *Virology* **47**:628.

Kramer, F. R., Mills, D. R., Cole, P. E., Nishihara, T., and Spiegelman, S., 1974, Evolution *in vitro:* Sequence and phenotype of a mutant RNA resistant to ethidium bromide, *J. Mol. Biol.* **89**:719.

Krueger, R. G., 1969, Serological relatedness of the ribonucleic acid–containing coliphages, *J. Virol.* **4**:567.

Landers, T. A., Blumenthal, T., and Weber, K., 1974, Function and structure of RNA phage Qβ replicase, *J. Biol. Chem.* **249**:5801.

Leipold, B., and Hofschneider, P. H., 1975, Isolation of an infectious RNA-A protein complex from the bacteriophage M12, *FEBS Lett.* **55**:50.

Lin, L., and Schmidt, J., 1972, Adsorption of a ribonucleic acid bacteriophage of *Pseudomonas aeruginosa, Arch. Microbiol.* **83**:120.

Lodish, H. F., 1975, Regulation of *in vitro* protein synthesis by bacteriophage RNA by RNA tertiary structure, in: *RNA Phages* (N. D. Zinder, ed.), p. 301, Cold Spring Harbor Laboratory, Cold Spring Harbor, NY.

Loeb, T., and Zinder, N. D., 1961, A bacteriophage containing RNA, *Proc. Natl. Acad. Sci. USA* **47**:282.

Looman, A. C., 1986, Effects of heterologous ribosomal binding sites on the expression of the *lacZ* gene of *E. coli,* PhD thesis, University of Leiden.

Looman, A. C., and Van Knippenberg, P. H., 1986, Effects of GUG and AUG initiation codons on the expression of *lacZ* in *E. coli, FEBS Lett.* **197**:315.

Lowary, P. T., and Uhlenbeck, O. C., 1987, An RNA mutation that increases the affinity of an RNA–protein interaction, *Nucleic Acids Res.* **15**:10483.

Mekler, P., 1981, Determination of nucleotide sequences of the bacteriophage Qβ genome: Organization and evolution of an RNA virus, PhD thesis, University of Zurich.

Meyer, F., 1978, Structure and function of Qβ replicase binding sites on Qβ RNA, PhD thesis, University of Zurich.

Meyer, F., Weber, H., and Weissmann, C., 1981, Interactions of Qβ replicase with Qβ RNA, *J. Mol. Biol.* **153**:631.

Mills, D. R., Peterson, R. L., and Speigelman, S., 1967, An extracellular Darwinian experiment with a self-duplicating nucleic acid molecule, *Proc. Natl. Acad. Sci. USA* **58**:217.

Mills, D. R., Kramer, F. R., and Spiegelman, S., 1973, Complete nucleotide sequence of a replicating RNA molecule, *Science* **180**:916.

Mills, D. R., Kramer, F. R., Dobkin, C., Nishihara, T., and Spiegelman, S., 1975, Nucleotide sequence of microvariant RNA: Another small replicating molecule, *Proc. Natl. Acad. Sci. USA* **72**:4252.

Mills, D. R., Bodkin, C., and Kramer, F. R., 1978, Template determined, variable rate of RNA chain elongation, *Cell* **15**:541.

Mills, D. R., Kramer, F. R., Dobkin, C., Nishihara, T., and Cole, P. E., 1980, Modification of cytidines in a Qβ replicase template: Analysis of conformation and localization of lethal nucleotide substitutions, *Biochemistry* **19**:228.

Min Jou, W., Haegeman, G., Ysebaert, M., and Fiers, W., 1972, Nucleotide sequence of the gene coding for the bacteriophage MS2 coat protein, *Nature (Lond.)* **237**:82.

Model, P., Webster, R. E., and Zinder, N. D., 1979, Characterization of *op3,* a lysis defective mutant of bacteriophage f2, *Cell* **18**:235.

Munson, L. M., Stormo, G. D., Niece, R. L., and Reznikoff, W. S., 1984, LacZ translation initiation mutations, *J. Mol. Biol.* **177**:663

Napoli, C., Gold, L., and Swebelius Singer, B., 1981, Translational reinitiation in the rIIB gene of phage T4, *J. Mol. Biol.* **149**:433.

Nishihara, T., Mills, D. R., and Kramer, F. R., 1983, Localization of the Qβ replicase recognition site in MDV-1 RNA, *J. Biochem.* **93**:669.

Nomura, M., Jinks-Robertson, S., and Miura, A., 1982, Regulation of ribosome biosynthesis in E. coli, in: *Interaction of Translational and Transcriptional Controls in the Regulation of Gene Expression* (M. Grunberg-Manago and B. Safer, eds.), Elsevier, New York.

Normark, S., Bergström, S., Edlund, T. Grundström, T., Jaurin, B., Lindberg, F. P., and Olsson, O., 1983, Overlapping genes, *Annu. Rev. Genet.* **17**:499.

Olson, R. H., and Thomas, D. D., 1973, Characteristics and purification of PRP1, an RNA phage specific for the broad host range *Pseudomonas* R1822 drug resistance plasmid, *J. Virol.* **12**:1560.

Oppenheim, D. S., and Yanofsky, C., 1980, Translational coupling during expression of the tryptophan operon of E. coli, *Genetics* **95**:785.

Paranchych, W., 1975, Attachment, ejection and penetration. Stages of the RNA phage infection process, in: *RNA Phages* (N. D. Zinder, ed.), p. 85, Cold Spring Harbor Laboratory, Cold Spring Harbor, NY.

Pley, C. W. A., Rietveld, K., and Bosch, L., 1985, A new principle of RNA folding based on pseudoknotting, *Nucleic Acids Res.* **12**:1717.

Remaut, E., De Waele, P., Marmenout, A., Stanssens, P., and Fiers, W., 1982, Functional expression of individual plasmid-coded RNA bacteriophage MS2 genes, *EMBO J.* **1**:205.

Robertson, H. D., 1975, Functions of replicating RNA in cells infected by RNA bacteriophages, in: *RNA Phages* (N. D. Zinder, ed.), p. 113, Cold Spring Harbor Laboratory, Cold Spring Harbor, NY.

Robertson, H. D., and Lodish, H. F., 1970, Messenger characteristics of nascent bacteriophage RNA, *Proc. Natl. Acad. Sci. USA* **67**:710.

Romaniuk, P. J., and Uhlenbeck, O. C., 1985, Nucleoside and nucleotide inactivation of R17 coat protein: Evidence for a transient covalent RNA-protein bond, *Biochemistry* **24**:4239.

Ryoji, M., Berland, R., and Kaji, A., 1981, Reinitiation of translation from the triplet next to the amber termination codon in the absence of ribosome-relasing factor, *Proc. Natl. Acad. Sci. USA* **78**:5973.

Schaffner, W., Ruegg, K. J., and Weissmann, C., 1977, Nanovariant RNA's: Nucleotide sequence and interaction with bacteriophage Qβ replicase, *J. Mol. Biol.* **117**:877.

Schmidt, B. F., Berkhout, B., Overbeek, G. P., Van Strien, A., and Van Duin, J., 1987, Determination of the RNA secondary structure that regulates lysis gene expression in bacteriophage MS2, *J. Mol. Biol.* **195**:505.

Schmidt, J. M., 1966, Observations on the adsorption of *Caulobacter* bacteriophages containing RNA, *J. Gen. Microbiol.* **45**:347.

Schmidt, J. M., and Stanier, R. Y., 1965, Isolation and characterization of bacteriophage active against stalked bacteria, *J. Gen. Microbiol.* **39**:95.

Schumperli, D., McKenney, K., Sobieski, D. A., and Rosenberg, M., 1982, Translational coupling at an intercistronic boundary of the E. coli galactose operon, *Cell* **30**:865.

Senear, A. W., and Steitz, J. A., 1976, Site-specific interaction of Qβ host factor and ribosomal protein S1 with Qβ and R17 bacteriophage RNA, *J. Biol. Chem.* **251**:1902.

Shapiro, L., and Bendis, I., 1975, RNA phages of bacteria other than E. coli, in: *RNA Phages* (N. D. Zinder, ed.), p. 397, Cold Spring Harbor Laboratory, Cold Spring Harbor, NY.

Shepherd, J. C. W., 1981, Periodic correlations in DNA sequences and evidence suggesting their evolutionary origin in a comma-less genetic code, *J. Mol. Evol.* **17**:94.

Shiba, R., 1975, Reconstitution of an infectious complex in RNA phages, *Proc. Mol. Biol. Meeting Jpn.*, p. 4

Shiba, T., and Miyake, T., 1975, New type of infectious complex of E. coli RNA phage, *Nature (Lond.)* **254**:157.

Shiba, R., and Suzuki, Y., 1981, Localization of A protein in the RNA-A protein complex of RNA phage MS2, Biochim. Biophys. Acta 654:249.

Steitz, J. A., 1969, Polypeptide chain initiation: Nucleotide sequences of the three ribosomal binding sites in bacteriophage R17, RNA, Nature 224:957.

Steitz, J. A., 1979, Genetic signals and nucleotide sequences in messenger RNA, in: Biological Regulation and Development, Vol. 1: Gene Expression (R. Goldberger, ed.), p. 349, Plenum Press, New York.

Subramanian, A. R., 1983, Structure and functions on ribosomal protein S1, Progr. Nucl. Acids Res. Mol. Biol. 28:101.

Sumper, M., and Luce, R., 1975, Evidence for de novo production of self-replicating and environmentally adapted RNA structure by bacteriophage Qβ replicase, Proc. Natl. Acad. Sci. USA 72:162.

Taniguchi, T., Palmieri, M., and Weissmann, C., 1978, Qβ DNA-containing hybrid plasmids giving rise to Qβ phage formation in the bacterial host, Nature 274:223.

Travers, A., Kamen, R., and Schleif, R., 1970, Factor necessary for ribosomal RNA synthesis, Nature 228:748.

Uhlenbeck, O. C., Carey, J., Romaniuk, P. J., Lowary, P. T., and Beckett, D., 1983, Interaction of R17 coat protein with its RNA binding site for translational repression, J. Biomol. Struct. Dyn. 1:539.

Van Dieyen, G., Van der Laken, C. J., Van Knippenberg, P. H., and Van Duin, J., 1975, Function of E. coli ribosomal protein S1 in translation of natural and synthetic messenger RNA, J. Mol. Biol. 93:351.

Van Dieyen, G., Van Knippenberg, P. H., and Van Duin, J., 1976, The specific role of ribosomal protein S1 in the recognition of native phage RNA, Eur. J. Biochem. 64:511.

Van Dieyen, G., Zipori, P., Van Prooyen, W., and Van Duin, J., 1978, Involvement of ribosomal protein S1 in the assembly of the initiation complex, Eur. J. Biochem. 90:571.

Van Duin, J., Overbeek, G. P., and Backendorf, C., 1980, Functional recognition of phage RNA by 30S ribosomal subunits in the absence of initiator tRNA, Eur. J. Biochem. 110:593.

Wahba, A. J., Miller, M. J., Niveleau, A., Landers, T. A., Carmichael, G., Weber, K., Hawley, D. A., and Slobin, L. J., 1974, Subunit I of Qβ replicase and 30S ribosomal protein S1 of E. coli, J. Biol. Chem. 249:3314.

Watanabe, M., Watanabe, H., and August, J. T., 1968, Replication of RNA bacteriophage R23, J. Mol. Biol. 33:1.

Weber, H., 1976, The binding site for coat protein on bacteriophage Qβ RNA, Biochim. Biophys. Acta 418:175.

Weber, K., and Konigsberg, W., 1975, Proteins of the RNA phages, in: RNA Phages (N. D. Zinder, ed.), p. 51, Cold Spring Harbor Laboratory, Cold Spring Harbor, NY.

Weber, H., Billeter, M. A., Kahane, S., Weissmann, C., Hindley, J., and Porter, A., 1972, Molecular basis for repressor activity of Qβ replicase, Nature (Lond.) 237:166.

Weiner, A. M., and Weber, K., 1971, Natural read-through at the UGA termination signal of Qβ coat protein cistron, Nature New Biol. 234:206.

Winter, R. B., and Gold, L., 1983, Overproduction of bacteriophage Qβ maturation protein leads to cell lysis, Cell 33:877.

Wittmann-Liebold, B., and Wittmann, H. G., 1967, Coat proteins of strains of two RNA viruses: Comparison of their amino acid sequences, Mol. Gen. Genet. 100:358.

Wulff, D. L., Mahoney, M., Shatzman, A., and Rosenberg, M., 1984, Mutational analysis of a regulatory region in bacteriophage that has overlapping signals for the initiation of transcription and translation, Proc. Natl. Acad. Sci. USA 81:555.

Yates, J. L., and Nomura, M., 1981, Feedback regulation of ribosomal protein synthesis in E. coli: Localization of the mRNA target sites for repressor action of ribosomal protein L1, Cell 24:243.

Yonesaki, T., and Aoyama, A., 1981, In vitro replication of bacteriophage GA RNA. Involvement of host factor(s) in GA RNA replication, J. Biochem. 89:751.

Yonesaki, T., and Haruna, I., 1981, In vitro replication of bacteriophage GA RNA. Subunit structure and analytic properties of GA replicase, J. Biochem. 89:741.

Yonesaki, T., Furuse, K., Haruna, I., and Watanabe, I., 1982, Relationships among four groups of RNA coliphages based on the template specificity of GA replicase, *Virology* **116:**379.

Zaug, A. J., and Cech, T. R., 1986, The intervening sequence RNA of *Tetrahymena* is an enzyme, *Science* **231:**470.

Zinder, N. D. (ed.), 1975, *RNA Phages,* Cold Spring Harbor Laboratory, Cold Spring Harbor, NY.

Zipper, P., Kratky, O., Herrmann, R., and Hohn, T., 1971, An X-ray small angle study of the bacteriophage fr and R17, *Eur. J. Biochem.* **18:**1.

CHAPTER 5

Phages with Protein Attached to the DNA Ends

Margarita Salas

I. INTRODUCTION

The finding of specific proteins covalently linked to the 5' ends of viral DNAs, the so-called terminal proteins, lead to the discovery of a new mechanism for the initiation of replication in which the primer, instead of being the 3' OH group of a nucleotide provided by RNA or DNA molecules, is the OH group of a serine, threonine, or tyrosine residue of the terminal protein.

Several phages are known to contain a terminal protein: the *B. subtilis* phage φ29, the *Streptococcus pneumoniae* phage Cp-1, and the *E. coli* phage PRD1. In addition, phages related to φ29 and Cp-1, respectively, have been shown to contain a terminal protein also.

Besides the role of the terminal protein as a primer in the initiation of replication (reviewed in Salas, 1983), the terminal protein of phage φ29 is required for the packaging of the phage DNA in the proheads.

In this review I will describe the characterization of the protein covalently linked to the DNA ends of φ29, Cp-1, and PRD1 and related phages, the role of these proteins as primers in the initiation of replication, and the role, in the case of phage φ29, in DNA packaging.

MARGARITA SALAS • Centro de Biología Molecular (CSIC-UAM), Universidad Autónoma, Canto Blanco, 28049 Madrid, Spain.

II. CHARACTERIZATION OF THE TERMINAL PROTEIN AT THE DNA ENDS OF THE PHAGE φ29, CP-1, AND PRD1 FAMILIES

A. Phage φ29 and Related Phages

The first evidence that suggested the existence of a protein at the ends of a DNA was obtained in the case of the *B. subtilis* phage φ29 with the finding of circular molecules and concatemers in the DNA isolated from the viral particles. These DNA structures were converted into unit-length linear DNA, of about 12×10^6 daltons (Sogo *et al.*, 1979), by treatment with proteolytic enzymes (Ortín *et al.*, 1971). In agreement with these findings, it was reported that φ29 DNA, after treatment with proteolytic enzymes, lost the capacity to transfect competent *B. subtilis* (Hirokawa, 1972) and that the DNA isolated from a *ts* mutant in gene 3 was thermolabile for transfection (Yanosfky *et al.*, 1976). Later on, it was found that a specific protein, p3, the gene 3 product (Salas *et al.*, 1978), of molecular weight 31,000 daltons, was covalently linked to the two 5' ends of the viral DNA (Salas *et al.*, 1978; Ito, 1978, Yehle, 1978).

The *B. subtilis* phages φ15, φ21, PZE, PZA, Nf, M2Y, B103, SF5, and GA-1 are morphologically similar to φ29 and have linear, double-stranded DNAs, of similar size, $11–13 \times 10^6$ daltons. As in the case of φ29, the transfection capacity of this DNAs is lost by treatment with proteolytic enzymes, again suggesting the existence of a terminal protein (Geiduschek and Ito, 1982; Fučik *et al.*, 1980). Indeed, a terminal protein of size similar to that of φ29 has been characterized in φ15, PZA, Nf, M2Y, B103, and GA-1 DNAs (Yoshikawa and Ito, 1981; Gutiérrez *et al.*, 1986b). In fact, the amino acid sequence of the terminal protein of phage PZA only has 6 amino acid changes (Pačes *et al.*, 1985) with respect to that of phage φ29 (Escarmís and Salas, 1982; Yoshikawa and Ito, 1982), of which five are replacements by similar amino acids. On the other hand, a comparison of the tryptic or chymotryptic peptides of the terminal proteins of phages φ29, φ15, PZA, Nf, M2Y, B103, and GA-1 has indicated that the terminal proteins of phages φ29, φ15, and PZA are similar; those of Nf, M2Y, and B103 are related to each other and less related to φ29, φ15, or PZA; and the one of phage GA-1 is not related to any other terminal protein (Yoshikawa and Ito, 1981; Gutiérrez *et al.*, 1986b).

B. Phage Cp-1 and Related Phages

The *S. pneumoniae* phage Cp-1 has a linear, double-stranded DNA of about 12×10^6 daltons. The first evidence for the existence of a terminal protein was obtained from the fact that the transfecting activity of Cp-1 DNA was lost after treatment with proteolytic enzymes (Ronda *et al.*, 1983). Indeed, a terminal protein of about 28,000 daltons was shown to be covalently linked to the two 5' ends of the DNA (García *et al.*, 1983a). By tryptic peptide analysis the terminal protein of phage Cp-1 is completely unrelated to that of phage φ29 (Escarmís *et al.*, 1985).

The Cp-1-related phages Cp-5 and Cp-7 also have a terminal protein covalently linked at the 5' ends of the DNA (López *et al.*, 1984). By tryptic peptide analysis, the terminal protein of phage Cp-5 was shown to be closely related to that of phage Cp-1 and to have some differences with respect to that of phage Cp-7 (Escarmís *et al.*, 1985).

C. Phage PRD1 and Related Phages

The lipid-containing phage PRD1 which infects *E. coli* and *Salmonella typhimurium* also has a linear, double-stranded DNA of about 14,500 base pairs (McGraw *et al.*, 1983). Evidence for the existence of a terminal protein linked to the ends of PRD1 DNA was obtained by labeling infected cells with ^{35}S-methionine and finding that a labeled protein remained associated with the DNA after phenol extraction and boiling with sodium dodecyl sulfate and was released from the DNA by nuclease treatment, migrating as the early phage-specific protein p8 (Bamford *et al.*, 1983).

There are other lipid-containing phages—PR4, PR5, PR722, and L17—closely related to PRD1. Although in none of these cases has a terminal protein been characterized, other properties of these phages (see later) make it very likely that all these phages also contain a terminal protein bound at the DNA ends.

III. LINKAGE BETWEEN THE TERMINAL PROTEIN AND THE DNA OF PHAGES φ29, Cp-1, AND PRD1

The linkage between the φ29 terminal protein p3 and the DNA was shown to be a phosphodiester bond between the OH group of a serine residue and 5' dAMP, the terminal nucleotide at both DNA ends (Hermoso and Salas, 1980). More recently, the serine residue involved in the linkage was determined as the serine at position 232 (Hermoso *et al.*, 1985), 266 being the total number of amino acids of the protein (Escarmís and Salas, 1982). Prediction of the secondary structure in the region around the linking site suggested that the above serine residue is located in a β turn, probably located on the external part of the molecule (Hermoso *et al.*, 1985).

In phage Cp-1, the linkage is a phosphodiester bond between threonine and 5'-dAMP (García *et al.*, 1986a), and in phage PRD1, the linkage between the terminal protein and the DNA is a phosphodiester bond between tyrosine and 5'-dGMP (Bamford and Mindich, 1984).

IV. NUCLEOTIDE SEQUENCE AT THE DNA ENDS OF THE φ29, Cp-1, AND PRD1 FAMILIES

A. φ29 Family

The complete nucleotide sequence of φ29 DNA, 19,285 base pairs long (Yoshikawa and Ito, 1982; Garvey *et al.*, 1985; Vlček and Pačes, 1986) and of the related phage PZA, 19,366 base pairs long (Pačes *et al.*, 1985, 1986a,b) is now known.

All the phages of the φ29 family have a short, inverted terminal repeat six nucleotides long (5' AAAGTA) for φ29, PZA, φ15, and B103 DNAs; eight nucleotides long (AAAGTAAG) for Nf and M2Y DNAs; and seven nucleotides long (AAATAGA) for GA-1 DNA. In addition, the sequence of the first 18 nucleotides at the left DNA end of φ29, PZA, φ15, Nf, M2Y, and B103 is identical, and there are homologies from nucleotides 19 to 50. The remaining known sequence in phages φ29, PZA, and φ15 is similar to each other and different from that of phages Nf, M2Y, and B103, which share a similar sequence. The sequence at the left end of GA-1 DNA is completely unrelated to that of the other phage DNAs except for the three terminal nucleotides. At the right end, the sequence of the first 13 nucleotides is very similar for the DNA of phages φ29, PZA, φ15, Nf, M2Y, and B103, and there is an identical sequence from nucleotide 27 to 38. The rest of the known sequence is very similar for the DNA of phages φ29, PZA, and φ15 and different from that of phages Nf, M2Y, and B103. The sequence at the right end of GA-1 DNA is also unrelated to that of the other phage DNAs except for the three terminal nucleotides and the sequence from nucleotide 29 to 41, which is almost identical to that present from nucleotides 27 to 38 in the other phage DNAs (Escarmís and Salas, 1981; Yoshikawa *et al.*, 1981, 1985; Pačes *et al.*, 1985, 1986b; Gutiérrez *et al.*, 1986b).

B. Cp-1 Family

Phage Cp-1 DNA has a very long, inverted terminal repeat, of 236 base pairs, being the following 116 base pairs 93% homologous at the two DNA ends (Escarmís *et al.*, 1984). The inverted terminal repeats of the related phages Cp-5 and Cp-7 are 343 and 347 base pairs long, respectively (Escarmís *et al.*, 1985). The homology between the inverted terminal repeats of the three DNAs is 84–92%, similar to the average homology calculated from restriction enzyme analysis (López *et al.*, 1984). The first 39 base pairs are the same for the three DNAs, and the differences are concentrated between nucleotides 74 and 98 and between nucleotides 229 and 253. On the other hand, Cp-5 DNA is more related to Cp-1 than Cp-7 DNA, in agreement with the degree of homology of the terminal proteins.

C. PRD1 Family

The closely related phages PRD1, PR4, PR5, PR722, and L17, isolated from different parts of the world, have long, inverted terminal repeats of 109–110 base pairs. The first 17 and the last 35 base pairs of the inverted terminal repeats are conserved in all the viruses except for a mismatch in L17. Between these conserved nucleotide sequences there is a variable area of 58 base pairs that enables to divide the phages into two groups; (1) phages PRD1 and PR5 from the United States and Canada, and (2) phages PR4, PR722, and L17 from Australia, South Africa, and the United Kingdom, respectively (Savilahti and Bamford, 1986).

The comparison of the DNA ends of the PRD1 family with those of the φ29 and Cp-1 families reveals that the conserved terminal sequence of the PRD1 family and the left end of the φ29 family is 17–18 base pairs long. Within this sequence, there is a homologous sequence CCCC$_A^T$CCC which is also found in the Cp-1 termini some nucleotides further from the DNA end (Savilathi and Bamford, 1986). The fact that the right end of φ29 does not have this sequence and that the left end of φ29 DNA is first packaged in the phage heads suggests that this sequence might be required for DNA packaging.

V. TRANSCRIPTION OF φ29 DNA

The φ29 early genes, located at the two ends of the DNA, are transcribed from right to left; the late genes are clustered in the middle of the genome and are transcribed from left to right (Mellado et al., 1976; Schachtele et al., 1973; Carrascosa et al., 1976) under the control of the viral gene 4 product (Sogo et al., 1979).

The initiation sites of the RNA transcripts synthesized in vivo in B. subtilis infected with phage φ29 or in vitro using φ29 DNA with the B. subtilis RNA polymerase holoenzyme have been mapped by S1 protection experiments (Barthelemy et al., 1986; Dobinson and Spiegelman, 1985; Mellado et al., 1986a,b). Nine transcription initiation sites were localized along the φ29 genome, close to previously reported RNA polymerase binding sites (Sogo et al., 1979). Eight of these sites correspond to early transcription, and only one to late transcription. The eight early promoters have consensus sequences that correspond to the ones used by the B. subtilis RNA polymerase with the σ^{43}subunit (TTGACA and TATAAT at the −35 and −10 regions, respectively). The late promoter has the TATAAT sequence at the −10 region, but no consensus sequence is present at the −35 region.

The main in vivo early and late transcription termination sites in φ29 DNA have been determined by S1 mapping experiments. Transcription of the early genes located at the left end of the genome terminates at the very end of the DNA and within the HindIII G fragment. Transcription termination of the early genes located at the right end of the genome, and that of the viral late genes, overlapped in a region within the EcoRI D fragment. Stem-loop structures followed by uridine-rich tails can be derived for early and late mRNAs, suggesting rho-independent transcription termination in ϕ29 DNA (Barthelemy et al., 1987b).

The ϕ29 gene 4, which controls ϕ29 late transcription (Sogo et al., 1979), has been cloned under the control of the P_L promoter of phage lambda, and protein p4 was overproduced (Mellado and Salas, 1982). The overproduced protein p4 has been purified and promotes the in vitro transcription from the late promoter when added to the B. subtilis RNA polymerase (Barthelemy et al., 1987a). Experiments aimed to elucidate whether protein p4 acts as a sigmalike factor or as an activator are under way.

The transcriptional maps of Cp-1 and PRD1 DNAs is not yet available.

VI. IN VIVO REPLICATION OF ϕ29, Cp-1, AND PRD1 DNAs

A. Phage ϕ29

Analysis of the in vivo ϕ29 replicative intermediates by electron microscopy showed that ϕ29 DNA replication starts at either end of the DNA, nonsimultaneously, and proceeds by a strand displacement mechanism (Inciarte et al., 1980; Harding and Ito, 1980). In agreement with a role of the terminal protein in the initiation of relication, protein was found at the ends of the parental and daughter DNA strands (Sogo et al., 1982).

By using ts and sus mutants of ϕ29 (Talavera et al., 1971; Moreno et al., 1974), five genes were shown to be involved in the viral DNA replication—genes 2, 3, 5, 6, and 17 (Talavera et al., 1972; Carrascosa et al., 1976; Hagen et al., 1976). By shift-up experiments using ts mutants available in genes 2, 3, 5, and 6, genes 2 and 3 were shown to be involved in an initiation process, whereas genes 5 and 6 were involved in an elongation step in ϕ29 DNA replication (Mellado et al., 1980). In this kind of experiment, the possibility that genes 5 and 6 are also involved in initiation cannot be ruled out. Gene 17, for which ts mutants are not available, gave rise to delayed DNA replication (Carrascosa et al., 1976). Recent transfection experiments using recombinant ϕ29 DNA molecules lacking gene 17 have shown that this gene is dispensable (C. Escarmís, D. Guirao, and M. Salas, in preparation).

In the ϕ29-related phage M2, three genes involved in the viral DNA replication have been characterized—G, E, and T. Genes G and E correspond to the ϕ29 genes 2 and 3, respectively (Matsumoto et al., 1983).

In vivo ϕ29 DNA replication does not require the bacterial DNA polymerases I or III, since replication occurs in polA⁻ mutants and in the presence of 6(p-hydroxyphenylazo)uracil (Peñalva and Salas, 1982; Talavera *et al.*, 1972). Development of ϕ29 also occurs normally in the *B. subtilis* replication mutants (reviewed in Henney and Hoch, 1980) *ts* dna B19, *ts* dna C30, *ts* dna D23, *ts* dna E20, *ts* dna F133, *ts* dna I102, and QB1506, whereas the burst size is reduced about 10-fold in the mutants *ts* dna G34 and *ts* dna H151, and no development takes place in mutant *ts* dna A13, involved in ribonucleotide reduction (Salas, unpublished results).

In vivo ϕ29 and M2 DNA replication is inhibited by aphidicolin (Hirokawa *et al.*, 1982), a known inhibitor of the eukaryotic DNA polymerase α (reviewed by Huberman, 1981). As will be shown later, the target of the aphidicolin inhibition is the ϕ29-induced DNA polymerase.

B. Phage Cp-1

Cp-1 DNA replication starts at the DNA ends by a protein-priming mechanism (see later). As in the case of phage ϕ29, Cp-1 replication is not inhibited by 6(p-hydroxyphenylazo)uracil, and it is inhibited by aphidicolin (García *et al.*, 1986b), suggesting a mechanism of replication very similar to that of phage ϕ29.

C. Phage PRD1

Viral DNA synthesis was determined in cells infected under restrictive conditions by PRD1 *sus* mutants. PRD1 DNA was not synthesized in class I and VIII mutants, which involve proteins p1 and p8, respectively. Mutants of class XII, involving protein p12, showed a decrease in the amount of the viral DNA synthesized. The same result was found with mutant *sus* 239, which either overproduces p17 or is also a mutant of p12 (Mindich *et al.*, 1982).

On the other hand, the host proteins rep, topoisomerase I, and topoisomerase II are not needed for PRD1 DNA replication (Davis and Cronan, 1983; Bamford and Mindich, 1984).

VII. *IN VITRO* REPLICATION OF ϕ29, Cp-1, AND PRD1 DNAs: INITIATION REACTION

A. Formation of a Covalent Complex between the Terminal Protein and 5′ dAMP in ϕ29 and M2

When extracts from ϕ29 or M2-infected *B. subtilis* were incubated with [α-^{32}P] dATP in the presence of the DNA-terminal protein complex, a labeled protein with the electrophoretic mobility of the terminal protein was found, which was not formed in the presence of antiterminal protein serum or when extracts from uninfected cells were used (Peñalva and Salas, 1982; Shih et al., 1982, 1984; Watabe et al., 1982, 1983; Matsumoto et al., 1983). Incubation of the ^{32}P-labeled protein with piperidine under conditions that hydrolyze the linkage between protein p3 and φ29 DNA released 5′ dAMP, indicating the formation of a protein p3-dAMP covalent complex. This complex could be elongated in vitro, indicating that it was indeed an initiation complex (Peñalva and Salas, 1982).

By using extracts from B. subtilis infected under restrictive conditions with φ29 mutants in genes 2, 3, 5, 6, and 17, it was shown that genes 2 and 3 are essential for the formation of the initiation complex (Blanco et al., 1983; Matsumoto et al., 1983). The products of genes 5, 6, or 17 were not needed for the initiation reaction. In agreement with the φ29 results, extracts from su⁻ B. subtilis infected with the M2 mutants susG or susE, equivalent to the φ29 sus2 or sus3 mutants, respectively, were also inactive in the formation of the M2 initiation complex (Matsumoto et al., 1983).

B. Formation of a Covalent Complex between the Terminal Protein and 5′ dAMP in Phage Cp-1

Incubation of extracts of Cp-1-infected S. pneumoniae with [α-^{32}P]dATP produced a labeled protein with the electrophoretic mobility of the Cp-1 terminal protein. Incubation of the ^{32}P-labeled protein with piperidine released 5′ dAMP, indicating that a covalent complex between the terminal protein and 5′ dAMP was formed in vitro. Addition of the four deoxynucleoside triphosphates to the incubation mixture produced a labeled complex of slower mobility in sodium dodecyl sulfate-polyacrylamide gels, indicating that the Cp-1 terminal protein-dAMP complex can be elongated and, therefore, it is an initiation complex (García et al., 1986a).

C. Formation of a Covalent Complex between the Terminal Protein and 5′ dGMP in Phage PRD1

When extracts from PRD1-infected S. typhimurium were incubated with [α-^{32}P]dGTP, a labeled protein band with the electrophoretic mobility of the terminal protein p8 was formed (Bamford et al., 1983). This reaction did not occur when extracts from uninfected cells were used. When the extract was made from cells containing cloned genes of classes I and VIII, coding for the replication proteins p1 and p8, respectively, the efficiency of the labeling of protein p8 greatly increased. On the other hand, when extracts from su⁻ cells infected with sus mutants of class I or VIII were used, no p8-dGMP complex was formed, although they complemented each other (Bamford and Mindich, 1984).

The involvement of host replication functions in p8-dGMP complex formation was tested by using extracts from *E. coli* cells temperature-sensitive for the replication genes dna B, dna E, dna A, dna C, dna G, dna Z, and dna P, each of them harboring a plasmid containing the viral essential genes coding for proteins p1 and p8. All mutants, except dna G, were temperature-sensitive for the formation of the initiation complex, suggesting that most of the host replication complex may be needed for the initiation reaction (Bamford and Mindich, 1984).

VIII. PURIFICATION AND CHARACTERIZATION OF THE φ29 PROTEINS p2, p3, p5, AND p6, INVOLVED IN DNA REPLICATION

A. Purification of Protein p2: Characterization of DNA Polymerase and $3' \rightarrow 5'$ Exonuclease Activities

Since the amount of protein p2 in φ29-infected *B. subtilis* is very low, gene 2 was cloned in an *E. coli* plasmid to overproduce the proteins to facilitate the purification. A φ29 DNA fragment containing gene 2 was cloned in plasmid pPLc28 under the control of the P_L promoter of phage lambda. A protein with the electrophoretic mobility expected for protein p2 was overproduced in the *E. coli* cells harboring the recombinant plasmid pLBw2, and the protein was active in the formation of the initiation complex when complemented with extracts from *sus* 2–infected *B. subtilis* (Blanco *et al.*, 1984). Protein p2 was highly purified in an active form from the above cells (Blanco and Salas, 1984).

The purified protein p2, in addition to catalyzing the initiation reaction when complemented with extracts from *sus* 2–infected cells or with purified protein p3 (see later), was shown to have DNA polymerase activity when assayed with a template primer such as poly dA-$(dT)_{12n-18}$ or on activated DNA (Blanco and Salas, 1984; Watabe *et al.*, 1984a). The DNA polymerase activity was shown by *in situ* gel analysis to be present in the protein p2 band (Blanco and Salas, 1984). In agreement with these results, a partially purified DNA polymerase isolated from *B. subtilis* infected with a *ts* mutant in gene 2 had greater heat lability than the protein p2 from wild-type-infected cells (Watabe and Ito, 1983).

In addition, protein p2 has a $3' \rightarrow 5'$ exonuclease activity on single-stranded DNA, which might provide a proofreading mechanism, but not $5' \rightarrow 3'$ nuclease (Blanco and Salas, 1985a; Watabe *et al.*, 1984b). The $3' \rightarrow 5'$ exonuclease activity was shown to be associated with the DNA polymerase and initiation activities of protein p2, since all these activities cosedimented in a glycerol gradient (Blanco and Salas, 1985a). Moreover, the DNA polymerase and $3' \rightarrow 5'$ exonuclease activities were heat-inactivated with identical kinetics (Blanco and Salas, 1985a).

B. Purification of Protein p3

Since in φ29-infected *B. subtilis* most of the protein p3 synthesized is used for priming replication and is present as a covalent complex with φ29 DNA, to overproduce free protein p3, a φ29 DNA fragment containing genes 3, 4, 5, and most of 6 was cloned in plasmid pKC30 under the control of the P_L promoter of phage lambda. A protein with the electrophoretic mobility of p3 that reacted with anti-p3 serum was overproduced in the *E. coli* cells harboring to recombinant plasmid pKC30 A1, and the protein was active in the formation of the initiation complex when complemented with extracts from *sus* 3–infected *B. subtilis* (García *et al.*, 1983b). Protein p3 has been highly purified in an active form from *E. coli* cells harboring the gene 3–containing recombinant plasmid (Prieto *et al.*, 1984; Watabe *et al.*, 1984a).

C. Activity of Purified Proteins p2 and p3 in the Formation of the p3-dAMP Initiation Complex and Its Further Elongation

When purified proteins p2 and p3 were incubated with [α-^{32}P] dATP in the presence of φ29 DNA-p3 as template, p3-dAMP initiation complex was formed. The reaction was greatly stimulated by addition of NH_4^+ ions (Blanco and Salas, 1985b). The stimulation is probably due to the formation of a complex between proteins p2 and p3 which is detected by glycerol gradient centrifugation in the presence but not in the absence of NH_4^+ ions (Blanco *et al.*, 1987).

The p3-dAMP initiation complex formed in the minimal system containing purified proteins p2 and p3 is elongated in the presence of dATP, dTTP, dGTP, and ddCTP to give rise to p3 linked to oligonucleotides 9 and 12 bases long (Blanco and Salas, 1984; Watabe *et al.*, 1984a), according to the sequence at the DNA ends. In addition, when the four dNTPs were present, this minimal system was able to elongate the p3-dAMP initiation complex to produce full-length φ29 DNA (Blanco and Salas, 1985b). The rate of elongation in this system was stimulated about threefold by addition of NH_4^+ ions (Blanco *et al.*, 1987). This result could be in agreement with those of Matsumoto *et al.* (1984) suggesting the participation of the φ29 terminal protein in elongation.

D. Effect of Aphidicolin and Nucleotide Analogues on the φ29 DNA Polymerase

Aphidicolin and the nucleotide analogues butylanilino dATP (BuAdATP) and butylphenyl dGTP (BuPdGTP), inhibitors of the eukaryotic DNA polymerase (Huberman, 1981; Khan *et al.*, 1984), inhibited the protein-primed replication of φ29 DNA-p3 using the minimal system with purified proteins p2 and p3. The aphidicolin effect was mainly on the elongation step, inhibiting the rate of elongation. The nucleotide analogues inhibited both the initiation and elongation steps. All the drugs inhibited polymerization on activated DNA and the $3' \rightarrow 5'$ exonuclease activities of protein p2, suggesting that the target of the drugs is the φ29 DNA polymerase itself (Blanco and Salas, 1986). This is in agreement with the finding that aphidicolin-resistant mutants map within gene 2 and they have an altered DNA polymerase that is less sensitive to aphidicolin than the wild-type enzyme (Matsumoto et al., 1986).

That aphidicolin inhibits mainly the elongation step of φ29 DNA- p3 replication whereas the nucleotide analogues inhibit both initiation and elongation suggests the existence of two active sites in protein p2—one for initiation for the covalent linkage of dAMP to the OH group of serine residue 232 in protein p3, and one for elongation for the linkage of dNMPs to the 3' OH group of a nucleotide (Blanco and Salas, 1986). The existence of two active sites in protein p2 is in agreement with the finding of a different dATP concentration requirement for initiation and elongation. Whereas initiation can occur at very low dATP concentration (0.1 μM), elongation requires a higher dATP concentration (> 1 μM) (Blanco et al., 1986).

E. Isolation of Mutants of Protein p3 by *in Vitro* Mutagenesis of Gene 3: Effect of the Mutations on the *in Vitro* Formation of the Initiation Complex

By *in vitro* manipulation of gene 3–containing recombinant plasmids, two p3 mutants with some residues changed at the carboxyl end were obtained; these mutants showed a reduced activity in the *in vitro* formation of the initiation complex (Mellado and Salas, 1982, 1983). To further study the importance of the carboxyl end in the activity of protein p3, deletion mutants were constructed. A deletion of four amino acids showed 50% of the activity of the wild-type protein. Deletions of 20 or more amino acids resulted in a small production of the protein and no priming activity. Addition of the last five carboxyterminal amino acids to the latter gave rise to a higher production of the proteins but no priming activity, suggesting that the region between residues 240 and 262 at the p3 carboxyl end, or part of it, may be essential for the normal function of the protein (Zaballos et al., 1986).

Deletion mutants at the amino end of protein p3 have also been constructed. Deletions of 5–13 amino acids resulted in a protein with higher activity than the wild-type p3. Deletions of 17–47 amino acids produced a protein with low activity, and a deletion of 54 amino acids resulted in an inactive protein (Zaballos et al., 1988). The finding that deletions at the amino end of protein p3 produce a protein more active than the wild-type p3 is interesting and might suggest some processing in protein p3 from ϕ29-infected *B. subtilis* that might not occur in the protein p3 synthesized in *E. coli*. Experiments aimed to elucidate this possibility are in progress.

To determine the specificity of the serine residue at position 232 of protein p3 involved in the linkage to dAMP, the above serine was changed to a threonine residue by site-directed mutagenesis. No detectable activity of the mutated p3 was obtained, indicating a very high specificity in the linking site (C. Garmendia, M. Salas, and J. M. Hermoso, in preparation).

F. Purification and Characterization of Protein p6

A ϕ29 DNA fragment containing gene 6 was cloned in plasmid pPLc28 under the control of the P_L promoter of phage lambda. A protein with the electrophoretic mobility of protein p6 (molecular weight ~12,000) was overproduced and highly purified. The apparent molecular weight of the purified protein p6 was 23,600, suggesting that the native form of the protein is a dimer (Pastrana *et al.*, 1985).

When purified protein p6 was added to the *in vitro* system with proteins p2 and p3 in the presence of ϕ29 DNA-p3 as template and 0.25 μM dATP, the formation of the p3-dAMP intiation complex was increased (Pastrana *et al.*, 1985). The stimulation by protein p6 was dependent on the dATP concentration and decreased about fivefold the K_m value for dATP (Blanco *et al.*, 1986).

Protein p6 also stimulated the limited elongation reaction in the presence of ddCTP in the minimal system with purified proteins p2 and p3 and ϕ29 DNA-p3 as template. The activities in protein p6 that stimulate the initiation and limited elongation reactions in ϕ29 DNA-p3 replication cosediment with the protein p6 peak in a glycerol gradient, indicating that the two activities are present in the same protein (Blanco *et al.*, 1986). The stimulation by protein p6 of elongation in ϕ29 DNA-p3 replication in the absence of ddCTP was similar to that in its presence, and no effect on the rate of elongation or on the K_m value for the dNTPs was obtained. A possible role for protein p6 in elongation might be to stimulate the incorporation of the first nucleotide(s) into the p3-dAMP initiation complex (Blanco *et al.*, 1986).

The effect of protein p6 on the elongation reaction requires the use of templates containing the ϕ29 DNA replication origin. In addition, the stimulation of elongation by protein p6, but not that of initiation, seems to be sequence-dependent, since the replication from the right end is preferentially stimulated over that from the left end (Blanco *et al.*, 1986).

To try to understand the mechanism by which protein p6 stimulates ϕ29 DNA-p3 replication, the binding of p6 to different DNA fragments was studied by DNAase I footprinting experiments. Binding to the 269

base pairs long HindIII L fragment from the right end of φ29 DNA or to a
307 base pairs long fragment from the left φ29 DNA end produced a
specific pattern of protected regions 22 nucleotides long, all along the
fragment, flanked by DNAase I hypersensitive sites located 24 nucleotides apart. A different pattern was found with an internal φ29 DNA
fragment or with a pBR322 DNA fragment, suggesting that protein p6
might recognize some signal at the ends of φ29 DNA (Salas *et al.*, 1986).
In addition, when relaxed circular DNA is incubated with protein p6 and
further treated with topoisomerase I, supercoiling of the DNA is obtained, suggesting that the binding of protein p6 at the ends of φ29 DNA
produces some conformational change, and this might facilitate replication (Prieto *et al.* 1988).

G. Purification and Characterization of Protein p5

The open-reading frame 10 in the sequence of Yoshikawa and Ito
(1982) was shown to be the one coding for protein p5 by sequencing a *ts5*
mutant (G. Martín, J. M. Lázaro, and M. Salas, in preparation).

A φ29 DNA fragment containing open-reading frame 10 was cloned
in plasmid pPLc28, under the control of the P_L promoter of phage lambda.
A protein of the expected molecular weight, about 13,000, was labeled
after heat induction. The protein has been highly purified and shown to
bind to single-stranded DNA (Martín *et al.*, in preparation). Experiments
to determine the activity of the purified protein in the *in vitro* φ29 DNA
replication system are under way.

IX. TEMPLATE REQUIREMENTS FOR THE FORMATION OF THE INITIATION COMPLEX

Protein p3-containing fragments from the left or right φ29 DNA
ends, but not internal fragments, were active as templates for the formation of the p3-dAMP initiation complex provided they had a minimal
size: a 26-base-pairs-long fragment was active, whereas a 10-base-pairs-
long one was essentially inactive. The activity of the latter was restored
by ligation of an unspecific DNA sequence, suggesting that the low activity of the 10-base-pairs-long fragment is due to its small size rather
than to the lack of a specific sequence (García *et al.*, 1984).

When φ29 DNA-protein p3 was treated with proteinase K, the resulting DNA was not active as a template for the initiation reaction. However, treatment with piperidine to remove the peptide that remains after
proteinase K digestion produced an active template for the initiation
reaction, although the activity was about 5–10 times lower than that
obtained with φ29 DNA-p3. The activity was due to the terminal sequences in φ29 DNA, since isolated piperidine-treated terminal frag-

ments HindIII B and L, but not internal fragments, were the active templates (Gutiérrez et al., 1986b).

To study the DNA sequence requirements for the initiation reaction further, the terminal fragments BclI C and HindIII L, 73 and 269 base pairs long, from the left and right ends of φ29 DNA, respectively, were cloned in plasmid pKK 223-3 in such a way that treatment of the recombinant plasmid pID13 with AhaIII released the φ29 DNA terminal sequences at the ends of two fragments. These fragments were active templates for the initiation and elongation reactions, although the activity was about 15% of that obtained with φ29 DNA-p3. No activity was obtained with the circular plasmid or with the plasmid linearized with HindIII that does not place the φ29 DNA terminal sequences at the DNA ends (Gutiérrez et al., 1986a). The fact that only terminal, but not internal, φ29 DNA fragments are active as templates suggests the existence of specific sequences at the φ29 DNA ends that allow the recognition for the initiation reaction.

The results obtained with the protein-free terminal fragments of φ29 DNA suggested that the parental terminal protein at the ends of φ29 DNA, although it was not an absolute requirement, was an important one, since, in its absence, the template activity was greatly reduced. To further show the relevance of the parental terminal protein, the DNA-terminal protein complex of the φ29-related phages PZA, φ15, Nf, B103, and GA-1, or the piperidine-treated DNAs, were used as templates for the initiation reaction with purified proteins p2 and p3 from φ29. The template activity of the terminal protein-DNA complex from phages PZA and φ15, with a terminal protein closely related to that of φ29, was similar to that obtained with the φ29 DNA-p3 complex, and the activity was similarly reduced after removal of the terminal protein by treatment with piperidine. The terminal protein-DNA complex from phages Nf, B103, or GA-1, with a terminal protein much less related (Nf and B103) or unrelated (GA-1) to that of φ29, was very little active (Nf and B103) or inactive (GA-1). Activity was restored to the level of that obtained with protein-free φ29 DNA after removal of the terminal protein (Gutiérrez et al., 1986b). These results are in agreement with an important role of the parental terminal protein, since the presence of a related one allows the initiation reaction to occur, probably by protein-protein interaction, whereas the presence of an unrelated one avoids such interaction and that with the DNA signals. Removal of the φ29-related parental terminal protein decreases the template activity, suggesting that protein-DNA interaction is less efficient than protein-protein interaction. Removal of the unrelated parental terminal protein gives rise to template activity, allowing the protein-DNA interaction to occur.

The important role of the parental terminal protein in φ29 DNA replication is in agreement with experiments of mixed infection of B. subtilis, at 42°C, with a ts 2 and a ts 3 mutant. Most of the progeny of the infection had the ts 2 genotype, suggesting that the heat-inactivated parental terminal protein in the *ts* 3 mutant cannot initiate replication (Salas *et al.*, 1978).

On the other hand, transfection of *B. subtilis* protoplasts with φ29 DNA molecules lacking one of the two parental terminal proteins give rise to neither phage progeny nor DNA replication, in agreement with a key role of the parental terminal protein in φ29 DNA replication (Escarmís *et al.*, in preparation).

X. POSSIBLE ROLE OF THE INVERTED TERMINAL REPEAT IN THE REPLICATION OF φ29, Cp-1, AND PRD1 DNAs

A possible role for the inverted terminal repeat in DNA replication is the formation of a panhandle structure of the parental DNA strand that could be displaced in the process of replication. This role has been suggested in the case of adenovirus DNA replication that occurs by a similar protein-priming mechanism and has an inverted terminal repeat 100 base pairs long (Lechner and Kelly, 1977). The inverted terminal repeat of phages Cp-1 and PRD1 is long enough to allow the formation of such a panhandle structure. However, the 6-base-pairs-long inverted terminal repeat of phage φ29 may be too short to allow the formation of such structure.

To test whether the φ29 parental DNA strand is displaced during replication and initiates replication after displacement, φ29 DNA molecules with only one parental terminal protein at the left or right ends were constructed, and *B. subtilis* protoplasts were transfected with such molecules. If the parental DNA strand is replicated after displacement, the φ29 DNA molecules with terminal protein at only one of the ends will have a normal displaced strand and therefore should replicate. No replication was obtained with such φ29 DNA molecules, suggesting that the parental strand does not start replication after displacement but probably starts replication at the opposite end before being displaced (Escarmís *et al.*, in preparation). The possibility that the parental DNA strand in phages Cp-1 and PRD1 is replicated after displacement will have to be determined.

From the fact that protein-free DNAs from phages φ29, PZA, φ15, Nf, B103, and GA-1 have template activity using the φ29 proteins p2 and p3, and taking into account that a common sequence for all the phage DNAs, except for GA-1, is the 6-base-pairs-long inverted terminal repeat, the latter could be the minimal sequence recognized for the initiation of replication. If this is the case, the template activity of GA-1 DNA, which only shares the sequence 5'AAA with the other phage DNAs (and this sequence is not enough to provide template activity) needs to be explained (Gutiérrez *et al.*, 1986a). Experiments aimed to elucidate which is the DNA sequence required for the initiation of φ29 DNA replication are under way.

XI. MODEL FOR THE PROTEIN-PRIMED REPLICATION OF φ29

Figure 1 shows our present knowledge of the protein-primed replication of φ29 and, most likely, the φ29, Cp-1, and PRD1 families. A free molecule of the terminal protein p3 interacts with the DNA polymerase p2 in the presence of NH_4^+ ions, and the p2–p3 complex is located at the ends of the φ29 DNA-p3 template by protein-protein and protein-DNA interaction. In the presence of dATP, the DNA polymerase catalyzes the formation of a covalent complex between the OH group of serine residue 232 in protein p3 and 5' dAMP. This reaction is stimulated by protein p6 that lowers the K_m value for dATP. Whether other factors are involved in this initiation reaction remains to be determined. The p3-dAMP initiation complex is further elongated by the DNA polymerase that probably remains associated with the DNA in a replication complex. The rate of elongation is increased about threefold by NH_4^+ ions. The viral protein p6 stimulates elongation, probably by coupling the initiation and elongation steps by stimulating the incorporation of the first nucleotide(s) to the p3-dAMP initiation complex. The role of the viral protein p5 could be to act as an *ssb* by interaction with the displaced parental strand. Further experiments are needed to determine the role of protein p5 and to establish the function of p17 and the possible involvement of other factors.

By using the purified system with proteins p2, p3, and p6, unit length φ29 DNA is obtained when the φ29 DNA-p3 template is used. If the parental strand is displaced before it initiates replication, formation of a

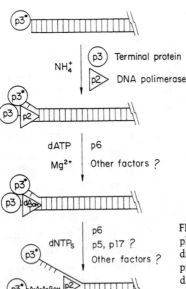

FIGURE 1. Protein-primed replication of bacteriophage φ29 DNA. The parental terminal protein is indicated by an asterisk to show the difference from free protein p3. Whether the terminal protein in the p3-dAMP complex resembles the parental or the free protein p3 remains to be determined.

panhandle structure through the 6-base-pairs-long inverted terminal repeat could provide a replication origin. The experimental evidence available does not support this possibility. Alternatively, the replication at the opposite end may start before the parental strand is completely displaced, thus using the replication origin on the double-stranded DNA.

On the other hand, although the model presented in Fig. 1 shows a linear φ29 DNA-p3 molecule as template, a possibility is the formation of circular DNA molecules by interaction of the terminal proteins at the DNA ends (Ortín et al., 1971). This alternative might explain why initiation of replication does not occur simultaneously at the two DNA ends in vivo. In addition, the existence of the displaced parental strand as a circle held by protein-protein interaction might provide a mechanism for the initiation of its replication.

XII. ROLE OF THE PARENTAL TERMINAL PROTEIN IN φ29 DNA PACKAGING

An in vitro φ29 DNA packaging system has been developed that depends on proheads, the gene 16 product, and φ29 DNA-p3 complex (Bjornsti et al., 1981, 1984; Guo et al., 1986). In this in vitro system, the φ29 DNA-p3 complex is encapsulated in an oriented way, the left end of φ29 DNA being packaged first (Bjornsti et al., 1983). Treatment of φ29 DNA-p3 with proteinase K avoids the packaging reaction, suggesting an essential role of the parental terminal protein in DNA packaging in addition to its role in the initiation of DNA replication.

The protein that forms the connector of φ29, p10, binds to DNA in a way that does not seem to be sequence-specific. However, the presence of the terminal protein at the ends of the DNA produces a significant increase of p10 molecules bound to the DNA ends, suggesting a terminal protein-mediated recognition of DNA ends by the phage connector (Herranz et al., 1986). It remains to be determined whether protein p16, the only free protein that seems to be essential for DNA packaging in vitro, provides the specificity to recognize the left φ29 DNA end.

It is likely that in the other phages of the φ29 family, as well as in the case of the Cp-1 and PRD1 families, the parental terminal protein is also essential for DNA packaging.

XIII. PROTEIN-PRIMED INITIATION OF REPLICATION: A GENERAL MECHANISM

In addition to the φ29, Cp-1, and PRD1 families reported in this review, adenovirus (Rekosh et al., 1977), the S1 and S2 mitochondrial DNA from maize (Kemble and Thompson, 1982), the linear plasmid pSLA2 from Streptomyces (Hirochika and Sakaguchi, 1982), and the linear plasmids pGKL1 and pGKL2 from yeast (Kikuchi *et al.*, 1984) have a terminal protein covalently linked at the 5' ends of the DNA. The protein-priming mechanism for the initiation of replication has been studied in great detail in the case of adenovirus (reviewed in Stillman, 1983), showing great similarities with the φ29 system. It is likely that the other reported cases of DNAs with a terminal protein also initiate replication by protein priming.

In addition, several RNA genomes of animal and plant viruses have a terminal protein covalently linked at the 5' end of the RNA (reviewed in Daubert and Bruening, 1984). Some evidence for the formation of a covalent complex between the terminal protein and the terminal nucleotide has been obtained for poliovirus and encephalomyocarditis virus RNAs (Morrow *et al.*, 1984; Vartapetian *et al.*, 1984). Therefore, the protein-priming mechanism is likely to be a general way to initiate replication in protein-containing nucleic acids.

It is also likely that, as happens in phage φ29, the parental terminal protein may be required for packaging of the nucleic acid into the viral heads.

ACKNOWLEDGMENTS. The most recent replication work from the Madrid laboratory was carried out by A. Bernad, L. Blanco, M. A. Blasco, C. Escarmís, C. Garmendia, D. Guirao, J. Gutiérrez, J. M. Hermoso, J. M. Lázaro, G. Martín, R. P. Mellado, I. Prieto, M. Serrano, L. Villar, and A. Zaballos. I am grateful to D. Anderson, D. Bamford, and V. Pačes for sending me unpublished information. This work has been supported by grant 5RO1 GM27242 from the National Institutes of Health and by grants from the Comisión Asesora para el Desarrollo de la Investigación Científica y Técnica and Fondo de Investigaciones Sanitarias.

REFERENCES

Bamford, D. H., and Mindich, L., 1984, Characterization of the DNA-protein complex at the termini of the bacteriophage PRD1 genome, *J. Virol.* **50:**309–315.

Bamford, D. H., McGraw, T., MacKenzie, G., and Mindich, L., 1983, Identification of a protein bound to the termini of bacteriophage PRD1 DNA, *J. Virol.* **47:**311–316.

Barthelemy, I., Salas, M., and Mellado, R. P., 1986, In vivo transcription of bacteriophage φ29 DNA. Transcription initiation sites, *J. Virol.* **60:**874–879.

Barthelemy, I., Lázaro, J. M., Méndez, E., Mellado, R. P., and Salas, M., 1987a, Purification in an active form of the phage φ29 protein p4 that controls the viral late transcription, *Nucleic Acid Res.* **15:**7781–7793.

Barthelemy, I., Salas, M., and Mellado, R. P., 1987b, In vivo transcription of bacteriophage φ29 DNA. Transcription termination, *J. Virol.* **61:**1751–1755.

Bjornsti, M. A., Reilly, B. E., and Anderson, D. L., 1981, In vitro assembly of the *Bacillus subtilis* bacteriophage φ29, *Proc. Natl. Acad. Sci. USA* **78:**5861–5865.

Bjornsti, M. A., Reilly, B. E., and Anderson, D. L., 1983, Morphogenesis of bacteriophage φ29 of *Bacillus subtilis:* Oriented and quantized in vitro packaging of DNA-gp3, *J. Virol.* **45:**383–396.

Bjornsti, M. A., Reilly, B. E., and Anderson, D. L., 1984, Bacteriophage ф29 proteins required for *in vitro* DNA-gp3 packaging, *J. Virol.* **50:**766–772.

Blanco, L., and Salas, M., 1984, Characterization and purification of a phage ф29-encoded DNA polymerase required for the initiation of replication, *Proc. Natl. Acad. Sci. USA* **81:**5325–5329.

Blanco, L., and Salas, M., 1985a, Characterization of a 3′→5′ exonuclease activity in the phage ф29-encoded DNA polymerase, *Nucleic Acids Res.* **13:**1239–1249.

Blanco, L., and Salas, M., 1985b, Replication of phage ф29 DNA with purified terminal protein and DNA polymerase: Synthesis of full-length ф29 DNA, *Proc. Natl. Acad. Sci. USA* **82:**6404–6408.

Blanco, L., and Salas, M., 1986, Effect of aphidicolin and nucleotide analogs on the phage ф29 DNA polymerase, *Virology* **153:**179–187.

Blanco, L., García, J. A., Peñalva, M. A., and Salas, M., 1983, Factors involved in the initiation of phage ф29 DNA replication *in vitro*: Requirement of the gene 2 product for the formation of the protein p3-dAMP complex, *Nucleic Acids Res.* **11:**1309–1323.

Blanco, L., Prieto, I., Gutierrez, J., Bernaod, A., Lazar, M., Hermoso, M., and Salas, M., 1987, Effect of NH₄⁺ ions on ф29 DNA-protein p3 replication: Formation of a complex between the terminal protein and the DNA polymerase, *J. Virol.* **61:**3983–3991.

Blanco, L., García, J. A., and Salas, M., 1984, Cloning and expression of gene 2, required for the protein-primed initiation of the *Bacillus subtilis* phage ф29 DNA replication, *Gene* **29:**33–40.

Blanco, L., Gutiérrez, J., Lázaro, J. M., Bernad, A., and Salas, M., 1986, Replication of phage ф29 DNA *in vitro*: Role of the viral protein p6 in initiation and elongation, *Nucleic Acids Res.* **14:**4923–4937.

Carrascosa, J. L., Camacho, A., Moreno, F., Jiménez, F., Mellado, R. P., Viñuela, E., and Salas, M., 1976, *Bacillus subtilis* phage ф29: Characterization of gene products and functions, *Eur. J. Biochem.* **66:**229–241.

Daubert, S. D., and Bruening, G., 1984, Detection of genome-linked proteins of plants and animal viruses, *Methods Virol.* **8:**347–379.

Davis, T. N., and Cronan, E. J. Jr., 1983, Nonsense mutants of the lipid-containing bacteriophage PR4, *Virology* **126:**600–613.

Dobinson, K. F., and Spiegelman, G. B., 1985, Nucleotide sequence and transcription of a bacteriophage ф29 early promoter, *J. Biol. Chem.* **260:**5950–5955.

Escarmís, C., García, P., Méndez, E., López, R., Salas, M., and García, E., 1985, Inverted terminal repeats and terminal proteins of the genomes of pneumococcal phages, *Gene* **36:**341–348.

Escarmís, C., Gómez, A., García, E., Ronda, C., López, R., and Salas, M., 1984, Nucleotide sequence at the termini of the DNA of *Streptococcus pneumoniae* phage Cp-1, *Virology* **133:**166–171.

Escarmís, C., and Salas, M., 1981, Nucleotide sequence at the termini of the DNA of *Bacillus subtilis* phage ф29, *Proc. Natl. Acad. Sci. USA* **78:**1446–1450.

Escarmís, C., and Salas, M., 1982, Nucleotide sequence of the early genes 3 and 4 of bacteriophage ф29, *Nucleic Acids Res.* **10:**5785–5798.

Fučik, V., Grunow, E., Grünnerová, H., Hostomský, Z., and Zadražyl, S., 1980, New members of *Bacillus subtilis* phage group containing a protein link in their circular DNA, in: *DNA: Recombination, Interactions and Repair* (S. Zadražyl and J. Sponar, eds.), pp. 111–118, Pergamon, New York.

García, E., Gómez, A., Ronda, C., Escarmís, C., and López, R., 1983a, Pneumococcal bacteriophage Cp-1 contains a protein tightly bound to the 5′ termini of its DNA, *Virology* **128:**92–104.

García, J. A., Pastrana, R., Prieto, I., and Salas, M., 1983b, Cloning and expression in *Escherichia coli* of the gene coding for the protein linked to the ends of *Bacillus subtilis* phage ф29 DNA, *Gene* **21:**65–76.

García, J. A., Peñalva, M. A., Blanco, L., and Salas, M., 1984, Template requirements for the initiation of phage ф29 DNA replication *in vitro*, *Proc. Natl. Acad. Sci. USA* **81:**80–84.

García, P., Hermoso, J. M., García, J. A., García, E., López, E., and Salas, M., 1986a, Formation of a covalent complex between the terminal protein of pneumococcal bacteriophage Cp-1 and 5'-dAMP, *J. Virol.* **58**:31–35.

García, E., Ronda, C., García, P., and López, R., 1986b, Studies on the replication of bacteriophage Cp-1 DNA in *Streptococcus pneumoniae, Microbiologia* **2**:115–120.

Garvey, K. J., Yoshikawa, H., and Ito, J., 1985, The complete sequence of the *Bacillus* phage ϕ29 right early region, *Gene* **40**:301–309.

Geiduschek, E. P., and Ito, J., 1982, Regulatory mechanisms in the development of lytic bacteriophages in *Bacillus subtilis,* in: *The Molecular Biology of the Bacilli* Vol. 1 (D. A. Dubnau, ed.), pp. 203–245, Academic Press, New York.

Guo, P., Grimes, S., and Anderson, D. L., 1986, A defined system for *in vitro* packaging of DNA-gp3 of the *Bacillus subtilis* bacteriophage ϕ29, *Proc. Natl. Acad. Sci. USA* **83**:3505–3509.

Gutiérrez, J., García, J. A., Blanco, L., and Salas, M., 1986a, Cloning and template activity of the origins of replication of phage ϕ29 DNA, *Gene* **43**:1–11.

Gutiérrez, J., Vinós, J., Prieto, I., Méndez, E., Hermoso, J. M., and Salas, M., 1986b, Signals in the ϕ29 DNA–terminal protein template for the initiation of phage ϕ29 DNA replication, *Virology* **155**:474–483.

Hagen, E. W., Reilly, B. E., Tosi, M. E., and Anderson, D. L., 1976, Analysis of gene function of bacteriophage ϕ29 of *Bacillus subtilis:* Identification of cistrons essential for viral assembly, *J. Virol.* **19**:501–517.

Harding, N. E., and Ito, J., 1980, DNA replication of bacteriophage ϕ29: Characterization of the intermediates and location of the termini of replication, *Virology* **104**:323–338.

Henney, D. J., and Hoch, J. A., 1980, The *Bacillus subtilis* chromosome, *Microbiol. Rev.* **44**:57–82.

Hermoso, J. M., Méndez, E., Soriano, F., and Salas, M., 1985, Location of the serine residue involved in the linkage between the terminal protein and the DNA of ϕ29, *Nucleic Acids Res.* **13**:7715–7728.

Hermoso, J. M., and Salas, M., 1980, Protein p3 is linked to the DNA of phage ϕ29 through a phosphoester bond between serine and 5'-dAMP, *Proc. Natl. Acad. Sci. USA* **77**:6425–6428.

Herranz, L., Salas, M., and Carrascosa, J. L., 1986, Interaction of the bacteriophage ϕ29 connector protein with the viral DNA, *Virology* **155**:289–292.

Hirochika, H., and Sakaguchi, R., 1982, Analysis of linear plasmids isolated from *Streptomyces:* Association of protein with the ends of the plasmid DNA, *Plasmid* **7**:59–65.

Hirokawa, H., 1972, Transfecting deoxyribonucleic acid of *Bacillus* bacteriophage ϕ29, *Proc. Natl. Acad. Sci. USA* **69**:1555–1559.

Hirokawa, H., Matsumoto, K., and Ohashi, M., 1982, Replication of *Bacillus* small phage DNA, in: *Microbiology—1982* (D. Schlessinger, ed.), pp. 45–46, American Society for Microbiology, Washington.

Huberman, J. A., 1981, New views of the biochemistry of eukaryotic DNA replication revealed by aphidicolin, an unusual inhibitor of DNA polymerase, *Cell* **23**:647–648.

Inciarte, M. R., Salas, M., and Sogo, J. M., 1980, Structure of replicating DNA molecules of *Bacillus subtilis* bacteriophage ϕ29, *J. Virol.* **34**:187–199.

Ito, J., 1978, Bacteriophage ϕ29 terminal protein: Its association with the 5' termini of the ϕ29 genome, *J. Virol.* **28**:895–904.

Khan, N. W., Wright, G. E., Dudycz, L. W., and Brown, N. C., 1984, Butylphenyl dGTP: A selective and potent inhibitor of mammalian DNA polymerase alpha, *Nucleic Acids Res.* **12**:3695–3706.

Kemble, R. J., and Thompson, R. D., 1982, S1 and S2, the linear mitochondrial DNAs present in a male sterile line of maize, possess terminally attached proteins, *Nucleic Acids Res.* **10**:8181–8190.

Kikuchi, Y., Hirai, K., and Hishinuma, F., 1984, The yeast linear DNA killer plasmids pGLK1 and pGLK2, possess terminally attached proteins, *Nucleic Acids Res.* **12**:5685–5692.

Lechner, R. L., and Kelly, T. J. Jr., 1977, The structure of replicating adenovirus 2 DNA molecules, *Cell* **12**:1007–1020.

López, R., Ronda, C., García, P., Escarmís, C., and García, E., 1984, Restriction cleavage maps of the DNAs of *Streptococcus pneumoniae* bacteriophages containing protein covalently bound to their 5′ ends, *Mol. Gen. Genet.* **197**:67–74.

Matsumoto, K., Kim, C. I., Urano, S., Ohashi, H., and Hirokawa, H., 1986, Aphidicolin-resistant mutants of bacteriophage ϕ29: Genetic evidence for altered DNA polymerase, *Virology* **152**:32–38.

Matsumoto, K., Saito, T., and Hirokawa, H., 1983, *In vitro* initiation of bacteriophage ϕ29 and M2 DNA replication: Genes required for formation of a complex between the terminal protein and 5′dAMP, *Mol. Gen. Genet.* **191**:26–30.

Matsumoto, K., Saito, T., Kim, C. I., Ando, T., and Hirokawa, H., 1984, Bacteriophage ϕ29 DNA replication *in vitro*: Participation of the terminal protein and the gene 2 product in elongation, *Mol. Gen. Genet.* **196**:381–386.

McGraw, T., Yang, H. L., and Mindich, L., 1983, Establishment of a physical and genetic map for bacteriophage PRD1, *Mol. Gen. Genet.* **190**:237–244.

Mellado, R. P., and Salas, M., 1982, High level synthesis in *Escherichia coli* of the *Bacillus subtilis* phage ϕ29 proteins p3 and p4 under the control of phage lambda P_L promoter, *Nucleic Acids Res.* **10**:5773–5784.

Mellado, R. P., and Salas, M., 1983, Initiation of phage ϕ29 DNA replication by the terminal protein modified at the carboxyl end, *Nucleic Acids Res.* **11**:7397–7407.

Mellado, R. P., Moreno, F., Viñuela, E., Salas, M., Reilly, B. E., and Anderson, D. L., 1976, Genetic analysis of bacteriophage ϕ29 of *Bacillus subtilis:* Integration and mapping of reference mutants of two collections, *J. Virol.* **19**:495–500.

Mellado, R. P., Peñalva, M. A., Inciarte, M. R., and Salas, M., 1980, The protein covalently linked to the 5′ termini of the DNA of *Bacillus subtilis* phage ϕ29 is involved in the initiation of DNA replication, *Virology* **104**:84–96.

Mellado, R. P., Barthelemy, I., and Salas, M., 1986a, *In vivo* transcription of bacteriophage ϕ29 DNA early and late promoter sequences, *J. Mol. Biol.* **191**:191–197.

Mellado, R. P., Barthelemy, I., and Salas, M., 1986b, *In vitro* transcription of bacteriophage ϕ29 DNA. Correlation between *in vitro* and *in vivo* promoters, *Nucleic Acids Res.* **14**:4731–4741.

Mindich, L., Bamford, D., Goldthwaite, C., Laverty, M., and MacKenzie, G., 1982, Isolation of nonsense mutants of lipid-containing bacteriophage PRD1, *J. Virol.* **44**:1013–1020.

Moreno, F., Camacho, A., Viñuela, E., and Salas, M., 1974, Suppressor-sensitive mutants and genetic map of *Bacillus subtilis* bacteriophage ϕ29, *Virology* **62**:1–16.

Morrow, C. D., Hocko, J., Navab, M., and Dasgupta, A., 1984, ATP is required for initiation of poliovirus RNA synthesis *in vitro*: Demonstration of tyrosine-phosphate linkage between *in vitro*–synthesized RNA and genome-linked protein, *J. Virol.* **50**:515–523.

Ortín, J., Viñuela, E., Salas, M., and Vásquez, C., 1971, DNA-protein complex in circular DNA from phage ϕ29, *Nature New Biol.* **234**:275–277.

Pačes, V., Vlček, C., Urbánek, P., and Hostomský, Z., 1985, Nucleotide sequence of the major early region of *Bacillus subtilis* phage PZA, a close relative of ϕ29, *Gene* **38**:45–46.

Pačes, V., Vlček, C., and Urbánek, P., 1986a, Nucleotide sequence of the late region of *Bacillus subtilis* phage PZA, a close relative of phage ϕ29, *Gene* **44**:107–114.

Pačes, V., Vlček, C., Urbánek, P., and Hostomský, Z., 1986b, Nucleotide sequence of the right early region of *Bacillus subtilis* phage PZA completes the 19366-bp sequence of PZA genome. Comparison with the homologous sequence of phage ϕ29, *Gene* **44**:115–120.

Pastrana, R., Lázaro, J. M., Blanco, L., García, J. A., Méndez, E., and Salas, M., 1985, Overproduction and purification of protein p6 of *Bacillus subtilis* phage φ29: Role in the initiation of DNA replication, *Nucleic Acids Res.* **13**:3083–3100.

Peñalva, M. A., and Salas, M., 1982, Initiation of phage φ29 DNA replication *in vitro:* Formation of a covalent complex between the terminal protein, p3, and 5'-dAMP, *Proc. Natl. Acad. Sci. USA* **79**:5522–5526.

Prieto, I., Lázaro, J. M., García, J. A., Hermoso, J. M., and Salas, M., 1984, Purification in a functional form of the terminal protein of *Bacillus subtilis* phage φ29, *Proc. Natl. Acad. Sci. USA* **81**:1639–1643.

Prieto, I., Serrano, M., Lazaro, J. M., Salas, M., and Hermoso, J. M., 1988, Interaction of the bacteriophage φ29 protein p6 with double-stranded DNA, *Proc. Natl. Acad. Sci. U.S.A.* **85** (in press).

Rekosh, D. M. K., Russell, W. C., Bellett, A. J. D., and Robinson, A. J., 1977, Identification of a protein linked to the ends of adenovirus DNA, *Cell* **11**:283–295.

Ronda, C., López, R., Gómez, A., and García, E., 1983, Protease-sensitive transfection of *Streptococcus pneumoniae* with bacteriophage Cp-1 DNA, *J. Virol.* **48**:721–730.

Salas, M., 1983, A new mechanism for the initiation of replication of φ29 and adenovirus DNA: Priming by the terminal protein, *Curr. Top. Microbiol. Immunol.* **109**:89–106.

Salas, M., Mellado, R. P., Viñuela, E., and Sogo, J. M., 1978, Characterization of a protein covalently linked to the 5' termini of the DNA of *Bacillus subtilis* phage φ29, *J. Mol. Biol.* **119**:269–291.

Salas, M., Prieto, I., Gutiérrez, J., Blanco, L., Zaballos, A., Lázaro, J. M., Martin, G., Bernad, A., Garmendia, C., Mellado, R. P., Escarmís, C., and Hermoso, J. M., 1986, Replication of phage φ29 DNA primed by the terminal protein, in: *Mechanisms of DNA Replication and Recombination,* UCLA Symposia on Molecular and Cellular Biology, New Series (T. Kelly and R. McMacken (eds.), Vol. 47, Alan R. Liss, New York.

Savilahti, H., and Bamford, D. H., 1986, Linear φDNA replication: Inverted terminal repeats of five closely related *Escherichia coli* bacteriophages, *Gene* **49**:199–205.

Schachtele, C. F., De Sain, C. V., and Anderson, D. L., 1973, Transcription during the development of bacteriophage φ29: Definition of early and late ribonucleic acid, *J. Virol.* **11**:9–16.

Shih, M. F., Watabe, K., and Ito, J. 1982, *In vitro* complex formation between bacteriophage φ29 terminal protein and deoxynucleotide, *Biochem. Biophys. Res. Commun.* **105**:1031–1036.

Shih, M. F., Watabe, K., Yoshikawa, H., and Ito, J., 1984, Antibodies specific for the φ29 terminal protein inhibit the initiation of DNA replication *in vitro*, *Virology* **133**:56–64.

Sogo, J. M., Inciarte, M. R., Corral, J., Viñuela, E., and Salas, M., 1979, RNA polymerase binding sites and transcription map of the DNA of *Bacillus subtilis* phage φ29, *J. Mol. Biol.* **127**:411–436.

Sogo, J. M., García, J. A., Peñalva, M. A., and Salas, M., 1982, Structure of protein-containing replicative intermediates of *Bacillus subtilis* phage φ29 DNA, *Virology* **116**:1–18.

Stillman, B. W., 1983, The replication of adenovirus DNA with purified proteins, *Cell* **35**:7–9.

Talavera, A., Jiménez, F., Salas, M., and Viñuela, E., 1971, Temperature-sensitive mutants of bacteriophage φ29, *Virology* **46**:586–595.

Talavera, A., Salas, M., and Viñuela, E., 1972, Temperature-sensitive mutants affected in DNA synthesis in phage φ29 of *Bacillus subtilis*, *Eur. J. Biochem.* **31**:367–371.

Vartapetian, A. B., Koonin, E. V., Agol, V. I., and Bogdanov, A. A., 1984, Encephalomyocarditis virus RNA synthesis *in vitro* is protein-primed. *EMBO J* **3**:2593–2598.

Vlček, C., and Pačes, V., 1986, Nucleotide sequence of the late region of *Bacillus* phage φ29 completes the 19285-bp sequence of φ29 genome. Comparison with the homologous sequence of phage PZA, *Gene* **46**:215–225.

Watabe, K., and Ito, J., 1983, A novel DNA polymerase induced by *Bacillus subtilis* phage φ29, *Nucleic Acids Res.* **11**:8333–8342.

Watabe, K., Shih, M. F., Sugino, A., and Ito, J., 1982, *In vitro* replication of bacteriophage φ29 DNA, *Proc. Natl. Acad. Sci. USA* **79**:5245–5248.

Watabe, K., Shih, M. F., and Ito, J., 1983, Protein-primed initiation of phage φ29 DNA replication, *Proc. Natl. Acad. Sci. USA* **80**:4248–4252.

Watabe, K., Leusch, M., and Ito, J., 1984a, Replication of bacteriophage φ29 DNA in vitro: The roles of terminal protein and DNA polymerase, *Proc. Natl. Acad. Sci. USA* **81**:5374–5378.

Watabe, K., Leusch, M., and Ito, J., 1984b, A 3′ to 5′ exonuclease activity is associated with phage φ29 DNA polymerase, *Biochem. Biophys. Res. Commun.* **123**:1019–1026.

Yanofsky, S., Kawamura, F., and Ito, J., 1976, Thermolabile transfecting DNA from temperature-sensitive mutant of phage φ29, *Nature* **259**:60–63.

Yehle, C. O., 1978, Genome-linked protein associated with the 5′ termini of bacteriophage φ29 DNA, *J. Virol.* **27**:776–783.

Yoshikawa, H., and Ito, J., 1981, Terminal proteins and short inverted terminal repeats of the small *Bacillus* bacteriophage genomes, *Proc. Natl. Acad. Sci. USA* **78**:2596–2600.

Yoshikawa, H., and Ito, J., 1982, Nucleotide sequence of the major early region of bacteriophage φ29, *Gene* **17**:323–335.

Yoshikawa, H., Friedmann, T., and Ito, J., 1981, Nucleotide sequences at the termini of φ29 DNA, *Proc. Natl. Acad. Sci. USA* **78**:1336–1340.

Yoshikawa, H., Garvey, K. J., and Ito, J., 1985, Nucleotide sequence analysis of DNA replication origins of the small *Bacillus* bacteriophages: Evolutionary relationships, *Gene* **37**:125–130.

Zaballos, A., Salas, M., and Mellado, R. P., 1986, Initiation of phage φ29 DNA replication by deletion mutants at the carboxyl end of the terminal protein, *Gene* **43**:103–110.

Zaballos, A., Mellado, P., and Salas, M., 1988, Initiation of phage φ29 DNA replication by mutants with deletions at the amino end of the terminal protein, *Gene* (in press).

CHAPTER 6

Phage Mu

Rasika M. Harshey

I. INTRODUCTION

Phage Mu was discovered accidentally when an *E. coli* strain was being tested for phage P1 lysogeny (Taylor, 1963). The first observation Taylor made on the new temperate phage was its ability to cause mutations, and hence the name Mu for *mutator*. By a series of genetic linkage tests, Taylor confirmed that the mutations resulted from an insertion of the phage genome into the host genes. He drew parallels between the ability of Mu to occupy many chromosomal sites and to suppress the phenotypic expression of genes, and that of "controlling elements" postulated to move between many sites in the maize chromosome and suppress the function of some genes (McClintock, 1956). Taylor's insight proved prophetic.

Because of the unique properties of Mu, a large body of Mu research has been focused on the extraordinary capacity of Mu to cause mutations as well as a gamut of gene rearrangements (reviewed by Couturier, 1976; Bukhari, 1976; Toussaint and Resibois, 1983; Mizuuchi and Craigie, 1986). These properties of Mu have been exploited in mobilizing and isolating genes and studying gene structure and function in many bacteria (see Faelen and Toussaint, 1976; Toussaint and Resibois, 1983; Casadaban and Cohen, 1979; Casadaban and Chou, 1984). More recently, Mu has been used in studying bacterial colony organization (Shapiro, 1984). As a consequence of the emphasis of studies on Mu as a transposon, other areas of phage biology have received less attention. I will attempt here to summarize our knowledge of these other areas. A general description of Mu was provided by Howe and Bade (1975) and is now the subject of an entire book (Symonds *et al.*, 1987).

RASIKA M. HARSHEY • Department of Molecular Biology, Research Institute of Scripps Clinic, La Jolla, California 92037.

II. VIRION MORPHOLOGY

The Mu phage particle consists of an icosahedral head and a contractile tail, similar to that of T-even phages (Fig. 1). The head is 540 Å in
diameter, and the contractile tail sheath is 1000 Å long and 180 Å wide.
The tail terminates in a base plate from which spikes and fibers emanate.
There is a knoblike structure at the junction of the head and tail (To et
al., 1966; Admiraal and Mellema, 1976; Grundy and Howe, 1984; Grundy
and Howe, 1985). The density of the phage particles is 1.468 g/ml (To et
al., 1966). Electron microscopy of phage DNA and of disrupted particles
has revealed that the DNA (MW ~ 26 × 10^6; Torti et al., 1970; Martuscelli et al., 1971) is encapsulated within the head by a headful packaging mechanism (Bukhari and Taylor, 1975) with the right end of the
genome adjacent to the attached tail in the finished virion (Breepoel et al.,
1976; Inman et al., 1976; Bade et al., 1977; Howe et al., 1977).

III. PHYSICAL AND GENETIC STRUCTURE OF MU DNA

The DNA in the phage heads is double-stranded and linear, and it has
a base composition of 51% G+C and a density of 1.71–1.712 gm/ml in
cesium chloride (Torti et al., 1970; Martuscelli et al., 1971). The adenine
residues carry an unusual, sequence-specific modification (see Section
VIII). There is a nonuniform distribution of bases in the DNA. This allows the two strands to be separated from each other by the technique of
poly(UG) binding and equilibrium centrifugation (Bade, 1972; Chaconas
et al., 1983), originally used for phage λ and T7 (Szybalski et al., 1971).
When the DNA is partially denatured, the nonuniform distribution of
bases produces a characteristic pattern in the electron microscope, the
most striking feature being the clustering of AT-rich, readily denaturable
segments, close to the ends of phage DNA (Inman et al., 1976; Bade et al.,
1977). This property has proved useful in the physical analysis of DNA
containing Mu genomes (Harshey et al., 1982; Resibois et al., 1982).

When phage DNA is denatured and reannealed, two peculiar features
are revealed (Hsu and Davidson, 1972, 1974; Daniell et al., 1973a,b). One
class of renatured molecules is linear, double-stranded with nonrenatured
single-stranded split ends (SE) at one end of the molecule, and the other
class has in addition a nonrenaturing bubble—the G bubble (Fig. 2). The
segments of the reannealed DNA are referred to as follows: the long
double-stranded segment originates at the left end and is called α; the
short one at the right end is called β; the single-stranded segments between α and β constitute the G bubble; and the nonreannealing variable
ends are called the split ends (also sometimes called VE—variable ends).
The origin of the G bubble and the function of the G segment will be
discussed in section VII. The split ends vary in length from 500 to 3000
bp. The left end was shown to have the same feature except that the

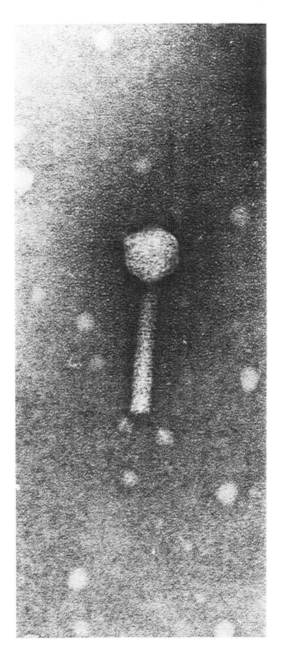

FIGURE 1. Electron micrograph of a Mu phage particle stained with uranyl acetate. The head is 540 Å in diameter, and the contractile tail 1000 Å long and 180 Å wide. Tail fibers can be seen emanating from the end of the tail.

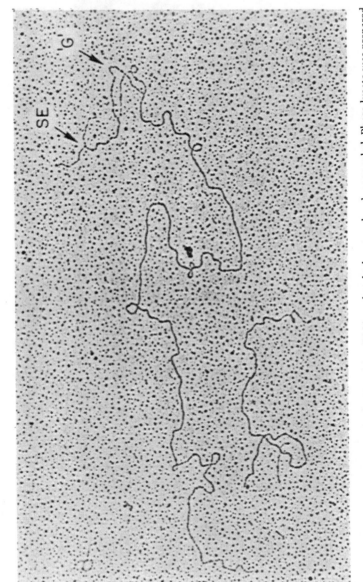

FIGURE 2. Electron micrograph of a Mu virion DNA molecule denatured and reannealed. Phage lysate was prepared by induction of a lysogen. The long double-stranded segment originates at the left end and is called α. It is followed by the nonreannealing G segment. The short double-stranded segment that follows is β. SE refers to the nonannealing variable "split ends."

length of the single strands in renatured molecules is too small (50–150 bp) to be seen in the electron microscope (Allet and Bukhari, 1975; George and Bukhari, 1981). These variable sequences at the left and right ends of Mu are derived from different segments of the host DNA (Daniell et al., 1975; Bukhari et al., 1976) and are acquired as a result of headful packaging of Mu DNA randomly integrated in the host genome (Bukhari and Taylor, 1975). These unusual physical features serve as landmarks on the Mu genome and have helped in physical mapping and orientation of Mu sequences (Daniell et al., 1973b; Waggoner et al., 1974).

A Mu prophage can be integrated in both possible orientations in a given operon (Boram and Abelson, 1973; Bukhari and Metlay, 1973; Howe, 1973a; Wijffelman et al., 1973; Zeldis et al., 1973). Zeldis et al. (1973) developed a technique for readily determining the orientation of any Mu prophage. Here one measures the direction of chromosome mobilization occurring as a result of recombination between a Mu prophage in the chromosome and a prophage on a F' episome. From a knowledge of this, as well as the orientation of either of these prophages, one can deduce the orientation of the second one. The Mu prophage is not inducible by UV light or other agents known to induce a λ prophage. Therefore one of the initial types of mutants isolated was that inducible by heat, the cts mutants (Howe, 1972; Waggoner et al., 1974). Since then, several hundred nonsense and temperature-sensitive mutants have identified 25 cistrons (see Howe et al., 1979a; Toussaint and Resibois, 1983). In addition to these, six semiessential or accessory functions have been identified. Mapping of Mu amber mutants during vegetative growth by two factor crosses resulted in a linear linkage map with genes A and S at opposite ends (Wijffelman et al., 1972, 1973; Couturier and Van Vliet, 1974). In such crosses recombination within Mu was found to be low (approxibately 1% between A and S) and completely dependent on the bacterial rec (recombination) function. Mu therefore appears not to encode an enzyme system that can efficiently catalyze this type of vegetative recombination. The genetic map of Mu has been established mostly by deletion mapping. These studies determined that the prophage map is the same for all Mu prophages and is the same as the vegetative map obtained from lytic crosses.

A system that has proved very useful in mapping Mu genes was first developed by Bukhari and Allet (1975; Allet and Bukhari, 1975). They heat-induced dilysogens in which Mu was inserted into the lac genes in λplac5cI857S7 and isolated λpMu phages from the population of rare plaque-forming phage that was produced. These hybrids carried deletions of Mu DNA. Subsequently several such λpMu's were isolated that carried deletions starting from the left or the right end of the Mu genome (Magazin et al., 1977), and later in vitro techniques were used for isolating λpMu's carrying DNA in the middle of the Mu genome (Moore et al., 1977). This collection of λpMu's covers the entire Mu genome and has been used to refine the genetic map of Mu (O'Day et al., 1979) (Fig. 3). To

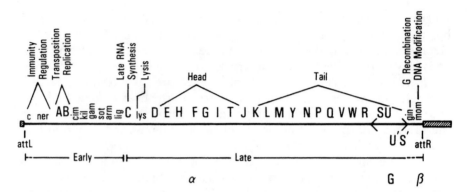

FIGURE 3. Schematic map of the Mu genome. The hatched boxes represent host sequences found at the ends of virion DNA. The attachment sites at the left (attL) and right (attR) ends are at the junction of the variable host and unique Mu sequences. Genes involved in various phage functions are indicated. "Early" and "Late" refer to the time in the phage lytic cycle during which the corresponding genes are expressed. Dashed lines denote genes known to be expressed during the lysogenic state.

date, the sequence of approximately 10 kb of DNA from the left end and 5 kb from the right end has been determined and can be found in the Mu book (Symonds et al., 1987).

A. The α Region

This region encompasses an area from the left end to the beginning of the G segment. The first gene at the left end is the immunity or repressor determinant, c. This end is therefore also called the immunity end. c and ner are the early regulatory functions involved in the lysis-lysogeny decision (Van Leerdam et al., 1982; Wijffelman et al., 1974). Both clear-plaque (c) and virulent (vir) mutants of Mu have been isolated (Howe, 1972; Van Vilet et al., 1978a). A and B are early functions involved in transposition and replication (Van de Putte and Gruijthuijsen, 1972; Toussaint and Faelen, 1973; Wijffelman and Lotterman, 1977; Faelen et al., 1978; O'Day et al., 1978). They are followed by six early semiessential or accessory functions—cim, kil, gam, sot, arm, and lig—identified on the basis of various phenotypes. cim, for control of immunity, was identified as a locus that abolished immunity when inactivated in phages lacking the Mu right end (Van de Putte et al., 1978). kil is responsible for killing even when Mu replication is blocked (Van de Putte et al., 1977a; Waggoner et al., 1984). Recent results suggest that cim and kil are most probably different phenotypes of the same gene (Goosen and Van de Putte, 1984). gam is a locus that complements the gam gene of phage λ (Van Vliet et al., 1978b). Unlike λ gam, which acts by binding to and inactivating ExoV (Karu et al., 1974), Mu gam acts by binding to DNA and shielding it against exonucleases (Akroyd et al., 1984). Sot, for stimulation of Mu transfection (Van de Putte *et al.*, 1977b) might be identical to *gam* (Akroyd *et al.*, 1984). *arm*, for amplified replication of *Mu*, is a locus that stimulates the replication of mini-Mu DNA (Waggoner *et al.*, 1981). Mu mutants that carry an insertion or deletion around this region make very small plaques (Goosen *et al.*, 1982). *lig* was identified as a gene that could substitute for *E. coli* and T4 ligase (Ghelardini *et al.*, 1979, 1980). It also complements the topoisomerase activity of T4*amG39* mutant (Ghelardini *et al.*, 1982). Isolation of *ts* mutants (Paolozzi *et al.*, 1980) and more recently amber mutants in this gene, which are conditional mutants for viral development, suggests that this is an essential function (Paolozzi and Ghelardini, 1986). Expression of *C* is required for transcription of the late genes, although the mechanism by which it positively regulates late transcription is not known (Van Meeteren, 1980). *lys* is required for host cell lysis (Faelen and Toussaint, 1973). Genes D through J are involved in head morphogenesis, and genes K through U are involved in tail morphogenesis (see Section IV.D and Toussaint and Resibois, 1983).

B. The G Segment

The S gene is found both in and out of the G segment. S, S', U, and U' are genes involved in synthesis of tail fibers (Howe *et al.*, 1979b; Grundy and Howe, 1984, 1985). The inversion of this segment controls the host range of the phage and will be discussed in Section VII.

C. The β Region

This region is also referred to as the right end. Two nonessential genes, *gin* and *mom*, are located here. *gin* is a recombination function required for G inversion (Chow *et al.*, 1977; Kamp *et al.*, 1978; see Section VII), and *mom* is responsible for modification of Mu DNA, which protects Mu from restriction (Allet and Bukhari, 1975; Toussaint, 1976; see Section VIII).

IV. LIFE CYCLE

A. Infection/Integration

Most studies with Mu have used *E. coli* K12 as the host, although Mu can infect a variety of gram-negative hosts (see Section VII). With *E. coli* K12 strains, some Mu-resistant mutants are also resistant to P1 and P2 (Howe, 1972). Resistance to P2 vir is now routinely used to select for Mu-resistant hosts. The rate of Mu adsorbtion to sensitive cells varies with the conditions. In medium containing 5–10 mM Ca^{2+} and Mg^{2+} adsorption is 80–95% complete within 15 min at 37°C (see Howe and Bade, 1975; for standard methods of growth, see Bukhari and Ljungquist, 1977). Transfection with Mu DNA is not possible, perhaps because the ends of the linear DNA, unlike those of λ or T4, are neither cohesive nor terminally redundant. A low transfection frequency of 10^{-7} per phage equivalent can be obtained with recBC strains (Kahmann et al., 1976). Transfection can be stimulated 100- to 1000-fold if a recBC sbcB host expresses the Mu sot function (Van de Putte et al., 1977b). Chase and Benzinger (1982) showed that if Mu particles are disrupted by freeze-thaw treatment and the DNA is purified on a CsCl gradient, a small class of highly infective DNA molecules can be recovered that transfect recBC strains 100-fold better than phenol-extracted DNA. The infective DNA is associated with a 65-kD protein probably present in the viron.

When Mu particles infect a sensitive bacterium, their DNA is injected right end first (Inman et al., 1976; Breepoel et al., 1976). Circular forms of infecting Mu DNA have been recovered from both sensitive and Mu-immune cells (Harshey and Bukhari, 1983). The circular infecting DNA still has the variable host sequences at the ends, which are held together by a 64-kD protein bound noncovalently to the ends (Fig. 4). The DNA protein complex is very efficient in the transfection of recBC spheroplasts and in vitro shows protection of Mu DNA ends by 5' and 3' exonucleases. Both Chase and Benzinger (1982) and Harshey and Bukhari (1983) inferred that in vivo the protein must provide at least one function in addition to protection against exonucleases. Perhaps the protein serves to bring the two Mu ends or att sites in close proximity so as to facilitate integration (transposition). Another possibility is that the protein blocks Mu replication, because, as we shall see later, the first integration event after Mu infection does not result in replication of Mu DNA. Circular forms of Mu DNA have also been seen after infection of minicells (Puspurs et al., 1983). These authors showed that a 64-kD protein is injected into the minicells along with Mu DNA. The Mu N gene has been shown to code for this protein (Gloor and Chaconas, 1986).

Infecting Mu may develop lytically to produce more phage, or it may form a lysogen, the major determinant of lysogeny being the repressor gene c. A unique feature of Mu DNA is that it is always found integrated in host DNA, whether in virions, when present as a prophage, or when replicating in the lytic phase of growth. As we shall see later, integration generates primers for Mu DNA replication. Upon infection, one sees two forms of integration—a high-frequency integration event which is observed within 5–10 min and leads to lytic growth, and a low-frequency event which results in the formation of stable lysogens and may occur over a long period of time (see Harshey, 1987). Genes A and B are required for the high-frequency integration (Chaconas et al., 1985a; see Toussaint and Resibois, 1983), whereas B mutants are only 10-fold reduced for lysogenization (O'Day et al., 1978). Lysogeny with Mu is not a very efficient process. It is estimated that in a single cycle of infection, the major-

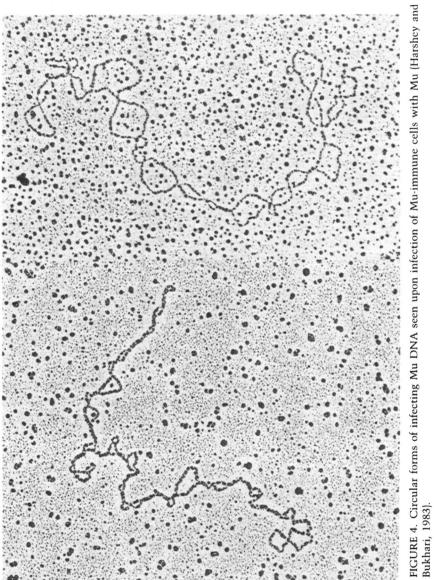

FIGURE 4. Circular forms of infecting Mu DNA seen upon infection of Mu-immune cells with Mu (Harshey and Bukhari, 1983).

ity of cells are killed with only 1–10% of the survivors being lysogens (see Howe and Bade, 1975). Isolation of drug-resistant derivatives of Mu has provided a convenient method for selection of lysogens (Leach and Symonds, 1979; Ross *et al.*, 1986).

Newly formed lysogens of a Mu *cts* prophage segretate Mu$^+$/Mu$^-$ clones, and it has not yet been distinguished if this is due to the *cts* mutation or a delay in the integration of potentially lysogenic Mu molecules, or whether newly integrated DNA excises at a high frequency if lytic growth does not ensue. Overnight incubation of phage-cell mixture can increase the proportion of lysogens to 100%. Multiple cycles of phage infection select for survival of lysogens that are immune to superinfection. Although most Mu lysogens contain only a single Mu prophage, polylysogenic strains containing two or more prophages at different sites are also found to occur. The frequency of such polylysogens is estimated to be high in a lysogenic population that is generated from infection at a high multiplicity (Howe and Bade, 1975). The variable host sequences present at the ends of infecting DNA are lost upon integration, since lysogens are found not to retain them (Hsu and Davidson, 1974; Allet and Bukhari, 1975; Allet, 1979; Kahmann and Kamp, 1979). Analysis of Mu lysogens has shown that Mu integration is associated with duplication of a 5-bp sequence originally present in the target DNA (Allet, 1979; Kahmann and Kamp, 1979). Such a duplication is characteristic of a transposable element and is presumed to arise by staggered single-stranded cuts within the target DNA during integration (Grindley and Sherratt, 1978). Although most Mu-induced mutation events cause no other alteration of target DNA, about 1–10% of mutations in a given gene are deletions (Howe and Zipser, 1974; Daniell *et al.*, 1972; Howe, personal communication).

Specific sequences at the ends of linear Mu DNA, called *att* sites, are used for Mu integration. Since linear insertion of Mu DNA, without a deletion of host sequences at the site of insertion, could best be imagined by the available Campbell model for phage λ integration (Campbell, 1971), one of the early problems that Mu workers grappled with was that of making Mu circles. Mu DNA ends had no feature that would allow them to circularize, and no covalently closed circles were detected after infection (Ljungquist and Bukhari, 1979). This paradox led to the seminal experiment by Ljungquist and Bukhari (1977; see next section) which showed that upon induction, a Mu prophage did not excise like λ but rather followed an integrate-replicate pathway, and therefore to the idea that Mu could integrate without having to make covalently closed circles. The models proposed for such integration involved making a replica of Mu DNA without the flanking host sequences, the ends of this replicated DNA then recombining with the target (Faelen *et al.*, 1975; Toussaint *et al.*, 1977).

Numerous other schemes, all involving replication of Mu followed by integration, were also proposed (see Bukhari, 1981). Another set of models invoked the idea of integration at Mu ends first, followed by replication (Shapiro, 1979; Harshey and Bukhari, 1981). All these models featured the known role of DNA replication in Mu transposition. However, Liebart *et al.* (1982) published a paper suggesting that during the first integration (transposition) event after infection, Mu may not replicate. Experiments have now confirmed that a high-frequency integration event takes place soon after infection (Chaconas *et al.*, 1983) and that Mu DNA does not replicate during this event (Harshey, 1984a). Details of the molecular mechanism of this integration are as yet unknown, and one of the current problems involves understanding why Mu DNA does not replicate during this integration. The features of infecting DNA that might block replication include (1) physical structure of the DNA, (2) 64-kd protein at the DNA ends, and (3) differential expression of replication-transposition functions. A study of the participating proteins and Mu DNA forms should help elucidate the mechanism of this nonreplicative DNA transposition.

B. Lysis–Lysogeny Decision

Very little is known about the regulation of Mu genes and the influence of the host on this regulation that allows the phage to develop lysogenically or lytically. Expression of the repressor gene *c* is required for establishment of lysogeny. *In vitro* transcription has shown that in contrast to the rest of the genome, the repressor gene is transcribed leftward on the 1-strand (Van Meeteren *et al.*, 1980). Fusions of the repressor promoter region to the structural *galK* gene and S1 nuclease mapping have indicated that the repressor is expressed from two promoters—P_{c-1} and P_{c-2} (Goosen *et al.*, 1984). P_{c-2} is much stronger than P_{c-1} and overlaps P_e. Two longer repressor transcripts have been observed *in vivo* in a Mu lysogen (Krause and Higgins, 1986). For Mu to be a successful lysogen, it must transcribe the early region without initiating DNA replication and transposition. The repressor binds to the promoter-operator region of the early operon, thus blocking its expression (Fig. 5). It binds specifically and cooperatively to three sites designated O_1, O_2, and O_3 between the repressor and *ner* genes (Craigie *et al.*, 1984; Krause and Higgins, 1984). A consensus repressor binding sequence covers 14 bp (Krause and Higgins, 1986). The number of such sequences per binding site correlates with binding strength. O_1 and O_2 are stronger binding sites than O_3. Binding to O_2 prevents transcription of the early replication transposition genes from the overlapping promoter P_e (Goosen *et al.*, 1984; Krause *et al.*, 1983). Repressor gene expression is autoregulated (Giphart-Gassler *et al.*, 1981b). This might be achieved by competition between repressor and RNA polymerase for binding O_3, since repressor bound at O_1 or O_2 does not impede leftward transcription (Krause and Higgins, 1986). Transcription of the repressor is also negatively regulated

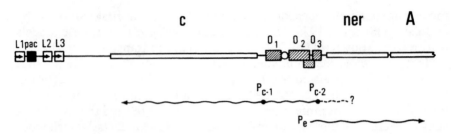

FIGURE 5. Regulatory region at the left end of the Mu genome. L1, L2, and L3 represent transposase (A) binding sites at attL. Between L1 and L2 lies the *pac* sequence containing signals for packaging of Mu DNA into virions. O_1, O_2, and O_3 are operator sites bound by the repressor c. The stippled box represents the DNA-binding site for Ner which overlaps O_2 and O_3. The circle between O_1 and O_2 contains a site bound by *E. coli* IHF. c and A can bind each others' sites *in vitro*. P_{c-1} and P_{c-2} are promoters for c transcription. ? represents start of larger transcripts found in a lysogen. P_e is a strong promoter at which early transcription initiates. Lysis-lysogeny decision is apparently controlled by the impact on transcription of the regulatory proteins competing for the various DNA-binding sites.

by the early gene *ner* (Van Leerdam *et al.*, 1982; Goosen and Van de Putte, 1984), whose binding site overlaps the repressor binding sites O_2 and O_3 (Goosen and Van de Putte, 1984; Tolias and DuBow, 1986). The Ner protein may serve a role analogous to the λ Cro protein (Ptashne *et al.*, 1980; Van Leerdam *et al.*, 1982).

The decision between lytic and lysogenic development is also controlled by two *E. coli* proteins. Some mutants of DNA gyrase and integration host factor (IHF; product of genes *himA* and *himD*) restrict the ability of Mu to develop lytically (Miller and Friedman, 1980; DiNardo *et al.*, 1982; Ghelardini *et al.*, 1984; Friedman *et al.*, 1984; Ross *et al.*, 1986). Compensating mutations in the phage that restore the ability for lytic growth in the host *himA* or *himD* mutants map in the regulatory region between the repressor and *ner* genes (Yoshida *et al.*, 1982; Goosen and Van de Putte, 1984) and suggest that the host proteins may act by regulating early transcription (Goosen *et al.*, 1984). Recent DNA footprinting experiments with IHF suggest that IHF binds to DNA between O_1 and O_2 and modulates Mu transcription on supercoiled DNA at both the opposing promoters P_c and P_{c-2} (Fig. 5) so as to favor the transcript from P_e over the transcript from P_{c-2}. In contradiction to these *in vitro* results, *in vivo* experiments using fusions of these promoters to the *galK* gene show that both transcripts are stimulated by IHF (Goosen *et al.*, 1984). As in phage λ, IHF may be involved in the lysis–lysogeny decision at more than one point. Thus Mu and other phages may have evolved a mechanism to gauge the physiological state of the cell and use it to decide whether to develop lytically or stay quiescent. It is not clear if the effect of the gyrase mutations are indirect, for example, by causing an alteration in the level of DNA supercoiling (see Gellert, 1981), or whether they reflect a direct role for DNA gyrase as an active component of Mu replication.

C. Lytic Growth

Mu infection of sensitive strains or heat induction of lysogens carrying a *ts* mutation in the repressor gene initiates lytic development which proceeds with a similar time course in both cases. At 42°C, phage production begins around 25 min, lysis is complete around 55–60 min and the burst size is 150–200 (Howe, 1972). The time course of Mu development varies subject to growth conditions and is particularly sensitive to cell concentration; therefore, the times may be increased or decreased in different experiments. A small number of the phage particles are generalized transducing particles (Howe, 1973b) which contain exclusively host DNA fragments (Teifel and Schmieger, 1981). The ratio of plaque-forming particles to total particles has not been determined. Mu lysates (with the exception of clear plaque mutants) are notoriously unstable, losing 90% of their titer in 1–2 weeks. Some investigators use 1 mM $CdCl_2$ to stabilize lysates (see Pato and Reich, 1982). For standard methods of cultivation and storage of Mu, see Bukhari and Ljungquist (1977).

1. Early Events

Early lytic transcription initiates at P_c (Fig. 5) and proceeds rightward to make a transcript that is 8.5 kb long (J. Engler, unpublished). Early transcription can be detected within a few minutes of infection or induction (Wijffelman and Van de Putte, 1974). The first early gene is *ner*. The *ner* gene product negatively regulates both early transcription and repressor transcription (see Goosen and Van de Putte, 1984). Two *ner* binding sites overlap P_c and P_{c-2} (Fig. 5; Goosen and Van de Putte, 1984; Tolias and DuBow, 1986). *ner* may play a role analogous to the λ cro protein in initiating lytic development by repressing the expression of Mu repressor (Van Leerdam *et al.*, 1982). Recently *ner* was shown to be an essential function, since *ner⁻* mutants could form plaques only when complemented for Ner (Goosen and Van de Putte, 1986). Mutation in the repressor protein did not abolish the need for Ner. However, when transcription of the repressor gene *c* was blocked, Ner was no longer essential for normal Mu development. The results suggest that transcription of the repressor gene blocks lytic transposition. This might occur by RNA polymerase molecules traversing the left attachment site (*attL*) of Mu, interfering with the action of Mu transposase at *attL*. Mu transposition (integration) can occur with or without Mu DNA replication (see Harshey, 1987). A and B genes are required for Mu replication-transposition both *in vivo* and *in vitro* (Wijffelman and Lotterman, 1977; Faelen *et al.*, 1978; Harshey, 1983; Mizuuchi, 1983). They are also required for the non-replicative integration of infecting Mu (O'Day *et al.*, 1978; Chaconas *et al.*, 1985a). Most mutants in gene *A* are completely deficient for integration. Mutations that map in the carboxyterminal region of *A*, however, are defective in replicative lytic transposition and in the high-frequency nonreplicative transposition observed early after Mu infection, but they show normal levels of lysogenization and low-frequency nonreplicative transposition (Harshey and Cuneo, 1986). Phages defective in *B* have a reduced level of lysogenization. Except for two amber mutants at the carboxyl terminus of B (deletion group 9.5; O'Day *et al.*, 1979), B mutants are defective in both replicative and nonrepliative transposition. The former two mutants are proficient only in the high-frequency, non-replicative transposition following Mu infection (Chaconas *et al.*, 1985a), suggesting a role for the carboxyl terminus of B in Mu DNA replication. A protein is viewed as the integrase or transposase, and B protein as a function that enhances transposition. It might well be, however, that A and B form the subunits of an active transposase.

The activity of A is regulated at more than one level. For example, the amino terminus of the A protein has considerable sequence homology with the amino terminus of the repressor (Harshey *et al.*, 1985). The two proteins also share binding sites on the DNA—i.e., bind both Mu *att*L and *att*R as well as the Mu operator, although with different affinities (Craigie *et al.*, 1984; see Fig. 5). *In vitro*, transposition can be inhibited by excess repressor. It is possible that *in vivo* the two proteins may interact in subtle ways to regulate transposition. The repressor may, for instance, remain bound to the *att* sites in the lysogenic state so that accidental transcription from P_e and expression of *A* might prevent A from acting on the ends. Alternatively, the two proteins may interact to form inactive heterooligomers.

A protein synthesis must also be regulated posttranscriptionally, since although both A and B proteins are translated from the same message, the downstream B protein is synthesized 20-fold more than A (Magazin *et al.*, 1978; Giphart-Gassler *et al.*, 1981b). Recent *in vitro* experiments suggest that this could be achieved by A-protein binding to its mRNA, thus blocking translation (Harshey, unpublished). A protein is unstable and active for just one round of replication-transposition; i.e., it is used stoichiometrically (Pato and Reich, 1982, 1984). The biochemical basis for this mechanism is not known. Increasing the levels of *A* causes an inhibition of *A*-activity (assayed by scoring for precise excision of Mu sequences inserted in *Lac*), suggesting that the phage uses more than one mechanism for regulating the activity and hence the mutagenic potential of *A* (Harshey and Cuneo, 1986). As discussed earlier, the function of the other early genes in the life cycle of Mu are not well understood, although they serve to enhance Mu replication. Fifteen minutes after Mu induction, transcription of gene *C* can be detected, the product of which positively regulates the transcription of late genes (Van Meeteren, 1980; Marrs, 1982). The precise role of C in regulating late transcription is not yet known. The need for the host RNA polymerase throughout Mu development (Toussaint and Lecocq, 1974) and the small size of this protein

(15 kD; Giphart-Gassler *et al.*, 1981a) make it unlikely that C is a new phage-encoded RNA polymerase. Based on amino acid homologies to other sigma factors, suggestions have been made that C may be a possible phage-specific sigma factor or a λ cII-like DNA-binding protein that activates transcription from Mu late promoters (Hattman *et al.*, 1985; Margolin and Howe, 1986).

2. Replication

Besides phage-encoded A and B and some accessory proteins in the early region, Mu replication is largely dependent on host proteins. *dnaA* and PolI mutants show a substantial effect on Mu replication (McBeth and Taylor, 1982, 1983). dnaB, dnaC, dnaE, dnaG, and gyrB products are required for Mu replication (Resibois *et al.*, 1984). Some experiments have suggested that Mu prefers to integrate near host replication forks (Fitts and Taylor, 1980; Paolozzi *et al.*, 1978, 1979), although more recent experiments have ruled out such a requirement (Nakai and Taylor, 1985). *In vitro* experiments have demonstrated a role for *E. coli* HU protein at an early step in Mu transposition (Craigie *et al.*, 1985). That replication and transposition are coupled processes became evident from observations that mutants defective in one event were defective in the other. Ljungquist and Bukhari (1977) induced a Mu prophage in parallel with a λ prophage and showed that unlike λ, Mu does not excise upon induction but remains integrated in the host DNA, while Mu replicas are made and integrated into new host sites. No "free" replicating Mu DNA was observed, suggesting strongly that replication and integration are coupled. Mu DNA synthesis is initiated 6–8 min after induction or infection (Waggoner *et al.*, 1977; Wijffelman and Lotterman, 1977). The bacterial nucleoid remains intact (Pato and Waggoner, 1981), and there is no amplification of host sequences adjacent to either end of the original prophage (Waggoner and Pato, 1978). Host DNA synthesis progressively decreases as Mu synthesis increases (Schroeder *et al.*, 1974; Pato and Reich, 1982). About 15–20 min after induction of the lytic cycle, rare, covalently closed supercoils of different sizes begin to appear. They all contain at least one copy of Mu (Waggoner *et al.*, 1974, 1977) and are products of replication-transposition (see Section V).

Electron-microscopic studies of DNA structures induced during Mu replication show that circular forms of DNA with tails called "key" structures are the most abundant species of DNA structures induced (Schroeder *et al.*, 1974; Harshey and Bukhari, 1981). The lengths of the circles and tails in the "key" structures vary and they have been shown to contain replicating and nonreplicating Mu DNA linked to host sequences (Harshey *et al.*, 1982; Resibois *et al.*, 1982). Structures conforming to "Shapiro" intermediates, where the two ends of Mu integrate in a concerted manner, generating potential replication forks at both ends, are rarely observed during *in vivo* replication but have been shown to occur during *in vitro* replication (Craigie and Mizuuchi, 1985). The absence of such Shapiro structures and the observation that "key" structures contain replicative Mu DNA led to the proposition that key structures were replication-transposition intermediates and that integration was asymmetric in that one end integrated first, followed by replication (Harshey and Bukhari, 1981; Harshey *et al.*, 1982). The subsequent demonstration that *in vitro* transposition proceeds via a Shapiro intermediate might mean that during DNA extraction from whole cells, such an intermediate may be particularly susceptible to breakage at one end, giving rise to a "key" structure. Why this would be so *in vivo* and not *in vitro* is not clear.

On the basis of hybridization of Okazaki fragments to separated Mu DNA strands, both *in vivo* and *in vitro* experiments suggest that the direction of Mu replication is predominantly from left to right on the Mu genome (Wijffelman and Van de Putte, 1977; Higgins *et al.*, 1983; Resibois *et al.*, 1984). Electron microscopy experiments suggest that replication can also be initiated at the right end, although the true frequency of this event cannot be assessed (Harshey *et al.*, 1982). Analysis of the first round of Mu replication after synchronization has also shown that although replication proceeds predominantly from the left end, a significant amount of initiation occurs from the right end (Reich *et al.*, 1984). Studies with mini-Mu's (internally deleted Mu phages) have shown that replication can start from either the left or the right end with equal efficiency (Harshey *et al.*, 1982; Resibois *et al.*, 1982, 1984). It is not clear whether efficient replication from the right end is a peculiar property of mini-Mu's. Recent experiments with maxi-Mu's (whose length ranged from 39.8 to 88.2 kb) indicate that transposition frequency decreases as the length of the prophage increases (Faelen *et al.*, 1986). Replication of the larger maxi-Mu's could not be detected. Electron microscopy experiments have indicated that Mu integrations tend to be clustered (Harshey *et al.*, 1982; Resibois *et al.*, 1984). *In vitro* experiments have shown that Mu DNA synthesis is semidiscontinuous and very similar to host chromosomal replication in many aspects (Higgins *et al.*, 1983). Not only does Mu require many host replication proteins, but the size and distribution of nascent Mu DNA are similar to those of host DNA. *In vitro* experiments confirm that after the initial strand exchange reactions during Mu integration are carried out by Mu proteins A and B, replication of Mu can be completed by the host proteins (Craigie and Mizuuchi, 1985). Thus *in vivo*, Mu must recruit the bacterial replisome for its own purpose. For Mu replication-transposition to occur, both Mu ends are required, correctly oriented and in cis (Chaconas *et al.*, 1981b; Howe and Schumm, 1980; Mizuuchi, 1983). Mu replication initiates around 6–8 min after induction, large-scale replication of Mu can be detected at 20–30 min, and packaging of Mu DNA into virions begins around 25 min (see Waggoner *et al.*, 1977).

4. Morphogenesis and DNA Packaging

Late RNA synthesis starts about 25 min after Mu induction or infection, rises sharply thereafter, and proceeds left to right (Wijffelman et al., 1974). Late proteins have been identified by analysis of minicells infected with amber mutants of Mu defective in various genes (Magazin et al., 1978; Giphart-Gassler et al., 1981a; Puspurs et al., 1983). Analysis of virion proteins has identified F and T as being major head proteins, L and Y as major tail proteins, and N, S, and U as minor tail proteins (Giphart-Gassler et al., 1981a; Shore and Howe, 1982). In vitro complementation assays for reconstitution of infective phage particles have helped assign most Mu late genes into two functional groups. Genes D, E, I, and J have been assigned to one group, and K, L, M, N, P, Q, R, and S to a second group (Giphart-Gassler et al., 1981a). Mutants defective in genes belonging to the same group do not complement. However, if the defects are in genes belonging to different groups, mixing two defective lysates produces viable phage whose genotype is that of the member of the second group. Thus the first group was categorized as possible head-defective mutants, since they can still serve as tail donors, and the second group as potential tail-defective mutants, because they can produce functional heads.

A recent electron microscopic study of extracts from cells growing Mu amber phages (Grundy and Howe, 1985) has revealed the following:

1. Mutations in lys, F, G, some H mutations, S, and U did not cause a visible change in the structure of the phage particle. The lys mutant particles were functionally normal, but the others in this group were defective for phage growth. By analogy with other phage systems, F, G, and H might be required for other stages such as DNA packaging, injection, DNA protection, replication, or DNA expression. G may be involved in tail morphogenesis (or joining of heads and tails), since Giphart-Gassler et al. (1981a) found that extracts defective in G function poorly as head donors and not at all as tail donors. Although F is a virion protein, its precise function is not yet known. The role of the H gene is also not clear, since two phenotypes are observed for H amber mutants. Promoter-proximal H mutants accumulate proheads, while promoter-distal ones make complete but defective phage particles. Genes S, U, S', and U' are involved in the determination of Mu host range and are required for the production of functional phage tail fibers (Van de Putte et al., 1980; Giphart-Gassler et al., 1982; Grundy and Howe, 1984). When G inversion is blocked, S and U mutants not only lack tail fibers but have contracted tails (Howe et al., 1979b; Grundy and Howe, 1984). The reason S and U mutants produced normal particles in the experiments by Grundy and Howe (1985) was because the G segment could invert, thus

expressing S' and U' which produced alternate tail fibers whose attachment could prevent tail sheath contraction.

2. Mutants defective in L, M, Y, N, P, Q, V, W, and R contained only head structures which appeared normal.

3. K defectives accumulated longer-than-normal free heads as well as free tails. The K gene product is needed for the production of normal length tails (Admiraal and Mellema, 1976). Electron microscopic analysis of polysheath structure from K amber mutants has led to a model for tail sheath protein organization (Admiraal and Mellema, 1976; Cremers et al., 1977).

Results 2 and 3 are similar to those obtained for λ, in which none of the tail-defective mutants other than those in Z or U produce tail-related structures (Mount et al., 1968; Kemp et al., 1968). Like the K defectives, long tails are produced by λ mutants lacking gene U. Both U and Z are believed to act late in λ tail assembly to produce tails that are competent for head attachment (Katsura, 1983). The Mu K protein may share a function analogous to λ U which terminates tail growth at the normal length. Grundy and Howe (1985) suggest that the unusual association of contracted tail sheath with the baseplate in K⁻ tails may implicate the K product in functioning at the head-proximal end of the tail to stabilize the sheath-core association, as does the gene 3 product of phage T4 (King, 1971; Berget and King, 1983). Thus, with the exception of K and L, the precise roles for M, N, P, Q, R, Y, V, and W genes in tail morphogenesis is undefined.

4. T-defective and some I-defective extracts had only tails, which appeared normal. T mutants are defective in the major head protein (Shore and Howe, 1982) and, as expected, lack all head-related structures. I mutants exhibited two phenotypes. Those at the left end of I accumulated small, thick-shelled, proheadlike structures similar to T7 proheads (Serwer, 1976; Roeder and Sadowski, 1977), whereas mutants at the right end produced no detectable head-related structures. By analogy to other phage systems (see reviews by Casjens and King, 1975; Murialdo and Becker, 1978; Wood and King, 1979; Earnshaw and Casjens, 1980; Black and Showe, 1983; Feiss and Becker, 1983; Georgopoulos et al., 1983). Grundy and Howe (1985) suggest that I may produce a scaffolding protein that is completely defective in prohead assembly in some mutants and blocked in prohead scaffolding protein loss in others (scaffolded proheads often lose the core protein as a result of protein cleavage). Low levels of complementation are observed between I mutants located at opposite ends of the gene (Howe et al., 1979a), suggesting either the existence of different proteins or a single protein whose different domains carry out distinct functional roles. In λ, Nu3 and C are overlapping genes involved in head morphogenesis, with Nu3 product acting as a scaffolding protein (Shaw and Murialdo, 1980). A similar overlap of analogous genes

in Mu might explain why centrally located *I* mutations cannot complement mutations at either end of the gene while those at the opposite ends do complement. *I*4037 mutants (at the left end of *I*) do not make significant amounts of mature Mu length DNA.

5. Free tails and empty heads accumulated in D-, E-, and some I- and H-defective extracts. These heads were smaller than normal heads. By analogy to other viral systems, these mutants may be defective in DNA processing and packaging. Consistent with this hypothesis, *E* mutants do not produce significant amounts of mature Mu-length DNA. *D* and *E* may act after *I* in the morphogenetic pathway.

6. Defects in *I* resulted in accumulation of unattached tails and full heads. This gene may therefore be involved in head-tail joining.

Mu morphogenesis appears to proceed via a branched system whereby heads and tails are synthesized independently and then join to form a complete particle. The tail fibers of Mu join with the tail independently of head attachment, in contrast to the tail fibers of T4 (King and Wood, 1969). Genes controlling the biosynthesis and assembly of head and tail structures are grouped according to function on the genetic map of Mu. Exceptions are *G*, which appears to be more defective in tail than in head synthesis (Giphart-Gassler *et al.*, 1981a), and possibly *F*, whose function is not identified.

Bukhari and Taylor (1975) proposed that Mu DNA is packaged by a headful mechanism, proceeding from left to right and terminating when the head is full. Because the head can accommodate approximately 40 kb of DNA and the Mu genome is only 37 kb long (see Marrs and Howe, 1983), packaging proceeds into the host sequences generating the variable host sequences at the Mu right end. This mechanism was tested by measuring the right end host sequences in phages that carry insertions or deletions of DNA, and it was shown that as predicted, less or more host sequences are packaged in the virion. The variable left end sequences must be generated by a mechanism that measures double helical turns in the DNA, since the length of these sequences varies by multiples of 11 bp, with a minimum length of 56 bp and a maximum of about 144 bp (George and Bukhari, 1981). A similar study of the variable right end has not been carried out. A site essential for DNA packaging has recently been identified between nucleotides 35 and 58 from the left end and lies between A-protein binding sites L1 and L2 (see Fig. 5; Groenen and Van de Putte, 1985).

V. TRANSPOSITION

This topic has been extensively reviewed (see Symonds *et al.*, 1987), so I will summarize here only some of the salient features. Mu transposition is accompanied by a variety of gene rearrangements, summarized in

Fig. 6. The most important and striking property of Mu is the frequency of transposition (Fig. 6.1). There are as many as 100 (or more) transposition events per cell. Compare this to other transposable elements that at best undergo 10^{-2} to 10^{-3} events per cell (for reviews see Kleckner, 1981; Grindley and Reed, 1985). The end products of Mu transposition are both nonreplicative simple insertions (Fig. 6.2), produced almost exclusively during the first integration event after Mu infection (Chaconas et al., 1983; Liebart et al., 1982; Harshey, 1984a, 1987), and cointegrates produced exclusively during the lytic phase of growth (Fig. 6.3; Faelen et al., 1971; Chaconas et al., 1981b). In the absence of B, as well as during

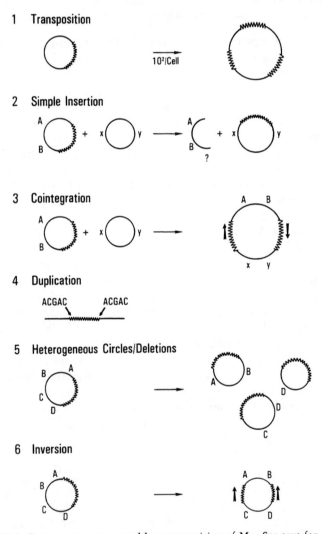

FIGURE 6. Rearrangements caused by transposition of Mu. See text for details.

lysogenisation, low-frequency transposition events can be observed. Both simple insertions and cointegrates are also the end products of these reactions. Both events generate a 5-bp duplication at the site of insertion (Fig. 6.4; Allet, 1979; Kahmann and Kamp, 1979; Szatmari *et al.*, 1986). Mu integrates into random target sites (Taylor, 1963). Bukhari and Zipser (1972) calculated that the specific target recognition site could be no more than 2–3 nucleotides.

During lytic growth one observes the formation of heterogeneous circles containing Mu linked to host sequences (Fig. 6.5; Waggoner *et al.*, 1977). Upon partial induction of a Mu prophage, host DNA deletions starting from one or the other end are observed (Faelen and Toussaint, 1978). Observation of a change in the orientation of DNA transfer in the bacteria upon Mu induction suggested that Mu causes inversions of DNA as well (Fig. 6.6; Faelen and Toussaint, 1980a). These were shown to be bracketed by two Mu prophages in opposite orientation. Many models were put forward to explain transposition and attendant DNA rearrangements (for reviews see Bukhari, 1981; Grindley and Reed, 1985). The prototype of the model that has been shown from *in vitro* experiments to be largely correct was put forward by Shapiro (1979) and is shown in Fig. 7. It explains how inversions are produced as the result of an intramolecular cointegration event, as are deletions. The heterogeneous circles would be the result of such an intramolecular deletion event. Recent results suggest that Mu transposition may be biased toward an inversion-type event (Pato and Reich, 1985). The replicating key structures observed during electron microscopy (Harshey *et al.*, 1982) could be explained by invoking a break in the chain at one end of the integrated Mu DNA, perhaps during DNA extraction (see Fig. 7.II). All these properties

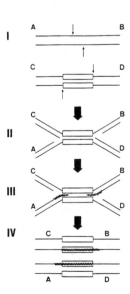

FIGURE 7. Replicative transposition model proposed by J. Shapiro. Open bars show parental transposon DNA, and wavy lines show newly synthesized DNA. Letters A through D represent genetic markers. If the two DNA molecules shown in I are circular species, the end product of this scheme (IV) is a cointegrate or replicon fusion. Site-specific recombination within the two transposon copies would resolve the cointegrate to generate simple insertions.

have been exploited exhaustively to isolate and study gene structure and function (see Toussaint and Resibois, 1983; Casadaban and Cohen, 1979; Van Gijsegem et al., 1987).

An in vitro cell-free system for Mu transposition was established by Mizuuchi (1983). The reaction requires a negatively supercoiled donor DNA carrying the two Mu ends in their proper relative orientation, A and B proteins and host factors. The requirement for a negatively supercoiled substrate with the ends of Mu correctly oriented can be eliminated provided the topology of the DNA energetically favors the same configuration of the pair of ends as is favored in the negatively supercoiled molecule. This suggests that the relative orientation of the Mu ends is needed only to energetically favor a particular configuration that the ends must adopt in a synaptic complex (Craigie and Mizuuchi, 1986). RNA synthesis by E. coli RNA polymerase is not required for the reaction. Both simple insertions and cointegrates are generated. The simple inserts do not seem to undergo a full round of DNA replication. An analysis of the distribution of newly synthesized DNA strands in these transposition products suggests that 3' ends of Mu DNA are joined to 5' protruding staggered ends of the target DNA (Mizuuchi, 1984). The resulting 3' -OH in the flanking target DNA is used as a primer for DNA replication into Mu. Under conditions that do not permit efficient initiation of DNA replication, a transposition intermediate with a structure predicted by Shapiro (1979) is the major product (Craigie and Mizuuchi, 1985). Mu proteins A and B, E. coli protein HU, ATP, and a divalent cation are required for the efficient formation of this intermediate (Craigie et al., 1985).

In a two-step reaction, the intermediate was isolated, stripped of bound proteins, and shown to produce either simple insertions or cointegrates in the presence of E. coli proteins alone (Craigie and Mizuuchi, 1985). Craigie and Mizuuchi suggest that both simple insertions and cointegrates are generated from a common Shapiro intermediate as proposed by Ohtsubo et al. (1980). The mechanism for generation of simple insertions in infecting Mu, however, is yet unknown. There must be something special about infecting DNA that blocks cointegration and channels the DNA exclusively to produce simple insertions (see Section IV.A). There is an alternate mechanism involving double-strand break and integration that can be invoked to explain simple insertions; it is depicted in Fig. 8.

The Mu A protein carries out the site-specific recognition of the Mu ends (Craigie et al., 1984). It binds to three sites on the left end and three on the right end. A consensus binding sequence covers 22 bp. This site-specific DNA binding property has been localized to the amino terminus of the protein (Nakayama et al., 1987). Mutations at the carboxyl terminus of the A protein suggest a role for this region in protein-protein interactions, especially with the B protein (Harshey and Cuneo, 1986). The A protein must also contain the enzymatic (nicking and strand exchange) activity, since transposition can occur in the absence of B, al-

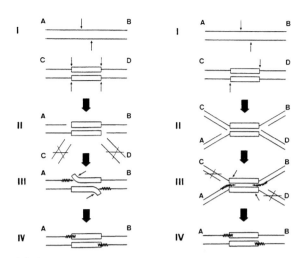

FIGURE 8. Models for conservative transposition. Symbols, as in Fig. 7. (Left panel) Double-stranded excised intermediate. (I) Double-stranded breaks occur at the ends of a transposon and a pair of staggered single-stranded nicks in the target DNA. (II) Extended single strands of the target DNA are joined to the transposon ends. Donor sequences flanking the transposon are destroyed. (III) Single-stranded gaps are repaired by primer extension from the target DNA, and the displaced transposon single strands are broken (perhaps by DNA polymerase 1). (IV) Ligation of newly synthesized DNA to transposon sequences completes transposition. (Right panel) Shapiro intermediate. (I) As in (left panel), except single-stranded nicks occur at the ends of the transposon. (II) As in (left panel), except a four-stranded Shapiro intermediate is generated. (III) As in (left panel). Donor sequences flanking the transposon are destroyed. (IV) As in (left panel).

though at a low frequency. Indeed, very recently it has been observed that in the presence of *E. coli* protein HU and Mg^{2+}, the A protein and Mu DNA ends form a stable complex (type I, Surette *et al.*, 1987). The 3' ends of the left and right Mu end sequences in this complex are nicked (Craigie and Mizuuchi, 1987). Conversion into a type II complex (transposition intermediate) then occurs with addition of target DNA, B protein, ATP, and Mg^{2+} (Surette *et al.*, 1987).

The role of the B protein is still not clear. It increases the overall efficiency of the transposition reaction. *In vitro*, it has an ATPase activity which is stimulated in the presence of A protein and DNA (Maxwell *et al.*, 1987), although ATP hydrolysis is not required for formation of the transposition intermediate (Harshey, unpublished). *In vivo* experiments show that in the absence of *B*, transposition events are preferentially intramolecular (Coelho *et al.*, 1982; Harshey, 1984b, 1987), as do *in vitro* experiments (Maxwell *et al.*, 1987), B binds DNA nonspecifically (Chaconas *et al.*, 1985b), and a putative DNA binding sequence has been identified on its amino-terminal region (Miller *et al.*, 1984). ATP stimulates B protein binding to DNA (Harshey, unpublished). The carboxyterminal region of the protein plays a key role in replicative transposition, since mutants in this region can integrate efficiently upon Mu infection to produce simple insertions but are blocked in subsequent replicative transposition (Chaconas *et al.*, 1985a). Thus the B protein must play a role in bringing distant DNA into the transposition complex as well as in sequestering the bacterial replisome onto Mu DNA. The HU protein has recently been shown to organize *E. coli* DNA into nucleosomelike structures (Broyles and Pettijohn, 1986) and may play a similar role in Mu transposition.

VI. EXCISION

Mu-induced mutations are stable (reversal frequency is less than 10^{-10}; Taylor, 1963; Bukhari and Taylor, 1971), completely polar (Jordan *et al.*, 1968; Toussaint, 1969; Boram and Abelson, 1971), and nonleaky (Daniell and Abelson, 1973). Bukhari (1975) discovered that under special circumstances one could detect a precise reversal (excision) of Mu integration. Precise and imprecise excision were later shown to be characteristic of transposable elements (see Kleckner, 1981). Mu excision, unlike that of Tn*10* (see Kleckner, 1981), requires *A* and a *rec*$^+$ host. Excision was first observed in MuX mutants (Bukhari, 1975). These can be readily obtained by plating Mu *cts* lysogens at 42–44°C, at which temperature most of the lysogenic cells are killed because of Mu induction (survival frequency is approximately 10^{-4}). A major class of survivors carries the X mutations, which are strong polar insertion mutations of IS elements in the *B* gene (Bukhari, 1975; Bukhari and Taylor, 1975; Bukhari and Froschauer, 1978; Chow and Bukhari, 1978; De Bruijn and Bukhari, 1978; Engler and Van Bree, 1981; Baker *et al.*, 1983). X mutants are unable to replicate their DNA or to express their morphogenetic functions. Excision of Mu *cts* X from the *lacZ* gene of *E. coli* has been monitored. It requires intact Mu ends (Bukhari, 1975; Khatoon and Bukhari, 1981) and is a low-frequency event (10^{-6}), with imprecise excision being 10–100 times more frequent than precise excision. Imprecise excision of Mu from *lacZ* was detected as relief of polarity on the distal *lacY* gene (Mel$^+$ phenotype). Mel$^+$ revertants exhibited four main classes of DNA rearrangements (based on Southern blots): (1) a predominant class that had no detectable Mu DNA left. They were presumed to retain the 5-bp duplication originally present at the prophage-host junction; these could further revert to *lacZ*$^+$; (2) a class that in addition to loss of Mu DNA had deletions that extended generally but not always to only one side of the prophage; (3) a class that still had Mu DNA but had deletions in the *lacZ* gene; and (4) a class that had Mu DNA but no deletions in *Z* and could revert further to *lacZ*$^+$, indicating that the *lacY* gene must have been turned on by rearrangements within Mu DNA. The behavior of Mu in excision seems more akin to that of transposable elements in plants in its requirement for a functional transposase (see Saedler and Nevers, 1985). Unlike the plant elements, however, there is no evidence that the excised element can reintegrate (Bukhari, 1975).

To study Mu DNA excision at the nucleotide level, Nag and Berg

(1987) determined the DNA sequences of products of excision of a Mu-derived phage from sites in the transcribed (but not translated) leader region of a tetracycline (*tet*) resistance gene. Most of these (17/21) were due to simple deletions, usually involving only the Mu prophage but in some cases extending into adjacent host sequences. Four excisants were more complex: each contained one transposed and inverted Mu end as well as a deletion of most of Mu plus some adjacent host sequences. The deletion breakpoints in most excisants occurred in short direct repeats, and one pair of repeats in the A protein binding sequence at the *att* sites was a deletion hot spot.

Precise excision has been proposed to be a reaction occurring during chromosome replication (Egner and Berg, 1981). In this view the element does not excise from its parent site, but rather a daughter missing the element is formed during DNA synthesis when the replication fork and its associated polymerase skip past the sequences of the transposon on the parental strand. The pairing of the long inverted repeats in large stem transposons might "blind" the replication fork to the presence of the transposon on the parental chromosome strand. Since Mu ends do not contain such long inverted repeats (Kahmann and Kamp, 1979), the A protein may be required to help the ends pair (Bukhari, 1975; Khatoon and Bukhari, 1981). It is not known if the enzymatic (nicking and strand exchange) activity of A is required for excision. Excision has also been described as an abortive transposition event (Bukhari, 1975; Harshey and Bukhari, 1981), and, interestingly, some of the excisants isolated by Nag and Berg appear to fit that definition.

VII. G INVERSION

The G segment is an invertible 3-kb region near the variable end of Mu DNA originally detected as a nonrenaturing loop in the electron microscope (Daniell *et al.*, 1973b; Hsu and Davidson, 1974). Daniell *et al.* (1973a) found that Mu DNA from phage grown by infection of *E. coli* K12 had the G segment primarily (99%) in one orientation (+), whereas DNA from phage grown by induction contained the G segment in both orientations (+ and −; 25–75% of each). They suggested that one orientation, G(+), might be preferred for lytic growth by infection. Deletion mutants of Mu missing various portions of G and β segments were found to be locked in the G(+) orientation and still viable (Chow *et al.*, 1977). For one of these mutants that could have the G segment locked in either the G(+) or G(−) orientation, lysogens of both produced phage particles, but only G(+) lysogens produced phages able to plaque on *E. coli* K12 (Kamp *et al.*, 1978). Complementation studies for inversion of the G segment revealed that a specific function, *gin* (*G in*version), was required for inversion and located in the β DNA segment adjacent to G. *gin* is expressed during the prophage state, and the frequency of inversion is such that during the growth of a lysogen in a culture of 10^8 cells/ml, the G(+)/G(−) ratio is close to 1. This accounts for the observation that induction of lysogens produces phage particles with equal numbers of + and − orientations of G (Symonds and Coelho, 1978).

Bukhari and Ambrosio (1978) found that in a mixed lysate of G(+) and G(−) particles, only the G(+) particles were efficient in adsorbing to *E. coli* K12, suggesting that the G region might encode functions involved in phage adsorption. The observation that genes *S* and *U* were located in the G segment (Howe *et al.*, 1979b) and that the map order of *S* and *U* relative to *R* (see Fig. 3) was inverted in strains deleted for portions of the G segment, suggested that inversion of G from *R−S−U* to *R−U−S* in plaque forming G(+) phage might result in inviable G(−) phages. Howe *et al.* (1979b) suggested that the promoter for *S* expression might be located in the noninverting α DNA segment or that the *S* gene might span the α− G boundary. All these observations led to the hypothesis that the G segment might encode two different host ranges, with inversion switching expression from one type to the other (see Howe, 1978, 1980).

That the host G segment determines the host specificity of Mu was demonstrated by Van de Putte *et al.* (1980), who showed that G(−) phages can grow on some isolates of *E. coli*, *Citrobacter freundii*, and *Shigella sonnei*. They proposed that the G segment encodes two sets of genes, *S− U* and *S'−U'* (Fig. 9). *S* and *U* are needed for adsorption to *E. coli* K12, and *S'* and *U'* serve analogous functions for adsorption to other strains. Giphart-Gassler *et al.* (1982) showed that the *S* and *S'* genes are located partially outside the G segment and share a common or constant (*Sc*) segment in the α region. This constant segment is joined by G inversion to the variable segment *Sv* or *S'v*, resulting in the synthesis of tail fiber

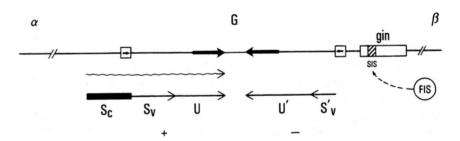

FIGURE 9. Components of the genetic switch controlling host-range specificity of phage Mu. A site-specific recombinase, Gin, acts on inverted repeats (boxed arrows) flanking the G segment to cause inversion of the segment. A *cis*-acting site, "sis," located within the coding region of Gin, along with an *E. coli* factor FIS, acts to enhance Gin-mediated recombination. The dark arrows within G are longer inverted repeats containing parts of the coding information of genes U and U'. Both sets of S and U genes are involved in tail-fiber biosynthesis. Transcription initiates in the noninverting α segment and spans the α−G boundary. In the + orientation of G, only S and U genes are expressed. When G is inverted and in the − orientation, the S gene gets spliced to express a new gene which shares its unique or constant N-terminal region with S (Sc) but contains a different or variable C-terminal segment (S'v). Thus S and U are expressed in the + state and S'U' in the − state. The + and − state determine two different sets of host-range specificities.

proteins S or S'. Grundy and Howe (1984) demonstrated that S and U are required for G(+) tail fiber biosynthesis and/or attachment. In the light of these results, it is understandable that host-range switching by Gin should be carried out at a low level, thus avoiding the phenotypic mixing of the two tail fiber genes (Symonds, 1982). Gin expression may be constitutive to ensure a constant presence of both types of lysogens, G(+) and G(−), in a cell population. G inversion is the first example of gene splicing by an inversion event and the first example of gene spicing within the prokaryotic cell. A phenomenom such as the assembly of immunoglobulin genes (Joho et al., 1983) appears to occur during the antigenic variation of gonococcal pili (Segal et al., 1986). An interesting observation appears to be that mutants of Mu that make slightly larger or smaller S proteins were found during the screening of amber mutants of Mu unrelated to S, indicating that like the somatic mutations in immunoglobulin genes, changes in S might occur with a rather high frequency (Van de Putte et al., 1980).

Besides E. coli K12, a his− strain of Salmonella typhi (Soberon et al. (1986) and a strain of Serratia marcescens (HY) have been found to plaque Mu G(+) (Harshey, unpublished). Strains sensitive to Mu G(−) include Enterobacter cloacae and some species of Erwinia besides those mentioned above (Kamp, 1981; Faelen et al., 1981). Recent studies have identified a lipopolysaccharide cell wall component as a receptor for Mu particles (Sandulache et al., 1984, 1985). Mu G(+) and G(−) particles differ, however, in their requirement for the type of glycosidic linkage of the terminal glucose residue in the lipopolysaccharide, which is $\alpha 1,2$ for G(+) and $\beta 1,6$ for G(−).

The G segment of Mu belongs to a family of closely related invertible segments. The first such homologue to be discovered was the C segment of phage P1 and P7 (Chow and Bukhari, 1976; Yun and Vapnek, 1977). Although Mu and P1 phages are very different, they share host-range specificity (Toussaint et al., 1978). Complementation studies have shown that the H segment of Salmonella typhimurium that determines variation of flagellar antigens and the P segment of a defective E. coli prophage e14, whose function is unknown, are closely related to the G and C segments of Mu and P1 (see Plasterk et al., 1983a). The products of the invertase genes in all these elements complement each other and share more than 60% homology at the amino acid level.

The analysis of the inversion mechanism has recently become possible with the development of in vitro assays for inversion (Plasterk et al., 1984a; Mertens et al., 1984). Recombination takes place between two identical, 34-pb-long sites which flank the G segment as inverted repeats. If the recombining sequences are artificially arranged as direct repeats, Gin-mediated deletion occurs at a 100-fold lower frequency than inversions between substrates arranged as inverted repeats (Plasterk et al., 1983b). A site in the β region, called sis (site for inversion stimulation), overlapping the coding sequence of gin, is required in cis for efficient inversion (Kahmann et al., 1985b). For full activity, purified Gin protein must be supplemented with a host factor termed *FIS* (*factor for iversion stimulation*) from *E. coli* K12, which acts on the site *sis* (Kahmann *et al.*, unpublished). The effect of *sis* is independent of orientation and distance from the inverted repeats and is reminiscent of enhancer elements involved in regulating eukaryotic gene expression (Khoury and Gruss, 1983). Gin and FIS together possess a sequence specific topoisomerase activity which changes the linking number of the inversion substrate in steps of 4 (Kahmann *et al.*, unpublished; Van de Putte *et al.*, unpublished).

VIII. THE *mom* GENE

The *mom* gene encodes a DNA modification function (reviewed by Kahmann, 1984). The gene is intriguing in two respects. First, the regulation of its expression follows an elaborate scheme without precedence in prokaryotes, and second, the modification introduced in DNA is not methylation.

The *mom* gene was first discovered as providing protection against restriction systems such as those elaborated by phage P1 and *E. coli* strains B and K (Allet and Bukhari, 1975; Toussaint, 1976; Toussaint *et al.*, 1980). Phage grown by lytic infection are 10- to 50-fold more sensitive to such restriction than those grown by lytic induction. Mom expression is dependent on the host *dam* gene (Toussaint, 1977; Khatoon and Bukhari, 1978). Mom-specific modification acts in *trans* to modify even cellular DNA sequences (Toussaint, 1976). The modification protects all *Hga*I, *Pvu*II, and *Sal*I sites from cleavage and allows only partial cleavage products to be formed with a host of other enzymes (see Kahmann, 1984). A comparison of the sequence specificity of all these enzymes allowed the derivation of a consensus sequence

$$5'-\begin{smallmatrix}C\\G\end{smallmatrix} A \begin{smallmatrix}G\\C\end{smallmatrix} NPy-3'$$

for mom modification. The modification is $N^6-(1\text{-acetamido})$-adenine (Swinton *et al.*, 1983) and alters about 15% of the adenine residues in phage DNA (Hattman, 1979, 1980). The modification is expressed as a "late" phage function and is deleterious to the host cell (Hattman and Ives, 1984; Kahmann *et al.*, 1985a).

The *mom* gene maps to the right of *gin* in the β region (Toussaint *et al.*, 1980; Kwoh *et al.*, 1980). The organization of this region is shown in Fig. 10. The region has been sequenced (Kahmann, 1983) and contains two open reading frames. The first of these codes for a 7.4-kD peptide and is called *com* (control of *mom*). The second one, *mom*, codes for a 28.3-kD protein. *mom* is transcribed separately from *gin*. Recent excitement in the regulation of *mom* comes from experiments that showed a link

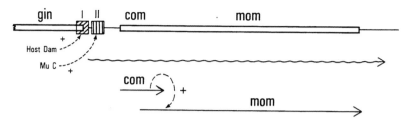

FIGURE 10. The *mom* region. Two DNA sites, I and II, are acted upon by the host Dam function and the Mu C function, respectively, to positively regulate transcription of the *mom* region. *com* codes for a small protein which positively regulates the expression of *mom* at a step after transcription. Expression of *mom* is lethal to the cells and occurs late in the lytic cycle.

between methylation of specific sequences upstream of the *mom* gene (3 GATC sites 5' to the promoter in region I) and its transcription (Kahmann, 1983; Hattman *et al.*, 1983; Plasterk *et al.*, 1983c). In the absence of the *E. coli dam* gene, there is 20-fold reduction in *mom* transcription (Hattman, 1982).

These results sparked speculation that secondary structure in the promoter region could be influenced by methylation which in turn could influence transcription from that promoter (Plasterk *et al.*, 1983c; Kahmann, 1983). Removal of the 5' GATC sequences makes *mom* expression *dam*-independent (Hattman and Ives, 1984; Kahmann, 1984; Plasterk *et al.*, 1983c, 1984b). This implies that whatever the role of the methylated DNA, it is not indispensable for transcription initiation. Besides *dam*, *mom* expression requires a second transactivation function encoded by Mu (Chaconas *et al.*, 1981a) and shown now to be *C* (Plasterk *et al.*, 1983c; Hattman *et al.*, 1985; Margolin and Howe, 1986). The site at which *C* acts is close to but distinct from the site of *dam* action (region II; Kahmann, 1984). This site is required for transcription of *mom*, and the need for transactivation, like that of *dam* methylation, is alleviated when *mom* is fused to a foreign promoter. In addition to this complex regulation, the *com* gene product is required for efficient expression of *mom* and is proposed to act as a positive regulator of *mom* expression at a posttranscriptional stage (Kahmann *et al.*, 1985b).

The function of the methylation by Dam in *E. coli* is not clear. It is known that Dam methylase affects different processes such as resistance to UV, spontaneous induction of phage lambda, and DNA mismatch correction; that double mutants of *dam* and *polA* or *recA* are nonviable; and that many Dam-methylated sequences are found in the origin of replication of *E. coli* (Marinus and Morris, 1975; Brooks *et al.*, 1983; Arraj and Marinus, 1983). The only well-established function for the Dam methylase is that it enables the mismatch correction enzymes to recognize the newly synthesized DNA strand (Wagner and Meselson, 1976). This fact, in combination with the observation that Mom is only expressed after replication of phage DNA, led to the hypothesis that the function of methylation dependence of Mom expression would be to make it depend on hemimethylated DNA and therefore on active phage replication. A recent finding (Kahmann, unpublished) that expression of *mom* is affected by *mutH*, a component of the methyl-directed mismatch repair system of *E. coli*, suggests a role for *dam* dependence of *mom* expression. *mom* expression in a *mutH⁻* strain becomes *dam*-independent. It is suggested that MutH protein functions as a repressor by binding to region I when the three GATC sites are hemi- or unmethylated. Methylation by *dam* would therefore serve to block the binding of MutH and allow *mom* expression.

High levels of mom modification are lethal to the host. Therefore it most certainly must shut down cellular functions. As mentioned before, *mom* is expressed "late" in the lytic cycle. Late expression is achieved in many ways. The dependence on *C* links *mom* expression to replication and therefore to lytic development. The late expression of *mom* might also be linked to its requirement for *dam* methylation, since during the early phase of replication it is most likely to be repressed by *mutH*. The requirement for *com* must in addition delay the onset of *mom* expression.

The biological role of the mom function is not fully clear. Although G(+) phage overcome host-controlled restriction, not a single host has been found in which the modification of G(−) phage improves their plating efficiency (Kahmann, 1984).

IX. MU-LIKE PHAGES

To date three Mu-like phages have been isolated, two of them from *Pseudomonas* (Krylov *et al.*, 1980a,b). These two generalized transducing phages, called D3112 and B3, have a genome size similar to Mu and variable sequences at their ends. D3112 has no mutagenic effect upon lysogenization of bacterial cells, suggesting that it might exist as a plasmid or be integrated into nonessential genes. Upon induction of *cts* phage lysogens, however, many different mutants could be found among the survivors. The two phages D3112 and B3 show regions of nonhomology at the right ends of their genomes. A third phage, D108, which has begun to receive attention lately, was first isolated by Mise (1971) as a generalized transducing phage and is closely related to Mu (Hull *et al.*, 1978). Heteroduplex analysis has shown D108 to be 90% homologous to Mu at the DNA level (Gill *et al.*, 1981), and the two phages behave similarly in catalyzing transposition and gene rearrangements (Faelen and Toussant, 1980b; Szatmari *et al.*, 1986).

There are three regions of nonhomology between the two phages. The first of these is the early region spanning the repressor, *ner* and beginning of the *A* gene. In agreement with the region of nonhomology detected at the immunity end, the two phages are heteroimmune; i.e., Mu grows in bacteria lysogenic for D108, and *vice versa*. The primary structures of the two repressors are similar toward the carboxyterminal halves but less so toward the amino-terminal halves (Mizuuchi *et al.*, 1986). A possible amino-terminal DNA-binding sequence has been identified for the Mu repressor (Harshey *et al.*, 1985) in a region that has considerable homology with the D108 repressor (Mizuuchi *et al.*, 1986). The primary structures of the Ner proteins are even more similar than the repressor proteins, although significant divergence has occurred (Tolias and Du-Bow, 1985; Mizuuchi *et al.*, 1986). The Ner proteins do not bind to each others' DNA-binding sites (Tolias and Dubow, 1985, 1986).

Differences in the manner of binding to their DNA sites have also been observed. The A proteins show some divergence in the first 65 amino acids (Toussaint *et al.*, 1983) but can function interchangeably to promote transposition (Craigie *et al.*, 1984). The repressors and A proteins from the two phages share homology at their amino-termini as well as share binding sites on DNA (Craigie *et al.*, 1984; Harshey *et al.*, 1985; Mizuuchi *et al.*, 1986). Suggestions have been made that the homologies at the amino-termini of the repressor and transposase act to control lysis-lysogeny decision by competing for DNA binding sites and/or by protein-protein interactions to form inactive heterooligomers. The second region of nonhomology between Mu and D108 is contained entirely within the third open reading frame after the *B* gene (Waggoner and Pato, unpublished), and the third nonhomologous region between the two phages resides in the β region close to right end, which is 0.5 kb longer in D108. The *mom* gene of D108 shows all the characteristics of Mu *mom* (Kahmann, 1984). The correspondence between the location of genes and sites in the control regions of D108 and Mu is reminiscent of the lambdoid family of phages (Hershey and Dove, 1971).

X. PERSPECTIVES

Mu is a unique phage in having so many fascinating features packed into so small a genome. At the left end, for example, are the variable host sequences that show an intriguing pattern of packaging, a repressor protein that shows structural and functional homology with the transposase, a regulatory system that has built-in elaborate sensing devices to control lysis-lysogeny decisions, and transposition proteins that catalyze two kinds of transposition with unprecedented frequencies. At the right end, there is the genetic switch that controls host range by a unique mechanism of gene splicing, a site-specific recombination system with an intriguing enhancer site and host factor, and a mysterious modification function with an intricate and uncommon control system. The two ends make up a fifth of the Mu genome. We have not even begun to tap the rest.

ACKNOWLEDGMENTS. I thank Martha Howe, Barbara Waggoner, Martin Pato, and Nora Goosen for useful comments on this review and Mina Finch for excellent secretarial assistance. My laboratory is supported by grants from the National Institutes of Health.

REFERENCES

Admiraal, G., and Mellema, J. E., 1976, The structure of the contractile sheath of bacteriophage Mu, *J. Ultrastruct. Res.* **56**:49.

Akroyd, J., Barton, B., Lund, P., Smith, S. M., Sultana, K., and Symonds, N., 1984, Mapping and properties of the *gam* and *sot* genes of phage Mu: Their possible roles in recombination, *Cold Spring Harbor Symp. Quant. Biol.* **49**:261.

Allet, B., 1979, Mu insertion duplicates a 5 base pairs sequence at the host insertion site, *Cell* **16**:123.

Allet, B., and Bukhari, A. I., 1975, Analysis of bacteriophage Mu and lambda-Mu hybrid DNAs by specific endonucleases, *J. Mol. Biol.* **92**:529.

Arraj, J. A., and Marinus, M. G., 1983, Phenotypic reversion in *dam* mutants of *E. coli* K12 by a recombinant plasmid containing the *dam*+ gene, *J. Bacteriol.* **153**:562.

Bade, E. G., 1972, Asymmetric transcription of bacteriophage Mu-1, *Virology* **10**:1205.

Bade, E. G., Delius, H., and Allet, B., 1977, Structure and packaging of Mu DNA, in *DNA Insertion Elements, Plasmids and Episomes* (A. I. Bukhari, J. A. Shapiro, and S. L. Adhya, eds.), pp. 315–318, Cold Spring Harbor Laboratory, Cold Spring Harbor, NY.

Baker, T. A., Howe, M. M., and Gross, C. A., 1983, MudX, a derivative of MudI (lacApr) which makes stable *lacZ* fusions at high temperature, *J. Bacteriol.* **156**:970.

Berget, P. B., and King, J., 1983, T4 tail morphogensis, in *Bacteriophage T4* (C. K. Matthews, E. M. Kutter, G. Mosig, and P. B. Berget, eds.), pp. 246–258, American Society for Microbiology, Washington.

Black, L. W., and Showe, M. K., 1983, Morphogenesis of the T4 head, in *Bacteriophage T4* (C. K. Matthews, E. M. Kutter, G. Mosig, and P. B. Berget, eds.), pp. 219–245, American Society for Microbiology, Washington.

Boram, W., and Abelson, J., 1971, Bacteriophage Mu integration: On the mechanism of Mu induced mutations, *J. Mol. Biol.* **62**:171.

Boram, W., and Abelson, J., 1973, Bacteriophage Mu integration: On the orientation of the prophage, *Virology* **54**:102.

Breepoel, H., Hoogendorp, J., Mellema, J. E., and Wijffelman, C., 1976, Linkage of the variable ends of the bacteriophage Mu DNA to the tail, *Virology* **74**:279.

Brooks, J. E., Blumenthal, R. M., and Gingeras, R., 1983, The isolation and characterization of the *E. coli* DNA adenine methylase (*dam*) gene, *Nucleic Acids Res.* **11**:837.

Broyles, S. S., and Pettijohn, D. E., 1986, Interaction of the *Escherichia coli* HU protein with DNA. Evidence for formation of nucleosome-like structures with altered DNA helical pitch, *J. Mol. Biol.* **187**:47.

Bukhari, A. I., 1975, Reversal of mutator phage Mu integration, *J. Mol. Biol.* **96**:87.

Bukhari, A. I., 1976, Bacteriophage Mu as a transposition element, *Annu. Rev. Genet.* **10**:389.

Bukhari, A. I., 1981, Models of DNA transposition, *Trends Biochem. Sci.* **6**:56.

Bukhari, A. I., and Allet, B., 1975, Plaque forming lambda-Mu hybrids, *Virology* **63**:30.

Bukhari, A. I., and Ambrosio, L., 1978, The invertible segment of bacteriophage Mu DNA determines the adsorption properties of Mu particles, *Nature* **271**:575.

Bukhari, A. I., and Froschauer, S., 1978, Insertion of a transposon for chloramphenicol resistance into bacteriophage Mu, *Gene* **3**:303.

Bukhari, A., I., and Ljungquist, E., 1977, Bacteriophage Mu: Methods for cultivation and use, in *DNA Insertion Elements, Plasmids and Episomes* (A. I. Bukhari, J. A. Shapiro, and S. L. Adhya, eds.), pp. 749–756, Cold Spring Harbor Laboratory, Cold Spring Harbor, NY.

Bukhari, A. I., and Metlay, M., 1973, Genetic mapping of prophage Mu, *Virology* **54**:109.

Bukhari, A. I., and Taylor, A. L., 1971, Genetic analysis of diaminopimelic acid and lysine requiring mutants of *E. coli*, *J. Bacteriol.* **105**:844.

Bukhari, A. I., and Taylor, A. L., 1975, Influence of insertions on packaging of host sequences covalently linked to bacteriophage Mu DNA, *Proc. Natl. Acad. Sci. USA* **72**:4399.

Bukhari, A. I., and Zipser, D., 1972, Random insertion of Mu-1 DNA within a single gene, *Nature New Biol.* **236:**240.

Bukhari, A. I., Froschauer, S., and Botchan, M., 1976, The ends of bacteriophage Mu DNA, *Nature* **264:**580.

Campbell, A., 1971, Genetic structure, in *The Bacteriophage* λ (A. D. Hershey, ed.), pp. 13–44, Cold Spring Harbor Laboratory, Cold Spring Harbor, NY.

Casadaban, M. J., and Cohen, S. N., 1979, Lactose genes fused to exogenous promoters in one step using a Mu-*lac* bacteriophage: *In vivo* probe for transcriptional control sequences, *Proc. Natl. Acad. Sci. USA* **76:**4530.

Casadaban, M. J., and Chou, J., 1984, *In vivo* formation of gene fusions encoding hybrid β-galactosidase proteins in one step with a transposable Mu-*lac* transducing phage, *Proc. Natl. Acad. Sci. USA* **81:**535.

Casjens, S., and King, J., 1975, Virus assembly, *Annu. Rev. Biochem.* **46:**555.

Chaconas, G., De Bruijn, F. S., Casadaban, M. J., Lupski, J. R., Kwoh, T. J., Harshey, R. M., DuBow, M. S., and Bukhari, A. I., 1981a, *In vitro* and *in vivo* manipulation of bacteriophage Mu DNA: Cloning of Mu ends and construction of mini-Mu carrying selectable markers, *Gene* **13:**37.

Chaconas, G., Harshey, R. M., Sarvetnick, N., and Bukhari, A. I., 1981b, The predominant end products of prophage Mu DNA transposition during the lytic cycle are replicon fusions, *J. Mol. Biol.* **150:**341.

Chaconas, G., Kennedy, D. L., and Evans, D., 1983, Predominant integration end products of infecting bacteriophage Mu DNA are simple insertions with no preference for integration of either Mu DNA strand, *Virology* **128:**48.

Chaconas, G., Giddens, E. B., Miller, J. L., and Gloor, G., 1985a, A truncated form of the bacteriophage MuB protein promotes conservative integration but not replicative transposition, of Mu DNA, *Cell* **41:**857.

Chaconas, G., Gloor, G., and Miller, J. L., 1985b, Amplification and purification of the bacteriophage Mu encoded B transposition protein, *J. Biol. Chem.* **260:**2662.

Chase, C. D., and Benzinger, R. H., 1982, Transfection of *Escherichia coli* spheroplasts with a bacteriophage Mu DNA-protein complex, *J. Virol.* **42:**176.

Chow, L. T., and Bukhari, A. I., 1976, The invertible DNA segments of coliphages Mu and P1 are identical, *Virology* **74:**242.

Chow, L. T., and Bukhari, A. I., 1978, Heteroduplex electron microscopy of phage Mu mutants containing IS1 insertions and chloramphenicol resistance transposons, *Gene* **3:**333.

Chow, L. T., Kahmann, R., and Kamp, D., 1977, Electron microscopic characterization of DNAs from non-defective deletion mutants of bacteriophage Mu, *J. Mol. Biol.* **113:**591.

Coelho, A., Maynard-Smith, S., and Symonds, N., 1982, Abnormal cointegrate structure mediated by gene B mutants of phage Mu: Their implications with regard to gene function, *Mol. Gen. Genet.* **185:**356.

Couturier, M., 1976, The integration and excision of bacteriophage Mu-1, *Cell* **7:**155.

Couturier, M., and Van Vliet, F., 1974, Vegetative recombination in bacteriophage Mu-1, *Virology* **60:**1.

Craigie, R., and Mizuuchi, K., 1985, Mechanism of transposition of bacteriophage Mu: Structure of a transposition intermediate, *Cell* **41:**967.

Craigie, R., and Mizuuchi, K., 1986, Role of DNA topology in Mu transposition: Mechanism of sensing the relative orientation of two DNA segments, *Cell* **45:**793.

Craigie, R., and Mizuuchi, K., 1987, Transposition of Mu DNA: Joining of Mu to target DNA can be uncoupled from cleavage at the ends of Mu, *Cell* **51:** 493.

Craigie, R., Mizuuchi, M., and Mizuuchi, K., 1984, Site-specific recognition of the bacteriophage Mu ends by the MuA protein, *Cell* **39:**387.

Craigie, R., Arndt-Jovin, D. J., and Mizuuchi, K., 1985, A defined system for the DNA strand transfer reaction at the initiation of bacteriophage Mu transposition: Protein and DNA substrate requirements, *Proc. Natl. Acad. Sci. USA* **82:**7570.

Cremers, A. F. M., Schepman, M. H., Visser, M. P., and Mellema, J. E., 1977, An analysis of the contracted sheath structure of bacteriophage Mu, *Eur. J. Biochem.* **80**:393.

Daniell, E., and Abelson, J., 1973, Lac messenger RNA in *lacZ* gene mutants of *Escherichia coli* caused by insertion of bacteriophage Mu, *J. Mol. Biol.* **76**:319.

Daniell, E., Roberts, R., and Abelson, J., 1972, Mutations in the lactose operon caused by bacteriophage Mu, *J. Mol. Biol.* **69**:1.

Daniell, E., Boram, W., and Abelson, J., 1973a, Genetic mapping of the inversion loop in bacteriophage Mu DNA, *Proc. Natl. Acad. Sci USA* **70**:2153.

Daniell, E., Abelson, J., Kim, J. S., and Davidson, N. 1973b, Heteroduplex structures of bacteriophage Mu DNA, *Virology* **51**:237.

Daniell, E., Kohne, D. E., and Abelson, J., 1975, Characterization of the inhomogeneous DNA in virions of bacteriophage Mu by DNA reannealing kinetics, *J. Virol.* **15**:739.

De Bruijn, F. J., and Bukhari, A. I., 1978, Analysis of transposable elements inserted in the genomes of bacteriophage Mu and P1, *Gene* **3**:315.

DiNardo, S., Voelkel, K. A., Sternglanz, R., Reynolds, A. E., and Wright, A., 1982, *Escherichia coli* DNA topoisomerase I mutants have compensatory mutations in DNA gyrase genes, *Cell* **31**:43.

Earnshaw, W. C., and Casjens, S. R., 1980, DNA packaging by the double-stranded DNA bacteriophage, *Cell* **21**:319.

Egner, C., and Berg, D. E., 1981, Excision of transposon Tn5 is dependent on the inverted repeats but not the transposon function of Tn5, *Proc. Natl. Acad. Sci. USA* **78**:459.

Engler, J. E., and Van Bree, M. P., 1981, The nucleotide sequence and protein-coding capability of the transposable element IS5, *Gene* **14**:155.

Faelen, M., and Toussaint, A., 1973, Isolation of conditional defective mutants of temperate phage Mu-1 and deletion mapping of the Mu-1 prophage, *Virology* **54**:117.

Faelen, M., and Toussaint, A., 1976, Bacteriophage Mu-1, a tool to transpose and to localize bacterial genes, *J. Mol. Biol.* **104**:525.

Faelen, M., and Toussaint, A., 1978, Stimulation of deletions in the *E. coli* chromosome by partially induced Mu *cts62* prophages. *J. Bacteriol.* **136**:477.

Faelen, M., and Toussaint, A., 1980a, Inversion induced by temperate bacteriophage Mu-1 in the chromosome of *E. coli* K12, *J. Bacteriol.* **142**:391.

Faelen, M., and Toussaint, A., 1980b, Temperate phage D108 induces chromosomal rearrangements, *J. Bacteriol.* **143**:1029.

Faelen, M., Toussaint, A., and Couturier, M., 1971, Mu-1 promoted integration of a lambda-gal phage in the chromosome of *E. coli*, *Mol. Gen. Genet.* **113**:367.

Faelen, M., Toussaint, A., and De Lafonteyne, J., 1975, Model for the enhancement of lambda-gal integration into partially induced Mu lysogens, *J. Bacteriol.* **121**:873.

Faelen, M., Huisman, O., and Toussaint, A., 1978, Involvement of phage Mu-1 early functions in Mu mediated chromsomal rearrangements, *Nature* **271**:580.

Faelen, M., Toussaint, A., Lefebvre, N., Mergeay, M., Braipson-Thiry, J., and Thiry, G., 1981, Certaines souches de *Erwinia* sont senibles au bacteriophage Mu, *Arch. Intern. Physiol. Biochim.* **89**:B55.

Faelen, M., Toussaint, A., Waggoner, B., Desmet, L. and Pato, M., 1986, Transposition and replication of maxi-Mu derivatives of bacteriophage Mu, *Virology* **153**:70.

Feiss, M., and Becker, A., 1983, DNA packaging and cutting, in *Lambda II* (R. W. Hendrix, J. W. Roberts, F. W. Stahl, and R. A. Weisberg, eds.), pp. 305–330, Cold Spring Harbor Laboratory, Cold Spring Harbor, NY.

Fitts, R., and Taylor, A. L., 1980, Integration of bacteriophage Mu at host chromosomal replication forks during lytic development, *Proc. Natl. Acad. Sci. USA* **77**:2801.

Friedman, D. I., Plantefaber, L. C., Olsen, E. J., Carver, D., O'Dea, M., and Gellert, M., 1984, Mutations in *gyrB* that are *ts* for λ site-specific recombination, Mu growth and plasmid maintenance, *J. Bacteriol* **157**:490.

Gellert, M., 1981, DNA topoisomerases, *Annu. Rev. Biochem.* **50**:879.

George, M., and Bukhari, A. I., 1981, Heterogenous host DNA attached to the left end of mature bacteriophage Mu DNA, *Nature* **292**:175.

Georgopoulous, C., Tilly, K., and Casjens, S., 1983, Lambdoid phage head assembly, in *Lambda II* (R. W. Hendrix, J. W. Roberts, F. W. Stahl, and R. A. Weisberg, eds.), pp. 279–304, Cold Spring Harbor Laboratory, Cold Spring Harbor, NY.

Ghelardini, P., Paolozzi, L., and Liebart, J. C., 1979, Restoration of DNA synthesis at nonpermissive temperature and UV resistance induced by bacteriophage Mu in *Escherichia coli lig*ts7, *Ann. Microbiol. Paris* **130B**:275.

Ghelardini, P., Paolozzi, L., and Liebart, J. C., 1980, Restoration of ligase activity in *E. coli, lig*ts7 strain by bacteriophage Mu and cloning of a DNA fragment harboring the Mu *lig* gene, *Nucleic Acids Res.* **8**:3157.

Ghelardini, P., Pedrini, A. M., and Paolozzi, L., 1982, The topoisomerase activity of T4 *am*G39 mutant is restored in Mu lysogens, *FEBS Lett.* **137**:49.

Ghelardini, P., Liebart, J. C., Marchelli, C., Pedrini, A. M., and Paolozzi, L., 1984, *E. coli* K12 *gyrB* gene product is involved in the lethal effect of the *lig*ts2 mutant of bacteriophage Mu, *J. Bacteriol.*, **157**:665.

Gill, G. S., Hull, A. C., and Curtis, R. III, 1981, Mutator bacteriophage D108 and its DNA: An electron microscopic characterization, *J. Virol.* **37**:420.

Giphart-Gassler, M., Wijffelman, C., and Reeve, J., 1981a, Structural polypeptides and products of late genes of bacteriophage Mu: Characterization and functional aspects, *J. Mol. Biol.* **145**:139.

Giphart-Gassler, M., Reeve, J., and Van de Putte, P. 1981b, Polypeptides encoded by the early region of bacteriophage Mu synthesized in minicells of *E. coli, J. Mol. Biol.* **145**:165.

Giphart-Gassler, M., Plasterk, R. H. A., and Van de Putte, P. 1982, G inversion in bacteriophage Mu: A novel way of gene splicing, *Nature* **297**:339.

Gloor, N., and Chaconas, G., 1986, The bacteriophage Mu*N* gene encodes the 64kDa virion protein which is injected with and circularizes infecting Mu DNA, *J. Biol. Chem.* **261**:16682.

Goosen, N., and Van de Putte, P., 1984, Regulation of Mu transposition. I. Localization of the presumed recognition site for HimD and Ner functions controlling bacteriophage Mu transposition, *Gene* **30**:41.

Goosen, N., and Van de Putte, P., 1986, Role of Ner protein in bacteriophage Mu transposition, *J. Bacteriol.* **167**:503.

Goosen, T., Giphart-Gassler, M., and Van de Putte, P., 1982, Bacteriophage Mu DNA replication is stimulated by nonessential early functions, *Mol. Gen. Genet.* **186**:135.

Goosen, N., Van Heuvel, M., Moolenaar, G. F., and Van de Putte, P., 1984, Regulation of Mu transposition. II. The *Escherichia coli* HimD protein positively controls two repressor promoters and the early promoter of bacteriophage Mu, *Gene* **32**:419.

Grindley, N. D. F., and Reed, R. R., 1985, Transpositional recombination in prokaryotes, *Annu. Rev. Biochem.* **54**:863.

Grindley, N. D. F., and Sherratt, D. J., 1978, Sequence analysis at IS1 insertion sites: Models for transposition, *Cold Spring Harbor Symp. Quant. Biol.* **43**:1257.

Groenen, M. A. M., and Van de Putte, P., 1985, Mapping of a site for packaging of bacteriophage Mu DNA, *Virology* **144**:520.

Grundy, F. J., and Howe, M. M., 1984, Involvement of the invertible G segment in bacteriophage Mu tail fiber biosynthesis, *Virology* **134**:296.

Grundy, F. J., and Howe, M. M., 1985, Morphogenetic structures present in lysates of amber mutants of bacteriophage Mu, *Virology* **143**:485.

Harshey, R. M., 1983, Switch in the transposition products of Mu DNA mediated by proteins: Cointegrates versus simple insertions, *Proc. Natl. Acad. Sci. USA* **30**:2012.

Harshey, R. M., 1984a, Transposition without duplication of infecting bacteriophage Mu DNA, *Nature* **311**:580.

Harshey, R. M., 1984b, Non-replicative DNA transposition: Integration of infecting bacteriophage Mu, *Cold Spring Harbor Symp. Quant. Biol.* **49**:273.

Harshey, R. M., 1987, Integration of infecting Mu DNA, in *The Phage Mu* (N. Symonds, A. Toussaint, P. Van de Putte, and M. M. Howe, eds.), pp. 111–135, Cold Spring Harbor Laboratory, Cold Spring Harbor, NY (in press).

Harshey, R. M., and Bukhari, A. I., 1981, A mechanism of DNA transposition, *Proc. Natl. Acad. Sci. USA* **78**:1090.

Harshey, R. M., and Bukhari, A. I., 1983, Infecting bacteriophage Mu DNA forms a circular DNA-protein complex, *J. Mol. Biol.* **167**:427.

Harshey, R. M., and Cuneo, S., 1986, Carboxyl-terminal mutants of phage Mu transposase, *J. Genet* **65**:159.

Harshey, R. M., McKay, R., and Bukhari, A. I., 1982, DNA intermediates in transposition of phage Mu, *Cell* **29**:561.

Harshey, R. M., Getzoff, E. D., Baldwin, D. L., Miller, J. L., and Chaconas, G., 1985, Primary structure of phage Mu transposase: Homology to Mu repressor, *Proc. Natl. Acad. Sci. USA* **82**:7676.

Hattman, S., 1979, Unusual modification of bacteriophage Mu DNA, *J. Virol.* **32**:468.

Hattman, S., 1980, Specificity of the bacteriophage Mu *mom* + -controlled DNA modification, *J. Virol.* **34**:277.

Hattman, S., 1982, DNA methyl transferase-dependent transcription of the phage Mu *mom* gene, *Proc. Natl. Acad. Sci. USA* **79**:5518.

Hattman, S., and Ives, J., 1984, S1 nuclease mapping of the phage Mu *mom* gene promoter: A model for the regulation of *mom* expression, *Gene* **29**:185.

Hattman, S., Goradia, M., Monaghan, C., and Bukhari, A. I., 1983, Regulation of the DNA modification function of bacteriophage Mu, *Cold Spring Harbor Symp. Quant. Biol.* **47**:647.

Hattman, S., Ives, J., Margolin, W., and Howe, M. M., 1985, Regulation and expression of the bacteriophage Mu *mom* gene: Mapping of the transactivation function (Dad) to the C region, *Gene* **39**:71.

Hershey, A. D., and Dove, W. D., 1971, Introduction to lambda, in *The Bacteriophage* λ (A. D. Hershey, ed.), pp. 3–11, Cold Spring Harbor Laboratory, Cold Spring Harbor, NY.

Higgins, N. P., Moncecchi, D., Manlapaz-Ramos, P., and Olivera, B. M., 1983, Bacteriophage Mu DNA replication *in vitro*, *J. Biol. Chem.* **258**:4293.

Howe, M. M., 1972, Genetic studies on bacteriophage Mu, Ph.D. thesis, Massachusetts Institute of Technology.

Howe, M. M., 1973a, Prophage deletion mapping of bacteriophage Mu-1, *Virology* **54**:93.

Howe, M. M., 1973b, Transduction of bacteriophage Mu-1, *Virology* **55**:103.

Howe, M. M., 1978, Invertible DNA in phage Mu, *Nature* **271**:608.

Howe, M. M., 1980, The invertible G segment of phage Mu, *Cell* **21**:605.

Howe, M. M., and Bade, E. G., 1975, Molecular biology of bacteriophage Mu, *Science* **190**:624.

Howe, M. M., and Schumm, J. W., 1980, Transposition of bacteriophage Mu. Properties of lambda phages containing both ends of Mu, *Cold Spring Harbor Symp. Quant. Biol.* **45**:337.

Howe, M. M., and Zipser, D., 1974, Host deletions caused by the integration of bacteriophage Mu-1, *Am. Soc. Microbiol. Abstr.* **208**:235.

Howe, M. M., Schnos, M., and Inman, R. B., 1977, DNA partial denaturation mapping studies of packaging of bacteriophage Mu DNA, in *DNA Insertion Elements, Plasmids and Episomes* (A. I. Bukhari, J. A. Shapiro, and S. L. Adhya, eds.), pp. 319–327, Cold Spring Harbor Laboratory, Cold Spring Harbor, NY.

Howe, M. M., O'Day, K. J., and Schultz, D. W., 1979a, Isolation of mutants defining five new cistrons essential for development of bacteriophage Mu, *Virology* **93**:303.

Howe, M. M., Schumm, J. W., and Taylor, A. L., 1979b, The *S* and *U* genes of bacteriophage Mu are located in the invertible G segment of Mu DNA, *Virology* **92**:108.

Hsu, M. T., and Davidson, N., 1972, Structure of inserted bacteriophage Mu-1 DNA and physical mapping of bacterial genes by Mu-1 DNA insertion, *Proc. Natl. Acad. Sci. USA* **69**:2823.

Hsu, M. T., and Davidson, N., 1974, Electron microscope heteroduplex study of the heterogeneity of Mu phage and prophage DNA, *Virology* **58**:229.

Hull, R. C., Gill, G. S., and Curtis, R. III, 1978, Genetic characterization of Mu-like bacteriophage D108, *J. Virol.* **27**:513.

Inman, R. B., Schnos, M., and Howe, M., 1976, Location of the "variable end" of Mu DNA within the bacteriophage particle, *Virology* **72**:393.

Joho, R., Nottenburg, C., Coffman, R. L., and Weissman, I. L., 1983, Immunoglobulin gene rearrangement and expression during lymphocyte development, *Curr. Top. Dev. Biol.* **18**:16.

Jordan, E., Saedler, H., and Starlinger, P., 1968, 0^0 and strong-polar mutations in the *gal* operon are insertions, *Mol. Gen. Genet.* **102**:353.

Kahmann, R., 1983, Methylation regulates the expression of a DNA-modification function encoded by bacteriophage Mu, *Cold Spring Harbor Symp. Quant. Biol.* **47**:639.

Kahmann, R., 1984, The *mom* gene of bacteriophage Mu, *Curr. Top. Microbiol. Immunol.* **108**:29.

Kahmann, R., and Kamp, D., 1979, Nucleotide sequences of the attachment sites of bacteriophage Mu DNA, *Nature* **280**:247.

Kahmann, R., Kamp, D., and Zipser, D., 1976, Transfection of *E. coli* by Mu DNA, *Mol. Gen. Genet.* **149**:323.

Kahmann, R., Seiler, A., Wulczyn, F. G., and Pfaff, E., 1985a, The *mom* gene of bacteriophage Mu: A unique regulatory scheme to control a lethal function, *Gene* **39**:61.

Kahmann, R., Rudt, F., Koch, C., Mertens, G., 1985b, G inversion in bacteriophage Mu DNA is stimulated by a site within the invertase gene and a host factor, *Cell* **41**:771.

Kamp, D., 1981, Invertible deoxyribonucleic acid: The G segment of bacteriophage Mu, in *Microbiology—1981* (D. Schlessinger, ed.), pp. 73–76, American Society for Microbiology, Washington.

Kamp, D., Chow, L. T., Broker, T. R., Kwoh, D., Zipser, D., and Kahmann, R., 1978, Site specific recombination in phage Mu, *Cold Spring Harbor Symp. Quant. Biol.* **43**:1159.

Karu, A., Sakaki, Y., Echols, H., and Linn, S., 1974, *In vitro* studies of the *gam* gene product of bacteriophage, in *Mechanisms of Recombination* (R. F. Grell, ed.), pp. 95–109, Plenum Press, New York.

Katsura, I., 1983, Tail assembly and injection, in *Lambda II* (R. W. Hendrix, J. W. Roberts, F. W. Stahl, and R. A. Weisberg, eds.), pp. 331–346, Cold Spring Harbor Laboratory, Cold Spring Harbor, NY.

Kemp, C. L., Howatson, A. F., and Siminovitch, C., 1968, Electron microscopic studies of mutants of lambda bacteriophage. I. General description and quantitation of viral products, *Virology* **36**:490.

Khatoon, H., and Bukhari, A. I., 1978, Bacteriophage Mu-induced modification of DNA is dependent on a host function, *J. Bacteriol.* **136**:423.

Khatoon, H., and Bukhari, A. I., 1981, DNA rearrangements associated with reversion of bacteriophage Mu induced mutations, *Genetics* **98**:1.

Khoury, G., and Gruss, P., 1983, Enhancer elements, *Cell* **33**:313.

King, J., 1971, Bacteriophage T4 tail assembly: Four steps in core formation, *J. Mol. Biol.* **58**:693.

King, J., and Wood, W. B., 1969, Assembly of bacteriophage T4 tail fibers: The sequence of gene product interaction, *J. Mol. Biol.* **39**:583.

Kleckner, N., 1981, Transposable elements in prokaryotes, *Annu. Rev. Genet.* **15**:341.

Krause, H. M., and Higgins, N. P., 1984, On the Mu repressor and early DNA intermediates of transposition, *Cold Spring Harbor Symp. Quant. Biol.* **49**:827.

Krause, H. M., and Higgins, N. P., 1986, Positive and negative regulation of the Mu operator by Mu repressor and *Escherichia coli* integration host factor, *J. Mol. Biol.* **261**:3744.

Krause, H. K., Rothwell, M. R., and Higgins, N. P., 1983, The early promoter of bacteriophage Mu: Definition of the site of transcript initiation, *Nucleic Acids Res.* **11**:5483.

Krylov, V. N., Bogush, V. G., and Shapiro, J., 1980a, *Pseudomonas aeroginosa* phages whose DNA structure is similar to Mu-1 phage DNA I, *Genetika USSR* **15**(5):824.

Krylov, V. N., Bogush, V. G., Yanenko, A. S., and Kirsanov, N. B., 1980b, *Pseudomonas aeroginosa* phages whose DNA structure is similar to Mu-1 phage DNA II, *Genetika USSR* **16**(6):975.

Kwoh, D. Y., Zipser, D., and Erdmann, D. S., 1980, Genetic analysis of the cloned genome of phage Mu, *Virology* **101**:419.

Leach, D., and Symonds, N., 1979, The isolation and characterization of a plaque forming derivative of bacteriophage Mu carrying a fragment of Tn3 conferring ampicillin resistance, *Mol. Gen. Genet.* **172**:179.

Liebart, J. C., Ghelardini, P., and Paolozzi, L., 1982, Conservative integration of bacteriophage Mu DNA into pBR322 plasmid, *Proc. Natl. Acad. Sci. USA* **79**:4362.

Ljungquist, E., and Bukhari, A. I., 1977, State of prophage Mu DNA upon induction, *Proc. Natl. Acad. Sci. USA* **74**:3143.

Ljungquist, E., and Bukhari, A. I., 1979, Behavior of bacteriophage Mu DNA upon infection of *E. coli*, *J. Mol. Biol.* **133**:339.

Magazin, M., Howe, M., and Allet, B., 1977, Partial correlation of the genetic and physical maps of bacteriophage Mu, *Virology* **77**:677.

Magazin, M., Reeve, J. N., Maynard-Smith, S., and Symonds, N., 1978, Bacteriophage Mu encoded polypeptides synthesized in infected mini-cells, *FEMS Microbiol. Lett.* **4**:5.

Margolin, W., and Howe, M. M., 1986, Localization and DNA sequence analysis of the C gene of bacteriophage Mu, the positive regulator of Mu late transcription, *Nucleic Acids Res.* **14**:4881.

Marinus, M. G., and Morris, N. R., 1975, Pleiotropic effects of a DNA adenine methylation mutation (*dam-3*) in *E. coli* K12, *Mutat. Res.* **28**:15.

Marrs, C. F., 1982, Transcription of bacteriophage Mu, Ph.D. thesis, University of Wisconsin, Madison.

Marrs, C. F., and Howe, M. M., 1983, AvaII and BglI restriction maps of bacteriophage Mu, *Virology* **126**:563.

Martuscelli, J., Taylor, A. L., Cummings, D. J., Chapman, V. A., Delong, S. S., and Canedo, L., 1971, Electron microscopic evidence for linear insertion of bacteriophage Mu-1 in lysogenic bacteria, *J. Virol.* **8**:551.

Maxwell, A., Craigie, R., and Mizuuchi, K., 1987, B protein of bacteriophage Mu is an ATPase that preferentially stimulates intermolecular DNA strand transfer, *Proc. Natl. Acad. Sci. USA,* **84**:699.

McBeth, D. L., and Taylor, A. L., 1982, Growth of bacteriophaage Mu in *Escherichia coli dnaA* mutants, *J. Virol.* **44**:555.

McBeth, D. L., and Taylor, A. L., 1983, Involvement of *Escherichia coli* K12 DNA polymerase I in the growth of bacteriophage Mu, *J. Virol.* **48**:149.

McClintock, B., 1956, Controlling elements and the gene, *Cold Spring Harbor Symp. Quant. Biol.* **21**:197.

Mertens, G., Hoffmann, A., Blocker, H., Frank, R., and Kahmann, R., 1984, Gin-mediated site-specific recombination in bacteriophage Mu DNA: Overproduction of the protein and inversion *in vitro*, *EMBO J.* **3**:2415.

Miller, H. I., and Friedman, D. I., 1980, An *E. coli* gene product required for lambda site-specific recombination, *Cell* **20**:711.

Miller, J. L., Anderson, S. K., Fujita, D. J., Chaconas, G., Baldwin, D., and Harshey, R. M., 1984, The nucleotide sequence of the B gene of bacteriophage Mu, *Nucleic Acids Res.* **12**:8627.

Mise, K., 1971, Isolation and characterization of a new generalized transducing bacteriophage different from P1 in *E. coli*, *J. Virol.* **7**:168.

Mizzuuchi, K., 1983, *In vitro* transposition of bacteriophage Mu: A biochemical approach to a novel replication reaction, *Cell* **35**:785.

Mizuuchi, K., 1984, Mechanism of transposition of bacteriophage Mu: Polarity of the strand transfer reaction at the initiation of transposition, *Cell* **39**:395.

Mizuuchi, K., and Craigie, R., 1986, Mechanism of bacteriophage Mu transposition, *Ann. Rev. Genet.,* **20**:385.

Mizuuchi, M., Weisberg, R. A., and Mizuuchi, K., 1986, DNA sequence of the control region of phage D108: The N-terminal amino acid sequences of repressor and transposase are similar both in phage D108 and in its relative phage Mu, *Nucleic Acids Res.* **14**:3813.

Moore, D. D., Schumm, J. W., Howe, M. M., and Blattner, F. R., 1977, Insertion of Mu DNA fragments in phage lambda *in vitro*, in *DNA Insertion Elements, Plasmids and Epi-*

somes (A. I. Bukhari, J. A. Shapiro, and S. L. Adhya, eds.), pp. 567–574, Cold Spring Harbor Laboratory, Cold Spring Harbor, NY.

Mount, D. W. A., Harris, A. W., Fuerst, C. R., and Siminovitch, L., 1968, Mutations in bacteriophage lambda affecting particle morphogenesis, *Virology* **35**:134.

Murialdo, H., and Becker, A., 1978, Head morphogenesis of complex double-stranded deoxyribonucleic acid bacteriophages, *Microbiol. Rev.* **42**:529.

Nag, D. K., and Berg, D. E., 1987, Specificity of bacteriophage Mu excision, *Mol. Gen. Genet.* **207**:395.

Nakai, H., and Taylor, A. L., 1985, Host DNA replication forks are not preferred targets for bacteriophage Mu transposition, *J. Bacteriol.* **163**:282.

Nakayama, C., Teplow, D., and Harshey, R. M., 1987, Structural domains in phage Mu transposase: Identification of the site-specific DNA binding domain, *Proc. Natl. Acad. Sci. USA* **84**:1809.

O'Day, K. J., Shultz, D. W., and Howe, M. M., 1978, A search for integration deficient mutants of bacteriophage Mu-1, in *Microbiology (1978)* (D. Schlessinger, ed.), pp. 48–51, ASM Publications, Washington.

O'Day, K. J., Schultz, D. W., Ericsen, W., Rawluk, L., and Howe, M. M., 1979, Correction and refinement of the genetic map of bacteriophage Mu, *Virology* **93**:320.

Ohtsubo, E., Zenilman, M., Ohtsubo, H., McCormick, M., Machida, C., and Machida, Y., 1980, Mechanism of insertion and cointegration mediated by IS1 and Tn3, *Cold Spring Harbor Symp. Quant. Biol.* **45**:283.

Paolozzi, L., and Ghelardini, P., 1986, General method for the isolation of conditional lethal mutants in any required region of the virus genome: Its application to the "semiessential" region of phage Mu, *J. Mol. Microbiol.* **132**:79.

Paolozzi, L., Jucker, R., and Calef, E., 1978, Mechanism of phage Mu-1 integration: Nalidixic acid treatment causes clustering of Mu-1 induced mutations near replication origin, *Proc. Natl. Acad. Sci. USA* **75**:4940.

Paolozzi, L., Ghelardini, P., Kepes, A., and Markovich, H., 1979, The mechanism of integration of bacteriophage Mu-1 in the chromosome of *Escherichia coli*, *Biochem. Biophys. Res. Commun.* **88**:111.

Paolozzi, L., Ghelardini, P., Liebart, J. C., Capozzoni, A., and Marchelli, C., 1980, Two classes of Mu *lig* mutants: The thermosensitive for integration and replication and the hyperproducers for ligase, *Nucleic Acid Res.* **8**:5859.

Pato, M. L., and Reich, C., 1982, Instability of transposase activity: Evidence from bacteriophage Mu DNA replication, *Cell* **36**:197.

Pato, M. L., and Reich, C., 1984, Stoichiometric use of the transposase of bacteriophage Mu, *Cell* **36**:197.

Pato, M. L., and Reich, C., 1985, *Genome Rearrangement*, pp. 27–35, Alan R. Liss, New York.

Pato, M. L., and Waggoner, B. T., 1981, Cellular location of Mu DNA replicas, *J. Virol.* **38**:249.

Plasterk, R. H. A., Brinkman, A., and Van de Putte, P., 1983a, DNA inversions in the chromosome of *Escherichia coli* and in bacteriophage Mu: Relationship to other site-specific recombination systems, *Proc. Natl. Acad. Sci. USA* **80**:5355.

Plasterk, R. H. A., Ilmer, T. A. M., and Van de Putte, P., 1983b, Site-specific recombination by Gin of bacteriophage Mu: Inversions and deletions, *Virology* **127**:24.

Plasterk, R. H. A., Vrieling, H., and Van de Putte, P., 1983c, Transcription initiation of Mu *mom* depends on methylation of the promoter region and a phage-coded transactivator, *Nature* **301**:344.

Plasterk, R. H. A., Kanaar, R., and Van de Putte, P., 1984a, A genetic switch *in vitro*: DNA inversion by gin protein of phage Mu, *Proc. Natl. Acad. Sci. USA* **81**:2689.

Plasterk, R. H. A., Vollering, M., Brinkman, A., and Van de Putte, P., 1984b, Analysis of the methylation-regulated Mu *mom* transcript, *Cell* **36**:189.

Ptashne, M., Jeffrey, A., Johnson, A. D., Maurer, R., Meyer, B. J., Pabo, C. O., Roberts, T. M., and Sauer, R. T., 1980, How the λ repressor and Cro work, *Cell* **19**:1.

Puspurs, A. H., Trun, N. J., and Reeve, J. N., 1983, Bacteriophage Mu DNA circularizes following infection of *Escherichia coli*, *EMBO J.* **2**:345.

Reich, C., Waggoner, B. T., and Pato, M. L., 1984, Synchronization of bacteriohage Mu DNA replicative transposition: Analysis of the first round after induction, *EMBO J.* **3**:1507.

Resibois, A., Colet, M., and Toussaint, A., 1982, Localication of mini-Mu in its replication intermediates, *EMBO J.* **1**:965.

Resibois, A., Pato, M., Higgins, P., and Toussaint, A., 1984, Replication of bacteriophage Mu and its mini-Mu derivatives, in *Proteins Involved in DNA Replication* (S. Spadari and U. Hubscher, eds.), p. 69–76, Plenum Press, New York.

Roeder, G. S., and Sadowski, P. D., 1977, Bacteriophage T7 morphogenesis: Phage-related particles in cells infected with wild type and mutant T7 phage, *Virology* **76**:263.

Ross, W., Shore, S. H., and Howe, M. M., 1986, Mutants of *Escherichia coli* defective for replicative transposition of bacteriophage Mu, *J. Bacteriol.* **167**:905.

Saedler, H., and Nevers, P., 1985, Transposition in plants: A molecular model, *EMBO J.* **4**:585.

Sandulache, R., Prehm, P., and Kamp, D., 1984, Cell wall receptor for bacteriophage Mu G(+), *J. Bacteriol.* **160**:299.

Sandulache, R., Prehm, P., Expert, D., Toussaint, A., and Kamp, D., 1985, The cell wall receptor for bacteriophage G(−) in *Erwinia* and *Escherichia coli* C, *FEMS Microbiol. Lett.* **28**:307.

Schroeder, W., Bade, E. G., and Delius, H., 1974, Participation of *E. coli* DNA in the replication of temperate bacteriophage Mu-1, *Virology* **60**:534.

Segal, E., Hagblom, P., Seifert, H. S., and So, M., 1986, Antigenic variation of gonococcal pilus involves assembly of separated silent gene segments, *Proc. Natl. Acad. Sci. USA* **83**:2177.

Serwer, P., 1976, Internal proteins of bacteriophage T7, *J. Mol. Biol.* **107**:271.

Shapiro, J. A., 1979, Molecular model for the transposition and replication of bacteriophage Mu and other transposable elements, *Proc. Natl. Acad. Sci. USA* **76**:1933.

Shapiro, J. A., 1984, The use of Mu*lac* transposons as tools for vital staining to visualize clonal and non-clonal patterns of organization in bacterial growth on agar surfaces, *J. Gen. Microbiol.* **130**:1169.

Shaw, J. E., and Murialdo, H., 1980, Morphogenetic genes *C* and *Nu3* overlap in bacteriophage λ, *Nature* **283**:30.

Shore, S. H., and Howe, M. M., 1982, Bacteriophage Mu *T* mutants are defective in synthesis of the major head polypeptide, *Virology* **120**:264.

Soberon, M., Gama, M. J., Richelle, J., and Martuscelli, J., 1986, Behaviour of temperate phage Mu in *Salmonella typhi*, *J. Mol. Microbiol.* **132**:83.

Surette, M. G., Buch, S. J., and Chaconas, G., 1987, Transpososomes: Stable protein–DNA complexes involved in the *in vitro* transposition of bacteriophage Mu DNA, *Cell* **49**:253.

Swinton, D., Hattman, S., Crain, P. F., Cheng, C.-S., Smith, D. L., and McCloskey, J. A., 1983, Purification and characterization of the unusual deoxynucleoside, α-N-(9-β-D-2′-deoxyribofuranosylpurin-6yl) glycinamide, specified by the phage Mu modification function, *Proc. Natl. Acad. Sci. USA* **80**:7400.

Symonds, N., 1982, A novel gene splice in bacteriophage Mu, *Nature* **297**:288.

Symonds, N., and Coelho, A., 1978, Role of the G-segment in the growth of phage Mu, *Nature* **271**:573.

Symonds, N., Toussaint, A., Van de Putte, P., and Howe, M. M., eds., 1987, *Phage Mu*, Cold Spring Harbor Laboratory, New York.

Szatmari, G. B., Kahn, J. S., and DuBow, M. S., 1986, Orientation and sequence analysis of right ends and target sites of bacteriophage Mu and D108 insertions in the plasmid pSC101, *Gene* **41**:315.

Szybalski, W., Kubinski, H., Hradecna, Z., and Summers, W. C., 1971, Analytical and preparative separation of the complementary DNA strands, *Methods Enzymol.* **21**:383.

Taylor, A. L., 1963, Bacteriophage-induced mutation in *E. coli.*, *Proc. Natl. Acad. Sci. USA* **50**:1043.

Teifel, J., and Schmieger, H., 1981, The origin of the DNA in transducing particles of bacteriophage Mu. Density gradient analysis of DNA, *Mol. Gen. Genet.* **184**:312.

To, C. M., Eisenstark, A., and Toreci, H., 1966, Structure of mutator phage Mu-1 of *Escherichi coli, J. Ujtrastruct. Res.* **14**:441.

Tolias, P. P., and DuBow, M. S., 1985, The cloning and characterization of the bacteriophage D108 regulatory DNA-binding protein ner, *EMBO J.* **4**:3031.

Tolias, P. P., and DuBow, M. S., 1986, The overproduction and characterization of the bacteriophage Mu regulatory DNA-binding protein ner, *Virology* **148**:293.

Torti, F., Barksdale, C., and Abelson, J., 1970, Mu-1 bacteriophage DNA, *Virology* **41**:567.

Toussaint, A., 1969, Insertion of phage Mu-1 within prophage lambda: A new approach for studying the control of the late functions in bacteriophage lambda, *Mol. Gen. Genet.* **106**:89.

Toussaint, A., 1976, The DNA modification of temperate phage Mu-1, *Virology* **70**:17.

Toussaint, A., 1977, The modification of bacteriophage Mu-1 requires both a bacterial and a phage function, *J. Virol.* **23**:825.

Toussaint, A., and Faelen, M., 1973, Connecting two unrelated DNA sequences with a Mu dimer, *Nature New Biol.* **242**:1.

Toussaint, A., and Lecocq, J.-P., 1974, Sensitivity of bacteriophage Mu-1 development to rifampicin and streptolydigin, *Mol. Gen. Genet.* **129**:185.

Toussaint, A., and Resibois, A., 1983, Phage Mu: Transposition as a life style, in *Mobile Genetic Elements* (J. A. Shapiro, ed.), pp. 105–158, Academic Press, New York.

Toussaint, A., Faelen, M., and Bukhari, A. I., 1977, Mu mediated illegitimate recombination as an integral part of the Mu life cycle, in *DNA Insertion Elements, Plasmids, and Episomes* (A. I. Bukhari, J. A. Shapiro, and S. L. Adhya, eds.), pp. 275–285, Cold Spring Harbor Laboratory, Cold Spring Harbor, NY.

Toussaint, A., LeFebvre, N., Scott, J. R., Cowan, J. A., De Bruijn, F., and Bukhari, A. I., 1978, Relationship between temperate phages Mu and P1, *Virology* **89**:146.

Toussaint, A., Desmet, L., and Faelen, M., 1980, Mapping of the modification function of temperature phage Mu-1, *Mol. Gen. Genet.* **177**:351.

Toussaint, A., Faelen, M., Desmet, L., and Allet, B., 1983, The products of gene *A* of related phages Mu and D108 differ in their specificities, *Mol. Gen. Genet.* **190**:70.

Van de Putte, P., and Gruijthuijsen, M., 1972, Chromosome mobilization and integration of F-factors in the chromosome of *recA* stains of *E. coli* under the influence of bacteriophage Mu-1, *Mol. Gen. Genet.* **118**:173.

Van de Putte, P., Westmaas, G. C., Giphart, M., and Wijffelman, C., 1977a, On the *kil* gene of bacteriophage Mu, in *DNA Insertion Elements, Plasmids, and Episomes* (A. I. Bukhari, J. A. Shapiro, and S. L. Adhya, eds.), pp. 287–294, Cold Spring Harbor Laboratory, Cold Spring Harbor, NY.

Van de Putte, P., Westmaas, G. C., and Wijffelman, C., 1977b, Transfection with Mu DNA, *Virology* **81**:152.

Van de Putte, P., Giphart-Gassler, M., Goosen, T., Van Meeteren, A., and Wijffelman, C., 1978, in *Integration and Excision of DNA Molecules* (P. Hofshneider and P. Starlinger, eds.), pp. 33–40, Springer-Verlag, Berlin.

Van de Putte, P., Cramer, S., and Giphart-Gassler, M., 1980, Invertible DNA determines host specificity of bacteriophage Mu, *Nature* **286**:218.

Van Gijsegem, F., Toussaint, A., and Casadaban, M., 1987, Mu as a genetic tool, in: *Phage Mu* (N. Symonds, A. Toussaint, P. Van de Putte, and M. M. Kowe, eds.), pp. 215–250, Cold Spring Harbor Laboratory, Cold Spring Harbor, NY.

Van Leerdam, E., Karreman, C., and Van de Putte, P., 1982, Ner, a Cro-like function of bacteriophage Mu, *Virology* **123**:19.

Van Meeteren, A., 1980, Transcription of bacteriophage Mu, Ph.D. thesis, Rijks Universiteit, Leiden, The Netherlands.

Van Meeteren, A., Giphart-Gassler, M., and Van de Putte, P., 1980, Transcription of bacteriophage Mu II: Transcription of the repressor gene, *Mol. Gen. Genet.* **179**:185.

Van Vliet, F., Couturier, M., Desmet, L., Faelen, M., and Toussaint, A., 1978a, Virulent mutants of temperate phage Mu-1, *Mol. Gen. Genet.* **160**:195.

Van Vliet, F., Couturier, M., de Lafonteyne, J., and Jedlicki, E., 1978b, Mu-1 directed inhibition of DNA breakdown in *Escherichi coli, recA* cells, *Mol. Gen. Genet.* **164**:109.

Waggoner, B. T., and Pato, M. L., 1978, Early events in the replication of Mu prophage DNA, *J. Virol.* **27**:587.

Waggoner, B. T., Gonzales, N. S., and Taylor, A. L., 1974, Isolation of heterogeneous circular DNA from induced lysogens of bacteriophage Mu-1, *Proc. Natl. Acad. Sci. USA* **71**:1255.

Waggoner, B. T., Pato, M. L., and Taylor, A. L., 1977, Characterization of covalently closed DNA molecules isolated after bacteriophage Mu induction, in *DNA Insertion Elements, Plasmids, and Episomes* (A. I. Bukhari, J. A. Shapiro, and S. L. Adhya, eds.), pp. 263–274, Cold Spring Harbor Laboratory, Cold Spring Harbor, NY.

Waggoner, B. T., Pato, M. L., Toussaint, A., and Faelen, M., 1981, Replication of mini-Mu prophage DNA, *Virology* **113**,379.

Waggoner, B. T., Marrs, C. F., Howe, M. M., and Pato, M. L., 1984, Multiple factors and processes involved in host cell killing by bacteriophage Mu: Characterization and mapping, *Virology* **136**:168.

Wagner, R. Jr., and Meselson, M., 1976, Repair tracts in mismatched DNA heterduplexes, *Proc. Natl. Acad. Sci. USA* **73**:4135.

Wijffelman, C. A., and Lotterman, B., 1977, Kinetics of Mu DNA synthesis, *Mol. Gen. Genet.* **151**:169.

Wijffelman, C. A., and Van de Putte, P., 1974, Transcription of bacteriophage Mu: An analysis of the transcription pattern in the early stage of the phage development, *Mol. Gen. Genet.* **135**:327.

Wijffelman, C. A., and Van de Putte, P., 1977, Asymmetric hybridization of Mu strands with short fragments synthesized during Mu DNA replication, in *DNA Insertion Elements, Plasmids, and Episomes* (A. I. Bukhari, J. A. Shapiro, and S. L. Adhya, eds.), pp. 329–333, Cold Spring Harbor Laboratory, Cold Spring Harbor, NY.

Wijffelman, C. A., Westmaas, G. C., and Van de Putte, P., 1972, Vegetative recombination of bacteriophage Mu-1 in *E. coli, Mol. Gen. Genet.* **116**:40.

Wijffelman, C. A., Westmaas, G. C., and Van de Putte, P., 1973, Similarity of vegetative map and prophage map of bacteriophage Mu-1, *Virology* **54**:125.

Wijffelman, C. A., Gassler, M., Stevens, W. F., and Van de Putte, P., 1974, On the control of transcription of bacteriophage Mu, *Mol. Gen. Genet.* **131**:85.

Wood, W. B., and King, J., 1979, Genetic control of complex bacteriophage assembly, *Comp. Virol.* **13**:581.

Yoshida, R. K., Miller, J. L., Miller, H. I., Friedman, D. I., and Howe, M. M., 1982, Isolation and mapping of Mu *nu* mutants which grow in *him* mutants of *E. coli, Virology* **120**:269.

Yun, T., and Vapnek, D., 1977, Electron microscopic analysis of bacteriophage P1, P1Cm, and P7, *Virology* **77**:376.

Zeldis, J. B., Bukhari, A. I., and Zipser, D., 1973, Orientation of prophage Mu, *Virology* **55**:289.

CHAPTER 7

Bacteriophage T1

Henry Drexler

I. BACKGROUND

Bacteriophage T1 is one of the many phages isolated by Milislav Demerec and was originally identified by the Greek letter α (Delbrück and Luria, 1942). In an attempt to focus the research of phage workers, Max Delbrück selected what he believed to be seven distinct phages from the Demerec collection and designated them T1 through T7, the "T" standing for "type." Among the many debts owed to Delbrück by the community of phage workers is that his designations for the T phages were symbols that appear on standard American and western European typewriters. It is now known that phages T2, T4, and T6 form a clan of related phages (the so-called "T-even" group) and T3 and T7 are related to each other. T1 and T5 are different from each other and the other T phages (Adams, 1959).

Phage lore says that one of Delbrück's criteria for the selection of the T phages was that they would form clear plaques, thus ensuring that they did not lysogenize cells. Apparently Delbrück believed that the confusion that at that time surrounded lysogeny would detract from the usefulness of phages as research tools. According to early terminology (which many investigators still use), the T phages were referred to as "virulent," in contrast to the lysogenizing phages, which were designated "temperate" (Adams, 1959); modern usage refers to the T phages as "lytic" and the lysogenizing phages as "lysogenic."

Although much work on T1 has been published, it remains the least-understood of the T phages. To a great extent, the avoidance of research

HENRY DREXLER • Department of Microbiology and Immunology, Wake Forest University Medical Center, Winston-Salem, North Carolina 27103.

with T1 stems from its almost legendary ability to contaminate stocks of other phages or cultures of *Escherichia coli*. Although the dangers of contamination with T1 have been greatly exaggerated, both its short latent period (Delbrück, 1945a) and its resistance to drying present challenges to careful workers and nightmares to those whose sterile technique is less than adequate. The unusually strong dependence of a lytic phage such as T1 on the DNA replicating and RNA transcription mechanisms of its host (Bourque and Christensen, 1980; Wagner *et al.*, 1977a) suggests that studies of the T1 life cycle will reveal new and interesting data about the organization and expression of the T1 genomes.

The goal of this review is to present a concise review of the physical nature and biological development of T1 together with a few miscellaneous facts we believe to be of interest. No attempt has been made to present an exhaustive review of the T1 literature or all the facts known about T1. The reader who wishes to learn more about T1 should consult several fine reviews that have been previously published (Hausmann, 1973; McCorquodale, 1975; Wagner *et al.*, 1983). Adams's (1959) classic book *Bacteriophages* remains a fine source of information about all the T phages with an especially enlightening, side-by-side comparison of many aspects of T-phage development.

II. BASIC CHARACTERISTICS AND METHODS OF STUDY

T1 infects most of the *E. coli* strains commonly used by researchers as well as many strains of *Shigella*. Plaque formation is usually studied with the well-known L media using 1% bottom agar and 0.65% overlays of 2–3 ml (Bertani, 1951; Drexler, 1967). Plaques are 3–4 mm in diameter with a clear, sharply defined center forming about one-third the total diameter and surrounded by a turbid, sharply defined halo. T1 will also plate on media less rich than L media provided its host grows well. Generally, quantitation of T1 by plaque formation can be done under a variety of temperatures with many types of media. However, where uniform plaques of maximum size are desired, investigators are advised to utilize L agar at 37°C.

High-titer lysates of T1 can easily be prepared in standard ways with plates covered with confluent plaques, cells growing in liquid, or through the use of a two-phase system (broth over solidified agar). Since NaCl interferes with rapid adsorption of T1, it is common to pellet lysates several times and resuspend phage particles in an NaCl-free medium such as T1RM (T1 resuspension medium; Drexler, 1970). Lysates may be sterilized by storage over chloroform or filtration. The titer of a properly stored T1 lysate may fall less than 1 order of magnitude over a period of 20 years. If lysates dry up owing to improper storage, it is nearly always possible to recover some viable phage by adding fresh broth to the dried stock.

The latent period of T1, in standard strains of *E. coli* growing exponentially in L broth at 37°C, is about 15–20 min, and the average burst size is 60–100 (Delbrück, 1945a; Borchert and Drexler, 1980; Roberts and Drexler, 1981a).

When T1 amber mutants are employed in experiments, it must be remembered that all amber mutations isolated by Michalke (1967), Figurski and Christensen (1974), Ritchie and Joicey (1980), Ritchie *et al.* (1983), and Liebeschuetz and Ritchie (1985, 1986) were isolated using *E. coli* KB-3 as the permissive strain (KB-3 carries *sup*E) and (usually) *E. coli* B as the nonpermissive strain. Results obtained with amber mutants may not be identical to data in previously reported work if permissive strains other than those with *sup*E (e.g., *sup*B or *sup*D; Borchert and Drexler, 1980) or other nonpermissive cells (e.g., the classic W3350 Su⁻ strain; Campbell, 1961; Drexler, unpublished observations) are employed.

III. THE VIRUS PARTICLE

T1 and phage λ particles are indistinguishable in electromicrographs (Maniloff and Christensen, 1967). T1 has a polyhedral head with a width and height of about 55–60 nm and a flexible, noncontractile tail about 150 nm long and 7 nm in diameter (Tikhonenko, 1970). The T1 head carries a single, unnicked length of double-stranded DNA about 16 μm long (Lang *et al.*, 1967; Bresler *et al.*, 1967). The molecular weight of T1's packaged DNA is about 31×10^6 daltons (Bresler *et al.*, 1967; MacHattie *et al.*, 1972), and the density of mature phage particles is about 1.499–1.505 gm/cm³ (Kyleberg *et al.*, 1975; MacHattie and Gill, 1977). T1 is believed to contain only the four common bases, and its G + C content is about 48% (Creaser and Tausigg, 1957; Brody *et al.*, 1967). The precise number of base pairs in a unit T1 genome has not been determined; the molecule packaged into a mature viron contains a unit genome plus a terminal repetition of 6.1% (Gill and MacHattie, 1976). The terminal repetition of packaged T1 DNA contains about 2800 bp with a variation of ± 530 bp (MacHattie *et al.*, 1972). Because of the similarity of its head size and density to λ and the slightly smaller G + C content of its DNA, it can be deduced that T1 probably packages about 49 kbp and a unit genome probably contains about 46 kbp.

IV. THE T1 MAP

The first successful attempts to isolate and map T1 markers were carried out by Carsten Bresch and co-workers. In addition to the host range marker (known variously as h, hr, or Hr) that permitted T1 to infect *E. coli* tonB, Bresh's laboratory isolated a number of mutants that affected halo size and (by use of a medium containing 2 dyes) halo color (Bresch,

1952, 1953; Bresch and Mennigmann, 1954; Bresch and Trautner, 1956). Color photographs of the plaques obtained through Bresch's unique system have been published (Bresch and Trautner, 1956). The color plaque mutants of T1 are rarely, if ever, used by modern phage workers and are not discussed in this review. For a detailed description of the color plaque mutants, the reader is referred to Bresch and Mennigmann (1954).

The climax of the work done by Bresch's group to obtain a T1 map came with the publication of the work of Michalke (1967). Michalke isolated 36 amber (am) and 12 temperature-sensitive (ts) mutants of T1 which, through complementation experiments, were placed in 19 cistrons. Representative mutants were then crossed against each other to create a map. As is customary among geneticists, Michalke numbered the am mutations in the order in which he discovered them; cistrons were numbered as they were studied.

Figurski and Christensen (1974) studied the functional nature of the T1 cistrons using both Michalke's mutants and ones they isolated. By means of biochemical tests, in vivo and in vitro complementation, and electron microscopy, Figurski and Christensen were able to classify T1 functional units according to whether they played a role in DNA synthesis or were essential for the production of competent tails or heads. For the most part, Figurski and Christensen confirmed Michalke's original ordering of the markers, although a few changes were made and the number of functional units was reduced from 19 to 18.

In common with other phages, it was found that the genes of T1 were organized topologically according to function. Figurski and Christensen suggested that the description of the T1 map follow the conventions set by Studier (1969) for phage T7 and that the genes specifying "early functions" (i.e., those essential to phage development prior to maturation as, for example, those whose products are essential for synthesis of phage DNA) be placed at the left end of the map and that the genes be numbered in order from left to right with genes discovered in the future to be given decimal values. Confusion with Michalke's system of numbering functional units was avoided by using the term "gene" instead of "cistron." Since gene 1 was located in the terminally redundant region of the T1 genome (MacHattie and Gill, 1977), it could not be properly mapped at either the left or right end of the map. Inasmuch as the gene product of gene 1 (gp1) was essential for phage DNA synthesis, it was placed, arbitrarily, at the left end of the T1 map. The T1 map shown in Fig. 1 was adapted from Figurski and Christensen (1974) and locates the original 18 genes plus 6 discovered since 1974. In addition, the map shown in Fig. 1 locates the pac site (see below) and two mutations (h and pip; see below), which have not been proved to identify separate genes. Table I is a compendium of T1 mutants.

Genes 1 and 2 are essential for phage DNA replication (Figurski and Christensen, 1974). Since synthesis of phage T1 DNA and degradation of the host chromosome are closely coupled events (Christensen et al.,

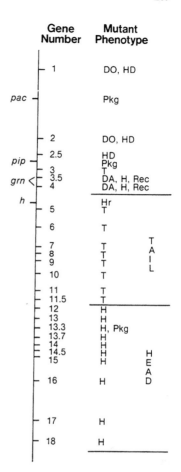

FIGURE 1. The map is adapted from Figurski and Christensen (1974), and except for gene 1 and *pac*, the relative distances between the sites or genes are based on recombination frequencies. Gene 1 is in the region of T1 DNA that becomes permuted during packaging and is placed arbitrarily at the gene 2 or left end of the map; *pac* is located between genes 1 and 2, but its exact location is unknown. The designations written to the left of the map identify the packaging initiation site (i.e., *pac*), two mutations not proved to identify unique genes (i.e., *pip* and *h*), and a system of cooperating genes (i.e., *grn*), which specifies a general recombination function. The derivations of symbols used to identify mutants are given in the footnotes to Table I. The derivation of symbols used to identify mutant phenotypes are as follows: DA, *D*NA *a*rrest (i.e., early shut off of phage DNA synthesis); DO, synthesis of phage *D*NA is zero; H, blocks formation of functional *h*eads; HD, *h*ost *D*NA is not degraded; Hr, *h*ost *r*ange; Pkg, affects *p*ac*k*a*g*ing of phage DNA; Rec, blocks phage mediated, general *rec*ombination; T, blocks formation of functional *t*ails.

1981), gp1 and gp2 are also essential for host DNA degradation. Mutations in gene 2.5 are able to block host DNA degradation without blocking phage DNA synthesis (Roberts and Drexler, 1981a,b). Curiously, gp3 appears to be essential for the production of functional phage tails even though it is located in the region of early genes (Figurski and Christensen, 1974; Ritchie and Joicey, 1980). Ritchie *et al.* (1983) showed that gene 3 mutants failed to produce a 75-kD protein. More recently, Liebeshuetz and Ritchie (1985) have shown that a cloned restriction fragment from the gene 3 region was able to fully complement gene 3 mutants, although the length of the fragment did not allow it to code for a protein larger than 45 kD; whether the activity of the cloned fragment resulted from a truncated gene 3 protein has not been established. However, the results of Liebeschuetz and Ritchie (1985), together with those of Ritchie *et al.* (1983) suggested the possibility that gp3 was a regulatory protein that was essential for the expression of tail genes. The products of genes 3.5 and 4 affect a variety of events during phage development. Either gp3.5 or gp4

TABLE I. Compendium of T1 Mutants[a]

Gene number	Amber (am) mutants	Other mutants
1	16, 20	
pac		
2	3, 5, 21, 26, 39, 221	ts120, ts252
2.5		tar1-tar7
—		pip
3	6, 41, 235, 274	
3.5	201	ts271
4	23	ts4, ts27, ts133, ts243
—		h
5	15, 203, 207, 273, 308	ts25
6	18, 31, 301	
7	35, 46	
8	1, 32, 102, 222, 233	
9	13, 232	
10	2, 19, 22, 27, 28, 40, 204, 227, 259	
11	9, 29	
11.5	304	
12	37	
13	10, 25, 260	
13.3	283	
13.7	216, 267	ts254
14	45	
14.5	246	ts3, ts7, ts27 ts28, ts30
15	11, 205	
16	4, 216, 247, 250	ts52, ts63
17	17, 219	ts49
18	30, 245, 248	

[a] The reference for all amber (am) and temperature-sensitive (ts) mutants identified with numbers less than 100 is Michalke (1967). References for other mutants are as follows: ts133 (Figurski, 1974); ambers 201, 208, 216, 246, and 283 (Ritchie and Joicey, 1980); ts254 and ts271 (Ritchie et al., 1980); am102 and ts120 (Christensen et al., 1981); ambers 221, 235, and 260 (Ritchie et al., 1983); ambers 203, 204, 205, 207, 216, 219, 222, 227, 232, 233, 245, 247, 248, 250, 259, 267, 273, 274, 301, 308 (Liebeschuetz and Ritchie, 1985, 1986); pac (MacHattie and Gill, 1977); tar (Roberts and Drexler, 1981a); pip (Drexler and Christensen, 1986); h (Drexler and Christensen, 1961; Figurski and Christensen, 1974). The derivations of the symbols used to identify mutants are as follows: am, amber; h, host range; pac, site at which initiation of phage DNA packaging is initiated; pip, affects packaging initiation and processive packaging; tar, transduces at an altered rate; ts, temperature-sensitive.

mutations cause an early shutoff of phage DNA synthesis, and both are required for synthesis of functional heads (Figurski and Christensen, 1974; Ritchie and Joicey, 1978). Further, both gp3.5 and gp4 are essential for the functioning of a phage-specified, general recombination system, termed the "grn" system (Ritchie et al., 1980).

The products of genes 5 through 11.5 are essential for the production of functional tails as measured by in vitro complementation; the success of this test is due to the ability of mature, competent phage heads and mature, competent phage tails to join *in vitro* to form infectious particles (Edgar and Wood, 1966; Weigle, 1966; Figurski and Christensen, 1974). Genes 12 through 18 are essential for the formation of functional heads as demonstrated by *in vitro* complementation (Figurski and Christensen, 1974; Ritchie and Joicey, 1980). Gene 13.3 mutants also fail to terminate head filling.

The *pac* site is the site at which T1 initiates packaging of T1 DNA and is located between genes 1 and 2 (MacHattie and Gill, 1977). The only host range mutation that has been accurately mapped with respect to nearby amber mutations (Figurski and Christensen, 1974) was located between *am23* (gene 4) and *am15* (gene 5) and is probably within gene 5. The *pip* mutation causes an increased initiation of DNA packaging at a non-T1 site (the *esp-λ* site located between the O and P genes of phage λ; Drexler, 1984) and probably a decrease in the efficiency with which T1 initiates DNA packaging at *pac*; *pip* is also defective in processive packaging; *pip* is located between *tar-1* (gene 2.5) and *am6* (gene 3), but it is not known whether it identifies a separate gene (Drexler and Christensen, 1986).

A number of restriction enzymes have been used to analyze variants of phage T1 (Ramsay and Ritchie, 1980; Ramsay and Ritchie, 1982). This discussion will pertain only to the variant of T1 that was used to select the bulk of Michalke's original *am* mutants (Michalke, 1967)—namely T1Ds^{++} ("Michalke wild type" or type 3 of Ramsay and Ritchie, 1982). T1Ds^{++} was formed by a cross of Bresch's T1 and the T1-related phage D20 (Trautner, 1960; Michalke, 1967). Of the various types of T1, T1Ds^{++} has been subjected to the most precise genetic analyses. Since the starting place for any understanding of the physical relationships between the known types of T1 is the work of Ramsay and Ritchie (1982) we would like to suggest that when it is necessary to distinguish T1Ds^{++} from other T1 variants it be referred to by the designation "T1 type 3."

T1 (or if you will T1 type 3) is more sensitive to *BglII* than to any of the restriction enzymes which recognize targets of six bases and which have thus far been reported to cut T1 (Ramsay and Ritchie, 1980, 1982) (Fig. 2). *BglII* cuts T1 16 times, and these cuts together with the packaging initiation cut at *pac* cause the DNA isolated from mature virions to form 17 precisely determined fragments, of which 15 are produced in equimolar amounts. The two precisely determined fragments of T1 that are submolar are fragments D and C. Fragment D is cleaved at one end by the initiation cleavage at *pac* and can be formed only from the first genome packaged from concatenated DNA. Fragment C is cut at both ends by *BglII* and contains the *pac* site. Fragment C can be cut only from genomes packaged processively from concatenated DNA. Since T1 headful packaging is imprecise and the number of molecules cut from a concatenate is limited (MacHattie *et al.*, 1972), a number of submolar, broad (or "fuzzy") bands appear on any agarose gel used to separate restriction enzyme fragments of T1 by gel electrophoresis (Ramsay and Ritchie, 1980).

BglI cuts six targets in T1 DNA to form (from DNA isolated from

mature virions) seven precisely determined fragments. All other restriction enzymes that recognize specific targets of six bases and which
have been reported to cut T1, cleave T1 DNA fewer times than *Bgl*I. For
example, *Eco*RI and *Pst*I make three cuts in T1 DNA, and *Hind*III makes
two cuts (Ramsay and Ritchie, 1980). Figure 2 shows the location of *Bgl*II
and *Eco*RI targets in T1.

In his work establishing a map for T1, Michalke (1967) noticed that
the markers in the center of the map appeared to be more tightly clustered than those at the ends. Michalke raised, but was unable to answer,
the question whether by chance he had isolated fewer markers at the ends
the T1 chromosome or whether there was actually more recombination
at the ends of the genome. After the discovery by MacHattie *et al.* (1972)
of limited permutation of packaged T1 DNA, the possibility arose that
markers located near the ends of the chromosome might appear to undergo more genetic segregation from nearby, more centrally located markers
because of their appearance in the terminally repetitious regions. That
gene 1 is genetically unlinked from the rest of the genome is undoubtedly
due to its location in the terminally redundant ends (Michalke, 1967;
Figurski and Christensen, 1974). It has also been demonstrated that gene
18 appears in the permuted region (P. Ooi and J. R. Christensen, personal
communication).

Liebeschuetz and Ritchie (1985) have used the size of restriction
fragments, locations of specific genes in restriction fragments, and the
size of proteins that can be assigned to particular genes to correlate the
physical and genetical maps. What Liebeschuetz and Ritchie found was
that genes 3.5 through 17 (about 80% of the known genes) occupied a
stretch of about 23,500 bases that accounted for about 50% of the T1 unit
genome. Furthermore, it was estimated that little, if any, space remained
in the gene 3.5 to gene 17 region for any additional, undiscovered, genes.
The region occupied by genes 1 through 3 (4 known genes) occupied 25%
of the genome. Even when the relatively large physical distances between
the known genes at the left end of the map were taken into account, it
appeared that more recombination (about 36%) took place than was expected from random crossovers. Likewise, Liebeschuetz and Ritchie
found that the right end of the map from gene 15 through 18 was more
recombinogenic than the center of the genome. Whether the ends of the
T1 chromosome are truly more recombinogenic than the center or other

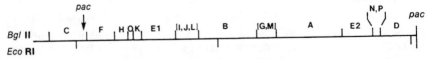

FIGURE 2. The map, showing the cut sites of *Bgl*II and *Eco*RI, is adapted from Ramsay and
Ritchie (1980) and P. Ooi and J. R. Christensen (personal communication). For the effect of
other restriction enzymes on T1 and T1-related phages, see Ramsay and Ritchie (1980,
1982), Liebeschuetz and Ritchie (1985).

genes, in addition to genes 1 and 18, are occasionally packaged in the permuted region, remains to be determined.

V. PHAGE DEVELOPMENT

A. Adsorption

Phage T1 adsorbs to host cells by a two-step process. The first step in adsorption is reversible and, since it is greatly influenced by the ionic strength of the adsorption medium, is probably electrostatic. The second, energy-requiring step is irreversible and apparently covalent (Garen and Puck, 1951; Christensen and Tolmach, 1955; Christensen, 1965). T1 seems to have adapted its adsorption mechanism to utilize surface proteins associated with the ferrichrome uptake system of the cell, and mutations in either an outer membrane function (tonA) or an inner membrane function (tonB) affect both T1 adsorption and uptake of ferric ions (Wagner et al., 1980). T1 shares the use of tonA and tonB with the lambdoid phage ϕ80.

It is interesting that although it is possible to isolate T1 mutants capable of adsorbing to tonB cells, no phage mutants capable of infecting tonA have ever been found. In T1-sensitive cultures that have been grown from cells obtained from a single, isolated colony, the ratio of spontaneous tonA mutants to tonB mutants is about 4 to 1. Since tonB is located near the tryptophan operon, selection for T1-resistant cells is frequently used as a tool to isolate deletions that penetrate the tryptophan operon.

Adsorption of T1 always takes longer and is less complete with K strains of *E. coli* than B strains (Drexler, unpublished observations), but good adsorption (i.e., >90%) can be obtained in 5–10 min with either strain, provided the cells have been washed several times to remove excess NaCl and resuspended in low-salt medium (Drexler and Christensen, 1961). Antiserum prepared against purified lysates of T1 and absorbed with host cell antigens will prevent T1 adsorption. However, T1 is one of those phages for which it is difficult to obtain high K values for antiserum (Adams, 1959). Therefore, complete inhibition of T1 adsorption by means of antiserum requires higher concentrations of antiserum and longer exposure times than for phages such as T4 or λ.

B. DNA Replication

It has been found (Bourque and Christensen, 1980) that T1, unlike certain other lystic phages (Wickner, 1978), depends on the host for some of the products known to be essential for the elongation of *E. coli* DNA. Products of the *polC*, *dnaG*, and *dnaZ* genes are all essential for the synthesis of T1 DNA. However, none of the four *dna*B mutations tested by Bourque and Christensen were able to prevent synthesis of T1 DNA. Cell products generally associated with the initiation and termination of DNA synthesis such as *dna*A, *dna*C, dnaI, *dna*P, and *dna*T are not essential to synthesis of T1 DNA.

Figurski and Christensen (1974) found that mutations in either gene 1 or gene 2 of T1 would completely block phage DNA synthesis. Results obtained with a *ts* mutation of gene 2 showed that gp2 was required for T1 DNA synthesis throughout the latent period (Walling and Christensen, 1981). Unsuppressed *am* mutations in genes 3.5 (Ritchie *et al.*, 1980) and gene 4 (Figurski and Christensen, 1974) caused an early shutoff of phage DNA synthesis. In addition, gp3.5 and gp4 resulted in an ATP-independent exonuclease activity and constituted a general recombination system (*grn*). Concatenates were absent from T1 infections in which the production of either gp3.5 or gp4 was blocked (Ritchie and Joicey, 1978; Pugh and Ritchie, 1984a); presumably the essential nature of *grn* arose from a requirement for general recombination to produce the concatenated phage DNA which served as a precursor for the phage DNA packaged into mature heads. The reason for early shutoff of T1 DNA synthesis when genes 3.5 or 4 are not functional remains unknown.

Although circular forms of T1 DNA have been observed (MacHattie *et al.*, 1972), they occurred in numbers much lower than 1 per cell (Ritchie and Joicey, 1978). The scarcity of T1 circles led Ritchie and Joicey (1978) to conclude that T1 did not replicate circularly and that concatenated T1 DNA was produced by end-to-end recombination rather that by rolling circles. The origin of replication has not yet been identified.

J. I. Mitchell and D. A. Ritchie (personal communication) found that upon infection of cells with T1, the parental phage DNA became associated with the inner membrane of the host. Association of the parental DNA with the membrane occurred even when synthesis of DNA or protein was blocked. Association did not depend on phage-coded functions. However, the observation that attachment of parental phage DNA was reduced in the presence of rifampicin suggested that attachment depended on the formation of transcriptional complexes. Labeled parental DNA did not become part of concatemers, which probably explains why only a small proportion of parental phage DNA was transferred to progeny phage. Concatenates 3–4 genomes in length were also membrane-associated, and release of DNA from the concatenates was blocked by any mutation that prevented assembly of phage heads. Mutations of gene products required for the completion of mature heads did not prevent phage DNA detachment. Therefore, Mitchell and Ritchie suggested that release of phage DNA from the membrane occurred in the form of immature heads.

The inability of the host's *recArecBrecC* general recombination system to substitute for *grn* suggested that either the production or activity of the *E. coli* exoV nuclease (which is specified by the *recBrecC* genes) was inhibited by a T1 product (see, e.g., the effect of gp*gam* of phage λ on exoV activity; Enquist and Skalka, 1973). It is interesting to note that either the *red* system of λ (Christensen, 1976) or the *recE* system of *E. coli* (Ritchie *et al.*, 1980; Pugh and Ritchie, 1984b) was able to substitute for the *grn* system. The activity of gp4 must have been exerted early, since concatenated phage DNA made soon after infection was packaged into mature particles even if a temperature-sensitive gp4 was raised to the nonpermissive temperature shortly after infection. The use of phages with *ts* mutations in gene 4 demonstrated that in contrast to gp2 (see above), gp4 became progressively dispensable during phage development (Walling and Christensen, 1981).

Recently C. L. A. Paiva and D. A. Ritchie (personal communication) isolated pseudorevertants of the phage carrying amber mutations in both gene 3.5 and gene 4; these *DNA arrest suppressors*, or "*das*," did not function by restoring either the normal products of gene 3.5 and gene 4 or the host RecBC pathway (blocked after T1 infection). In infections by T1 carrying both *das* and g3.5⁻ or g4⁻ (i.e., a mutation in gene 3.5 or gene 4 or both), concatenate formation was partially restored (thus leading to viable phage production), but the premature shutoff of phage DNA synthesis characteristic of infections with T1 g3.5⁻ or T1 g4⁻ phages was not suppressed. A low-molecular-weight protein, referred to by Paiva and Ritchie as polypeptide X, was found to be present at a higher levels in infections with g3.5⁻ or g4⁻ phages than infections with wild-type T1. However, during infections with g4⁻ *das* phages, polypeptide X was present in normal (i.e., wild-type) levels. Paiva and Ritchie suggested the possibility that in addition to its ATP-independent, exonuclease activity (see above), the *grn* system specified by gp3.5 and gp4 may have expressed a protease activity that cleaved polypeptide X.

Figurski (1974) and Roberts and Drexler (1981b) demonstrated that although host DNA degradation depended on phage DNA synthesis (Figurski and Christensen, 1974), the synthesis of T1 DNA did not depend on host DNA degradation. Hydroxyurea (HU) blocks DNA replication that depends on *de novo* synthesis of deoxyribonucleoside diphosphates (Sinha and Snustad, 1972). Ordinarily, T1 derives about two-thirds of its deoxyribonucleotides from degradation of the host chromosome (Labaw, 1951, 1953). Therefore, T1 phages capable of degrading the bacterial chromosome show almost normal development in the presence of HU. Figurski (1974) showed that even at permissive temperatures, the T1 mutant *ts*120 (gene 2) was 100 times more sensitive to HU than wild-type T1. Although *ts*120's sensitivity to HU demonstrated a significant defect in host DNA degradation by the mutant, T1 *ts*120 produced yields that were 60% those of wild-type T1. Roberts and Drexler (1981b) showed that although gene 2.5 mutations blocked degradation of host DNA (T1 gene 2.5, mutations would not produce progeny in the presence of HU), phage with mutations in gene 2.5 produced yields 30–50% those of wild type.

Ritchie and Joicey (1980) categorized T1 genes into five groups according to the amounts of phage DNA produced as well as the types of DNA structures made and accumulated during phage development. Group I contained genes 1 and 2 whose products were essential for synthesis of T1 DNA. Group IIA was made up of genes 3.5 and 4, whose functions were essential for the production of concatenated DNA; the absence of gp 3.5 and gp4 also caused an early shutoff of phage DNA synthesis. Group II contained all the genes known to be essential for the production of mature, competent tails (i.e., genes 3, 5, 6, 7, 8, 9, 10, 11, and 11.5). Phage DNA synthesis in cells infected with mutants of any of the genes of group II did not differ from synthesis in infections by wild-type T1. Mutations in the head genes, which make up group III (i.e., genes 12, 13, 13.7, 14, 14.5, 15, 16, 17, and 18), did not affect phage DNA synthesis, but infection with group III mutants showed an accumulation of concatenated phage DNA. Cells infected with T1 having a mutation in gene 13.3, which was the sole member of group IV, showed a pattern of development intermediate between group II and group III.

It is a peculiarity of T1 infection that degradation of the bacterial chromosome appears to require ongoing synthesis of phage DNA. Infection of amber-nonpermissive cells with amber mutants of either genes 1 or 2 of T1 totally blocked trichloracetic acid solubilization of host DNA. Further, if 5, 10, or 15 min after infecting and incubating cells at 30°C with ts120 (gene 2) the temperature was elevated to 40°C there was an immediate cessation of both phage DNA synthesis and host DNA degradation. Likewise, addition of naladixic acid promptly stopped both phage DNA synthesis and degradation of host DNA (Christensen et al., 1981).

C. Protein and mRNA Synthesis

Three laboratories independently studied the synthesis of phage-specific proteins during T1 infections, and their work was published within a few months of each other (Toni et al., 1976; Martin et al., 1976; Wagner et al., 1977a). Because different schemes for the purification and measurement of the proteins were used by the different investigators, it is difficult to make precise comparisons of the results obtained by the three laboratories. Nevertheless, the results of the several investigations were, as one might expect, quite similar. Since differences between the results do exist, the interested reader is advised to consult the original reports.

Thirty-one phage-specific proteins have been detected, and their molecular weights range from 9000 to 150,000 daltons (Table II). The sum total of the molecular weights of all T1 proteins is about 1.6×10^6 daltons and corresponds to about 80–100% of the coding capacity of the T1 genome. The T1 virions contain about 14 proteins whose total molecular weight is about 60% of the total coding required for all of T1's proteins. Wagner et al. (1977a) and Ritchie et al. (1983) have been able to

TABLE II. Proteins Specified by T1 DNA[a]

Protein number	Molecular weight (kD)
1	150
2	140
3	125
4	120
5	81
6	80
7	73
8	70
9	64
10	62
11	60
12	51
13	45
14	44
15	41
16	40
17	38
18	36
19	34
20	33
21	31
22	26
23	24
24	20
25	19
26	17
27	15
28	14
29	12
30	10
31	9

[a] Adapted from Wagner et al. (1977a).

associate polypeptide products with 18 of the 24 known genes of T1. There is no doubt that certain T1 proteins are subjected to posttranslational processing. For example, Ritchie et al. (1983) showed that a major constituent of the T1 head (the P7 protein listed in Table III) was derived from the precursor protein P7p. Both gp13 and gp14 were required for the production of P7, but mutations in genes 13 or 14 did not affect the production of P7p; furthermore, it was demonstrated that an am mutation in gene 14 resulted in the absence of a protein other than P7 or P7p.

Male and Christensen (1970) showed that phage specific mRNA synthesis began shortly after infection. Gawron et al. (1980) have demonstrated that most of the mRNA of T1 was transcribed from the same strand of phage DNA. It has been found by P. Ooi and J. R. Christensen

TABLE III. Proteins in Mature T1 Particles[a]

Protein band	Molecular weight (kD)	Percent total particle protein	Number of molecules per particle
P1	152	1.5	3
P2	117	1.0	3
P3	103	1.0	3
P4	57	1.2	7
P5	50	1.5	10
P6	42.5	0.7	6
P7	33	49.6	505
P8	29.5	1.5	17
P9	29	3.5	41
P10	26	16.3	211
P11	16	20.1	422
P12A	14 ⎫		
P12B	14 ⎬	2.1	52
P13	13.5 ⎭		

[a] Adapted from Martin et al. (1976).
Molecular weights of P11, P12A, P12B, and P13 are only approximations.

(personal communication) that the synthesis of mRNA that hybridized with restriction fragments known to code for gene products involved in DNA metabolism reached a peak early in the phage's latent period and later declined. mRNA that hybridized to restriction fragments whose gene products were essential for capsid formation reached a peak only after the peak of early mRNA began to decline. Furthermore, the intensity of hybridization by late RNA messages to restriction fragments was greater for some fragments than others. Even though synthesis of host mRNA continued throughout the latent period, the inducible enzyme β-galactosidase could not be induced (Male and Christensen, 1970; Jiresova and Jenecek, 1977). Expression of T1 genes did not require phage DNA replication, and T1 did not appear to make its own RNA polymerase (Wagner et al., 1977a).

T1 proteins can be categorized according to the time of their appearance and the persistence of their synthesis as (1) "early," which means their synthesis begins within 4–8 min after infection and shuts off at about 11 min; (2) "early-continuous," which appear within 4–8 min after infection and continue to be synthesized throughout the latent period although with diminishing rates; and (3) "late," which do not appear until about 8 min after infection and then are produced with increasing rates until lysis. Host protein synthesis was rapidly inhibited after infection by T1 (Toni et al., 1976; Wagner et al., 1977a); the block occurred at the level of mRNA translation and occurred *before* the synthesis of any viral protein could be detected (Wagner et al., 1977b).

In summary, the potential coding capacity of T1 suggests that the 31 phage-specific proteins that have been observed to be synthesized during T1 infection account for all, or nearly all, the proteins that can be coded for by T1. Proteins specifically associated with 18 genes known to be essential for phage development have been identified. Although the expression of T1 genes appears to follow an orderly, regulated program, no parts of the regulatory mechanism are yet known, nor is it known what part of the regulation occurs at the level of transcription or translation.

D. Maturation

As mentioned in Section IV (see also Fig. 1 and Table I), Figurski and Christensen (1974) were able to classify certain T1 genes as "head" (H) or "tail" (T) according to their ability to make mature, competent heads or tails. Ramsay and Ritchie (1984) have examined and compared the various headlike structures produced by infections with, respectively, T1, a representative unsuppressed *am* tail mutant, and unsuppressed, *am* mutation lysates of all known head genes. Ramsay and Ritchie concluded that the assembly of T1 heads led first to the formation of a "prohead" whose major component was the 33,000-dalton virion protein P7 (Table III). The prohead also contained smaller amounts of the protein P5 and probably contained P1, P3, P8, and P9. Filled heads also contained substantial amounts of the 16,000-dalton virion protein P11, which probably stabilized filled heads.

The 26,000-dalton P10 protein was a major constituent of the tail, which also contained smaller amounts of P4. The virion protein P2 was probably a part of the tail (Ramsay and Ritchie, 1984).

Packaging T1 DNA into heads required the formation of concatenated phage DNA (Ritchie and Joicey, 1978). Packaging was initiated at the *pac* site (see Fig. 1), proceeded unidirectionally toward gene 1, and terminated by means of a headful mechanism which resulted in about 6.1% terminal redundancy (MacHattie and Gill, 1977; Gill and MacHattie, 1975, 1976; Ramsay and Ritchie, 1980). After the first headful was packaged from a concatenate, packaging continued processively; about three headfuls were packaged from a single initiation event, and if the first headful is referred to as a class I molecule, the second as class II, the third as class III, etc., the proportions of classes were 0.4 : 0.4 : 0.2 (Gill and MacHattie, 1976). Since headful termination was imprecise and led to a slight variability in the amount of DNA in each packaged molecule, the various classes of packaged molecules formed closely related families rather than families of precisely determined lengths of DNA (Ritchie and Malcolm, 1970; MacHattie and Gill, 1977).

In addition to *pac*, packaging initiation is controlled by at least one other function—the *pip* function. The *pip* mutation (Drexler and Christensen, 1986) caused an increase in the efficiency with which T1 initiated packaging of DNA at the *esp-λ* site (located between the O and P genes of phage λ). Since the *pip* mutant had a relatively low burst size of 8 to 10, it was thought that the mutation led to a reduction in the efficiency of recognition of *pac* compared to *pip*⁺. In addition, T1 *pip* packaged predominantly class I molecules—that is, molecules whose packaging was initiated at *pac*. Therefore, *pip* was defective in processive packaging.

Gene 13.3 (Ritchie and Joicey, 1980) is located in the region of the T1 chromosome whose genes code for functions essential for the production of heads. It was demonstrated that during infections of cells with gene 13.3 mutants of T1, cleavage at the *pac* site occurred, but the maturation of T1 concatenates into packaged monomers did not occur. In other words, gp13.3 appeared to be required for the termination of head filling (Ramsay and Ritchie, 1983).

Studies of *in vitro* packaging of T1 and non-T1 DNA by extracts isolated from T1-infected cells have shed considerable light on the mechanism of initiation of DNA packaging by T1. Davison *et al.* (1984) found that when extracts of cells infected with T1g16$^-$ (lack of gp16 leads to the production of unfilled phage heads) were mixed with extracts of cells infected with T1g13$^-$ (gp13 is essential for prohead assembly), endogenous DNA was packaged into plaque-forming units with an efficiency of 1.5×10^3 PFU/μg DNA. When extracts of cells infected with T1g1$^-$g16$^-$ (gp1 is essential for replication of phage DNA) were mixed with extracts from cells infected with T1g1$^-$g13$^-$ and supplied with either concatenated T1 DNA isolated from cells infected with T1g13.3$^-$ (gp13.3 is essential for the termination of the headfilling reaction and its absence leads to a pileup of concatenated T1 DNA) or T1 monomers isolated from mature virions, the efficiency of PFU formation was 1×10^4/μg DNA. Davison *et al.* (1984) also found that the formation of PFU by infected cell extracts required ATP, Mg^{2+}, and spermidine.

Liebeschuetz *et al.* (1985) demonstrated that the *in vitro* packaging of T1 DNA extracted from virions was significantly enhanced when the extracts were able to convert the phage DNA into concatenates; specifically, it was shown that extracts from cells infected with T1g1$^-$ were able to convert virion DNA into PFUs 30- to 60-fold more efficiently than extracts from cells infected with either T1g1$^-$g3.5$^-$ or T1g1$^-$g4$^-$ (gp3.5 and gp4 are both essential for the formation of concatenated phage DNA). Indeed, Liebeschuetz *et al.* (1985) were able to show directly that virion DNA was efficiently converted to concatemers by the T1g1$^-$ extracts but not the T1g1$^-$g3.5$^-$ or T1g1$^-$g4$^-$ extracts.

Hug *et al.* (1986) examined the *in vitro* packaging of virion DNA isolated from more than 40 T1-like phages as well as phages thought to be totally unrelated to T1 (e.g., T3, T7, and λ). DNA from the T1-like phages was found to fall into three categories according to the efficiency with which the T1 type 3 (i.e., wild-type) extracts were able to convert the DNA into PFU. The DNA of one group of phages apparently carried the same (or a very similar) *pac* site as T1 and was packaged as efficiently as T1 DNA. The DNA of a second group seemed to have a *pac* site that was similar, but not identical, to T1 DNA and was packaged with 10–30% the efficiency of T1. The DNA of the third group appeared to have a *pac* site unrelated to T1, and their DNA was packaged with an efficiency of about 1% that of wild-type T1.

When extracts of T1-infected cells were supplied with DNA isolated from T3 or T7 virions, infectious particles with the antigenic specificity of T1 and the genome of T3 (or T7) were detected, but the efficiency of infectious particle formation was less than 1% that of wild-type T1 PFU. Hug *et al.* (1986) tested the ability of extracts from T1-infected cells to package monomers and concatemers of either unit length or "short" (i.e., 80% unit length) λ DNA. The packaging of λ DNA into T1 particles was unaffected by multimerization. The "short" λ DNA was packaged into PFU about 40 times more efficiently than the unit length λ genomes.

The observations by Liebeschuetz *et al.* (1985) that T1 DNA monomers were not only converted to concatenates in an *in vitro* system but that the conversion increased the efficiency of packaging strongly suggested that *in vitro* packaging mimicked *in vivo* packaging. Furthermore, the demonstration of specificity in packaging by Hug *et al.* (1986) supported the conclusions of Liebeschuetz *et al.* (1985) by indicating that the normal *pac* specificity operated during *in vitro* packaging. The ability of the *in vitro* packaging system to form PFU from monomers of T1, T3, or T7, etc., suggested that T1 has more than one mechanism of initiating DNA packaging and that in addition to the mechanism that initiated packaging at *pac* located on a concatenate, T1 was able to less efficiently package precut lengths of DNA probably through initiation at a "free end" of DNA. Certain results obtained through the study of transduction (see below) are also useful in understanding the initiation of head filling by T1.

VI. MISCELLANEOUS

A. Restriction and Modification

T1 is not known to specify a restriction modification system. T1 is affected by a number of restriction modification systems. The effects of restriction and modification on T1 have been studied the most with the restriction modification system specified by phage P1. In general, the effect of the P1 restriction modification system on T1 is similar to the effect of other systems on other phages and can be summarized as follows: the damaging effects of P1 restriction of T1 DNA (Lederberg, 1957; Drexler and Christensen, 1961), as well as the modification of genomes that escape damage, were shown to be mediated by a gene product(s) specified by phage P1 (Christensen, 1961). Multiple infection of restricting cells by unmodified T1 overcame restriction through cooperation between T1 genomes (Freshman *et al.*, 1968). Modification of T1 in P1-lysogenic cells was the result of methylation of T1 DNA (Klein, 1965).

Curiously, T1 was unaffected by the restriction-modification systems mediated by the K and B strains of *E. coli* (Lederberg, 1957). Wagner *et al.* (1979) presented evidence that T1 infections may inhibit host methyltransferase while at the same time T1 induced its own methyltransferase. It is possible that T1 specified a product that modified (i.e., protected) some of the sites in its own DNA that served as targets for certain restriction enzymes. Indeed, Auer and Schweiger (1984) demonstrated that during T1 infection of *E. coli dam* strains (which lack 5'-GATC-3' methyltransferase activity), the 5'-GATC-3' sites present in T1 showed almost complete methylation. Thus, as was true for many phages (see e.g., Toussaint, 1976; Krüger and Bickle, 1983), T1 appeared to encode enzymes that enabled it to modify its own DNA.

B. Coinfection and Exclusion

T1 and T2 were among the earliest phages studied in coinfections (Delbrück and Luria, 1942). When T1 and T2 infected a cell at the same time, T1 was always excluded. Only if T1 infection proceeded T2 by 6–7 min were coinfected cells able to produce T1. During the simultaneous infection of cells with T1 and T7 (Delbrück, 1945b), T7 usually prevailed, and only a 4-min head start by T1 enabled it to exclude T7 regularly. Several mechanisms for the exclusion of T1 by either T2 or T7 have been proposed (Delbrück, 1945b) but there is no experimental data bearing on this subject.

T1 is not always excluded in mixed infections (Adams, 1959; Geiman *et al.*, 1974). For example, in their study of coinfection of cells by T1 and λ, Geiman *et al.* (1974) showed that although T1 and λ have indistinguishable morphologies (see Section III), phenotypic mixing did not occur, at least in part, because T1 excluded λ. The lack of phenotypic mixing during coinfection by T1 and λ was not the result of an inability by T1 to package λ DNA, since it was shown that T1 was able to package entire λ prophages (Drexler, 1977; Drexler and Christensen, 1979). During coinfection of cells with T1 *am*23 (gene 4) and λ, it was discovered that the λ *red* system was able to substitute for the defective T1 function and that expression of both *red*X and *red*B were required for the complementation to take place. T1 was unable to complement the growth of λ *red* in *pol*A hosts (Christensen, 1976).

In cells lysogenic for λ, the λ *rex* function caused premature lysis of T1-infected cells, and as a result the average burst size of T1 was reduced from about 100 to 17 (Christensen and Geiman, 1973). Christensen *et al.* (1978) found that the growth of λ in a cell prior to superinfection with T1 resulted in the exclusion of T1, and, as a result of further experiments, they were able to show that exclusion was due to both an "early" and a "late" mechanism. The employment of a variety of λ *bio* phages and various mutants of λ enabled Christensen *et al.* (1978) to show that early exclusion of T1 by λ was caused directly by gpN rather than a λ function whose expression depended on gpN. In the absence of gpN, T1 was excluded by a later-developing λ function which was either gpQ or a product whose expression depended on gpQ. An interesting, but so far unexplained, observation by Christensen *et al.* (1978) was that if T1 carried a gene 4 mutation, it would be excluded by the late exclusion mechanism of λ but not the early one.

C. Transduction

The existence of transduction by a phage with a strictly lytic mode of development (e.g., T1) was once considered unlikely (Garen and Zinder, 1955; Luria *et al.*, 1960). At the time transduction by T1 was discovered (Drexler, 1970), it was generally believed that only phages that initiated the packaging of DNA at random sites (i.e., phages with completely permuted genomes) were capable of being generalized transducers (Ozeki and Ikeda, 1968), whereas T1 was (at that time) thought to have a unique, unpermuted genome (Michalke, 1967). Based on present knowledge, a more likely generality is that most (perhaps all) generalized transducers found in nature initiate the packaging of their own DNA at (or near) a specific site and terminate packaging by a headful mechanism. The ability to observe transduction by T1 required a technique that permitted the survival of transductants, and this was accomplished through the propagation of phages with *am* mutations on amber-permissive cells and the transduction of amber-nonpermissive cells.

It was shown that T1 preferred to initiate unidirectional packaging of non-T1 DNA at specific sites (Drexler, 1977, 1984). Therefore, T1 seemed to initiate the packaging of non-T1 DNA by the same mechanism it utilized to initiate the packaging of its own DNA. That transducing particles had the same density as T1 pfu suggested that termination of the packaging of non-T1 DNA was by the headful mechanism (Kyleberg *et al.*, 1975). T1 was able to package bacterial DNA (Drexler, 1970), integrated prophages (Bendig and Drexler, 1977; Drexler, 1977; Drexler and Christensen, 1979), and plasmids (Drexler, 1984).

Since transduction by T1 is highly reproducible, it is a useful tool to study the characteristics of generalized transduction (Drexler and Kyleberg, 1975; Kyleberg *et al.*, 1975; Bendig and Drexler, 1977; Drexler and Christensen, 1979; Borchert and Drexler, 1980). One of the most valuable contributions of T1 transduction to the study of the phage is that transduction can be used to isolate mutants that define gene functions that are important in the phage development. For example, transduction led to the discovery of the *tar* mutants that block the degradation of the host DNA (see Sections IV and V.B; Roberts and Drexler, 1981a,b). Likewise, transduction was used to discover the *pip* mutation which affects packaging initiation and processive packaging (see Section V.D; Drexler and Christensen, 1986).

As mentioned above, Liebeschuetz and Ritchie (1985) succeeded in cloning many T1 restriction fragments into plasmids. Similar to results obtained by Orbach and Jackson (1982), Liebescheutz and Ritchie (1986)

observed that the transduction of a multicopy plasmid by T1 was significantly increased when the plasmid contained any fragment of T1. The
efficiency of transduction of plasmids whose cloned T1 DNA did not
contain the *pac* site was dependent on the length of the cloned DNA and
a functioning T1 *grn* system in the donor. Since T1 interferes with the
RecBC recombination system of the *E. coli* host, the requirement for
expression of *grn* for efficient transduction of T1 DNA-containing
plamids that lacked *pac* suggested that efficient packaging of the plasmids depended on cointegration of the plasmid and transducing phage
through general recombination between homologous lengths of T1 DNA.
Resolution of the transduced plasmid DNA to form viable, circular plasmids in recipients could be mediated by either RecBC or *grn*.

Similar to phage P22 (Schmidt and Schmieger, 1984), Liebeschuetz
and Ritchie (1986) found that the presence of T1 *pac* in a cloned fragment
increased the efficiency of transduction of a plasmid above that of plasmids whose T1 fragment did not have *pac*. Furthermore, when *pac* was
present in a plasmid, efficient transduction of the plasmid did not require
expression of *grn* in the donor, although general recombination was still
required to establish a transduced plasmid in the recipient. Liebeschuetz
and Ritchie (1986) also confirmed the observation of Drexler (1984) that
the presence of a *pac*-like site, namely *esp*-λ, increased the ability of T1 to
transduce a plasmid compared to one with no *pac*-like site.

If the DNA of *pac*-containing plasmids was extracted from transducing particles and digested with a restriction enzyme that recognized only
one site in the plasmid, plasmid molecules of unit length were produced
(Liebeschuetz and Ritchie, 1986). Therefore plasmid DNA must be packaged by T1 as head-to-tail multimers at least 2 units in length; in fact, all
the data were consistent with the idea that *pac*-containing plasmid DNA
was packaged into T1 heads as a continuous head-to-tail multimer containing about seven copies of the plasmid. How T1 is able to "create"
such a multimeric plasmid is unknown.

Silberstein and Cohen (1987) have reported that inactivation of the
RecBCD nuclease (i.e., exoV) coupled with increased activity by either
RecF or RecE leads to a significant diversion of replication by a plasmid
such as pBR322 from the circular mode to the production of linear multimers; perhaps T1 is able to reproduce such conditions during infection.

VII. SUMMARY

T1 is a lytic (i.e., virulent) phage which appears to have certain characteristics of both lytic and lysogenic (i.e., temperate) phages. In common
with lytic phages and T1's short latent period and its degradation of the
host chromosome. Similar to lysogenic phages is T1's strong dependence
on the host for its DNA replication and transcription.

A comparison of the functions so far identified in Fig. 1 and Table I
with the number of proteins identified in Table II suggests that 5–10
protein-specifying T1 genes remain to be discovered. No T1 function that

interacts directly with host functions has been specifically identified, although either gp1 or gp2 (or both) seems certain to interact with one or more host functions during early phage development. Probably a T1 product interacts with exoV and (perhaps) another with the function specified by the host cell's *dam* gene.

T1 is representative of certain phages that initiate unidirectional packaging of their DNA at a specific site, terminate packaging by the headful mechanism, and package processively (e.g., phage P22; Tye *et al.*, 1974; Jackson *et al.*, 1978). The study of DNA packaging by T1 will probably contribute valuable information about the T1 and P22 mechanism of packaging and, at the same time, the packaging of host DNA by generalized transducing phages.

Despite the sizable amount of information available about T1, some of the most critically important facts needed to explain phage development are missing. Especially conspicuous by their absence are facts about the control of phage mRNA transcription by T1. Likewise, replication of phage DNA by T1 is understood only in the most primitive way.

ACKNOWLEDGMENTS. I dedicate this review to my friend and mentor Dr. James Roger Christensen, whose honest and idealistic approach to science has always been an inspiration to me. I am deeply grateful to P. Ooi and J. R. Christensen, J. Liebeschuetz and D. A. Ritchie, J. I. Mitchell and D. A. Ritchie, and C. L. A. Paiva and D. A. Ritchie for making their unpublished results available to me. J. R. Christensen read this review and made a number of helpful suggestions.

REFERENCES

Adams, M. H., 1959, *Bacteriophages*, Interscience, New York.

Auer, B., and Schweiger, M., 1984, Evidence that *Escherichia coli* virus T1 induces a DNA methyltransferase, *J. Virol.* **49**:588.

Bendig, M. M., and Drexler, H., 1977, Transduction of bacteriophage Mu by bacteriophage T1, *J. Virol.* **22**:640.

Bertani, G., 1951, Studies on lysogenesis. I. The mode of phage liberation by lysogenic *Escherichia coli*, *J. Bacteriol.* **62**:293.

Borchert, L. D., and Drexler, H., 1980, T1 genes which affect transduction, *J. Virol.* **33**:1122.

Bourque, L. W., and Christensen, J. R., 1980, The synthesis of coliphage T1 DNA: Requirement for host *dna* genes involved in elongation, *Virology* **102**:310.

Bresch, C., 1952, Unterscheidung verschiedener Bakteriophagentypen Durch Farbindikator *Nährböden, Zentral. Bakteriol. Parasitenkd. Infectionskr. Hyg. Abt 1 Orig.* **159**:47.

Bresch, C., 1953, Genetical studies on bacteriophage T1, *Ann. Inst. Pasteur (Paris)* **84**:157.

Bresch, C., and Mennigmann, H.-D., 1954, Weitere Untersuchungen zur Genetik von T1 Bacteriophagen, *Z. Naturforsch.* **9b**:212.

Bresch, C., and Trautner, T., 1956, Die Bedeutung des Zweifarb-Nährbodens für Bacteriophagen T1, *Z. Indukt. Abstamm. Vererbungs.* **87**:590.

Bresler, S. E., Kiselev, N. A., Manjakov, V. F., Mosevitsky, M. I., and Timhovsky, A. L., 1967, Isolation and physicochemical investigations of T1 bacteriophage DNA, *Virology*, **33**:1.

Brody, E. N., Mackal, R. P., and Evans, E. A. Jr., 1967, Properties of infectious T1 deoxyribonucleic acid, *J. Virol.* **1**:76.

Campbell, A., 1961, Sensitive mutants of bacteriophage λ, *Virology* **14**:22.

Christensen, J. R., 1961, On the process of host-controlled modification of bacteriophage, *Virology* **13**:40.

Christensen, J. R., 1965, The kinetics of reversible and irreversible attachment of bacterio-phage T1, *Virology* **26**:727.

Christensen, J. R., 1976, The red system of bacteriophage λ complements the growth of a bacteriophage T1 gene 4 mutant, *J. Virol.* **17**:713.

Christensen, J. R., and Geiman, J. M., 1973, A new effect of the *rex* gene of phage λ: Premature lysis after infection by phage T1, *Virology* **56**:285.

Christensen, J. R., and Tolmach, L. J., 1955, On the early stages of infections of *Escherichia coli* B by bacteriophage T1, *Arch. Biochem. Biophys.* **57**:195.

Christensen, J. R., Gawron, M. C., and Halpern, J., 1978, Exclusion of bacteriophage T1 by bacteriophage λ. I. Early exclusion requires λ N gene product and host factors involved in N gene expressions, *J. Virol.* **25**:527.

Christensen, J. R., Figurski, D. H., and Schreil, W. H., 1981, The synthesis of coliphage T1 DNA: Degradation of the host chromosome, *Virology* **108**:373.

Creaser, E. H., and Tausigg, A., 1957, The purification and chromotography of bacterio-phages on anion-exchange cellulose, *Virology* **4**:200.

Davison, P. J., Ramsay, N., and Ritchie, D. A., 1984, *In vitro* packaging of exogenous phage T1 DNA by extracts of phage-infected *Escherichia coli, FEMS Microbiol. Lett.* **21**:71.

Delbrück, M., 1945a, The burst size distributions in the growth of bacterial viruses, *J. Bacteriol.* **50**:131.

Delbrück, M., 1945b, Interference between bacterial viruses. III. The mutual exclusion effect and the depressor effect, *J. Bacteriol.* **50**:151.

Delbrück, M., and Luria, S. E., 1942, Interference between bacterial viruses: I. Interference between two bacterial viruses acting on the same host, and the mechanism of virus growth, *Arch. Biochem.* **1**:111.

Drexler, H., 1967, The kinetics of yielder cell formation in host-controlled modification caused by P1 lysogenic cells, *Virology* **33**:674.

Drexler, H., 1970, Transduction by bacteriophage T1, *Proc. Natl. Acad. Sci. USA* **66**:1083.

Drexler, H., 1977, Specialized transduction of the biotin region of *Escherichia coli* by phage T1, *Mol. Gen. Genet.* **152**:59.

Drexler, H., 1984, Initiation by bacteriophage T1 of DNA packaging at a site between the P and Q genes of bacteriophage λ, *J. Virol.* **49**:754.

Drexler, H., and Christensen, J. R., 1961, Genetic crosses between restricted and unre-stricted phage T1 in lyogenic and nonlysogenic hosts, *Virology* **13**:31.

Drexler, H., and Christensen, J. R., 1979, Transduction of bacteriophage lambda by bacterio-phage T1, *J. Virol.* **30**:543.

Drexler, H., and Christensen, J. R., 1986, T1 *pip*: A mutant which affects packaging initia-tion and processive packaging of T1 DNA, *Virology* **150**:373.

Drexler, H., and Kyleberg, K. J., 1975, Effect of UV irradiation on transduction by coliphage T1, *J. Virol.* **16**:263.

Edgar, R. S., and Wood, W. B., 1966, Morphologies of bacteriophage T4 in extracts of mutant infected cells, *Proc. Natl. Acad. Sci. USA* **55**:498.

Enquist, L., and Skalka, A., 1973, Replication of bacteriophage λ DNA dependent on the function of host and viral genes I. Interaction of *red, gam*, and *rec, J. Mol. Biol.* **75**:185.

Figurski, D. H., 1974, Genes of phage T1 involved in DNA metabolism, Ph.D. thesis, University of Rochester, Rochester, NY.

Figurski, D. H., and Christensen, J. R., 1974, Functional map of the genes of bacteriophage T1, *Virology* **59**:397.

Freshman, M., Wannag, S. A., and Christensen, J. R., 1968, Cooperative infection of P1-lysogenic bacteria by restricted phage T1, *Virology* **35**:427.

Garen, A., and Puck, T. T., 1951, The first two steps of the invasion of host cells by bacterial viruses, *J. Exp. Med.* **94**:177.

Garen, A., and Zinder, N. D., 1955, Radiological evidence for partial genetic homology between bacteriophage and host bacteria, *Virology* **1**:347.

Gawron, M. C., Christensen, J. R., and Shoemaker, T. M., 1980, Exclusion of bacteriophage T1 by bacteriophage λ. II. Synthesis of T1-specific macromolecules under N-mediated excluding conditions, *J. Virol.* **35**:93.

Geiman, J. M., Christensen, J. R., and Drexler, H., 1974, Interaction between the vegetative states of phages λ and T1, *J. Virol.* **14:**1430.

Gill, G. S., and MacHattie, L. A., 1975, Oriented extrusion of DNA from coliphage T1 particles, *Virology* **65:**297.

Gill, G. S., and MacHattie, L. A., 1976, Limited permutations of the nucleotide sequence in bacteriophage T1 DNA, *J. Mol. Biol.* **104:**505.

Hausmann, R., 1973, The genetics of T-odd phages, *Annu. Rev. Microbiol.* **27:**51.

Hug, H., Hausmann, R., Liebeschuetz, J., and Ritchie, D. A., 1986, In vitro packaging of foreign DNA into heads of T1, *J. Gen. Virol.* **67:**333.

Jackson, E. N., Jackson, D. A., and Deam, R. J., 1978, Eco RI analysis of bacteriophage P22 DNA packaging, *J. Mol. Biol.* **118:**365.

Jiresova, M., and Jenecek, J., 1977, Inhibition of β-galactosidase synthesis in *E. coli* after infections with different DNA and RNA phages, *Folia Microbiol. (Praha)* **22:**173.

Klein, A., 1965, Mechanismin der Wirtskontrollierten Modifikation des Phagen T1, *Z. Vererbrungsl.* **96:**346.

Krüger, D. H., and Bickle, T. A., 1983, Bacteriophage survival: Multiple mechanisms for avoiding the deoxyribonucleic acid restriction systems of their hosts, *Microbiol. Rev.* **47:**345.

Kyleberg, K. J., Bendig, M. M., and Drexler, H., 1975, Characterization of transduction by bacteriophage T1: Time of production and density of transducing particles, *J. Virol.* **16:**854.

Labaw, L. B., 1951, The origin of phosphorus in *Escherichia coli* bacteriophage, *J. Bacteriol.* **62:**169.

Labaw, L. B., 1953, The origin of phosphorus in T1, T5, T6 and T7 bacteriophage of *E. coli*, *J. Bacteriol.* **66:**429.

Lang, D., Bujard, H., Wolff, B., and Russell, D., 1967, Electron microscopy of size and shape of viral DNA in solutions of different ionic strength, *J. Mol. Biol.* **23:**163.

Lederberg, S., 1957, Suppression of the multiplication of heterologous bacteriophages in lysogenic bacteria, *Virology* **3:**496.

Liebeschuetz, J., and Ritchie, D. A., 1985, Correlation of the genetic and physical maps of phage T1, *Virology* **143:**175.

Liebeschuetz, J., and Ritchie, D. A., 1986, Phage T1-mediated transduction of a plasmid containing the T1 *pac* site, *J. Mol. Biol.* **192:**681.

Liebeschuetz, J., Davison, P. J., and Ritchie, D. A., 1985, A coupled in vitro system for the formation and packaging of concatemeric T1 DNA, *Mol. Gen. Genet.* **200:**451.

Luria, S. E., Adams, J. N., and Ting, R. C., 1960, Transduction of lactose-utilizing ability among strains of *E. coli* and *S. dysenteriae* and properties of the transducing phage particles, *Virology* **12:**348.

MacHattie, L. A., and Gill, G. S., 1977, DNA maturation by the "headful" mode in bacteriophage T1, *J. Mol. Biol.* **110:**441.

MacHattie, L. A., Rhodes, M., and Thomas, C. A. Jr., 1972, Large repetition in the nonpermuted nucleotide sequence of bacteriophage T1, *J. Mol. Biol.* **72:**645.

Male, C. J., and Christensen, J. R., 1970, Synthesis of messenger ribonucleic acid after bacteriophage T1 infection, *J. Virol.* **6:**727.

Maniloff, J., and Christensen, J. R., 1967, Electron microscopy of T1 structures and infection, *Biophysl. Soc. Abstr.* p. 134.

Martin, D. T. M., Adair, C. A., and Ritchie, D. A., 1976, Polypeptides specified by bacteriophage T1, *J. Gen. Virol.* **33:**309.

McCorquodale, D. J., 1975, The T-odd bacteriophage, *CRC Crit. Rev. Microbiol.* **4:**101.

Michalke, W., 1967, Erhöhte Rekombinationshäufigkeit an den Enden der T1-Chromosoms, *Mol. Gen. Genet.* **99:**12.

Orbach, M. J., and Jackson, E. N., 1982, Transfer of chimeric plasmids among *Salmonella typhimurium* strains by P22 transduction, *J. Bacteriol.* **149:**985.

Ozeki, H., and Ikeda, H., 1968, Transduction mechanisms, *Annu. Rev. Genet.* **2:**245.

Pugh, J. C., and Ritchie, D. A., 1984a, The structure of replicating bacteriophage T1DNA: Comparison between wild type and DNA-arrest mutant infections, *Virology* **135:**189.

Pugh, J. C., and Ritchie, D. A., 1984b, Formation of phage T1 concatmers by the *recE* recombination pathway of *Escherichia coli, Virology* **135**:200.

Ramsay, N., and Ritchie, D. A., 1980, A physical map of the permuted genome of bacteriophage T1, *Mol. Gen. Genet.* **179**:669.

Ramsay, N., and Ritchie, D. A., 1982, Restriction endonuclease analysis of variants of phage T1: Correlations of the physical and genetic maps, *Virology* **121**:420.

Ramsay, N., and Ritchie, D. A., 1983, Uncoupling of initiation site cleavage from subsequent headful cleavages in bacteriophage T1 DNA packaging, *Nature* **301**:264.

Ramsay, N., and Ritchie, D. A., 1984, Phage head assembly in bacteriophage T1, *Virology* **132**:239.

Ritchie, D. A., and Joicey, D. H., 1978, Formation of concatemeric DNA as an intermediate in the replication of bacteriophage T1 DNA molecule, *J. Gen. Virol.* **41**:609.

Ritchie, D. A., and Joicey, D. H., 1980, Identification of some steps in the replication of bacteriophage T1 DNA, *Virology* **103**:191.

Ritchie, D. A., and Malcolm, F. E., 1970, Heat stable and density mutants of phages T1, T3, and T7, *J. Gen. Virol.* **9**:35.

Ritchie, D. A., Christensen, J. R., Pugh, J. C., and Bourque, L. W., 1980, Genes of coliphage T1 whose products promote general recombination, *Virology* **105**:371.

Ritchie, D. A., Joicey, D. H., and Martin, D. T. M., 1983, Correlation of genetic loci and polypeptides specified by bacteriophage T1, *J. Gen. Virol.* **64**:1355.

Roberts, M. D., and Drexler, H., 1981a, Isolation and genetic characterization of T1-transducing mutants with increased transduction frequency, *Virology* **112**:662.

Roberts, M. D., and Drexler, H., 1981b, T1 mutants with increased transduction frequency are defective in host chromosome degradation, *Virology* **112**:670.

Schmidt, C., and Schmeiger, H., 1984, Selective transduction of recombinant plasmids with cloned *pac* sites by *Salmonella* phage P22, *Mol. Gen. Genet.* **196**:123.

Silberstein, Z., and Cohen, A., 1987, Synthesis of linear multimers of oriC and pBR322 derivatives in *Escherichia coli* K12: Role of recombination and replication functions, *J. Bacteriol.* **169**:3131.

Sinha, N., and Snustad, D., 1972, Mechanism of inhibition of deoxyribonucleic acid synthesis in *Escherichia coli* by hydroxyurea, *J. Bacteriol.* **112**:1321.

Studier, F. W., 1969, The genetics and physiology of bacteriophage T7, *Virology* **39**:562.

Tikhonenko, A. S., 1970, *Ultrastructure of Bacterial Viruses* (translated from Russian by B. Haigh), p. 30, Plenum Press, New York.

Toni, M., Conti, G., and Schito, G. C., 1976, Viral protein synthesis during replication of bacteriophage T1, *Biochem. Biophys. Res. Commun.* **68**:545.

Toussaint, A., 1976, The DNA modification function of phage Mu-1, *Virology* **70**:17.

Trautner, T. A., 1960, Genetische und physiologische Beziehungen zwischen den Bacteriophagen T1 and D20, *Z. Vererbungsl.* **91**:317.

Tye, B.-K., Huberman, J. A., and Botstein, D., 1974, Non-random circular permutation of phage P22 DNA, *J. Mol. Biol.* **85**:501.

Wagner, E. F., Ponta, H., and Schweiger, M., 1977a, Development of *E. coli* virus T1: The pattern of gene expression, *Mol. Gen. Genet.* **150**:21.

Wagner, E. F., Ponta, H., and Schweiger, M., 1977b, Development of *Escherichia coli* virus T1: Repression of host gene expression, *Eur. J. Biochem.* **80**:255.

Wagner, E. F., Auer, B., and Schweiger, M., 1979, Development of *E. coli* virus T1: Escape from host restriction, *J. Virol.* **29**:1129.

Wagner, E. F., Ponta, H., and Schweiger, M., 1980, Development of *Escherichia coli* virus T1: Role of proton motive force, *J. Biol. Chem.* **255**:534.

Wagner, E. F., Auer, B., and Schweiger, M., 1983, *Escherichia coli* virus T1: Genetic controls during virus infection, *Curr. Top. Microbiol. Immunol.* **102**:131.

Walling, L., and Christensen, J. R., 1981, The synthesis of coliphoge T1 DNA: Studies on the roles of T1 genes 1, 2, and 4, *Virology* **114**:309.

Weigle, J., 1966, Assembly of phage lambda *in vitro, Proc. Natl. Acad. Sci. USA* **55**:1462.

Wickner, S., 1978, DNA replication proteins of *Escherichia coli, Annu. Rev. Biochem.* **47**:1163.

CHAPTER 8

The T7 Group

RUDOLF HAUSMANN

I. INTRODUCTION

In this chapter I shall not try to review in detail the genetics and molecular biology of phage T7, since this topic has been covered repeatedly (Studier, 1972; Hausmann, 1976; Rabussay and Geiduschek, 1977; Studier and Dunn, 1983). Besides, Dunn and Studier (1983) have exhaustively analyzed the genome structure of T7, giving the full sequence of the 39,936 base pairs of its DNA and a complete survey of its 50-odd genes and its functions (where known), in addition to a full analysis of its signal sequences, such as the terminal repeats, promoters, terminators, processing sequences for RNase III, and origins of DNA replication. Rather, I shall here recapitulate some of the basic findings with regard to structure and life cycle of T7 in order to allow the reader to follow my thought with regard to the main problem I have in mind (but which I am unable to solve): How did this surprising biological entity come about? Unless we find satisfactory answers to the questions regarding origins and evolution of this phage, we cannot really claim to have understood it, notwithstanding the full grasp we might come to have of all molecular details of the workings of that self-replicating machine.

Since phages have left no fossils, the only way to monitor their phylogenetic past is by means of comparative anatomy and physiology. Thus, several related phages ought to be compared with each other in a search for fundamental similarities which hopefully would give us clues as to the genome structure of a more primitive ancestor of present-day complexity.

By comparing similar biological structures, one inevitably becomes

RUDOLF HAUSMANN • Institut für Biologie III der Universität, 78 Freiburg, Federal Republic of Germany.

entangled with the problems of systematics and taxonomy. With regard to phages, these problems are particularly acute in many respects (see e.g., Matthews, 1982, 1985; Reanney and Ackermann, 1982; Fraenkel-Conrat, 1985), and I would like to steer clear of them, as far as possible, by choosing the noncommittal term "T7 group" for encompassing all phages which, as I shall try to show, may be considered as sharing a common phylogenetic origin with T7. In particular, I want to avoid such terms as "family," "genus," and "species," since these have connotations of certain well-defined degrees of phylogenetic relatedness and therefore, at the present stage of our knowledge, might be applied improperly.

Since phage T7 is by far the best-investigated representative of its group, other group members will primarily be compared to this prototype, which therefore will be dealt with first. Then other phages, independently isolated from nature but growing on *Escherichia coli* B (the host strain on which T7 grows), will be described, followed by phages growing on other *E. coli* strains. Finally, phages growing on non-*E. coli* enterobacteria or hosts other than enterobacteria will be considered.

II. THE T7 PROTOTYPE

The original source of phage T7 is not known with certainty; the stock was present under another denomination in a laboratory collection and was included in an arbitrary series of seven virulent phages growing on the host strain *E. coli* B (Demerec and Fano, 1945). This host is still the usual laboratory host for T7, which is not able to grow on most *E. coli* strains isolated from nature, although it will grow on some *Shigella* and *Yersinia* strains (Hausmann, unpublished).

A. The T7 Virion

The T7 virion is made up of a linear genome of double-stranded DNA with a length of 12 μm, enclosed within an icosahedral head (with a diameter of about 60 nm), to which a short tail (about 20 nm in length and 10 nm wide) is attached; tenuous tail fibers have also been described (Fig. 1) (see Hausmann, 1976; Matsuo-Kato *et al.*, 1981). At least three proteins enter the tail (1 protein constitutes the tail fiber), and six proteins are found in the head (Roeder and Sadowski, 1977). Although most of these proteins are represented by only a few molecules per capsid, each such capsid has about 460 copies of the main coat protein (Adolph and Haselkorn, 1972).

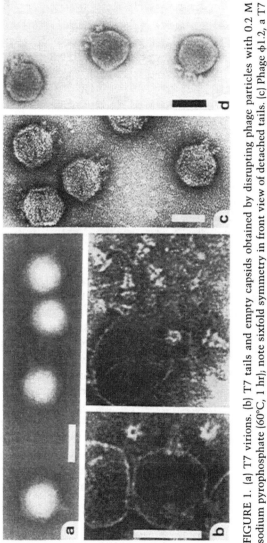

FIGURE 1. (a) T7 virions. (b) T7 tails and empty capsids obtained by disrupting phage particles with 0.2 M sodium pyrophosphate (60°C, 1 hr); note sixfold symmetry in front view of detached tails. (c) Phage φ1.2, a T7 relative specific for the encapsulated E. coli strain K235. (d) Klebsiella phage K11, another capsule-specific T7 relative. The bars represent 50 nm. Negative staining with phosphotungstic acid (a,b) or uranyl acetate (c,d). Unpublished electron micrographs by O.-G. Issinger (a,b), E. Boschek (c), and E. Freund-Mölbert (d).

B. The T7 Growth Cycle

T7 was the object of many pioneer experiments in phage biology (see, e.g., Adams, 1959), but no real insights into the molecular mechanisms of phage progeny formation resulted from those early investigations. The breakthrough came with the availability of amber mutants in the 1960s. Studier (1969) published a genetic map of 19 essential genes (defined as yielding amber mutants which are lethal on a nonsuppressor host). The map turned out to be a linear linkage group of about 200 morgan units, and the genes were numbered 1 through 19, from left to right. Additional new genes were then given fractional numbers, according to their map position (Studier, 1969, 1972).

By now, about 50 genes are located on this map with no, or only small, intergenic regions and basically without overlaps (Fig. 2)(Dunn and Studier, 1983). At the ends there are two direct repeats (Ritchie et al., 1967) of 160 base pairs (Dunn and Studier, 1981, 1983), which play a crucial, if ill-defined, role in the replication and maturation of phage DNA. Although the genes are all transcribed in the same direction, from left to right, on the genetic map, the patterns of expression differ, and these differences have been the basis for dividing the T7 genome into three regions (Studier, 1969, 1972; Studier and Maizel, 1969): (1) an early region, covering the leftmost 19% of the genome, coding for the so-called early, or class I, proteins, expressed from the onset until about 8 min after infection (all times refer to conditions at 30°C, unless otherwise stated); (2) the region coding for class II proteins, ranging from units 19 to 46 of the map (the T7 map is divided into 100 T7 units) and expressed from about 6 to 15 min after infection; and finally (3) the region coding for class III proteins, to the right of map position 46, expressed from about 8 min after infection until lysis. A functional clustering is observed, as far as class I proteins prepare the intracellular environment for phage multiplication (see below), whereas class II proteins are responsible for phage DNA synthesis, and class III genes code for coat proteins or are otherwise involved in the assembly of mature progeny particles. A schematic representation of the T7 genetic map and the patterns of its expression, as summarized in the following text, is given in Fig. 2; the patterns of T7-directed protein synthesis, as visualized by polyacrylamide gel electrophoresis (PAGE) autoradiography, are represented in Fig. 3a.

1. Expression of Early (Class I) Genes

Upon infection, the host RNA polymerase recognizes three closely linked strong promoters located near the left end of the T7 DNA molecule (Chamberlin and Ring, 1972; Bordier and Dubochet, 1974; Stahl and Chamberlin, 1977; Dunn and Studier, 1983), and from these all class I genes are transcribed before the E. coli RNA polymerase is thrown off the template with an efficiency of about 75% (McAllister and Barrett, 1977) at a rho-independent (Peters and Hayward, 1974) terminator at position 19 (Dunn and Studier, 1980, 1981; O'Hare and Hayward, 1981). The efficiency of termination at this site is highly enhanced by a newly identified host protein, the so-called tau factor (Briat and Chamberlin, 1984). Other terminators to the right have been described (Garner *et al.*, 1985). The early functions thus expressed are crucial to the problems the phage is confronted with while diverting the synthesizing capacity of the cell to the exclusive production of phage progeny. An immediate first task is the maintenance of the integrity of the genome by avoiding the action of restriction endonucleases. This function is exerted by the first T7 gene to be expressed, the leftmost on the map, gene *0.3*, whose product, the antirestriction protein, inactivates the restriction endonucleases *Eco*B and *Eco*K by tightly binding to these enzymes (Bandyopadhyay *et al.*, 1985). As a second step to take over the host machinery, RNA synthesis is redirected to the exclusive transcription of the 81% of the phage DNA located beyond (i.e., to the right) of the terminator for the host RNA polymerase. This is achieved by a dual mechanism of (1) inactivation of the host RNA polymerase, and (2) synthesis of a new, phage-coded RNA polymerase which recognizes specific promoters located on the phage DNA and which are absent from the host DNA (Chamberlin *et al.*, 1970). This newly synthesized RNA polymerase, the product of T7 gene *1*, is a single polypeptide of 883 amino acids, with a calculated molecular weight of 98,856 (Moffatt *et al.*, 1984). Inactivation of the host RNA polymerase is brought about by the gene *0.7* product, a protein kinase (Rothman-Denes *et al.*, 1973; Rahmsdorf *et al.*, 1974) and by the gene *2* product which acts by simple binding (Hesselbach and Nakada, 1975, 1977a,b). (It is worth mentioning here that gene *0.7* is transcribed by the host RNA polymerase whereas gene *2* is transcribed by the T7 RNA polymerase. The reasons for these two genes with a similar physiological effect being expressed differently are not obvious and might be simple evolutionary accidents.) One other important class I gene is that for ligase (gene *1.3*), whose function can normally be replaced by the host ligase (Studier, 1973).

2. Expression of Class II Genes

Class II genes, which are transcribed by the gene *1* RNA polymerase from so-called class II promoters, are collectively responsible for phage DNA metabolism. Although some clear-cut *in vitro* enzymatic activity has been assigned to most of class II proteins, their precise *in vivo* roles are often less well defined (Sadowski *et al.*, 1984). Nevertheless, gene *3* and gene *6*, coding for an endonuclease and an exonuclease, respectively, are certainly responsible for the breakdown of host DNA to acid-soluble material (Sadowski and Kerr, 1970), which is reincorporated into phage DNA by the action of a new DNA polymerase activity, exerted by a protein whose quaternary structure is built up from the product of T7

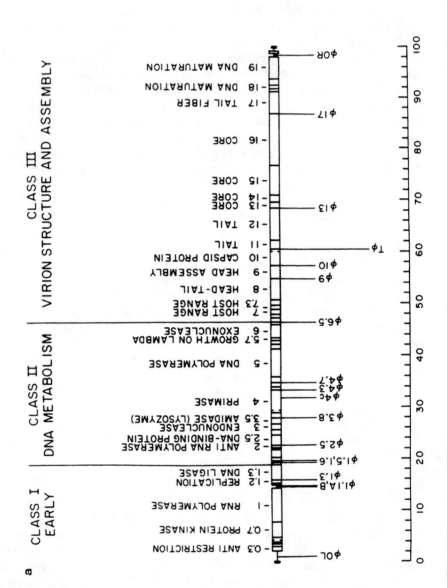

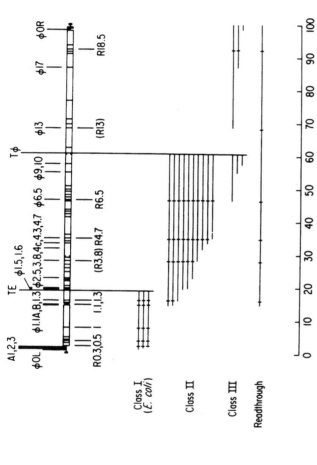

FIGURE 2. The T7 genome and its expression. (a) The main genes (open boxes drawn to scale) are characterized by their numbers and by a reference to their functions. The denominations of the promoters for the T7 RNApolymerase initiate with φ; φ1.1 through φ4.7 are class II promoters; φ6.5 through φ17 are class III promoters; φOL and φOR are promoters that characterize secondary origins of DNA replication. Tφ represents a terminator for the T7 RNA polymerase; this terminator is not fully efficient, thus allowing a certain rate of read-through transcription. The terminal repeats of T7 DNA are represented by the small filled boxes at the ends. In (b) there is also a representation of the three promoters, A1, A2, and A3, for the *E. coli* RNA polymerase; the terminator, TE, for this RNA polymerase; the RNase III cleavage sites, R 0.3 through R 18.5 (the parentheses indicate less efficient cleavage sites); the individual primary transcripts (horizontal lines) with the locations of their processing sites. From Studier and Dunn (1983), with the permission of the authors and of Cold Spring Harbor Laboratory.

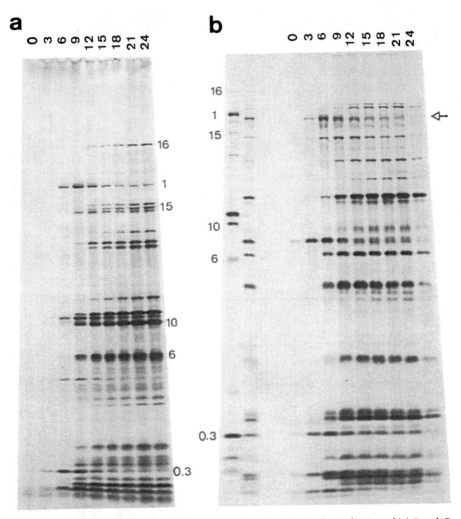

FIGURE 3. Time course of synthesis of phage-coded proteins after infection of (a) *E. coli* B by phage T7, (b) *E. coli* 09 : K31 by phage K31. Host-coded protein synthesis was inhibited by UV irradiation of bacteria growing in a medium without methionine. These bacteria were infected with phage and incubated at 30°C. At regular intervals after infection, 3-min pulses of [^{35}S]methionine were given to different samples of the infected cultures. The cells were harvested and disrupted by heating for 2 min in a buffer containing 2% sodium dodecyl sulfate. Samples of the disrupted cells were subjected to electrophoresis (200 V, 12 hr) in polyacrylamide gradient gels (a: 8–18%; b: 10–18%) containing sodium dodecyl sulfate. The gels were dried and autoradiographed for selective visualization of bands of phage-coded proteins. Numbers at the top of the tracks refer to the times after infection (in minutes) at which [^{35}S]methionine was added. Numbers in columns indicate corresponding T7 genes. The arrow at the right points to the phage K31–coded early protein of molecular weight of about 95,000 which presumably is homologous to the T7 gene *1* RNA polymerase. The first track of (b) refers to T7, the second track to K31; these tracks contain samples of continuously labeled proteins and are included for comparison. (a) is from Mertens and Hausmann (1982), and (b) from Dietz *et al.* (1986), with permission of the *Journal of General Virology*.

gene 5 and the bacterial thioredoxin (Modrich and Richardson, 1975; Adler and Modrich, 1983; Tabor and Richardson, 1985). The T7 gene 5 protein by itself has no detectable activity, as shown by means of thioredoxin mutants of *E. coli*, unable to reproduce T7 (Holmgren et al., 1978, 1981). A further requirement for T7 DNA synthesis is the presence of the gene 4 product, a protein that synthesizes RNA primers (Romano and Richardson, 1979) and also shows helicase activity (Matson et al., 1983). On the other hand, the function of the gene 2.5 protein, binding to single-stranded DNA, can be vicariously exerted by a corresponding host protein (Araki and Ogawa, 1981, 1982).

3. T7 DNA Synthesis

The first round of T7 DNA replication involves individual linear genomes. These first give rise to so-called eye forms by bidirectional DNA replication, initiated at an origin of DNA replication sequence (at map position 15); later, Y-shaped replication intermediates are observed (Dressler et al., 1972). Similar structures have also been produced with purified proteins *in vitro* (Wever et al., 1980; Fuller and Richardson, 1985). For further rounds of replication, the formation of concatemers (linear tandem arrays) of individual genomes is required (Schlegel and Thomas, 1972; Paetkau et al., 1977; Serwer et al., 1982). These concatemers form by a mechanism that is not yet well understood, but it is based on the complementary base-pairing involving the terminal repeats of different T7 DNA copies, accompanied by digestion of the surplus single strands. Thus, according to Watson (1972), the problem of RNA primer-dependent DNA synthesis at the 5' ends of the two strands of an individual genome is circumvented (except, of course, for the 5' end pieces at the extremes of a concatemer). It is not clear, however, how the terminal repeats of a single T7 molecule are prevented from annealing (this would result in T7 DNA circle formation—which has never been observed).

It is also not clear why newly synthesized T7 DNA is not normally broken down to acid-soluble material, which occurs in certain nonpermissive hosts as *Shigella sonnei* (Hausmann et al., 1968; Beck et al., 1986) and which is the normal fate of the host DNA.

Genetic recombination within the pool of replicating T7 DNA is independent of host recombination enzymes; the role of phage genes in recombination has been repeatedly investigated, and genes 3, 4, 5, and 6 were reported to be involved (Powling and Knippers, 1974; Miller et al., 1976; Lee and Sadowski, 1981; Stone and Miller, 1984). De Massy et al. (1984, 1987) found that one of the functions of the gene 3 endonuclease is that of cleaving the branched concatemeric complexes that had been described by Paetkau et al. (1977) and Tsujimoto and Ogawa (1977, 1978). However, it is not known how these branched recombinatorial intermediates are formed, and, in general, our views of the molecular events within the vegetative T7 DNA pool are not very precise. The further development of *in vitro* systems for genetic recombination of T7 DNA, such as described by Sadowski *et al.* (1984), will certainly help to clarify the picture.

4. Expression of Class III Genes

With T7 DNA synthesis and concatemer formation vigorously going on, after a few minutes of expression of class II genes, class III promoters become active, and thus class III genes begin to be expressed. [However, expression of class III proteins is not dependent on phage-directed DNA synthesis (Studier and Maizel, 1969).] These class III promoters, similarly to class II promoters, are recognized only by the gene *1* RNA polymerase. Class III promoters are much stronger than class II promoters. This, on the one hand, explains the fact that transcription of class II genes is depressed after the onset of transcription of class III genes, apparently by competition for nucleotide building blocks. On the other hand, the question arises of why class III genes are not expressed at all in the period between about 6 and 8 min, during which the weaker class II promoters are functional. The answer, according to McAllister *et al.* (1981), may simply lie in the directional and relatively slow entry of the T7 genome into the cell. As is known to occur with phage T5 (Lanni, 1969), T7 DNA would always be injected with one specific end first (in the case of T7, the left end). It has been suggested that the driving force for the entry of the whole T7 genome might be the act of transcription itself (Zavriev and Shemyakin, 1982).

Most class III genes code for coat proteins. Important exceptions are gene *9*, which codes for a protein with a scaffolding function during capsid assembly but is not a constituent part of infectious particles (Studier, 1972; Roeder and Sadowski, 1977), and genes *18* and *19*, which are essential for DNA packaging, coupled with the cutting of the concatemers to individual genome sizes (Studier, 1972). This cutting, in turn, requires repair DNA synthesis in the region of the terminal repeat which is present only once in the junction region between two concatemeric genomes, but which is present as a full double strand at both ends of an individual T7 DNA molecule (Langman *et al.*, 1978).

5. Maturation of Progeny Particles and Lysis

The appearance of mature T7 particles (Fig. 4) starts abruptly about 9 min after infection at 37°C (corresponding to about 18 min at 30°C), several minutes after the onset of synthesis of class III proteins (Hausmann and Härle, 1971). The reasons for this delay are not clear, but once initiated, maturation proceeds at a constant rate until shortly before lysis (Hausmann and Härle, 1971). The process of packaging has been repeatedly analyzed *in vitro* (Masker *et al.*, 1978; Roeder and Sadowski,

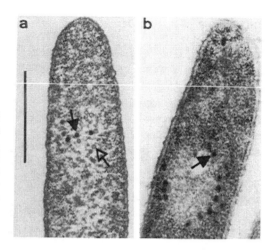

FIGURE 4. Thin sections of *E. coli* B cells fixed at 10 min (a) and 12 min (b) after infection (37°C) with T7. Empty proheads (open arrow) and filled particles (black arrows) are seen associated with the pool of vegetative T7 DNA. Thin sections negatively stained with uranyl acetate. Bar represents 1 μm. From A. Kuhn, unpublished.

1979; Masker and Serwer, 1982). Apparently, the gene 8 product is the only minor particle protein essential for DNA packaging (Roeder and Sadowsky, 1977), and since it is the only core constituent that remains stably associated with the T7 tail structure upon disruption of mature particles (Serwer, 1976), this protein has tentatively been assigned the role of a portal protein (Bazinet and King, 1985). *In vivo*, T7 DNA packaging proceeds apparently at the periphery of a pool of T7 DNA concatemers, similar in appearance to the DNA nucleoid of uninfected bacteria (Fig. 4); at no time are there membrane-associated phage-related structures (Kuhn *et al.*, 1982). This is in contrast to what has been observed for phages such as T4, where capsid assembly is membrane-bound (for review see Bazinet and King, 1985; Cajsens and Hendrix, this volume). *In vivo* packaging of T7 DNA seems to depend on specific host charactersistics, since in certain nonpermissive hosts, such as *E. coli* M (Kuhn *et al.*, 1982) and *E. coli* tsn B (Serwer and Watson, 1985), there is a specific accumulation of empty proheads and T7 concatemers.

The final step in the T7 growth cycle, lysis of the host cell, also starts rather abruptly, at about 30 min after infection. The regulation of this phenomenon is ill understood, but it is independent of the gene 3.5 lysozyme (which is a class II protein whose synthesis starts about 8 min after infection). Gene 17.5 has recently been associated with cell lysis; its mechanism of action is not clear, but it induces no lysozyme activity (Yamada *et al.*, 1986).

6. mRNA Processing and Principles of Regulation

An interesting feature of T7 genome expression is the efficient processing of transcripts by RNase III of *E. coli*. This RNase recognizes segments of double-stranded RNA (hairpin structures) formed by complementary base-pairing within adjacent sequences of an RNA strand, and it introduces specific cuts at these sites (Dunn and Studier, 1973, 1983; Gross and Dunn, 1987). Since T7 DNA transcripts (early as well as late) have several such sites, located between genes, most transcription units are cut into several individual mRNAs, each of which has its own ribosome-binding site (Dunn and Studier, 1983) (Fig. 2b). The different efficiencies of ribosome binding, which possibly depend on whether the mRNAs are processed or not (Dunn and Studier, 1975; Saito and Richardson, 1981), together with the efficiency of transcription, as determined by the strength and number of promoters from which a given T7 DNA segment is transcribed (see Fig. 2), constitute a set of variables that can be combined to regulate within wide limits the relative amounts of the individual gene products synthesized.

This overview of the T7 reproductive cycle shows a coordinated sequence of events aimed at producing a high yield of progeny virus within a few minutes. These events are very specific for T7, and studies with other phage types have disclosed quite different situations. This sum total of peculiar genetic, biochemical, and physiological observations that characterize the reproductive cycle of a virus is what has become known as its strategy of infection (Wolstenholme and O'Connor, 1971). Thus, the strategy of infection of T7 is characterized essentially by the transcription of only a small segment of the phage genome by the RNA polymerase of the host, thus enabling the early synthesis of a phage-coded single peptide RNA polymerase (with a molecular weight of about 100,000) which only recognizes promoters situated exclusively on the phage DNA and which is responsible for the selective and efficient transcription of the rest of the phage genome, while transcription of the host genome is stopped by the phage-directed inactivation of the *E. coli* RNA polymerase.

I argue that such a complex strategy evolved only once and that therefore other phages following the same strategy are to be considered as phylogenetic relatives of T7. Here I tentatively define a phage as belonging to the T7 group if, in general outline, it follows the T7 strategy.

III. THE T7 RELATIVES

A. T7 Relatives Growing on *E. coli* B

1. Phage T3

That one member of the T series, phage T3, was related to T7 was established by serological cross-reactions of these phages: T3 was inactivated by anti-T7 serum with about 10% the homologous rate (Adams and Wade, 1954). The relationship was confirmed by the observation that genetic recombination (Hausmann *et al.*, 1961), and phenotypic mixing of coat components (Issinger *et al.*, 1973) occurs between these phages.

However, the recombination rates in T3 × T7 crosses were several orders of magnitude lower than in homologous crosses (Beier and Hausmann, 1973, 1974), and *in vitro* phenotypic mixing (association of T3 tail fibers to fiberless particles of T7) occurred at only about 1/10 the homologous rate (Issinger *et al.*, 1973). In addition, host cells mixedly infected with T3 and T7 displayed the phenomenon of mutual exclusion; i.e., only 2–5% of these cells produced a mixed burst while the others produced exclusively either T3 or T7. The molecular nature of this intriguing phenomenon is far from clear, but expression of gene *1* seems to be decisive (Hausmann and Gomez, 1967). From these findings it is evident that T3 and T7 do not belong to a naturally interbreeding phage population, and since they grow on the same host, the question arises of how the genetic isolation that must underlie their divergent evolution was possible (see Section III.B).

Comparative studies with T3 and T7 took a new twist with a detailed electron microscopical heteroduplex DNA analysis by Davis and Hyman (1971). This analysis revealed that T3 and T7 share extensive sequences of partial homology, which is particularly accentuated in the right third of the genome (genes *15*, *16*, and *17*) and in a segment corresponding to gene *1*. For this gene, the base sequence homology between T3 and T7 was estimated at about 80%, a value that is in good agreement with that derived from comparing the base sequences of gene *1* of T7 (Moffatt *et al.*, 1984) and T3 (McGraw *et al.*, 1985). The heteroduplex analysis of Davis and Hyman (1971) revealed no homology in the regions of the terminal repeats of T3 and T7 (see, however, Dietz *et al.*, 1985), which might be one of the causes for the intracellular incompatibility between these phages.

One important, distinctive feature between T3 and T7 lies in the different specificities of their RNA polymerases which *in vitro* transcribe the heterologous DNA at a reduced rate (Dunn *et al.*, 1971; Hausmann and Tomkiewicz, 1976) and which *in vivo* are apparently unable to recognize the heterologous promoters (Morris *et al.*, 1986).

2. Close T7 Relatives

Besides T3, over the years a series of other T7 relatives was isolated from different locations of the globe, the main strains being the following: φI, from Milan sewage (Dettori *et al.*, 1961); φII, from Paris sewage (Monod and Wollman, 1947); W31, from Tokyo sewage (Watanabe and Okada, 1964); H, from California sewage (Cavanaugh and Quan, 1953; Molnar and Lawton, 1969); Y, from a collection at the Institut Pasteur, Paris (Girard, 1943; Hertman, 1964), and A1122, from the blood of a plague patient (Lazarus and Gunnison, 1947).

Hyman *et al.* (1974) compared phage φI, φII, W31, and H with T7 and T3 with regard to the PAGE-autoradiographic patterns of phage-directed protein synthesis and the early patterns of phage-directed RNA synthesis.

These authors also did an electron microscopic analysis of heteroduplices of T7 DNA with DNAs of W31 or φI, in order to complete similar previous comparisons among T7, T3, φII, and H by Hyman *et al.* (1973) and Brunovskis *et al.* (1973). These investigations indicated that all these additional isolates were closely related to T7 and less so to T3. In addition, it was shown that phages H and φII were nearly identical (Brunovskis *et al.*, 1973). This near identity was confirmed by Studier (1979), who subjected these strains, as well as T7, φI, W31, and T3, to restriction fragment analysis after digestion with *Hpa*I. This author proposed the use of *Hpa*I restriction patterns as easily testable and precisely informative identifying characteristics for T7-related phages. Following this suggestion, we reproduce here (Fig. 5b) a scheme of these patterns obtained for a series of T7-related phages growing on *E. coli*B, together with their PAGE-autoradiographic patterns of phage-coded proteins (Fig. 5a).

The phage collection of Fig. 5 includes the classical group representatives T7 and T3, as well as some of the already described phages. [Among these, φII and H were not included, because they have already been extensively analyzed and shown to be similar to φI (Hyman *et al.*, 1974; Studier, 1979).] The rest of the phages were isolated in our laboratory. They were chosen for this overview among some 50-odd isolates from Freiburg sewage or from the Rhine river at Cologne, on the basis of their clearly individual *Hpa*I restriction patterns.

Although no DNA homology studies were made with the new isolates of Fig. 5, I am confident that in spite of the clear differences in restriction patterns, they are—with the exception of BA14, BA127, and BA156—closely related to T7 and less so to T3. This conclusion is based on the following evidence from our laboratory: (1) crosses with amber mutants of these phages show efficient genetic recombination with T7 amber mutants, but recombination with T3 is depressed by 1–3 orders of magnitude, as compared to T3 × T3 crosses; (2) the RNA polymerases (gene *1* products) of these phages show the same template specificity as T7 RNA polymerase (although they vary in temperature stability and specific activity); (3) all these phages abortively infect the host strain *Shigella sonnei* $D_2$371–48, displaying the same pattern of breakdown of newly synthesized phage DNA (see Hausmann *et al.*, 1968; Beck *et al.*, 1986), while T3 grows normally on this host.

T7 and its close relatives have repeatedly been characterized as female-specific—i.e., unable to grow on male strains (Williams and Meynell, 1971). However, this characteristic is not typical for all isolates and is probably due to a single allelic difference. In T3, which normally grows on F^+ hosts, mutants in gene *1.2* had lost this ability (Molineux and Spence, 1984).

3. Phages BA14, BA127, and BA156

Phages BA14 (Mertens and Hausmann, 1982) and BA127 and BA156 (first described here) constitute a group apart, as judged by their inability to recombine either with T7 or with T3, and by the inability of their RNA polymerases to transcribe the DNAs of T7 and T3. On the other hand, these three phages recombine readily with each other, and they code for RNA polymerases with the same template specificity.

Thus, based on the criterion of genetic isolation, we distinguish three groups of T7-related phages infecting *E. coli* B: one rather large group (with over 50 independent isolates distinguishable by *Hpa*I restriction pattern analysis); T3 as the only representative of the second group; and finally BA14, BA127, and BA156, constituting the third group (see Fig. 5).

B. T7 Relatives Growing on *E. coli* Strains Other Than B

No attempt has been made up to now to systematically isolate T7-related phages for the large variety of coli strains found in nature, as characterized by Whittam *et al.* (1983) and Ochman and Selander (1984). In the few instances where T7 relatives were looked for on *E. coli* strains different from the usual laboratory hosts B and K12, several phages that follow the T7 strategy were isolated. They can be placed in at least two new groups, according to the template specificity of their RNA polymerases (Dietz *et al.*, 1986): phage A16 and CK235 on the one hand and K31 and ϕ1.2 on the other hand. A16 and CK235 do not transcribe the DNAs of K31 and ϕ1.2, and vice versa. None of these phages transcribes T7 or T3 DNA. The PAGE-autoradiographic patterns of phage-coded proteins (Figs. 3b and 6) strongly suggest that the new RNA polymerase activities (which appear according to the pattern of early protein synthesis in T7) are associated with the early proteins of molecular weights of about 100,000, whose bands are seen on the autoradiograms. It is characteristic of many of the other bands of the proteins coded by these phages that they cannot be easily homologized with T7 bands. Parallel to this observation go the findings that ϕ1.2 DNA and K31 DNA do not hybridize with T7 DNA and that A16 DNA and CK235 DNA hybridize with T7 DNA to a very limited extent (Fig. 7). Thus, the question arises of whether the detailed genome structure is indeed similar to that of T7, although the grand design of the infection strategy (selective transcription of the phage genome by a phage-coded RNA polymerase) is obviously the same.

It is interesting to note that the two groups established on the basis of RNA polymerase template specificity are not identical with the grouping according to host range: while phage ϕ1.2 and CK235 grow on the capsulated *E. coli* strain K235, phage A16 grows on *E. coli* E112, and phage K31 grows on *E. coli* K31. Although A16 and CK235 grow on different hosts, they show clear similarities in their PAGE-autoradiographic patterns, and thus they are clearly more closely related to each other than ϕ1.2 and CK235, which grow on the same host. Since no host range mutants for different hosts could be isolated from these phages, the occurrence of a divergent multistep change of host range during

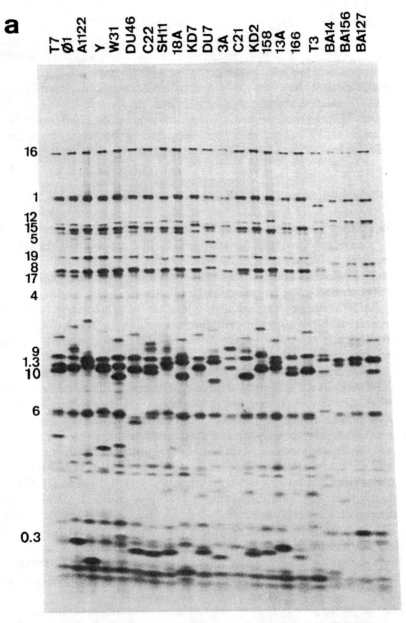

FIGURE 5. (a) Banding patterns of intracellular proteins coded by T7-related phages grow-ing on *E. coli* B. The experimental layout was as described in the legend of Fig. 3, except that [35S]methionine was present throughout the whole infectious cycle, and the cells were harvested 20 min after infection. One sample of disrupted cells infected with the phage indicated at the top was placed on a track of a polyacrylamide gradient gel of 8–18%. Numbers at the left refer to corresponding T7 genes. (b) Scheme of *Hpa*I restriction fragment patterns of the phages represented in (a). The fragment pattern of T7 (see Studier, 1979) is used as molecular weight reference; the numbers at the left indicate the lengths of some of these fragments (in kilobases).

b

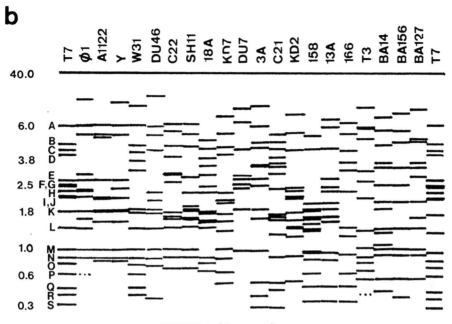

FIGURE 5. (Continued)

the evolution of these phages is quite obvious. This point is also emphasized by phage K09 (obtained from S. Stirm, University of Giessen), whose PAGE-autoradiographic pattern turned out to be very similar to that of T7 (Fig. 6) and which showed the same RNA polymerase specificity as T7, although it will not grow on *E. coli* B and T7 will not grow on the host of K09, the capsulated *E. coli* K09, nor does T7 produce any mutants able to do so. These observations are also relevant with regard to the divergent evolution of T3 and T7, which could be accounted for if the respective ancestors did not always share the same host strain.

Similarly to what is observed for T7 (Fig. 5), phage A16 seems to have a large entourage of closely related but clearly distinguishable kin strains. Contrary to what is observed for T7, though, in some cases there are clearly measurable differences in the template specificities of the RNA polymerases (Lützelschwab and Hausmann, unpublished.)

C. T7-Related Phages with Non-*E. coli* Enterobacterial Hosts

The phages considered in this section have all been previously described: phage IV, infecting *Serratia marcescens* (Wassermann and Seligmann, 1953); phage VilII, infecting *Citrobacter* spp. Ci23 (Kwiatkowski *et al.*, 1975); phage K11, infecting *Klebsiella* sp. 390 (Bessler *et al.*, 1973), and phage SP6, infecting *Salmonella typhimurium* (Butler and Cham-

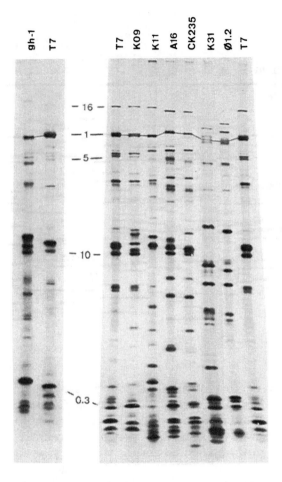

FIGURE 6. Banding patterns of intracellular proteins coded by some T7-related phages growing on hosts on which T7 does not grow. T7 is included for comparison. See legend of Fig. 5a. The bands connected by lines with the T7 gene *1* band were identified by pulse-labeling as corresponding to early proteins, presumably responsible for the early phage-directed RNA polymerase activity observed during infection with these phages. (Diffuse and multiple bands for some of these proteins are possibly due to proteolytic cleavage or electrophoretic artifact; the kinetics of early appearance are the same for all these bands.)

berlin, 1982; Kassavetis *et al.*, 1982). All code for RNA polymerases with strict template specificities for their own DNAs; i.e., they show no appreciable heterologous transcription with any other DNAs (Korsten *et al.*, 1979; Hausmann, unpublished) and thus, on the basis of this finding, each phage would be clearly set apart from all the others. Nevertheless, by the Cot technique of DNA-DNA hybridization (Britten and Kohne, 1968), Korsten *et al.* (1979) detected a relatively high base sequence homology of T3 with *Serratia* phage IV (about 25%), whereas T7 showed much less homology to that phage. The only other T7-related non-*E. coli* phage whose DNA showed a clear-cut, albeit low, hybridization with T7 DNA was *Klebsiella* phage K11 (Korsten *et al.*, 1979). Strong additional evidence for a common evolutionary origin of T7 and K11 has come from DNA base sequence comparisons in the regions of the terminal repeats (Dietz *et al.*, 1985) and in the region of gene *1* (Dietz, 1985).

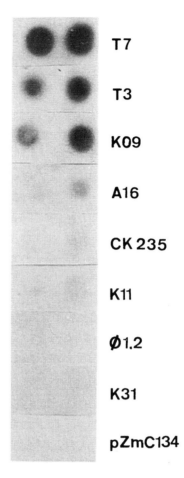

T7

T3

K09

A16

CK 235

K11

Ø1.2

K31

pZmC134

FIGURE 7. Dot hybridization of ³²P-labeled T7 DNA with the DNAs of various representatives of the T7 group. Two different amounts of DNA (right: 0.5 μg; left: 0.1 μg) of the phages indicated at the right were dotted on a nitrocellulose sheet, denatured, and incubated with nick-translated radioactive T7 DNA under nonstringent conditions. After washing, the extent of hybridization was visualized by autoradiography. DNA from pZmC134, a plasmid with cloned *Zea mays* sequences, was used as a control.

D. T7-Related Phages for Nonenterobacterial Strains

That phages following the same strategy of infection as T7 were also growing on nonenterobacterial hosts was made evident by Towle *et al.* (1975), who described the phage-coded RNA polymerase of gh-1, a phage infecting *Pseudomonas putida*. This new RNA polymerase, although not recognizing T7 DNA as a template, has several characteristics in common with the T7 RNA polymerase; (1) a high specificity for the DNA of the phage which directed its synthesis; (2) resistance to rifampicin; and (3) it is a single polypeptide with a molecular weight of 98,000, which is close to that of the T7 RNA polymerase (Towle *et al.*, 1975). *In vivo*, the gh-1 RNA polymerase is responsible for late transcription. In contrast to what is observed in T7, though, the host RNA polymerase is not inactivated in the course of the infectious cycle (Jolly, 1979). Further evidence for a phylogenetic relationship between gh-1 and T7 was provided by Korsten *et al.* (1979), who showed that the PAGE pattern of intracellular proteins coded for by gh-1 has many features in common with that of T7; the same applies to a comparison of the coat proteins (Fig. 8). However, no hybridization between gh-1 DNA and the DNAs of T7 or T3 could be detected.

A further relative of T7 seems to have been found in the form of *Caulobacter crescentus* phage φCdl. Besides being morphologically similar to T7 (West *et al.*, 1976), φCdl directs the early synthesis of a rifampicin-resistant RNA polymerase responsible for late transcription

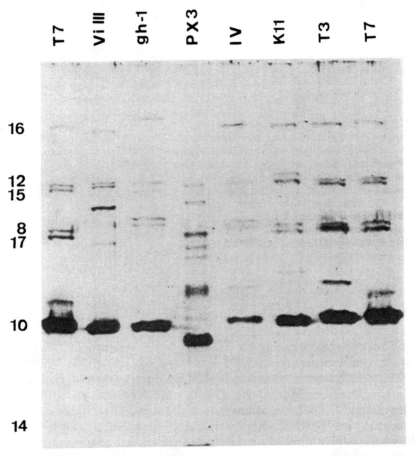

FIGURE 8. Banding patterns of main coat proteins of some phages morphologically similar to T7. Purified suspensions of particles were heated in the presence of sodium dodecyl sulfate and then subjected to electrophoresis (150 V, 3 hr) in a polyacrylamide gradient gel (10–18%) containing sodium dodecyl sulfate. The protein bands were stained with Coomassie blue. Numbers at the left indicate corresponding T7 genes. From Korsten *et al.* (1979), with permission of the *Journal of General Virology*.

(Amemiya et al., 1980). This phage also directs the early synthesis of a new protein kinase, able to phosphorylate at least some 40 host proteins, including the β' subunit of the C. crescentus RNA polymerase (Hodgson et al., 1985). But it has not been checked yet whether this phosphorylation leads to the functional inactivation of the host RNA polymerase.

A relationship to T7 has also been proposed for blue-green algae virus LPP-1 (Sherman and Haselkorn, 1970); the hypothesis is based on suggestive similarities in the capsid structure (Luftig and Haselkorn, 1968; Adolph and Haselkorn, 1972) and in the PAGE-autoradiographic patterns of intracellular virus-directed proteins (Sherman and Haselkorn, 1970). No hybridization between LPP-1 DNA and T7 RNA was observed (Luftig and Haselkorn, 1968). It has not been investigated whether LPP-1 codes for an RNA polymerase specific for the respective virus DNA, although the appearance of an early protein with a molecular weight of about 100,000 was observed within the infected host cells of *Plectonema boryanum*. Thus, a relationship of LPP-1 with T7 can only be envisaged on a tentative basis; an effort to clarify this point seems quite worthwhile.

IV. FURTHER COMPARISONS AND GENERAL REMARKS

The sum of the observations reported here makes it clear that the T7 strategy of infection is common to a large number of phages infecting hosts belonging to different families of prokaryotes. Although most known members of that group, which I here chose to call the T7 group, infect enterobacteria, especially *E. coli*, this fact may simply reflect the bias of phage investigators rather than what is prevalent in nature. Considering that a comparatively small number of phages has been genetically and physiologically investigated to some extent, it is also clear that the known representatives of the group represent only the tip of the iceberg, with regard to both the number and the degree of variation of phages following the T7 strategy.

All those phages are structurally characterized by an icosahedral head with a diameter of about 60 nm and a short tail of about 15–20 nm. Capsule-specific phages such as φ1.2 (Kwiatkowski et al., 1982) (Fig. 1) show a tail structure morphologically different from that of T7, due to the attachment of the capsule-lytic virion-bound hydrolases (Rudolph et al., 1975; Kwiatkowski et al., 1975, 1982). Whatever the differences in structural details might be, all phages here described, by virtue of their short, noncontractile tails, belong to what the International Commission for the Nomenclature of Viruses (ICNV) defined as *Podoviridae* (see Frankel-Conrat, 1985). This raises the question of the weight to be given to morphological similarity when evaluating systematic categories. Although all T7-related phages are to be classified as *Podoviridae*, the reverse is certainly not true: not all stubby-tailed phages are related to T7 by the criterion of sharing with this phage the strategy of infection based on the synthesis of a phage-coded RNA polymerase for "late" transcription. Thus, the temperate phage P22, a typical Podovirus, has a genome structure and an infection strategy that bear no resemblance to T7 (see Poteete, 1988); the same can be said for *Pseudomonas* phage PX3, which is morphologically very similar to T7, although the capsid proteins (Fig. 8) and the intracellular phage-coded proteins display a PAGE pattern that is quite different from that of T7-related phages, and no phage-directed RNA polymerase activity was detected during infection (Korsten *et al.*, 1979).

Although not all phages that qualify as *Podoviridae* appear to follow the T7 strategy, it is to be expected that many will turn out to do so. This outlook is especially interesting in view of the fact that *Podoviridae* have been described for dozens of genera, including, for instance, *Acinetobacter*, *Azotobacter*, *Brucella*, *Hydrogenomonas*, *Proteus*, *Rhizobium* (cf. Fraenkel-Conrat, 1985), and *Vibrio* (see Chatterjee and Maiti, 1984).

In spite of the lack of a comprehensive view of the distribution and variability of phages of the T7 group, some firm conclusions regarding some of the group's characteristics can be drawn from our present observations: First, in the case of T7 proper, there is a strong evidence of a vigorous diversification, possibly based on clonal evolution which is rather recent, as shown by the relative homogeneity of new isolates belonging to what I would like to call the "T7 superclone" (Fig. 5) and by the full genetic compatibility in crosses between its members. Particularly noteworthy with regard to this group is the great similarity between some different isolates from different places of the globe at different times (for instance: ϕI, ϕII, and H). The genomes of some of these seem to have diverged by only a few percent of base substitutions, as judged by the similarity of restriction patterns. Since about 5–10% of base substitutions would practically erase all similarity of the corresponding restriction fragment patterns, it is evident that such strains as W31, ϕII, and H are very closely related to T7, a conclusion reached already by Hyman *et al.* (1974). On the other hand, there is what appears to be a continuum of increasing divergence from the standard type (Fig. 5).

Phage T3, another phage for *E. coli* B, for which no close relatives have been described, represents a quantum jump for its distinctive features which set it aside from the multiplicity of phages of the T7 superclone. Nevertheless, in view of its genetic incompatibility with T7 and the close T7 relatives, the overall genome organization of T3 is surprisingly similar to that of T7 (Beier and Hausmann, 1973; Yamada *et al.*, 1986). Greatly similar also are the respective assembly and DNA maturation pathways (Serwer *et al.*, 1983; Yamagishi *et al.*, 1985; Hamada *et al.*, 1986). But there are some marked differences. Thus, as already mentioned, their RNA polymerases recognize different promoters (Dunn *et al.*, 1971; Hausmann and Tomkiewicz, 1976; Bailey *et al.*, 1983; Morris *et al.*, 1986). Further, in the early region there are at least three nonessential

T3 genes with no counterpart in T7, and three nonessential T7 genes have no counterpart in T3; and the T3 primary origin of replication does not occupy the same map position as that of T7 (Schmitt *et al.*, 1987). In an *in vitro* packaging system with concatemers, T3 proheads package T3 DNA much more efficiently than T7 DNA, and *vice versa* (Yamagishi *et al.*, 1985). (However, if mature DNA is used, T3 proheads will package T3 DNA and T7 DNA with the same efficiency (Hamada *et al.*, 1986).) Another difference between T3 and T7 lies in the enzymatic properties, if not the physiological functions, of their gene *0.3* products: whereas in T7 the gene *0.3* protein inactivates the *E. coli* restriction endonucleases *Eco*B and *Eco*K by complex formation (Spoerel *et al.*, 1979; Bandyopadhyay *et al.*, 1985), in T3 this protein also displays *S*-adenosyl-methionin hydrolase (SAMase) activity (Spoerel *et al.*, 1979). Recently, a base sequence analysis of gene *0.3* of T3 revealed no homology with gene *0.3* of T7; thus, functional similarity may have arisen by convergent evolution (Hughes *et al.*, 1987).

SAMase activity has also been detected early upon infection with *Serratia* phage IV, with *Klebsiella* phage K11 (Korsten *et al.*, 1979), and with coliphage BA14 (Mertens and Hausmann, 1982). These findings can be seen as indicating a close relationship among these phages, setting them apart from those unable to induce SAMase synthesis upon infection. Indeed, in the study of Korsten *et al.* (1979), *Serratia* phage IV revealed an overall base sequence homology of about 25% in relation to T3 but much less in relation to T7. Also, a comparison of the terminal repeats of T7 (Dunn and Studier, 1983), T3 (Fujisawa and Sugimoto, 1983), and K11 (Dietz *et al.*, 1985), and of portions of gene *1* of K11 (Dietz, 1985) with homologous sequences of T7 (Moffatt *et al.*, 1984) and of T3 (McGraw *et al.*, 1985) reveals a closer base sequence homology between K11 and T3 than between K11 and T7. However, postulating a closer relationship between T3 and K11 is questionable in view of the fact that the overall base sequence homology of K11 on the one hand, and T3 or T7 on the other, is much lower than that between T3 and T7 (Korsten *et al.*, 1979) (see Fig. 7). These findings could be reconciled by the hypothesis of modular evolution (Botstein 1980; Campbell and Botstein, 1983). According to this hypothesis, new phage genomes could arise by recombination of functional units of linked genes (modules) of phages which already diverged to a considerable extent. Information in this regard may come from hybridization studies using individual restriction fragments or (especially in those cases where no heterologous hybridization is observed) from base sequence analyses.

The varying promoter specificities of many of the RNA polymerases coded for by the phages here described deserve special attention. In those instances where promoters have been sequenced, namely for T7 (Rosa, 1979; Carter and McAllister, 1981; Dunn and Studier, 1983), T3 (Basu *et al.*, 1984; McGraw *et al.*, 1985), SP6 (Melton *et al.*, 1984; Brown *et al.*, 1986), and K11 (Dietz, 1985), it turned out that within a 23-bp promoter region there is a difference of only a few base pairs to account for the totally different specificities. Similar differences are likely to be the bases for all other specific RNA polymerase-promoter interactions within phages of the T7 group. (Altogether there are at least 10 such specificities known by now: those of T7, T3, BA14, SP6, K11, *Serratia* phage IV, *Citrobacter* phage ViIII, gh-1, ϕ1.2, and A16.)

In spite of the low or nonexistent heterologous interactions of the different phage RNA polymerases and their promoters, these enzymes obviously belong to the same protein family, as judged by their very similar molecular weights, their rifampicin resistance, their identical physiological roles, and (where known) their homologies in amino acid sequence and by recognizing similar promoters. Thus, these enzymes must have evolved from one ancestral RNA polymerase by successive mutational steps, accompanied by a coevolution of the respective promoters. This implies that, in the course of evolution, deleterious mutations in enzyme specificity were not lost by selection so fast as to preclude adaptive promoter mutations, and *vice versa*. Further comparative studies of promoters of T7-related phages are clearly needed to possibly arrive at a pedigree of molecular evolution of their base sequences. Thus, the promoter sequence of the patriarch RNA polymerase could perhaps be deduced.

Similar comparisons, extended to entire genomes, could conceivably yield a reasonably detailed picture of the very founding ancestor of the group. However, by comparing all extant forms that are recognizably related to T7, we might realize that a paradox persists, namely that the evolutionary origin of one of the most simple self-replicating structures is one of the most mysterious. Although we might be able to reconstruct to some extent the archetypal T7 ancestor from its present-day descendants by means of comparative genome analysis and base sequence studies, this patriarch phage will turn out to be just as complex with regard to genome size and strategy pursued as all its diverse descendants, since these all have about the same degree of complexity. Thus, unlike what we take for granted when comparing higher organisms, we are not going to find a chain of increasing complexity like that of the vertebrates, which display a series of evolutionary stages, from the lowly amphioxus to the most evolved mammals. Rather, the putative archetypal T7 ancestor is suddenly there, full-blown in its complex molecular design, not giving us any hints as to how it might have reached this complexity in the first place. It seems obvious, however, that this patriarchal structure was, in its turn, the product of a long evolution shrouded in complete mystery.

Some of the questions that arise in this regard are as follows: Could a comparative study of other phage groups reveal genome organizations and infection strategies reminiscent of T7 in a way that one could imagine even more remote phylogenetic relationships to T7, observable in phages that do not follow the T7 strategy *sensu strictu* (i.e., by coding for an early protein with an RNA polymerase activity specific for promoters of the phage DNA)? It seems that none of the better-described coliphages is a candidate for such a hypothetical far-off T7 relative. Did the forerunners of the potentially well definable T7 ancestor arise from genome segments of host bacteria, and at which stage of biological evolution did this occur? Or do these structures stem from parasitic DNA whose origin goes straight back to the primordial soup? Posing these questions more precisely, one would ask about the origin of the individual proteins, the individual genes composing the primitive phage. Considering that modern studies of molecular evolution make it clear that new proteins arise from ancient proteins (see Goodman, 1982), what is the origin, for instance, of the key regulatory protein in T7, the gene 1 RNA polymerase? Sequence analyses with other proteins have not yet been carried out in this direction. This, however, could be a promising endeavor, as judged by comparative studies which disclosed a domain of sequence homology between DNA polymerase I of *E. coli* and the T7 DNA polymerase subunit, the gene 5 product (Ollis *et al.*, 1985).

It is possible that further analyses regarding the evolutionary origins of phage genome organization and evolution of individual phage proteins might contribute to answer general questions regarding molecular evolution, the origin of proteins, and the origin of self-replicating structures— i.e., of life itself.

ACKNOWLEDGMENTS. Work from our laboratory was done with the skillful technical assistance of Ms. Monika Messerschmid and was financially supported by the Deutsche Forschungsgemeinschaft. I thank B. Boschek, E. Freund-Mölbert, O. G. Issinger, and A. Kuhn for providing unpublished electron micrographs.

REFERENCES

Adams, M. H., 1959, *Bacteriophages*, Interscience Publishers, New York.

Adams, M. H., and Wade, E., 1954, Classification of bacterial viruses: The relationship of two *Serratia* phages to coli-dysentery phages T3, T7, and D44, *J. Bacteriol.* **68**:320.

Adler, S., and Modrich, P., 1983, T7-induced DNA polymerase, *J. Biol. Chem.* **258**:6956.

Adolph, K. W., and Haselkorn, R., 1972, Comparison of the structures of blue-green algal viruses LPP-IM, LPP-2, and bacteriophage T7, *Virology* **47**:701.

Amemiya, K., Raboy, B., and Shapiro, L., 1980, Involvement of host RNA polymerase in the early transcription program of *Caulobacter crescentus* bacteriophage φCdl DNA, *Virology* **104**:109.

Araki, H., and Ogawa, H., 1981, A T7 amber mutant defective in DNA-binding protein, *Mol. Gen. Genet.* **183**:66.

Araki, H., and Ogawa, H., 1982, Novel amber mutants of bacteriophage T7, growth of which depends on *Escherichia coli* DNA-binding protein, *Virology* **118**:260.

Bailey, J. N., Klement, J. F., and McAllister, W. T., 1983, Relationship between promoter structure and template specificities exhibited by the bacteriophage T3 and T7 RNA polymerses, *Proc. Natl. Acad. Sci. USA* **80**:2814.

Bandyopadhyay, P. K., Studier, F. W., Hamilton, D. L., and Yuan, R., 1985, Inhibition of the type I restriction-modification enzymes EcoB and EcoK by the gene 0.3 protein of bacteriophage T7, *J. Mol. Biol.* **182**:567.

Basu, S., Sarkar, P., Adhya, S., and Maitra, U., 1984, Locations and nucleotide sequences of three major class III promoters for bacteriophage T3 RNA polymerase on T3 DNA, *J. Biol. Chem.* **259**:1993.

Bazinet, C., and King, J., 1985, The DNA translocating vertex of dsDNA bacteriophage, *Annu. Rev. Microbiol.* **39**:109.

Beck, P. J., Condreay, J. P., and Molineux, I. J., 1986, Expression of the unassembled capsid protein during infection of *Shigella sonnei* by bacteriophage T7 results in DNA damage that is repairable by bacteriophage T3, but not T7, DNA ligase, *J. Bacteriol.* **167**:251.

Beier, H., and Hausmann, R., 1973, Genetic map of bacteriophage T3, *J. Virol.* **12**:417.

Beier, H., and Hausmann, R., 1974, T3 × T7 phage crosses leading to recombinant RNA polymerases, *Nature* **251**:538.

Bessler, W., Freund-Mölbert, E., Knüfermann, H., Rudolph, C., Thurow, H., and Stirm, S., 1973, A bacteriophage-induced depolymerse active on *Klebsiella* K11 capsular polysaccharide, *Virology* **56**:134.

Bordier, C., and Dubochet, J., 1974, Electron microscopic localization of the binding sites of *Escherichia coli* RNA polymerase in the early promoter region of T7 DNA, *Eur. J. Biochem.* **44**:617.

Botstein, D., 1980, A theory of modular evolution for bacteriophage, *Ann. N.Y. Acad. Sci.* **354**:484.

Briat, J. F., and Chamberlin, M. J., 1984, Identification and characterization of a new transcriptional termination factor from *Escherichia coli*, *Proc. Natl. Acad. Sci. USA* **81**:7373.

Britten, R. J., and Kohne, D. E., 1968, Repeated sequences in DNA, *Science* **161**:529.

Brown, J. E., Klement, J. F., and McAllister, W. T., 1986, Sequences of three promoters for the bacteriophage SP6 RNA polymerase, *Nucleic Acids Res.* **14**:3521.

Brunovskis, L., Hyman, R. W., and Summers, W. C., 1973, *Pasteurella pestis* bacteriophage H and *Escherichia coli* bacteriopage φII are nearly identical, *J. Virol.* **11**:306.

Butler, E. T., and Chamberlin, M. J., 1982, Bacteriophage SP6-specific RNA polymerase I. Isolation and characterization of the enzyme, *J. Biol. Chem* **257**:5772.

Campbell, A., and Botstein, D., 1983, Evolution of the lambdoid phages, in:*Lambda II* (R. W. Hendrix, J. W. Roberts, F. W. Stahl, and R. Weisberg, eds.), p. 365, Cold Spring Harbor Laboratory, Cold Spring Harbor, New York.

Carter, A. D., and McAllister, W. T., 1981, Sequence of three class II promoters for the bacteriophage T7 RNA polymerase, *J. Mol. Biol.* **153**:825.

Cavanaugh, D. C., and Quan, S. F., 1953, Rapid identification of *Pasteurella pestis*, *Am. J. Clin. Pathol.* **23**:619.

Chamberlin, M., McGrath, J., and Waskell, L., 1970, New RNA polymerase from *Escherichia coli* infected with bacteriophage T7, *Nature* **228**:227.

Chamberlin, M., and Ring, J., 1972, Studies of the binding of *Escherichia coli* RNA polymerase to DNA. V. T7 RNA chain initiation by enzyme-DNA complexes, *J. Mol. Biol.* **70**:221.

Chatterjee, S. N., and Maiti, M., 1984, Vibriophages and vibriocins: Physical, chemical, and biological properties, *Adv. Virus Res.* **29**:263.

Davis, R. W., and Hyman, R. W., 1971, A study in evolution: The DNA base sequence homology between coliphages T7 and T3, *J. Mol. Biol.* **62**:287.

De Massy, B., Studier, F. W., Dorgai, L., Appelbaum, E., and Weisberg, R. A., 1984, Enzymes and sites of genetic recombination: Studies with gene-3 endonuclease of phage T7 and with site-affinity mutants of phage lambda, *Cold Spring Harbor Symp. Quant. Biol.* **49**:715.

De Massy, B., Weisberg, R. A., and Studier, F. W., 1987, Gene 3 endonuclease of bacteriophage T7 resolves conformationally branched structures in double-stranded DNA, *J. Mol. Biol.* **193**:359.

Demerec, M., and Fano, U., 1945, Bacteriophage-resistant mutants in *Escherichia coli*, *Genetics* **30:**119.

Dettori, R., Maccacaro, G. A., and Piccinin, G. L., 1961, Sex-specific bacteriophages of *Escherichia coli* K-12, *G. Microbiol.* **9:**141.

Dietz, A., 1985, Untersuchungen zur Evolution von Verwandten des Phagen T7 durch vergleichende DNA-Basensequenzanalyse, Ph.D. Thesis, University of Freiburg, Freiburg, F.R.G.

Dietz, A., Andrejauskas, E., Messerschmid, M., and Hausmann, R., 1986, Two groups of capsule-specific coliphages coding for RNA polymerases with new promoter specificities, *J. Gen. Virol.* **67:**831.

Dietz, A., Kössel, H., and Hausmann, R., 1985, On the evolution of the terminal redundancies of *Klebsiella* phage No. 11 and of coliphages T3 and T7, *J. Gen. Virol.* **66:**181.

Dressler, D., Wolfson, J., and Magazin, M., 1972, Initiation and reinitiation of DNA synthesis during replication of bacteriophage T7, *Proc. Natl. Acad. Sci. USA* **69:**998.

Dunn, J. J., and Studier, F. W., 1973, T7 early RNAs are generated by site-specific cleavages, *Proc. Natl. Acad. Sci. USA* **70:**1559.

Dunn, J. J., and Studier, F. W., 1975, Effect of RNAase III cleavage on translation of bacteriophage T7 messenger RNAs, *J. Mol. Biol.* **99:**487.

Dunn, J. J., and Studier, F. W., 1980, The transcription termination site at the end of the early region of bacteriophage T7 DNA, *Nucleic Acids Res.* **10:**2119.

Dunn, J. J., and Studier, F. W., 1981, Nucleotide sequence from the genetic left end of bacteriophage T7 DNA to the beginning of gene 4, *J. Mol. Biol.* **148:**303.

Dunn, J. J., and Studier, F. W., 1983, Complete nucleotide sequence of bacteriophage T7 DNA and the location of T7 genetic elements, *J. Mol. Biol.* **166:**535.

Fraenkel-Conrat, H., 1985, *The Viruses. Catalogue, Characterization, and Classification*, Plenum Press, New York.

Fuller, C. W., and Richardson, C. C., 1985, Initiation of DNA replication at the primary origin of bacteriophage T7 by purified proteins. Site and direction of initial DNA synthesis, *J. Biol. Chem.* **260:**3185.

Garner, I., Cromie, K. D., Marson, E. A., and Hayward, R. S., 1985, Transcription termination regions of coliphage T7 DNA: The effects of *nusA1, Mol. Gen. Genet.* **200:**295.

Girard, G., 1943. Sensibilité des bacilles pesteux et pseudotuberculeux aux bacteriophages, *Ann. Inst. Pasteur* **69:**52.

Gross, G., and Dunn, J. J., 1987, Structure of secondary cleavage sites of *E. coli* RNAaseIII in A3t RNA from bacteriophage T7, *Nucl. Acids Res.* **15:**431.

Hamada, K., Fujisawa, H., and Minagawa, T., 1986, A defined *in vitro* system for packaging of bacteriophage T3 DNA, *Virology* **151:**119.

Hausmann, R., 1976, Bacteriophage T7 genetics, *Curr. Top. Microbiol. Immunol.* **75:**77.

Hausmann, R., and Gomez, B., 1967, Amber mutants of bacteriophages T3 and T7 defective in phage-directed deoxyribonucleic acid synthesis, *J. Virol.* **1:**779.

Hausmann, R., Gomez, B., and Moody, B., 1968, Physiological and genetic aspects of abortive infection of a *Shigella sonnei* strain by coliphage T7, *J. Virol.* **2:**335.

Hausmann, R., and Härle, E., 1971, Expression of the genomes of the related bacteriophages T3 and T7, in: *Proceedings of the First European Biophysics Congress* (E. Broda, ed.), pp. 467–488, Wiener Medizinische Akademie, Vienna.

Hausmann, R., and Tomkiewicz, C., 1976, Genetic analysis of template specificity of RNA polymerases (gene 1 products) coded by phage T3 × T7 recombinants within gene 1, in: *RNA Polymerase (Monograph)* (R. Losick and M. Chamberlin, eds.), pp. 731–743, Cold Spring Harbor Laboratory, Cold Spring Harbor, NY.

Hausmann, R. L., Almeida-Magalhães, E. P., and Araujo, C., 1961, Isolation and characterization of hybrids between bacteriophages T3 and T7, *An. Microbiol. Univ. Brasil* **10:**35.

Hertman, I., 1964, Bacteriophage common to *Pasteurella pestis* and *Escherichia coli, J. Bacteriol.* **88:**1002.

Hesselbach, B. A., and Nakada, D., 1975, Inactive complex formation between *E. coli* RNA polymerase and an inhibitor protein purified from T7 phage infected cells, *Nature* **258**:354.

Hesselbach, B. A., and Nakada, D., 1977a, "Host shutoff" function of bacteriophage T7: Involvement of T7 gene 2 and gene 0.7 in the inactivation of *Escherichia coli* RNA polymerase, *J. Virol.* **24**:736.

Hesselbach, B. A., and Nakada, D., 1977b, I protein: Bacteriophage T7-coded inhibitor of *Escherichia coli* RNA polymerase, *J. Virol.* **24**:746.

Hodgson, D., Shapiro, L., and Amemiya, K., 1985, Phosphorylation of the β' subunit of RNA polymerase and other host proteins upon ϕCdl infection of *Caulobacter crescentus*, *J. Virol.* **55**:238.

Holmgren, A., Kallis, G. B., and Nordström, B., 1981, A mutant thioredoxin from *Escherichia coli tsnC7007* that is nonfunctional as subunit of phage T7 DNA polymerase, *J. Biol. Chem.* **256**:3118.

Holmgren, A., Ohlsson, I., and Grankvist, M. L., 1978, Thioredoxin from *Escherichia coli*. Radioimmunological and enzymatic determinations in wild type cells and mutants defective in phage T7 DNA replication, *J. Biol. Chem.* **253**:430.

Hughes, J. A., Brown, L. R., and Ferro, A. J., 1987, Nucleotide sequence and analysis of the coliphage T3 S-adenosylmethionine hydrolase gene and its surrounding ribonuclease III processing sites, *Nucl. Acids Res.* **15**:717.

Hyman, R. W., Brunovskis, I., and Summers, W. C., 1973, DNA base sequence homology between coliphages T7 and ϕII and between T3 and ϕII as determined by heteroduplex mapping in the electron microscope, *J. Mol. Biol.* **77**:189.

Hyman, R. W., Brunovskis, I., and Summers, W. C., 1974, A biochemical comparison of the related bacteriophages T7, ϕI, ϕII, W31, H, and T3, *Virology* **57**:189.

Issinger, O. G., Beier, H., and Hausmann, R., 1973, *In vivo* and *in vitro* "phenotypic mixing" with amber mutants of phages T3 and T7, *Mol. Gen. Genet.* **122**:81.

Jolly, J. F., 1979, Program of bacteriophage gh-1 DNA transcription in infected *Pseudomonas putida*, *J. Virol.* **30**:771.

Kassavetis, G. A., Butler, E. T., Roulland, D., and Chamberlin, M. J., 1982, Bacteriophage SP6-specific RNA polymerase. II. Mapping of SP6 DNA and selective *in vitro* transcription, *J. Biol. Chem.* **257**:5779.

Korsten, K. H., Tomkiewicz, C., and Hausmann, R., 1979, The strategy of infection as a criterion for phylogenetic relationships of non-coli phages morphologically similar to phage T7, *J. Gen. Virol.* **43**:57.

Kuhn, A. H. U., Moncany, M. I. J., Kellenberger, E., and Hausmann, R., 1982, Involvement of the bacterial *groM* gene product in bacteriophage T7 reproduction, *J. Virol.* **41**:657.

Kwiatkowski, B., Beilharz, H., and Stirm, S., 1975, Disruption of Vi bacteriophage III and localization of its deacetylase activity, *J. Gen. Virol.* **29**:267.

Kwiatkowski, B., Boschek, B., Thiele, H., and Stirm, S., 1982, Endo-*N*-acetylneuraminidase associated with bacteriophage particles, *J. Virol.* **43**:697.

Langman, L., Paetkau, V., Scraba, D., Miller, R. C. Jr., Roeder, G. S., and Sadowski, P. D., 1978, The structure and maturation of intermediates in bacteriophage T7 DNA replication, *Can. J. Biochem.* **56**:508.

Lanni, Y. T., 1969, Function of two genes in the first-step-transfer DNA of bacteriophage T5, *J. Mol. Biol.* **44**:173.

Lazarus, A. S., and Gunnison, J. B., 1947, The action of *Pasteurella pestis* bacteriophage on strains of *Pasteurella*, *Salmonella*, and *Shigella*, *J. Bacteriol.* **53**:705.

Lee, D., and Sadowski, P., 1981, Genetic recombination of bacteriophage T7 *in vivo* studied by use of a single physical assay, *J. Virol.* **40**:839.

Luftig, R., and Haselkorn, R., 1968, Comparison of blue-green algae virus LPP-1 and the morphologically related viruses GIII and coliphage T7, *Virology* **34**:675.

Masker, W. E., Kuemmerle, N. B., and Allison, D. P., 1978, *In vitro* packaging of bacteriophage T7 DNA synthesized *in vitro*, *J. Virol.* **27**:149.

Masker, W. E., and Serwer, P., 1982, DNA packaging *in vitro* by an isolated bacteriophage T7 procapsid, *J. Virol.* **43**:1138.

Matson, S. W., Tabor, S., and Richardson, C. S., 1983, The gene 4 protein of bacteriophage T7, *J. Biol. Chem.* **258**:14017.

Matsuo-Kato, H., Fujisawa, H., and Minagawa, T., 1981, Structure and assembly of bateriophage T3 tails, *Virology* **109**:157.

Matthews, R. E. F., 1982, Classification and nomenclature of viruses, *Intervirology* **17**:1.

Matthews, R. E. F., 1985, Viral taxonomy for the nonvirologist, *Annu. Rev. Microbiol.* **39**:451.

McAllister, W. T., and Barrett, C. L., 1977, Hybridization mapping of restriction fragments from the early region of bacteriophage T7 DNA, *Virology* **82**:275.

McAllister, W. T., Morris, C., Rosenberg, A. H., and Studier, F. W., 1981, Utilization of bacteriophage T7 late promoters in recombinant plasmids during infection, *J. Mol. Biol.* **153**:527.

McGraw, N. J., Bailey, J. N., Cleaves, G. R., Dembinski, D. R., Gocke, C. R., Joliffe, L. K., MacWright, R. S., and MacAllister, W. T., 1985, Sequence and analysis of the gene for bacteriophage T3 RNA polymerase, *Nucleic Acids Res.* **18**:6753.

Melton, D. A., Krieg, P. A., Rebagliati, M. R., Maniatis, T., Zinn, K., and Green, M. R., 1984, Efficient *in vitro* synthesis of biologically active RNA and RNA hybridization probes from plasmids containing a bacteriophage SP6 promoter, *Nucleic Acids Res.* **12**: 7035.

Mertens, H., and Hausmann, R., 1982, Coliphage BA14: A new relative of phage T7, *J. Gen. Virol.* **62**:331.

Miller, R. C. Jr., Lee, M., Scraba, D. G., and Paetkau, V., 1976, The role of bacteriophage T7 exonuclease (gene 6) in genetic recombination and production of concatemers, *J. Mol. Biol.* **101**:223.

Modrich, P., and Richardson, C. C., 1975, Bacteriophage T7 deoxyribonucleic acid replication *in vitro*. Bacteriophage T7 DNA polymerase: An enzyme composed of phage- and host-specific subunits, *J. Biol. Chem.* **250**:5515.

Moffatt, B. A., Dunn, J. J., and Studier, F. W., 1984, Nucleotide sequence of the gene for bacteriophage T7 RNA polymerase, *J. Mol. Biol.* **173**:265.

Molineux, I. J., and Spence, J. L., 1984, Virus-plasmid interactions: Mutants of bacteriophage T3 that abortively infect plasmid F–containing (F⁺) strains of *Escherichia coli*, *Proc. Natl. Acad. Sci. USA* **81**:1465.

Molnar, D. M., and Lawton, W. D., 1969, *Pasteurella* bacteriophage sex specific in *Escherichia coli*, *J. Virol.* **4**:896.

Monod, J., and Wollman, E., 1947, L'inhibition de la croissance et de l'adaptation enzymatique chez les bactéries infectées par le bacteriophage, *Ann. Inst. Pasteur* **73**:937.

Morris, C. E., Klement, J. F., and McAllister, W. T., 1986, Cloning and expression of the bacteriophage T3 RNA polymerase gene, *Gene* **41**:193.

Ochman, H., and Selander, R. K., 1984, Evidence for clonal population structure in *Escherichia coli*, *Proc. Natl. Acad. Sci. USA* **81**:198.

O'Hare, K. M., and Hayward, R. S., 1981, Termination of transcription of the coliphage T7 "early" operon *in vitro*: Slowness of enzyme release and lack of any role for sigma, *Nucleic Acids Res.* **9**:4689.

Ollis, D. L., Kline, C., and Steitz, T. A., 1985, Domain of *E. coli* DNA polymerase I showing sequence homology to T7 DNA polymerase, *Nature* **313**:818.

Paetkau, V., Langman, L., Bradley, R., Scraba, D., and Miller, R. C. Jr., 1977, Folded concatenated genomes as replication intermediates of bacteriophage T7 DNA, *J. Virol.* **22**:130.

Peters, G. G., and Hayward, R. S., 1974, Transcriptional termination *in vitro*: The 3'-terminal sequence of coliphage T7 "early" RNA, *Biochem. Biophys. Res. Commun.* **61**:809.

Poteete, A. R., 1988, Bacteriophage P22, in: *The Bacteriophages*, Vol. 2 (R. Calendar, ed.), pp. 647–682, Plenum Press, New York.

Powling, A., and Knippers, R., 1974, Some functions involved in bacteriophage T7 genetic recombination, *Mol. Gen. Genet.* **134:**173.

Rabussay, D., and Geiduschek, E. P., 1977, Regulation of gene action in the development of lytic bacteriophages, in: *Comprehensive Virology 8* (H. Fraenkel-Conrat and R. R. Wagner, eds.), pp. 1–196, Plenum Press, New York.

Rahmsdorf, H. J., Pai, S. H., Ponta, H., Herrlich, P., Roskoski, R. Jr., Schweiger, M., and Studier, F. W., 1974, Protein kinase induction in *Escherichia coli* by bacteriophage T7, *Proc. Natl. Acad. Sci. USA* **71:**586.

Reanney, D. C., and Ackermann, H. W., 1982, Comparative biology and evolution of bacteriophages, *Adv. Virus Res.* **27:**205.

Ritchie, D. A., Thomas, C. A. Jr., MacHattie, L. A., and Wensink, P. C., 1967, Terminal repetition in non-permuted T3 and T7 bacteriophage DNA molecules, *J. Mol. Biol.* **23:**365.

Roeder, G. S., and Sadowski, P. D., 1977, Bacteriophage T7 morphogenesis: Phage-related particles in cells infected with wild-type and mutant T7 phage, *Virology* **76:**263.

Roeder, G. S., and Sadowski, P. D., 1979, Pathways of recombination of bacteriophage T7 DNA *in vitro, Cold Spring Harbor Symp. Quant. Biol.* **43:**1023.

Romano, L. J., and Richardson, C. C., 1979, Characterization of the ribonucleic acid primers and the deoxyribonucleic acid product synthesized by the DNA polymerase and gene 4 protein of bacteriophage T7, *J. Biol. Chem.* **254:**10483.

Rosa, M. D., 1979, Four T7 RNA polymerase promoters contain an identical 23 bp sequence, *Cell* **16:**815.

Rothman-Denes, L. B., Muthukrishnan, S., Haselkorn, R., and Studier, F. W., 1973, A T7 gene function required for shut-off of host and early T7 transcription, in: *Virus Research* (C. F. Fox and W. S. Robinson, eds.), pp. 227–239, Academic Press, New York.

Rudolph, C., Freund-Mölbert, E., and Stirm, S., 1975, Fragments of *Klebsiella* bacteriophage No. 11, *Virology* **64:**236.

Sadowski, P. D., and Kerr, C., 1970, Degradation of *Escherichia coli* B deoxyribonucleic acid-defective amber mutants of bacteriophage T7, *J. Virol.* **6:**149.

Sadowski, P. D., Lee, D. D., Andrews, B. J., Babineau, D., Beatty, L., Morse, M. J., Proteau, G., and Vetter, D., 1984, *In vitro* systems for genetic recombination of the DNAs of bacteriophage T7 and yeast 2-micron circle, *Cold Spring Harbor Symp. Quant. Biol.* **49:**789.

Saito, H., and Richardson, C. C., 1981, Processing of mRNA by ribonuclease III regulates expression of gene 1.2 of bacteriophage T7, *Cell* **27:**533.

Schlegel, R. A., and Thomas, C. A. Jr., 1972, Some special structural features of intracellular bacteriophage T7 concatemers, *J. Mol. Biol.* **68:**319.

Schmitt, M. P., Beck, P. J., Kearney, C. A., Spence, J. L., DiGiovanni, D., Condreay, J. P., and Molineaux, I. J., 1987, Sequence of a conditionally essential region of bacteriophage T3 including the primary origin of DNA replication, *J. Mol. Biol.* **193:**479.

Serwer, P., 1976, Internal proteins of bacteriophage T7, *J. Mol. Biol.* **107:**271.

Serwer, P., Grennhaw, G. A., and Allen, J., 1982, Concatemers in a rapidly sedimenting, replicating bacteriophage T7 DNA, *Virology* **123:**474.

Serwer, P., and Watson, R. H., 1985, Alterations of the bacteriophage T7 and T3 DNA packaging pathway in *Escherichia coli* mutant tsnB, *Virology* **140:**80.

Serwer, P., Watson, R. H., Hayes, S. J., and Allen, J. L., 1983, Comparison of the physical properties and assembly pathways of the related bacteriophages T7, T3, and ϕII, *J. Mol. Biol.* **170:**447.

Sherman, L. A., and Haselkorn, R., 1970, LPP-1 infection of the blue-green alga *Plectonema boryanum.* III. Protein synthesis, *J. Virol.* **6:**841.

Spoerel, N., Herrlich, P., and Bickle, T. A., 1979, A novel bacteriophage defence mechanism: The anti-restriction protein, *Nature* **278:**30.

Stahl, S. J., and Chamberlin, M. J., 1977, An expanded transcriptional map of T7 bacteriophage. Reading of minor T7 promoter sites *in vitro* by *Escherichia coli* RNA polymerse, *J. Mol. Biol.* **112:**577.

Stone, J. C., and Miller, R. C. Jr., 1984, Plasmid-phage recombination in T7 infected *Escherichia coli*, *Virology* **137**:305.

Studier, F. W., 1969, The genetics and physiology of bacteriophage T7, *Virology* **39**:562.

Studier, F. W., 1972, Bacteriophage T7, *Science* **176**:367.

Studier, F. W., 1973, Genetic analysis of non-essential bacteriophage T7 genes, *J. Mol. Biol.* **79**:227.

Studier, F. W., 1979, Relationships among different strains of T7 and among T7-related bacteriophages, *Virology* **95**:70.

Studier, F. W., and Dunn, J. J., 1983, Organization and expression of bacteriophage T7 DNA, *Cold Spring Harbor Symp. Quant. Biol.* **47**:999.

Studier, F. W., and Maizel, J. V. Jr., 1969, T7-directed protein synthesis, *Virology* **39**:575.

Tabor, S., and Richardson, C. C., 1985, A bacteriophage T7 RNA polymerase/promoter system for controlled exclusive expression of specific genes, *Proc. Natl. Acad. Sci. USA* **82**:1074.

Towle, H. C., Jolly, J. F., and Boezi, J. A., 1975, Purification and characterization of bacteriophage gh-1-induced deoxyribonucleic acid-dependent ribonucleic acid polymerase from *Pseudomonas putida*, *J. Biol. Chem.* **250**:1723.

Tsujimoto, Y., and Ogawa, H., 1977, Intermediates in genetic recombination of bacteriophage T7 DNA, *J. Mol. Biol.* **103**:423.

Tsujimoto, Y., and Ogawa, H., 1978, Intermediates in genetic recombination of bacteriophage T7 DNA. Biological activity and the roles of gene 3 and gene 5, *J. Mol. Biol.* **125**:255.

Watanabe, T., and Okada, M., 1964, New type of sex factor-specific bacteriophage of *Escherichia coli*, *J. Bacteriol.* **87**:727.

Watson, J. D., 1972, Origin of concatemeric T7 DNA, *Nature (New Biol.)* **239**:197.

West, D., Lagenaur, C., and Agabian, N., 1976, Isolation and characterization of *Caulobacter crescentus* bacteriophage φCdl, *J. Virol.* **17**:568.

Wever, G. H., Fischer, H., and Hinkle, D. C., 1980, Bacteriophage T7 DNA replication *in vitro*. Electron micrographic analysis of T7 DNA synthesized with purified proteins, *J. Biol. Chem.* **255**:7965.

Whittam, T. S., Ochman, H., and Selander, R. K., 1983, Multilocus genetic structure in natural populations of *Escherichia coli*, *Proc. Natl. Acad. Sci. USA* **80**:1751.

Williams, L., and Meynell, G. G., 1971, Female-specific phages and F-minus strains of *Escherichia coli* K12, *Mol. Gen. Genet.* **113**:222.

Wolstenholme, G. E. W., and O'Connor, M. (eds.), 1971, *Strategy of the Viral Genome* Ciba Foundation Symposium, Churchill Livingstone, London.

Yamada, M., Fujisawa, H., Kato, H., Hamada, K., and Minagawa, T., 1986, Cloning and sequencing of the genetic right end of bacteriophage T3 DNA, *Virology* **151**:350.

Yamagishi, M., Fujisawa, H., and Minagawa, T., 1985, Isolation and characterization of bacteriophage T3/T7 hybrids and their use in studies on molecular basis of DNA-packaging specificity, *Virology* **144**:502.

Zavriev, S. K., and Shemyakin, M. F., 1982, RNA polymerase-dependent mechanism for the stepwise T7 phage DNA transport from the virion into *E. coli*, *Nucleic Acids Res.* **10**:1635.

CHAPTER 9

Bacteriophage P1

Michael B. Yarmolinsky and Nat Sternberg

I. INTRODUCTION

We preface this review with a brief chronology of seminal P1 studies in order to illuminate the circumstances that molded the idiosyncratic development of P1 biology. The account that follows this brief historical preface deals first with P1 structure, second with successive stages in the life cycle of the phage, and third with comparative studies. It covers information available to us prior to November 1986 but additional material received throughout 1987 has also been inserted. Topics in P1 biology that have been treated in recent reviews include: the P1 genomic map (Yarmolinsky, 1987), transduction (Margolin, 1987; Masters, 1985), restriction–modification (Yuan, 1981; Krüger and Bickle, 1983), immunity to superinfection (Sternberg and Hoess, 1983; Scott, 1980), site-specific recombinations (Sadowski, 1986; Plasterk and Van de Putte, 1984; Sternberg and Hoess, 1983; Simon and Silverman, 1983), maintenance of the plasmid prophage (Scott, 1984; Sternberg and Hoess, 1983), and methylation-regulated gene expression and DNA processing (Sternberg, 1985; Marinus, 1984, 1987).

The isolation of bacteriophage P1 was reported by G. Bertani in 1951, the same year in which Esther Lederberg reported the discovery of bacteriophage λ. P1 is a temperate phage that differs from λ in a number of fundamental and interesting ways, but whereas λ is the subject of two substantial monographs or "testaments," much of P1 lore has been confined to oral tradition. The reasons for this difference are partly biological, partly historical.

MICHAEL B. YARMOLINSKY • Laboratory of Biochemistry, National Cancer Institute, National Institutes of Health, Bethesda, Maryland 20892.   NAT STERNBERG • E. I. DuPont de Nemours & Co., Central Research and Development Department, Experimental Station, Wilmington, Delaware 19898.

P1 was recognized as one of three temperate phages harbored by the lysogenic *Escherichia coli* of Lisbonne and Carrère that Bertani had received from Joshua Lederberg. The three phages, named by Bertani P1, P2, and P3, were characterized by their plaque sizes on a strain of *Shigella dysenteriae*, respectively: small, large, and variable. The phages were released spontaneously at low frequency from individual lysogenic bacteria in pure bursts of one or another of three serologically distinguishable types. A significant serological cross-reactivity was noted at the time, but it does not appear to have been further explored. [Further study of the relationship is likely to resume, given the current knowledge of and interest in the genetic determinant in P1 of serum blocking power (see Section III.B).] Of the three phages, P3 has received the least attention because preparations of it appeared markedly unstable. It is rather closely related to P2, with which it forms viable recombinants (see Bertani and Six, 1988). Bertani and his colleagues concentrated their attention on P2 for the reason that its plaque characteristics on *E. coli* C made it amenable to study away from *Shigella*, unlike the original isolate of P1. Lennox (1955) selected a P1 mutant, P1*kc*, that made somewhat clearer plaques on *E. coli* K12 than the original isolate, and it is from this strain that P1 phages in current circulation are derived. Simultaneously published experiments of Jacob (1955) made use of what is probably an independent isolate of P1, called 363, described in Jacob and Wollman (1961) but not heard of since.

P1 might have remained as obscure as is P3* were it not for the discovery by Lennox (1955) of P1-mediated generalized transduction between strains of *Escherichia* and *Shigella* and the discovery by Seymour Lederberg (1957) and Arber and Dussoix (1962) of P1-mediated modification and restriction. Inasmuch as studies by Werner Arber that exploited P1 helped to establish the fundamental biology of restriction–modification (see Section V.A), P1 was instrumental in bringing about the age of genetic engineering, even if the P1 restriction nuclease has not been useful to genetic engineers. P1 as a transducing phage continues to be a workhorse of gross genetic manipulations in bacteria even at this writing.

One of the first uses to which P1-mediated transduction was put was to demonstrate a chromosomal location of determinants of λ, 434, and 82 lysogeny (Jacob, 1955). However, in bacterial mating experiments, the determinant of P1 lysogeny could not be localized to any specific locus on the host chromosome (Jacob and Wollman, 1959). Studies of P1-mediated transduction initiated in Luria's laboratory in the late 1950s revealed that P1 could effect specialized as well as generalized transduction. The convenient *lac* marker that had been inserted into the P1 genome made the resultant P1*dl* prophage easy to follow in exconjugants, but Boice and

---

*P3 was nearly rescued from obscurity in 1976, when, during the spring, Nancy Kleckner considered using this phage for her 1977 class experiments at Harvard. According to oral tradition, she was dissuaded by political considerations, since P3 experiments (in the sense of requiring high-level physical containment) were explicitly forbidden by the Cambridge City Council at the time.

Luria (1963) were likewise unsuccessful in localizing P1*dl* prophage to one or a few specific chromosomal sites.* The reason for the failures became clear in 1968, when Ikeda and Tomizawa (who had also been studying P1-mediated transduction) demonstrated that P1 prophage exists extrachromosomally as a plasmid. They further showed that P1 DNA does not integrate within the bacterial chromosome even for as brief a period as would be required to allow its passive replication. It must therefore be actively replicated. The low copy number and stability of the plasmid implied that P1 is also actively partitioned between daughter cells rather than passively (and randomly) distributed by diffusion.

At about the time that Ikeda and Tomizawa in Tokyo recognized P1 to be a plasmid in the prophage state, Ravin and Shulga in Moscow were independently showing that another coliphage, N15, also lysogenizes as a plasmid. The Soviet workers, having been denied access to the phage Mu, proceeded to isolate phages that alter the phenotype of lysogenized hosts, hoping to isolate a mutator phage of their own (E. Golub, personal communication). In 1964 they isolated a phage, which they named N15, that renders *E. coli* T1-resistant. They subsequently found that this altered phenotype was not due to mutagenesis by integration. It was instead the result of a lysogenic conversion by N15 prophage (Ravin and Golub, 1967). The prophage was stably maintained without chromosomal integration (Ravin and Shulga, 1970). The 1970 report presented evidence that N15 exhibits considerable homology to phage λ. It also described mutants in N15 affecting plasmid maintenance more than a decade before any such mutants were obtained in P1. Had N15 been granted an exit visa from the Soviet Union, the attention devoted to it would likely have been considerable. As it is, N15 appears not even to have been tested for assignment to an incompatibility group.

After a considerable latent period, the findings of Ikeda and Tomizawa provided a new impetus and a new direction to P1 biology. Progress was not rapid at first, in part because P1 did not appear to be a convenient biological material with which to work. P1 has a complex genome (twice the size of the λ genome), and two genetically marked P1 prophages fail to coexist together stably, a feature of P1 that can interfere with tests of genetic complementation and dominance. Luria *et al.* (1960) had noted that attempts to form stable lysogens of two differently marked phages result in the displacement of one prophage by another. The instability of polylysogens of P1 is attributable to incompatibility, the inability of two related plasmids to coexist stably. Incompatibility determinants are key

---

*The possibility that "P1 can persist as an episome without chromosome-associated phase" was considered by Luria *et al.* (1960). In the absence of evidence that prophages in lysogens are integrated, Bertani (1951) had assumed that P1 prophage *is* extrachromosomal and reasonably, but mistakenly, favored the idea that "the number of prophages must be so high that their random distribution at cell division does not cause any important shift in the average number of prophages per cell" over the alternative that P1 employs "a mechanism . . . that is somehow connected with the process of cell division . . . to assure equal distribution of prophages to daughter cells."

elements in the machinery of plasmid maintenance. Their study, although hampered by incompatibility, has been particularly rewarding for the geneticist interested in the plasmid way of life (see Sections VII.A.2 and VII.A.3).

Systematic genetic studies of P1 were initiated in the late sixties by J. R. Scott. She isolated a large number of mutations affecting plaque clarity, widely scattered over the genome. Some are implicated in immunity, others in unrelated or unidentified functions. Insight into the complex immunity system of P1 was gained with the discovery (Smith, 1972) of ΦAmp (now called P7), a phage largely homologous to P1 but of differing immunity specificity.* Differences between P1 and P7 have been particularly informative not only in the study of immunity (see section VI.B) but also in the study of plasmid maintenance (see Section VII.A.3).

A possible connection of immunity regulation in P1 to plasmid replication control was suggested by previous studies of the λdv plasmid in which replication is under control of the *cro* immunity repressor. However, an analysis of IS*1*-generated deletions in P1 showed that it is possible to dissociate plasmid replication control from the immunity system of P1 (Austin *et al.*, 1978). This result was the first indication that the prophage and vegetative replicons of P1 are probably separate, and it was the first step in the eventual dissection of plasmid maintenance genes into a series of functionally discrete cassettes that can fulfill their role even when incorporated into unusual DNA contexts. Ironically, it now appears that the formal aspects of the λdv circuitry of plasmid replication control may have relevance for the control of P1 plasmid replication, but we defer (until section VII.B.2) further discussion of this topic.

The genetic map that was generated by Scott (1968) and extended by the Walkers (1975, 1976b) was based on the frequencies of recombination between markers along the cyclically permuted viral DNA. It was, surprisingly, a linear rather than a circular map. However, by marker rescue from prophage deletion strains, circularity of the prophage map was demonstrated, and the possibility that a recombinational hot spot might account for the discrepancy was proposed (Walker and Walker, 1975). The existence of a recombinational hot spot that defines the ends of the genetic map was demonstrated by Sternberg (1978). The site-specific recombination at this locus is determined by a phage-encoded enzyme. The system plays varied, and normally minor, roles in the life of P1, but owing to its intrinsic interest has been the object of intensive study, as has a separate recombinational system that inverts a 4.2-kb segment of P1 DNA (see Section III.C). The latter system was discovered as a result of a heteroduplex analysis of P1 DNA by electron microscopy (Lee *et al.*, 1974). The biological role of DNA inversion in varying host range (Iida,

---

*The plasmid nature of P7 prophage, and hence its possible homology with P1, was deduced from the sensitivity of P7 prophage to acridine orange, a reagent known to cure bacteria of plasmids such as F. Paradoxically, attempts to eliminate P1d*l* with acridine orange had been previously reported to fail (P. Amati and N. C. Franklin, cited in Boice and Luria, 1963).

1984) was first appreciated in phage Mu, a phage that carries a homologous invertible DNA segment but appears otherwise unrelated to P1. As further discussed in Section X.C, the relationship of P1 to Mu is one of several indications of the chimeric nature of the phage-plasmid P1.

With the advent of genetic engineering technology, research into various areas of P1 biology has been altered in pace and in nature. Some features of the phage are being revealed in molecular detail, whereas other features have yet to be discerned even in outline. The reader may be surprised to learn how many elementary questions about P1 as an organism remain to be addressed. One of our aims in writing this review is to raise these questions.

II. PHYSICAL AND GENETIC STRUCTURE

A. Phage Particles

1. Morphology

The structure of infectious P1 particles as visualized by the electron microscopy of Walker and Walker (1983) is shown in Fig. 1A. The head or capsid has icosahedral symmetry, a symmetry that is generally preferred in viral architecture as in the construction of geodesic domes (Caspar and Klug, 1962). An inflexible tail is attached to one vertex of the head via a head-neck connector and a hollow neck. The tail consists of a tail tube (which may be slightly thinner than the neck) and a contractile sheath. The sheath has a cap at the end proximal to the head and, at its distal end, a baseplate to which are attached kinked tail fibers, probably six of them (Walker and Anderson, 1970).

P1 lysates normally contain particles with similar tails but with mature isometric heads of different widths: 85 ± 2 nm, 65 ± 2 nm, and 47 ± 2 nm (Anderson and Walker, 1960; Ikeda and Tomizawa, 1965c; Walker and Anderson, 1970).* The largest particles, P1B (big), are infectious. The particles of P1S (small) (Fig. 1B), and possibly the much rarer P1M (minute) can adsorb and inject their DNA but are nonproductive on single infection; they lack the capacity to hold a complete P1 genome.

---

*The ratios of the head widths are consistent with the principles of viral architecture elucidated by Caspar and Klug (1962). The capsids are built of identical capsomeres, measured to be 5 nm thick but idealized as two-dimensional figures, distributed with their centers d units apart on the surface of an icosadeltahedron of diameter D. (An icosahedron has 20 equilateral triangular faces, and an icosadeltahedron has 20 T facets, where T is the triangulation number.) D is less than the observed width of a phage head by the thickness of a capsomere. The values that D can assume are related to d, the lattice constant, and to the triangulation number T (which characterizes the limited number of capsomere distributions) by the formula $D = 1.62 \, dT^{\frac{1}{2}}$ where, for all icosahedral viruses, $T = f^2$ or $3f^2$ and f is an integer. The values of D for P1B, P1S, and P1M (making allowance for the capsomere thickenss) are about 80, 60, and 42 nm, respectively. These values are in the ratio 4:3:2 within experimental error and satisfy the formula for D with $T = 4^2$, 3^2, and 2^2 and a lattice constant of 12 to 13 nm (Walker and Anderson, 1970).

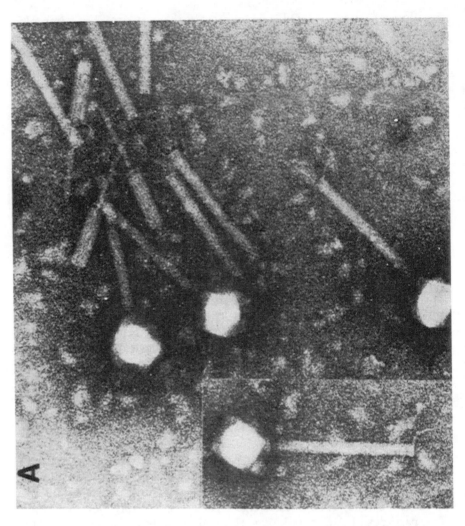

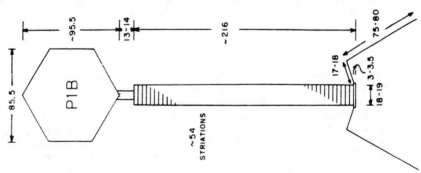

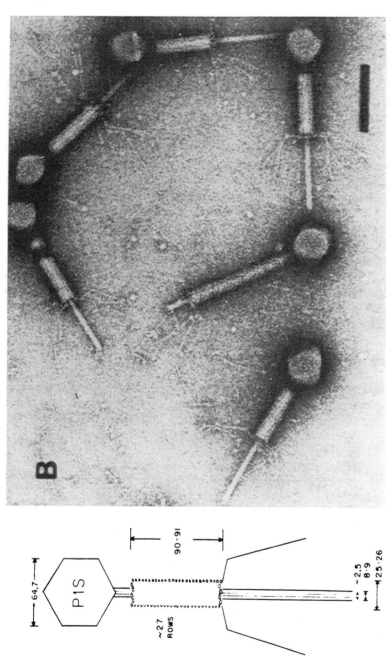

FIGURE 1. Micrographs showing two morphological variants of P1. (A) Infectious P1B particles and tails. In this micrograph the majority of tails have extended sheaths. (B) P1S particles. In this micrograph the majority of tails have contracted sheaths, as diagrammed; baseplates and tail fibers are readily seen. The black bar represents 100 nm. Micrographs after Walker and Walker (1983). The accompanying diagrammatic scale drawings of P1B and P1S particles and of a P1M (minute) head are after Walker and Anderson (1970).

Burst sizes of P1 as high as 400–500 pfu per bacterium have been reported for P1 grown in *Shigella dysenteriae* or in *E. coli* (Bertani and Nice, 1954; Rosner, 1972), although, at least in *E. coli*, figures around 100–200 may be more common (Scott, 1970b; Mise and Arber, 1976; Takano and Ikeda, 1976). The relative abundance of the P1 head size variants in the burst is influenced by both host and phage. Normally, P1S particles constitute one-sixth of the yield (Walker *et al.*, 1979), but, depending on the host, the proportion of P1S can vary considerably (Table I). The basis for this variation remains to be studied. The proportion of P1B to P1S particles is also under control of the phage *vad* (viral architecture determinant) gene localized to *Eco*RI-12 (Fig. 2A). An amber mutation affecting *vad* expression has been localized to this fragment (D. H. Walker Jr. and J. T. Walker, personal communication). Deletion derivatives of P1 that lack *vad* produce as many P1S as P1B particles (Iida and Arber, 1977, 1980). A deletion that removes about 10% of the P1 genome clockwise from the IS1 at map coordinate 24 eliminates an operon that includes *vad*. This operon carries late genes for four internal proteins which would otherwise be found in association with P1 heads (Walker and Walker, 1981; Iida *et al.*, 1987; S. Iida, J. Meyer, and W. Arber, personal communication) (see Sections VIII.D.1 and IX.B.2.b).

2. Buoyant Density

Plaque-forming particles can be separated from various classes of defective particles on CsCl density gradients. The buoyant density of the plaque-forming particles is about 1.476 g cm^{-3} (Kondo and Mitsuhashi,

TABLE I. Characteristics of P1 Morphological Variants[a]

Class	Relative abundance (% of total)	Head width (nm)	DNA content (genome equivalents)	Buoyant density (g cm^{-3})
P1B	~ 83 (i)	85.5 ± 1.8 (e)	1.12 (d) 1.09 (h)	1.482 (a) 1.47 (c) 1.476 (b,g) 1.473 ± 0.002 (e) 1.473 (f)
P1S	6–27 (e) ~ 17 (i)	64.7 ± 1.5 (e)	0.45 (c) 0.40 (i)	1.438 (a) 1.42 (c) 1.433 ± 0.002 (e)
P1M	0.2–3.6 (e)	46.7 ± 2.0 (e)	—	—

[a]The data are compiled from the following sources: (a) Ting (1962); (b) Kondo and Mitsuhashi (1964); (c) Ikeda and Tomizawa (1965c); (d) Ikeda and Tomizawa (1968); (e) Walker and Anderson (1970); (f) Karamata (1970); (g) Mise and Arber (1976); (h) Yun and Vapnek (1977); (i) Walker *et al.* (1979).
The existence in P1 lysates prepared on *Shigella* of cosedimenting P1S and defective P1B particles, each with a calculated DNA content of 60% of the P1B viral DNA, was reported by Walker and Anderson (1970), but has not been noted in other published particle profiles (Ting, 1962; Ikeda and Tomizawa, 1965c; Walker *et al.*, 1979).

1964; Mise and Arber, 1976). Values on either side of this figure have also been reported (Table I). Five density classes of P1 particles (ρ = 1.473–1.422 g cm^{-3}), representing various tetrameric combinations of P1B, P1S, and phage tails adhering at their baseplates, were reported by Karamata (1970). Aggregates of these kinds might account for a phenotypic variation in density (around a mean of 1.482) reported by Ting (1962). Aggregation might also account for the small size of P1 plaques, claimed by Karamata to be smaller than expected for a phage with the burst size of P1 and with the diffusion constant of an individual P1B particle.

P1S particles, which carry about 40–45% of the P1 genome, are 0.04–0.05 g cm^{-3} less dense than P1B particles (Table I). The permutations of DNA in such populations of P1S particles represent the entire genome found in P1B particles and are presumably similarly generated (Walker et al., 1979). This result is in conformity with the earlier findings of Ikeda and Tomizawa (1965c) that purified P1S particles at high multiplicities of infection can produce infectious particles by recombination (multiplicity reactivation).

3. Viral Proteins

In addition to the major head protein (44 kD), 14 other head proteins (4 of which are dispensable) are revealed by SDS–PAGE analysis of P1 particles (Walker and Walker, 1981). Nine tail proteins and four proteins of undetermined provenance are also seen. The major tail proteins have M_rs of 21.4 and 72 kD and, by analogy with other phages, are presumed to be the subunits of tube and sheath, respectively. An unusual feature of the band pattern is the presence of more than 30 additional bands regularly spaced at intervals corresponding to increments of 1–2 kD. The multiple bands appear to arise by addition, through disulfide bonds, of multimers of a short, discrete polypeptide to at least one major head protein and one major tail protein.

4. Viral DNA

The DNA of plaque-forming particles is double-stranded, linear (Ikeda and Tomizawa, 1965a), terminally redundant, and cyclically permuted (Ikeda and Tomizawa, 1968). Its molecular weight is about 66 MD (Yun and Vapnek, 1977), and its buoyant density, ρ, is 1.706 g cm^{-3}, corresponding to about 100 kb of DNA with a G+C mole fraction of 46% (Ikeda and Tomizawa, 1965a). For comparison, the G+C mole fraction of Escherichia coli DNA and of Shigella dysenteriae DNA is 51% (Schildkraut et al., 1962). The extent of terminal redundancy was estimated by Ikeda and Tomizawa (1968) as 12% of the viral DNA and by Yun and Vapnek (1977) as 7.9%. The very closely related phage, P7, similar to P1 in DNA content, but of larger genome size, was found to have only 0.6% redundancy (Yun and Vapnek, 1977). The 25.5 MD of DNA in P1S particles lacks terminal redundancy but is cyclically permuted (Walker et al., 1979).

Ikeda and Tomizawa (1968) proposed that recombination between terminally redundant ends of viral DNA permit cyclization. In support of this claim, they showed that the nonredundant DNA has the length of prophage. Preparations of viral DNA denatured and reannealed in vitro contained "artificial circles" carrying two big single-stranded "bushes." The length of these "artificial circles" corresponded within experimental error to the length of the circular DNA of P1 prophage. Careful analysis of restriction fragments of DNA isolated from viral particles (Bächi and Arber, 1977) revealed in each enzymatic digest one additional band not obtained from digested plasmid DNA. These extra bands were of relatively low intensity. Certain other viral bands were found to be slightly more intense than expected for equimolar fragments. The overrepresented fragments were shown to be derived from DNA that is situated in the genome to one side of the DNA of the novel bands. Bächi and Arber (1977) interpreted these results to mean that packaging of P1 from DNA concatamers starts from a unique site which generates the novel bands. It proceeds unidirectionally and continues sequentially by the headful. DNA adjacent to the site at which packaging starts will be slightly overrepresented in the average phage population because of the terminal redundancy. The details of packaging and its control are covered in section VIII.E.

One consequence of the extensive terminal redundancy that permits cyclization is tolerance for large insertions. A large insertion can diminish terminal redundancy without preventing the complete viral genome from being fully represented in each viral particle.* Plaque-forming P1 carrying multiple antibiotic resistance markers of the same kind (Meyer and Iida, 1979) or of different kinds (Mise and Arber, 1976; Iida and Arber, 1977; Scott et al., 1982) have been selected; other plaque formers carrying substantial insertions, such as the entire gal operon of E. coli (Austin et al., 1982), have been constructed. Several sites exist at which insertions can be made without compromising the capacity to form a plaque (Arber et al., 1978; Iida et al., 1985a). At one of these sites is the insertion element IS1 which Iida et al. (1978) showed to be a natural constituent of the P1 genome. In the P1Cm of Kondo and Mitsuhashi (1964) the cat (chloramphenicol acetyltransferase) gene is flanked by two direct IS1 re-

---

*An insertion that exceeds the length of the terminally redundant region, provided it does not replace or interrupt essential operons of the phage, results in a prophage that yields a genetically inhomogeneous burst upon induction. The burst is a population of defective particles from which viable phage and defective particles carrying the insertion may possibly be reconstituted by recombination in multiply infected cells (Rae and Stodolsky, 1974; Rosner, 1975; cf. Tye et al., 1974b). Specialized transducing prophages that require helper to produce a burst arise when essential genes of the phage are replaced or interrupted by the acquired material. The generation of specialized transducing P1 of these several kinds is discussed in Section IX.A.

peats at this same locus (MacHattie and Jackowski, 1976; Iida and Arber, 1977). The IS*1* is also convenient for making insertions *in vitro,* because the IS*1* of P1 is the locus of a unique *Pst*I cleavage site (Bächi and Arber, 1977; Ohtsubo and Ohtsubo, 1978; Alton and Vapnek, 1979).

B. Transducing Particles

A few particles in every lysate, about 0.3–0.5% of the total, carry only (>95%) bacterial DNA (Ikeda and Tomizawa, 1965a; Harriman, 1972). The frequency and composition of transducing particles in a lysate can be manipulated experimentally (see Sections IX.B.1.c and IX.B.2.b). Deliberately selected HTF and LTF mutants of P1 that respectively raise and lower the proportion of transducing particles in phage bursts have been isolated (Wall and Harriman, 1974; Yamamoto, 1982). A fortuitously observed increase in the proportion of transducing particles occurs in deletion mutants of P1 that lack the operon of which *vad* is a member. The phenotype is named *gta* (generalized transduction affected) (Iida *et al.,* 1987).

Bacterial DNA is found in both P1B and P1S particles but, uniquely in P1B transducing particles, the DNA appears linked (probably at a terminus) to protein (Ikeda and Tomizawa, 1965b). Based on its contribution to buoyant density, the amount of protein associated with the DNA of a transducing particle is estimated to be 5×10^2 kD. The protein does not appear in association with the DNA of plaque-forming P1 particles. A plausible biological role for the protein is discussed in Section IX.B.2.b.

C. Prophages

1. Size and Copy Number

P1 plasmid DNA in the form of supercoils and open circles was identified in extracts of endonuclease-negative lysogenic *E. coli* by sedimentation rate and by visualization in the electron microscope (Ikeda and Tomizawa, 1968). The molecular weights of the plasmid DNA have been estimated from electron microscopic measurements of DNA contour length (Abelson and Thomas, 1966; Ikeda and Tomizawa, 1968; Yun and Vapnek, 1977; Meyer et al., 1981) and by summing the lengths of restriction fragments as deduced from their electrophoretic mobilities (Bächi and Arber, 1977). If we average the commonly cited estimates of Yun and Vapnek (60.9 ± 1.7 MD, or 91.8 kb) and of Bächi and Arber (58.6 ± 1.2 MD, or 88.4 kb), we arrive at approximately 60 MD, or 90 kb, for the size of the P1 genome. Conventionally, this is subdivided into 100 map units.

The content of P1-specific DNA in lysogenic *E. coli* was estimated by

Ikeda and Tomizawa (1968), in agreement with independent measurements by Inselberg (1968), to be about 1.0 per bacterial chromosome. For this reason, P1 prophage is referred to as a unit copy plasmid or a plasmid under stringent control (Clowes, 1972).* Both host and plasmid functions contribute to copy number control, as is evident from the existence of E. coli mutants (Cress and Kline, 1976) and P1 mutants (Sternberg et al., 1978; Scott et al., 1982; Froehlich et al., 1986) with elevated copy number. Measurements of P1 DNA relative to DNA of prophages Mu (near ilv) and ϕ80 (at att80)—markers for bacterial origin and terminus, respectively—were made in 1977 by Prentki et al. The copy number of wild-type prophage was shown to decrease from about 1.9 to 1.0 per bacterial replication terminus as the cell doubling time, τ, increases from 24 to 215 min. Over the same range of τ, the copy number increases from about 0.6 to 0.9 per bacterial replication origin, reaching a plateau at τ >90 min. These results imply that P1 plasmid replication is not linked to either the initiation or termination of chromosomal replication; if a locus must be chosen for standardization, one near the origin will show less fluctuation with growth rate than one near the terminus. Various methods have been used to determine (and define) the copy number of P1 plasmids. The simplest assay is based on the proportionality of antibiotic resistance to copy number of plasmids bearing a suitable antibiotic resistance determinant. An improper choice of antibiotic-resistance determinant, however, can be misleading (Scott et al., 1982). DNA-DNA hybridization methods, in particular the slot-blot technique using external standards, provides at present the most direct and accurate determinations (Pal et al., 1986; Shields et al., 1986).

The isolation of a thermoinducible P1Cm by Rosner (1972)† permitted the selection of rare descendants of lysogenic bacteria that had lost the potentially lethal prophage. The frequency with which such "cured" cells arise per bacterial generation was shown to be independent of whether the bacteria had been transiently and reversibly induced. The frequency observed was about 2×10^{-5} cells per generation in Rosner's experiments and lower by almost another order of magnitude in subsequent experiments (Austin et al., 1981). In the latter experiments, more than half the survivors could be shown to carry prophage with deletions that appeared to be generated by the chloramphenicol resistance trans-

---

*Stringent control was subsequently redefined to be used in reference to plasmid replication that is obligatorily coupled to chromosome replication (Novick et al., 1976). In this sense of the term, P1 is not under stringent control. P1 prophage can undergo several cycles of replication in a dnaAts host in which chromosome initiation has been inhibited (Abe, 1974).

†Rosner's mutant P1Cm0 cl.100 (originally called P1Cm c1r-100) is widely used in bacterial transductions, because the availability and convenience of this phage were made known by Miller's popular laboratory manual (Miller, 1972). The number after the Cm, normally omitted, designates the particular location of the chloramphenicol resistance transposon; zero designates the site of the IS1.

poson, Tn*9*, of the parental prophage. These experiments show that P1 as a plasmid appears as stably maintained as those prophages that are covalently integrated within the chromosome of their host. Mechanisms that contribute to this remarkable extrachromosomal stability are considered in Section VII.

2. Localization: Plasmid and Integrated States

The nonrandom distribution of P1 prophages at cell division has been taken to imply that at some time in the cell cycle P1 must be anchored to a cellular structure that has the same segregation pattern as the bacterial chromosome. The concept that stable, low-copy-number genetic elements, the bacterial chromosome included, are independently attached to specific components of the cell surface was incorporated into the replicon model posited in the classic contribution of Jacob *et al.* (1963). Only recently have the tools been developed to determine the genetic basis of plasmid anchorage and hence to specify the structures to which plasmids are attached (see Section VII.C.4).

Instances of P1 and P7 integrated into the *E. coli* chromosome have been described. The integrated prophage, P1 cryptic, is a derivative of P1Cm that has suffered a deletion of two-thirds of its genome (Scott, 1970a). The prophage is cryptic in the sense of not expressing immunity; it does retain restriction and modification functions. The deletion appears to have initiated from the chloramphenicol resistance transposon. Integration is in the *lac-pro* region of the bacterial chromosome. It is not clear how the integration occurred.

By requiring that bacterial survival depend on replication initiated from an integrated plasmid, Chesney and Scott (1978) were able to select bacteria carrying P1 and P7 integrated in various ways. The frequency of such integrations was about 2×10^{-5} in a *rec*$^+$ host and about twofold less in a *recA* host. At least nine different chromosomal integration sites were used, corresponding, in at least two cases, to sites at which plasmid–*E. coli* homology could be detected (Chesney *et al.*, 1978, 1979). Six of 10 sites of P1–*E. coli* homology have been identified as IS*1* elements and appear to be relatively common sites for P1 integrations (Chesney *et al.*, 1979; Hansen and Yarmolinsky, 1986). Many integrated prophages were also shown to have sustained deletions that render them defective. Chesney *et al.* (1979) provided evidence that deletions in P7 were generated during the course of integration by the ampicillin resistance transposon (Tn*902*) that P7 carries and were catalyzed, at least in part, by the host recombination system. The structure of insertions that occur in *recA* hosts suggested that in the absence of bacterial generalized recombination functions, P1 or P7 integration is almost exclusively by means of a site-specific recombination mechanism acting at a specific chromosomal site. The reader is referred to Section IX.A.1 for a detailed discussion of the alternative modes of P1 integration. The several makeshift

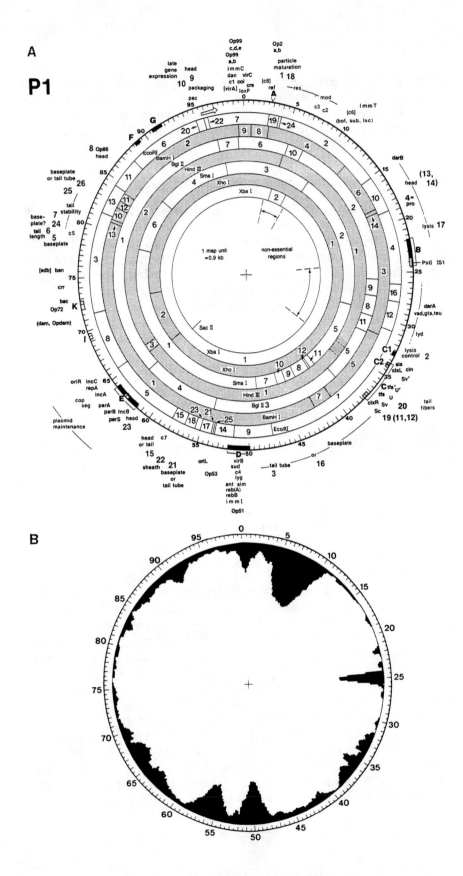

FIGURE 2. Physical and genetic maps of P1. (A) Genetic and restriction map. The present figure is reprinted from Yarmolinsky (1987), to which the reader is referred for gene definitions and relevant references. Allele assignments are given in an earlier version of the map (Yarmolinsky, 1984). The following conventions are used here: Genes are indicated outside the circles by boldface numbers, acronyms, or numbers and acronyms where both are in current use. Bracketed symbols designate genes that have been identified only in P7 but are assumed to correspond to similarly positioned homologues in P1 (Scott *et al.*, 1977a). Each c1 protein-binding site or operator (Op) sequence is assigned the integral number portion of its map position. Where there is more than one site in any interval of a map unit, we assign letters to the sites in alphabetical and clockwise order. Op *dam* refers to the operator of the *dam* analogue, not yet mapped precisely. The approximate positions of the principle nonhomologies between P1 and P7 are based on data from Yun and Vapnek (1977), Iida and Arber (1979), Iida *et al.* (1985b), and Meyer *et al.* (1986). Insertions relative to P1 (A, the ampicillin resistance transposon Tn902; F) and substitutions (B, C1, D, G, E) are indicated by boldface letters and black boxes; the invertible segment of P1 (and P7) is designated C. The C1 nonhomology region is separated from the *cin* structural gene by about 100 bp (Meyer *et al.*, 1986) and is probably related to a segment called X by Chow *et al.* (1978). The stippled bars flanking C represent inverted repeats within one of which a deletion relative to the P1 sequence (Iida *et al.*, 1985b) is marked as a black line labeled C2. IS1 (part of B) is also stippled. Additional regions of apparent partial nonhomology between P1 and P7, within E and at positions labeled I and K (Meyer *et al.*, 1983, 1986), are indicated by open boxes. The cleavage sites of restriction enzymes EcoRI, BamHI, BglII, HindIII, and the unique sites of SacII and PstI (in the IS1) were mapped by Bächi and Arber (1977), XhoI by Iida *et al.* (1985a), SmaI and XbaI by Mattes (1985). An additional unique site within IS1, for Tth111I, was also found (S. Iida, personal communication) but is not marked here. No cleavage site was found for SalI (Bächi and Arber, 1977). Gene designations S_c, S_v, S'_v, U, and U' (also called *tfs* and *tfs'* (tail fiber specificity) by Kamp *et al.*, 1984) are derived from names of homologues in phage Mu (Iida *et al.*, 1984). Gene 19 is a synonym for S. Several cistron designations (numbers or letters) are no longer in use, although allele numbers formerly associated with them are retained and have been reassigned in many cases (Walker and Walker, 1976b). Cistron numbers 11 through 16 of Scott (1968) were reassigned by Razza *et al.* (1980), as was 17 by Walker and Walker (1975). Also, the map coordinates have undergone a minor change. At a time when the position of *loxP* (now coordinate 0/100) was not correlated precisely with the physical map of P1, the unique PstI cleavage site was assigned map unit 20 so as to fix the physical and genetic maps in approximate alignment (Yarmolinsky, 1977). Linkage cluster boundaries (Walker and Walker, 1975, 1976b) are not included in the present map. We also omit coordinates of deletion prophages, some of which were represented in early maps (Walker and Walker, 1975; Yarmolinsky, 1977). Marker rescue from deletion prophages has until recently provided the most convenient and reliable technique for mapping P1 mutations (Walker and Walker, 1975; Scott and Kropf, 1977; Razza *et al.*, 1980), but the availability of cloned restriction fragments for this purpose and of methods of systematic deletion justifies the present omission of deletion end points. The DNA sequences of selected regions of P1 have been determined: from *loxP* through *cre* (Sternberg *et al.*, 1986), the *ref* gene (D. Lu *et al.*, in preparation; Windle and Hays, 1986), the *res* and *mod* genes (Humberlin *et al.*, 1988), part of the *lyd* gene and the region between *lyd* and *cin* (S. Iida, personal communication), the *cin* gene and its substrates (Heistand-Nauer and Iida, 1983; Iida *et al.*, 1984) from *cixL* through the BglII:5–3 junction (Kamp *et al.*, 1984; W. Ritthaler, in preparation) from the BglII:5–3 junction to the BamHI:5–7 junction (Sengstag, *et al.*, 1983), from the c4 gene in EcoRI-9 (Baumstark and Scott, 1987) through the oriL region, i.e. from the EcoRI9–14 junction to within EcoRI-17 (E. B. Hansen, personal communication; N. Sternberg, unpublished), the entire *rep* (Abeles *et al.*, 1984) and *par* (Abeles *et al.*, 1985) regions required for stable plasmid maintenance, a 2 kb region that symmetrically flanks the EcoRI-3,8 junction (J. Coulby and N. Sternberg, unpublished), the region from which DNA packaging initiates (Sternberg and Coulby, 1987b) and the region from *pac* to *loxP*, i.e., from mid EcoRI-20 up to 1 kb into EcoRI-7 (N. Sternberg, unpublished) and from the BglII:1–7 junction to *loxP* (Eliason and Sternberg, 1987), within which lies the c1 gene which has been sequenced independently (F. Osborne, S. R. Stovall, and B. Baumstark, personal communication).

(B) Denaturation map. The length of each bar provides an approximate measure of A + T richness. The map was prepared for this chapter through the kindness of Jürg Meyer, based on data of Meyer *et al.* (1981).

mechanisms by which P1 can integrate at low frequencies, although of considerable interest to the experimentalist, may be of only marginal biological significance. P1 prophage is not an episome in the sense in which Campbell (1969) uses this term—namely, to mean an added and dispensable element that has evolved a specialized integration/excision apparatus.

3. Genome Organization

The coding capacity of P1 is about 100 genes. Of these, less than half have been defined by mutations. The first genetic map of P1 was established by Scott (1968). Subsequently, 91 amber mutations affecting plaque formation were isolated by Walker and Walker (1975, 1976b) and, with 12 from Scott, were classified into about 18 cistrons. These were assigned a map order by deletion mapping and two- and three-factor crosses. Some additional mutations remained unassigned to specific cistrons because of ambiguity in the complementation tests. Ninety-four amber mutations affecting plaque formation were isolated by Yamamoto (1982) and classified into 31 presumptive complementation groups, but no more about this classification, beyond the fact of its existence, has been published.

The physical cartography of P1 is largely due to Arber and his colleagues. A thorough restriction mapping of P1 was undertaken by Bächi and Arber (1977). Restriction and heteroduplex maps allow P1 to be compared with its relatives P7 and P15B (Yun and Vapnek, 1977; Iida and Arber, 1979; Meyer et al., 1983, 1985) and a denaturation map of P1 (Meyer et al., 1981) provides an additional frame of reference (Fig. 2B). Heteroduplex mapping provided the initial evidence for alternative orientations of a 4.2-kb region of the prophage genome (Lee et al., 1974). The mechanism and significance of site-specific DNA inversion in P1 is discussed in Section III.C.

A correlated genetic and physical map of P1 is presented in Fig. 2. We present the map in circular form, although (as already noted in the Introduction) the recombinational hot spot created by a P1-determined site-specific recombination system effectively masks continuity of the genetic map. The locus of crossover in the phage (loxP) defines the ends of the linear map. When loxP is deleted, markers that would otherwise be assigned to opposite ends of the map become genetically linked (Sternberg and Hoess, 1983). The pac site at which packaging of P1 initiates (at map unit 95.5) is seen to be situated relative to loxP (at map unit 0/100) such that the first headful of P1 DNA packaged into P1B particles will contain a second loxP centrally within the terminal redundant region. The significance of this organization for DNA cyclization is discussed in section IV.B.3.

In addition to the loss of genetic linkage across loxP, other genetic inhomogeneities have been noted by Walker and Walker (1975, 1976b), leading them to subdivide the genetic map into 11 "linkage clusters." Crosses between cloned P1 fragments in λ vectors that were tested have failed to confirm the presence of particular recombinational hot spots (N. Sternberg, unpublished), but it is possible that an additional P1 function is required to detect them.

An obvious feature of the organization of the P1 genome is the dispersion of related functions, both those related to morphogenesis and those related to immunity. Head genes and tail genes are each found in at least three well-separated regions of the genome. Three widely separated regions are also involved in the control of immunity. The coordination of dispersed genes in the course of phage development is discussed in Section VIII.C.

III. ADSORPTION AND INJECTION

A. The Receptor

One of the first observations made about the P1 receptor was that its synthesis is dependent on a functional UDP-glucose pyrophosphorylase gene (galU). Franklin (1969) showed that P1 could not adsorb to, or kill, E. coli K12 strains with a galU mutation. Since this mutation prevents the addition of glucose to the lipopolysaccharide (LPS) core of the bacterial outer membrane, it was suggested that glucose residues in that core are essential components of the receptor. This conclusion is consistent with the observation that P1 mutants that plaque on galU strains derived from P1Cm (Kondo and Mitsuhashi, 1964) and P1Km (Takano and Ikeda, 1976) contain host range mutations that map in the P1 tail fiber genes and alter serological cross-reactivity (Toussaint et al., 1978).*

The recent experiments of Sandulache et al. (1984, 1985) more precisely identify the P1 receptor as a terminal glucose of the LPS core. In E. coli K12, that core is composed of long-chain phosphate-containing heteropolymers that are associated with the heptose sugar 2-keto-3-deoxy octonate (KDO) and with the hexose sugars D-glucose and D-galactose (Luderitz et al., 1968; Prehm et al., 1976) (Fig. 3). Evidence for the importance of the terminal glucose for P1 adsorption comes primarily from an analysis of mutant E. coli K12 strains with various incomplete cores (Sandulache et al., 1984). Both plaque formation and LPS inactivation assays indicate that a Glc-Hep-Hep-KDO core lacking both the terminal and the penultimate glucose residue and the branched galactose residue

---

*Although Toussaint et al. (1978) confirmed the plating efficiency defect of P1 in galU strains, they also noted that in liquid medium P1 adsorbs to, and injects its DNA into, isogenic galU+ and galU strains with equal efficiency. Moreover, the burst size of P1 was the same in both strains in liquid medium. An explanation for these seemingly contradictory results was not provided. Aberrant results due to media contamination with trace amounts of sugar have been noted (Sandulache et al., 1984).

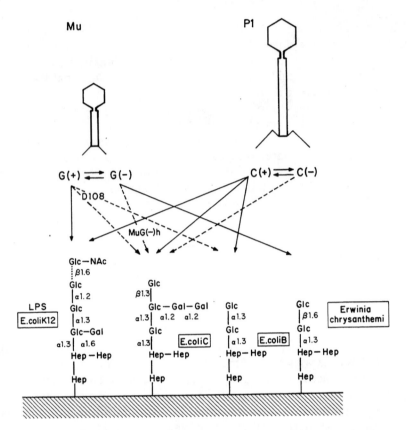

FIGURE 3. The bacterial receptors of bacteriophages P1, Mu, and D108. This figure was kindly prepared by D. Kamp. It shows the polysaccharide portion of the LPS core structures of the P1, Mu, and D108 receptors of *E. coli* K12, *E. coli* C, *E. coli* B, and *Erwinia chrysanthemi*. A solid arrow indicates that the particular LPS core structure is the receptor for the tail fibers produced by that particular alternative state of the fiber genes. A dashed arrow indicates that only a mutant form of the tail fiber genes in the configuration shown specifies fibers that adsorb to the receptor. Glc, glucose; Gal, galactose; Hep, L-glycero-D-mannoheptose; Glc-NAc, N-acetylglucosamine. β1.6, β1.3, α1.3, α1.2 refer to bonds between sugar residues. The dotted β1.6 linkage in the *E. coli* K12 LPS core indicates that the C6 position of the terminal glucose is partially substituted by N-acetylglucosamine.

still functions as a P1 receptor. In contrast, a Hep-Hep-KDO core lacking all of its glucose residues is without receptor activity. Moreover, P1 appears not to be able to distinguish between a terminal 1,2-glycosidic linkage or a terminal 1,3-glycosidic linkage, since it adsorbs to mutant *E. coli* K12 strains with both types of terminal residues. It also can adsorb to *E. coli* B and *E. coli* C, both of which contain terminal 1,3 linkages, but not to *Erwinia chrysanthemi*, which contains a terminal 1,5-glycosidic linkage (Sandulache *et al.*, 1984, 1985) (Fig. 3).

A comparison of the P1 receptor with that of other related phages indicates that, like P1, phage Mu adsorbs to *E. coli* K12, but not to a mutant in which the terminal glucose residue of the *E. coli* K12 LPS core is missing (Sandulache *et al.*, 1984). It also cannot adsorb to *E. coli* C or *E. coli* B (Sandulache *et al.*, 1985). These results suggest that the Mu receptor is a terminal glucose with a 1,2-glycosidic linkage. Mu-related phage D108 and P1-related phage P7 have the same receptor as P1 (Sandulache *et al.*, 1984).

In *Salmonella typhimurium*, the polysaccharide side chain of the LPS core contains polymerized oligosaccharide repeating units that carry the O-antigenic specificity necessary for the adsorption of phage P22. P1 does not adsorb to wild-type *Salmonella typhimurium* but will adsorb to rough derivatives that have lost part of their polysaccharide side chains owing to a mutation in either the galactose epimerase gene (*galE*), which results in galactose-deficient LPS, or a mutation in the *galU* gene, which results in glucose-deficient LPS (Ornellas and Stocker, 1974). Presumably the polysaccharide side chains characteristic of *Salmonella* interfere with the ability of P1 to contact its LPS core receptor. Unless the *galU* mutation used by Ornellas and Stocker is leaky, the results obtained with that mutant would suggest that glucose may not be part of the P1 receptor in *Salmonella*. Conversion from a smooth to a rough stage is not necessary for P1 adsorption in all cases, since the virus can adsorb to smooth versions of *Shigella flexneri* (Godard *et al.*, 1971).

It has been suggested that a protein component of the outer membrane of *E. coli* is also part of the P1 receptor, based on the observation that a protein-rich fraction isolated from outer membranes appeared to enhance the adsorption of P1 to *E. coli* (Tomás *et al.*, 1984). Moreover, an analysis of the outer membrane of *E. coli* and *Klebsiella pneumoniae* suggested that following P1 lysogeny there is a dramatic change in both the protein content of the outer membrane and the LPS core structure such that either the adsorption or injection of superinfecting λ or P1 is prevented (Tomás and Kay, 1984; Tomás *et al.*, 1984). It is difficult to take this study seriously because of deficiencies in the rigor of the experimental protocols and because the conclusions are contradicted by a significant body of data. Most compelling is the ability of λ to grow in a P1 lysogen if the P1 restriction enzyme is absent and of P1 to grow in a P1 lysogen if either one of two prophage repressor proteins is inactivated. Dramatic changes in the composition of the cell envelope appear both unreasonable and unnecessary.

P1 shows little ability to adsorb to or kill *E. coli* cells in the absence of Ca^{2+} (Franklin, 1969). While the role of Ca^{2+} in the adsorption process is still not clear, the isolation of host range mutants of P1 that show less of a requirement for Ca^{2+} than does the parent P1 (Franklin, 1969) suggests that Ca^{2+} is involved in the proper alignment of phage tail fibers and cell receptors. Ca^{2+} also appears to play a role in the stability of P1 in phage lysates, especially in the presence of chloroform (D. Touati, personal communication). The adsorption rate constants of P1 to *Shigella* ($k = 5 \times 10^{-10} \, min^{-1}$) and of P1*k* mutants to *E. coli* K12 strains ($k = 5 \times 10^{-10}$ to $5 \times 10^{-11} \, min^{-1}$) have been determined (Luria *et al.*, 1960).

B. Host Range Extension

Among the gram-negative bacteria, it is primarily in *E. coli* K12 that techniques for the construction of episomes, and for the generation of specialized transducing phage, are well developed. Moreover, it is in this host that generalized transduction, largely mediated by phage P1, has been a significant contributor to genetic manipulation and chromosomal mapping. If one could expand the host range of P1, then it should be possible to transfer genes from a variety of gram-negative bacteria into *E. coli* K12, where those genes could be more easily manipulated and characterized, and then back into their natural host, so that the effects of those manipulations could be studied. Although the host range of P1 is rather broad when compared to that of other phages such as λ, T4, and P22, it is quite narrow when compared to that of broad host range plasmids such as RK2 and RSF1010. Besides *E. coli* B, C, and K (Bertani, 1958), P1 can grow in *Shigella* and in rough forms of *Salmonella typhi* and *S. typhimurium*. P1 can also adsorb to, and lysogenize, but does not grow in, *Yersinia* (formerly *Pasteurella*) *pestis* and *Y. pseudotuberculosis* (Lawton and Molnar, 1972). P1 adsorbs to and injects into, but does not lysogenize, *Myxococcus xanthus* (Kaiser and Dworkin. 1975).

The host range of P1 has been expanded by the isolation of either phage or bacterial mutants. Among the phage variants is a mutant (P1*k*) able to plate with elevated efficiency on *E. coli* K12 (Lennox, 1955), two mutants (*h* and *hhf*) that allow P1 to grow on P1-resistant derivatives of *Shigella dysenteriae* strain Sh and *Shigella flexneri* strain 2a, respectively (Luria et al., 1960), and, as previously noted, mutants that grow on *E. coli* K12 *galU* strains.

The procedure used to isolate bacterial mutants with extended host range rests on the ability of P1c1.100 Km to transfer its DNA into bacteria at 30°C if those bacteria are permissive for P1 adsorption, injection, cyclization, and plasmid maintenance or integration (Table II). Among the kanamycin-resistant cells produced by this procedure there should be variants that are more permissive for P1 than is the population as a whole. To determine if any of the kanamycin-resistant cells can support the vegetative growth of P1, they were induced by heat [the c1.100 mutation makes the c1 repressor protein thermosensitive (Rosner, 1972)], and the yield of phage was measured. Phage production proves that the cells are lysogens and are permissive for P1 growth (Table II). The absence of detectable phage after induction (e.g., in the cases of *Flavobacterium* sp. M64, *Agrobacterium tumefaciens*, and *Alcaligenes faecalis*) indicates either that the phage growth defect is more severe in these bacteria than it is in others or that the kanamycin-resistant cells lack a complete complement of essential P1 genes. All attempts to select kanamycin-resistant colonies of gram-positive bacteria belonging to various genera were unsuccessful (Murooka and Harada, 1979).

A striking feature of the data of Murooka and Harada (1979) is the

TABLE II. Host Range of P1[a]

Bacterium	P1 DNA injection	P1 Phage production	References[b]
Escherichia coli K12, C, B	+	+	2
Shigella dysenteriae	+	+	1
Shigella flexneri	+		6
Salmonella typhimurium	(−) +	+	(4),5,9,11
Salmonella typhi and abony	+		4
Klebsiella aerogenes	+	+	8,11
Klebsiella pneumoniae	+	+	7,8,11
Citrobacter freundii	+	+	8,11
Serratia marcescens	+	(−) +	(3),11
Enterobacter aerogenes	+	+	8,11
Enterobacter liquefaciens and cloacae	+	+	8
Erwinia carotovora	+	+	11
Erwinia amylovora	+	+	8
Yersinia pestis and pseudotuberculosis	+	−	7
Proteus mirabilis, vulgaris and inconstans	+	+	11
Pseudomonas putida	−		8
Pseudomonas aeruginosa	+	(−) +	(3),11
Pseudomonas amyloderamosa	(−) +	+	(8),11
Flavobacterium sp. M64	+	−	11
Agrobacterium tumefaciens	+	−	11
Acetobacter suboxydans	−		11
Alcaligenes faecalis	+	−	11
Myxococcus xanthus	+	−	10

[a]Symbols + and − indicate that P1 injection or phage production was detected and not detected, respectively, in either wild type members of the bacterial species or in mutants of that species. The assay for injection was generally based on detection of antibiotic-resistant colonies following exposure of the bacteria to P1Cm or P1Km. The efficiency of acquisition of antibiotic resistance by the tested strains relative to E. coli was as low as 10^{-8} (in Myxococcus xanthus). In the experiments of reference 11, phage production was based on the detection of particles capable of plaque formation on E. coli following exposure of bacteria having acquired kanamycin resistance from P1Kmc1.100 to an "inducing" heat pulse and a further incubation to allow phage maturation. The phage yield from P1 lysogens of the tested Serratia, Proteus, and Pseudomonas species was no more than about 10^{-4} the yield obtained with E. coli.

[b]References: (1) Bertani, 1951; (2) Bertani, 1958; (3) Amati, 1962; (4) Kondo and Mitsuhashi, 1966; (5) Okada and Watanabe, 1968; (6) Godard et al., 1971; (7) Lawton and Molnar, 1972; (8) Goldberg et al., 1974; (9) Ornellas and Stocker, 1974; (10) Kaiser and Dworkin, 1975; (11) Murooka and Harada, 1979. See also Tominaga and Enomoto, 1986.

low efficiency of transfer of the kanamycin-resistance gene by P1Km into members of various genera of gram-negative bacteria, relative to E. coli. Moreover, there is generally a low yield of P1 following the presumptive induction of lysogens other than those of E. coli, Shigella dysenteriae, and Klebsiella aerogenes. The low efficiency of transfer of antibiotic resistance could reflect any one of several physiological defects including problems in adsorption or injection, a sensitivity to restriction systems present in the host, a failure to cyclize P1 DNA after injection, an inability to express plasmid maintenance functions, etc. Similarly, the low yields on "induction" could have many causes.

C. Recombinational Control of Host Range: The *cix-cin* System

1. Organization of *cix-cin* and Homologies with Other Systems

The genome of bacteriophage P1 carries an invertible 4.2-kb segment of DNA, called the C segment, that consists of a 3-kb unique sequence flanked by 0.62-kb inverted repeats (Iida *et al.*, 1982; Hiestand-Nauer and Iida, 1983) (Fig. 4). The C segment contains genes involved in tail fiber assembly and controls P1 host range by varying its orientation in the phage genome (Iida, 1984). Walker and Walker (1983) reported that an amber mutation in P1 gene 19, which has been mapped to the C segment, fails to make a tail fiber protein whose molecular weight is 105 kD. Moreover, the wild-type alleles of amber mutations in phage Mu tail fiber genes *S* and *U* can be rescued from the P1 C segment, and the wild-type alleles of P1 amber mutations 33, 72, and 66, which interfere with P1 tail fiber assembly, can be rescued from an analogous invertible segment, the G segment, of Mu (Toussaint *et al.*, 1978). Mu and P1 are also serologically cross-reactive. The electron microscopic studies of Chow and Bukhari (1976) reveal remarkable similarity between the P1 C loop and the Mu G loop. Heteroduplexes between the two loops reveal no observable non-homologous regions (the limit of resolution is 50–100 bp) over the entire 3-kb length of the duplex. These results suggest that the organization of the P1 tail fiber genes in the C segment is like that of the Mu tail fiber genes in the G segment (Kamp *et al.*, 1978; Iida *et al.*, 1982). These expectations appear to be justified (Iida *et al.*, 1984). The P1 C segment contains the analogous genes, *U* and *U'*, and the two C-terminal parts, *Sv* and *Sv'*, of analogous genes *S* and *S'* (Figs. 2A, 4). The common N-terminal portion of the *S* gene analogue and the promotor for both *U* and *S* gene analogues are outside of the C segment on the *imm*I-proximal side of that segment. The direction of transcription is toward the C segment. The 105-kD gene 19 protein has been identified with the *S* gene product (Iida, 1984; Iida *et al.*, 1984). This identification is consistent with results obtained from a construction in which the *S* gene promoter and the constant portion of the *S–S'* gene (*Sc*) have been replaced by the λP$_L$ promoter. Activation of that promoter generates a 22.5-kD protein that is presumed to be the analogue of the Mu *U* or *U'* gene product (Huber *et al.*, 1985a).

Support for the hypothesis that C segment inversion affects tail fiber production and P1 host range comes from an analysis of insertion and deletion mutations that span the C segment (Iida, 1984). In the orientation shown in Fig. 4, termed C(+), mutations in the *imm*I-distal portion of the C segment (containing *Sv'–U'*) do not affect the ability of P1 to plaque on *E. coli* K12 or *E. coli* C. In contrast, mutations in the *imm*I-proximal portion of the C segment (containing *Sv–U*) block the ability of P1 to plaque on these hosts. Evidence that the *Sv'–U'* portion of the C segment contains tail fiber genes comes from the isolation of *E. coli* C

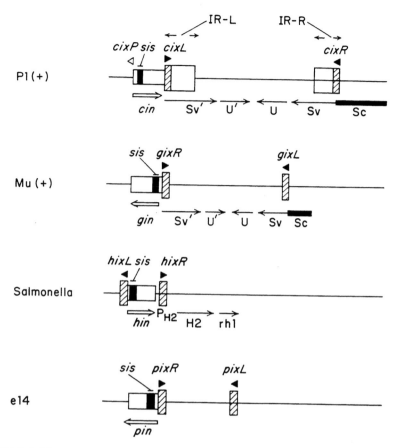

FIGURE 4. The *cix-cin* region and its homologues. The figure is adapted from Iida *et al.* (1984) and shows the organization of the inversion systems of P1(+), Mu(+), *Salmonella*, and e14. The recombination sites (*cix, gix, hix,* and *pix*), the recombinase genes (*cin, gin, hin,* and *pin*), the *sis* enhancer sites, and the genes whose synthesis is controlled by inversion are all described in Section III.C. Arrows indicate the direction of transcription, arrowheads the orientation of recombination sites (cross-hatched), and *sis* the location of enhancer sequences (filled rectangles) within recombinase genes (open narrow rectangles).

mutants that are permissive for P1 phage containing insertion mutations in the *Sv–U* region of the C segment, provided the C segment is in the C(−) orientation (Iida, 1984). Both P1C(−) and P1C(+) phages produce plaques on this *E. coli* C mutant.* In summary, the P1 C segment carries

---

*Tominaga and Enomoto (1986) have shown that the inability of P1C(−) to plaque on a wild-type strain of *E. coli* C (and on *Shigella sonnei*) can be remedied by the presence of Mg^{2+}. These workers postulate that Mg^{2+} mediates a strong interaction between lipopolysaccharides that exposes a hidden receptor. The candidates proposed for the P1C(−) receptor are the terminal galactose or subterminal side-chain glucose (β1,3) that is characteristic of the LPS core of *E. coli* C and *Sh. sonnei* (see Fig. 3).

two sets of genes for host range specificity, and C inversion alters the P1 host range through activation of one set of genes.

The essential components of the system that inverts the P1 C segment are a recombinase protein encoded by the phage *cin* gene and two imperfect, inverted, repeat sequences at the outside ends of the C segment (*cix* sites) in which crossover occurs during inversion (Hiestand-Nauer and Iida, 1983). The *cin* gene is located on the *immI*-distal side of the C segment. It encodes a 21-kD protein and is transcribed in the clockwise direction on the P1 map, toward the C segment (Fig. 4). The *cin* termination codon, TAA, is within the *immI*-distal *cix* site, *cixL*, of the C segment (Hiestand-Nauer and Iida, 1983; Huber *et al.*, 1985a). The *cixL* and *cixR* sites differ only in the sequence of the *cixL* termination codon, presumably reflecting the need to terminate the *cin* open reading frame in *cixL* before it progresses too far into the C segment. The repeats at the end of the C segment are about 600 bp long and contain at their outside ends either 9-bp (*cixL*) or 13-bp (*cixR*) regions of imperfect dyad symmetry flanking a 2-bp –TT– spacer region (Fig. 5A) (Hiestand-Nauer and Iida, 1983). Based on the sequence of hybrid sites generated by recombination between a *cixL* site and *cixQ* sites (quasi *cix* sites present in pBR322 DNA), the actual position of the crossover could be localized to the 2-bp –TT– spacer region (Iida *et al.*, 1984). Some of the *cixQ* sites contain only one T residue in the spacer, and when they recombine with *cixL*, conversion can occur at the mismatched position. This indicates that Cin generates a 2-bp staggered cut flanking the –TT– spacer and that reciprocal strands exchange within that 2-bp sequence (Iida and Hiestand-Nauer, 1986).

The *cin-cix* system of P1 is one of a family of DNA inversion systems that also includes the Mu *gin-gix* system, the *Salmonella hin-hix* system and the *E. coli* e14 *pin-pix* system (Fig. 4). While a biological role for the *pin-pix* system has yet to emerge (Plasterk *et al.*, 1983), the *hin-hix* system, like the *cin-cix* and *gin-gix* systems, serves as a biological switch for gene expression. DNA inversion at *hix* sites controls flagellin phase variation by connecting and disconnecting the flagellin promoter P_{H2}, located in the H-invertible segment, and the H2 operon, located outside of the H-invertible segment (Simon and Silverman, 1983; Zieg *et al.*, 1977). The H2 operon consists of two genes—the upstream H2 gene, which encodes the phase 2 flagellin protein, and the rh1 gene, which specifies a repressor of the phase 1 flagellin gene, H1.

The four inversion systems show much functional and structural homology. The recombinases (Cin, Gin, Hin, and Pin) encoded by the four systems can complement each other for inversion and exhibit more than 60% identity at the amino acid level (Kutsukake and Iino, 1980a; Plasterk and Van de Putte, 1984). Significant similarity also exists between these invertases and the resolvases of the Tn3 and γδ transposons, although no complementation between the invertases and the resolvases has been detected (Simon *et al.*, 1980). The Mu G segment and the *Salmo-*

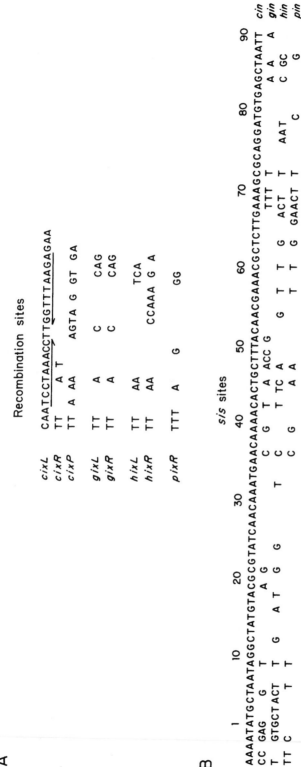

FIGURE 5. Sequences of recombination and enhancer sites within the *cin*, *gin*, *hin*, and *pin* inversion systems. (A) Recombination sites. The sequence of the *cixL* recombination site is shown in the top line. The 9-bp imperfect inverted repeat with the 2-bp –TT– spacer is indicated by the arrows below the *cixL* sequence. Only the bases that differ from those in the *cixL* sequence are shown for the other recombination sites. (B) Enhancer sites. The sequence of the *sis* (*cin*) enhancer site is shown on the top line. Only the bases that differ from those in the *sis* (*cin*) sequence are shown for *sis* (*gin*), *sis* (*hin*), and *sis* (*pin*).

nella H segment are flanked by 34-bp and 14-bp perfect inverted repeats, respectively, that contain the sites (*gixL* and *gixR* for Mu and *hixL* and *hixR* for *Salmonella*) in which crossover occurs during the inversion reaction (Plasterk *et al.*, 1983; Kahmann *et al.*, 1985; Johnson and Simon, 1985). The *gixL*, *gixR*, and *hixL* crossover sites consist of 26- to 30-bp segments of DNA containing 12- to 14-bp regions of imperfect dyad symmetry that bear considerable homology with similar regions of the *cixL* and *cixR* sites (Fig. 5A) and that flank 2-bp –TT– or –TC– spacers (Hiestand-Nauer and Iida, 1983). The *hixR* site contains only one of the 12-bp symmetry regions present in *hixL* and shares the –TT– spacer with *hixL*. Johnson and Simon (1985) have shown that a synthetic 26-bp sequence containing the *hixL* crossover site is sufficient for Hin-mediated inversion. Recombination between *cix* and *hix* sites has also been demonstrated (Iida *et al.*, 1984). *In vitro* reactions with purified recombinase proteins indicate that supercoiled circular DNA is recombined much more efficiently than is either relaxed circular DNA or linear DNA, and that recombination between sites in inverted orientation is much more efficient than recombination between sites that are in direct orientation (Huber *et al.*, 1985b; Kahmann *et al.*, 1985; Johnson and Simon, 1985).

The recombinase genes in all four inversion systems lie adjacent to one of the recombination sites (Fig. 4). The *hin* gene is located between the sites, and the *cin*, *gin*, and *pin* genes are located just to one side of the sites. *cin* is transcribed toward the invertible region, and *gin* and *pin* are transcribed away from it (Zieg *et al.*, 1977; Kamp *et al.*, 1978; Plasterk *et al.*, 1983). The *hin*, *pin*, and *gin* promoter elements overlap the *hix*, *pix*, and *gix* sites, respectively, and the *cin* promoter overlaps a quasi-*cix* site, called *cixP*, which is opposite in orientation to *cixL*. Additional quasi-*cix* sites have been identified, including four within *cin* itself (Iida and Hiestand-Nauer, 1987). The significance of this organization of transcription and recombination elements in regulating the inversion reaction will be discussed below.

2. Enhancer-Mediated Bacterial Control of Inversion

An exciting development in the regulation of site-specific recombination in recent years is the discovery of *cis*-acting recombination enhancer elements, *sis* (sequence for inversion stimulation) sites, in the *cin*, *gin*, and *hin* inversion systems (Huber *et al.*, 1985b; Kahmann *et al.*, 1985; Johnson and Simon, 1985). The *sis* sites consist of 60- to 72-bp segments of DNA (Fig. 5B) that are contained within the amino-terminal portion of the recombinase genes and that enhance recombination of adjacent inversion segments 15- to 20-fold in the *hin* and *gin* systems and as much as 100-fold in the *cin* system. The *sis* (*cin*), *sis* (*hin*), and *sis* (*gin*) elements show much sequence homology (Huber *et al.*, 1985b; Fig. 5B) and *sis* (*hin*) can replace *sis* (*cin*) for *cix* inversion (Huber *et al.*, 1985b). The *sis* elements behave like true eukaryotic enhancer elements in that their activity is largely independent of orientation with respect to the crossover sites. The *sis* (*gin*) enhancer can function as far as 3 kb from the G segment (Kahmann *et al.*, 1985). The *sis* (*hin*) enhancer consists of at least two noncontiguous sequence domains whose relative orientation, but not precise spacing, with respect to each other is important; i.e., a 2- to 10-bp insertion in the middle of *sis* (*hin*), between the domains, does not interfere with enhancer activity (Johnson and Simon, 1985). While *sis* can be far from the recombination sites, it cannot be too close. If *sis* (*hin*) is placed 48 bp from the center of one of the recombination sites, it fails to function (Johnson and Simon, 1985).

In vitro studies indicate that in addition to the recombinase, the *hin* system requires two host factors, a 12-kD protein and the *E. coli* HU protein (Johnson and Simon, 1985; Johnson *et al.*, 1986), and the *gin* system requires a 12-kD host protein, called Fis (*factor for inversion stimulation*) (Kahmann *et al.*, 1985). Since both Fis and the 12-kD *hin*-host protein specifically bind to the enhancer sequence, the two proteins are probably one and the same. The difference in the HU requirement in the two systems may be one of degree rather than substance. Johnson *et al.* (1986) noted that the HU requirement varies with the distance of the *sis* (*hin*) element from the *hix* sites, and R. Kahmann (personal communication) recently showed that an HU requirement can be generated in the *gin* system if *sis* (*gin*) is moved further than it normally would be from the *gix* sites.

The failure of *sis* to act when located very close to the recombination sites suggests that there are steric limitations on *sis* action. Perhaps the DNA between *sis* and recombination sites needs to bend so that the proteins bound to each of those two DNA segments can be properly aligned. When the sites are too close, bending is prevented. A similar type of mechanism has been proposed for the action of *gal*, *ara*, and λ repressors (Majumdar and Adhya, 1984; Dunn *et al.*, 1984; Hochschild and Ptashne, 1986).

3. Potential Autoregulatory Circuits in Inversion

The overlap between the recombination sites and the *gin* and *hin* recombinase gene promoters offers a mechanism for autoregulation of gene expression. In addition to promoting inversion, binding of the recombinase to its recombination site might also block access of the promoter to RNA polymerase at the recombinase gene. Reduction in the level of recombinase transcription and consequent synthesis would in turn reduce binding to the site and permit enhanced transcription of the recombinase gene. An interesting variation on this theme is provided by the P1 *cin* gene. Since *cin* is located to one side of the invertible C segment and is transcribed toward that DNA, its promoter does not overlap one of the *cix* sites normally used in the inversion. Rather, it overlaps a quasi-*cix* site. Since that site rarely, if ever, takes part in recombination

(Hiestand-Nauer and Iida, 1983; Iida et al., 1984) but presumably binds Cin protein, cin expression can be regulated as a function of Cin concentration independent of recombination. Whether the autoregulatory circuits outlined above actually operate to ensure a constant frequency of switching remains to be determined.

Two other features of the P1 cin system might also function in regulating inversion: the presence of the cin termination codon within cixL could affect transcription at the very end of the cin gene, and convergent transcription of cin and C segment tail fiber genes might interfere with each other and, therefore, with the synthesis of either Cin or tail fibers or both. Huber et al. (1985a) tested the termination codon hypothesis by fusing a cin gene, containing an upstream strong promoter (lacUV5), to the E. coli galactokinase gene. In one construction, cin contains the wild-type cixL site at its C-terminal end, and in a second construction it contains a cixL site with a deletion that renders the site inactive for recombination but leaves the cin termination codon intact. No significant effect of the deletion on galK expression was observed, indicating that binding of Cin to cixL does not affect transcription of the C terminus of cin. Apparently, an actively transcribing RNA polymerase can remove any Cin protein bound to a cix site ahead of it. Moreover, neither transcription from the weak cin promoter nor transcription from the stronger lacUV5 promoter through the cixL site affects the ability of that site to recombine in vivo. In contrast, transcription from the very strong phage λ P_L promoter reduced cixL recombination function drastically. Thus elevated levels of transcription could act as a regulator of inversion. Whether such transcription levels are ever achieved during P1 growth is not known.

4. A Physiological Consequence of the Frequency of C and G Segment Inversion

For the case of C and G segments of phages P1 and Mu, another factor operates to regulate the distribution of invertible segments in the phage population. If a phage stock is prepared by induction of E. coli K12 lysogens, both (+) and (−) configurations are present in the stock with almost equal frequency. In contrast, if a stock is prepared by infection, the phages exhibit predominantly the orientation that produces tail fibers capable of adsorbing to the host used in the experiment. For E. coli K12, that is the (+) orientation (Fig. 3) (Kamp et al., 1978; Iida et al., 1982). These results are due to the improbability of inversion within the time needed for either P1 or Mu to go through a single burst cycle (Symonds and Coelho, 1978; Iida, 1984). Inversion rarely occurs in a single phage cycle, but it is likely to have occurred repeatedly during growth of a lysogenic colony. If a phage stock is prepared by induction of a lysogen, the burst from any one cell will reflect the orientation of the prophage at the moment of induction and that orientation will have an equal probability of being either (+) or (−). When a stock prepared by induction is in turn used to infect *E. coli* K12 only, the (+) phage will produce successful infections, little inversion will occur during a single burst cycle, and the stock will be predominantly of the (+) type. Since P1 generates a higher percentage of (−) phage in an infection of *E. coli* K12 than does Mu (Iida, 1984), the amount of C inversion in a single P1 cycle must be greater than the amount of G inversion in a single Mu cycle.

D. Injection

To our knowledge, the process has not been studied in P1.

IV. CYCLIZATION

A. The Need for Cyclization

The DNA encapsidated in a P1 virion is a terminally redundant linear molecule of about 100 kb. When that DNA is injected into a sensitive bacterium, one of the first events that occurs is a conversion to a closed circular form. Segev *et al.* (1980) showed that circular P1 molecules could be detected as early as 5 min after injection, well before the DNA is replicated. The importance of the cyclization event is accentuated by the almost complete failure of P1 to grow lytically or establish lysogeny in the absence of recombination systems necessary for the cyclization reaction (Sternberg *et al.*, 1986). Presumably if the injected P1 DNA is not cyclized, it is rapidly degraded from its ends by cellular nucleases such as the product of the host's *recBCD* genes, exonuclease V (Smith, 1983; Amundsen *et al.*, 1986). Other viruses, such as λ and T4, deal with this threat to their existence by encoding an inhibitor of the *recBCD* nuclease (Sakaki, 1974). It has been reported that P1 also encodes such an inhibitor (Sakaki, 1974). However, if this report is correct, the inhibition must be less than complete, since three different observations suggest that the *recBCD* nuclease is still active in P1-infected cells. First, the lysogenization frequency of P1 is reduced about 10-fold in a *recB* mutant (Rosner, 1972). Second, the burst size of P1 is reduced about 20-fold in a *recBC* host when compared to that observed in the isogenic *recBC*+ host (Zabrovitz *et al.*, 1977). Third, DNA ends generated during P1 infection by cleavage at the packaging site (*pac*) are very sensitive to the degradative function of the RecBCD nuclease (Sternberg and Coulby, 1987a). Thus, the RecBCD nuclease remains quite active during P1 infection, at least during the initial stages, and this activity can account for the need to cyclize P1 DNA rapidly.

What, then, are the recombination systems responsible for cyclization? First we will consider the *lox-cre* site-specific recombination system, whose involvement in this process is well documented, and then we will discuss alternative cyclization modes.

B. *lox-cre*-Mediated Site-Specific Recombination

As noted previously, the existence of a site-specific recombination system in P1 was predicted from the observation that the genetic map of P1 is linear despite the fact that virion DNA is cyclically permuted. Phages such as P22 and T4, which also have cyclically permuted DNA, but which lack site-specific recombination systems, have circular genetic maps. The ends of the P1 genetic map define the locus, *loxP* [locus of crossing over (X) in P1], at which the site-specific recombination system operates. The recombination system has been well characterized with regard to both the necessary components and their mechanism of action.

1. Organization of the *lox-cre* System

The *loxP* site is a 34-bp sequence that consists of two perfect 13-bp inverted repeats separated by an 8-bp spacer region (Fig. 6) (Hoess *et al.*, 1982). Recombination between two sites is mediated by a single phage-encoded protein, Cre (*cyclization recombination* protein), of 343 amino acids (Sternberg *et al.*, 1986). The Cre structural gene, *cre*, is located 434 bp clockwise from *loxP* (Figs. 2A, 6) and is transcribed in the clockwise direction. Between *loxP* and *cre* is a 73-codon open reading frame, desig-

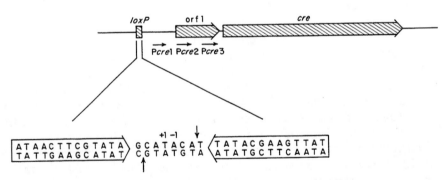

FIGURE 6. The *lox-cre* region. The top line indicates the relative position of *loxP*, orf-1, and *cre*. The position of three promoters (Pcre1, Pcre2, and Pcre3) is shown below the line. The transcript specified by Pcre1 includes both orf-1 and *cre*, and its synthesis is sensitive to Dam methylation (Sternberg *et al.*, 1986). The sequence of the 34-bp *loxP* site is shown at the bottom of the figure. It consists of 13-bp inverted repeats (enclosed within the broad horizontal arrows) and an 8-bp spacer. The bases are numbered from the center of the spacer leftward (+) and rightward (−). The positions of the cleavages during recombination (between +3 and +4 on the upper strand and between −3 and −4 on the lower strand) are indicated by the thin vertical arrows. The mechanism of *loxP* recombination is discussed in Section IV.B.2.

nated orf-l, and a set of three weak promoters that have been implicated in orf-l and *cre* expression. One of the promoters, Pcre1 (formerly pR1), is upstream of both orf-l and *cre* and, therefore, could produce a transcript of both genes. The other two promoters, Pcre2 and Pcre3 (formerly pR2 and pR3), are located within orf-l. All three of these promoters are relatively weak, with activities ranging from 7% to 10% that of the galactose operon promoter, and they are active in P1 lysogens. Thus *cre* expression appears to be insensitive to repressors made by a P1 prophage. Constitutive expression permits the *lox-cre* system to play a role in P1 plasmid maintenance (see Section VII.C.3). The only evidence to date for regulation of *cre* expression is based on the observation that transcription from Pcre1 is sensitive to methylation by the *E. coli* Dam protein. That enzyme methylates the adenine in the sequence GATC at the N^6 position (Hattman *et al.*, 1978; Sternberg, 1985). The Pcre1 promoter contains two Dam methylation sites in its −35 region and is five times more active in a *dam* host than it is in a *dam*$^+$ host (Sternberg *et al.*, 1986). A mutational analysis of the *cre* gene suggests that a carboxyl terminus of the protein is involved in the cleavage of DNA and an amino terminal domain is involved in recognition of the binding site (Wierzbicki *et al.*, 1987).

2. Mechanism of *lox-cre* Recombination

The effects of DNA deletion mutations in the region of *loxP* and the results of DNase I footprinting experiments with purified Cre protein (Hoess *et al.*, 1984; Hoess and Abremski, 1984) indicate that sequences outside of the 34-bp *loxP* site are not required for recombination and they are not in contact with the recombination protein. Moreover, half-sites (one 13-bp repeat and 4-bp of the spacer) bind Cre just as efficiently as does the entire site and are fully protected from DNase I by Cre binding. These results suggest that each site consists of two binding domains that are nonidentical by virtue of the different sequences in the *loxP* spacer region. The asymmetry of the spacer sequence accounts for the asymmetry in the recombination reaction. Thus, recombination between sites that are in directly repeated orientation results in the excision of DNA between those sites, and recombination between sites that are in inverted orientation relative to each other inverts the DNA between them (Abremski *et al.*, 1983). If the spacer region is made symmetric, then 50% of the recombination events are of the excision type and 50% are of the inversion type (Hoess *et al.*, 1986).

Unlike many other recombination systems, the *lox-cre* system is quite flexible as to the substrate that it can utilize. We have already alluded to the fact that it can recombine sites on the same molecule that are either in direct or inverted orientation relative to each other. In addition, it shows no preference for DNA that is linear or circular and works both on relaxed and supercoiled circles (Abremski *et al.*, 1983). It will also promote recombination between separate DNAs containing *loxP* sites, although with reduced efficiency relative to recombination between sites on the same molecule. The latter observation suggests that distance between sites on any one DNA molecule is not a limiting element in the recombination reaction—an important point, because *loxP* sites in the terminally redundant regions of P1 virus DNA are about 100 kb apart.

The location of the cleavage sites associated with *loxP* recombination have been mapped using an efficient *in vitro* system (Hoess and Abremski, 1985). During recombination the cuts are made six base pairs apart on opposite strands of the spacer and result in a 5′ protruding terminus (Fig. 6). At the point of cleavage, Cre becomes covalently attached to a 3′ phosphate and produces a free 5′ hydroxyl. Analysis of point mutations in the *loxP* spacer region indicates that homology between the spacer regions of two sites is a requirement for efficient recombination (Hoess *et al.*, 1986). Thus, a *loxP* with a C → T change at position +3 of the spacer (Fig. 6) will not recombine well with a wild-type site but will recombine well with an identical mutant site.

An interesting consequence of recombination between *loxP* sites on a circular substrate is that the products of the reaction are not topologically linked after the reaction. Recombination between directly repeated sites on a highly supertwisted circle generates two supertwisted circular products that are either free of each other or present as simple catenanes (Abremski *et al.*, 1983; Abremski and Hoess, 1985). The simplest interpretation of this result is that the events associated with the recombination process are not initiated until the sites come together in a way that physically separates the DNA domains between the sites. For plectonemically twisted supercoiled circles containing two *loxP* sites, this could occur if the Cre molecules bound to *loxP* sites interact primarily, but not necessarily exclusively, when the sites are brought close to each other by the sliding of DNA along itself. Presumably, only then does a productive recombination event occur.

3. Role of *lox-cre* Recombination in the Cyclization of P1 DNA

The *lox-cre* system appears to play an important role in at least three different stages of the P1 life cycle: the cyclization of injected P1 DNA, the maintenance of the P1 plasmid state, and the vegetative growth of the virus. We defer discussion of the latter two *lox-cre* functions until later sections of this review and concentrate here on the role of site-specific recombination in cyclization.

The properties of P1 *cre* mutants lend support to the proposed role of the *lox-cre* system in the cyclization reaction. These mutants lysogenize *recA* bacteria 10–25 times less efficiently than does a *cre*[+] phage, but they are indistinguishable from *cre*[+] phage in their ability to lysogenize *recA*[+] bacteria (Schultz *et al.*, 1983; Sternberg *et al.*, 1986). Since *cre* lysogens are no less immune than *cre*[+] lysogens (N. Sternberg, unpublished), the reduced lysogenization frequency in the *recA* cells probably reflects the failure of the *cre* DNA to cyclize. Its ability to cyclize in *recA*$^+$ bacteria indicates that other systems dependent on RecA function can replace *lox-cre* for cyclization.

More direct evidence of a role for *lox-cre* in cyclizing P1 DNA comes from the experiments of Segev and Cohen (1981) and Hochman *et al.* (1983). By following the conversion of linear ^{3}H-labeled P1 DNA to a closed circular form after injection, these authors showed that cyclization could occur in sensitive *recA*$^+$ and *recA* bacteria as well as in *recA*$^+$ and *recA* P1 lysogens but could not occur if a sensitive host was treated with either chloramphenicol or rifampicin prior to the infection. However, if the chloramphenicol-treated cells had previously accumulated Cre protein from a functioning *cre* gene, the ability to cyclize was restored. Two conclusions emerge from these experiments: (1) the *lox-cre* recombination system can cyclize at least some of the injected P1 DNA molecules, and (2) the host generalized recombination system, at least by itself, cannot. In these experiments, however, no more than 1–2% of the labeled DNA was converted to a circular form. The expected value is 25–30%, because only one of every three or four P1 DNAs is expected to have *loxP* sites in its terminally redundant region, and only that DNA can be cyclized by *lox-cre* recombination. A better accounting of the labeled P1 DNA in the above experiments needs to be achieved before we can be sure of the conclusions drawn from those experiments.

The supposition that only one of every three or four plaque-forming P1 virions contains *loxP* sites in the terminally redundant region derives from the following observations. First, the packaging site (*pac*) is about 5 kb counterclockwise from *loxP*, and the direction of packaging is from *pac* toward *loxP* (Bächi and Arber, 1977; Sternberg and Coulby, 1987a) (see section VIII.E). Since the amount of DNA that can be packaged into a P1 head is about 100 kb and the size of the P1 genome is only 90 kb, the terminal redundancy is about 10 kb. Packaging proceeds for an average of about 3–4 headfuls on any one concatemer (Bächi and Arber, 1977), generating a corresponding number of classes of viral DNAs. Only class I DNA, the first headful, will have terminally redundant *loxP* sites that allow it to cyclize by the *lox-cre* system. This feature may be important to the virus, because it permits cyclization of the P1 DNA that is packaged first without consideration for the recombination potential of the host cell into which that DNA is injected.

In considering what might control the cyclization process, it is obvious that if there is a competition between the degradation of linear P1 DNA injected into cells and its cyclization, it would be wise to maximize Cre synthesis during the short period immediately after injection. The regulation of *cre* transcription by adenine methylation provides an opportunity to do just that. When DNA is replicated, the newly synthesized strand is not immediately methylated by the host's Dam methylase (Lyons and Schendel, 1984; Marinus, 1976). Consequently, newly replicated DNA is transiently hemimethylated. Moreover, during rapid viral replication, newly synthesized DNA exceeds the methylation capacity of the cell, resulting in fully unmethylated DNA (Marinus, 1984). Thus, replication provides a way of increasing the expression of genes that are transcribed more efficiently from either hemi- or unmethylated DNA than they are from fully methylated DNA. As indicated above, *cre* is such a gene. If one makes the assumption that *cre* is rapidly replicated following injection of P1 DNA, then hemimethylated, and perhaps even unmethylated, *cre* DNA would be produced, causing elevated *cre* expression at a time when its function is critical for the cyclization reaction. What could be the source of this proposed replication? One possibility is that there exists an origin of vegetative replication near *cre*. Conceivably the origin detected by Froehlich *et al.* (1986) fills this function. An alternative possibility is that injected P1 DNA is replicated from its ends. Since *cre* would be closer to an end in class I DNA than in the other three classes of DNAs, this mode of replication would ensure overproduction of Cre in just those cells in which that protein could be used to cyclize P1 DNA.

C. Alternative Modes of Cyclizing P1 DNA

The most obvious alternative to the *lox-cre* system for cyclizing P1 DNA is the general recombination system of the host. Theoretically, it could generate P1 circles by recombination at any position within the terminally redundant region of an injected P1 DNA molecule. The first indication that this alternative mode might not function, at least by itself, in the cyclization reaction comes from the previously discussed experiment in which chloramphenicol-treated cells were shown not to be able to cyclize injected P1 DNA despite the fact that they should still be recombination proficient (Segev and Cohen, 1981). This result is consistent with one of three possibilities: (1) The P1 *lox-cre* system is the only one that can cyclize P1 DNA, and, therefore, only class I DNAs are cyclized. (2) P1 encodes a general recombination system that can work at the redundant ends of all classes of packaged DNA. (3) P1 encodes a protein that enhances bacterial generalized recombination and permits injected P1 DNA to cyclize before that DNA is degraded by the exonuclease V. The first possibility cannot be correct, since, as noted above, P1 *cre* mutants lysogenize bacterial strains that are proficient for general recombination no less efficiently than does P1 wild type (Schultz *et al.*, 1983; Sternberg *et al.*, 1986). The second possibility is highly unlikely in light of the fact that P1 *cre* mutants lysogenize *recA* bacteria 10- to 25-fold less efficiently than does P1 wild type (Schultz *et al.*, 1983; Sternberg *et al.*, 1986). Moreover, the latter result, in combination with the experiments of Segev and Cohen (1981), supports the third model. That model is also consistent with the two- to fivefold reduction in P1 *cre*⁺ lysogenization frequency caused by a *recA* mutation in the host (Rosner, 1972;

Segev and Cohen, 1981; Sternberg *et al.*, 1986). The reduced efficiency is presumably a consequence of the failure to cyclize all but the class I packaged DNAs.

A candidate for a gene that might encode a general recombination enhancement function involved in cyclizing P1 DNA is the P1 *ref* gene (Windle and Hays, 1986; D. Lu, S. Lu, and M. Gottesman, personal communication). It is located just clockwise from *cre* (Fig. 2A), and it encodes a protein that enhances the host's general recombination system, at least for certain recombination events. Unfortunately, it appears to fail the critical test as far as P1 cyclization is concerned, since P1 *cre ref* phages lysogenize recombination-proficient bacteria no less efficiently than do P1 *cre ref*$^{+}$ phages (Windle and Hays, 1986). A P1 cyclization enhancement gene has yet to be found.

Although we have excluded the possibility that P1 encodes a general recombination system capable, by itself, of cyclizing P1 DNA, it should be noted that Hertman and Scott (1973) reported that recombinants between P1 *am* mutants could be produced at reasonably high levels in *recA* bacteria. They concluded that P1 encodes a general recombination function. Sternberg *et al.* (1980) showed that this was not the case and that the data of Hertman and Scott could be accounted for by *lox-cre* recombination in the *recA* host, packaging of recombined molecules to generate terminally redundant heterozygotes, and resolution of those heterozygotes by the bacterial general recombination system during cyclization in the *recA*$^{+}$ host used to assay the recombinants.

Another recombination system that could function in the cyclization reaction is the P1 *cin-cix* system. Iida *et al.* (1982) showed that Cin could promote recombination between *cix* sites on separate DNA molecules, indicating that it probably could also function to cyclize P1 DNA containing the invertible C segment in the redundant region. However, three points argue against the actual involvement of *cin-cix* in cyclization. First, the recombination is relatively inefficient when compared to the *lox-cre* system (Iida *et al.*, 1982) and therefore might not be able to work fast enough to prevent the degradation of the linear P1 molecule. Second, *in vitro* and *in vivo* studies indicate that the preferred substrate for the *cin-cix* system is a supercoiled molecule containing sites in inverted orientation (Iida *et al.*, 1984; Huber *et al.*, 1985b). The cyclization reaction would require that it work on a linear molecule with directly repeated sites. More pertinent, however, is the position of the *cin-cix* system in the P1 genome. Its location, about 34–38 kb from *loxP* and *pac*, would preclude its presence in the redundant region of P1 DNA except for the rare class V-packaged P1 DNAs.

Until this point we have considered the cyclization reaction only as it concerns P1. The case of P7 is worth mentioning here, as it represents an interesting paradox. Despite the fact that the size of the terminally redundant region is much smaller for P7 DNA (0.5–1 kb) than it is for P1 DNA (about 10 kb), P7 lysogenizes recombination proficient hosts as

efficiently as does P1 (N. Sternberg, unpublished). Surprisingly, P7 can do
this in the absence of terminally redundant *loxP* sites, since the small
size of the P7 redundant region should preclude the generation of phage
with two *loxP* sites. Thus, the cyclization of P7 DNA would appear to
depend solely on systems other than *lox-cre*. The P7 data also suggest
that the 10-fold difference in homology at the ends of P1 and P7 DNA is
not limiting for cyclization when viral DNA is injected into a sensitive
bacterium. This could be the case if DNA ends greatly facilitate the
recombination of adjacent homologous sequences. To appreciate the extent of this increase, we remind the reader that homologous recombination between two adjacent integrated prophages (each about 100 times
the length of the terminally redundant region of P7) occurs so infrequently as to excise a prophage in no more than 1% of the bacterial
population per cell generation (Arber, 1960a).

V. RESTRICTION-MODIFICATION

A. Organization and Function of the Restriction-Modification Genes

   When P1 lysogenizes *E. coli*, the lysogens acquire P1-specific modification and restriction activities. In a elegant set of experiments, Arber
and Dussoix (1962) and Dussoix and Arber (1962) showed that a P1 prophage could cause the phenotypic modification of λ DNA or the destruction of unmodified λ DNA. The studies were extended, using Hfr × F⁻
conjugation and P1 transduction, to show that also the bacterial chromosome is sensitive to P1 restriction and modification (Arber and Morse,
1965; Inselberg, 1966). Indeed, if P1 DNA itself is not modified, owing to
a mutation that inactivates the phage modification function (Glover *et
al.*, 1963), then it too is subject to restriction when it enters a P1 lysogen.
   The P1 restriction-modification system provides a means of protect-
ing lysogenic cells from the deleterious effects of foreign viral and cellular
genes. It is more general in its effect than the P1 immunity system, since
the DNA sequence recognized by the restriction enzyme is only 5 bp long
(see below) and therefore likely to be present in nearly all DNAs that are
larger than 2 kb.
   The genes encoding the P1 restriction (*res*) and modification (*mod*)
functions are located between coordinates 3 and 9 on the P1 genetic map
(Fig. 2A) and are contained within a 5-kb segment of DNA at the *loxP*-
proximal end of *Bam*HI-4 (Heilmann *et al.*, 1980b; Iida *et al.*, 1983). Since
*Eco*RI-2 expresses *mod* but not *res* function (Sternberg, 1979; Heilmann
et al., 1980b), *mod* must be located on the *loxP*-distal side of *res*. The
precise location of the genes encoding both of these activities was determined by generating transposon-induced mutations of the *res* and *mod*
genes (Heilmann *et al.*, 1980b; Iida *et al.*, 1983). The mutations were

mapped by electron microscopy and restriction site analysis, and their effects on *res* and *mod* functions were determined. Two classes of mutations were identified—those that inactivate restriction function, and those that inactivate both restriction and modification function. The former were shown to be in the *res* gene, and the latter in the *mod* gene. This analysis localizes *res* to a 2.8-kb segment of DNA that spans the *loxP*-proximal end of *Eco*RI-2 and at least a portion of *Eco*RI-24. The *mod* gene is about 2 kb long and is located within *Eco*RI-2.

The direction of *res-mod* transcription and the nature of the transcripts were assessed by electron microscopic mapping of *in vitro* transcripts (Iida *et al.*, 1983) and by the sizing of truncated proteins generated by insertion mutations (Heilmann *et al.*, 1980b). These studies indicate that both the *res* and *mod* genes have their own promoters and that in both cases transcription is counterclockwise. Transcripts were also found that start at the *mod* promoter and span the *res* and *mod* genes, indicating that both Res and Mod polypeptides may also be made from the same transcript.

B. The P1 Restriction and Modification Enzyme and Its Mode of Action

It is now well established that the *Eco*P1 restriction enzyme consists of two subunits of 106–110 and 73–77 kD (Heilmann *et al.*, 1980b; Hadi *et al.*, 1983). The sizes of the subunits agree quite well with those predicted from the size of the genes that encode those subunits (Hümbelin *et al.*, 1988). Brockes (1973) identified a second modification subunit of 45 kD not seen in the subsequent studies and concluded that the P1 modification enzyme was a dimer of two unlike subunits. Since the enzyme purified by Brockes was isolated following induction of a P1 prophage whereas the enzyme isolated in the subsequent studies was either from a cloned fragment or from an uninduced lysogen, it is possible that the 45-kD modification protein is not part of the *Eco*P1 enzyme but is a protein encoded by a gene expressed only after P1 induction. We discuss supporting evidence for this hypothesis below.

The DNA site methylated by the *Eco*P1 enzyme is AGA$\overset{*}{\text{A}}$C^C_T, where A is N^6-methyladenine (Hattman *et al.*, 1978; Bächi *et al.*, 1979). This site is asymmetric, with enzyme methylating only the strand shown (Bächi *et al.*, 1979). The complementary strand lacks an adenine residue and is not methylated. In its restriction mode, the enzyme recognizes the unmethylated 5-bp sequence and cuts the DNA 25-27 bp in the 3' direction from the methylation site. The cut leaves staggered ends of 2–4 bp. The enzyme prefers to cleave the DNA on the 5' side of a thymidine residue. The failure to find such a residue at the appropriate position in the DNA may account both for the variability in the stagger at the cut site and for the variable rate of cleavage at different sites (Bächi *et al.*, 1979). As previously indicated, the results of transposon mutagenesis of *res* and *mod* demonstrate that insertion in the *res* gene produces a Res⁻Mod⁺ phenotype, whereas insertion in the *mod* gene produces a Res⁻Mod⁻ phenotype (Iida *et al.*, 1983). These results were interpreted to mean that the Mod subunit contains both the modification and sequence-recognition function of the enzyme. Inactivation of that subunit thus creates a completely inactive enzyme. The existence of Res⁻Mod⁺ insertions indicates that the Mod subunit can function in the absence of the Res subunit to modify DNA. Support for these conclusions comes from the purification of Mod protein from a Res⁻Mod⁺ bacterium. That Mod subunit is similar to *Eco*P1 in its ability to modify DNA (Hadi *et al.*, 1983). In contrast, the purified Res subunit exhibits neither restriction nor modification function.

Before the *Eco*P1 modification sequence was determined by Hattman *et al.* (1978) and Bächi *et al.* (1979), an alternative modification sequence, AGATCT, was proposed by Brockes *et al.* (1974). This was based on the methylation of two tetranucleotide sequences, GATC and ATCT, and one trinucleotide sequence, AGA, by their preparation of modification enzyme. However, the above hexanucleotide cannot be the *Eco*P1 methylation site. That sequence is a *Bgl*II restriction site, and P1 methylation does not follow the pattern of *Bgl*II restriction; e.g., plasmid pBR322 contains no *Bgl*II site but is subject to P1 restriction (N. Sternberg, unpublished data). Moreover, as noted above, methylation occurs on only one strand, because the site is asymmetric. The *Bgl*II site is symmetric and should therefore be methylated in both strands.

How, then, do we interpret these results? The simplest hypothesis is that the purified preparation isolated by Brockes *et al.* (1974) contained two different modification proteins. The first is the 77-kD protein, the modification subunit of *Eco*P1, and is responsible for modifying the sequence AGA. The second is the 45-kD protein, responsible for the methylation of the sequences GATC and ATCT, and presumably produced after P1 infection. J. Coulby and N. Sternberg (unpublished) have recently shown that P1 encodes an analogue of the host Dam methylase. Since that enzyme methylates the sequence GATC, it could account for the two methylated tetranucleotides identified by Brockes *et al.* (1974).

*Eco*P1 is a type III restriction enzyme; i.e., it is intermediate in its properties between type I restriction enzymes, such as *Eco*K and *Eco*B (Linn and Arber, 1968; Yuan and Meselson, 1968), and the type II restriction enzymes that are the tools of modern molecular biology (Zabeau and Roberts, 1978). The class II enzymes require no factor other than Mg^{2+}, recognize a short nucleotide sequence on the DNA, and cleave either that sequence or just a few base pairs away from it. The type I enzymes are very large protein complexes, as much as 900 kD, and have a strict requirement for both ATP and S-adenosyl methionine (Ado-Met) in addition to Mg^{2+}. The type I enzyme can both methylate and cleave DNA, and it consists of three different subunits—an endonuclease, a methylase, and a recognition subunit. Type I enzymes recognize specific sequences but cleave the DNA at sites that can be thousands of nucleotides away from the recognition site (Horiuchi and Zinder, 1972). The cleavage reaction is accompanied by massive ATP hydrolysis.

In its native state, EcoP1 is a homotetramer of about 300 kD (D. Hornby and T. Bickle, personal communication) that consists of two subunits—an endonuclease and a methylase-recognition subunit. It requires Mg^{2+} and ATP for cleavage, but unlike type I enzymes, it does not catalyze extensive ATP hydrolysis. In the absence of ATP, the enzyme can carry out methylation but not cleavage (Haberman, 1974). Although type I enzymes show an absolute requirement for Ado-Met, EcoP1 restriction activity is only stimulated by Ado-Met (Haberman, 1974). Thus, in the absence of Ado-Met, the enzyme exclusively restricts DNA. When both ATP and Ado-Met are present, cleavage and methylation occur in competition. A comparison of the methylation abilities of the purified Mod subunit and the entire enzyme indicates that the Mod subunit is relatively inefficient when compared to the native enzyme (Hadi et al., 1983). Also modification of DNA by the Mod subunit is unaffected by ATP, whereas modification by the native enzyme is stimulated by ATP (Haberman, 1974).

C. Timing of Restriction and Modification Following P1 Infection

It seems reasonable to expect that for either lysogenization or a productive lytic cycle to occur following P1 infection, the expression of modification must precede the expression of restriction. Were this not the case, infection would almost invariably be lethal to both the bacterium and the phage itself. Despite the soundness of this logic, P1 mutants that restrict but do not modify do exist, and their existence very likely depends on a timing mechanism that delays the expression of restriction until, under normal conditions, modification has been assured. We begin this subsection with a consideration of such mutants and then discuss the nature of the timing mechanisms.

Among the clear plaque mutants of P1 isolated by Scott (1968, 1970b) were mutants that appeared to be altered at closely linked cistrons named c2 and c3. These mutations were mapped to the region of the P1 chromosome containing the res and mod genes. The first indication that the res-mod system might be involved in the lysogenization defect exhibited by these mutants came from the observation that they also exhibit a 100-fold reduced plating efficiency in a P1 cryptic lysogen when compared to a nonlysogen (Scott, 1970b). The P1 cryptic lysogen has P1 modifying and restricting ability but not P1 immunity (Scott, 1970a). Taking these results as a clue, Rosner (1973) showed that c2 and c3 mutants were deficient in P1-modifying ability and could complement each other for the expression of P1 modification, although neither of them could complement a Res⁻Mod⁻ insertion mutant. Rosner presumed that c2 and c3 mutants are restriction-proficient, but this was not directly shown. He suggested that these modification-deficient mutants are unable to lysogenize bacteria, because the action of their restriction endonuclease on the unmodified host chromosome eventually destroys that chromosome and kills cells that are in the process of being lysogenized. Normally, c2 and c3 phages are not subject to self-restriction during lytic growth because the P1 restriction function is not expressed until late in the P1 infection cycle (Arber et al., 1975). The inability of c3 mutants to grow at low temperature (Scott, 1968, 1970b) might indeed reflect a situation where self-restriction is occurring. The c2-c3 complementation studies prompted Rosner to suggest that there were two modification subunits in the EcoP1 enzyme. Considering that we now know that there is only one Mod subunit, a more likely explanation is that intragenic complementation occurs—i.e., different mutant proteins assemble to yield an active enzyme. It should also be noted that as the DNA sequence recognition domain of the Mod subunit is essential for restriction activity (Hadi et al., 1983), the c2 and c3 mutations, unlike the mod insertion mutations, must not affect the recognition domain.

Arber et al. (1975) and C. Levy, S. Iida, and T. Bickle (personal communication) have shown that Mod function is detectable within a few minutes after P1 infection, whereas Res function is detectable after about 1 h and does not become fully established until 3–4 h after infection. This pattern of expression is clearly a desirable one, but how is it achieved? A transcriptional model is ruled out by the observation that both Mod and Res subunits can be detected as early as 30 min after P1 infection. Instead, it appears that the synthesis of functional Res protein is delayed relative to the synthesis of functional Mod protein, because the former process depends on translational suppression of a stop codon. This possibility was first suggested by the observation that certain streptomycin-resistant strains of E. coli are phenotypically Res⁻Mod⁺ (Arber and Dussoix, 1962). These streptomycin-resistant strains carry a mutation in one of the ribosomal proteins that greatly reduces the frequency of spontaneous misreading.

The hypothesis that ribosomal misreading of a stop codon is necessary for expression of P1 restriction was confirmed by the demonstration that two other kinds of mutations that affect ribosomal misreading also affect the expression of P1 restriction (C. Levy, S. Iida, and T. Bickle, personal communication). One of these is a ribosomal mutation that is temperature-sensitive for the suppression of UGA stop codons and is also streptomycin-resistant (Engelberg-Kulka, 1979). P1 lysogens of this strain are Res⁺Mod⁺ at the permissive temperature (when UGA is suppressed) but are Res⁻Mod⁺ at the nonpermissive temperature. In the presence of streptomycin, which stimulates misreading, they are Res⁺ even at the nonpermissive temperature. The res termination codon is UGA. The second mutant that was tested is an miaA mutant, which contains a defective tRNA-modifying enzyme (Petrillo et al., 1983). miaA mutants fail to make an isopentenyl modification which is normally found 3' to the anticodon of all E. coli tRNAs that recognize codons beginning with U.

The phenotype associated with *miaA* mutations is a drastically decreased level of misreading by the nonmodified tRNAs. P1 lysogens of a *miaA* mutant have a Res⁻Mod⁺ phenotype. Based on these findings, Levy, Iida, and Bickle propose that if protein synthesis stops at the UGA codon in *res*, the resulting protein is inactive in restriction. Active protein is produced only if the UGA codon is suppressed and protein synthesis is terminated downstream. Since the UGA codon is suppressed in wild-type strains of *E. coli* about 1–2% of the time, this would ensure that active Mod would be produced almost 100 times faster than active Res.

If the P1 methylation site is asymmetric and therefore methylated on only one strand, how does the system avoid restricting its own DNA after replication? Replication of the unmethylated strand should generate fully unmethylated DNA, which should be subject to the competing activity of the restriction and modification functions. However, modified P1 DNA does replicate without being restricted by EcoP1. To determine whether modification alters both strands of P1 modified DNA, hybrid DNA molecules were generated *in vivo* by single-cycle infection of *E. coli* with density-labeled λ·P1 phage. Progeny phage containing only one P1-modified DNA strand still appeared to be resistant to P1 restriction *in vivo* (Arber *et al.*, 1963). Hybrid DNA was also generated *in vitro* by denaturing and reannealing a mixture of λ and λ·P1 DNAs. Both of the hybrid species so generated were resistant to EcoP1 restriction (Hattman *et al.*, 1978). These results imply that both strands of DNA containing an EcoP1 modification site are rendered, by the P1 modification function, insensitive to P1 restriction. Thus, replication of P1 modified DNA does not produce a substrate that is even transiently sensitive to restriction. How restriction is averted is not clear.

D. Defense against Restriction (*darA* and *darB*)

Following its injection into *E. coli*, the DNA of phage P1 is much less sensitive to type I restriction enzymes than it is to type II and type III enzymes. This insensitivity has been attributed to the presence of two proteins within the phage head, the products of the nonessential phage genes *darA* and *darB* (Krüger and Bickle, 1983; Iida *et al.*, 1987). *darA* is located between P1 map coordinates 25 and 30, and *darB* is located near coordinate 15 (Fig. 2A). Mutants defective in *darA* lack both DarA and DarB functions; i.e., they are sensitive to the EcoK, EcoB, and EcoA restriction enzymes. In contrast, mutants defective in *darB* lack DarB but retain DarA function. They are restricted by EcoK and EcoB but not by EcoA. The product of the *darA* gene is a 68-kD protein, and the product of the *darB* gene is a 200-kD protein (Iida *et al.*, 1987; Streiff *et al.*, 1987). Both proteins are absent from P1 *darA* mutant phage particles, whereas only the 200-kD protein is absent from *darB* mutant phage. Most likely the *darB* protein cannot be incorporated into phage head precursors in the absence of the *darA* protein. The Dar proteins will protect any DNA packaged into a P1 capsid but will not work in *trans* to protect DNA injected by another phage. Thus λ DNA packaged into a P1 head is protected, but λ DNA packaged into its own head is not protected even if it is coinjected with P1 DNA. Dar⁻ P1 phages can be protected against restriction if they are first grown on strains carrying a cloned and functional *dar* gene. The exact mechanism of Dar protection is unknown, but the above results clearly suggest that the proteins are injected into host cells along with the phage DNA and remain associated with it.

VI. LYSOGENIZATION VERSUS LYTIC GROWTH: IMMUNITY AND VIRULENCE

A. Physiological Influences on the Decision Process

P1 DNA that has been injected into a sensitive bacterium, that has circularized by recombination, and that has foiled the restriction enzymes of its new host and those of its own making, arrives at a critical decision: to lysogenize or not. If conditions are propitious, prophage establishment will occur. However, the prophage state is only as stable as the conditions that favor it. The established prophage is continuously poised to reverse the prior decision in the light of changing circumstances and initiate a lytic cycle. Very likely, many factors enter into the decision process, since the host range of P1 is broad and several different symptoms of bacterial distress merit interpretation. Eventually it may be possible to justify in these terms the formidable complexity of the immunity systems of phages P1 and P7 that we describe below.

Physiological factors that influence the decision to establish immunity were studied in *Shigella* (Bertani and Nice, 1954; Luria *et al.*, 1960) and in *E. coli* (Jacob, 1955; Rosner, 1972). In both genera, P1 was found to lysogenize more efficiently at 20°C than at 37°C. The lysogenization frequency, 80–90% at 20°C in *S. dysenteriae* strain Sh (Bertani and Nice, 1954) and ~ 30% at 32°C in *E. coli* (Rosner, 1972), was reported to be independent of multiplicity of infection (*moi*), although Luria *et al.* (1960), at variance with Bertani and Nice, found that the frequency of lysogenization of *Shigella* increased from 15% on single infection to 90% at an *moi* of 5 to 10. The discrepancy may possibly be due to the fact that by 1960 P1*kc* had replaced the P1 wild type of Bertani as the P1 of laboratory use.

A striking feature of lysogenization by P1, irrespective of whether the host is *Shigella* or *E. coli* and independent of *moi* and of postinfection temperature, is that bacteria that survive infection by P1 give rise almost exclusively to pure lysogenic clones (Adams and Luria, 1958). Two explanations can be offered for the absence of abortive lysogens: either the prophage is prompt in establishing the machinery for its own maintenance, or it is prompt in blocking the formation of viable progeny of the infected cell until such machinery is in operation. We defer discussion of the latter possibility until Section VII.B.5.

Reports of the degree to which P1 is inducible by ultraviolet (UV) irradiation of lysogens are surprisingly contradictory. Bertani (1951) was unable to induce P1 by UV, as was Jacob (using 363) (Jacob and Wollman, 1961); Rosner's attempts at P1 induction by UV or mitomycin C treatments were also unsuccessful (personal communication). Luria et al. (1960) found that P1 lysogens were weakly UV inducible. With optimum UV doses, up to 20% of the lysogens could be induced to yield phage. Kondo et al. (1970) reported a maximum of 30% of the lysogens. On the other hand, Mise and Arber (1976) reported that optimally UV irradiated P1 lysogens normally yield more than 300 pfu per irradiated bacterium within a 5- to 6-h incubation period. Published induction curves of Kondo et al. (1970) make it unlikely that UV markedly prolongs the latent period for P1 development and thereby masks inducibility (as was the case for phage 186, originally assumed to be noninducible (Woods and Egan, 1974; Hooper et al., 1981)). More probably, P1 variants with different susceptibilities to UV are in circulation. Possibly the host strain is the critical factor. Neither the physiology nor the genetics of UV inducibility of P1 appear to have been systematically studied.

B. The Bipartite Nature of the Immunity System

To identify the genetic elements in P1 involved in lysogeny, clear plaque mutants of the phage were isolated, and the mutations were classified into cistrons (originally 5) by cross-streak complementation tests and mapped to scattered locations (originally 4) (Scott, 1968, 1970b, 1972; Scott and Kropf, 1977). Clear plaque mutations in P7 were subsequently isolated and characterized, resulting in the identification of three additional cistrons with counterparts in P1 (Scott et al., 1977a). The functions of cistrons $c5$, $c6$, $c7$, and $c8$ are obscure. The $c6$ cistron is discussed briefly in Section VI.C, the $c7$ cistron in the first footnote of Section V.II, and the $c8$ cistron in the footnote to Section VI.D.2. We accord $c5$ a footnote here.* We indicate in Fig. 7A the locations of those cistrons that are implicated in the determination of immunity to superinfection.

*The $c5$ cistron was mapped by Scott (1972) to a region containing genes since recognized to be involved in tail morphogenesis. The originally isolated P1 $c5$ts mutant was generated by nitrosoguanidine mutagenesis that appears to have created a confusing array of mutations unrelated to plaque clarity: a $seg5$ mutation that is responsible for a rapid, temperature-dependent loss of prophage; a ban mutation; and a thermosensitive copy number control mutation, cop-5 (Jaffé-Brachet and D'Ari, 1977; Sternberg et al., 1981b). Among thermosensitive clear plaque mutations obtained in P7 by hydroxylamine mutagenesis (Scott et al., 1977a), some are located near $c5$ and are complemented weakly by P1$c5$, perhaps as a result of intracistronic complementation (Scott, 1980). Attempts to isolate stable lysogens of either the P1$c5$ts mutant or the P7$c5$ts mutants were uniformly unsuccessful (Scott et al., 1977a).

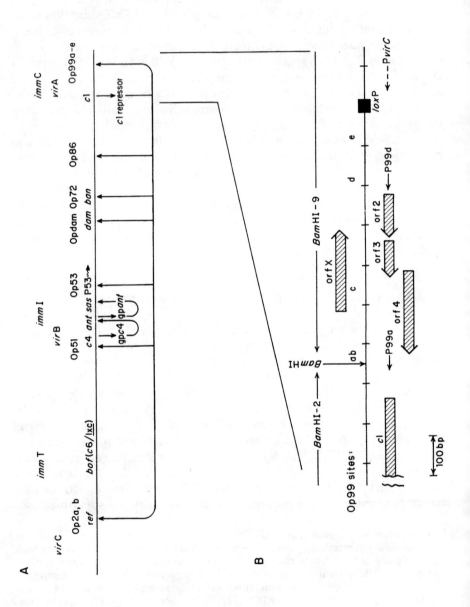

FIGURE 7. The tripartite immunity system of P1. (A) Immunity circuitry. This correlated physical and genetic map of elements involved in immunity circuitry (not drawn to scale) indicates the arrangement of immunity functions and of operator sites (Op) that bind or are presumed to bind the *c1* repressor. These sites are assigned the integral number portion of their map position; additional letters indicate more than one per map unit. Op *dam* is located close to Op 72 but is not precisely mapped. Virulent mutations are found at three locations. Op Suppressors of *virB* and *virC* have been identified. The suppressor of *virB* is *ant* (*reb*). The control of *ant* (*reb*) by *c4* and the complex behavior of *ant* (*reb*) are discussed in Section VI.E. The suppressor of *virC* is *coi* (located on the opposite side of *loxP* from *virC*, but not shown). It is discussed in Section VI.C. The *c8* cistron, identified in P7 (Scott and Kropf, 1977), is not *bof*, *lxc*, and *c6* are discussed in Section VI.C. The *c8* cistron, identified in P7 (Scott and Kropf, 1977), is not included here, but its map position (on the clockwise (right) side of *loxP*) suggests a possible relationship to the circuitry of *c1* regulation discussed in Section VI.D.3. The genes *ref*, *dam*, and *ban* are included as genes that are under immunity control.

(B) The *immC* region. The essential component of the expanded *immC* region is the *c1* gene. It is separated from the rest of the region by the *Bam*HI site located at the junction of *Bam*HI-9 and *Bam*HI-2. Within *Bam*HI-9 are the five Op99a–e sites and at least four open reading frames (orfs 2, 3, 4, and X). Presumed promoters for orfs 2, 3, 4, and c1 (P99d, P*vir*C, P99a) are also shown. P99d is presumed to produce a transcript of orf 2, 3, 4, and c1. Since P99d overlaps Op99d, its activity is likely to be negatively regulated by repressor. P*vir*C is a promoter generated by the *vir*C mutation. It presumably results in the transcription of the same genes as are transcribed from P99d, but it is not regulated by repressor. The *vir*C virulent phenotype is due to the constitutive synthesis of Coi, a protein that is probably encoded by orf3. P99a is probably the maintenance promoter for c1. We have postulated that its activity is regulated in both a positive and a negative sense by repressor. Orf-X is read in the opposite direction from the other orfs. Its significance, if any, remains unclear. Further details can be found in Section VI.D.3.

Among the clear mutants isolated, some are thermosensitive. The cistron designated c1 was the first cistron to be specifically implicated in immunity maintenance, because thermosensitive mutations in this cistron conferred thermoinducibility (Scott, 1970b; Rosner, 1972).

Two lines of evidence suggested that the expression of the c1 gene, although necessary for expression of immunity, is not by itself sufficient. One line of evidence came from a study of a defective hybrid between P1 and an R factor that had been isolated by Delhalle (1973). The defective prophage, P1d91tet, failed to confer P1 immunity, but Scott (1975) showed that it possessed a functional c1 gene. A clts allele crossed into P1d91tet rendered the viability of the lysogen thermosensitive. A second line of evidence for an additional component to immunity determination came from a study of the basis for the immunity difference between P1 and P7. P1 and P7 are reciprocally heteroimmune: each will grow on a lysogen of the other, but the two phages are very closely related (Chesney and Scott, 1975; Walker and Walker, 1976a; Wandersman and Yarmolinsky, 1977).* Electron microscopy of DNA heteroduplexes showed that 90% of the P1 genome is homologous to the P7 genome (Yun and Vapnek, 1977). Surprisingly, the c1 genes of P1 and P7 were found to be functionally identical; they can be exchanged without altering immunity specificity (Chesney and Scott, 1975; Wandersman and Yarmolinsky, 1977). For example, a thermoinducible P1 crossed with P7 yields, among the progeny, thermoinducible hybrids with P7 immunity characteristics: able to grow on P1 lysogens, unable to grow on P7 lysogens. The codon sequences of the c1 genes of the two phages are almost identical (F. Osborne, S. R. Stovall, and B. Baumstark, personal communication). The interchangeability of the c1 genes led to the search for the locus that determines the immunity difference between P1 and P7, the site of a second immunity determinant.

The immunity difference between P1 and P7 was mapped by Chesney and Scott (1975) to the region of cistron c4 (defined by Scott, 1968) and vir^s (a mutation to immunity insensitivity isolated by S. Sarkar in Luria's laboratory). The c4-vir^s region is seen to be within the P1–P7 nonhomology loop D of Yun and Vapnek (1977). Mutations in c4 are unique in being the only clear plaque mutations of P1 that P7 fails to complement. The search for thermoinducible mutants was widened with the isolation by Scott et al. (1977a) of thermosensitive clear-plaque P7 mutants. Thermoinducibility was found to be characteristic of P7c4ts

---

*The unit efficiency with which P7 plates on lysogens of P1 provided immediate reinforcement for the idea that the two phages are in some measure homologous. The simplest explanation for the ability of P7 to resist restriction by P1, namely that P7 modifies its DNA as does P1, was shown to be correct (Chesney and Scott, 1975; Wandersman and Yarmolinsky, 1977). Marker rescue spot tests showed that each of 107 amber mutants of P1 tested could be rescued by a nonsuppressing E. coli lysogenic for P7. The marker rescue was shown to be due largely to recombination, although a role for complementation was not excluded (Walker and Walker, 1976a).

mutants as well as of P7clts mutants.* These results were interpreted to mean that immunity is under bipartite (c1 and c4) control.

Bipartite control over immunity in phage P22 (Levine et al., 1975; Botstein et al., 1975) provided a model and a nomenclature for the immunity systems of P1 and P7. Immunity in P22 requires the simultaneous expression of two repressors. The primary repressor, encoded in a region called immC, represses lytic functions. The secondary repressor is needed to repress an antirepressor that would otherwise inactivate the primary repressor. Both the antirepressor gene (ant) and its repressor are encoded in a region called immI. Inducing virulent mutations that were mapped to immI express the antirepressor constitutively. By analogy with P22, the c1 cistrons of P1 and P7 were said to belong to a region named immC, and the c4 cistrons and closely linked vir mutations (designated virB by West and Scott, 1977) were said to belong to a region named immI. The prediction was made that virulence due to virB mutations (such as vir^s) and the heteroimmunity of P1 and P7 are due to the expression of antirepressor by the superinfecting phage. The phage specificity of the c4 genes was taken to mean that P1 prophage prevents the expression of the P1 antirepressor but fails to block antirepressor expression by a superinfecting P7 phage. By the same reasoning, the failure of a P7 lysogen to block P1 growth was explained (Chesney and Scott, 1975; Wandersman and Yarmolinsky, 1977; Scott et al., 1978b).

A compelling verification of the model was provided by the isolation of phages with the expected properties of antirepressor deficient mutants. They were isolated in three different ways: as lysogenization-proficient derivatives of c4 mutants or of virB mutants or as mutants of wild-type phages unable to grow on bacteria carrying cryptic $c1^+\Delta immI$ prophages (P1cry and P1d91tet are in this category) (Wandersman and Yarmolinsky, 1977; Scott et al., 1978b). These various ant mutants are sensitive to the immunity exerted by P1 or P7 or $c1^+\Delta immI$ prophages, presumably because they lack the means to inactivate c1 repressor. On nonlysogenic hosts they plaque poorly, if at all, unless also c1-deficient. This result suggests that the establishment of immunity is normally influenced by the expression of antirepressor activity. Nonsense mutations in the antirepressor were isolated by Wandersman and Yarmolinsky (1977) and Scott et al. (1978b). The antirepressor, like the repressors determined by c1 and c4, was therefore deduced to be a protein. As prophage, antirepressor-deficient mutants derived from P1c4 differ from those derived from P1 vir^s in that only the latter are immune to P1 superinfection. Presumably the vir^s mutant continues to make the c4 repressor.

---

*Productive induction does not occur upon induction of lysogens carrying the known P1c4ts mutants for reasons that remain obscure, although the presence of secondary mutations in the P1 that block phage production has not been excluded. We note here also the existence of the mutation lyg (Scott, 1974), tightly linked to c4 and possibly an allele of it (Scott, 1980). The lyg mutation causes a lysogenization defect, but not plaque clarity, perhaps because of residual $c4^+$ activity.

C. The Tripartite Nature of the Immunity System

Before considering the modes of action of *imm*C and *imm*I proteins, we consider what is known of a third, less clearly defined component of immunity. When P7 thermosensitive clear plaque mutants were tested for thermoinducibility, one mutant, affected in a cistron outside of *imm*C and *imm*I, was found to yield 4–10 pfu per cell following a temperature shift (Scott *et al.*, 1977a). The weak thermoinducibility of this single P7*c6*ts mutant suggested the possibility that immunity is under the tripartite control of *c1*, *c4*, and *c6*. In the same region of the viral genome as *c6*, near map unit 10, are found two highly pleiotropic mutations, *bof-1* and *1xc*, which probably alter the same prophage function and possibly belong to the *c6* cistron.

The amber mutation *bof*-1 was discovered in a search for mutations that would conditionally decrease the expression of *ban* (an analogue of the bacterial *dnaB* gene) in a mutant (P1*bac*-1) that expresses the *ban* gene constitutively (D'Ari *et al.*, 1975). Instead of obtaining an amber mutation in *ban* (near map unit 75), as expected, the distant *bof*-1 (amber) mutation was isolated.* The effect of *bof*-1 on *ban* expression may be specific for mutants in which *ban* is not under normal *c1* repression. The feature of *bof*-1 that appears most clearly relevant to immunity is that *bof*-1 lowers the temperature threshold for P1 *c1*.100 thermal induction. It has also been noted that P1Cm *bof*-1 and P1Km *bof*-1 confer resistance to concentrations of chloramphenicol or kanamycin about twofold higher than is obtained with their *bof*+ counterparts. One explanation for the increased antibiotic resistance is that *bof*-1 increases P1 gene dosage by relaxing slightly the normal immunity repression of vegetative phage replication. Paradoxically, *bof*-1 also increases immunity to superinfection. These results of Touati-Schwartz (1979b) have been interpreted by her to mean that the function of the wild-type *bof* gene is to increase both the efficacy of *c1* repressor and its sensitivity to antirepressor. This interpretation does not account for the phenotype of *bof*-1 that led to its discovery, but multiple or indirect effects of *bof* on gene expression are not excluded.

P1*bof*-1 prophage is derepressed for two additional P1 functions normally under *c1* control (Windle and Hays, 1986). These are a recombination enhancement function, determined by *ref* at about map unit 3, and a single-stranded binding protein that can complement an *ssbA* defect in *E. coli* and that is determined by an as yet unmapped, unnamed gene. Mu-

---

*The etymology of *bof*-1 is as follows: *bof* was chosen as an acronym for *ban off*, because *ban*-1 was originally isolated as a mutant with this phenotype. When it was pointed out that the gene name should indicate the gene's function in the wild type, the name was rationalized as an acronym for *ban on function*. When it became apparent that the function of the *bof* gene was not specific for *ban*, but more probably connected to immunity, the suggestion from one of us that the name be changed was rebuffed with a curt and disdainful "*bof*." The name stands as a fine example of Gallic *je m'en foutisme*.

tants of P1 that express the *ssbA* analogue as prophage were isolated by
Johnson (1982) and named *lxc* (an abbreviation of *lexC*, an alternative
name for *ssbA*). Several P1 mutants (isolated by J. L. Rosner) that permit
the survival of *E. coli* *ssbA*ts at 42°C were shown by Windle and Hays
(1986) to be derepressed for *ref* as well. These several observations lead to
the tentative conclusion that *bof*, *lxc*, and *c6* designate a single gene (or
closely linked set of genes) that influences immunity expression. They
may do this in more than one way. Studies on the control of *ref* indicate
that *ref* repression can be relieved by either a *c1* or a *bof*/*lxc* defect
(Windle and Hays, 1986). A λ-P1:*Bam*HI-8 *ref*⁺ can be repressed for *ref*
expression by an intact P1, but not by cloned *c1* or *bof* genes that are each
capable of complementing mutations of the corresponding genes in P1
(Windle, 1986). These results imply that *bof* acts in a concerted manner
with *c1*, but the crucial reconstruction experiments to test this hypothesis remain to be performed. We designate as *imm*T the tertiary immunity region that includes *bof*, *lxc*, and *c6* (see Fig. 7A).

To facilitate further analysis of the tripartite immunity system, the
component elements were separately cloned into λ vectors. Use of integrated λ-P1 hybrid prophages enabled Sternberg *et al.* (1978) to dissect and
reconstruct the P1 immunity system without interference from incompatibility barriers or the distortion of gene dosage relationships that is
inherent in the use of multicopy vectors. This study provided a more
precise localization of components of the *imm*I region and a deeper understanding of interactions among the three immunity regions. We summarize the results below.

A resident λ prophage bearing a P1 DNA fragment that expresses *c1*⁺
(abbreviated λ-P1:*c1*⁺) was found to confer immunity against infection by
P1 or P7 *c1*⁺ phages, not against infection by *c1*⁻ phages. This result
suggested that the antirepressor made by infecting phage is able to overcome the primary *c1*⁺ repressor made by the incoming phage, but only if it
does not also have to contend with additional preexisting repressor in the
infected cell. Similarly, a resident λ prophage bearing a P1 DNA fragment
that expresses *c4*⁺ was found to confer immunity to P1, not to P1 *c1*⁻, P1
*vir*ˢ, or P7. This result supports the conclusion that the control of antirepressor is critical to the attainment of the concentration of *c1* repressor
needed for lysogeny during the establishment phase. In a third experiment,
λ-P1 hybrids expressing both *c1*⁺ and *c4*⁺ were found to confer immunity
that extended to P7 as well as to P1, unless *Eco*RI-2, bearing *imm*T, was
also present. These results are consistent with the previously noted conclusion that a product of *imm*T, presumably *bof*, renders *c1* repressor more
sensitive to antirepressor. An *imm*T-mediated diminution of the immunity conferred by *imm*C is also seen from the observation that immunity
to P1*c1*⁺ conferred by a λ-P1:*c1*⁺ prophage is eliminated if a λ-P1:*Eco*RI-2
prophage is also present. This additional result of Sternberg *et al.* (1978)
reveals that the ability of P1*c1*⁺ to plate on bacteria harboring either of the
c1⁺Δ*imm*I prophages P1*cry* or P1*d91tet* as well as the ability of P1*c1*⁺ to

plate on bacteria harboring the various c4*ant* prophages (the P1*c4reb* of Scott *et al.*, 1978b) depends not only on the lack of *immI* but also on the presence, in each of these lysogens, of *imm*T.

D. A Closer Look at *imm*C

The *imm*C region contains the *c*1 repressor gene and its upstream regulatory region (Fig. 7B). The region includes the *c*1 promoter and genes that encode small proteins that apparently affect *c*1 function. Also present in this region is the *vir*C mutation, which, in addition to conferring a virulent phenotype, results in constitutive synthesis of at least one of the small proteins (Heilmann *et al.*, 1980a). In this section we discuss how these components might interact to regulate *c*1 protein synthesis and function. We also describe data from several laboratories that tentatively identify DNA sequences to which the *c*1 protein binds, and we present models to account for the regulation of the synthesis of that critical protein following infection.

1. A Multiplicity of *c*1 Protein-Binding Sites in the P1 Genome

There are now seven regions of the P1 chromosome known to interact with the *c*1 repressor (Fig. 7A). These regions have been identified either physiologically as containing repressible promoters or biochemically as containing sites that bind the repressor *in vitro*. Where DNA sequence data are available, each of these regions has been shown to contain similar sequences which we presume to be the *c*1 protein-binding sites. Several of the regions contain more than one presumptive binding site. The DNA sequence of each site and a derived consensus sequence is shown in Table III. Note that the sequence is not symmetric and hence has directionality.

The P1 *ref* gene (at map unit 2.5 to 3) is repressed in the prophage state by the *c*1 repressor (Windle and Hays, 1986). Analysis of the *ref* gene sequence (Windle, 1986) reveals two putative binding sites, Op2a and Op2b* (J. Eliason, personal communication). Op2a overlaps one of several promoter sequences located upstream of *ref* and is oriented (as shown in Table III) in a direction defined as clockwise on the P1 map. Op2b is located at the C-terminal end of the *ref* structural gene and is in the same orientation as Op2a. The repressor binding site in *EcoRI-9* (Op51, at map unit 51.4) was discovered in an *in vitro* binding assay (Baumstark *et al.*, 1987) and is located about 100 bp upstream of the promoter for the c4

---

*To simplify subsequent discussion, each region containing a *c*1 protein binding site or operator (Op) sequence is assigned the integral number portion of its map position. Where there is more than one site in any one map unit, we assign letters to the sites in alphabetical and clockwise order.

TABLE III. cl Repressor-Binding Sites[a]

Operator site	DNA sequence	Map orientation
Consensus	A T T G C T C T A A T A A A T T T	
Op2a	A T T G C T C T A A T (T)(G) A T T (G)	c
Op2b	(C)(A) T G C (A) C T A A T A A A T (A) T	c
Op51	A (A) T G C T C T A A T A A A T T T	c
Op53	(T) T T G C T C T A A T A A A T T T	c
Op72	A T T G C T C T A A T A A A T T T	c
Op86	A T T G C T C T A A T A A A (A)(A)(A)	ND
Op99a	A (A) T G C (A) C T A A T A A A T (C) T	cc
Op99b	(G)(A)(C)(T) C (A) C (O) A A T A A A T (G)(C)	cc
Op99c	A T T G C T C T A A (C)(G)(C)(T) T T (A)	cc
Op99d	A T T G C T C T A A T A A A T T (A)	cc
Op99e	A T T G C T C T A A T A A (T) T (C) T	c

[a]Symbols used are: c, clockwise; cc, counterclockwise; ND, not determined. Circled bases deviate from the consensus sequence. Operator sites Op53, Op72, Op86, and Op99a,c,d,e have been identified by DNase footprinting experiments (Velleman et al. 1987; Eliason and Sternberg, 1987). The Op51 site was first identified by repressor binding to specific DNA fragments and then further localized by comparison of the sequence of those fragments with the consensus repressor-binding site (B. Baumstark, personal communication; Velleman et al., 1987). The localization of Op2a and Op2b sites to the P1 ref gene was initially based on the sensitivity of ref expression to cl repression (Windle and Hays, 1986). The positions of the operator sequences were identified by comparing the ref sequence to that of the consensus repressor-binding sequence. The existence of Op99b is inferred from its similarity to the consensus site and from its proximity to Op99a. Velleman et al. (1987) have not detected DNase protection of Op99b by repressor under conditions in which Op99a is protected by repressor.

gene (B. Baumstark, personal communication), which is also on this fragment (Baumstark and Scott, 1987). Op53 is in EcoRI-14 about 550 bp from its junction with EcoRI-25 and overlaps a strong promoter (P53), which directs transcription toward EcoRI-25. When P53 is fused to the galK gene, production of galactokinase is comparable to that observed with the gal operon promoter. In the presence of cl repressor (made from a single-copy λ-P1:EcoRI-7 prophage), transcription from P53 is reduced about 20-fold (Sternberg et al., 1986). The orientation of Op53 is clockwise. The existence of Opdam is inferred from the observation that the P1 dam gene is repressed in the prophage state but is induced upon inactivation of repressor (J. Coulby and N. Sternberg, unpublished). Presumably there is a repressor-binding site upstream of dam. The dam gene is located in a region of the P1 map that overlaps the EcoRI-8,3 junction (J. Coulby and N. Sternberg, unpublished). Op72 (between map units 72 and 73) is a site at which the repressor acts to inhibit expression of the P1 dnaB gene analogue, ban (D'Ari et al., 1975; Ogawa, 1975; Sternberg et al., 1978). Its orientation is also clockwise (Velleman et al., 1987). Op86 is located centrally within EcoRI-11, and its existence is inferred from in vitro binding experiments with purified repressor (Velleman et al., 1987). Its orientation is unknown. Between map units 99.3 and 99.9 are five binding sites, Op99a–e. These are located upstream of cl in the 640-bp seg-map units 99.3 and 99.9 are five binding sites, Op99a–e. These are located upstream of $c1$ in the 640-bp segment of DNA between the $c1$ structural gene and $loxP$ (Baumstark et al., 1987; J. Eliason and N. Sternberg, unpublished; H. Schuster, personal communication). Op99a and Op99b are adjacent to each other, overlapping by 5 bp, and are located 615 bp and 606 bp, respectively, from $loxP$.* The Op99c, Op99d, and Op99e binding sites are located 444, 151, and 48 bp, respectively, from $loxP$. DNA fragments containing Op99a, Op99b and Op99c, or Op99d and Op99e bind to nitrocellulose filters in the presence of $c1$ protein specifically (J. Eliason, personal communication). Op99a, Op99b, Op99c, and Op99d are all oriented in the same direction (counterclockwise on the P1 map), and Op99e is oriented in the opposite direction.

2. Genes and Regulatory Elements in immC

The DNA sequences of the $c1$ genes of P1 and P7 have been determined and are nearly identical (F. Osborne, S. R. Stovall, and B. Baumstark, personal communication). The P1 repressor was partially purified by Baumstark and Scott (1980) and to near homogeneity by Dreiseikelmann et al. (1988). Its molecular weight was estimated to be 33 kD from its relative mobility on SDS–PAGE (Heilmann et al., 1980a). This estimate is in good agreement with the value calculated from the 283-codon open reading frame identified in the laboratory of B. Baumstark. The positions of two amber mutations, $c1.55$ and $c1.169$, have also been assigned. Inspection of the DNA sequence of the $c1$ gene (B. Baumstark and J. Eliason, personal communications) suggests that the $c1$ protein of P1 does not possess a helix-turn-helix structure that is characteristic of a number of other repressor and DNA-binding proteins (Pabo and Sauer, 1984). Moreover, the P1 repressor is unusual in that the binding sites that it recognizes are not symmetrical (Velleman et al., 1987; Eliason and Sternberg, 1987).

The $c1$ structural gene is contained within BamHI-2, is read in a counterclockwise direction on the P1 map, and begins about 80 bp from the BamHI-2,9 junction (Fig. 7B). Deletion mutations that start within BamHI-9 and extend 70 bp into BamHI-2 do not express $c1$ unless fused to an appropriate promoter, such as λP_L. In contrast, deletions that extend 100 bp into BamHI-2 irreversibly destroy $c1$ function, because they remove a portion of the $c1$ structural gene (J. Eliason, personal communication). The promoter for $c1$ has not been precisely defined, but the most promosing candidate is a sequence contained within a region about 70–90 bp upstream of the structural gene. That sequence contains a perfect −10 region but lacks a −35 region. The two $c1$ protein-binding sites, Op99a and Op99b, are located just upstream of where the −35 region should be. The importance of this region is revealed by the properties of

---

*The distances between $loxP$ and Op99 sites are from the central base of the $loxP$ spacer to the central base of each Op consensus sequence.

deletion mutants that dissect it (N. Sternberg, unpublished). A mutant retaining 174 bp upstream of c1 produces normal levels of repressor. If that mutant is present on a multicopy plasmid, then the cell containing it is superimmune; it is immune not only to P1 but also to P1virB. We presume that the extra copies of c1 result in an elevated level of repressor synthesis. If the mutant DNA extends only up to the BamHI site 80 bp upstream of c1, then a reduced level of repressor is synthesized. A cell containing that mutant plasmid is still immune but no longer superimmune. This deletion mutation leaves the −10 region of the c1 promoter intact but removes all other upstream sequences. If the −10 region is removed (e.g., by a deletion that leaves only 60 bp upstream of c1), then cells containing that mutant plasmid no longer express immunity. These results suggest that sequences on either side of the BamHI site are essential for c1 expression and that all the essential elements are contained within the upstream 174 bp. Presumably the important element within BamHI-2 is the −10 promoter sequence, and the important elements in BamHI-9 are the c1 protein-binding sites, Op99a and Op99b. We assume that c1 protein binding to one or both of these Op sites acts as a regulator of c1 transcription (see below).

Upstream of the c1 promoter region are four small open reading frames (orfs) (B. Baumstark, personal communication; J. Eliason and N. Sternberg, unpublished). All four orfs are contained within the 640-bp region between loxP and the BamHI-9,2 junction (Fig. 7B). Orfs 2, 3, and 4 are read counterclockwise on the P1 map and are 37, 31, and 69 codons, respectively (J. Eliason and N. Sternberg, unpublished). The order of the orfs from loxP to c1 is orf2, orf3, and orf4, with orfs3 and -4 overlapping by 10 codons. The P1 virC mutation leads to the increased synthesis of a 3.5-kD protein, which is probably the product of orf3 (Heilmann et al., 1980a). The predicted molecular weight of the protein encoded by orf3 is 3.7 kD. There are ribosome-binding sites appropriately positioned to help initiate translation of all three open reading frames, although the ribosome-binding site upstream of orf2 is in least agreement with the consensus.

There is also a promoter (P99d) about 35 bp upstream of orf2 that could be used to initiate transcription of all three open reading frames. The Op99d repressor-binding site is located between the −10 and −35 regions of P99d. Presumably repressor binding to this site blocks transcription from the promoter. The virC mutation results in elevated synthesis of the orf3 gene product, because it creates a more effective −35 region in a promoter sequence located about 250 bp upstream of P99d (N. Sternberg, unpublished). This places virC on the opposite side of loxP from c1, in agreement with Scott et al. (1977b), who mapped virC to the opposite end of the P1 map from c1. Transcription from the virC promoter must pass through loxP to reach orfs2–4. The virC promoter contains no c1 protein-binding site, is insensitive to repressor, and promotes the constitutive synthesis of the orf3 product. The fourth open reading frame (orfX), in the 640 bp loxP-c1 region, was identified by Baumstark et al. (1987), is read clockwise on the P1 map (Fig. 7B), and has 64 codons. It extends from about the center of orf4 toward *loxP* and terminates near the end of orf2. Since only the orf3 protein has been identified, we will limit our speculations as to the function of this protein but note that similar arguments may be made for orf2 and orf4 products as well.

Several results suggest that the orf3 protein inhibits *c*1 repressor function. P1 *virC*, which synthesizes the orf3 product constitutively, grows on a P1 or a P7 lysogen and, when it does so, induces the prophage (West and Scott, 1977; Scott *et al.*, 1977b).* Since *virC* does not reduce the synthesis of *c*1 repressor in the cell (Heilmann *et al.*, 1980a), it must affect its function. The simplest hypothesis is that *virC* results in the constitutive overproduction of the orf3 protein or another protein encoded by one of these reading frames and that this protein inhibits *c*1 repressor activity. Support for this hypothesis comes from the characterization of mutations that suppress the virulence of the *virC* mutation (Scott, 1980). The mutations define a gene called *coi* that is located in the *immC* region closely linked to *c*1. Since *coi* mutations are recessive to *coi*+, *coi* probably inactivates a gene product that is responsible for the virulent phenotype of P1*virC*. The size of the protein overproduced by P1*virC* suggests that orf3 is *coi*, although the effect of insertions into a cloned fragment that interferes with P1 lysogeny points to orf4 as *coi* (B. Baumstark, personal communication).

3. The Regulation of *c*1 Repressor

How are *c*1 protein synthesis and function regulated after sensitive cells are infected by P1? The data presented in this section allow us to propose two speculative models. We defer until the next section consideration of how the *immI* region and its products participate in these processes. The role of *c*1 protein in the repression of phage lytic growth is discussed in Sections VIII.B and VIII.C.

We propose that when P1 infects a sensitive host, there is a race between the synthesis of the repressor encoded by *c*1 and the synthesis of the repressor antagonists—the product of orf3, encoded at *immC*, as well as *ant*, encoded at *immI*. The fate of the cell is determined, at least in part, by the outcome of the race at *immC*. Since there is no repressor in the cell when it is infected by P1, P99d should become active soon after infection, resulting in the synthesis of the protein products of orf3 and *c*1. The orf3 protein should act to inhibit *c*1 repressor function and drive the infection into a lytic mode. On the other hand, the repressor should bind to Op99d, blocking any further transcription. This would favor a lysogen-

---

*The unique *c*8 mutation, originally isolated in P7, on which it confers a thermosensitive clear plaque phenotype but not thermoinducibility (Scott *et al.*, 1977a), has been mapped to the vicinity of *virC* (Scott and Kropf, 1977). The *c*8 mutation might possibly be functionally similar to, but less effective than, *virC*.

ic response. For lysogenization to occur, two other conditions need to be met. First, the orf3 repressor antagonist must be unstable, so that blocking its synthesis effectively halts the inhibition of repressor function. Second, an alternative promoter for repressor synthesis must be available to maintain the level of repressor normally present in the lysogen. The promoter just upstream of $c1$ (P99a) is such an alternative.

The synthesis of $c1$ protein in the lysogenic state appears to be autoregulated. This conclusion is deduced from a demonstration that variation in the capacity of a cell to translate $c1$ repressor mRNA has little effect on the amount of $c1$ repressor in the cell (N. Sternberg, unpublished).

The extent of suppression of a $c1.55$ (amber) mutation in a P1 prophage was varied over a threefold range. This variation was achieved by use of the $supD$ts strain MX399 of Oeschger and Wiprud (1980) grown at 25–34°C. The preincubated cultures were shifted to 40°C, which abolished suppression. Continued cell growth at 40°C diluted the previously synthesized $c1$ repressor. When its concentration decreased to a critical threshold, prophage induction occurred. Whereas induction by a shift to 40°C of a control culture in which the P1 prophage carried the thermoinducible $c1.100$ mutation caused a rapid increase in free phage (beginning 30 min after the shift and reaching a plateau at about 70 min), thermal induction of the cultures of MX399 (P1$c1.55$) required an additional 40–50 min. The induction of MX399 (P1$c1.55$) lysogens was equally delayed regardless of the prior growth temperature. This result argues that the extent of amber suppression does not affect the concentration of repressor in the cell and therefore that the synthesis of $c1$ repressor must be autoregulated. Consistent with this conclusion are the observations that (1) multiple copies of the repressor-sensitive promoter in EcoRI-14, P53, can be almost fully repressed by a single copy of the $c1$ gene (Sternberg et al., 1985), and (2) the introduction of multiple copies of Op99-binding sites into a cell containing a single copy P1 prophage does not induce the prophage (B. Baumstark, personal communication).

The structure of the P99a $c1$ maintenance promoter conforms to the expectation that it include autoregulatory sites. The promoter lacks a typical -35 region and contains instead, just upstream of -35, two $c1$ repressor-binding sites. The lack of a -35 region suggests that the promoter activity might depend on a diffusible activator of transcription. The requirement for autoregulation suggests that the $c1$ protein acts as a repressor of transcription. It is possible that the $c1$ protein could act in both capacities, as activator and as autorepressor, the particular activity depending on the site or sites to which the protein binds. Op99a is likely to be the stronger of the two binding sites, as its DNA sequence conforms more closely to the consensus binding site sequence given in Table III. We propose, therefore, that the $c1$ protein binds first to Op99a and that this binding activates transcription of the $c1$ maintenance promoter. As the concentration of $c1$ protein in the cell increases, the Op99b site is also filled. Filling of both a and b sites creates a complex that represses, rather than activates, transcription.

Alternatively, activation and autorepression could be accomplished by different proteins. In this model, the two cl protein-binding sites would both be involved in autorepression. What might be the activator? We suggest that the *bof* gene product is a reasonable candidate. Some of the properties conferred by a *bof* mutation (see section VI.C) fit with a role for Bof in the activation of cl transcription: the *bof* mutation lowers the temperature threshold for P1cl.100 induction, raises the plasmid copy number, and derepresses the *ssbA* analogue and *ref* genes; moreover, putative *bof* alleles (c6 mutations) cause plaque clarity. However, the decrease in *ban* expression and increase in immunity associated with a *bof* mutation remain unexplained. The first of the two possibilities described above implies that the promoter we have referred to as the cl maintenance promoter is also regulated in such a way as to favor immunity establishment.

E. A Closer Look at *immI*

How similar is the immunity circuitry of P1 and P7 to that of phage P22? A close look suggests that the similarity is in part illusory. As expected of phages that express an antirepressor constitutively, P1 *virB* and P7 *virB* are inducing virulents: the yield of phages of the prophage type in homoimmune infections is at least as great as the yield of virulent phages. However, a heteroimmune phage (*virB* or *vir*+) fails to induce a resident prophage (West and Scott, 1977; Wandersman and Yarmolinsky, 1977; Scott *et al.*, 1978b; Sternberg *et al.*, 1979). This latter result is inconsistent with expectations based on the P22 analogy, according to which antirepressor expression should inactivate, with equal efficiency, the functionally equivalent *immC* repressors of homo- and heteroimmune prophages alike. We are forced to the conclusion that P1 and P7 repression antagonists, unlike the *ant* gene product of P22, must act indirectly. Repression antagonism does not necessarily mean direct repressor inactivation.

What, then, is the mediator of repression antagonism? On what does the inducibility of P1 prophage in heteroimmune infections depend? The following experiments suggest that the critical element is a cis-acting locus within *EcoRI-14*, a restriction fragment within which also lies at least part of the putative antirepressor gene (Sternberg *et al.*, 1979; Sternberg and Hoess, 1983). The requirements for the antirepressor gene to derepress prophage are deduced from experiments in which derepression is measured by the yield of P1 *virB ant*+ *ban*− phage. The infected bacterium is a *dnaBts E. coli* carrying a λ-P1:*EcoRI-3*(*ban*+) prophage as a source of the essential *ban* gene product and a P1*cry*(cl+) or λ-P1:*EcoRI-7*(cl+) prophage as a source of the repressor that blocks *ban* expression. The *ant* of the superinfecting *virB* phage was found to allow an appreciable phage burst at a temperature nonpermissive for *dnaB* only if an *Eco*RI-14 fragment was positioned in *cis* to the *ban* gene. The *Eco*RI-14 fragment could be variously oriented and could be separated from *ban* by as much as a λ prophage. These results were taken as evidence that *Eco*RI-14 bears a loading site, designated *sas* (site of *ant* specificity) (Sternberg *et al.*, 1979), at which the homologous repressor antagonist acts. From this site, the antagonist might diffuse along the DNA to operators at which it displaces or inactivates the *c*I repressor. Alternatively, it might alter the superhelicity of the DNA or otherwise change its structure over a considerable distance.

The loading site hypothesis, in its simplest form, retains an important feature of the P22 model—namely, that the antagonist of repression does eventually interact with the *imm*C repressor, even if it requires the intervention of a special site to do so. Evidence for interaction of these gene products has been claimed from a study of suppressor mutations (D'Ari, 1977). A mutation, *dan-1*, tightly linked to *c*I.100, renders P1 *c*I.100 no longer thermoinducible. The P1*c*I.100 *dan-1* phage is also defective for plaque formation at 32°C, and at 42°C it is insensitive to the repressor antagonist expressed by P1 *vir*^s or P7. These results suggest that the *dan-1* mutation alters the structure or abundance of the mutant repressor such that it exerts a stronger repression than is achieved by the wild-type repressor. The plating defect of P1*c*I.100 *dan-1* is presumably due to excess repression, since it is suppressed by an external suppressor of *dan*, *sud-2*, which has been mapped to the *imm*I region.* This suppressor might result in the constitutive low-level synthesis of the wild-type *ant* protein or might produce a mutant *ant* protein that can antagonize the *dan-1* repressor. The second hypothesis assumes a basal level of *ant* expression by P1 prophage. The dependence of the effectiveness of the presumptive antirepressor on mutations affecting *c*I structure or abundance was taken by D'Ari to be convincing evidence that the corresponding proteins do interact with one another. Although we find this evidence to be merely suggestive, we retain the term *ant* for the repression antagonist genes in the *imm*I regions of P1 and P7, recognizing that considerable differences among the *ant* genes of P22, P1, and P7 may be masked by use of this generic term.

As an alternative to the antirepressor model, Scott (1980) proposed that repression antagonism is mediated by a replicative *r*epressor *b*ypass and accordingly offered the name *reb* for the genes that we have termed *ant*. She proposed that *reb* gene products are predominantly *cis*-acting

---

*P1*c*I.100 *sud-2* segregates nonlysogens at a high frequency (D'Ari, 1977; Jaffé-Brachet and D'Ari. 1977). The lysogens grow slowly and form filaments. Recall that filament formation on thermal derepression of P1*c*4ts mutants had been noted by Scott (1970b). The P1*c*I.100 *sud-2* phage is unable to lysogenize a *recA* host. The phenotypes associated with *sud-2* are no longer manifest if an *ant* mutation is introduced. These results implicate in the stability of plasmid maintenance a presently obscure factor that is under immunity control.

proteins that enable an associated replicon to initiate rounds of replication despite the presence of *imm*C repressor. Scott attributed the *imm*I-specific induction of prophage in *virB* infections to *c*4 repressor titration by the superinfecting phage. This titration would have to outpace *c*4 repressor synthesis from the newly replicated *c*4 gene copies and not seriously delay extensive replication of the induced prophage. Support for the repressor bypass hypothesis was particularly vulnerable, because it relied on evidence of a negative nature, the failure to obtain specific experimental findings. Specifically, it was based on the apparent inability of *reb*+ phages to complement *reb*− phages in coinfection experiments (Scott *et al.*, 1978b) and the apparent inability of P1 *virB* to induce a prophage when its own replication was dependent on that induction (West and Scott, 1977). An inducing action of *reb/ant* in *trans* was observed in the experiments of Sternberg *et al.* (1979) and Sternberg and Hoess (1983) that we have outlined earlier in this section. Moreover, the induction they observed was found to be *independent* of prior replication. These results undermine the experimental support for the repressor bypass hypothesis. However, evidence of a different nature (see Sections VII.B.3 and VIII.B.2) does suggest that clockwise transcription from *imm*I is capable of activating a lytic P1 replicon.

The possibility that the *cis*-acting "loading site," *sas*, detected in *Eco*RI-14, mediates the action of *ant* by serving as an origin of replication, was tested directly. A λ-P1:*Eco*RI-14 lysogen was infected with P1 *vir*s, and markers flanking the *Eco*RI-14 fragment were assayed by DNA hybridization. No evidence for replication originating from the cloned P1 fragment was detected (N. Sternberg and M. Zeise, unpublished experiments).

Further progress in understanding immunity circuitry in P1 is likely to follow promptly from the isolation and characterization of immunity region gene products and the sequencing of the relevant genes and regulatory sites. Both the *c*1 gene (F. Osborne, S. R. Stovall, and B. Baumstark, personal communication) and the *c*4 gene (Baumstark and Scott, 1987) have been sequenced. The *c*4 gene lies within *Eco*RI-9 (Sternberg *et al.*, 1978), is transcribed clockwise toward the *ant/reb* genes, and, on the basis of the location of two *c*4 mutations within an open reading frame, encodes a 69 amino acid protein (Baumstark and Scott, 1987). The *vir*s mutation that renders *ant* constitutive has been mapped to *Eco*RI-9, and mutations in or affecting *ant* have been mapped to both sides of the *Eco*RI-9,14 junction (Sternberg *et al.*, 1978). These locations suggest that *ant* transcription is also clockwise. The products of two genes, either or both of which may correspond to *ant*(*reb*), are detected in minicells infected with P1*vir*s or P1*c*4 phage, but not with P1*reb* mutants. These 27.5 kD and 40 kD polypeptides have been designated the gene products of *reb*A and *reb*B respectively (Heilmann *et al.*, 1980a). *Eco*RI fragments 14 and 25 have recently been sequenced, revealing two open reading frames that could encode the two *reb* polypeptides (E. Hansen, personal commu-

nication; N. Sternberg, unpublished). Since the P53/Op53 promot-
er/operator in *Eco*RI-14 is located between these two open reading frames
this organization may be indicative of a two way communication between *imm*C and *imm*I. Alternatively, the two *reb* proteins could be
encoded by the single open reading frame between *c4* and Op53, with the
smaller 27 kD polypeptide initiating at an in-frame start codon about
one-third of the way into the 40 kD open reading frame.

We conclude discussion of the *imm*I region with mention of *sim*, a
gene that perhaps deserves inclusion here. Fragment *Eco*RI-9, when
cloned in a high copy number vector, was fortuitously discovered to
confer an unexpected protection against an infecting P1 *c*1 phage. The
gene responsible for this protection, *sim*, is closely linked to *c4* but transcribed from an independent promoter (Devlin *et al.*, 1982). Most of the
bacteria carrying pBR322:*sim* survive P1 *c*1 infection, but the infected
cells do not become lysogens. Infection by all P1 clear plaque and virulent
derivatives tested and by P7 is also effectively aborted. The protection,
conceived to be an extended immunity, led to the naming of the gene,
sim, for super immunity. The blockage by *sim* is selective for lytic functions; prophage maintenance functions are not affected. When present in
cells at low copy number, *sim* is without detectable effect and is therefore
unlikely to particpate in immunity maintenance. Moreover, a *sim* mutation does not alter the lysogenization frequency of P1Cm. Whether *sim*
should be considered an immunity function under normal physiological
conditions is therefore open to question. The study of P1 mutants that
overcome the superimmunity conferred by *sim* (Devlin, Soper, and Scott,
referred to in Devlin *et al.*, 1982) offers a promising route to understanding the target of *sim* action and the possible role of *sim* in the life of P1.

F. Bacterial Functions Implicated in Lysogenization

Mutants of *E. coli* that do not permit P1Cm to lysogenize were selected by Takano (1971, 1977) and identified as mutants in *lon*, the determinant of a specific heat shock-inducible protease (Neidhardt *et al.*,
1984). The capacity to lysogenize *E. coli lon* at normal frequency is conferred on P1Cm by *ant* mutations (Yarmolinsky and Stevens, 1983) and
by unmapped mutations in P1 designated *p* or *pla* by Takano (1971,
1977).* The lysogeny defect of P1 in a *lon*-negative host can also be
overcome by a *lexA3* mutation (Yarmolinsky and Stevens, 1983) or by *sul*
(*sfi*) mutations (Gottesman and Zipser, 1978; Huisman *et al.*, 1980). The
lexA3 mutation renders the set of SOS functions normally induced by
DNA damage no longer inducible. The SOS functions include the *recA*
protease (which mediates the UV induction of prophages) and the *sul* (*sfi*)-

---

*The *pla* mutations of Takano (1977) are reported to be dominant and include both *cis* and
trans dominants. These characteristics suggests that they might not be simple *ant*
defectives.

determined inhibitor of septation (which is responsible for the UV induction of cellular filamentation). Cellular filamentation is also induced when *ant* is derepressed by raising the temperature of a P1c4ts lysogen (Scott, 1970). Derivatives of P1c4ts that fail to cause this filamentation have been isolated by Scott. These observations suggest possible connections among components of the P1 immunity system and regulatory functions of *E. coli* that respond to various environmental stimuli. It is presumably via such connections that selective forces operate to ensure the retention of the complex immunity circuitry that P1 possesses.

VII. THE PROPHAGE STATE

A. General Considerations

1. The Extent of Prophage Genome Expression

The fraction of the P1 genome that is transcribed in the prophage state is large, although the average rate of transcription is modest (see Fig. 5 of Sternberg and Hoess, 1983, after Mural, 1978). Of the total mRNA in a P1 lysogen, about 0.2% is P1-specific (Mural, 1978), whereas the ratio of P1 to *E. coli* DNA is an order of magnitude greater.

We have already noted that several scattered genes of P1 which confer a selective advantage to P1 lysogens are expressed in the prophage state. Full immunity requires the expression of genes located at *imm*C, *imm*I, and *imm*T. The *cre* recombinase and the genes *res* and *mod* are also expressed constitutively, whereas *ref*, between *cre* and *res*, is not. One plasmid-encoded replication gene (*rep*A) and two plasmid-encoded partition genes (*par*A, *par*B) are known to be expressed from the P1 prophage (Chattoraj *et al.*, 1985a; Austin and Abeles, 1985). It has been suggested that the *c7* cistron, mapped to *Eco*RI-15 (Mural *et al.*, 1979) and therefore separated by one or more head genes from the *par* region (Fig. 2A), is also expressed by prophage (Scott *et al.*, 1977a, 1978a).* Given the small number of P1 genes that have been studied, we expect, on statistical grounds, that several other P1 genes will also be found to function in the prophage state. The prophage state of P1, unlike that of λ, is not a condition of passivity, nor is there reason to assume that the several prophage genes that do function are continuously expressed throughout the cell cycle.

---

*The basis for the suggestion that the *c7* cistron is expressed by prophage is the observation that a thermosensitive P7c7 mutant (P7c7.106) appears to be thermosensitive for plasmid partitioning as well as plaque clarity. The rate of loss of a P7c7.106 prophage at 42°C approximates the expected rate of loss of a randomly partitioned plasmid or a plasmid with a partial replication defect. The latter possibility is inconsistent with the finding of an unchanged rate of thymidine incorporation into plasmid following a thermal shift (Scott *et al.*, 1978a). We cannot dismiss entirely the possibility that a fortuitous and misleading secondary mutation in P7c7.106 accounts for the surprisingly paired traits in this phage.

For a discussion of the expression of *vegetative* functions in the prophage state, the reader is referred to Section VIII.C.1.

2. Relation of Plasmid Incompatibility to Mechanisms of Maintenance

A characteristic that P1 shares with plasmids in general is the expression of and sensitivity to plasmid incompatibility. Luria *et al.* (1960) noted the failure of two genetically marked P1 prophages to coexist stably in the same cell, and Hedges *et al.* (1975) reported the failure of P1 to coexist stably with certain other plasmids. Both phenomena are examples of plasmid incompatibility. A third phenomenon that is possibly due to P1 incompatibility was noted by Mise and Arber (1976), Meurs and D'Ari (1979), and Jaffé-Brachet and Briaux-Gerbaud (1981). These authors observed a low-frequency curing of P1 lysogens by P1 superinfection. The cause of this plasmid destabilization has not been established, but it might be explained by an incompatibility exerted by the presence, among the superinfecting phage, of defective P1 phages (e.g., P1S particles) that are themselves incapable of forming stable lysogens.

The basis of incompatibility was recognized by Jacob *et al.* (1963) to be competition for the machinery of plasmid maintenance. These authors proposed that all cellular genetic elements are attached to rare specific sites of the cell surface that are rate-limiting for both replication and partition. Despite the seminal importance of the concepts set forth in this article, both parts of the proposal are incorrect: the part that implies a necessary connection between the specific partitioning and replication machinery of the plasmid and the part that assumes an essentially passive control* of plasmid replication. First of all, we now know that replication and partition are essentially independent processes in P1, as they are in other plasmids of comparable copy number. Deletion of the partition region of a miniP1 plasmid does not interfere with its replication, although the plasmid is no longer as stably inherited (Austin *et al.*, 1982). Moreover, *rep* and *par* regions of differing provenance can be exchanged and function in their new contexts apparently as well as they did in the original plasmids and even when their relative orientations are reversed (Austin and Abeles, 1983a). Secondly, the proposal that plasmid replica-

---

*Our use of the term "passive control" in connection with the membrane-site model of Jacob *et al.* (1963) derives from Pritchard (1978), who introduced the term to refer to the theoretical case in which a system external to the plasmid (a host factor) sets the maximum level of replication. Jacob *et al.* had introduced the unfortunate term "positive regulation" in connection with the proposal that every unit capable of independent replication (replicon) carries a gene that determines the structure of a specific initiator. Initiators are elements of replicons that act in a positive fashion, but, as Nordström (1985b) points out, positive regulation simply cannot exist. A replicon that is limited in the rate of its replication solely by an initiator of its own is an unregulated runaway replicon. If positive control is nonsensical, then negative control is pleonastic. We avoid both terms henceforth. The distinction between active and passive control, on the other hand, is a useful one.

tion is limited by host factors that are the determinants of plasmid incompatibility implies an unreasonable degree of complexity. Each incompatibility group would have to correspond to a different host factor that is rate-limiting for the replication of the members of that group. It is difficult to imagine that plasmids with similar requirements for replication would belong to as many different (in)compatibility groups* as have been distinguished.

We owe to Pritchard et al. (1969) the insight that plasmids are actively, rather than passively, controlled. Plasmids themselves determine the means to correct deviations from their characteristic copy numbers. They do this by a negative loop which assures that most of the time the plasmid is inhibited from replicating. Only some plasmids specify initiator proteins, but all plasmids studied so far control their own replication by specifying a negative loop (Nordström, 1985a).

The major determinants of incompatibility are the plasmid-determined elements that limit the availability of a rate-limiting component in the initiation of replication. The mechanism of their action can be understood by reference to Fig. 8 (compare B and A). Each of the generic low-copy-number plasmids of the figure possesses such an element.

Each of the plasmids in the figure also possesses a partitioning element without which a low-copy-number plasmid would be unstable. If such elements limit the availability of a rate-limiting component in partitioning, they will also contribute to incompatibility (Novick and Hoppenstaedt, 1978), as shown in Fig. 8 (compare C and A).

Certain plasmids possess a third system that adds to the stability with which they are inherited, the *ccd* (coupled cell division, or control of cell death) system. In F, in which *ccd* functions were first described (Ogura and Hiraga, 1983), the system leads to loss of viability of plasmid-free segregants (Jaffé et al., 1985; Hiraga et al., 1986). Under exceptional conditions, this system too can contribute an incompatibility determinant (Ogura and Hiraga, 1983; Hiraga et al., 1985) (compare b and a in Fig. 8).

The machinery of plasmid maintenance was first illumined by studies of incompatibility. We therefore preface the sections on P1 plasmid replication and partitioning with a discussion of the localization and preliminary characterization of incompatibility determinants.

3. Localization of P1 *inc* Determinants

The determinant of P1 incompatibility expression was first localized to a deletion derivative of P1, the P1ΔN19 of Austin et al. (1978), then to

---

*The terms "compatibility group" and "incompatibility group" are interchangeable, like "flammable" and "inflammable." An attempt to decree "incompatibility group" as standard usage (Novick et al., 1976) appears not to have been fully successful (see Nordström, 1985b).

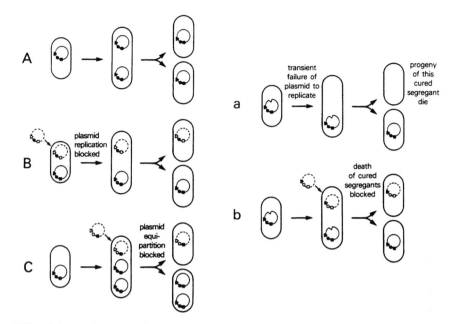

FIGURE 8. Mechanisms of incompatibility. (A) Unperturbed plasmid maintenance. The diagram depicts the replication and segregation through one cell cycle of a stable plasmid maintained at a low copy number (shown as 1 copy per baby cell). An element of the plasmid that controls replication frequency is represented by a small filled circle, one that controls partitioning of plasmid replicas to daughter cells is represented by a small filled square, and one that antagonizes the *ccd* inhibitor of cell division is represented by a small filled triangle. (a) Plasmid maintenance unperturbed by a transient failure of the plasmid to replicate. The diagram shows the consequences for cell viability of a temporary block in plasmid replication. The plasmid fails to replicate owing to damage (represented by a blip in the circle) that is repaired only after a segregant cured of plasmid has been born. Functioning of the *ccd* genes determined by the damaged plasmid assures that the progeny of the cured segregant bacterium will die. (B) Replication-control-mediated incompatibility. A plasmid (dashed circle) bearing a replication control element functionally identical to that of the resident plasmid is introduced prior to replication and competes with the resident plasmid for replication. The competition is possible, because replicons in general (see Nordström, 1985b) and P1 in particular (Abe, 1974) are chosen at random for replication (rather than solely on the basis of a feature of replicon history such as the time at which the replicon last engaged DNA polymerase). Consequently in some bacteria carrying both plasmids the introduced plasmid will exclude the replication of the resident plasmid. In the subsequent generation, cells cured of plasmids of the resident type will be born, as shown in the figure. (b) *ccd*-mediated incompatibility. A plasmid (dashed circle) identical to the resident plasmid at an element that antagonizes the lethal *ccd* protein is introduced into a cell with an abnormally low content of plasmids. As in panel (a), the resident plasmid has failed to replicate owing to reparable damage. Were it not for the introduced element, cell division would have produced largely nonviable segregants and would thereby have reduced the proportion of cells cured of plasmids of the resident type. (C) Partition-mediated incompatibility. A plasmid (dashed circle) bearing a partitioning element identical to that of the resident plasmid is introduced following plasmid replication and competes with the resident plasmids for partitioning. By a line of reasoning parallel to the one given in (B) above, the competition will occasionally generate cells cured of plasmids of the resident type, as shown in the figure. The randomness with which plasmids are assorted in pairs for partitioning to daughter cells is an essential requirement for this kind of partition-mediated incompatibility.

the P1-derived plasmid, pIH1972 of Shafferman *et al.* (1979), of which the relevant portion was shown to be within *Eco*RI-5 (Sternberg *et al.*, 1981b). Subsequently, the dual nature of incompatibility exerted by P1 was recognized (Cowan and Scott, 1981), and incompatibility determinants were localized to two separate regions of *Eco*RI-5, named *incA* and *incB* (Austin *et al.*, 1982). A single extra copy of either of these elements is sufficient to destabilize P1, but only *incA* of P1 will destabilize P7. We recognize that the locus of group Y incompatibility expression common to P1 and P7 is *incA*, a segment of DNA of about 0.3 kb (Abeles *et al.*, 1984). If we define IncY as the group of plasmids sensitive to *incA*, then it follows that, since P1 and P7 also possess incompatibility determinants of differing specificity, a plasmid cannot be unambiguously classified as a member of IncY solely on the basis of its incompatibility with one or the other prophage alone.

The presumption that incompatibility determinants are implicated in plasmid maintenance led to the search for a stable plasmid replicon centered on the *Eco*RI fragment that contains both *incA* and *incB*.* Such a replicon was generated by integration of λ-P1:*Eco*RI-5 into P1 followed by an illegitimate excisive recombination event that extended the P1 DNA of the cloned fragment "rightwards" (Sternberg and Austin, 1983). The resultant λ-P1:5R phages lysogenize by plasmid formation and cannot integrate, because they are deleted for the λ *attP* site. The copy number of the hybrid plasmid is similar to that of the parental P1; its stability, although somewhat less [0.04% loss per generation from a *recA* strain of *E. coli* in rich medium (Austin *et al.*, 1982)], is due to an active partitioning mechanism.

Deletion analysis of λ-P1:5R and of comparable λ-miniF plasmids led to the conclusion that the primary plasmid replication and partition regions of P1 and F are similarly organized (Chattoraj *et al.*, 1985a). P1 and F possess a replication control element immediately downstream of a gene that determines an essential plasmid replication protein. This element in P1 is the incompatibility determinant *incA* (Austin *et al.*, 1982; Chattoraj *et al.*, 1984). The replication control element also adjoins the partition

---

*It was initially assumed that *Eco*RI-5 would confer the capacity for autonomous replication, since *Eco*RI-5 was found to cover all of the P1 DNA common to two different P1-derived plasmids: pIH1972 and P1ΔN19 (Sternberg *et al.*, 1981b). The paradoxical inability of λ-P1:*Eco*RI-5 to replicate as a plasmid was explained by the discovery that extension of *Eco*RI-5 to include P1 DNA normally present on either one or the other side of *Eco*RI-5 allowed for autonomous replication (Sternberg and Austin, 1983). Extensions that include part of *Eco*RI-8 are called R replicons. L replicons contain DNA to the other side of *Eco*RI-5 and need not contain any portion of *Eco*RI-5 (Sternberg and Austin, 1983; Sternberg and Hoess, 1983). We treat these replicons in the next section. It has been reported that pIH1972 exerts a weak incompatibility towards P7 (I. Hertman, personal communication, cited in Cowan and Scott, 1981), although the maintenance of pIH1972 is unlikely to involve functions homologous to those that maintain P7. The incompatibility is presumably due to *incA* DNA present in pIH1972. The element can exert incompatibility, although we presume that pIH1972 is itself *incA*-insensitive.

region which, in each plasmid, encodes two *trans*-acting proteins (*parA* and *parB*) required for plasmid stability (Abeles *et al.*, 1985; Austin and Abeles, 1985; Friedman and Austin, 1988). The partition regions of P1 and F each terminate in a component of the plasmid partitioning apparatus that necessarily acts in *cis* to the DNA undergoing partitioning. In P1 this DNA sequence is designated *parS* (Friedman *et al.*, 1986; Martin *et al.*, 1987). (The regions of F corresponding to *parA, B, S* have been named *sopA, B, C* by Ogura and Hiraga, 1983b.) The *parS* region is capable of exerting incompatibility toward P1 when present in *trans*. This region is therefore given the alternative designation *incB*. Stabilization in *cis* and destabilization in *trans* are properties that we expect of a centromere analogue.

A portion of the basic plasmid replicon of P1 (part of *oriR*) can also exert incompatibility toward P1. Although this element exerts little incompatibility toward P1 when both are present at low copy numbers, the incompatibility becomes obvious when the *oriR* region is cloned into a vector of higher copy number than P1. The relevant portion of *oriR* has been named *incC* and has the specificity of *incA* (Abeles *et al.*, 1984).* The relationship between *incA* and *incC* will be considered in the following subsection.

B. Plasmid Replication and Its Control

1. Organization of the R Replicon

P1 plasmid replication in *E. coli* is normally driven by the R replicon, the basic P1 replicon of λ-P1:5R. This conclusion is supported by the observation that the associated incompatibility element, *incA*, acts as an inhibitor of replication of both the isolated R replicon and intact P1 (Austin *et al.*, 1982). Extra copies of *incA* in a P1 lysogen cure the bacteria of the P1. Bacteria are rendered nonviable by extra copies of *incA* under special conditions that make bacterial survival dependent upon the initiation of DNA replication from the R replicon of an integrated P1 prophage (Chattoraj *et al.*, 1984). Additional evidence for a critical role of the R replicon comes from localization of mutations affecting P1 copy number to the essential *repA* gene of the R replicon (Scott *et al.*, 1982; Baumstark *et al.*, 1984). The R replicon is a basic P1 replicon in the sense that it replicates with the same average copy number as intact P1 (Sternberg and Austin, 1983).

---

*A curious result follows from these features of P1. One might expect that deletion of *incA* from an IncY plasmid would lead to an increase in plasmid copy number that would place the altered plasmid in a different incompatibility group determined by a new rate-limiting factor. The result of *incA* deletion from a miniP1 is a nonlethal increase in plasmid copy number with, *mirabile dictu*, no alteration in incompatibility specificity (Pal *et al.*, 1986). The increased gene dosage of the latent incompatibility element, *incC*, allows retention of the original incompatibility phenotype.

The basic R replicon is contained within 1.5 kb of DNA and consists of an origin (oriR) of no more than 245 bp (Chattoraj et al., 1985b), the gene (repA) for an essential replication protein of 286 amino acids (Abeles et al., 1984) and the adjoining 285 bp of the replication control element incA (Abeles et al., 1984; Chattoraj et al., 1984). The control element is completely dispensable for replication (Pal et al., 1986).

The basic R replicon of P1 is compared in Figs. 9 and 10 with the basic replicons of a few low-copy-number plasmids with which P1 is

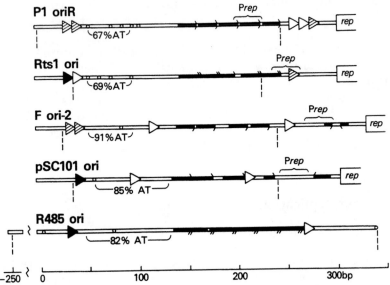

FIGURE 9. Similarly organized origins of replication in plasmids of different incompatibility groups. Functional origins lie between the vertical dashed lines. Solid, hatched, and open triangles represent dnaA boxes that conform to the consensus sequence TTAT(C/A)CACA for DnaA binding sites (Fuller et al., 1984) perfectly, deviating by one or by two bases, respectively. Each origin requires a specific plasmid-encoded rep gene product transcribed from the indicated promoters labeled Prep. Black bars with angled lines at their ends represent repeated sequences (some incomplete) to which the different Rep proteins bind specifically. Circles designate GATC sequences. P1 and Rts1 exhibit extensive DNA base sequence homology with each other (Kamio et al., 1984), whereas the other plasmids do not. Although it is possible to obtain a minimal Rts1 ori that lacks the pair of adjacent dnaA boxes, origin function is much reduced relative to that of an ori that carries at least one dnaA box of the pair at the left in the figure (Itoh et al., 1987). Similar results with ori-2 miniF have been obtained by Murakami et al., 1987). The stability of a miniRts1 and of a miniF that includes all the dnaA boxes shown in the figure is greatly reduced by a dnaA null mutation (Hansen and Yarmolinsky, 1986). R485 belongs to the incompatibility group X, of which the more thoroughly studied R6K is a member. Although R485 has not been tested, replication of oriγ of R6K is markedly reduced in a dnaA null mutant host (D. Bastia, personal communication). The sources used are as follows: For P1: Abeles et al., 1984; Chattoraj et al., 1985b. For Rts1: Kamio et al., 1984; Itoh et al., 1987. For F: Murotsu et al., 1984; Søgaard-Anderson et al., 1984. For pSC101: Linder et al., 1985; Churchward et al., 1983; Yamaguchi and Masamune, 1985; K. Yamaguchi, personal communication. For R485: Stalker and Helinski, 1985.

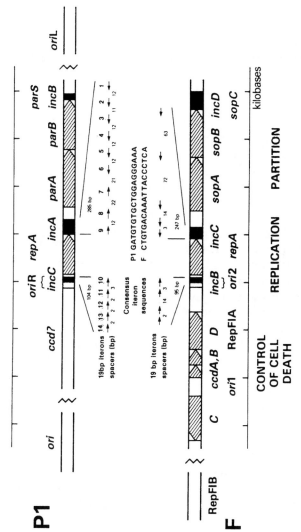

FIGURE 10. Organization of *rep-par* regions of P1 and F. The hatched arrows represent open reading frames of the indicated structural genes. Black boxes represent segments that can exert incompatibility. Small arrows indicate the organization of the 19-bp iterons. The distance from *oriR* to *oriL* in P1 is between 10 and 11 kb (E. B. Hansen, personal communication), and the distance from *oriR* to the additional *ori* of Froehlich *et al.* (1986) is at least 25 kb (B. Froehlich and J. Scott, personal communication). The P1 coordinates of the replication region are from Abeles *et al.* (1984), and those of the partition region are from Abeles *et al.* (1985). The distance from *ori*-1 to the origin present in RepFIB of F is about 10 kb (Lane and Gardner, 1979). The F coordinates of the RepFIA region of Bergquist *et al.* (1986) (from *C* to *incD*) are taken from the review by Kline (1985). Enlargements of the *oriR* region of P1 and *ori*-2 region of F are shown in Fig. 9.

compatible. DNA sequence homology can be detected within the origin regions at sites that are identical or similar to consensus binding sites for the bacterial DnaA protein (Abeles *et al.*, 1984; Hansen and Yarmolinsky, 1986). These dnaA boxes are often found in association with GATC seqtences—e.g., in *oriC* regions of various gram-negative bacteria (Zyskind *et al.*, 1983) and in the *dnaA* gene promoter (Hansen *et al.*, 1982; Braun and Wright, 1986). The pair of *dnaA* boxes within *oriR* of P1 can be assumed to bind DnaA protein, because replication from this origin, as from *oriC* of *E. coli*, has been shown to depend directly upon *dnaA* function (Hansen and Yarmolinsky, 1986). It is uncertain what significance to attach to sequences that resemble vestigial dnaA boxes found in the basic replicons of P1, Rts1, F, and R485 upstream of the open reading frame of the replication protein in each.

With the exception of Rts1, the other plasmids listed exhibit little DNA sequence homology with P1 elsewhere in the plasmid maintenance region (Abeles *et al.*, 1984; Mori *et al.*, 1986). There are, however, striking organizational homologies.

In *oriR* of P1 and in the plasmid origins depicted with it, the presumptive DnaA protein-binding site is adjacent to an A+T-rich region that includes one or more GATC sequences. In each plasmid the origin region is completed with a set of repeated DNA sequences (iterons) of comparable length. There are five 19-bp iterons in *oriR* of P1. The *oriR* iterons have the same orientation and are spaced close together such that similar portions are two turns of the DNA helix apart. In P1, Rts1, and F, iterons similar to those in the respective origin regions are also present immediately downstream of the gene for the initiator. In P1 the set of nine iterons downstream of *repA* are designated *incA*. The set of five iterons within *oriR* has been designated *incC* (Abeles *et al.*, 1984). We refer to them here as the iterons of *oriR*, because they have a more significant role than simply in the generation of incompatibility.

The *incA* iterons, like those of *oriR*, are spaced so as to present similar segments when viewed from one face of the helix, but the iterons are not as close to each other, and the proximal three are opposite in orientation to the distal six (Abeles *et al.*, 1984). Comparison of the number, orientations, and spacing of the *incA* iterons of P1 with the similarly positioned sets of iterons in Rts1 and in F suggests that the organization of the iterons within *incA* may not be of critical importance for their function (see Fig. 10). Iterons within a given plasmid differ little from each other and are invariant at several positions. Seven of the 19 positions of the 14 P1 iterons are invariant. From one plasmid to another of the set P1, F, pSC101, and R485, there is no clear homology among the consensus sequences of the iterons.

With the development of an *in vitro* system capable of replicating an *oriR* miniplasmid, the determination of the direction of replication of this miniP1 has been achieved. Replication initiates preferentially within the origin and proceeds in most cases unidirectionally away from the *repA* gene in theta-structure intermediates (Wickner and Chattoraj, 1987). Note that both the similarly organized *ori-2* of F and the *ori* of pSC101 have also been shown to replicate *in vivo* in the opposite direction from the direction of transcription of the essential replication protein (Eichenlaub *et al.*, 1981; Yamaguchi and Yamaguchi, 1984).

Transcription initiated from within *oriR* of P1 is detected when the *oriR* region is cloned into a multicopy vector so as to form an operon fusion to *galK* (Chattoraj *et al.*, 1985a,b). The major promoter activity is directed toward *repA*. A minor activity, about twofold over background, is detected when the cloned fragment is inverted. Both activities are strongly repressed by RepA, as assayed with the *galK* vector. The extent of repression is about the same whether the cloned P1 DNA includes only the iterons of *oriR* or a much larger region of more than 1400 bp upstream of *repA* (D. K. Chattoraj, personal communication). It therefore seems unlikely that there exists an upstream promoter that can override a repressed P*repA*.

The promoter in *oriR* that is directed toward *repA* is the promoter for the *repA* transcript (Chattoraj *et al.*, 1985b). This conclusion is based on the properties of a genetic construct in which the N-terminal 62 codons of *repA* are fused in frame to *lacZ*. The protein fusion places β-galactosidase activity under control by the RepA protein. The extent of repression of β-galactosidase activity from the protein fusion is identical to the extent of repression of galactokinase activity from the operon fusion described in the previous paragraph when the source of RepA in each experiment is λ-P1:5R. When the *repA-lacZ* protein fusion is present on a unit copy vector (integrated λ), the extent of *lacZ* repression by RepA from λ-P1:5R is greater than 99% (Chattoraj *et al.*, 1985c). We conclude that expression of the *repA* gene is severely limited by an autoregulatory circuit.

The location of the *repA* promoter, P*repA*, has been specified precisely. The −35 and −10 regions of P*repA* have been localized to the iteron junction positions shown in Fig. 9. In *lacZ* protein fusions to *repA*, β-galactosidase activity could be reduced to background levels by insertion of a terminator near the end of the *oriR* iterons proximal to *repA* or by a deletion (*rep-11*) that removes the proximal two iterons from the promoter region. On the basis of additional operon and protein fusions, the *repA* promoter was localized to within three *oriR* iterons (D. K. Chattoraj, personal communication). The 5' end of *in vivo repA* transcripts was localized to the middle of iteron 10 (see Fig. 10). The −35 and −10 regions indicated in the figure are the best fit to the consensus promoter sequence (Hawley and McClure, 1983) from which *repA* transcription could be initiated (Chattoraj *et al.*, 1985b). Note that the promoter of *repA* in P1 is unlike the corresponding promoter in F or pSC101, which lies outside the minimal origin (Fig. 9).

Only one protein, RepA, is encoded within the basic R replicon as far as is known. This protein is essential for plasmid replication from *oriR* (Abeles *et al.*, 1984; Austin *et al.*, 1985). RepA has been overproduced from an expression vector and purified to 90% homogeneity (Abeles, 1986). The amino-terminal sequence and amino acid composition of the purified protein indicate that the entire *repA* open reading frame is translated. The electrophoretic mobility of the denatured protein on SDS–PAGE, M_r 32 kD, is in agreement with the calculated value of 32,224 daltons (Abeles, 1986). Gel filtration of the native protein through a Sephadex G100 column indicates an M_r of 66 kD, implying that the protein exists largely as a dimer (J. Swack and D. K. Chattoraj, personal communication). In respect to size and net charge (calculated from the amino acid composition to be +14), the *RepA* protein of P1 is typical of other plasmid initiator proteins that have been characterized (Chattoraj *et al.*, 1985a).

Specific binding of RepA protein to the iterons of *oriR* and *incA* is suggested by the strategic disposition of iterons on either side of *repA* in regions necessary for replication and replication control. Experiments on replication control, discussed in the subsection that follows, also lead to the prediction that the iterons are binding sites for RepA. This prediction was verified in studies of DNA-protein binding measured by retardation of the electrophoretic migration in polyacrylamide gels of linearized DNA segments (Chattoraj *et al.*, 1985a; Austin *et al.*, 1985; Abeles, 1986). Binding is specific for iteron DNA that is double-stranded (A. Abeles, personal communication). The binding of RepA to groups of identically oriented iterons, whether they derive from *oriR* or *incA*, results in a number of distinct retarded species. There appears to be a direct relationship between the number of iterons occupied and the degree of retardation. Binding to a DNA segment bearing a single iteron has also been demonstrated (Pal *et al.*, 1986). The binding of RepA to intact *incA* results in a complex pattern of bands and smears that is presumably due to the complexity of *incA* structure (Abeles, 1986). Under the conditions of these experiments, there is no evidence of cooperativity in the binding process nor of any striking differences in the affinity of particular iterons for the protein (Abeles, 1986).

A more intimate look at RepA-iteron interaction, by DNase I footprinting, does reveal some differences among the iterons of *oriR* (A. Abeles, personal communication). Although RepA protected all the iterons of *oriR* from digestion, the protection of iterons 10 and 11 was different from the protection of iterons 12–14 (see Fig. 10). Differences in the extent to which RepA caused an enhanced susceptibility to DNase I of certain bases centrally within the iterons of both strands also suggests that the interaction of RepA with iterons 10 and 11 is not the same as its interaction with the other three iterons. Evidence that RepA can displace RNA polymerase from the iterons of *oriR* has also been obtained (A. Abeles, personal communication).

2. Replication Control

The first evidence that P1 plasmid actively controls its own replication was the finding of P1 mutants with increased copy number (Sternberg et al., 1981b; Scott et al., 1982). The recognition that the incompatibility element incA is a regulator of both P1 and miniP1 replication (Austin et al., 1982; Abeles et al., 1984; Chattoraj et al., 1984) focused attention on the mechanism by which this unusual structure might exert its effect on the plasmids derived from P1.

Several lines of evidence indicate that incA exerts control over P1 plasmid replication by binding the essential RepA protein and making it unavailable for the initiation of DNA synthesis at oriR. (1) The concentration of RepA in bacteria that carry low-copy-number miniP1 plasmids, such as λ-P1:5R, is sufficiently low relative to the concentration of RepA-specific iterons that the binding reaction can dramatically alter the concentration of free RepA in the cell. Immunological (Western blot) assays of RepA in extracts of λ-P1:5R lysogens indicate that the number of RepA dimers per prophage is approximately 20 (Chattoraj et al., 1985c; Swack et al., 1987). Each of the 14 iterons of λ-P1:5R is a binding site for RepA protein (Abeles, 1986). (2) The copy number of miniP1 (oriR) plasmids is inversely related to the number of iterons present in the same cell (Chattoraj et al., 1984, 1985a). Even a single intact iteron is adequate to exert incompatibility (Pal et al., 1986), although its effect is considerably larger when the iteron is present in cis to the tested replicon than when it is on a separate, otherwise compatible vector (Pal et al., 1986). The significant result is that incA behaves as the sum of individual units no larger than the individual RepA binding sites of which it is composed. This is compelling evidence that incA exerts control directly without making a product. (3) The copy number of oriR plasmids supplied with RepA in trans is linearly related to the supply of RepA protein over a physiological range of concentrations (Chattoraj et al., 1985b,c; Pal and Chattoraj, 1986). (4) The effects of RepA and incA have been shown to be mutually antagonistic, at least under certain conditions. The inhibition of replication by incA is relieved by an increased supply of RepA. This relief, in turn, can be prevented by additional copies of incA (Chattoraj et al., 1985a). (5) Evidence from studies of a miniF, which is similarly organized to miniP1, provides independent support for the proposal that iterons control copy number by sequestering a rate-limiting, plasmid-determined initiator (Tsutsui et al., 1983; Tokino et al., 1986).

In order to understand how incA participates in the replication control circuitry of P1 or miniP1, it is necessary to distinguish the different functions that RepA serves and to determine how sequestration by incA affects each of them. The repA gene serves at least three functions in miniP1, one positive and two negative. Its primary gene product, RepA, plays a positive role in the initiation of replication (Austin et al., 1985) and can fulfill this function for a separate oriR in trans (Chattoraj et al.,

1985a,b). A major negative role of RepA at physiological concentrations is to modulate expression of the repA gene. The dramatic repression of repA transcription by low concentrations of RepA has already been noted in part B.1 of this section.

A hitherto unrecognized product of the repA DNA, when at relatively high concentrations, also plays an inhibitory role in the initiation of replication (Chattoraj et al., 1985b,c; Pal and Chattoraj, 1986). The repA gene has been placed under control of a pBR322 promoter weakened to various extents by insertions of terminators or under a lacZ promoter induced to various extents. As transcription of the repA gene is increased, the copy number of an oriR plasmid increases to a maximum value of about 8 relative to λ-P1:5R. The RepA concentration that results in maximum copy number is about sixfold the concentration that is present in a λ-P1:5R lysogen (S. Pal and D. K. Chattoraj, personal communication). A further increase in transcription causes the copy number to fall.* When transcription is sufficient to furnish RepA at about 40-fold its normal concentration, replication from oriR is effectively and specifically blocked. However, when RepA is supplied to an in vitro P1 replication system in 100-fold excess over its optimal concentration, the inhibition is no more than 25% (Wickner and Chattoraj, 1987). Purified RepA protein appears to lack inhibitory activity. This finding is consistent with evidence from the studies of the repA gene altered by site-directed mutagenesis (K. Muraiso and D. K. Chattoraj, personal communication). Alteration of the repA initiation codon appears to prevent RepA synthesis but not the synthesis of the inhibitor of oriR replication. The precise nature of this inhibitor is being determined. The utility for P1 of the replication inhibition exerted when repA is overexpressed in vivo is presumably to provide an additional restraint on runaway plasmid replication.

In preceding paragraphs we have outlined compelling arguments for control of RepA concentration by binding to incA on the one hand and control of RepA concentration by an autoregulatory circuit on the other. These alternatives would appear to be mutually exclusive. The autoregulatory circuit would be expected to replace RepA molecules as they become bound to the iterons at incA. The addition or removal of incA iterons would be expected to cause at most a transient change in plasmid copy number instead of the permanent change that is observed. What prevents the autoregulatory circuit from negating the effect of RepA sequestration by incA?

In principle, the dilemma can be resolved by making the assumption that sequestration of RepA by incA differentially affects the positive (initiator) and negative (repressor) roles of RepA. If binding of RepA at incA were to block the initiator action of RepA specifically, but not its re-

---

*Specific mutagenesis of the repA gene, followed by selection for mutants that are no longer as inhibitory for replication when repA is overexpressed, has yielded a mutation within repA that allows an oriR plasmid to reach a copy number about 30-fold higher than that of λ-P1:5R (S. Pal and D. K. Chattoraj, personal communication).

pressor action, then autoregulation would be unable to compensate for RepA sequestration. Two ways have been envisaged for this to happen.

Trawick and Kline (1985) have proposed that the forms of RepA responsible for initiation and for repression exist as separate species, of which one is derived from the other by an irreversible reaction. If the repressor form irreversibly generates the initiator form, then the removal of some of the initiator form of RepA will not diminish the pool of the repressor form (Fig. 11A). The repressor is effectively insulated from the initiator. The model can be formulated in mathematical terms and suitable values for the several parameters chosen so as to allow simulations of the response of plasmid copy number to various experimental situations. These simulations show that the model is consistent with the actual pattern of inheritance of miniF, the plasmid for which the model was originally designed (Womble and Rownd, 1987). The model should be equally satisfactory for P1.

Resolution of the autoregulation-sequestration dilemma does not require that RepA exist as a separate species in the free state. Different sites for initiator and repressor functions might reside in one molecular species, but sequestration would have to affect them differently. Chattoraj *et al.* (1988) suggest that sequestration of RepA by *incA* might specifically interfere with the initiator function of RepA without impairing its repressor function. This is perhaps simplest to imagine if we assume that the RepA dimer is asymmetric (either inherently or as a result of DNA binding) such that in the dimer one moiety has affinity for any iteron and the other moiety has a greater affinity for the iterons associated with the *repA* promoter. A RepA dimer bound to *incA* would be unavailable for initiation, but looping of the DNA could bring the moiety with repressor

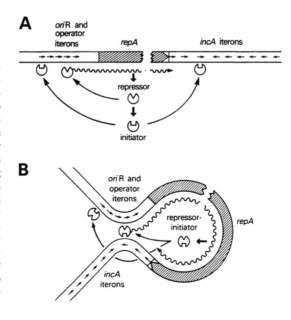

FIGURE 11. Alternative models that reconcile autoregulation of *rep* and initiation control by sequestration of Rep protein at *inc* iterons. (A) Proposal of Trawick and Kline (1985): Repressor and initiator are separate molecular species with the indicated binding specificities. The conversion of repressor to initiator is an irreversible reaction. (B) Proposal of Chattoraj *et al.* (1988). Repressor and initiator functions are carried out by different facets of a single molecular species capable of binding to *inc* and operator sites simultaneously.

activity to its operator target (Fig. 11B). Formally the two models are nearly identical.

Both models predict that it should be possible by mutation to affect the initiator and repressor form of the RepA protein differentially so as to cause an increase (or decrease) in P1 copy number. The finding of single base substitution mutations within *repA* that increase P1 copy number up to eightfold (Scott *et al.*, 1982; Baumstark *et al.*, 1984) is consistent with this expectation. According to the hypothesis of Trawick and Kline, repressor and initiator are separate molecular species of which only the latter has affinity for the repeats at the control locus, *incA* (Fig. 11A). An experimental test of this aspect of the model does not support it. Repeats at the origin region and at *incA* were comparable competitors for the repressor. The alternative hypothesis, according to which DNA participates more actively, is supported by the direct visualization of predicted structures. Electron microscopy of appropriately prepared material revealed RepA-mediated DNA loops between the *repA* promoter (*oriR*) region and the *incA* region of isolated miniP1 DNA (Chattoraj *et al.*, 1988). The autoregulation-sequestration dilemma appears to be more satisfactorily resolved by according *incA* the capacity to loop than by according it the capacity to discriminate between hypothetical initiator and repressor forms of free RepA.

3. Alternative Replicons

The R replicon is not the only replicon of P1 that, under suitable conditions, is capable of supporting plasmid replication. There is substantial evidence for at least two additional P1 replicons with this capability.

One of the first miniP1 plasmids to be studied, pIH1972 (Shafferman *et al.*, 1978, 1979) is a stable, low-copy-number plasmid (I. Hertman, personal communication) that lacks both *oriR* and *repA* but carries *par* and about 9 additional kb of contiguous P1 DNA. It is maintained by the replicon referred to as L (Sternberg and Austin, 1983; Sternberg and Hoess, 1983). The L replicon, defined as the region of P1 DNA between *par* and *immI* with the capacity for plasmid replication, is normally under control of P1 immunity repressors, although such control must be lacking in pIH1972. Plasmids λ-P1:L-4 and pIH1972 carry roughly equivalent regions of P1 DNA in which the minimal L replicon is centrally situated, but the copy number of λ-P1:L-4 is about 10 times the copy number of pIH1972.

Evidence for L control by the distant *cl* gene was provided first by Sternberg and Hoess (1983). Lysogeny by λ-P1:L-10, which lacks *Eco*-RI-14, is *cl*-insensitive, whereas lysogeny by λ-P1:L-11, which includes this DNA, is *cl*-sensitive.* The site at which *cl* exerts its control appears

---

*These experiments were performed in *recA* hosts. The λ-P1:L phages studied appear to be poorly maintained in *recA*+ *E. coli*, perhaps owing to an SOS response.

to be at a c1-sensitive promoter (P53) that has been detected in *EcoRI*-14 and that is directed toward the L replicon (Sternberg and Hoess, 1983). There is also evidence for control of L by the neighboring *c4* gene.

The ability of P1 (integrated at *loxB*) to integratively suppress the heat sensitivity of a *recA dnaAts E. coli* is not blocked if excess *incA* is furnished to prevent functioning of the R replicon of P1 (Chattoraj *et al.*, 1985a). Presumably a P1 replicon other than R is capable of driving chromosomal replication. This replicon is largely under the control of the c1 repressor protein; however, the residual amount of replication that remains when the c1 protein is present can be modulated by the c4 protein. When *incA* and *c4* are cloned together and introduced into a *recA dnaAts* (P1) host, the resultant strain is thermosensitive (D. K. Chattoraj, personal communication). The proximity of *c4* to the minimal L replicon was taken as suggestive evidence that the replicon under *c4* control is L. This conjecture is supported by the map position of a Tn5 insertion that makes integrative suppression by P1 largely *incA*-sensitive. The insertion is located in *EcoRI*-17 (E. B. Hansen, personal communication). Presumably normal *c4* expression in a lysogen carrying integrated P1 is insufficient to assure full L replicon repression.

As integration at *loxB* may alter the level of P1 immunity, it is possible that the L replicon is expressed to a greater or lesser extent in free P1 plasmid than it is in integrated P1 plasmid (N. Sternberg, unpublished). The L replicon must be normally less capable of plasmid maintenance than the R replicon, since extra copies of *incA* destabilizes P1, whereas deletion of a region we now know to include *oriL* does not (Austin *et al.*, 1978). However, recent experiments on the effect of blocking R replicon function in various ways reveal that the L replicon of P1 can substitute for the R replicon to assure continued plasmid maintenance (M. Yarmolinsky and S. Jafri, unpublished). The lytic replicon serves this purpose particularly effectively in lysogens of P1 c1.100 at 32°C, presumably because the lytic replicon is partially released from c1 repression under these conditions.

Possibly the activity of the L replicon that has been detected in P1 prophages is due to the incomplete immunity of the prophages studied, which are all derivatives of the relatively clear plaque mutant originally selected by Lennox (1955). We think it probable that the expression of *ref* at 30°C by P1Cm c1.100 is partly due to incomplete repression of *ref* by the thermosensitive c1 repressor and partly due to the plasmid copy number being slightly elevated as a consequence of a modest activation of the L replicon. This hypothesis accounts for the observation by Windle (1986) that *ref* expression from a P1c1.100 plasmid is eliminated by wildtype c1 repressor and slightly reduced by a Tn5 insertion (180° from *ref* on the P1 map) at a site in *EcoRI*-17 adjacent to *oriL* (see Section VIII.B.2).

Although the L replicon does appear to function in a repressed P1 prophage, it is in the absence of immunity repression, upon infection or following induction, that the L replicon comes into its own. We discuss in Section VIII.B.2 evidence that the L replicon is involved in lytic replication and summarize these recent studies on the precise localization and control of this key element.

Another replicon, derived from a region distant from the R and L origins, is responsible for the high level of resistance to ampicillin and chloramphenicol of certain mutants of a P1Cm-P7 hybrid that have undergone rearrangements and deletions (Froehlich *et al.*, 1986). Further discussion of it can be found in Section VIII.B.2.

We conclude with mention of a putative replicon, the existence of which is based purely on analogy. There is indirect evidence (to be discussed in Section VII.B.5) to suggest that the organizational similarity between P1 and F that we have illustrated in Figs. 9 and 10 extends toward the region where F has an origin of replication designated *ori*-1. The situations in which activity of miniF *ori*-1 is observed suggest that this origin responds to conditions that threaten continued maintenance of F, although the evidence for this proposal is only circumstantial at present (H. E. D. Lane, personal communication). For the sake of completeness, we mention the possibility that P1 may resemble F in possessing a replicon of this kind in addition to the replicons we have mentioned above.

4. Host Participation in Plasmid Replication

In various *dnaA*ts mutant strains of *E. coli*, the replication of P1 (Abe, 1974; Chesney *et al.*, 1978, 1979) and of the miniP1, λ-P1:5R (Chattoraj *et al.*, 1984), has been shown to continue under conditions nonpermissive for the bacterial host. These results imply only that P1 and its R replicon are less sensitive to a reduction in the function of DnaA protein than is *E. coli*. Experiments with *dnaA* null mutants of *E. coli* (in which bacterial replication proceeds from a *dnaA*-independent origin such as an integrated miniR1 plasmid) have permitted the determination of whether the P1 replicons can dispense with *dnaA* function completely (Hansen and Yarmolinsky, 1986). These experiments show that intact P1 makes plaques of normal size on a *dnaA* null mutant and is capable of integratively suppressing such a mutant. However, the R replicon of P1 appears completely dependent on *dnaA*. It can neither maintain itself as a free plasmid nor, in the integrated state, replace *oriC* function if the *host* carries a *dnaA* null mutation. Additional experiments both *in vivo* (Hansen and Yarmolinsky, 1986) and *in vitro* (Wickner and Chattoraj, 1987) indicate that the role played by *dnaA* in the replication of the R replicon is probably direct, as it appears to be in the initiation of bacterial replication at *oriC*. The DnaA requirement cannot be bypassed by providing additional RepA protein. It can be seen from Fig. 9 that the R replicon of P1 belongs to a class of similarly organized replicons bearing *dnaA* boxes similarly positioned at a border of the origin. Each replicon is largely, if not completely, *dnaA*-dependent, although only pSC101 appears as sensitive as *E. coli* to a depletion of DnaA function (see legend to Fig. 9).

Whereas several plasmids appear to be less sensitive to a DnaA deficiency than is *E. coli*, the reverse is true for a DNA gyrase deficiency. Bacteria carrying heat-sensitive gyrase mutations can be cured of a number of different plasmids by subjecting them to sublethal elevated temperatures. Wild-type bacteria can be cured of plasmids by poisoning the bacterial gyrase, the *gyrB* subunit in particular, with various specific inhibitors.* This sensitivity has been attributed to a preferential inhibition of plasmid replication (Wolfson *et al.*, 1983; Uhlin and Nordström, 1985). P1 and F are a pair of plasmids in which a study of plasmid sensitivity to gyrase inhibition should be particularly illuminating since, despite the similarity in the organization of their principal plasmid replicons, the plasmids exhibit a marked difference in their sensitivity to gyrase alterations. Mutations in the *gyrB* gene that are thermosensitive for maintenance of F (and pBR322) do not diminish the stability of P1 at temperatures that lead to a rapid loss of F (Friedman *et al.*, 1984). Differences in the sensitivity of P1 and F to the curing agent acridine orange, an intercalating dye, have also been noted (Amati, 1965). A reinvestigation of these reported differences using miniplasmids would be desirable.

In the study of P1 plasmid replication, the exploitation of conditional host mutations in genes affecting DNA synthesis (other than *dnaA* and *gyrB*) has been limited to an unpublished survey of a few *dna* mutations undertaken in 1975 (Nainen, 1975; Scott and Vapnek, 1980). The results of this effort indicate that prophage replication is dependent on the products of genes *dnaB*, *dnaC*, and *dnaG*, although there is a preferential replication of P1 DNA compared to chromosomal DNA at the restrictive temperature in *dnaB* and *dnaG* mutants. In *dnaE*486 a stimulation of P1 DNA synthesis was observed at a temperature restrictive for chromosomal DNA synthesis; however, the P1 DNA made was not in covalently closed form. P1 (Stephens and Vapnek, cited in Scott and Vapnek, 1980) and miniP1 (*oriR*) plasmids (Chattoraj *et al.*, 1984) are replication-proficient in *polA* mutants.

S. Wickner (Wickner and Chattoraj, 1987) has developed an *in vitro* system that replicates P1 miniplasmid DNA and promises to resolve a

---

E. coli DNA gyrase consists of subunits determined by widely separated genes *gyrA* and *gyrB*. The A subunit participates directly in the passage of one DNA strand through another and, in the process, attaches covalently to the DNA. The B subunit bears an ATP binding site and participates in energy transduction (Gellert, 1981). To explain why gyrase inhibitors that act on the B subunit should be more effective curing agents than those that act on the A subunit, the following, readily testable hypothesis has been suggested by Hooper *et al.* (1984). These authors suggest that with either type of poison the initiation of plasmid replication is blocked, but since poisons of the A subunit additionally lead to DNA damage (Gellert, 1981), they trigger an SOS response. Cell division is inhibited, and plasmid loss is consequently blocked.

number of questions concerning the roles of the various participants. Replication is dependent on purified P1 RepA protein and a partially purified protein fraction from uninfected E. coli similar to that described by Fuller et al. (1981) for the replication of oriC plasmid DNA. The P1 system replicates supercoiled DNA containing the P1 plasmid origin region. Replication is preferentially from the origin region and in the direction away from the repA promoter. The requirement for DnaA protein confirms the deductions of Hansen and Yarmolinsky (1986) from their genetic experiments. An inhibition of replication by novobiocin indicates an involvement of DNA gyrase.

The in vitro replication of P1 miniplasmid is also blocked by rifampicin, an inhibitor of RNA polymerase. An RNA polymerase dependency has been noted in the in vitro replication of oriC plasmids, although the requirement can be removed under certain conditions, such as in the absence of the histonelike bacterial protein HU (Ogawa et al., 1985). In the case of F (and RK2), in vivo experiments suggest the possibility that the sigma 32 protein, required for induction of heat-shock proteins in E. coli, is directly required for plasmid replication (Wada et al., 1986). The significance of these loosely connected observations must await the further development of the comparative biochemistry of plasmid DNA replication.

A possible role for the dam methylase of E. coli in the replication of the plasmids of Fig. 9, as well as in the replication of E. coli and other enteric bacteria, is suggested by the presence of one or more GATC sequences in the ori regions. There are eight conserved GATC sequences in the oriC regions of five related gram-negative bacteria (Zyskind et al., 1983), five in the oriR of P1 (Abeles et al., 1984) and a single GATC sequence in the corresponding ori-2 of F (Murotsu et al., 1984). The dam methylase catalyzes the N^6 methylation of adenine residues in GATC sequences (reviewed by Marinus, 1984, 1987). There is evidence in the case of E. coli that unmethylated oriC DNA is inefficiently utilized in the initiation of replication (Hughes et al., 1984; Messer et al., 1985; Smith et al., 1985). On the basis of their experiments, Messer et al. (1985) propose that the rate at which GATC sequences within oriC are methylated determines the length of a refractory period between successive replications.*

The failure of attempts to lysogenize a dam mutant of E. coli with λ-P1:5R has been noted (M. Yarmolinsky, unpublished; Austin et al., 1985), and recently investigated in some detail (Abeles and Austin, 1987). The

---

*In the case of those plasmids that have been tested (Nordström, 1985b), including P1 (Abe, 1974), Meselson–Stahl density shift experiments show that the replication of an individual plasmid can occur much earlier or later than one bacterial generation after its formation. Although the selection of plasmids for replication is nearly random, the spread in replication times is limited on the low side. The refractory period for R1 has been calculated to be about 20% of a generation time (Nordström, 1983); for P1, it remains to be determined.

recent studies demonstrate that *oriR* of P1 cannot function in DNA adenine methylase deficient strains, even when this origin is present in a miniF-driven plasmid that is already established in a *dam* mutant. The dependency of miniP1 plasmid replication upon methylation has also been demonstrated in an *in vitro* replication system. These results suggest that postreplicational methylation may limit the interval between successive plasmid replications in P1. The finding that λ-miniF carrying *ori-2* as its sole functional origin can be maintained readily in a *dam* mutant of *E. coli* implies that there is a significant difference in the way similarly organized replicons with differing numbers of GATC sequences are controlled by their host.

5. Plasmid Participation in Control of Its Host

P1 and other low copy number plasmids coordinate their replication with the cell cycle of their bacterial host. The controls operate in both directions. The first indication that P1 participates in controlling its host came from studies of a phenomenon called indirect induction (Rosner *et al.*, 1968). Indirect induction is the induction of an SOS response in an undamaged cell by introduction of a damaged replicon. It was originally described as the induction of prophage λ by the conjugal transfer of F from an irradiated F$^+$ donor to an unirradiated λ lysogen (Borek and Ryan, 1958). The observation that those replicons that can cause indirect induction are low-copy-number plasmids that are faithfully maintained, P1 among them, led to the hypothesis that the mechanism responsible for indirect induction is also implicated in plasmid maintenance (Rosner *et al.*, 1968). A damaged plasmid that elicits SOS responses, such as the induction of DNA repair enzymes and the temporary inhibition of cell division, would be expected to increase the fidelity with which the plasmid is maintained.

Plasmid-encoded genes that contribute to the fidelity of plasmid maintenance and to the capacity to elicit indirect induction have been described in F (Ogura and Hiraga, 1983a; Mori *et al.*, 1984; Bex *et al.*, 1983; Karoui *et al.*, 1983; Miki *et al.*, 1984a,b; Bailone *et al.*, 1984; Brandenburger *et al.*, 1984; Sommer *et al.*, 1985; Jaffé *et al.*, 1985; Hiraga *et al.*, 1986). These genes were named *ccdA* and *ccdB* by Ogura and Hiraga (1983a). The term *ccd* was originally an acronym for "coupled cell division," but the initials now stand for "control of cell death." When replication or partition of a *ccd*$^+$ plasmid is impaired and plasmid-free segregants appear, these segregants undergo a few residual cell divisions, elongate, and die (Jaffé *et al.*, 1985; Hiraga *et al.*, 1986). The *ccdB* gene encodes a protein that leads to the loss of viability of the plasmid-free segregants. The *ccdB* gene product is normally prevented from acting by the *ccdA* gene product, but when the plasmid and the *ccdA* gene it carries are lost, the functioning of the *ccdB* gene product becomes uninhibited. The fidelity of plasmid maintenance is enhanced in the sense that cured cells are rendered unable to compete with those that retain the plasmid. This primary function of the *ccd* mechanism is not dependent on *recA* function.*

Certain other effects of *ccdB* expression, however, are attributed to secondary SOS responses that depend on a functional *recA* gene. Among such SOS responses is the induction of λ in a small proportion of a population of λ lysogens from which a *ccd*+ miniF is being lost as a result of impaired replication or partition (Mori *et al.*, 1984; cf. Bailone *et al.*, 1984, and Brandenburger *et al.*, 1984). Indirect induction by damaged P1, like indirect induction by damaged F, is quite possibly the consequence of the expression of a P1-determined *ccdB* analogue.†

Evidence of a more convincing nature than the analogous behavior of P1 and F, although still indirect, suggests that P1 does encode a *ccd* mechanism. MiniP1 replicons that include part of *Eco*RI-5 and all of the adjacent *Eco*RI-8 (rather than only a portion of it, as in λ-P1:5R) have been constructed and found to induce SOS functions (Capage and Scott, 1983). An *Eco*RI-5,8 plasmid with an additional pBR322 origin of replication failed to induce SOS functions. SOS induction could also be prevented by the presense of *Eco*RI-6 of P1, acting in *trans*. One interpretation of these results is that *Eco*RI-8 has a *ccdB*-like gene whose expression is restrained by two functions—one expressed from *Eco*RI-8 and gene dosage-dependent, and the other expressed from *Eco*RI-6. An alternative interpretation is that the expression of *ccd* functions from *Eco*RI-8 is elicited when the DNA is replicated from *oriR* and not when it is replicated from the pBR322 *ori*. If *ccd* genes do reside on *Eco*RI-8 of P1, then the organiza-

---

*The function of the *ccd* operon of F is not only in control of cell death. The third member of the operon, the D gene, appears to be a site-specific resolvase with a site of action at *ori*-1 (O'Connor *et al.*, 1986; Lane *et al.*, 1984). It is also a repressor of a promoter within *ori*-1 (H. E. D. Lane, personal communication). Moreover, the protein product of *ccdB* (G or F6), along with the multifunctional products of genes C (F3 or *pifC*—Miller and Malamy, 1983; Tanimoto and Iino, 1984) and E (F4 or rep(A)—Lane *et al.*, 1984), is required for replication from *ori*-1 (H. E. D. Lane, personal communication). We also note that the two *ccd* genes may each give rise to two polypeptides in ratios modified by the bacterial genetic background (Bex *et al.*, 1983). With this degree of complexity, it would not be surprising if F could exhibit the kind of discrimination we normally associate with intellect.

†The relation of DNA damage to SOS induction by F and P1 is complex. The expression of *ccdB* can cause SOS induction even in the absence of DNA damage, as when F is lost owing to a partition defect. SOS induction mediated by the *ccd* system of an undamaged replicon, the *gratuitous induction* of Brandenburger *et al.* (1984), may be responsible for a slight SOS induction detected following infection by undamaged P1 (D'Ari and Huisman, 1982). Damage to the infecting DNA augments the induction observed, provided the host permits replication to be initiated. It has been suggested that when replication initiates but aborts, the process generates allosteric effectors of the RecA protease that activate it for SOS induction (D'Ari and Huisman, 1982; Mori *et al.*, 1984). Indirect induction by UV-damaged P1 has been found to be largely inhibited by the simultaneous presence of extra copies of *incA* and *c4* (Yarmolinsky and Stevens, 1984). This result is possibly attributable to a failure to generate activators of RecA when replication is blocked at the initiation stage by *incA* at *oriR* and by *c4* repressor at *oriL*.

tional homology between P1 and F might extend beyond the border of *oriR* (see Fig. 10).

The hypothesis that P1 encodes a *ccd* mechanism leads to the expectation that cell growth will be inhibited if P1 plasmid replication is specifically blocked. This expectation appears to be justified. Lysogens of P1 mutants that segregate cured cells at abnormally high frequency include many elongated cells (Jaffé-Brachet and D'Ari, 1977). Support of a more direct nature comes from the growth characteristics of a *supD*ts *E. coli* carrying P1Cm with a suppressor-sensitive *repA* mutation. This lysogen exhibits a prompt reduction in growth rate upon shifting to temperatures nonpermissive for P1 replication from *oriR* (M. Yarmolinsky and S. Jafri, unpublished). Growth ceases if replication from the alternative *oriL* is additionally inhibited by the presence of excess *c1* or *c4* repressor. Moreover, Rts1, a plasmid that enjoys extensive DNA sequence homology with P1 in the region of the R replicon (Kamio *et al.*, 1984) and which is naturally thermosensitive for replication (Terawaki *et al.*, 1967), was found to interfere with the growth of its hosts at temperatures nonpermissive for Rts1 replication (Terawaki *et al.*, 1968; Terawaki and Rownd, 1972).

C. Plasmid Partition and Its Control

1. Organization and Control of the *par* Region

The partition region of P1 is a discrete DNA sequence of about 2.7 kb (Abeles *et al.*, 1985) that can function to stabilize associated *par*-defective replicons, including a plasmid (pBR322) that is normally not endowed with a *par* mechanism (Austin and Abeles, 1983a; Austin *et al.*, 1986). The rate of loss of P1Cm carrying the unsuppressed nonsense mutation *seg*-101 (a *par* mutation) is about 10% per generation in rich medium (Som *et al.*, 1981). Under similar conditions of rapid growth, the rates of loss of λ-P1:5R plasmids that carry a *par* defect are only slightly, if at all, higher (Austin and Abeles, 1983b).

The DNA sequence of the *par* region (Fig. 10) reveals the existence of two large open reading frames that could encode proteins of molecular weight 44 and 38 kD, both read in the same direction (Abeles *et al.*, 1985), toward the *cis*-acting centromerelike site designated *parS*. The protein product of *parA*, a gene that corresponds to the first open reading frame, is a protein of molecular weight 44 kD. Of two possible ATG start points, separated by 15 codons, the first is favored because the second lacks a suitable ribosome-binding site. Upstream of the. *parA* coding sequence lies a highly A+T-rich region containing a 20-bp imperfect palindrome that is flanked by the ribosome-binding sequence on one side and a unique potential *parA* promoter on the other side. The protein product of *parB*, a gene that corresponds to the second open reading frame, is a protein with an anomalously high M_r of 45 kD on SDS–PAGE. The stop codon of parA and the only suitable ATG start codon of parB are separated by 16 bp. A probable promoter for parB lies just within the terminus of the parA open reading frame, although it is likely that parB can also be transcribed from the parA promoter. Both parA and parB have been shown to be essential for the function of the partition mechanism (Austin and Abeles, 1983b; Friedman et al., 1986). Both proteins have been shown to be effective when supplied in trans. The ParB protein appears to be produced in larger amounts than the ParA protein. This may be due in part to the presence of two promoters with about equal activity from which parB is transcribed and only one promoter from which parA is transcribed, and in part to a difference in translation efficiency (S. Austin, personal communication). Progress is being made on the purification of both proteins.

The parAB operon appears to be regulated by ParA protein and, more effectively, by ParA and ParB proteins in concert (Friedman and Austin, 1988). The target of the parA and parB proteins involved in autoregulation is presumably an A+T-rich palindrome that is situated adjacent to the putative parA promoter. A region that encompasses this palindrome has been cloned into pBR322. It fails to express incompatibility under conditions that permit a similarly cloned incB region to be effective (Abeles et al., 1985). However, the autoregulatory circuit does appear to be responsible for an unanticipated source of incompatibility to which we alluded earlier in this section (part A.3). Plasmids that contain an isolated parA gene or a major portion of the parA open reading frame together with its promoter efficiently displace a resident λ-P1:5R. The specificity of this incompatibility is identical to that of a cloned incB element. Abeles et al. (1985) propose that this incompatibility is due to a gross imbalance in the ratio of ParA to ParB that interferes with the proper functioning of these proteins. The amber mutation parA17 eliminates this incompatibility.

A more dramatic interference with partitioning is exerted by excess ParB protein on replicons that bear parS (Funnell, 1988; compare SopB of F: Kusukawa et al., 1987). The defect is even more severe than that due to complete elimination of par function. The binding of excess ParB to parS appears to reduce markedly the number of partitionable units, presumably by aggregation, such that the rate of plasmid loss exceeds what could be expected for individual plasmids distributed at random. Neither plasmid copy number nor host cell growth rate is significantly affected. These results suggest that par regulation is critical for the participation of parB in the appropriate pairing of plasmids.

The location of incB was shown to coincide approximately with location of parS, the cis-acting recognition site required for plasmid partitioning (Austin and Abeles, 1983b). Plasmids that carry a 175 bp restriction fragment derived from parS exert incompatibility against par+ miniP1 plasmids (Abeles et al., 1985, see erratum). The DNA of this fragment is 63% A+T and contains a 13 bp perfect inverted repeat followed by a stretch of C and T residues and the sequence AAAAAAATAAAAAAA.

In order to determine which portion of the *parS* region is the site required in *cis* for partitioning, fragments of the *parS* region were cloned into pBR322, and their potential for stabilizing a low-copy-number mini-P1 or miniF replicon was determined in a "pickup" assay. In this assay, cells were first cotransformed with pBR322 carrying the plasmid replication functions of P1 or F and with a second pBR322 carrying the *parS* fragment to be tested. Recombination between pBR322 sequences produces a small proportion of mixed dimers. The DNA was isolated and used to transform a *polA* strain, selecting for the two different drug resistance markers carried by the two parental plasmids. As pBR322 origins of replication are nonfunctional in a *polA* strain, only mixed dimers can transform, and replication proceeds from the low-copy-number origin supplied. The *polA* strain also carried a compatible plasmid that supplied the *parA* and *parB* proteins in *trans* (Martin *et al.*, 1987). Stabilization of the mixed-dimer plasmid was achieved with fragments of the *parS* region 109 and 112 bp in length that have in common an interval of only 49 bp. The striking feature of this 49-bp region is the presence of the 13-bp perfect inverted repeat, which is part of a 20-bp imperfect inverted repeat. Perhaps a plasmid segregation apparatus in which Par proteins directly or indirectly participate attaches at this site to bring about plasmid partitioning. In the case of F, the analogue of P1's *parB*, a protein of molecular weight 37 kD, has been shown to bind to the analogue of P1's *parS* (Hayakawa *et al.*, 1985).

An astonishing result obtained by Martin *et al.* (1987) is that the specificity of *parS* incompatibility can be altered by a deletion within *parS* that does not eliminate, or even reduce, *parS* functionality. The simplest interpretation of this result is that *parS* regions of different size, in conjunction with the appropriate partitioning proteins, can form homologous but not heterologous pairs, each of which is an adequate substrate for partitioning.

2. Mode of Action and Control of the *par* Region

How does the partitioning apparatus operate? In each cell cycle the *par* region promotes the equipartitioning of at least one and possibly all of the available P1 plasmid pairs in the cell. It can be assumed that partitioning involves attachment of plasmids undergoing partitioning to cellular structures that themselves segregate into separate daughter cells. The attachment is unlikely to be a permanent one, because of the ease with which a resident plasmid is displaced by an incompatible plasmid or an appropriate incompatibility determinant cloned from such a plasmid. The attachment site could possibly be *par* incompatibility group-specific, but the several known *par* specificities would need to be separately and differently accommodated.

Taking these considerations into account, Austin and Abeles (1983b) proposed an alternative in which partitioning involves cycles of plasmid attachment and detachment to and from one another and to and from cell envelope sites as follows. The Par proteins mediate the parS-specific formation of daughter plasmid pairs that each attach to any one of several equivalent membrane sites. Each of these sites subsequently splits and segregates to separate cells, in the process detaching the paired plasmids from each other. Detachment of the separated plasmids from their membrane sites could be triggered by plasmid replication or by an alteration of the membrane site in the course of the cell cycle. In this proposal the par-specific reaction is a pairing of daughter plasmids that *precedes* membrane attachment. The existence of this pairing reaction should be testable *in vitro* when pure Par proteins are available.

As originally formulated, the proposal of Austin and Abeles accounted for the ability of an incompatibility determinant such as *incB* of P1 to displace a resident plasmid by the formation of a heterologous plasmid pair, of which the components would be partitioned to separate cells, leaving the unpaired daughter adrift. To account for the observation by Austin (1984) that plasmids carrying two widely separated centromere analogues are stable and faithfully partitioned (unlike dicentric chromosomes of yeast or maize), it appeared necessary to make the additional assumption that duplicate *parS* regions on the same plasmid do not compete with each other. It was therefore proposed that plasmid DNA is compacted in such a way as to make it sterically impossible for two partitioning sites on the same plasmid, even if 50-kb pairs apart, to be simultaneously accessible to the membrane sites.

The *incB* element and its analogue in F, *incD*, are possibly able to confer incompatibility by more than one means. In addition to displacing a resident plasmid by pairing with it and thus blocking the association with its normal partner, these incompatibility elements might be capable of acting at a distance from the segregating plasmids. When in sufficiently high concentration the *inc* elements might sequester the specific factors (presumably the corresponding Par proteins themselves) that are unique to one or the other type of partitioning system (Austin, 1984). This elaboration of the partitioning model was introduced to explain the observation that low-copy-number plasmid composites carrying *par* regions of both P1 and F were found to be insensitive to pBR322 clones of either *incB* of P1 or *incD* of F. That observation cannot be explained by a model in which the *inc* elements can act only by forming inappropriate partnerships that deactivate any alternative partitioning site on the same plasmid. Were that the case, each double *par* composite when confronted with an excess of a cloned *inc* element would tend to pair with that element and thereafter behave as if it possessed only the partitioning site of the challenging *inc* element. It would therefore be sensitive to the challenge with either *inc* element, in contradiction to the results of experiment. In the modified model, the *inc* determinants would render the homologous partitioning sites on the composite useless without making the alternative partitioning sites inaccessible. The composite would then be free to pair with its partner via the remaining partitioning site, allowing for equipartitioning of the members of the pair.*

Whether or not the details of the tentative partitioning model described above are correct, the essential features of the partitioning apparatus as described initially by Austin and Abeles (1983) for P1 and by Ogura and Hiraga (1983b) for F are both simple and similar. In each case, a pair of plasmid-determined proteins, in concert with proteins of the host, act following plasmid replication to promote ordered movement of plasmid copies to daughter cells. A centromerelike site has been defined for each plasmid. The structure of the entire complex—or "partisome," as it has been called (Hiraga et al., 1985)—remains to be determined, as does the precise sequence of events by which partitioning is accomplished.

3. A Role for Site-Specific Recombination

The plasmid λ-P1:5R is at least an order of magnitude more stable in recA than in rec+ bacteria (Austin et al., 1981). This observation is not surprising. Under rec+ conditions, homologous recombination between sister plasmids should occasionally generate plasmid dimers, thereby effectively reducing the number of partitionable units in the bacterial population. For a plasmid of low copy number, reduction in the number of partitionable units greatly increases the risk of plasmid loss in the course of cell division.

The remarkable stability of P1 is no greater in recA than in rec+ bacteria (Austin et al., 1981). This result suggests that homologous recombination is prevented from adversely affecting the partitioning of the intact prophage. How is the adverse effect of homologous recombination overcome? There is no evidence for a P1-determined inhibitor of recombination, but we have already described here a highly efficient, site-specific recombinase, effective in the prophage state, that can be expected to keep P1 dimers and monomers in rapid equilibrium.

A role for the cre-lox site-specific recombination system in preventing homologous recombination from destabilizing P1 has been directly demonstrated (Sternberg et al., 1980; Austin et al., 1981). A cre mutation

*This explanation would appear to create another dilemma because of evidence from parA-lacZ protein fusions that the low level transcription of parA and parB from the parA promoter is subject to a tight autoregulation by the products of both genes (Austin and Abeles, 1985). Sequestration and autoregulation are incompatible, as already discussed (Section VII.B.2), without further ad hoc assumptions. The dilemma is avoided simply by assuming that autoregulation cannot cope with the high rate at which high copy numbers of incB sequester Par proteins. This assumption appears justified in view of evidence that high-copy-number clones of the incB region exert a stronger incompatibility than clones that also carry parA and parB (A. L. Abeles, personal communication; Austin et al., 1986). It is not unreasonable that supplies of additional components of the partitioning machinery become exhausted as the concentration of a competitor for partitioning is increased.

decreases the stability of P1 about 40-fold in a rec^+ host and is without significant effect in a $recA$ host. Conversely, the stability of λ-P1:5R in a rec^+ host can be improved by about as many fold if the plasmid carries the recombination site $loxP$ and is supplied in $trans$ with a source of the Cre recombinase.

Plasmid specified site-specific recombination systems that play a role in the stabilization of other plasmids have since been described (CloDF13: Hakkaart et al., 1984; ColE1: Summers and Sherratt, 1984; pSC101: Tucker et al., 1984; F: S. Austin, personal communication, and see O'Connor et al., 1986; Lane et al., 1984). In each case, the harm to plasmid stability done by a necessarily sluggish generalized recombination system can be overcome by a recombination system that is efficient, but to avoid widespread perturbations of both host and plasmid chromosomes that would result if generalized recombination were augmented, the efficient recombination is confined to a plasmid-specific site.

4. Host Participation in Plasmid Partitioning

Attempts to learn how the host participates in plasmid partitioning were first directed toward localizing plasmid DNA within the cell and examining its possible association with chromosomal DNA. Early studies suggested that P1 plasmid does not freely diffuse to all parts of the cytoplasm. A minicell-producing E. coli lysogenic for P1 generates, by septum formation at its poles, minicells that are free of P1 (L. R. Kass, cited in Kass and Yarmolinsky, 1970). To a slight extent, F does segregate into minicells (Kass and Yarmolinsky, 1970). Studies of cosedimentation of various plasmids with rapidly sedimenting bacterial nucleoids suggested that P1, F, and other low-copy-number plasmids (which had been noted not to segregate freely into minicells) are largely found in association with the nucleoid. The specificity of this association was indicated by its lack of dependence upon plasmid size or the presence of bacterial membrane in the preparations (Kline et al., 1976). Moreover, a study in which the localization of P1 prophage was determined in intact bacteria by autoradiography showed that P1 prophage that had been damaged by UV irradiation assumes a central position in infected cells (MacQueen and Donachie, 1977).

Only recently, with the identification and characterization of plasmid partition functions, have the tools required to identify host components of the partitioning mechanism become available. Protein complexes with DNA of the partitioning site have been detected in miniF. The proteins present, in addition to one of the Par proteins (SopB), are of molecular weight 75 and 33 kD. Both are probably of E. coli origin (Hayakawa et al., 1985). Progress has also been made in identifying bacterial genes specifically implicated in the functioning of the F par region (Hiraga et al., 1985; Mori et al., 1986; Hiraga, 1986). Studies of a similar nature are being initiated with P1. We anticipate that bacterial functions involved in partitioning P1 and other similarly organized plasmids will turn out to play important and interesting roles in cell division.

VIII. VEGETATIVE GROWTH

In this section we first review what we know about the location and organization of vegetative genes on the P1 chromosome and then discuss four aspects of the P1 vegetative growth cycle: vegetative replication, the timing of gene expression, phage morphogenesis, and *pac* cleavage.

A. Organization of the Genes for Vegetative Functions

1. Early Genes

The limited evidence available indicates that the genes determining early functions are located at four regions on the P1 map. Just clockwise of *lox*P, at about 2.5 map units, is the inducible gene *ref*, which has been discussed in sections IV.C, VI.C, and VI.D. The region between map coordinates 53 and 55 contains what is believed to be the origin of lytic replication and putative replication gene(s) that are under control of *c1* and *c4* immunity repressors (see Sections VII.B.3 and VIII.B.2). The region between map coordinates 67 and 77 is transcribed soon after induction (Mural, 1978, reviewed in Sternberg and Hoess, 1983) and probably contains an analogue of the bacterial *dam* gene as well as *ban*, an analogue of the bacterial *dnaB* gene (D'Ari *et al.*, 1975; Ogawa, 1975). The *ban* gene is under direct *c1* control. Only in the absence of *c1* repressor is the *ban* gene expressed by *Eco*RI-3 of P1 (Sternberg *et al.*, 1978, 1979; Sternberg, 1979). The *ban* gene is located centrally within that fragment at map coordinates 74.8 to 76.5 (Heisig *et al.*, 1987). The P1 *dam* gene, discussed in parts C2 and E3 of this section, has been mapped to the region of the *Eco*RI-3,8 junction (J. Coulby and N. Sternberg, unpublished). The *dam* gene is repressed in a lysogen and is rapidly induced after prophage induction, reaching maximal levels within 15–20 min. Finally, the region encompassing *Eco*RI-6 (map coordinates 87–95) is also transcribed early (Mural, 1978). At one end of this region (map coordinates 94–96) lie genes 9 and 10. The gene 10 product is necessary for the expression of most, if not all, P1 late genes (Walker and Walker, 1980, 1983). The gene 9 product is the protein that cleaves P1 DNA at the packaging (*pac*) site (Sternberg and Coulby, 1987a). Although packaging is a late function, gene 9 must be expressed early in the vegetative cycle, since *pac* cleavage can be detected about 15 min after phage infection at 37°C and lags only slightly behind the onset of phage DNA synthesis.

2. Late Genes

The late genes of P1 are those involved in phage tail and head morphogenesis, in cell lysis (Walker and Walker, 1980, 1983), and in certain functions not essential for phage production. These comprise most of the remainder of the P1 chromosome. The organization of morphogenesis genes in P1 is similar to that of T-even phages in that clusters of head and tail genes are scattered throughout the viral chromosome and are separated from each other by clusters of genes not involved in morphogenesis. This differs from the organization exhibited by phages such as λ, P22, and P2, in which all the head genes are present in a single cluster and all the tail genes are present in an adjacent cluster.

The tail morphogenesis genes are located in three separate regions of the P1 genome, at approximate map positions 34–50, 59–61, and 77–86. Within these regions, genes invoved in similar functions are clustered. The first region contains the alternative pairs of tail fiber genes (see Section III.C) and at least four tail tube or baseplate genes (3, 16, 20). The second region contains one tail sheath gene (22) and one baseplate or tail tube gene (21). The third region contains genes involved in tail length determination (6), tail stability (7, 24, 25), and baseplate or tail tube production (5, 26).

The known head genes of P1 have been localized to four regions of the map: 10–20, 55–61, 78–93, and 96–98. The first region includes genes 4, 13, and 14 (Walker and Walker, 1983; Razza et al., 1980). Gene 4 mutants produce normal phage tails and separate unfilled heads and cleave *pac* normally (Sternberg and Coulby, 1987a). Gene 4 has been renamed *pro*, as it is probably the structural gene for a maturation protease (Streiff et al., 1987) (see part D.1). Mutations in the second and third head regions (in genes 15, 23, and 8) result in tails with head-tail connectors and either no heads (23) or empty heads (8). The fourth head region, as noted above, contains gene 9, which encodes the *pac* cleavage protein (Sternberg and Coulby, 1987a).

Besides head and tail genes, P1 also encodes a protein whose absence results in the production of noninfectious particles that appear morphologically normal. The gene for this protein (gene 1) is probably completely contained within the 0.92 kb *Eco*RI-19 fragment (map coordinates 3–4), since the wild-type alleles of six amber mutations with this phenotype can be rescued from *Eco*RI-19 (cited in Walker and Walker, 1983). The function of the gene 1 protein is not clear.

There are at least three host cell lysis functions. Gene 17 mutants are located just counterclockwise from the IS*1* element (map coordinate 20–24) and are completely defective for cell lysis. Since the production of phages following infection with gene 17 mutants ceases at about the same time as one would normally expect the cells to lyse, and since the cells are not lysed by chloroform treatment (Walker and Walker, 1980), gene 17 would appear to be an analogue of the λ *R* gene. It probably encodes an endolysin. A second lysis function, encoded by the phage *lyd* gene (map coordinate 30–32), is analogous to the λ *S* gene (Iida and Arber, 1977). Lyd⁻ mutants make tiny plaques and exhibit a delayed lysis phenotype. Infected cells that have not lysed can be lysed with chloroform and continue to produce phage beyond the normal lysis time. The third gene involved in phage lysis (gene 2) is located just clockwise of *lyd* and appears to be a regulator of the time of lysis. Gene 2 mutants lyse cells prematurely (Walker and Walker, 1980) and therefore produce very few phage particles (Walker and Walker, 1983).

The region of the P1 chromosome between map coordinates 25 and 31 contains an operon that is dispensible for phage growth (Iida and Arber, 1977). It is probably expressed from a promoter located near map unit 32 that becomes active only 30 min after prophage induction (C. Senstag and W. Arber, personal communication). The operon contains the *lyd* gene, the defense against restriction *darA* gene (see Section V.D) and one or more genes involved in viral architecture determination (the *vad* gene) and in determination of the frequency of generalized transduction (see Sections IX.B.1.c and IX.B.2.b).

B. Vegetative Replication

1. The Nature of Vegetative Replication

P1 DNA starts to be synthesized by the fifth minute after infection of *E. coli* by P1 at 37°C. The rate of synthesis increases to a level of about 95% that of total bacterial DNA synthesis at about 30 min after infection (Segev *et al.*, 1980). Electron microscopy of replication intermediates from lytically infected cells indicates that theta forms and sigma forms are present in equal numbers during the early stages of replication (15–30 min after infection) but that sigma forms dominate the scene later in infection (Cohen, 1983). By an examination of sigma forms generated in cells that carried replicating P1 and P1 d*lac* phages, Bornhoeft and Stodolsky (1981) were able to show that concatemeric sigma forms do not contain phages of both types in the same molecule. This implies that sigma forms are rarely, if ever, generated by recombination but rather arise from rolling circle replication. These results suggest that the vegetative growth of P1, like that of λ, entails at least two modes of DNA replication.

Where is the origin of P1 vegetative replication? Is the same origin used for theta and sigma replication? What are the relevant replication functions? How is lytic replication controlled? Only recently have these elementary questions begun to be answered. However, it has been apparent for some time that the lytic origins are not identical to or even functionally connected with the origin of plasmid replication. The initial evidence for this conclusion came from Tn9-generated P1Cm deletion mutants obtained as chloramphenicol-resistant survivors of thermal induction. The deletions, covering about 30% of the P1 genome, eliminate immunity (specifically *immI*) and vegetative replication but leave plasmid replication unaffected (Austin *et al.*, 1978). Conversely, interference with plasmid replication (by introduction of multicopy *incA* or by incorporation of a *repA* mutation into P1) leaves vegetative replication unaffected (M. Yarmolinsky, unpublished). Moreover, host mutations affect vegetative and plasmid replication in different ways. Whereas *recA* mutations markedly reduce sigma vegetative replication, they do not diminish the efficacy of plasmid replication (Rosner, 1972; see also part B.3 of this section). Conversely, a *dnaA* null mutation blocks P1 plasmid replication, whereas vegetative P1 replication is normal (Hansen and Yarmolinsky, 1986). It can be concluded that the plasmid and vegetative replication systems of P1 are independent of each other.

The genes involved in lytic replication of P1 have been singularly elusive. From among more than 150 conditional mutations affecting plaque-forming capacity (Walker and Walker, 1975, 1976; Razza *et al..* 1980), no single mutation that blocks vegetative replication could be identified. Possible reasons for this failure include (1) the small size of the target, (2) the existence of alternative replicons, and (3) an excessively stringent screening procedure. What has been learned about the primary lytic replicon lends some credence to each of these possibilities.

2. The Nature of the Lytic Replicon(s)

For reasons that we state below, we believe that the primary vegetative replicon is the L replicon, which we briefly characterized in section VII.B.3. The study of this replicon was at first hampered by reliance upon the initially isolated λ clones that contained it. Several were defective λ-P1:L phages. They lysogenized unstably with very low frequency and could not be readily recovered from lysogens.

Compelling evidence that L is the primary lytic replicon comes from characterization of kanamycin-resistant survivors of thermal induction of *E. coli* carrying a P1 prophage with a Tn5 transposon inserted into *Eco*RI-17 at about map coordinate 54.5 (Windle, 1986). Whereas chloramphenicol-resistant survivors of *E. coli* carrying P1Cmc1.100 (Austin *et al.*, 1978) are mutants in which deletions extend clockwise from the Tn9 transposon at map unit 24 into or somewhat beyond *immI*, the deletions found among the kanamycin-resistant survivors extend counterclockwise from the Tn5 transposon (E. B. Hansen, personal communication). Some are very short, leaving *immI* intact. Two strains without detectable deletions were studied further and were shown to have acquired IS50 insertions within *Eco*RI-14. (Inverted IS50 elements flank the kanamycin resistance determinant of Tn5 and are themselves capable of independent transposition.) The IS50 insertions in *Eco*RI-14 are in slightly different positions and in different orientations, the nearest being separated from the Tn5 by only 1.8 kb. When these novel P1 lysogens were heated to

42°C, immunity was lost, but P1 DNA synthesis was maintained for at least 50 min in the same proportion to bacterial DNA synthesis as before the temperature shift and the cells remained viable. Similar properties are exhibited by P1 phage containing Tn5 insertions at four different positions within the 400 bp of EcoRI-14 proximal to the small fragments EcoRI-21,25 (N. Sternberg, unpublished). We conclude that the capacity to survive thermal induction in the mutant P1 lysogens is due to a marked inhibition of vegetative DNA synthesis.

The L replicon was proposed to be activated by a transcript from the cl repressor-sensitive promoter (P53) situated in EcoRI-14 (Sternberg and Hoess, 1983; Sternberg et al., 1986; G. Cohen, personal communication) since multicopy replication from λ-P1:L phage containing EcoR-14 is inhibited by the cl repressor, whereas multicopy replication from λ-P1:L phage lacking that fragment is not. Presumably, the IS50 and Tn5 insertions in EcoRI-14 prevent this activation. Direct inactivation of L is excluded because L replicon activity does not require the presence of EcoRI-14 DNA. The characterization of L-driven miniplasmids recently derived by E. B. Hansen from P1 EcoRI-17::Tn5 indicate that the only region common to all L replicon-driven plasmids is a 1.1-kb region that spans the EcoRI-17,21 junction within which lies an open reading frame of about 198 codons in clockwise orientation.* This open reading frame encodes a protein that is likely to be essential for oriL functioning. The gene is designated repL (E. B. Hansen; G. Cohen and N. Sternberg, personal communications). In the same operon as repL and immediately upstream of repL is an additional open reading frame of 270 codons that is appropriate in size to encode the product of rebA, the 27.5 kD protein recognized by Heilmann et al. (1980a) (see p. 349). Sequence analysis indicates that the 27 kD protein is similar at its C-terminal end to the 40 kD protein whose gene largely fills the interval between Op53 and c4. Expression of the smaller gene is lethal to E. coli and so is expression of the larger gene for recA but not recA+ bacteria. Neither gene is essential for repL function, but they are probably both involved in the lysis/lysogeny decision. The DNA common to each of the L replicons is unable to replicate without the addition of an adjacent DNA segment, e.g., part of EcoRI-14 or of EcoRI-9, that appears necessary to furnish a promoter (E. B. Hansen; G. Cohen and N. Sternberg, personal communications). The susceptibility of the L replicon to regulation by immunity repressors depends on the nature of these adjacent sequences.

P1 normally relies exclusively on the L replicon to assure vegetative replication. This conclusion derives from the lack of induced P1 DNA

---

*Among the apparently L-driven λ-P1 hybrids described by Sternberg and Austin (1983) is one (λ-P1:5L-3) that carries EcoRI-5 of P1 and contiguous DNA counterclockwise to it on the P1 map, terminating just short of the unique Bg1II site in EcoRI-17. DNA deemed essential to L replicon function (as determined by successive deletions into a much simpler L-driven plasmid) is absent. The nature of the replicon that drives λ-P1:5L-3 is currently in question.

replication during the first 50 min following the induction of L-defective P1 prophages (G. Cohen and N. Sternberg, E. B. Hansen, personal communications). Nevertheless, prolonged thermal induction of L-defective P1Cmc1.100 prophage produces a considerable yield of phage particles that can be titrated, not as plaque-formers, but as particles that are able to transduce chloramphenicol resistance. Restriction analysis of the DNA in these particles shows that the *pac*-containing fragment is present at less than stoichiometric amounts, indicating that concatemeric DNA is the substrate for packaging, as it is normally (E. B. Hansen, personal communication).

The possibility that P1 possesses additional replicons is not entirely excluded, although their functioning in *E. coli* must be normally masked. We noted in Section VII.B.3 the report by Froehlich *et al.* (1986) of a replicon distant from both *oriL* and *oriR* on the P1 map. That replicon can be used to maintain a plasmid about half the size of its parental plasmid (a P1Cm-P7 hybrid) at about eight times its copy number. The replicon has not been precisely localized, but the limiting map coordinates are from 95 clockwise to 24 (B. Froehlich and J. Scott, personal communication).

3. Host Functions and Their P1-Determined Analogues in Vegetative Replication

Among the early studies of P1 vegetative replication were attempts to determine which of the genes of *E. coli* that are directly involved in bacterial DNA replication are also involved in P1 vegetative replication. P1 not only replicates in a *dnaB*ts mutant of *E. coli* at a temperature nonpermissive for *E. coli* replication (Lanka and Schuster, 1970) but, in the process, permits *E. coli* DNA synthesis to recover transiently (Beyersmann and Schuster, 1971). This result was explained with the discovery that P1 determines a *dnaB* analogue that can complement a *dnaB* defect (D'Ari *et al.*, 1975; Ogawa, 1975). The gene, *ban*, is normally repressed. However, in mutants such as P1*bac*-1 (D'Ari *et al.*, 1975), *ban* expression is largely *c*1-insensitive, but it appears to depend on the functioning of the distant *bof* gene (Touati-Schwartz, 1979b) (here subsumed under *imm*T; see Section VI.C).

Genetic experiments (D'Ari *et al.*, 1975; Touati-Schwartz, 1979a) and subsequently biochemical experiments (Lanka *et al.*, 1978b; Edelbluth *et al.*, 1979) showed that Ban protein can form active, thermally stabilized heteromultimers with the DnaB protein of *dnaB*ts *E. coli* mutants. Negative complementation between a *ban* and a *dnaB* mutant has also been noted (D'Ari *et al.*, 1975). Ban protein also appears capable of substituting completely for DnaB protein in the replication of *E. coli*, since the presence of P1 *bac*-1 prophage permits construction of an otherwise nonviable *E. coli* bearing the unsuppressed amber mutation *dnaB*266 (D'Ari *et al.*, 1975). Such constructs are cryosensitive. Cryoresistant derivatives carry a mutation (*crr*) that is tightly linked to *bac*-1 (Touati-Schwartz, 1978). The *crr* mutants markedly overproduce the 56-kD Ban protein (Reeve *et al.*, 1980). The properties of Ban protein and interactions with DnaB protein are the subject of several additional biochemical studies (Schuster *et al.*, 1975, 1978; Lanka *et al.*, 1978a,b,c; Edelbluth *et al.*, 1979). The physiological advantage that the *ban* gene confers on P1 is not apparent.

Another *E. coli* gene involved in bacterial DNA synthesis for which P1 encodes an analogue is the *ssbA* gene (Johnson, 1982). Whereas expression of the P1 *ban* gene is activated (in P1 *bac*-1) by the product of *imm*T gene *bof/lxc*, the same product appears to act as a repressor of the P1 *ssbA* analogue (see Section VI.C). In possessing an *ssbA*-complementing gene, P1 superficially resembles a number of conjugative plasmids (Golub and Low, 1985), but hybridization studies suggest that the P1 *ssbA*-analogue is unrelated to the others (E. I. Golub, personal communication). The physiological significance of this seemingly redundant function is unknown.

Several studies have indicated that P1 can dispense with *dnaA* function for vegetative replication (e.g., Lanka and Schuster, 1970; Beyersmann and Schuster, 1971; Abe, 1974; Nainen, 1975; Hay and Cohen, 1983), but only recently has the validity of this conclusion been affirmed with a *dnaA* null mutant (Hansen and Yarmolinsky, 1986). There is no indication that P1 encodes a *dnaA* analogue. In contrast to *dnaA* and *dnaB* functions, the bacterial *dnaC* function appears to be essential for P1 vegetative replication (Nainen, 1975; Hay and Cohen, 1983). This was shown to be true of both the initiation and elongation components of the *dnaC* function.

Measurements of P1 DNA synthesis following infection of sensitive cells, or following induction of lysogens, indicate that genes involved in bacterial DNA elongation—*dnaE, dnaZ* (which determine subunits of polymerase III holoenzyme), and *dnaG* (which determines DNA primase)—are necessary for productive vegetative replication (Nainen, 1975; Hay and Cohen, 1983). Elongation-deficient polymerase I mutations appear not to affect replication of P1 as a plasmid or as a phage (D. K. Chattoraj, personal communication).

We discussed the possible role of adenine methylation by the products of the bacterial *dam* gene and its P1-determined analogue in the control of prophage replication in Section VII.B.4 and in the regulation of vegetative functions in Sections VIII.C.2 and VIII.C.3. Measurement of DNA synthesis in vegetative growth reveals that functioning of the host *dam* gene is not essential. Whether this result reflects the absence of a requirement for adenine methylation for vegetative replication or the synthesis of a P1-determined *dam* analogue has not yet been determined.

In addition to bacterial replication functions, bacterial recombination functions also play an important role in P1 vegetative replication. It has long been known that P1 grows poorly in *recA* bacteria (Rosner, 1972). As previously discussed, part of this defect is due to the failure to cyclize much of the P1 DNA injected into these bacteria. However, this cannot account for the entire defect, since the yield of phage following induction of a *recA* lysogen is only 5–10% that obtained following induction of a *recA*$^+$ lysogen. To understand this defect, the nature and extent of vegetative replication were studied in both *recA*$^+$ and *recA* bacteria (Segev *et al.*, 1980; Segev and Cohen, 1981; Cohen, 1983; Bornhoeft and Stodolsky, 1981). Closed circular DNA is produced at about the same rate in a *recA* host as it is in a *recA*$^+$ host during the first 15–30 min after P1 infection. Its accumulation then abruptly ceases in the *recA*$^+$ host, whereas it continues in the *recA* host (Segev and Cohen, 1981). Since total P1 DNA synthesis is about 10 times higher in a *recA*$^+$ host than it is in a *recA* host at late times during P1 infection (Segev and Cohen, 1981), the disappearance of circular forms in the *recA*$^+$ host must reflect their conversion to other replicative forms, such as sigma forms, at this time in the phage cycle (Cohen, 1983). In *recA* bacteria, circular forms continue to accumulate, presumably because these cells are blocked in the conversion of theta to sigma forms. The electron microscopic studies of Cohen (1983) indicate that *recA* cells have no more than a tenth as many sigma forms as do *recA*$^+$ cells. The role of RecA in the initiation and/or propagation of sigma structures is still unclear. Moreover, there is no good evidence for a precursor-product relationship between theta and sigma forms during the P1 life cycle, since these two replication intermediates are present in nearly equal amounts at early times after infection of *recA*$^+$ cells (Cohen, 1983).

C. Regulation and Timing of Gene Expression

1. Expression of Vegetative Functions in the Prophage State

Regions of the P1 genome concerned with vegetative growth are expressed in the prophage state. The transcription studies of Mural (1978) (summarized in Sternberg and Hoess, 1983) indicate that as much as 80% of the P1 chromosome is transcribed in lysogens, suggesting that vegetative promoters are not completely repressed in the prophage. Support for this contention comes from studies on the regulation of P1 promoters fused to a promoterless *galK* gene (N. Sternberg, unpublished). Two *Eco*RI fragments (*Eco*RI-6 and *Eco*RI-17) and one *Bam*HI fragment (*Bam*HI-10) were found to cause a *galK* expression that appears to be insensitive to prophage repressors; i.e., *galK* was expressed even in a lysogen of a multicopy P1 mutant. The direction of P1 transcription was determined for the two *Eco*RI fragments and is clockwise on the circular P1 map (Fig. 2A). In both cases, transcription is directed into a region of the phage genome containing vegetative functions: transcription from *Eco*RI-17 is toward the adjacent tail genes 21 and 22, and transcription from *Eco*RI-6 is toward the early genes 9 and 10. The direction of transcription from

*Bam*HI-10 was not determined, but it seems likely that the *Bam*HI-10 promoter is involved in the synthesis of tail proteins, since the region contains only tail genes. Presumably other promoters of P1 are similar to these three. It is not known whether the proteins from these transcripts are translated.

The above considerations indicate that the repression strategy for P1 vegetative functions may be quite different from that employed by phages such as λ, P22, P2, or Mu. In the latter phages, repression is primarily regulated at transcription, with the repressor acting to block the action of one or two critical promoters (often closely linked to the repressor gene) that control much of the transcription capacity of the virus. Accordingly, very little of the prophage genome is transcribed. While some P1 genes are repressed in the prophage state, examples being *ref, ban,* and *ant* (see Section VI.D.1), several are not. Moreover, there is no evidence to suggest the existence of master control promoters adjacent to the P1 *c*1 gene. Rather it would appear that the P1 chromosome is composed of independent blocks of genes, of which some are repressed and others are constitutive in lysogens.

2. Replicative Induction of Early Vegetative Functions: A Model

The degree to which vegetative functions are expressed in the prophage state and the sensitivity of certain P1 promoters to methylation lead us to propose the following hypothesis to explain how some (perhaps most) early vegetative functions become expressed following infection or induction. In the absence of *c*1 repressor, the efficient P53 promoter in *Eco*RI-14 that faces *oriL* initiates transcription that brings about lytic replication of the P1 genome. The expression of genes necessary for vegetative growth is stimulated in two ways—first by the increased gene dosage, second by the transient increase in the strength of those promoters that, like the *cre* promoter (see Section IV.B.1), are most active in the absence of Dam-dependent methylation. When the genes that specify early vegetative functions become no longer necessary for phage growth, expression of those genes can be shut off by methylation. Shutoff is presumably accelerated midway through the vegetative growth cycle, when the P1-encoded *dam* methylase is made in significant amounts (J. Coulby and N. Sternberg, unpublished). The survival of thermally induced lysogens in which a transposon-generated deletion has removed the region of *oriL* (Austin *et al.,* 1978; E. B. Hansen, personal communication) or in which *oriL* is separated from its activating promoter in *Eco*RI-14 by an IS*50* or Tn*5* insertion (E. B. Hansen, personal communication; N. Sternberg, unpublished) is consistent with the idea that the expression of early vegetative functions, including those that lead to the eventual expression of late lethal functions, can be regulated solely by a replicative mechanism.

3. Control of Late Vegetative Functions

P1, like λ and T4 and unlike T7, continues to depend on a rifampicin-sensitive (presumably bacterial) RNA polymerase throughout lytic development (Meyers and Landy, 1973). Whether P1 modifies host RNA polymerase or specifies production of specific transcription factors remains unknown, although the recognition that the product of phage gene 10 is necessary for the production of late viral proteins (Walker and Walker, 1980, 1983) opens the problem for further study. Cells infected with gene 10 mutants fail to lyse, do not produce serum-blocking power (tail fibers), and produce no tail or head structures detectable in thin sections of unlysed cells (Walker and Walker, 1983). However, they appear to be normal for early protein synthesis, because phage DNA replication and *pac* cleavage functions are unaffected by the lack of the gene 10 product (Sternberg and Coulby, 1987a). C. Senstag and W. Arber (personal communication) have recently identified two promoters that are inactive in uninduced P1 lysogens but become active about 30 min after prophage induction, reaching efficiencies comparable to that of the galactose operon promoter. It is tempting to speculate that these late promoters are turned on by the gene 10 protein. One is present near map coordinate 40 and is probably involved in tail protein synthesis; the other is present near map coordinate 32 and is probably a promoter of the *darA* operon. It is proposed that transcription from both of the promoters is counterclockwise. This agrees with the direction of tail fiber gene transcription in the invertible C segment (Huber *et al.*, 1985a).

Heilmann *et al.* (1980a) showed that minicells infected with any one of three gene 10 amber mutants were missing a protein of 64 kD. If this is the product of gene 10, then that gene must be about 2 kb. Since the wild-type alleles of all ten gene 10 amber mutations are present in the 0.62-kb *Eco*RI-20 fragment, at least part of the gene must be in that fragment (Walker and Walker, 1976, 1983). However, as 250 bp of the N-terminal portion of gene 9 is also present in *Eco*RI-20 (J. Coulby and N. Sternberg, unpublished), there is coding capacity for only about 15 kD of gene 10 protein in *Eco*RI-20. Thus it is somewhat surprising that all the gene 10 mutations map within this fragment. Based on these considerations, the remainder of the gene 10 protein must be encoded by the fragment adjacent to *Eco*RI-20 on the counterclockwise side of the map, *Eco*RI-6. Since we already know that there is a repressor-insensitive promoter within *Eco*RI-6 that can initiate the synthesis of a transcript that spans the *Eco*RI-6,20 junction, that transcript must contain gene 10, and gene 10 must be read in a clockwise direction from *Eco*RI-6 to *Eco*RI-20 (Fig. 2A). Since gene 9 is also transcribed in that direction (J. Coulby and N. Sternberg, unpublished), it is likely that genes 9 and 10 are part of an operon whose promoter is located in *Eco*RI-6 (Fig. 12A).

Recently, evidence has been adduced for the role of protein processing in the expression of certain late phage functions (Streiff *et al.*, 1987).

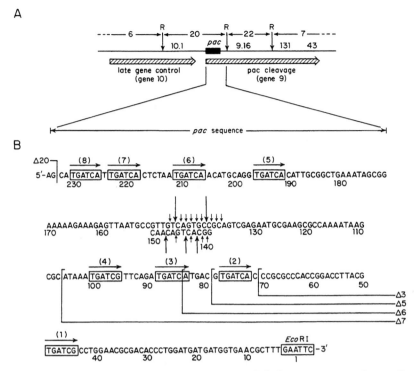

FIGURE 12. Organization of the P1 packaging region. (A) The organization of genes flanking *pac*. The gene that controls late P1 expression (gene 10) is located just to the left
(counterclockwise) of *pac*. It is contained within *Eco*RI-6 and *Eco*RI-20 and is transcribed in
a clockwise direction (see Section VIII.C.3). The *pac* cleavage gene (gene 9) contains the *pac*
sequence within its amino-terminal portion and extends from *Eco*RI-20 into *Eco*RI-7. Like
gene 10, it is transcribed in a clockwise direction. The location of amber mutations (10.1,
9.16, 131, 43) is also shown. (B) The DNA sequence of *pac*. The sequence of a 234-bp
segment of *Eco*RI-20 containing *pac* is shown. A minimal *pac* is contained between bp
positions 71 and 232. The location of *pac* cleavage termini is indicated by the arrows
between bp positions 137 and 148. The large arrows indicate preferred cleavage points.
Seven hexanucleotide elements (TGATCĄ) flank the cleavage termini and have been shown
by an analysis of deletion mutants Δ3–Δ7 and Δ20 to be essential for normal *pac* function
(see Section VIII.E). We propose that these elements are binding sites of the *pac* cleavage
protein. They are also Dam methylation sites.

The full extent of this mode of controlling gene expression in P1 remains
to be assessed.

D. Morphogenesis

1. Head Morphogenesis

There are 15–19 proteins of the P1 head as revealed by SDS–PAGE
analysis of P1 particles, the major head protein having a molecular weight

of 44 kD (Walker and Walker, 1981). Only a few head genes have been described. Mutations that result in head defects have not been assigned roles in the determination of specific structural proteins other than two internal proteins (DarA and DarB—Iida et al., 1987). However, different mutant phenotypes have been distinguished (Walker and Walker, 1976b, 1983). Gene 1 mutants are said to be defective in particle maturation. They produce either apparently complete, but noninfective, particles, or particles in which heads are mostly empty and tails are in various stages of contraction, suggesting the absence of a stabilizing component. The unique mutations that define genes 8 and 23 result in the production of normal amounts of tails and relatively few unattached empty heads. Since the tails end in a neck and a head-neck connector or in an aberrant head structure resembling one, it has been suggested that the defect exhibited by these mutants may be in the stabilization of DNA-filled heads (Walker and Walker, 1983). For gene 23 mutants, the heads presumably fall apart after being filled, whereas for gene 8 mutants, the DNA is lost without destruction of the head. These mutants might produce unstable heads owing to the absence of a minor head protein. Gene 9 mutants produce unattached empty heads that appear morphologically mature. Unattached empty heads are also produced by gene 4 mutants, but clumping and association with cell debris has prevented determination of whether these are mature heads or proheads. [Proheads are head precursors that are usually smaller than mature heads and are never filled with DNA, although they may be filled with scaffolding or other morphogenetic proteins (Murialdo and Becker, 1978).]

It was suggested by Walker and Walker (1983) that the gene 4 defect possibly prevents the release of proheads from the bacterial membrane, a defect characteristic of mutants in T4 gene 21 (a maturation protease (Black and Showe, 1983)). The astuteness of this comparison is now apparent; genetic studies reveal that gene 4 is likely to be the structural gene for such a protease of P1 (Streiff et al., 1987). Accordingly, the gene has been renamed pro. The protease is involved in the processing of the nonessential internal protein DarA and in the processing of other phage proteins that are essential.

Particles with small heads (P1S), depicted in Fig. 1B, usually represent 5–10% of the yield in P1 lysates. We have mentioned (Section II.A.1) the existence of a deletion derivative of P1 that produce as many P1S particles (which are nonproductive on single infection) as infectious P1B particles (Iida and Arber, 1977, 1980). The deletion (about 10% of the P1 genome, clockwise from the IS1) eliminates an operon that determines the expression of late genes for four internal proteins (200, 68, 47.5, and 10 kD) that would otherwise be found in association with P1 heads (Walker and Walker, 1981; Iida et al., 1980, 1987; Streiff et al., 1982). In addition to vad (the viral architecture determinant), the operon determines other dispensable functions, which are briefly discussed in Sections IX.B.1.c and IX.B.2.b. The vad gene itself lies within EcoRI-12

(Fig. 2A). An amber mutation affecting *vad* expression has been localized to this fragment (D. H. Walker and J. T. Walker, personal communication). Although the ratio of P1B (normal-size heads) to P1S particles varies with the presence or absence of the *vad* gene, the morphology of the two particles is similar. Moreover, based on SDS–PAGE, the protein composition of the normal and small-headed phage from the *vad* deletion mutant is similar. This suggests that the Vad protein is not found in the finished P1 particles (D. H. Walker and J. T. Walker, personal communication). Using cloned fragments of the *vad* region that had been isolated by S. Iida and M. B. Streiff, D. H. Walker and J. T. Walker (personal communication) found that Vad could act in *trans* to change the ratio of P1B to P1S particles. Their analysis of sectioned cells infected with a P1 mutant lacking the *darA* operon indicates that both P1B and P1S are produced in the same cell and that the ratio of P1B to P1S particles does not change with the time after infection. Thus the increased production of P1S particles is not due to the accumulation of an essential protein needed for P1B production that could result from the lysis delay phenotype associated with the deletion mutation.

2. Tail Morphogenesis

Nine tail proteins have been identified in phage particles. The two major tail proteins have molecular weights of 21.4 and 74 kD and are presumed to be proteins of the tail tube and tail sheath, respectively (Walker and Walker, 1981). Since 12–17 P1 tail genes have been identified and mutants exist that affect each tail component, it could well be that all or nearly all of the P1 tail genes have been identified. It remains now to determine which genes code for structural proteins and which for nonstructural proteins and how the assembly process occurs. Since tail fibers are often seen attached to headless tails, P1 fibers, unlike those of phage T4, must be able to attach to tails before the heads and tails join.

E. DNA Packaging and Cleavage of the P1 *pac* Site

1. Processive Headful Packaging

The packaging of P1 DNA into viral capsids occurs by a process that preferentially chooses viral chromosomes for encapsidation from a pool of DNA that consists mostly of bacterial DNA sequences. For the large, double-stranded DNA viruses such as λ, P22, and P1, a common feature of DNA packaging is the interaction of precursor head structures or proheads with virus DNA in a process mediated by phage-encoded DNA recognition and cleavage proteins (Murialdo and Becker, 1978; Botstein *et al.*, 1973). The site on the viral DNA at which these proteins bind is called *cos* in the case of λ (Feiss *et al.*, 1983a,b) and *pac* in the case of P1 or

P22 (Bächi and Arber, 1977; Casjens and Huang, 1982; Backhaus, 1985). For phages such as P1, in which the DNA in plaque-forming viruses is terminally redundant and cyclically permuted, the best explanation of how packaging occurs is by processive "headful" packaging starting with a concatemeric DNA substrate (Streisinger et al., 1967; Tye et al., 1974b). The substrate, consisting of tandemly repeated units of the phage genome arranged in a head-to-tail configuration, is packaged in four steps: (1) cleavage of the concatemeric DNA at the unique pac site; (2) unidirectional packaging of the DNA from the cut end into the prohead until no more DNA can be inserted; (3) cleavage of the packaged DNA from the rest of the concatemer (the "headful" cut); and (4) reinitiation of a second round of DNA packaging from the end generated by the previous "headful" cut. Generally, the pac cut is precise and the "headful" cut is imprecise [the latter can be spread over a 2-kb region of DNA (Casjens and Huang, 1982)]. Therefore, the only unique end in the population of packaged DNA is the end generated by the pac cut, and it is detected in restriction enzyme digests of viral DNA as contributing a nonstochiometric, but well-defined fragment, because it is represented in only the first "headful" in each concatemer. The ratio of the pac-generated end fragment to any of the equimolar restriction fragments is inversely proportional to the number of processive "headfuls" packaged from a concatemer. Bächi and Arber (1977) were able to localize pac by mapping the position of the pac-cut end on the nonequimolar fragments of viral DNA produced by digesting that DNA with various restriction enzymes. The nature and abundance of those fragments also allowed them to predict the direction of packaging and the processivity of the reaction. Their results place pac at a position about 5 kb on the counterclockwise side of the map from loxP (Fig. 12A). The predicted direction of packaging is clockwise, and the degree of packaging processivity is 3–4 "headfuls"; i.e., the relative molarity of the pac fragments is 0.25–0.33.

To determine whether the packaging process is similar for P1B and P1S phage particles, Walker et al. (1981) isolated DNA from small-headed variants produced by the deletion mutant that overproduces these particles and compared restriction digests of that DNA to the digests of DNA from wild-type phage. They concluded that P1S particles, as a population, contain the entire complement of DNA present in P1B particles. Moreover, based on the fact that pac fragments from P1S DNA are the same as those from P1B DNA, both in size and abundance relative to other fragments, they could also conclude that the DNA present in P1B and P1S particles is derived from a packaging process that recognizes the same pac site in both cases, packages DNA in the same direction, and packages an amount of DNA from each concatemer that is independent of the ratio of P1B to P1S particles produced by the cell. The latter result suggests that the processivity of packaging is not limited by the number of "headfuls" produced by each concatemer but rather by the physical length of the P1 concatemer. This conclusion is supported by results described below.

2. Localization of *pac* and *pac*-Cleavage Functions

To localize *pac* precisely, P1 fragments flanking the segment of P1 chromosome predicted by Bächi and Arber (1977) to contain *pac* were cloned into phage λ DNA (Sternberg and Coulby, 1987a). It could be shown that the 0.62-kb *EcoRI*-20 fragment contains a functional *pac* site. When a λ phage containing that fragment (λ-P1:*EcoRI*-20) was integrated into the bacterial chromosome as a prophage, and the resulting lysogens were then infected with P1, the chromosomal *pac* site could act as a starting point for the generation of tranducing particles that contained bacterial DNA as far as 10–20 min (5–10 "headfuls") from *pac*. The yield of transducers was increased 60- to 80-fold for a distance of about 10 min on one side of *pac*. If the orientation of *pac* was inverted, then markers on the opposite side of *pac* were transduced with elevated frequencies. These results confirm the unidirectionality of packaging and are consistent with the direction proposed by Bächi and Arber. However, the processivity of the reaction is 2–3 times greater than that predicted in the Bächi and Arber study and suggests, in agreement with the data of Walker *et al.* (1981), that a limiting element in the processive packaging of P1 DNA is the size of the P1 concatemer.

A biochemical assay for *pac* cleavage that is independent of DNA packaging or phage particle formation has been devised (Sternberg and Coulby, 1987a). Bacteria containing a *pac* site located either on an integrated λ prophage or on a resident multicopy plasmid (pBR322-P1:*Eco-RI*-20) were infected with P1, and total cellular DNA was isolated from the infected cells at various times after infection. The DNA was then digested with a restriction enzyme that cleaves on both sides of *pac*. That DNA was then analyzed by agarose gel electrophoresis and Southern transfer hybridization using a non-P1 DNA probe containing sequences between the cloned *pac* and one of the flanking restriction sites. If *pac* is not cleaved during the P1 infection, one should detect a single *pac*-containing restriction fragment. If *pac* is cleaved, two new fragments should appear, reflecting the extra cleavage event.

The results obtained with this assay indicate the following: (1) P1 *am* mutants that are unable to produce gene 9 protein cannot cleave *pac* (Fig. 12A). In contrast, mutants with defects in tail genes (genes 3, 5, 6, and 7), a particle maturation gene (gene 1), head genes (genes 4, 8, 23), and the late gene regulator (gene 10) were able to cleave *pac* with the same kinetics as P1 wild type. (2) *pac* cleavage is detectable as early as 10–15 min after P1 infection. This is shortly after the onset of P1 DNA synthesis in these cells. By 60 min after infection, as much as 80% of the *pac*-containing DNA is cleaved. These results, in combination with the observation that a defect in gene 10 does not affect *pac* cleavage, suggest that the gene 9 protein is expressed early during the phage growth cycle, that it is not under normal late regulation, and that *pac* sites can be cleaved at early times. (3) Of the two *pac* cleavage fragments that should be detected in the assay, one is always present about 10 times more abundantly than the other. The abundant fragment is always the one containing the *pac* end that would normally be packaged into a phage head *in vivo*. The simplest interpretation of this result is that the abundant fragment is detected in higher yields because it is protected from cellular nucleases after cleavage by being packaged into a phage head. The other fragment is not packaged and is presumably destroyed by those nucleases. This interpretation cannot be correct, since the same asymmetric recovery of fragments is observed after infection with a P1 gene 10 mutant, which makes no heads. An alternative model is one in which the *pac* cleavage protein cleaves *pac*, remains bound, protects the DNA end that will be packaged, and eventually provides the specific interaction needed to bring that DNA into the phage head. The other end is not protected. It has recently been shown that the unprotected *pac* end is subject to degradation in the cell by at least two nucleases (N. Sternberg, unpublished). One is the product of the host's *recBCD* genes, which degrades large segments of DNA (>5 kb) from the unprotected end very rapidly. The other nuclease is of unknown origin and degrades DNA more slowly.

The sequence of one strand of a segment of *Eco*RI-20 DNA containing *pac* is shown in Fig. 12B (Sternberg and Coulby, 1987b). The position of the 5' and 3' *pac* termini are indicated by the arrows above and below the double-stranded region at positions 140–150 bp from the *Eco*RI site that separates *Eco*RI-20 from *Eco*RI-22. Those termini were determined by end-labeling procedures using purified *pac* fragments obtained from P1 virion DNA. The *pac* termini present in the population of virion DNAs are distributed over about one turn of the DNA helix with preferred cleavage sites (large arrows) separated by a half-turn of the helix on the strand that generates the 5' end. On the other strand, preferred cleavage sites are separated by 2 bp. Also present in the *pac* sequence are eight hexanucleotide elements (TGATC$_C^A$), four of which are on one side of the cleavage sites and four on the other.

To determine the region necessary for *pac* cleavage, deletion mutations were constructed that penetrate the sequenced region from either side. Plasmids containing each of the deletions were assayed for *pac* cleavage as described above. The results indicate that a completely normal site is contained on a 161-bp segment of *Eco*RI-20 located between bp position 71 and bp position 232 from the center of the *Eco*RI site that separates *Eco*RI-20 from *Eco*RI-22 (Fig. 12B, the DNA between deletion mutations Δ3 and Δ20). Centrally located in this region are the *pac* termini. This region contains seven of the eight hexanucleotide elements; the only one missing is element 1. The importance of the hexanucleotide elements is evident from the properties of *pac* deletion mutants. A deletion of just one of those elements from either side of the minimal site (e.g., deletion Δ5; Fig. 12B) reduces cleavage about 10-fold. Moreover, the elements appear to be multiplicative in their effect on cleavage; removal of 1½ elements or all three elements from the right side of *pac* (deletions

Δ6 and Δ7) reduces the efficiency of *pac* cleavage 100-fold and more than 1000-fold, respectively. We believe that the hexanucleotide elements are binding sites for the *pac* recognition protein, presumably the gene 9 product. Increasing the number of binding sites on any one side of the minimal *pac* sequence presumably increases binding and subsequent cleavage.

3. Regulation of *pac* Cleavage

Each of the hexanucleotide elements present in the minimal *pac* site contains the sequence GATC. Since the adenine residue in that sequence is normally methylated by the bacterial *dam* gene product, the role of methylation in *pac* cleavage was assessed by measuring the kinetics of the reaction in *dam* and *dam*$^+$ bacteria (N. Sternberg and J. Coulby, unpublished). The results of that experiment indicate that the onset of *pac* cleavage in a *dam* host is delayed about 20 min relative to that observed in a *dam*$^+$ host, but by 70–80 min after infection the level of cleavage is comparable in both hosts. The delayed cleavage in the *dam* host is not due to a general effect of the mutation on P1 expression, since P1 replication is normal in the *dam* host. More likely, cleavage requires methylation, but at late times the source of that methylation is the phage-encoded methylase. To prove this point, an *in vitro pac* cleavage assay was developed in which the cleavage of exogenous *pac*-containing DNA is dependent on the addition of an extract from induced P1 lysogens (N. Sternberg, unpublished). It could be shown that unmethylated plasmid DNA containing *pac* (DNA prepared in a *dam* host) is not a substrate in the *in vitro* assay and that the *pac* site can be converted to a substrate if it is methylated by the *E. coli* Dam enzyme *in vitro* prior to being incubated with the *pac*-cleaving extract. The level of *in vitro* methylation and the ability of *pac* to be cleaved were measured at various times after incubation with the Dam methylase. The results indicate that not all of the 14 GATC sequences (there are 2 in each of the hexanucleotide elements in the minimal *pac*) need to be methylated for *pac* to be cleaved. It could also be shown that a hemimethylated *pac* site (1 strand fully methylated and the other fully unmethylated) is not cleaved. This is particularly important, because hemimethylated *pac* DNA is generated, if only transiently, every time *pac* is replicated in a wild-type host.

The properties of the *pac* cleavage reaction described above suggest a way in which *pac* cleavage might be regulated in the cell. However, before describing that regulatory model, we need to consider why it is important to regulate *pac* cleavage. For P1 we believe such regulation is vital for the normal production of infectious phage particles. Several observations lead to this conclusion. First, terminal redundancy in virion DNA is essential if P1 DNA is to be cyclized after it is injected into a sensitive host (see Section IV). In the absence of cyclization, infectivity is lost (Sternberg *et al.*, 1986). To generate infectious phages with terminally redundant DNA, the packaging system must avoid cutting each *pac* site on the concatemer, as that would generate monomer-length DNA. Ideally, the system should cut the concatemer once at *pac*, and packaging should proceed down the concatemer from that point, bypassing uncleaved *pac* sites, and packaging DNA by the "headful." Moreover, to maximize the efficiency of packaging, the *pac* site on the concatemer chosen for cleavage should be at one or the other end of the concatemer, rather than in the middle, since the *pac* terminus not destined to be packaged is rapidly degraded in the cell (Sternberg and Coulby, 1987a). Without a way of regulating *pac* cleavage, P1 will have difficulty establishing such a regimen, since *pac* cleavage functions are synthesized early during the phage growth cycle, probably well before phage heads are available for packaging. If those functions were to act at that time, it is difficult to imagine how each *pac* site on a concatemer could avoid being cut before the phage could package DNA. Clearly, *pac* cleavage needs to be inhibited at early times during phage infection, when both substrate and enzyme are present in the cell.

A model that can account for the regulation of P1 *pac* cleavage is based on the methylation requirement of that reaction. According to this model, *pac* sites present on P1 rolling circle concatemers early during infection are not substrates for the cleavage reaction. We propose that early during infection, when *pac* is replicated onto the tail of the rolling circle, it becomes transiently hemimethylated and must remain so; otherwise it would be cut. To accomplish this, the action of the host Dam methylase at *pac* must be blocked. We postulate that hemimethylated *pac* sites bind to the *pac* cleavage enzyme more avidly than they do to the bacterial Dam methylase. The *pac* cleavage enzyme remains bound to the site, does not cut it, but prevents the methylase from activating the site. If this process were to continue throughout the phage growth cycle, very little P1 DNA would be packaged. Thus we must assume that a change in the pattern occurs at late times during the infection.

That change is produced by the synthesis of a new methylase, the P1-encoded enzyme. This enzyme should become quite abundant in the cell late during infection, since there are many copies of the gene in the cell by then. We propose that it competes favorably with the *pac* cleavage enzyme for the newly replicated *pac* sites. Thus as the infection proceeds, the probability that a newly replicated *pac* site will be methylated, and thereby activated before it encounters the *pac* cleavage enzyme, increases. This model makes three predictions: (1) the direction of lagging strand DNA synthesis along the tail of the rolling circle must be the same as the direction of packaging; (2) hemimethylated *pac* sites bind the *pac* cleavage protein; and (3) uncleaved *pac* DNA isolated from P1 virions is a poor substrate for *in vitro* cleavage, because it is undermethylated. The first two predictions have not been tested; the third has, and the results are in agreement with the prediction (N. Sternberg, unpublished).

IX. TRANSDUCTION

Transduction, along with conjugation and transformation, is one of three methods by which genes are transferred in nature from one bacterium to another. The encapsidated DNA is transferred into the recipient cell by injection. There it either undergoes recombination with the host chromosome, producing a stable transductant, or, more frequently, remains extrachromosomal and generates an abortive transductant. In the latter case, the extrachromosomal DNA is not replicated and is passaged to only one of the two daughter cells at cell division.

Transduction can be classified into two types, specialized and generalized, based on the way the transducing particles are generated and the way the transducing DNA is incorporated into the genome of the recipient cell. Specialized transducing phages contain bacterial DNA physically linked to phage DNA. For viruses such as λ, these transducing phages are produced by aberrant excision of the prophage DNA and therefore contain chromosomal sequences located adjacent to the integrated prophage. For phage P1, the host DNA present in specialized transducing particles probably comes from sequences adjacent to sites used by the phage when it infrequently integrates into the chromosome (see below). Alternatively, bacterial sequences may insert directly into P1 DNA. When DNA from specialized transducing phage is injected into recipient cells, transduction occurs by prophage lysogeny and is usually independent of the generalized recombination system of the recipient cell. The cells so generated are diploid for the transduced DNA.

Generalized transducing particles differ from specialized transducing particles in that they contain only host chromosomal sequences. They originate during the phage vegetative growth cycle by the packaging of host DNA either randomly or from *pac*-like sites in the host chromosome. Stable transduction following injection of DNA from generalized transducing particles is generated by replacement of homologous recipient cell DNA with transduced DNA and depends on the bacterial generalized recombination system.

A. Specialized Transduction

P1 specialized transducing phages contain segments of bacterial DNA inserted into the P1 genome. In some cases, this insertion process is accompanied by the loss of P1 DNA adjacent to the site of the insertion. How this process occurs will be the subject of this section. We will first discuss the recombination systems used to insert DNA into P1, or to integrate P1 DNA into the bacterial chromosome, and then we will consider how P1 specialized transducing phages are generated using P1*argF* as a model system.

1. Modes of Integrative Recombination

P1 and P7 can integrate into the chromosome of a host bacterium at low frequency. This was first recognized with the discovery of P1 specialized transducing phages containing either the E. coli lac operon (P1dlac) (Rae and Stodolsky, 1974; Schultz and Stodolsky, 1976), the E. coli proA gene (P1dpro) (Stodolsky, 1973; Rosner, 1975), or the E. coli argF gene (Schultz and Stodolsky, 1976; York and Stodolsky, 1981, 1982a,b), and it was most convincingly demonstrated by showing that P1 or P7 prophages could suppress the temperature-sensitive replication defect of E. coli dnaAts mutants by integrating into the host chromosome (Chesney and Scott, 1978). Presumably, at elevated temperatures in such strains, chromosomal replication is driven by the plasmid replication system and initiates at the plasmid origin of replication.

In addition to recombination with the bacterial chromosome, P1 can also recombine with the DNA of plasmids such as pBR322 and NR1 to generate either cointegrate structures (Iida et al., 1981) or phages with resistance-determinant elements (Iida and Arber, 1980). If the recombinant DNA does not exceed the "headful" packaging limit of P1, then that DNA can be faithfully propagated as a specialized transducing phage. Phage-carried, site-specific recombination systems (lox-cre, cin-cix, IS1, and Tn902) appear responsible for most, but not all, cases of integrative recombination involved in the generation of specialized transducing phage.

a. Integrative Recombination Mediated by lox-cre

lox-cre is generally the most frequently used recombination system in the integration of P1 or P7 DNA. In recA$^+$ cells, 10 of 22 integration events scored were found to have occurred at the loxP site, and in recA cells, 16 of 17 events occurred at that site (Chesney et al., 1979). Moreover, the unique bacterial site (loxB)* at which the integration occurs bears a strong resemblance to the phage loxP site (Sternberg et al., 1981a; Hoess et al., 1982). That site is located between the E. coli dnaG and tolC genes at minute 66 on the standard E. coli map (Bachmann, 1983). The loxB site has been cloned and sequenced and found to have inverted repeats similar to those of loxP (Hoess et al., 1982). However, the loxB spacer region is 9 bp rather than 8 bp and is completely different in sequence from that of loxP. Since homology within the spacer is critical for efficient recombination between loxP sites (Hoess et al., 1986), it is not surprising that integration at loxB is rather inefficient. In vitro recombination between loxB and loxP sites is undetectable (K. Abremski, personal communication), and in vivo it generates integratively suppressed cells at a frequency of 10^{-5} to 10^{-6}. Moreover, as the direction of

*The designation attP1,P7 (Bachmann, 1983) is synonymous with loxB.

loxP recombination is determined by homologous pairing within the spacer regions of the two recombining sites, it is also not surprising that *loxP* × *loxB* recombination can integrate DNA in both possible orientations with only a 3:1 preference of one orientation over the other (Sternberg *et al.*, 1981a). Integration of P1 DNA at *loxB* generates two new sites at the junctions of the prophage and host sequences. These hybrids between *loxB* and *loxP* are designated *loxR* and *loxL* (Sternberg *et al.*, 1981a). *loxL* has essentially the *loxP* spacer region and recombines nearly as efficiently with *loxP* as does wild-type *loxP*. In contrast, *loxR* has essentially the *loxB* spacer region and recombines with either *loxP* or *loxL* poorly (about 100-fold less efficiently than *loxP* recombines with *loxP*) (Sternberg *et al.*, 1981a).

Because *loxL* recombines more efficiently with *loxP* than does *loxB*, the chromosome of cells containing these hybrid *lox* sites flanking an integrated prophage is a more efficient substrate for the subsequent integration of *loxP*-containing prophage than is the original chromosome. However, the integrated DNA can be maintained in the chromosome only if the resident prophage is lost (Sternberg *et al.*, 1981a). The efficiency of this integration by replacement can be readily understood in terms of the efficiencies of recombination of the various pairs of *lox* sites. Thus, the integration of a *loxP*-containing prophage at a chromosomal prophage flanked by *loxL* and *loxR* will occur primarily at the *loxL* site, since *loxP* × *loxL* recombination is much more efficient than *loxP* × *loxR* recombination. This generates a transient intermediate containing both the resident and the newly integrated prophage in the configuration *loxL*– new prophage–*loxP*–resident prophage–*loxR*. Since *loxL* × *loxP* recombination is efficient, the newly integrated prophage is readily excised. However, if *loxP* recombines with *loxR*, then the resident prophage DNA is excised, and the newly integrated DNA is stabilized in the chromosome, because the flanking *lox* sites (*loxL* and *loxR*) cannot recombine efficiently with each other.

b. Integrative Recombination Mediated by cix-cin

The P1-P7 C-loop inversion system has been implicated in at least one integrative suppression event (Chesney *et al.*, 1979) and in the formation of P1–pBR322 cointegrates. In the latter case, the recombination is promoted by Cin and occurs between a P1 *cix* site and quasi *cix* sites on pBR322 DNA (Iida *et al.*, 1982). The hybrid sites formed at the junctions of the cointegrate are also functional *cix* sites. A particular cell line that is integratively suppressed by P7 was found to be missing part of the prophage DNA that probably includes one of the *cix* pseudo-*cix* junctions of the prophage. This structure suggests that integration was Cin-mediated (Chesney *et al.*, 1979). Additional support for the role of the *cin-cix* system in P1 recombination with the bacterial chromosome comes from structural analysis of P1d*lac* DNA (Schulz and Stodolsky, 1976). The

DNA of that phage is specifically missing a segment of P1 DNA that includes contiguous EcoRI fragments 1, 12, and 16. Since the cin-cix system is contained within EcoRI-1, that system is implicated in the generation of this phage. It is interesting that in both P1dlac and the cin-mediated, integratively suppressed P7 lysogen, one of the junctions generated by the proposed integration event is missing. Perhaps such a deletion is necessary to prevent excision of the integrated P1-P7 DNA.

c. Integrative Recombination Mediated by IS1 and Tn902

P1 contains an IS1 element at map coordinate 24, which can recombine with IS1 elements in the E. coli chromosome. The recombination depends on the host's recA function and is therefore probably a homology-dependent, rather than a site-specific, process (Chesney et al., 1979; York and Stodolsky, 1981, 1982a). Two of the 14 P1 integratively suppressed recA+ dnaAts mutants studied by Chesney et al. (1979) contained P1 prophages integrated by means of IS1 recombination. Moreover, 13 of 13 P1Cm lysogens selected in a strain in which P1 could not replicate as a plasmid (a dnaA null mutant driven by an integrated mini-R1 plasmid) are the result of integration of the P1Cm DNA via IS1-mediated recombination (Hansen and Yarmolinsky, 1986). IS1-mediated recombination has also been implicated in the generation of P1argF, P1raf, and P1CmSmSu phage (York and Stodolsky, 1981; Schmitt et al., 1979; Iida and Arber, 1980). In all three of these cases, the DNA inserted into the P1 chromosome (the E. coli argF gene, the raf gene of plasmid pRSD2, and the NR1 plasmid r-det element) is flanked by IS1 elements both in its original context and in P1 (Hu and Deonier, 1981; Schmitt et al., 1979; Iida and Arber, 1980). The generation of P1dpro phage (Rosner, 1975) probably also involves integration of P1 DNA into the bacterial chromosome by means of homologous recombination between IS1 elements in P1 and those near proA. This is inferred from the structure of the P1dpro DNA, in which the bacterial sequences lie adjacent to the Tn9 transposon of phage P1Cm c1.100. That transposon contains a cat gene flanked by IS1 elements and, like the single P1 IS1 element, is located at map coordinate 24.

Three of 10 integratively suppressed recA+ dnats strains isolated by Chesney et al. (1979) that became auxotrophs as a consequence of the integration of P7 DNA and 6 of 7 integratively suppressed strains that contain an integrated P7 prophage with internal deletions appear to have integrated at or near the Tn902 element. In several of these prophages, one of the deletion boundaries is near the Tn902 element, implicating that element also in the deletion process.

d. Other Modes of Integrative Recombination

Integratively suppressed strains have been isolated in which the DNA of either P1 or P7 has recombined with the host chromosome at regions of the phage genome not known to contain site-specific recombination elements or obvious homology with the host chromosome. The two regions involved contain the *res-mod* genes (class d of Chesney *et al.*, 1979) and the phage packaging gene (class b of Chesney *et al.*, 1979). How these regions recombine with the bacterial chromosome is still unclear; however, it is tempting to speculate that, in the second case, the phage packaging site (*pac*) is involved. Generation of double-stranded breaks in P1 DNA at *pac* could act to initiate the insertion of P1 DNA into the bacterial chromosome by a double-strand break repair mechanism (Szostak *et al.*, 1983), provided there is at least some homology between *pac* and a region of the host chromosome.

2. IS1-Mediated Generation of a Specialized Transducing Phage: P1*argF*

How are specialized transducing phages generated? We confine most of our discussion of this question to P1*argF*, since it is the most extensively studied of the P1 specialized transducing phages (York and Stodolsky, 1981, 1982a,b). This is so largely because the *argF* gene is transduced by P1 in the specialized transduction mode much more efficiently than are other genes; *argF* is transduced into a *recA* host with an efficiency of 10^{-4} to 10^{-5} per plaque-forming phage, whereas for *pro, lac,* and *raf* the efficiencies are 10^3- to 10^4-fold lower (York and Stodolsky, 1981, 1982a; Schmitt *et al.*, 1979). Why this should be so is not clear. Several experimental results (York and Stodolsky, 1982a) suggest that recombination between IS1 elements, or IS1-mediated transposition, is involved in the generation of P1*argF* transducing phage: (1) The *argF* gene is flanked by IS1 elements in both the bacterial chromosome and in P1. (2) The *argF* gene in P1*argF* is located at the same place in the P1 genome as is the IS1 element. (3) The ability of phages to transduce *argF* is directly proportional to their IS1 content. Thus P7, which does not have an IS1 element, rarely transduces *argF*, whereas P1CmO (Iida *et al.*, 1978), which contains two IS1 elements flanking the *cat* gene at map coordinate 24, exhibits a high frequency of *argF* transduction. P1, with a single IS1, is intermediate in its ability to transduce *argF*. (4) P1*argF* phages generated by infection of sensitive cells with P1CmO frequently lose the *cat* gene. Although this result suggests that the pickup of *argF* and the loss of *cat* are occurring in the same process, it was not ruled out that the two events occur separately and that *cat* is lost because the P1CmO*argF* DNA is too big to be packaged. York and Stodolsky (1982a) also showed that the yield of P1*argF* phage/infective phage is reduced about 100-fold in a *recA* host, relative to that observed in a *recA*$^+$ host, and that the union of P1 and *argF* was stimulated during the phage lytic cycle in the donor cell. In the case where the phage lysate was prepared by prophage induction, few of the transducing phages could be accounted for by the union of *argF* with the prophage DNA prior to induction.

When the frequency of *argF* transduction was measured in a recipient host deleted for the *argF* region (to avoid detection of generalized transducing phage) at a low multiplicity of phage infection (<0.1 phage per cell), a linear relationship could be shown to exist between the multiplicity of infection and the number of *argF* transductants. This result indicates that one infecting virion is sufficient to generate a P1*argF* transductant in the recipient and contrasts with the results for transduction of *lac* by P1*dlac* phage (Rae and Stodolsky, 1974), where infection with at least two phages is necessary for transduction. When the *argF* transductants were examined, they all contained a P1*argF* prophage. Induction of those P1*argF* lysogens produced a normal yield of plaque-forming phages and a high-frequency *argF* transducing lysate (York and Stodolsky, 1982a). Presumably the 11-kb *argF*-containing insert into P1 DNA eliminates most but not all of the terminal redundancy required for plaque-forming capacity. If the recipient *argF*-deleted host is infected at a phage multiplicity of greater than 0.2 phage per cell, then the yield of transductants, plotted against the multiplicity of infection, has a slope of 2. This indicates that simultaneous infection with a second virion facilitates P1*argF* prophage formation. If the recipient host is a P1 lysogen, then the slope is 1 at both high and low multiplicities. Finally, at all multiplicities, the *argF* transduction frequency is increased if the prophage in the recipient cell is transiently induced.

These observations prompted York and Stodolsky (1982a) to propose three possible pathways for the generation of P1*argF* specialized transducing phage. In the first pathway, the *argF* gene excises from the bacterial chromosome by recombination between flanking IS1 elements. The resulting circle then integrates by homologous recombination into the natural IS1 element in P1. In the second pathway, P1 integrates into the chromosome adjacent to *argF* by a homologous recombination event between the P1 IS1 element and one of the IS1 elements flanking *argF*. Subsequent excision of the P1 DNA by IS1-mediated recombination would frequently transfer the *argF* gene to the P1 replicon, leaving behind a single IS1 in the chromosome. If the P1 used contains, instead of a single IS1, two IS1 elements flanking the *cat* gene (P1CmO), excision of the P1 with *argF* would frequently leave the *cat* gene in the chromosome flanked by IS1 elements. The observation that P1*argF* transduction is facilitated either by double infection or by lysogeny of the recipient strain suggests a third pathway of generating the P1*argF* prophage. In this third pathway, recombination between IS1 elements inserts P1 DNA into the *E. coli* chromosome near *argF*, as proposed for the second pathway. However, this is followed by "headful" packaging during the P1 lytic cycle, producing a transducing particle containing both P1 DNA and bacterial DNA, including the *argF* gene. By itself, that DNA would not be able to cyclize after injection into the recipient cell. However, if it were to recombine with a resident P1 prophage or with a second infecting phage at the IS1 locus, a DNA molecule could be produced that contains a full complement of P1 genes, the *argF* gene flanked by IS1 elements, and terminal redundancy to permit cyclization.

B. Generalized Transduction

In this section we will discuss three aspects of the P1 transduction process: the generation of transducing particles, the fate of transduced DNA when it enters the recipient cell, and the biological effects associated with transductional recombination in the recipient cell. Where pertinent, we will compare the P1 results with those obtained from studies of the other major generalized transducing phage, P22.

1. Formation of Generalized Transducing Particles

a. Evidence from Physical Experiments

Much of what we know about the genesis of transducing particles comes from the density-labeling experiments of Ikeda and Tomizawa (1965a,b). P1 lysates were prepared on cells that were labeled with bromouracil before P1 infection and then with either bromouracil or thymine after P1 infection, and the lysates were then analyzed by CsCl density-gradient centrifugation. It was shown that transducing particles carrying the wild-type alleles of *pro, his, leu, arg,* and *trp* markers had the same density whether the postinfection medium had bromouracil or thymine and that the density of particles prepared in cells whose DNA was labeled with bromouracil prior to infection was 0.010 g cm^{-3} greater than the density of transducing particles prepared on cells whose DNA was not labeled with bromouracil prior to infection. When phages were prepared in bromouracil-labeled bacteria in media containing thymine and $^{32}PO_4$, radioactivity was not found in the fractions of the CsCl gradient containing the transducing particles. These results were interpreted to mean that transducing particles lack significant amounts of phage DNA (<5%) and contain only fragments of the bacterial chromosome that existed prior to the time of phage infection. Ikeda and Tomizawa (1965a) were also able to show that transducing particles comprise 0.3% of the total phage particles and that the molecular weight of the DNA in infective and transducing phages is the same (60 MD). A surprising observation was the demonstration that the density of DNA in transducing particles was 0.004 g cm^{-3} less than that of bacterial DNA but could be increased to that of the bacterial DNA by treatment with proteolytic enzymes (Ikeda and Tomizawa, 1965b). The density of DNA from infective phage particles was not affected by treatment with those enzymes. These results were interpreted to mean that the DNA of transducing particles, but not infective particles, contains an associated protein. Further studies indicated that the protein was covalently linked to the DNA, was not evenly distributed over the DNA molecule, and was not associated with DNA derived from small-headed transducing particles (Ikeda and Tomizawa, 1965b,c). It was estimated that the total amount of protein bound to DNA in normal-size transducing particles was 500 kD.

b. Evidence from Physiological Experiments

The next major advance in our understanding of how transducing particles are produced in the donor cell came from single-burst analyses (Harriman, 1972). Using prophages (λ, 21, and 186) as chromosomal markers to increase the efficiency of scoring the transductants, Harriman could show that transducing particles containing any one of the three markers were produced in 0.5% of the P1-infected cells and that those cells also released infective P1 particles. Moreover, among cells producing one class of transducing particles, the likelihood that they also produced another was elevated. Thus among cells producing λ transducers, 12% produced 21 transducers, and 5% produced 186 transducers. However, among cells producing both λ and 186 transducers, the frequency of 21 transducers (14.5%) was not significantly greater than in cells producing transducers of λ but not 186. When the λ marker was located on an extrachromosomal F factor and the 21 marker on the bacterial chromosome, the percentage of cells producing both λ- and 21-transducing particles dropped from 12% to 3%.

These results were interpreted to mean that regions on the order of 20% of the E. coli chromosome were packaged into phage particles by the processive "headful" packaging mechanism. This would explain the higher frequency of coproduction of λ and 21 markers (separated by 10% of the chromosome) than λ and 186 markers (separated by 40% of the chromosome) or 21 and 186 markers (separated by 30% of the chromosome). It would also explain the low coproduction of λ and 21 markers when physically unlinked. The estimate of processivity proposed by Harriman is in good agreement with that determined by Sternberg and Coulby (1987a, referred to in Section VIII.E.1). Since the λ and 186 markers are further apart than the distance covered in a single sequence of sequential packaging events, one could not account for the higher coproduction frequencies of these markers or of unlinked markers, such as λ(F) and 21, by assuming that only one sequence of processive packaging occurs per cell. Rather, Harriman was forced to conclude that, in cells producing transducing particles, there are frequently several independent processive packaging events; i.e., cells producing transducing particles are a special class of cells in the population that are prone to initiate processive packaging of the bacterial chromosome. Harriman estimates that if one-tenth of the infected cells contained several regions of localized packaging of 20% of the chromosome, this could explain the genetic results reported above. It is not clear what distinguishes a cell producing transducing particles from one that is not. We discuss some of the possibilities below.

The frequencies with which bacterial markers are transduced by P1 can vary by as much as 30-fold (Masters, 1977), ranging from 3×10^{-4} transductants per infective phage particle, for markers near the origin of bacterial replication, to 10^{-5} transductants per infective phage particle, for markers near the terminus of replication. Nearly all of these differences can be attributed either to variability in events that lead to stabilization of transduced DNA in the recipient cell (Newman and Masters, 1980; Masters *et al.*, 1984) or to differences in gene concentration in the donor cell. The latter effect was evident from a comparison of the transducing abilities of lysates prepared on donors growing at different rates. The frequency of transduction of origin and terminus markers was proportional to the differences in the marker copy numbers at the various growth rates (Masters, 1985). Moreover, direct measurements by Southern blot hybridization of the relative abundance of bacterial DNAs in phage heads indicates that, except for DNA from the 2-min region of the *E. coli* chromosome (including the *leu* gene), all sequences are present in approximately equal amounts. Markers near the *leu* gene are overrepresented in transducing particles about threefold (M. Hanks and M. Masters, personal communication). These studies show that the packaging of chromosomal DNA into P1 capsids is nonselective; i.e., if one excludes copy number differences for various markers, any segment of chromosomal DNA is just as likely as any other to be packaged. Moreover, since we have already noted that a P1 packaging site (*pac*) in the bacterial chromosome can increase the transduction frequency of adjacent markers as much as 80-fold (Sternberg and Coulby, 1987a), it is reasonable to conclude that the encapsidation of bacterial DNA into P1 transducing particles is not initiated at *pac* or pseudo-*pac* sites. This conclusion is supported by the inability to detect cross-hybridization between P1 *pac* DNA and *E. coli* DNA (N. Sternberg, unpublished; M. Hanks and M. Masters, personal communication).

These results contrast with those for phage P22, where the difference in transduction frequencies for different chromosomal markers (10^{-9} to 10^{-6} transductants per infective phage particle) is much greater than it is for P1 and cannot be explained simply by differential stabilization of transduced DNA in the recipient cell or differences in gene dosage in the donor. Rather it would appear that in this case, selective packaging of regions of the bacterial chromosome is occurring. This is supported by the following observations: (1) Chelala and Margolin (1974) showed that a deletion of chromosomal DNA could alter the cotransduction frequencies of pairs of markers distal to the deletion even when the deletion itself is not cotransduced with those markers. This finding is readily explained if the deletion alters the register of "headfuls" that starts from a chromosomal *pac* site and proceeds through the deletion. (2) High-frequency transducing (HFT) mutants of P22 have been isolated that increase the proportion of transducing particles in the P22 lysate from 1–5% in wild-type lysates to as much as 50% in the mutant (Schmeiger, 1972; Masters, 1985). The mutations responsible for this HFT phenotype have been localized to the P22 gene whose product is thought to be involved in the recognition of *pac* (the gene 3 protein) (Raj *et al.*, 1974). These mutants transduce all chromosomal markers with a similar high frequency and package phage DNA with less site specificity (Schmieger, 1972). (3) Phage lysates prepared by the infection of cells with P22 HFT mutants exhibit altered cotransduction frequencies when compared to lysates prepared using P22 wild type (Schmieger and Backhaus, 1976). These results are interpreted in terms of a processive "headful" packaging process that initiates at various *pac* sites in the bacterial chromosome with specificities that are altered in the HFT mutants.

c. Evidence from Genetic Experiments

High-frequency transducing mutants of P1 have been isolated. The first set of mutants studied was isolated by Wall and Harriman (1974). The mutants transduce all chromosomal markers about five- to tenfold more efficiently than does the parental P1. Since the frequency of transducing particles in the mutant P1 lysate (2%) is about 10 times higher than in the parental P1 lysate (Sandri and Berger, 1980a), it would appear that much of the increased transduction is due to increased packaging of bacterial DNA in the donor cell. Moreover, as one of the mutants (an amber) is suppressible, the elevated transduction frequency would appear to be due to a loss of phage function. Unfortunately, none of the mutants have been mapped, so it has not yet been possible to correlate the mutant phenotype with any known P1 packaging functions. A second class of P1 mutants affecting transduction maps within the *darA* operon (map coordinates 25–30; Fig. 2A) and results in increased marker transduction frequency. The mutation with this phenotype is a deletion of the entire *darA* operon, and the function responsible has been designated *gta* (Iida *et al.*, 1987). It has not been determined whether the increase in transduction frequency correlates with an increased frequency of transducing particles in the phage lysate or whether it is the result of events occurring in the recipient cell after DNA injection.

d. A Speculative Proposal for the Formation of Generalized Transducing Particles

If P1 generalized transducing particles are not produced by packaging events that select preferred initiation sites on the bacterial chromosome, how is bacterial DNA packaged into P1 heads? The hypothesis that best fits the results presented above is one in which packaging of bacterial DNA into P1 capsids is initiated at a chromosomal end and proceeds progressively down that chromosome for 5–10 "headfuls" just as it does when packaging is initiated by cutting a *pac* site of P1 that has been inserted within the bacterial chromosome. If the ends from which packaging is initiated are randomly distributed in the bacterial chromosome, this would account for the near-random packaging of chromosomal DNA into transducing particles (Masters, 1977; Newman and Masters, 1980; Masters *et al.*, 1984). The processivity of the reaction would account for the coproduction of transducing phage in the same cell (Harriman, 1972).

Whether P1 can efficiently initiate packaging at an end has not been determined, but it must be able to carry out a reaction like that after each "headful" cut. In the case of processive packaging from a cleaved P1 *pac* site, the protein that cleaves *pac* and remains bound to the end that is destined to be packaged could be transferred to the nonpackaged free end generated by the "headful" cut. This would permit packaging of that DNA and would allow the reaction to proceed down the concatemer.

For the packaging of bacterial DNA, one need simply assume that the *pac* cleavage protein can associate with a free DNA end, at least to some degree, without having previously cleaved a *pac* site. The HFT mutants of Wall and Harriman could affect any one of several steps in the above process. They might produce a mutant *pac*-cleavage protein that more efficiently associates with free DNA ends. Alternatively, the mutant might generate DNA ends in bacterial DNA more efficiently than they are generated in wild-type infections. This could occur if the mutant encodes a more efficient endonuclease than does its parent or if the mutant has a *pac*-cleavage protein with less specificity. In this regard, Wall and Harriman (1974) note that one of the properties of cells infected with the HFT P1 mutant 607H is an accelerated breakdown of the *E. coli* chromosome.

One of the outstanding questions not addressed above is the nature of the protein associated with the ends of transducing DNA. It is tempting to speculate that this protein is the *pac*-cleavage protein. It is not associated with infectious particles, because it is transferred from one "headful" to another as the packaging process marches down the P1 concatemer. Since none of the infectious P1 particles have protein associated with their DNA, we must conclude that the last "headful" packaged is noninfectious or loses its *pac*-cleavage protein before the tail is put on the filled head. To account for the presence of protein on the DNA of transducing particles, one must assume that processive packaging from a preexisting end (as opposed to processive packaging from a cleaved *pac* site) does not involve a protein transfer step. Rather, the protein binds to the end and brings that DNA into an empty phage capsid. When the capsid is filled, the "headful" cut occurs, separating the chromosome from the packaged DNA, but the protein is not transferred to the free DNA end created by the "headful" cut. Rather, it remains associated with the bacterial DNA it helped to package. To initiate the next "headful" packaging event, additional protein must become associated with the free DNA end.

e. Transduction of Plasmid DNA by P1

The transduction of plasmids by P1 (first reported by Arber, 1960b) deserves special consideration because the transduced DNA is frequently smaller in size than a P1 "headful." Two pathways have been proposed to account for the production of generalized transducing particles containing plasmid DNA (O'Connor and Zusman, 1983). One possibility is that

P1 DNA can recombine with plasmid multimers in the cell to generate cointegrates whose size exceeds the "headful" limit of P1. That DNA is packaged from the normal P1 *pac* site but lacks terminal redundancy. When injected into the recipient cell, it cannot cyclize like DNA from P1 specialized transducing phage, but it can generate plasmid circles if the host's generalized recombination system excises a plasmid monomer from the injected linear DNA. An alternative pathway is the direct packaging of plasmid DNA into phage capsids in much the same way as P1 encapsidates chromosomal DNA. The process is limited by the need for the plasmid to attain a size that completely fills a P1 head. This could be achieved if either the plasmid to be packaged were large enough as a monomer or if a multimeric form of a smaller plasmid were packaged.

It should be noted that as small P1 heads contain only 40% of the normal P1 DNA, packaged plasmid DNA need only attain a size of about 40 kb. When injected into recipient cells, linear multimers can give rise to plasmid monomers by recombinational excision. Support for the second of these pathways comes from the observation that the efficiency of transduction of a plasmid is proportional to its size. Thus, plasmids in the 45–100 kb range are transduced with frequencies of 10^{-3} to 10^{-6} (Watanabe *et al.*, 1964; Kondo and Mitsuhashi, 1964; O'Connor and Zusman, 1983) and plasmids in the 6–7 kb range are transduced with a frequency of 10^{-6} to 10^{-7} (Iida *et al.*, 1981). These results are interpreted to mean that the smaller the plasmid, the higher the multimeric state it must attain in order to be packaged.

2. Alternative Fates of Transducing DNA in the Recipient

To study the fate of DNA transduced by P1 after it is injected into *E. coli*, Sandri and Berger (1980a,b) extracted DNA from ³H-labeled bacteria that were infected with ³²P-5-bromouracil-labeled transducing particles. Cesium chloride density-gradient analysis of that DNA indicated that it could recombine with, and thereby stably insert into, the bacterial chromosome, or it could be maintained extrachromosomally as a nonreplicating element that would be transmitted unilinearly to daughter cells at cell division. The former process leads to stable transductants, and the latter to abortive transductants.

a. Stable Transduction

Sandri and Berger (1980a) showed that about 7–15% of the ³²P-labeled DNA in P1 transducing particles becomes associated with recipient cell DNA in an alkali-resistant complex within 1 h of being injected into *recA⁺* or *recA* bacteria. However, only in the *recA⁺* cells could appreciable amounts (9–15%) of this ³²P label be released by sonication or shearing as fragments that are greater than 500 bp in size. Thus, only 1–2% of the ³²P-labeled DNA injected into a *recA⁺* host is ever incorporated within the bacterial chromosome as an intact fragment of at least 500 bp. These results confirm genetic experiments that indicate that a functional RecA protein in the recipient is necessary for stable transduction (Sandri and Berger, 1980a). The presence of ^{32}P label in chromosomes of *recA* bacteria is presumably due to degradation of injected ^{32}P DNA followed by reincorporation of the label into chromosomal DNA. Sandri and Berger (1980a) also showed the following. (1) All of the ^{32}P label incorporated into bacterial DNA was incorporated within the first hour after injection. (2) The DNA inserted into the bacterial chromosome was generally about 10 kb in size, substantially less than the size of the bacterial DNA encapsidated into each P1 head. (3) UV irradiation of the phage lysate increased the incorporation of transducing DNA into bacterial DNA but did not alter the size of the incorporated fragment, which confirms the observation that the frequency of transduction goes up when the phage lysate is exposed to low doses of UV light (Wall and Harriman, 1974; Newman and Masters, 1980). (4) P1 transducing DNA was integrated into bacterial DNA as double-stranded fragments.

Thus, ^{32}P-labeled fragments released from total bacterial DNA extracted 1 h after P1 infection banded in CsCl gradients at a "heavy" position. The latter result contrasts with the results of physiological experiments that seem to be most simply explained by assuming that stable transduction results from a single strand replacement (Hanks and Masters, 1987). It should also be noted that if only 10 kb of transduced DNA is inserted into the bacterial chromosome, there should be little cotransduction of markers that are more than 0.25 min apart. Since markers that are 0.5 min apart are cotransduced with a frequency of 50% (Harriman, 1971), there is reason to doubt the 10-kb upper limit for DNA insertion.

The frequency of P1 transduction can vary over about a 30-fold range (Masters, 1977). As noted in the previous section, except for gene dosage effects dependent upon the distance of the marker from the origin of *E. coli* replication, much of the difference in transduction frequency is due to events that occur in the recipient cell. An elevated frequency of transduction of markers very close to the bacterial replication origin is due to an ability of the transduced DNA containing that origin to replicate in the recipient cell, increasing its copy number and thereby affording that DNA a greater chance for subsequent integration in the chromosome (Masters, 1977; M. Masters, personal communication). Variability in transduction frequency for markers that are not near the origin of replication is due to variation in the efficiency of stably inserting different segments of chromosomal DNA into the bacterial chromosome.

What is responsible for the marker discrimination in the recipient of transducing DNA? To answer this question, physical and genetic factors affecting the recombinational uptake of DNA have been studied. If phage lysates are treated with UV light, or if the recombination pathway used in the recipient cell is the *recF* pathway, then frequency differences for non-origin-linked markers are virtually eliminated. This occurs by increases in the transduction frequencies of the poorly transduced markers (Newman and Masters, 1980; Masters et al., 1984). One mechanism proposed for the selective stimulation of transduction by irradiation rests on the assumption that recombination for poorly transduced markers is limited by the concentration of functional RecA protein in the recipient cell. Selective stimulation is then produced by the induction of RecA activity after the introduction of damaged DNA into cells (Little and Mount, 1982). However, this is probably not the mechanism at work, because elevation of the level of RecA protein in the cell or introduction of irradiated P1 DNA into a cell at the time of transduction is not sufficient to stimulate transduction or eliminate marker discrimination (Newman and Masters, 1980; Masters et al., 1984). The more likely possibility is that DNA damage renders the DNA a better substrate for recombination (Benbow et al., 1974; Konrad, 1977). Consistent with this interpretation is the observation that much of the recombination deficiency of a recBCD null mutant is suppressed if the substrate DNA is irradiated (Porter et al., 1978; Hays and Boehmer, 1978; Masters et al., 1984).

Since the RecBCD nuclease appears to promote recombination by nicking DNA at chi sites (Stahl, 1979; Smith, 1983), and since chi sites influence the distribution of marker exchanges along the DNA in P1-mediated transduction events (Dower and Stahl, 1981), it is tempting to speculate that the discrimination between markers during transduction in recBCD+ bacteria is due to the presence or absence of chi sites near those markers. Markers transduced with high frequency would be expected to contain nearby chi sites, and those transduced with low frequency would be expected to be far from chi sites. UV damage in DNA probably stimulates recombination for markers without nearby chi sites by providing an alternative to the recBCD pathway for generating nicks in DNA (Smith and Hays, 1985).

b. Abortive Transduction

In genetic studies with phages P1 and P22, a majority of the transductants are found to be abortive; i.e., the transduced DNA is not integrated into the recipient cell genome but remains extrachromosomal and is transferred to only one daughter cell at cell division (Gross and Englesberg, 1959; Ozeki, 1956, 1959; Lederberg, 1956; Hertman and Luria, 1967). Abortively transduced DNA markers apparently do not replicate, since the number of abortive transductants remains constant with increasing growth of recipient cultures (Lederberg, 1956; Ozeki, 1959). In confirmation of these points, Sandri and Berger (1980b) showed that about 75% of the total ^{32}P-5-bromouracil-labeled P1 transducing DNA injected into recA+ cells remains unassociated with the chromosome and fully heavy for as long as 5 h after infection. Since stable transductions do not increase after the first hour of infection, this extrachromosomal, nonreplicating, abortively transduced DNA must only rarely give rise to stable transductants. When Sandri and Berger (1980b) lysed recipient bacteria with a nonionic detergent procedure, they discovered that as much as 60% of the abortively transduced DNA was in a circular form. Since that DNA was sensitive either to pronase treatment or to SDS treatment at 70°C, the DNA is presumed to be held in a ring by association with a protein. This accounts for its long-term stability.

What is the protein that so efficiently cyclizes DNA transduced by P1? The obvious candidate is the protein detected by Ikeda and Tomizawa (1965b). We have postulated that this protein is the same protein that cleaves *pac* and is found associated with transducing DNA because of its affinity for DNA ends (see Section VIII.E). We suggest here that when the transducing DNA is injected into the recipient cell, this protein, bound to one end of the DNA molecule, associates with the other to generate a ring. Less frequently, this association does not occur, and then the DNA either is rapidly degraded from its unprotected end or is stably inserted into the bacterial chromosome.

Two functions that map in the *darA* operon of P1 have been implicated in the transduction process, since a deletion of that operon increases the frequency of stable transduction (the *gta* function) and at the same time makes it insensitive to stimulation by UV light (the *teu* function). Yamamoto (1982) described a P1 amber mutant, *sus50*, that has these same two phenotypes as the *darA* operon deletion mutant. Moreover, Yamamoto also showed that the increase in stable transduction exhibited by the *sus50* mutant is associated with a decrease in the frequency of abortive transduction. Since the *sus50* mutation maps within or near the *darA* operon, it is likely that the *sus50* mutation and the *darA* deletion impair the same function. What is that function? Iida *et al.* (1987) have shown that the proteins of the *darA* and *darB* genes are DNA-binding proteins located within the phage head (see section V.D). It is unlikely, however, that either of these proteins is the product of the gene with the *sus50* mutation, as they are both found in purified heads isolated from a P1*sus50* lysate. Moreover, neither DarA nor DarB is the protein that Ikeda and Tomizawa (1965b) found covalently attached to the end of bacterial DNA in transducing particles, because both DarA and DarB proteins are found in particles containing phage DNA, and because they seem to be easily removed from DNA by phenol treatment (Iida *et al.*, 1987). The relationship between the internal proteins encoded by the *darA* and *darB* genes and the transduction phenotypes of the *gta, teu,* and *sus50* mutations is still obscure.

3. The Physiology of Bacterial Recipients of Transducing DNA

Transduction experiments in *E. coli* have shown that there is a long lag after P1 infection before the number of stable transductants increases. For exponentially growing cells with a generation time of 30 min, total cells in the population show a delay in growth of about one generation after P1 infection, whereas stable transductants show a delay of six generations (Sandri and Berger, 1980a). For P22 transductants, the growth delay of *Salmonella typhimurium* is about three generations (Ebel-Tsipis *et al.*, 1972). The P1 phenomenon has been most extensively studied by Bender and Sambucetti (1983) in *Klebsiella aeroginosa*. As in the *E. coli* study, a six-generation lag is also observed for the *Klebsiella* transductants. Moreover, the effect could be shown to be independent of the marker tested, despite 50-fold differences in transduction frequencies. Despite the long lag before the increase of transductants, Bender and Sambucetti (1983) could show that transduced cells become sensitive to penicillin within 1–2 generations after P1 infection, and, thus, they must continue to grow. These results are interpreted to mean that the transductants are growing as long, nondividing filaments and that the effect of recombinational transduction is to suppress cell division. Support for this interpretation comes from the demonstration that transductional recombinants were 1000 times more frequent among filamentous cells isolated by gradient centrifugation after P1 infection than they were among normal cells. Bender and Sambucetti (1983) speculate that the suppression of division is an SOS response (Little and Mount, 1982) triggered by the degraded DNA not incorporated in the final recombinant.

Attempts to repeat the experiments of Bender and Sambucetti (1983) in *E. coli* have failed to demonstrate a significant enrichment for stable transductants among filaments (Hanks and Masters, 1987). While filaments were 10-fold enriched for transductants at the time of P1 infection in these experiments, the proportion of transduced filaments did not increase with further incubation; i.e., transduced cells were not being selectively recruited into the filamenting population. Hanks and Masters interpret the initially increased transduction frequency of filaments as simply due to the increased surface of the filaments. Larger cells are better targets for phage adsorption and so more likely to receive a transducing particle.

X. COMPARATIVE BIOLOGY

In the course of this review, we have indicated some striking structural and functional relationships linking P1 to other phages, to other plasmids, and to the bacterial hosts within which P1 lives. In this section we summarize these relationships and introduce others that we have not had occasion to mention previously. The eventual understanding of how such an organism as P1 came to be can emerge only by seeing P1 in relation to its contemporaries. We admit at the outset, however, that the question whence P1, like many other questions raised here, cannot be answered properly now. For a general discussion of the issues involved in phage evolution and speciation, the reader is referred to Chapter 1.

A. P1, the Plasmid

P1 is both a plasmid and a necessarily more complex structure, a bacteriophage. The critical element in a plasmid, the replicon capable of independent extrachromosomal existence, can be simple. The major plasmid replicon of P1, at least in *E. coli,* is the R replicon. Despite the existence of regulatory circuits that place phage replication under control of a separate plasmid replicon or *vice versa* (German and Syvanen, 1982; Miller and Malamy, 1983; Burck *et al.,* 1984), there is no evidence at present to suggest that the R replicon of P1 is controlled by specifically viral functions. We therefore feel justified in considering P1, the plasmid, separately from P1, the phage. The assumption that the particular plasmid replicon associated with the remainder of the P1 genome is of little importance to P1 as a phage is supported by the characteristics of D6 (line 8, Table IV). This phage is similar to P1 in several respects, but it is heterologous at the plasmid *rep* region.

What organisms carry replicons that resemble the R replicon of P1? The closest resemblance is presumably among plasmids that belong to incompatibility group Y (the first 5 entries in Table IV). However, it would be a mistake to use incompatibility as the sole criterion of relatedness among replicons. Slight differences can render similar replicons, such as miniP1 and miniRts1, completely compatible (Abeles *et al.,* 1984; Mori *et al.,* 1986). Therefore comparative studies of basic replicons that rely on hybridization tests for DNA sequence homologies are becoming increasingly important supplements to purely physiological studies (see, e.g., Bergquist *et al.,* 1986). The relationship of the incY to the incT plasmid Rts1 (entry 6 in Table IV) was discovered fortuitously, and, to our knowledge, there has been no systematic attempt to determine by hybridization techniques the relatedness of P1 to plasmids of various other incompatibility groups or to determine whether the homology between P1 and Rts1 extends beyond the basic replicon.

Although two of the six known members of IncY are not active phages, p15B is clearly a defective prophage. The remaining member of IncY, pIP231, appears likely to have acquired its replicon from a phage, because the wild-type alleles of P1 amber mutations 135 and 180, probably affecting phage baseplate or tube and phage sheath, respectively (Walker and Walker, 1983), can be rescued from pIP231 (Briaux *et al.,* 1979).

It seems reasonable to postulate that a viral ancestor of P1 acquired an R-like replicon and then gave rise to phages of differing immunity and, by loss of viral functions, to plasmids such as p15B and pIP231. Perhaps even

TABLE IV. Organisms in Which the Plasmid Maintenance Region
or Immunity Region Is Related to That of P1

Name	Description	References
(1) P7	A generalized transducing phage from *E. coli* H, incompatible with P1. Its 95–98 kb of DNA is homologous to 90% of the P1 genome. It is reciprocally heteroimmune to P1 owing to non-homology at *immI* and is nonhomologous at *par* and lacks IS*1*. P7 is *ref*-defective possibly owing to interruption of *ref* by Tn902, an ampicillin resistance transposon.	Smith (1972); Chesney and Scott (1975); Walker and Walker (1976a); Wandersman and Yarmolinsky (1977); Yun and Vapnek (1977); Iida and Arber (1979); Windle and Hays (1986); Ludtke and Austin (1987)
(2) φW39	A generalized transducing phage from *E. coli* W39, incompatible with P1. Its plasmid molecular weight and host range are the same as for P1, and its DNA exhibits partial homology to the DNA of P1. Like P1, it carries a normally repressed *dnaB* analog. φW39 is reciprocally heteroimmune to P1/P7 and nonreciprocally to D6, which plates on a φW39 lysogen. The phage DNA probably contains a long terminal redundancy.	Yoshida and Mise (1984)
(3) j2	A generalized transducing phage from *Salmonella typhi*, incompatible with P1, though less strongly than is P1 with P1. The molecular weights of prophages j2 and P1 are similar. P1 and j2 are heteroimmune. Despite some serological cross-reactivity, the phages differ in host range.	Mise *et al.* (1981, 1983)
(4) p15B	A defective prophage in *E. coli* 15T⁻, incompatible with P1. Of its 94 kb, about 80% is homologous to P1, with which p15B forms viable hybrids. It expresses no immunity to P1, and its Res-Mod functions are distinct from those of P1, although functionally related. It carries a 3.5-kb invertible segment with apparently six crossover sites, no homology to the C segment of P1, but some to the P1 *cin* gene. It lacks an IS*1* at map unit 24.	Ikeda *et al.* (1970); Arber and Wauters-Willems (1970); Meyer *et al.* (1983, 1986); Iida *et al.* (1983); H. Sandmeier, S. Iida, J. Meyer, and W. Arber (personal communication)
(5) pIP231	A conjugative plasmid isolated from an *E. coli* strain. pIP231 is highly incompatible with P1 and P7 and is homologous to 15% of the P1 genome. It carries wild-type alleles of certain P1 mor-	Stoleru *et al.* (1972); Briaux *et al.* (1979); Briaux-Ger-

(Continued)

TABLE IV. (Continued)

Name	Description	References
	phogenetic mutations rescuable by recombination. It carries IS1 and has homology to Tn902. It exhibits extensive homology to F and carries an ssb gene similar to one in F and several other conjugative plasmids. This ssb gene lacks obvious homology to the P1 gene, which also encodes an ssbA-complementing protein. pIP231 bears a determinant of H_2S production and Tn1523, a tetracycline resistance transposon.	baud et al. (1981); Capage et al. (1982); Golub and Low (1985, 1986); E. Golub, personal communication
(6) Rtsl and R401	Drug resistance transfer factors isolated from Proteus vulgaris and P. rettgeri, respectively. The rep region of P1 exhibits evident DNA sequence homology with the nearly identical rep regions of Rtsl and R401, but the R factors, which grow in E. coli, belong to IncT.	Coetzee et al. (1972); Kamio et al. (1984) (see Fig. 9)
(7) F	A sex factor identified in Escherichia (B.) coli. Slight amino acid sequence homologies and obvious organizational similarity between F and P1 are detected in the rep-par region. Both plasmids are composites of disparate replicons. F belongs to IncFI.	Hayes (1953); Mori et al. (1986) (see Figs. 9, 10)
(8) D6	A generalized transducing phage from Salmonella oranienburg also active on E. coli. D6 is non-reciprocally heteroimmune with P1 and P7, plaquing with low efficiency on E. coli (P1). D6 and P1 prophages are similar in molecular weight, and the phages are morphologically similar and related by tests for serological cross-reactivity and for DNA homology. Nevertheless, complementation and recombination tests fail. D6 is compatible with P1 and P7 and lacks a res-mod system.	Mise and Suzuki (1970); Watkins and Scott (1981)

the plasmid replicon of the Epstein-Barr virus (EBV) can be traced to an ancestor of P1, although an apparent organizational similarity in the minimal region required for plasmid maintenance (Rawlins et al., 1985) can also be ascribed to parallel evolution.* This postulated course of evolution, in which a prophage adopts a plasmid mode of propagation by incorporation of a replicon already adapted to controlled low-copy-number maintenance, may be contrasted with the known genesis of λdv. The λvir-derived plasmid utilizes a viral replicon that is inappropriately regulated for maintenance of a plasmid. Consequently that maintenance is rela-

---

*Within a 1.8-kb region of EBV required for plasmid maintenance is a cluster of tandem iterons 30 bp apart and paired dyad symmetry sites 21 bp apart. These iterons bind the EBNA (Epstein-Barr nuclear antigen) protein. It has been suggested that the tandem repeat cluster of EBV is the analog of P1's incA and acts like a "sink" for controlling the rate of binding of EBNA protein (a RepA analogue) to the dyad symmetry sites at the putative replication origin of the EBV plasmid (Rawlins et al., 1985).

tively unstable (Matsubara, 1981; Womble and Rownd, 1986). In this respect, the plasmid form of P4*vir* may also resemble λdv (see Bertani and Six, 1988).

The putative donor of the R-like replicon was probably a low-copy-number plasmid, a progenitor of the closely related Rts1 and the more distantly related F (see Fig. 9). The plasmid, in turn, may have originated from a bacterial chromosome. The presence of dnaA boxes in the origin regions that are similarly organized to that of P1 (as depicted in Fig. 9) links these replication origins to those of both gram-negative and gram-positive bacteria (Zyskind *et al.*, 1983; Moriya *et al.*, 1985). The impor-tance of *dam* methylation sites in the P1 origin and elsewhere in the genome (see Sections VII.B.4 and VIII) suggests that P1 originated within a relatively recent branch of the bacterial evolutionary tree that possesses the *dam* gene (Barbeyron *et al.*, 1984) and in which we still find P1 and its relatives.

B. P1, the Phage

Whether or not the plasmid replicon of P1 is ultimately of bacterial provenance, it is likely that some, if not most, genes of P1 are derived from bacterial genes. Sequence comparisons between the dam genes of P1 and *E. coli* reveal considerable homology at the proximal end although the P1 gene is considerably longer (J. Coulby and N. Sternberg, unpublished).

With respect to specifically viral structures, the capsid and the injection machinery, P1 resembles a number of other phages. In its morphology (Fig. 1), P1 resembles T4. The molecular weights of the major sheath and tube proteins of T4 (71,160 and 18,460, respectively) (F. Arisaka, T. Nakaka, M. Takahashi, and S. Ishii, personal communication) are similar to the molecular weights (72 and 21.4 kD) of major tail proteins in P1 that are assumed to have the corresponding functions (Walker and Walker, 1981). Cyclically permuted, terminally redundant DNA is characteristic of T4 as well as P1, and both phages have evolved anti-restriction mechanisms. Although it is not known whether P1 and T4 have evolved similar head and tail structures independently, the tail fibers of P1 and those of phages P7, Mu, D108, P2, and D6 must have a common ancestry (see Section III.C.1 and Table IV). They are known to be closely related on the basis of serological cross-reactivity and DNA sequence homology.

The recombinational alternation of host range is accomplished by machinery that exhibits significant conservation of DNA sequence in bacteria, phages, and nonviral plasmids. The conservation appears to be greater for the site-specific recombination functions, both the *cis*-acting sites and the *trans*-acting recombination proteins, than for the regions controlled by recombination. In p15B, for example, evidence from DNA hybridization shows the existence of homology to the *cin* gene of P1, but the invertible segment of P1 is replaced by nonhomologous genetic material of unknown function and the number of crossover sites has been increased (H. Sandmeier, S. Iida, J. Meyer, and W. Arber, personal communication). A clustered inversion region in certain plasmids with IncI type pili (R64, CoIIb, and R621a) (Komano *et al.*, 1986) and in *Herpes simplex* virus (Roizman, 1979) has been reported, but it remains to be determined whether these systems are in any way related.

As mentioned in Section VII.C.3, several plasmids specify site-specific recombination systems that play a role in their stabilization. A curious difference between P1 and F is that in P1 the recombination site *loxP* is located far from the origin of plasmid replication, whereas the recombination site that is presumed to play an analogous role in F is intimately connected to the plasmid replication origin (*ori-1*). Although we do not know whether the Cre recombination protein exhibits homology with the recombination protein of F (the D gene product), striking similarities are exhibited by the recombination sites. Each is a palindromic sequence interrupted by a short core sequence, an organization also found at the *att* sites of λ.

The *lox-cre* and the *cix-cin* inversion systems of P1 do not appear to be related, but an inversion system in the 2-micron plasmid of yeast uses a site (F*lox*) that shows considerable homology to *loxP* (Vetter *et al.*, 1983). We do not know the origins of the site-specific recombination systems present in P1, but we can attest to their widespread success.

It seems likely that favored hosts of P1 or of P1's immediate ancestors are responsible for the paucity of certain restriction sites in the DNA of the phage (see Fig. 2A). P1 DNA lacks a *Sal*I site (Bächi and Arber, 1977) and possesses only two *Xho*I sites. Its unique *Pst*I site and unique *Tth*111I site are located in the IS*1* that P1 may have acquired late in the course of evolution. (In P7, 3 of the 4 *Pst*I sites are located in Tn*902* (Iida and Arber, 1979), presumably a late acquisition as well.) P1 has a unique *Sac*II site and only two *Xba*I sites, whereas in 90 kb of DNA the expected number is 22 sites for each of these restriction enzymes. At present we do not know enough about the distribution in nature of restriction enzymes to say which bacterial genera might have selected against the particular underrepresented sequences.

The overall base composition of P1 DNA is perhaps more suggestive. The 46% G+C content of P1 (Ikeda and Tomizawa, 1965a) is between the values reported by Schildkraut *et al.* (1962) for *E. coli* (51%) and for *Proteus vulgaris* and *Pr. rettgeri* (39% and 42%). As seen in Table IV, P1 and most of its known relatives were isolated from *E. coli*, but the related plasmids Rts1 (with a G+C content of 45% (Shapiro, 1977)) and R401 were isolated from species of *Proteus*. If we look more closely at the base composition of P1 so as to see the inhomogeneities (Fig. 2B), we note that the region of the restriction-modification genes is singularly A+T-rich, suggesting a separate derivation, as might be expected for a region of DNA with a discrete and nonessential function.

The disposition of the two major immunity regions, *imm*C and *imm*I, 180° apart on the circular map, raises the possibility that P1 is a composite of two phages, each with its own immunity system. However, it is likely, in our view, that the complexity of immunity regulation and the dispersal of repressor-sensitive operators represents a primitive state of affairs rather than a result of the fusion of entire phage genomes. The centralized control we see in phages such as λ is probably the result of pruning and consolidation.

C. P1, the Pastiche

Comparison of P1 with other phages, plasmids, and their bacterial hosts reveals the modular nature of P1 construction. As with the lambdoid bacteriophages, the acquisition and exchange of modules appears to have played an important role in the origin of P1 and its relatives. However, insofar as P1 is an organism—and there is general agreement that it is—the pastiche is harmonious, and the elements of it are imbedded in a relatively fixed matrix.

What elements of the pastiche that we call P1 could have contributed to the genetic reassortments that occurred in the course of P1's evolutionary history? A list of such elements must include the site-specific recombination systems *cix-cin* (see Section III.C) and *lox-cre* (see Section IV.B), the recombinogenic element IS1, Ref and its special targets (Windle and Hays, 1986), and the hot spots in P1 for IS2 insertion described by Sengstag and Arber (1983, 1987) and for recombination during vegetative growth described by Walker and Walker (1975, 1976b). The system responsible for indirect induction by damaged P1 (see section VII.B.5) may also be assigned to this category. A detailed study of the members of Table IV might provide evidence as to which, if any, of these elements have been significant contributors to the shuffling of modules.

Are there other elements in P1 that enhance genetic exchanges of evolutionary importance? In the case of phage P22, it has been suggested that the antirepressor is such an element because it can induce any of a variety of prophages present in a new bacterial host (Susskind and Botstein, 1978). The product of the *ant* gene of P1 does not appear to have this property. It cannot even induce the prophage of P1's closest relative, P7. A rationale for the complex immunity system of P1 does not appear to come so easily.

Transcriptional control of very large operons is characteristic of the lambdoid phages. It has been suggested, again by Susskind and Botstein (1978), that this organization facilitates replacement of one segment by another in the course of evolution while maintaining proper regulation. Transcription of the late genes of P1 (with the possible exception of those in much of the fourth quadrant of the P1 map) occurs in relatively short units. This conclusion is based on a rudimentary transcription map of P1 (Mural, 1978) and on the degree to which genes expressed in the plasmid state and those expressed in the lytic cycle of phage development are interspersed. It is also based on the finding that *c*1 repressor binding sites are scattered over the genome rather than clustered uniquely near the $c1$ gene itself (see Fig. 7A). It is further supported by the existence of a few insertions (Yun and Vapnek, 1977; Fortson et al., 1979; deBruijn and Bukhari, 1978; Meyer et al., 1986) and a large inversion [from about map unit 24 to map unit 50 (Iida et al., 1985a)] that do not interfere with phage development. Perhaps the lack of a single center from which all late transcripts are controlled means that the evolution of P1 has been somewhat more conservative than has been the case with the lambdoid phages.

Arber et al. (1978, 1985) have presented evidence that under conditions that mimic those encountered by bacteria in their natural environment—stationary phases of long duration alternating with periods of active growth—the kinds of mutations that prevail in P1 are IS-mediated deletions and transpositions. The mutations scored were limited to those that block vegetative phage growth. The same conclusions were reached when the P1 was replaced with a P1-P15 hybrid that lacks IS1 and that is presumed to carry no other transposable element (Arber et al., 1980). However, considerable clustering of mutations was observed in both the P1 and the P1-P15 hybrid experiments, and therefore the conclusions drawn may be relevant to a limited number of phage genes. The molecular basis of this clustering is being studied further (Sengstag and Arber, 1983, 1987; Sengstag et al., 1983). Whatever the limitations of these experiments, they serve to remind us that the optimal growth conditions preferred in the laboratory are not the conditions under which natural selection normally operates.

D. Coda

Apart from any evolutionary perspective that might be gained from an appreciation of the many and varied connections of P1 to near and distant relatives, there are other satisfactions. We are reassured that P1 is of more than parochial interest, that its study has relevance for the understanding of widely disseminated and even ubiquitous vital processes. The study of P1 itself offers, within a small compass, an occasional vista of a beautiful and complex organization. Opportunities to glimpse this vista provide a motivation for our research and for the present undertaking, even if our efforts serve largely to bear out the truth in Sydney Brenner's words, "anything that is produced by evolution is bound to be a bit of a mess" (cited in Lewin, 1984).

ACKNOWLEDGMENTS. Members of the community of P1 workers have generously informed us of unpublished results and provided constructive criticism and encouragement. We thank particularly our co-workers during the protracted writing of this review: Kenneth Abremski, Dhruba Chattoraj, Gerald Cohen, James Eliason, Barbara Funnell, Norman

Grover, Egon Hansen, Ronald Hoess, and Subrata Pal, Richard Putscher, and colleagues Ann Abeles, Werner Arber, Fumio Arisaka, Stuart Austin, Barbara Baumstark, Giuseppe Bertani, Thomas Bickle, Allan Campbell, Michael Davis, Stanley Friedman, Barbara Froehlich, Efim Golub, Max Gottesman, John Hays, Shigeru Iida, Dietmar Kamp, Nancy Kleckner, H. E. David Lane, Douglas Ludtke, Kathy Martin, Millicent Masters, Jürg Meyer, Rüdiger Schmitt, Heinz Schuster, June Scott, Daniel Vapnek, Donald and Jean Walker, Sue Wickner, and Kazuo Yamaguchi. For cheerfully bearing the heavy burden of manuscript preparation, we thank Beverly Miller and Penelope Lockhart.

REFERENCES

Abe, M., 1974, The replication of prophage P1 DNA, *Mol. Gen. Genet.* **132**:63–72.

Abeles, A. L., 1986, P1 plasmid replication: Purification and DNA-binding activity of the replication protein, RepA, *J. Biol. Chem.* **261**:3548–3555.

Abeles, A. L., and Austin, S. J., 1987, P1 plasmid replication requires methylated DNA, *EMBO J.* **6**:3185–3189.

Abeles, A. L., Snyder, K. M., and Chattoraj, D. K., 1984, P1 plasmid replication: Replicon structure, *J. Mol. Biol.* **173**:307–324.

Abeles, A. L., Friedman, S. A., and Austin, S. J., 1985, Partition of unit-copy miniplasmids to daughter cells: III. The DNA sequence and functional organization of the P1 partition region, *J. Mol. Biol.* **185**:261–272. Erratum corrected in *J. Mol. Biol.* **189**:387.

Abelson, J., and Thomas, C. A. Jr., 1966, The anatomy of the T5 bacteriophage DNA molecule, *J. Mol. Biol.* **18**:262–291.

Abremski, K., and Hoess, R., 1983, Bacteriophage P1 site-specific recombination: Purification and properties of the Cre recombinase protein, *J. Biol. Chem.* **259**:1509–1519.

Abremski, K., and Hoess, R., 1985, Phage P1 Cre-*loxP* site-specific recombination: Effects of DNA supercoiling on catenation and knotting of recombinant products, *J. Mol. Biol.* **184**:211–220.

Abremski, K., Hoess, R., and Sternberg, N., 1983, Studies on the properties of P1 site-specific recombination: Evidence for topologically unlinked products following recombination, *Cell* **32**:1301–1311.

Adams, J. N., and Luria, S. E., 1958, Transduction by bacteriophage P1. Abnormal phage function in transducing particles, *Proc. Natl. Acad. Sci. USA* **44**:590–594.

Alton, N. K., and Vapnek, D., 1979, Nucleotide sequence of analysis of the chloramphenicol resistance transposon Tn9, *Nature* **282**:864–869.

Amati, P., 1962, Abortive infection of *Pseudomonas aeruginosa* and *Serratia marcescens* with coliphage P1, *J. Bacteriol.* **83**:433–434.

Amati, P., 1965, A case of P1*dl* prophage curable by acridine orange treatment, *Atti Assoc. Genet. It. Pavia* **10**:79–85.

Amundsen, S. K., Taylor, A. F., Chaudhury, A. M., and Smith, G. R., 1986, *recD*, the gene for an essential third subunit of exonuclease. V. *Proc. Natl. Acad. Sci. USA* **83**:5558–5562.

Anderson, T. F., and Walker, D. H. Jr., 1960, Morphological variants of the bacteriophage P1, *Science* **132**:1488.

Arber, W., 1960a, Polylysogeny for bacteriophage lambda, *Virology* **11**:250–272.

Arber, W., 1960b, Transduction of chromosomal genes and episomes in *Escherichia coli*, *Virology* **11**:273–288.

Arber, W., and Dussoix, D., 1962, Host specificity of DNA produced by *Escherichia coli*. I. Host controlled modification of bacteriophage λ, *J. Mol. Biol.* **5**:18–36.

Arber, W., and Morse, M. L., 1965, Host specificity of DNA produced by *Escherichia coli*. VI. Effects on bacterial conjugation, *Genetics* **51**:137–148.

Arber, W., and Wauters-Willems, D., 1970, Host specificity of DNA produced by *Escherichia coli*. XII. The two restriction and modification systems of strain 15T⁻, *Mol. Gen. Genet.* **108**:203–217.

Arber, W., Hattman, S., and Dussoix, D., 1963, On the host-controlled modification of bacteriophage λ, *Virology* **21**:30–35.

Arber, W., Yuan, R., and Bickle, T. A., 1975, Post-synthetic modification of macromolecules, *FEBS Proc.* **34**:3–22.

Arber, W., Iida, S., Jütte, H., Caspers, P., Meyer, J., and Hänni, C., 1978, Rearrangements of genetic material in *Escherichia coli* as observed on the bacteriophage P1 plasmid, *Cold Spring Harbor Symp. Quant. Biol.* **43**:1197–1208.

Arber, W., Hümberlin, M., Caspers, P., Reif, H. J., Iida, S., and Meyer, J., 1980, Spontaneous mutations in the *Escherichia coli* prophage P1 and IS-mediated processes, *Cold Spring Harbor Symp. Quant. Biol.* **45**:38–40.

Arber, W., Sengstag, C., Caspers, P., and Dalrymple, B., 1985, Evolutionary relevance of genetic rearrangements involving plasmids, in: *Plasmids in Bacteria* (D. R. Helinski, S. N. Cohen, D. B. Clewell, D. A. Jackson, and A. Hollaender, eds.), pp. 21–31, Plenum Press, New York.

Austin, S., 1984, Bacterial plasmids that carry two functional centromere analogs are stable and are partitioned faithfully, *J. Bacteriol.* **158**:742–745.

Austin, S., and Abeles, A. L., 1983a, The partition of unit-copy mini-plasmids to daughter cells. I. P1 and F mini-plasmids contain discrete, interchangeable sequences sufficient to promote equipartition, *J. Mol. Biol.* **169**:353–372.

Austin, S., and Abeles, A. L., 1983b, The partition of unit-copy mini-plasmids to daughter cells. II. The partition region of mini-P1 encodes an essential protein and a centromere-like site at which it acts, *J. Mol. Biol.* **169**:373–387.

Austin, S., and Abeles, A., 1985, The partition functions of P1, P7 and F mini-plasmids, in: *Plasmids in Bacteria* (D. R. Helinski, S. N. Cohen, D. B. Clewell, D. A. Jackson, and A. Hollaender, eds.), pp. 215–226, Plenum Press, New York.

Austin, S., and Wierzbicki, A., 1983, Two mini-F encoded proteins are essential for equipartition, *Plasmid* **10**:73–81.

Austin, S., Sternberg, N., and Yarmolinsky, M., 1978, Miniplasmids of bacteriophage P1. I. Stringent plasmid replication does not require elements that regulate the lytic cycle, *J. Mol. Biol.* **120**:297–309.

Austin, S., Ziese, M., and Sternberg, N., 1981, A novel role for site-specific recombination in maintenance of bacterial replicons, *Cell* **25**:729–736.

Austin, S., Hart, F., Abeles, A., and Sternberg, N., 1982, Genetic and physical map of a P1 miniplasmid, *J. Bacteriol.* **152**:63–71.

Austin, S., Mural, R., Chattoraj, D., and Abeles, A., 1985, *Trans-* and *cis*-acting elements for the replication of P1 miniplasmids, *J. Mol. Biol.* **83**:195–202.

Austin, S., Friedman, S., and Ludtke, D., 1986, The partition functions of three unit-copy plasmids can stabilize maintenance of plasmid pBR322 at low copy number, *J. Bacteriol.* **168**:1010–1013.

Bächi, B., and Arber, W., 1977, Physical mapping of *Bgl*II, *Bam*HI, *Eco*RI, *Hind*III and *Pst*I restriction fragments of bacteriophage P1 DNA, *Mol. Gen. Genet.* **153**:311–324.

Bächi, B., Reiser, J., and Pirrotta, V., 1979, Methylation and cleavage sequences of the *Eco*P1 restriction-modification enzyme, *J. Mol. Biol.* **128**:143–163.

Bachmann, B., 1983, Linkage map of *Escherichia coli* K12, edition 7, *Microbiol. Rev.* **47**:180–230.

Backhaus, H., 1985, DNA packaging initiation of *Salmonella* bacteriophage P22: Determination of cut sites within the DNA sequence coding for gene 3, *J. Virol.* **55**:458–465.

Bailone, A., Brandenburger, A., Lévine, M., Pierre, M., Dutreix, M., and Devoret, R., 1984, Indirect SOS induction is promoted by ultraviolet light-damaged miniF and requires the miniF *lynA* locus, *J. Mol. Biol.* **179**:367–390.

Barbeyron, T., Kean, K., and Forterre, P., 1984, DNA adenine methylation of GATC sequences appeared recently in the *Escherichia coli* lineage, *J. Bacteriol.* **160**:586–590.

Baumstark, B. R., and Scott, J. R., 1980, The c1 repressor of bacteriophage P1. I. Isolation of the c1 protein and determination of the P1 DNA region to which it binds, *J. Mol. Biol.* **140**:471–480.

Baumstark, B. R., and Scott, J. R., 1987, The c4 gene of phage P1, *Virology* **156**:197–203.

Baumstark, B. R., Lowery, K., and Scott, J. R., 1984, Location by DNA sequence analysis of cop mutations affecting the number of plasmid copies of prophage P1, *Mol. Gen. Genet.* **194**:513–516.

Baumstark, B. R., Stovall, S. R., and Ashkar, S., 1987, Interaction of the P1c1 repressor with P1 DNA. Localization of repressor binding sites near the c1 gene, *Virology,* **156**:404–413.

Benbow, R. M., Zuccarelli, A. J., and Sinsheimer, R. L., 1974, A role for single stranded breaks on bacteriophage ϕX174 genetic recombination, *J. Mol. Biol.* **88**:629–665.

Bender, R. A., and Sambucetti, L. C., 1983, Recombination-induced suppression of cell division following P1-mediated generalized transduction in *Klebsiella aerogenes, Mol. Gen. Genet.* **189**:263–268.

Bergquist, P., Saadi, S., and Maas, W. K., 1986, Distribution of basic replicons having homology with RepFIA, RepFIB, and RepFIC among IncFI group plasmids, *Plasmid* **15**:19–34.

Bertani, L. E., and Six, E. W., 1988, The P2-like phages and their parasite, P4, in: *The Bacteriophages* (R. Calendar, ed.), Vol. 2, pp. 73–143, Plenum, New York.

Bertani, G., 1951, Studies in lysogenesis. I. The mode of phage liberation by lysogenic *Escherichia coli, J. Bacteriol.* **62**:293–300.

Bertani, G., 1958, Lysogeny, *Adv. Virus Res.* **5**:151–193.

Bertani, G., and Nice, S. J., 1954, Studies on lysogenesis. II. The effect of temperature on the lysogenization of *Shigella dysenteriae* with phage P1, *J. Bacteriol.* **67**:202–209.

Beyersmann, D., and Schuster, H., 1971, DNA synthesis in P1 infected *E. coli* mutants temperature-sensitive in DNA replication, *Mol. Gen. Genet.* **114**:173–176.

Bex, F., Karoui, H., Rokeach, L., Drèze, P., Garcia, L., and Couturier, M., 1983, Mini-F encoded proteins: Identification of a new 10.5 kilodalton species, *EMBO J.* **2**:1853–1861.

Black, L. W., and Showe, M. K., 1983, Morphogenesis of the T4 head, in: *Bacteriophage T4* (C. K. Mathews, E. M. Kutter, G. Mosig, and P. B. Berget, eds.), pp. 219–245, American Society for Microbiology, Washington.

Boice, L. B., and Luria, S. E., 1963, Behavior of prophage P1 in bacterial matings. I. Transfer of the defective prophage P1dl, *Virology* **20**:147–157.

Borek, E., and Ryan, A., 1958, The transfer of irradiation-elicited induction in a lysogenic organism. *Proc. Natl. Acad. Sci. USA* **44**:374–377.

Bornhoeft, J. W., and Stodolsky, M., 1981, Lytic cycle replicative forms of bacteriophages P1 and P1dl concatemer forms, *Virology* **112**:518–528.

Botstein, D., Waddell, C. H., and King, J., 1973, Mechanism of head assembly and DNA encapsidation in *Salmonella* phage P22. I. Genes, proteins, structures, and DNA maturation, *J. Mol. Biol.* **80**:669–695.

Botstein, D., Lew, K., Jarvik, V., and Swanson, C. Jr., 1975, Role of antirepressor in the bipartite control of repression and immunity by bacteriophage P22, *J. Mol. Biol.* **91**:439–462.

Brandenburger, A., Bailone, A., Lévine, A., and Devoret, R., 1984, Gratuitous induction, *J. Mol. Biol.* **179**:571–576.

Braun, R. E., and Wright, A., 1986, DNA methylation differentially enhances the expression of one of the two *E. coli* dnaA promoters *in vivo* and *in vitro, Mol. Gen. Genet.* **202**:246–250.

Briaux, S., Gerbaud, G., and Jaffé-Brachet, A., 1979, Studies of a plasmid coding for tetracycline resistance and hydrogen sulfide production incompatible with the prophage P1, *Mol. Gen. Genet.* **170**:319–325.

Briaux-Gerbaud, S., Gerbaud, G., and Jaffé-Brachet, A., 1981, Transposition of a tetracycline-resistance determinant (Tn1523) and cointegration events mediated by the pIP231 plasmid in *Escherichia coli, Gene* **15**:139–149.

Brockes, J. P., 1973, The DNA modification enzyme of bacteriophage P1: Subunit structure, *Biochem. J.* **133**:629–633.

Brockes, J. P., Brown, P. R., and Murray, K., 1974, Nucleotide sequences at the sites of action of the deoxyribonucleic acid modification enzyme of bacteriophage P1, *J. Mol. Biol.* **88**:437–443.

Burck, C., Shapiro, J. A., and Hauer, B., 1984, The pλCM system: Phage immunity-specific incompatibility with IncP-1 plasmids, *Mol. Gen. Genet.* **194**:340–342.

Campbell, A. M., 1969, *The Episomes*, Harper and Row, New York, pp. 13, 162.

Capage, M. A., and Scott, J. R., 1983, SOS induction by P1 Km miniplasmids, *J. Bacteriol.* **155**:473–480.

Capage, M. A., Goodspeed, J. K., and Scott, J. R., 1982, Incompatibility group Y member relationships: pIP231 and plasmid prophages P1 and P7, *Plasmid* **8**:307–311.

Casjens, S., and Huang, W. M., 1982, Initiation of sequential packaging of bacteriophage P22 DNA, *J. Mol. Biol.* **157**:287–298.

Caspar, D. L. D., and Klug, A., 1962, Physical principles in the construction of regular viruses, *Cold Spring Harbor Symp. Quant. Biol.* **27**:1–24.

Chattoraj, D., Cordes, K., and Abeles, A., 1984, Plasmid P1 replication: Negative control by repeated DNA sequences, *Proc. Natl. Acad. Sci. USA* **81**:6456–6460.

Chattoraj, D. K., Abeles, A. L., and Yarmolinsky, M. B., 1985a, P1 plasmid maintenance: A paradigm of precise control, in: *Plasmids in Bacteria* (D. R. Helinski, S. N. Cohen, D. B. Clewell, D. A. Jackson, and A. Hollaender, eds.), pp. 355–381, Plenum Press, New York.

Chattoraj, D. K., Abeles, A. L., and Yarmolinsky, M. B., 1985a, P1 plasmid maintenance: A paradigm of precise control, in: *Plasmids in Bacteria* (D. R. Helinski, S. N. Cohen, D. B. Clewell, D. A. Jackson, and A. Hollaender, eds.), pp. 355–381, Plenum Press, New York.

Chattoraj, D. K., Mason, R. J., and Wickner, S. H., 1988, Mini-P1 plasmid replication: The autoregulation-sequestration paradox, *Cell* **52** (in press).

Chattoraj, D. K., Snyder, K. M., and Abeles, A. L., 1985b, P1 plasmid replication: Multiple functions of RepA protein at the origin, *Proc. Natl. Acad. Sci. USA* **82**:2588–2592.

Chattoraj, D. K., Pal, S. K., Swack, J. A., Mason, R. J., and Abeles, A. L., 1985c, An auto-regulatory protein is required for P1 replication, in: *Sequence Specificity in Transcription and Translation*, UCLA Symposia on Molecular and Cellular Biology, New Series (R. Calendar and L. Gold, eds.), Vol. 30, pp. 271–280, Aan R. Liss, New York.

Chelala, C. A., and Margolin, P., 1974, Effects of deletions on co-transduction linkage in *Salmonella typhimurium*. Evidence that bacterial chromosome deletions affect the formation of transducing DNA fragments, *Mol. Gen. Genet.* **131**:97–112.

Chesney, R. H., and Adler, E., 1982, Chromosomal location of attP7, the *recA-* independent P7 integration site used in the suppression of *Escherichia coli dnaA* mutations, *J. Bacteriol.* **150**:1400–1404.

Chesney, R. H., and Scott, J. R., 1975, Superinfection immunity and prophage repression in phage P1. II. Mapping of the immunity difference and ampicillin resistance loci of phage P1 and φAMP, *Virology* **67**:375–384.

Chesney, R. H., and Scott, J. R., 1978, Suppression of a thermosensitive *dnaA* mutation of *Escherichia coli* by bacteriophage P1 and P7, *Plasmid* **1**:145–163.

Chesney, R. H., Vapnek, D., and Scott, J. R., 1978, Site-specific recombination leading to the integration of phages P1 and P7, *Cold Spring Harbor Symp. Quant. Biol.* **43**:1147–1150.

Chesney, R. H., Scott, J. R., and Vapnek, D., 1979, Integration of the plasmid prophages P1 and P7 into the chromosome of *Escherichia coli*, *J. Mol. Biol.* **130**:161–173.

Chow, L. T., and Bukhari, A. I., 1976, The invertible DNA segments of coliphage Mu and coliphage P1 are identical, *Virology* **74**:242–248.

Chow, L. T., Broker, T. R., Kahmann, R., and Kamp, D., 1978, Comparison of the G DNA inversion in bacteriophages Mu, P1 and P7, in: *Microbiology—1978* (D. Schlessinger, ed.), pp. 55–56, American Society for Microbiology, Washington.

Churchward, G., Linder, P., and Caro, L., 1983, The nucleotide sequence of replication and maintenance functions encoded by plasmid pSC101, *Nucleic Acids Res.* **11**:5645–5659.

Clowes, R. C., 1972, Molecular structure of bacterial plasmids, *Bacteriol. Rev.* 36:361–405.

Coetzee, J. N., Datta, N., and Hedges, R. W., 1972, R factors from *Proteus rettgeri*, *J. Gen. Microbiol.* **72**:543–552.

Cohen, A., and Clark, A. J., 1986, Synthesis of linear plasmid multimers in *Escherichia coli,* *J. Bacteriol.* **167**:327–335.

Cohen, G., 1983, Electron microscopy study of early lytic replication forms of bacteriophage P1 DNA, *Virology* **131**:159–170.

Cowan, J. A., and Scott, J. R., 1981, Incompatibility among group Y plasmids, *Plasmid* **6**:202–221.

Cress, D. E., and Kline, B. C., 1976, Isolation and characterization of *Escherichia coli* chromosomal mutants affecting plasmid copy number, *J. Bacteriol.* **125**:635–642.

D'Ari, R., 1977, Effects of mutations in the immunity system of bacteriophage P1, *J. Virol.* **23**:467–475.

D'Ari, R., and Huisman, O., 1982, DNA replication and indirect induction of the SOS response in *Escherichia coli,* *Biochimie* **64**:623–627.

D'Ari, R., Jaffé-Brachet, A., Touati-Schwartz, D., and Yarmolinsky, M., 1975, A *dnaB* analog specified by bacteriophage P1, *J. Mol. Biol.* **94**:341–366.

deBruijn, F. J., and Bukhari, A. I., 1978, Analysis of transposable elements inserted in the genomes of bacteriophages Mu and P1, *Gene* **3**:315–331.

Delhalle, E., 1973, Restriction et modification de bactériophages par *Escherichia coli* K12 impliquant un prophage P1 cryptique associé à différents plasmides, *Ann. Microbiol. (Paris)* **124A**:173–178.

Devlin, B. H., Baumstark, B. R., and Scott, J. R., 1982, Superimmunity: Characterization of a new gene in the immunity region of P1, *Virology* **120**:360–375.

Dower, N. A., and Stahl, F. W., 1981, χ activity during transduction-associated recombination, *Proc. Natl. Acad. Sci. USA* **78**:7033–7037.

Dreiseikelmann, B., Velleman, M., and Schuster, H., 1988, The c1 repressor of bacteriophage P1. Isolation and characterization of the repressor protein, *J. Biol. Chem.* **263**:1391–1397.

Dunn, T., Hahn, S., Ogden, S., and Schlief, R., 1984, An operator at −280 base pairs that is required for repression of *araBAD* operon promoter: Addition of DNA helical turns between the operator and promoter cyclically hinders repression, *Proc. Natl. Acad. Sci. USA* **81**:5017–5020.

Dussoix, D., and Arber, W., 1962, Host specificity of DNA produced by *Escherichia coli* II. Control over acceptance of DNA from infecting phage λ. *J. Mol. Biol.* **5**:37–49.

Ebel-Tsipis, J., Fox, M. S., and Botstein, D., 1972, Generalized transduction by bacteriophage P22 in *Salmonella typhimurium.* II. Mechanism of integration of transducing DNA, *J. Mol. Biol.* **71**:449–469.

Edelbluth, C., Lanka, E., Von der Hude, W., Mikolajczyk, M., and Schuster, H., 1979, Association of the prophage P1*ban* protein with the *dnaB* protein of *Escherichia coli.* Overproduction of *ban* protein by a P1*bac crr* mutant, *Eur. J. Biochem.* **94**:427–435.

Eichenlaub, R., Wehlmann, H., and Ebbers, J., 1981, Plasmid mini-F encoded functions involved in replication and incompatibility, in: *Molecular Biology, Pathogenicity and Ecology of Bacterial Plasmids* (S. B. Levy, E. L. Koenig, and R. C. Clowes, eds.), pp. 327–336, Plenum Press, New York.

Eliason, J. L., and Sternberg, N., 1987, Characterization of the binding sites of c1 repressor of bacteriophage P1: evidence for multiple asymmetric sites, *J. Mol. Biol.,* **198**:281–293.

Engelberg-Kulka, H., 1979, The requirement of nonsense suppression for the development of several phages, *Mol. Gen. Genet.* **170**:155–159.

Feiss, M., Widner, W., Miller, G., Johnson, G., and Christiansen, S., 1983a, Structure of bacteriophage λ cohesive end site: Location of the sites of terminase binding (*cosB*) and nicking (*cosN*), *Gene* **24**:207–218.

Feiss, M., Kobayashi, I., and Widner, W., 1983b, Separate sites for binding and nicking of bacteriophage λ DNA by terminase, *Proc. Natl. Acad. Sci. USA* **80**:955–959.

Fortson, M. R., Scott, J. R., Yun, T., and Vapnek, D., 1979, Map location of the kanamycin resistance determinant in P1Km$_o$, *Virology* **96**:332–334.

Franklin, N. C., 1969, Mutation in *galU* gene of *E. coli* blocks phage P1 infection, *Virology* **38**:189–191.

Friedman, D. I., Plantefaber, L. C., Olson, E. J., Carver, D., O'Dea, M. H., and Gellert, M., 1984, Mutations in the DNA *gyrB* gene that are temperature sensitive for lambda site-specific recombination, Mu growth, and plasmid maintenance, *J. Bacteriol.* **157**:490–497.

Friedman, S. A. and Austin, S. J., 1988, The P1 plasmid-partition system synthesizes two essential proteins from an autoregulated operon, *Plasmid* (in press).

Friedman, S., Martin, K., and Austin, S., 1986, The partition system of the P1 plasmid, in: *Banbury Report 24. Antibiotic Resistance Genes: Ecology, Transfer, and Expression*, Cold Spring Harbor Laboratory, Cold Spring Harbor, New York, pp. 285–294.

Froehlich, B. J., Tatti, K., and Scott, J. R., 1983, Evidence for positive regulation of plasmid prophage P1 replication: Integrative suppression by copy mutants, *J. Bacteriol.* **156**:205–211.

Froehlich, B. J., Watkins, C., and Scott, J. R., 1986, IS1-dependent generation of high copy number replicons from P1ApCm as a mechanism of gene amplification, *J. Bacteriol.* **166**:609–617.

Fuller, R. S., Funnell, B., and Kornberg, A., 1984, The DNA protein complex with the *E. coli* chromosomal replication origin (*oriC*) and other DNA sites, *Cell* **38**:889–900.

Fuller, R. S., Kaguni, J. M., and Kornberg, A., 1981, Enzymatic replication of the origin of the *Escherichia coli* chromosome, *Proc. Natl. Acad. Sci. USA* **78**:7370–7374.

Funnell, B. E., 1988, Mini-P1 plasmid partitioning: Excess parB protein destabilizes plasmids containing the centromere *parS*, *J. Bacteriol.* **170**:954–960.

Gellert, M., 1981, DNA topoisomerases, *Annu. Rev. Biochem.* **50**:879–910.

German, M., and Syvanen, M., 1982, Incompatibility between bacteriophage λ and the sex factor F, *Plasmid* **8**:207–210.

Gill, G. S., Hull, R. C., and Curtiss, R. III, 1981, Mutator bacteriophage D108 and its DNA: An electron microscopic characterization, *J. Virol.* **37**:420–430.

Glover, S. W., and Colson, C., 1969, Genetics of host controlled restriction and modification in *Escherichia coli*, phage P1 transduction, *Genet. Res.* **13**:227–240.

Glover, S. W., Schell, J., Symonds, N., and Stacey, K. A., 1963, The control of host-induced modification by phage P1, *Genet. Res.* **4**:480–482.

Godard, C., Beumer-Jochmans, M. P., and Beumer, J., 1971, Apparition des sensibilités aux phages T et à des colicines chez *Shigella flexneri* F6S survivant à l'infection par un phage Lisbonne. 1. Modification des propriétés biologiques de surface, *Ann. Inst. Pasteur* **120**:475–489.

Goldberg, R. B., Bender, R. A., and Streicher, S. L., 1974, Direct selection of phage P1-sensitive mutants of enteric bacteria, *J. Bacteriol.* **118**:810–814.

Golub, E. I., and Low, K. B., 1985, Conjugative plasmids of enteric bacteria from many different incompatibility groups have similar genes for single-stranded DNA-binding proteins, *J. Bacteriol.* **162**:235–241.

Golub, E. I., and Low, K. B., 1986, Unrelated conjugative plasmids have sequences which are homologous to the leading region of the F factor, *J. Bacteriol.* **166**:670–672.

Gottesman, S., and Zipser, D., 1978, Deg phenotype of *E. coli lon* mutants. *J. Bacteriol.* **133**:844–851.

Gross, J., and Englesberg, E., 1959, Determination of the order of mutational sites governing L-arabinose utilization in *Escherichia coli* B/r by transduction with phage P1bt, *Virology* **9**:314–331.

Haberman, A., 1974, The bacteriophage P1 restriction endonuclease, *J. Mol. Biol.* **89**:545–563.

Hadi, S. M., Bächi, B., Iida, S., and Bickle, T. A., 1983, DNA restriction-modification enzymes of phage P1 and plasmid p15B, *J. Mol. Biol.* **165**:19–34.

Hakkaart, M. J. J., Van den Elzen, P. J. M., Veltkamp, E., and Nijkamp, H. J. J., 1984, Maintenance of multicopy plasmid CloDF13 in *E. coli* cells: Evidence for site-specific recombination at parB, *Cell* **36**:203–209.

Hanks, M. C. and Masters, M., 1987, Transductional analysis of chromosome replication time, *Mol. Gen. Genet.* **210**:288–293.

Hansen, E. B., and Yarmolinsky, M. B., 1986, Host participation in plasmid maintenance:

Dependence upon *dnaA* of replicons derived from P1 and F, *Proc. Natl. Acad. Sci. USA* **83**:4423–4427.

Hansen, F. G., Hansen, E. B., and Atlung, T., 1982, The nucleotide sequence of the *dnaA* gene promoter and of the adjacent *rpmH* gene, coding for the ribosomal protein L34, of *Escherichia coli, EMBO J.* **1**:1043–1048.

Harriman, P., 1971, Appearance of transducing activity in P1 infected *Escherichia coli, Virology* **45**:324–325.

Harriman, P. D., 1972, A single-burst analysis of the production of P1 infectious and transducing particles, *Virology* **48**:595–600.

Hattman, S., Brooks, J. E., and Masurekar, M., 1978, Sequence specificity of the P1 modification methylase (M.*Eco*P1) and the DNA methylase (M.*Eco*dam) controlled by the *Escherichia coli dam* gene, *J. Mol. Biol.* **126**:367–380.

Hawley, D. K., and McClure, W. R., 1983, Compilation and analysis of *Escherichia coli* promoter DNA sequences, *Nucleic Acids Res.* **11**:2237–2255.

Hay, N., and Cohen, G., 1983, Requirement of *E. coli* DNA synthesis functions for the lytic replication of bacteriophage P1, *Virology* **131**:193–206.

Hayakawa, Y., Murotsu, T., and Matsubara, K., 1985, Mini-F protein that binds to a unique region for partition of mini-F plasmid DNA, *J. Bacteriol.* **163**:349–354.

Hayes, W., 1953, Observations on a transmissible agent determining sexual differentiation in *Bact. coli, J. Gen. Microbiol.* **8**:72–88.

Hays, J. B., and Boehmer, S., 1978, Antagonists of DNA gyrase inhibit repair and recombination of UV-irradiated phage lambda, *Proc. Natl. Acad. Sci. USA* **75**:4125–4129.

Hedges, R. W., Jacob, A. E., Barth, P. T., and Grinter, N. J., 1975, Compatibility properties of P1 and φAMP prophages, *Mol. Gen. Genet.* **141**:263–267.

Heilmann, H., Reeve, J. N., and Puhler, A., 1980a, Identification of the repressor and repressor bypass (antirepressor) polypeptides of bacteriophage P1 synthesized in infected minicells, *Mol. Gen. Genet.* **178**:149–154.

Heilmann, H., Burkardt, H. J., Puhler, A., and Reeve, J. N., 1980b, Transposon mutagenesis of the gene encoding the bacteriophage P1 restriction endonuclease. Co-linearity of the gene and gene product, *J. Mol. Biol.* **144**:387–396.

Heisig, A., Severin, I., Seefluth, A.-K., and Schuster, H., 1987, Regulation of the *ban* gene containing operon of prophage P1, *Mol. Gen. Genet.* **206**:368–376.

Hertman, I., and Luria, S. E., 1967, Transduction studies on the role of a *rec⁺* gene in the ultraviolet induction of prophage lambda, *J. Mol. Biol.* **23**:117–133.

Hertman, I., and Scott, J. R., 1973, Recombination of phage P1 in recombination deficient hosts, *Virology* **53**:468–470.

Hiestand-Nauer, R., and Iida, S., 1983, Sequence of the site-specific recombinase gene *cin* and of its substrates serving in the inversion of the C segment of bacteriophage P1, *EMBO J.* **2**:1733–1740.

Hiraga, S., 1986, Mechanisms of stable plasmid inheritance, *Adv. Biophys.* **21**:91–103.

Hiraga, S., Ogura, T., Mori, H., and Tanaka, M., 1985, Mechanisms essential for stable inheritance of mini-F plasmid, in: *Plasmids in Bacteria,* (D. R. Helinski, S. N. Cohen, D. B. Clewell, D. A. Jackson, and A. Hollaender, eds.), pp. 469–487, Plenum Press, New York.

Hiraga, S., Jaffé, A., Ogura, T., Mori, H., and Takahashi, H., 1986, F plasmid *ccd* mechanism in *Escherichia coli, J. Bacteriol.* **166**:100–104.

Hochman, L., Segev, N., Sternberg, N., and Cohen, G., 1983, Site-specific recombinational circularization of bacteriophage P1 DNA, *Virology* **131**:11–17.

Hochschild, A., and Ptashne, M., 1986, Cooperative binding of λ repressors to sites separated by integral turns of the DNA helix, *Cell* **44**:681–687.

Hoess, R., and Abremski, K., 1984, Interaction of the bacteriophage P1 recombinase Cre with the recombining site *loxP, Proc. Natl. Acad. Sci. USA* **81**:1026–1029.

Hoess, R., and Abremski, K., 1985, Mechanism of strand cleavage and exchange in the Cre-*lox* site-specific recombination system, *J. Mol. Biol.* **181**:351–362.

Hoess, R. H., Ziese, M., and Sternberg, N., 1982, P1 site-specific recombination: Nucleotide sequence of the recombining sites, *Proc. Natl. Acad. Sci. USA* **79**:3398–3402.

Hoess, R. H., Abremski, K., and Sternberg, N., 1984, The nature of the interaction of the P1 recombinase Cre with the recombining site *loxP*, *Cold Spring Harbor Symp. Quant. Biol.* **49**:761–769.

Hoess, R. H., Wierzbicki, A., and Abremski, K., 1986, The role of the *loxP* spacer region in P1 site-specific recombination, *Nucleic Acids Res.* **14**:2287–2300.

Hooper, D. C., Wolfson, J. S., McHugh, G. L., Swartz, M. D., Tung, C., and Swartz, M. N., 1984, Elimination of plasmid pMG110 from *Escherichia coli* by novobiocin and other inhibitors of DNA gyrase, *Antimicrob. Agents Chemother.* **25**:586–590.

Hooper, I., Woods, W. H., and Egan, B., 1981, Coliphage 186 replication is delayed when the host cell is UV irradiated before induction, *J. Virol.* **40**:341–349.

Horiuchi, K., and Zinder, N., 1972, Cleavage of bacteriophage f1 DNA by the restriction enzyme of *Escherichia coli* B, *Proc. Natl. Acad. Sci. USA* **69**:3220–3224.

Hu, M., and Deonier, R. C., 1981, Mapping of IS*1* elements flanking the *argF* gene region on the *Escherichia coli* K12 chromosome, *Mol. Gen. Genet.* **181**:222–229.

Huber, H. E., Iida, S., and Bickle, T. A., 1985a, Expression of the bacteriophage P1 *cin* recombinase gene from its own and heterologous promoters, *Gene* **34**:63–72.

Huber, H. E., Iida, S., Arber, W., and Bickle, T. A., 1985b, Site-specific DNA inversion is enhanced by a DNA sequence element in *cis*, *Proc. Natl. Acad. Sci. USA* **82**:3776–3780.

Hughes, P., Squali-Houssaini, F.-Z., Forterre, P., and Kohiyama, M., 1984, *In vitro* replication of a *dam* methylated and non-methylated *ori*-C plasmid, *J. Mol. Biol.* **176**:155–159.

Huisman, O., D'Ari, R., and George, J., 1980, Inducible *sfi* dependent division inhibition in *Escherichia coli*, *Mol. Gen. Genet.* **177**:629–636.

Hümberlin, M., Suri, B., Rao, D. N., Hornby, D. P., Eberle, H., Pripfl, T., and Bickle, T. A., 1988, The type III DNA restriction and modification systems EcoP1 and EcoP15. Nucleotide sequence of the EcoP1 operon and of the EcoP15 gene, *J. Mol. Biol.* (in press).

Iida, S., 1984, Bacteriophage P1 carries two related sets of genes determining its host range in the invertible C segment of its genome, *Virology* **134**:421–434.

Iida, S., and Arber, W., 1977, Plaque forming specialized transducing phage P1: Isolation of P1CmSmSu, a precursor of P1Cm, *Mol. Gen. Genet.* **153**:259–269.

Iida, S., and Arber, W., 1979, Multiple physical differences in the genome structure of functionally related bacteriophages P1 and P7, *Mol. Gen. Genet.* **173**:249–261.

Iida, S., and Arber, W., 1980, On the role of IS*1* in the formation of hybrids between bacteriophage P1 and the R plasmid NR1, *Mol. Gen. Genet.* **177**:261–270.

Iida, S., and Hiestand-Nauer, R., 1986, Localized conversion at the crossover sequences in the site-specific DNA inversion system of bacteriophage P1, *Cell* **45**:71–79.

Iida, S., and Hiestand-Nauer, R., 1987, Role of the central dinucleotide at the crossover sites for the selection of quasi sites in DNA inversion mediated by the site-specific Cin recombinase of phage P1, *Mol. Gen. Genet.* **208**:464–468.

Iida, S., Meyer, J., and Arber, W., 1978, The insertion element Is*1* is a natural constituent of coliphage P1 DNA, *Plasmid* **1**:357–365.

Iida, S., Meyer, J., and Arber, W., 1981, Cointegrates between bacteriophage P1 DNA and plasmid pBR322 derivatives suggest molecular mechanisms for P1-mediated transduction of small plasmids, *Mol. Gen. Genet.* **184**:1–10.

Iida, S., Meyer, J., Kennedy, K. E., and Arber, W., 1982, A site-specific conservative recombination system carried by bacteriophage P1. Mapping the recombinase gene *cin* and the crossover sites *cix* for the inversion of the C segment, *EMBO J.* **1**:1445–1453.

Iida, S., Meyer, J., Bächi, B., Stålhammer-Carlemalm, M., Schrickel, S., Bickle, T. A., and Arber, W., 1983, DNA restriction-modification genes of phage P1 and plasmid p15B. Structure and *in vitro* transcription, *J. Mol. Biol.* **165**:1–18.

Iida, S., Huber, H., Hiestand-Nauer, R., Meyer, J., Bickle, T. A., and Arber, W., 1984, The bacteriophage P1 site-specific recombinase Cin: Recombination events and DNA recognition sequences, *Cold Spring Harbor Symp. Quant. Biol.* **49**:769–777.

Iida, S., Meyer, J., and Arber, W., 1985a, Bacteriophage P1 derivatives unaffected in their growth by a large inversion or by IS insertions at various locations, *J. Gen. Microbiol.* **131:**129–134.

Iida, S., Hiestand-Nauer, R., Meyer, J., and Arber, W., 1985b, Crossover sites *cix* for inversion of the invertible DNA segment C on the bacteriophage P7 genome, *Virology* **143:**347–351.

Iida, S., Streiff, M. B., Bickle, T. A., and Arber, W., 1987, Two DNA anti-restriction systems of bacteriophage P1, *darA* and *darB:* Characterization of darA-phage, *Virology,* **157:**156–166.

Ikeda, H., and Tomizawa, J.-I., 1965a, Transducing fragments in generalized transduction by P1. I. Molecular origin of the fragments, *J. Mol. Biol.* **14:**85–109.

Ikeda, H., and Tomizawa, J.-I., 1965b, Transducing fragments in generalized transduction by phage P1. II. Association of DNA and protein in the fragments, *J. Mol. Biol.* **14:**110–119.

Ikeda, H., and Tomizawa, J.-I., 1965c, Transducing fragments in generalized transduction by phage P1. III. Studies with small phage particles. *J. Mol. Biol.* **14:**120–129.

Ikeda, H., and Tomizawa, J.-I., 1968, Prophage P1, an extrachromosomal replication unit, *Cold Spring Harbor Symp. Quant. Biol.* **33:**791–798.

Ikeda, H., Inuzuka, M., and Tomizawa, J.-I., 1970, P1-like plasmid in *Escherichia coli, J. Mol. Biol.* **50:**457–470.

Inselberg, J., 1966, Phage P1 modification of bacterial DNA studied by generalized transduction, *Virology* **30:**257–265.

Inselberg, J., 1968, Physical evidence for the integration of prophage P1 into the *Escherichia coli* chromosome, *J. Mol. Biol.* **31:**553–560.

Itoh, Y., Kamio, Y., and Terawaki, Y., 1987, The essential DNA sequence for the replication of Rts1, *J. Bacteriol.,* **169:**1153–1160.

Jacob, F., 1955, Transduction of lysogeny in *Escherichia coli, Virology* **1:**207–220.

Jacob, F., and Wollman, E. L., 1959, The relationship between the prophage and the bacterial chromosome in lysogenic bacteria, in: *Recent Progress in Microbiology* (G. Tunevall, ed.), pp. 15–30. Charles C. Thomas, Springfield, IL.

Jacob, F., and Wollman, E. L., 1961, *Sexuality and the Genetics of Bacteria,* Academic Press, New York.

Jacob, F., Brenner, S., and Cuzin, F., 1963, On the regulation of DNA replication in bacteria, *Cold Spring Harbor Symp. Quant. Biol.* **28:**329–347.

Jaffé, A., Ogura, T., and Hiraga, S., 1985, Effects of the *ccd* function of the F plasmid on bacterial growth, *J. Bacteriol.* **163:**841–849.

Jaffé-Brachet, A., and Briaux-Gerbaud, S., 1981, Curing of P1 prophage from *Escherichia coli* K-12 *recA*(P1) lysogens superinfected with P1 bacteriophage, *J. Virol.* **37:**854–859.

Jaffé-Brachet, A., and D'Ari, R., 1977, Maintenance of bacteriophage P1 plasmid, *J. Virol.* **23:**476–482.

Johnson, B. F., 1982, Suppression of the *lexC* (*ssbA*) mutation of *Escherichia coli* by a mutant of bacteriophage P1, *Mol. Gen. Genet.* **186:**122–126.

Johnson, R. C., Bruist, M. F., and Simon, M., 1986, Host protein requirements for *in vivo* site-specific DNA inversion, *Cell* **46:**531–539.

Johnson, R. C., and Simon, M. I., 1985, Hin-mediated site-specific recombination requires two 26 bp recombination sites and a 60 bp recombinational enhancer, *Cell* **41:**781–791.

Kahmann, R., Rudt, F., Koch, C., and Mertens, G., 1985, G inversion in bacteriophage Mu DNA is stimulated by a site within the invertase gene and a host factor, *Cell* **41:**771–780.

Kaiser, D., and Dworkin, M., 1975, Gene transfer to a myxobacterium by *Escherichia coli* phage P1, *Science* **187:**653–654.

Kamio, Y., Tabuchi, A., Itoh, Y., Katagiri, H., and Terawaki, Y., 1984, Complete nucleotide sequence of mini-Rts1 and its copy mutant, *J. Bacteriol.* **158:**307–312.

Kamp, D., Kahmann, R., Zipser, D., Broker, T. R., and Chow, L. T., 1978, Inversion of the G DNA segment of phage Mu controls phage infectivity, *Nature* **271:**577–580.

Kamp, D., Kardas, E., Ritthaler, W., Sandulache, R., Schmucker, R., and Stern, B., 1984, Comparative analysis of invertible DNA in phage genomes, *Cold Spring Harbor Symp. Quant. Biol.* **49:**301–311.

Karamata, D., 1970, Multiple density classes of phage P1 due to tetramer formation, *Mol. Gen. Genet.* **107**:243–255.

Karoui, H., Bex, F., Drèze, P., and Couturier, M., 1983, *Ham*22, a mini-F mutation which is lethal to host cell and promotes *recA*-dependent induction of lambdoid prophage, *EMBO J.* **2**:1863–1868.

Kass, L. R., and Yarmolinsky, M. B., 1970, Segregation of functional sex factor into minicells, *Proc. Natl. Acad. Sci. USA* **66**:815–822.

Kennedy, K. E., Iida, S., Meyer, J., Stålhammer-Carlemalm, M., Hiestand-Nauer, R., and Arber, W., 1983, Genome fusion mediated by the site-specific DNA inversion system of bacteriophage P1, *Mol. Gen. Genet.* **189**:413–421.

Kline, B. C., 1985, A review of mini-F plasmid maintenance, *Plasmid* **14**:1–16.

Kline, B. C., Miller, J. R., Cress, D. E., Wlodarszyk, M., Manis, J. J., and Otten, M. R., 1976, Non-integrated plasmid-chromosome complexes in *Escherichia coli*, *J. Bacteriol.* **127**:881–889.

Komano, T., Kubo, A., Kayanuma, T., Furuichi, T., and Nisioka, T., 1986, Highly mobile DNA segment of IncIα plasmid R64: A clustered inversion region, *J. Bacteriol.* **165**:94–100.

Kondo, E., and Mitsuhashi, S., 1964, Drug resistance of enteric bacteria. IV. Active transducing bacteriophage P1CM produced by the combination of R-factor with bacteriophage P1, *J. Bacteriol.* **88**:1266–1276.

Kondo, E., and Mitsuhashi, S., 1966, Drug resistance of enteric bacteria. VI. Introduction of bacteriophage P1*CM* into *Salmonella typhi* and formation of P1*dCM* and F-CM elements, *J. Bacteriol.* **91**:1787–1794.

Kondo, E., Haapala, D. K., and Falkow, S., 1970, The production of chloramphenicol acetyltransferase by bacteriophage P1CM, *Virology* **40**:431–440.

Konrad, E. B., 1977, Method for the isolation of *Escherichia coli* mutants with enhanced recombination between chromosomal duplications, *J. Bacteriol.* **130**:167–172.

Krüger, D. H., and Bickle, T. A., 1983, Bacterial survival: Multiple mechanisms for avoiding the deoxyribonucleic acid restriction systems of their hosts, *Microbiol. Rev.* **47**:345–360.

Kuner, J. M., and Kaiser, D., 1981, Introduction of transposon Tn5 into *Myxococcus* for analysis of development and other nonselectable mutants, *Proc. Natl. Acad. Sci. USA* **78**:425–429.

Kusukawa, N., Mori, H., Kondo, A., and Hiraga, S., 1987, Partitioning of the F plasmid: overproduction of an essential protein for partition inhibits plasmid maintenance, *Mol. Gen. Genet.* **208**:365–372.

Kutsukake, K., and Iino, T., 1980, Inversions of specific DNA segments in flagellar phase variation of *Salmonella* and inversion systems of bacteriophages P1 and Mu, *Proc. Natl. Acad. Sci. USA* **77**:7338–7341.

Lane, D., and Gardner, R. C., 1979, Second *Eco*RI fragment of F capable of self-replication, *J. Bacteriol.* **139**:141–151.

Lane, D., Hill, D., Caughey, E., and Gunn, P., 1984, The mini-F primary origin. Sequence analysis and multiple activities, *J. Mol. Biol.* **180**:267–282.

Lanka, E., and Schuster, H., 1970, Replication of bacteriophages in *E. coli* mutants thermosensitive in DNA synthesis, *Mol. Gen. Genet.* **106**:279–285.

Lanka, E., Edelbluth, C., Schlicht, M., and Schuster, H., 1978a, *Escherichia coli dnaB* protein, *J. Biol. Chem.* **253**:5847–5851.

Lanka, E., Mikolajczyk, M., Schlicht, M., and Schuster, H., 1978b, Association of the prophage P1*ban* protein with the *dnaB* protein of *Escherichia coli*, *J. Biol. Chem.* **253**:4746–4753.

Lanka, E., Schlicht, M., Mikolajczyk, M., Geschke, B., Edelbluth, C., and Schuster, H., 1978c, Suppression of *E. coli dnaB* mutants by prophage P1*bac*: A biochemical approach, in: *DNA Synthesis: Present and Future* (I. Molineux and M. Kohiyama, eds.), pp. 669–682, Plenum Press, New York.

Lawton, W. D., and Molnar, D. M., 1972, Lysogenic conversion of *Pasteurella* by *Escherichia coli* bacteriophage P1*CM*, *J. Virol.* **9**:708–709.

Lederberg, J., 1956, Linear inheritance in transductional clones, *Genetics* **41**:845–871.

Lederberg, S., 1957, Suppression of the multiplication of heterologous bacteriophages in lysogenic bacteria, *Virology* **3**:496–513.

Lee, H.-J., Ohtsubo, E., Deonier, R. C., and Davidson, N., 1974, Electron microscope heteroduplex studies of sequence relations among plasmids of *Escherichia coli*. V. *ilv*⁺ deletion mutants of F14, *J. Mol. Biol.* **89**:585–597.

Lennox, E. S., 1955, Transduction of linked genetic characters of the host by bacteriophage P1, *Virology* **1**:190–206.

Leonard, A. C., Hucul, J. A., Helmstetter, C. E., 1982, Kinetics of mini-chromosome replication in *Escherichia coli* B/r, *J. Bacteriol.* **149**:499–507.

Levine, M., Truesdale, S., Ramakrishnan, T., and Bronson, M. J., 1975, Dual control of lysogeny by phage P22, *J. Mol. Biol.* **91**:421–438.

Lewin, R., 1984, Why is development so illogical? *Science* **224**:1327–1329.

Lindberg, A. A., 1973, Bacteriophage receptors, *Ann. Rev. Genet.* **27**:205–241.

Linder, P., Churchward, G., Yi-Yi, X. G. Y., and Caro, L., 1985, An essential replication gene, *repA*, of plasmid pSC101 is autoregulated, *J. Mol. Biol.* **181**:383–393.

Linn, S., and Arber, W., 1968, Host specificity of DNA produced by *Escherichia coli* X. In vitro restriction of phage fd replicative form, *Proc. Natl. Acad. Sci. USA* **59**:1300–1306.

Little, J. N., and Mount, D. W., 1982, The SOS regulatory system of *Escherichia coli*, *Cell* **29**:11–13.

Luderitz, O., Jann, K., and Wheat, R., 1968, Somatic and capsular antigens of gram-negative bacteria, *Comprehensive Biochem.* **26A**:105–228.

Ludtke, D. N., and Austin, S. J., 1987, The plasmid-maintenance functions of P7 prophage, *Plasmid*, **18**:93–98.

Luria, S. E., Adams, J. N., and Ting, R. C., 1960, Transduction of lactose-utilizing ability among strains of *E. coli* and *Sh. dysenteriae* and the properties of the transducing phage particles, *Virology* **12**:348–390.

Lyons, S. M., and Schendel, P. F., 1984, Kinetics of methylation in *Escherichia coli* K12, *J. Bacteriol.* **159**:421–423.

MacHattie, L. A., and Jackowski, J. B., 1976, Physical structure and deletion effects of the chloramphenicol-resistance element, Tn9, in phage lambda, in: *DNA Insertion Elements, Plasmids and Episomes* (A. I. Bukhari, J. A. Shapiro, and S. L. Adhya, eds.), pp. 219–228, Cold Spring Harbor Laboratory, Cold Spring Harbor, NewYork.

MacQueen, H. A., and Donachie, W. D., 1977, Intracellular localization and effects on cell division of a plasmid blocked in deoxyribonucleic acid replication, *J. Bacteriol.* **132**:392–397.

Majumdar, A., and Adhya, S. L., 1984, Demonstration of two operator elements in *gal*: In vitro repressor binding studies, *Proc. Natl. Acad. Sci. USA* **81**:6100–6104.

Margolin, P., 1987, Generalized transduction, in: *Escherichia coli and Salmonella typhimurium. Cellular and Molecular Biology* (F. C. Neidhardt, ed.), American Society for Microbiology, Washington, Vol. 2, pp. 1154–1168.

Marinus, M. G., 1984, Methylation of prokaryotic DNA, in: *DNA Methylation* (A. Razin, M. Cedar, and A. Riggs, eds.), pp. 81–109, Springer-Verlag, New York.

Marinus, M. G., 1987, DNA methylation in *Escherichia coli*, *Ann. Rev. Genet.* **21**:113–131.

Martin, K. A., Friedman, S. A., and Austin, S. J., 1987, The partition site of the P1 plasmid, *Proc. Natl. Acad. Sci., USA* **84**:8544–8547.

Masters, M., 1977, The frequency of P1 transduction of the genes of *Escherichia coli* as a function of chromosomal position: Preferential transduction of the origin of replication, *Mol. Gen. Genet.* **155**:197–202.

Masters, M., 1985, Generalized transduction in: *The Genetics of Bacteria* (J. G. Scaife, D. Leach, and A. Galizzi, eds.), pp. 197–215, Academic Press, New York.

Masters, M., Newman, B. J., and Henry, C. M., 1984, Reduction of marker discrimination in transductional recombination, *Mol. Gen. Genet.* **196**:85–90.

Matsubara, K., 1981, Replication control system in lambda dv, *Plasmid* **5**:32–52.

Mattes, R., 1985, *Habilitationsschrift*, University of Regensburg, Regensburg, F.R.G., 1985.

Meselson, M., and Yuan, R., 1968, DNA restriction enzyme from *E. coli, Nature* **217**:1110–1114.

Messer, W., Bellekes, U., and Lother, H., 1985, Effect of *dam* methylation on the activity of the *E. coli* replication origin, *oriC, EMBO J.* **4**:1327–1332.

Meurs, E., and D'Ari, R., 1979, Prophage substitution and prophage loss from superinfected *Escherichia coli recA* (P1) lysogens, *J. Virol.* **31**:277–280.

Meyer, J., and Iida, S., 1979, Amplification of chloramphenicol resistance transposons carried by phage P1Cm in *Escherichia coli, Mol. Gen. Genet.* **176**:209–219.

Meyer, J., Stålhammar-Carlemalm, M., and Iida, S., 1981, Denaturation map of bacteriophage P1 DNA, *Virology* **110**:167–175.

Meyer, J., Iida, S., and Arber, W., 1983, Physical analysis of the genomes of hybrid phages between phage P1 and plasmid p15B, *J. Mol. Biol.* **165**:191–195.

Meyer, J., Stålhammar-Carlemalm, M., Streiff, M., Iida, S., and Arber, W., 1986, Sequence relations among the IncY plasmids p15B, P1 and P7 prophages, *Plasmid* **16**:81–89.

Meyers, D. E., and Landy, A., 1973, The role of host RNA polymerase in P1 phage development, *Virology* **51**:521–524.

Miki, T., Chang, Z. T., and Horiuchi, T., 1984a, Control of cell division by sex factor F in *Escherichia coli*. II. Identification of genes for inhibitor protein and trigger protein on the 42.84–43.6 F segment, *J. Mol. Biol.* **174**:627–646.

Miki, T., Yoshioka, K., and Horiuchi, T., 1984b, Control of cell division by sex factor F in *Escherichia coli*. I. The 42.84–43.6 F segment couples cell division of the host bacteria with replication of plasmid DNA, *J. Mol. Biol.* **174**:605–625.

Miller, J. F., and Malamy, M. H., 1983, Identification of the *pifC* gene and its role in negative control of F factor *pif* gene expression, *J. Bacteriol.* **156**:338–347.

Miller, J. H., 1972, *Experiments in Molecular Genetics*, Cold Spring Harbor Laboratory, Cold Spring Harbor, NY.

Mise, K., 1971, Isolation and characterization of a new generalized transducing bacteriophage different from P1 in *Escherichia coli, J. Virol.* **7**:168–175.

Mise, K., 1980, New recombinant prophages between bacteriophage P1 and the R plasmid NR1, in: *Antibiotic Resistance. Transposition and Other Mechanisms* (S. Mitsuhashi, L. Rosival, and V. Krcmery, eds.), pp. 77–81, Springer-Verlag, Berlin.

Mise, K., and Arber, W., 1975, Bacteriophage P1 carrying drug resistance genes of the R factor NR1, in: *Microbial Drug Resistance* (S. Mitsuhashi and H. Hashimoto, eds.), pp. 165–167, University Park Press, Baltimore.

Mise, K., and Arber, W., 1976, Plaque-forming transducing bacteriophage P1 derivatives and their behaviour in lysogenic conditions, *Virology* **69**:191–205.

Mise, K., and Suzuki, K., 1970, New generalized transducing bacteriophage in *Escherichia coli, J. Virol.* **6**:253–255.

Mise, K., Kawai, M., Yoshida, Y., and Nakamura, A., 1981, Characterization of bacteriophage j2 of *Salmonella typhi* as a generalized transducing phage closely related to coliphage P1, *J. Gen. Microbiol.* **126**:321–326.

Mise, K., Yoshida, Y., and Kawai, M., 1983, Generalized transduction between *Salmonella typhi* and *Salmonella typhimurium* by phage j2 and characterization of the j2 plasmid in *Escherichia coli, J. Gen. Microbiol.* **129**:3395–3400.

Mori, H., Ogura, T., and Hiraga, S., 1984, Prophage λ induction caused by mini-F plasmid genes, *Mol. Gen. Genet.* **196**:185–193.

Mori, H., Kondo, A., Ohshima, A., Ogura, T., and Hiraga, S., 1986, Structure and function of the F plasmid genes essential for partitioning, *J. Mol. Biol.* **192**:1–15.

Moriya, S., Ogasawara, N., and Yoshikawa, H., 1985, Structure and function of the region of the replication origin of the *Bacillus subtilis* chromosome. III. Nucleic acid sequence of some 10,000 base pairs in the origin region, *Nucleic Acids Res.* **13**:2251–2265.

Murakami, Y., Ohmori, H., Yura, T., and Nagata, T., 1987, Requirement of the *E. coli dnaA* gene function for ori-2 dependent mini-F plasmid replication, *J. Bacteriol.* **169**:1724–1730.

Mural, R. J., 1978, Transcription of bacteriophage P1, PhD thesis, University of Georgia, Athens.

Mural, R. J., Chesney, R. H., Vapnek, D., Kropf, M. M., and Scott, J. R., 1979, Isolation and characterization of cloned fragments of bacteriophage P1 DNA, *Virology* **93**:387–397.

Murialdo, H., and Becker, A., 1978, Head morphogenesis of complex double-stranded deoxyribonucleic acid bacteriophages, *Microbiol. Rev.* **42**:529–576.

Murooka, Y., and Harada, T., 1979, Expansion of the host range of coliphage P1 and gene transfer from enteric bacteria to other gram-negative bacteria, *Appl. Environ. Microbiol.* **38**:754–757.

Murotsu, T., Tsutsui, H., and Matsubara, K., 1984, Identification of the minimal essential region for the replication origin of miniF plasmid, *Mol. Gen. Genet.*, **196**:373–378.

Nainen, O., 1975, Ph.D. thesis, University of Georgia, Athens.

Neidhardt, F. C., VanBogelen, R. A., and Vaughn, V., 1984, The genetics and regulation of heat-shock proteins, *Annu. Rev. Genet.* **18**:295–329.

Newman, B. J., and Masters, M., 1980, The variation in frequency with which markers are transduced by phage P1 is primarily a result of discrimination during recombination, *Mol. Gen. Genet.* **180**:585–589.

Nordström, K., 1933, Replication of plasmid R1: Meselson-Stahl density shift experiments revisited, *Plasmid* **9**:218–221.

Nordström, K., 1985a, Chairman's introduction: Replication, incompatibility and partition, in: *Plasmids in Bacteria* (D. R. Helinski, S. N. Cohen, D. B. Clewell, D. A. Jackson, and A. Hollaender, eds.), pp. 119–123, Plenum Press, New York.

Nordström, K., 1985b, Control of plasmid replication: Theoretical considerations and practical solutions, in: *Plasmids in Bacteria* (D. R. Helinski, S. N. Cohen, D. B. Clewell, D. A. Jackson, and A. Hollaender, eds.), pp. 189–214, Plenum Press, New York.

Novick, R. P., and Hoppenstaedt, F. C., 1978, On plasmid incompatibility, *Plasmid* **1**:421–434.

Novick, R. P., Clowes, R. C., Cohen, S. N., Curtis, R. III, Datta, N., and Falkow, S., 1976, Uniform nomenclature for bacterial plasmids: A proposal, *Bacteriol. Rev.* **40**:168–189.

O'Connor, K. A., and Zusman, D. R., 1983, Coliphage P1-mediated transduction of cloned DNA from *Escherichia coli* to *Myxococcus xanthus*: Use for complementation and recombinational analyses, *J. Bacteriol.* **155**:317–329.

O'Connor, M. B., Kilbane, J. J., and Malamy, M. H., 1986, Site-specific and illegitimate recombination in the *oriV1* region of the F factor, *J. Mol. Biol.* **189**:85–102.

Oeschger, M. P., and Wiprud, G. T., 1980, High efficiency temperature-sensitive amber suppressor strains of *Escherichia coli* K12: Construction and characterization of recombinant strains with suppressor-enhancing mutations, *Mol. Gen. Genet.* **178**:293–299.

Ogawa, T., 1975, Analysis of the *dnaB* function of *Escherichia coli* strain K12 and the *dnaB*-like function of P1 prophage, *J. Mol. Biol.* **94**:327–340.

Ogawa, T., Baker, T. A., Van der Ende, A., and Kornberg, A., 1985, Initiation of enzymatic replication at the origin of the *Escherichia coli* chromosome: Primase as the sole priming enzyme, *Proc. Natl. Acad. Sci. USA* **82**:3562–3566.

Ogura, T., and Hiraga, S., 1983a, Mini-F plasmid genes that couple host cell division to plasmid proliferation, *Proc. Natl. Acad. Sci. USA* **80**:4784–4788.

Ogura, T., and Hiraga, S., 1983b, Partition mechanism of F plasmid: Two plasmid gene-encoded products and a *cis*-acting region are involved in partition, *Cell* **32**:351–360.

Ohtsubo, H., and Ohtsubo, E., 1978, Nucleotide sequence of an insertion element, IS*1*, *Proc. Natl. Acad. Sci. USA* **75**:615–619.

Okada, M., and Watanabe, T., 1968, Transduction with phage P1 in *Salmonella typhimurium*, *Nature* **218**:185–187.

O'Regan, G. T., Sternberg, N. L., and Cohen, G., 1987, Construction of an ordered, overlapping library of bacteriophage P1 DNA in λD69, *Gene* **60**:129–135.

Ornellas, E. P., and Stocker, B. A. D., 1974, Relation of lipopolysaccharide character to phage P1 sensitivity in *Salmonella typhimurium*, *Virology* **60**:491–502.

Ozeki, H., 1956, Abortive transduction, *Carnegie Inst. of Washington, Yearbook* **55**:302–303.

Ozeki, H., 1959, Chromosome fragments participating in transduction in *Salmonella typhimurium*, *Genetics* **44**:457–470.

Pabo, C. O., and Sauer, K. T., 1984, Protein-DNA recognition, *Annu. Rev. Biochem.* **53:**293–321.

Pal, S. K., and Chattoraj, D. K., 1986, RepA is rate limiting for P1 plasmid replication, in: *Mechanisms of DNA Replication and Recombination* UCLA Symposium on Molecular and Cellular Biology, New Series (T. Kelly and R. McMacken, eds.), Vol. 47, pp. 441–450, Alan R. Liss, New York.

Pal, S. K., Mason, R. J., and Chattoraj, D. K., 1986, P1 plasmid replication: Role of initiator titration in copy number control, *J. Mol. Biol.* **192:**275–285.

Petrillo, L. A., Gallagher, P. J., and Elseviers, D., 1983, The role of 2-methylthio-N6-isopentenyladenosine in readthrough and suppression of nonsense codons in *Escherichia coli*, *Mol. Gen. Genet.* **190:**289–294.

Plasterk, R. H. A., Brinkman, A., and Van de Putte, P., 1983, DNA inversions in the chromosome of *Escherichia coli* and in bacteriophage Mu: Relationship to other site-specific recombination systems, *Proc. Natl. Acad. Sci. USA* **80:**5355–5358.

Plasterk, R. H. A., and Van de Putte, P., 1984, Genetic switches by DNA inversions in prokaryotes, *Biochim. Biophys. Acta* **782:**111–119.

Porter, R. D., McLaughlin, T., and Low, B., 1978, Transduction versus "conjugation." Evidence for multiple roles for exonuclease V in genetic recombination in *Escherichia coli*, *Cold Spring Harbor Symp. Quant. Biol.* **43:**1043–1048.

Prehm, P., Schmidt, G., Jann, B., and Jann, K., 1976, The cell wall lipopolysaccharide of *Escherichia coli* K12, *Eur. J. Biochem.* **56:**41–55.

Prentki, P., Chandler, M., and Caro, L., 1977, Replication of the prophage P1 during the cell cycle of *Escherichia coli*, *Mol. Gen. Genet.* **152:**71–76.

Pritchard, R. H., 1978, Control of DNA replication in bacteria, in: *NATO Advanced Study Series A: Life Sciences. DNA Synthesis: Present and Future* (I. Molineux and M. Kohiyama, eds.), pp. 1–26, Plenum Press, New York.

Pritchard, R. H., Barth, P. T., and Collins, J., 1969, Control of DNA synthesis in bacteria, *Symp. Soc. Gen. Microbiol.* **19:**263–297.

Rae, M. E., and Stodolsky, M., 1974, Chromosome breakage, fusion and reconstruction during P1*dl* transduction, *Virology* **58:**32–54.

Raj, A. S., Raj, A. Y., and Schmieger, H., 1974, Phage genes involved in the formation of generalized transducing particles in *Salmonella*-phage P22, *Mol. Gen. Genet.* **135:**175–184.

Rashtchian, A., Brown, S. W., Reichler, J., and Levy, S. B., 1986, Plasmid segregation into minicells is associated with membrane attachment and independent of plasmid replication, *J. Bacteriol.* **165:**82–87.

Ravin, V. K., and Golub, E. I., 1967, Study of phage conversion in *Escherichia coli*. I. Acquisition of resistance to bacteriophage T1 as a result of lysogenization, *Genetika* 3:113–121 (in Russian), English translation available in *Soviet Genetics*.

Ravin, V. K., and Shulga, M. G., 1970, Evidence for extrachromosomal location of prophage N15, *Virology* **40:**800–807.

Rawlins, D. R., Milman, G., Hayward, S. D., and Hayward, G. S., 1985, Sequence-specific DNA binding of the Epstein-Barr virus nuclear antigen (EBNA-1) to clustered sites in the plasmid maintenance region, *Cell* **42:**859–868.

Razza, J. B., Watkins, C. A., and Scott, J. R., 1980, Phage P1 temperature-sensitive mutants with defects in the lytic pathway, *Virology* **105:**52–59.

Reeve, J. N., Lanka, E., and Schuster, H., 1980, Synthesis of P1 ban protein in minicells infected by P1 mutants, *Mol. Gen. Genet.* **177:**193–197.

Roizman, B., 1979, The structure and isomerization of herpes simplex virus genomes, *Cell* **16:**481–494.

Rosenfeld, S. A., and Brenchley, J. E., 1980, Bacteriophage P1 as a vehicle for Mu mutagenesis of *Salmonella typhimurium*, *J. Bacteriol.* **144:**848–851.

Rosner, J. L., 1972, Formation, induction and curing of bacteriophage P1 lysogens, *Virology* **48:**679–689.

Rosner, J. L., 1973, Modification-deficient mutants of bacteriophage P1. I. Restriction by P1 cryptic lysogens, *Virology* **52:**213–222.

Rosner, J. L., 1975, Specialized transduction of *pro* genes by coliphage P1: Structure of a partly diploid P1-*pro* prophage, *Virology* **67**:42–55.

Sadowski, P., 1986, Site-specific recombinases: Changing partners and doing the twist, *J. Bacteriol.* **165**:341–347.

Sakaki, Y., 1974, Inactivation of the ATP-dependent DNase of *Escherichia coli* after infection with double-stranded DNA phages, *J. Virol.* **14**:1611–1612.

Sandri, R. M., and Berger, H., 1980a, Bacteriophage P1-mediated generalized transduction in *Escherichia coli:* Fate of transduced DNA in *rec⁺* and *recA⁻* recipients, *Virology* **106**:14–29.

Sandri, R. M., and Berger, H., 1980b, Bacteriophage P1-mediated generalized transduction in *Escherichia coli:* Structure of abortively transduced DNA, *Virology* **106**:30–40.

Sandulache, R., Prehm, P., and Kamp, D., 1984, Cell wall receptor for bacteriophage Mu G(+), *J. Bacteriol.* **160**:299–303.

Sandulache, R., Prehm, P., Expert, D., Toussaint, A., and Kamp, D., 1985, The cell wall receptor for bacteriophage Mu G(−) in *Erwinia* and *E. coli* C, *FEMS Microbiol. Lett.* **28**:307–310.

Schildkraut, C. L., Marmur, J., and Doty, P., 1962, Determination of the base composition of deoxyribonucleic acid from its buoyant density in CsCl, *J. Mol. Biol.* **4**:430–443.

Schmieger, H., 1972, Phage P22 mutants with increased or decreased transduction abilities, *Mol. Gen. Genet.* **119**:75–88.

Schmieger, H., and Backhaus, H., 1976, Altered co-transduction frequencies exhibited by HT-mutants of *Salmonella* phage P22, *Mol. Gen. Genet.* **143**:307–309.

Schmitt, R., Mattes, R., Schmid, K., and Altenbuchner, J., 1979, RAF plasmids in strains of *Escherichia coli* and their possible role in enteropathogeny, in: *Plasmids of Medical, Environmental and Commercial Importance in Development in Genetics* (K. N. Timmis and A. Puhler, eds.), Vol. 1, pp. 199–210, Elsevier-North Holland, Amsterdam.

Schulz, G., and Stodolsky, M., 1976, Integration sites of foreign genes in the chromosome of coliphage P1: A finer resolution, *Virology* **73**:299–302.

Schulz, D. W., Taylor, A. F., and Smith, G. R., 1983, *Escherichia coli* RecBC pseudorevertants lacking Chi recombinational hotspot activity, *J. Bacteriol.* **155**:664–680.

Schuster, H., Mikolajczyk, M., Rohrschneider, J., and Geschke. B., 1975, φX174 DNA-dependent DNA synthesis *in vitro:* Requirement for P1 *ban* protein in *dnaB* mutant extracts of *Escherichia coli, Proc. Natl. Acad. Sci. USA* **72**:3907–3911.

Schuster, H., Lanka, E., Edelbluth, C., Geschke, B., Mikolajczyk, M., Schlicht, M., and Touati-Schwartz, D., 1978, A *dnaB*-analog DNA-replication protein of phage P1, *Cold Spring Harbor Symp. Quant. Biol.* **43**:551–557.

Scott, J. R., 1968, Genetic studies on bacteriophage P1, *Virology* **36**:564–574.

Scott, J. R., 1970a, A defective P1 prophage with a chromosomal location, *Virology* **40**:144–151.

Scott, J. R., 1970b, Clear plaque mutants of phage P1, *Virology* **41**:66–71.

Scott, J. R., 1972, A new gene controlling lysogeny in phage P1, *Virology* **48**:282–283.

Scott, J. R., 1973, Phage P1 cryptic II. Location and regulation of prophage genes, *Virology* **53**:327–336.

Scott, J. R., 1974, A turbid plaque forming mutant of phage P1 that cannot lysogenize *Escherichia coli, Virology* **62**:344–349.

Scott, J. R., 1975, Superinfection immunity and prophage repression in phage P1, *Virology* **65**:173–178.

Scott, J. R., 1980, Immunity and repression in bacteriophages P1 and P7, *Curr. Top. Microbiol. Immunol.* **90**:49–65.

Scott, J. R., 1984, Regulation of plasmid replication, *Microbiol. Rev.* **48**:1–23.

Scott, J. R., and Kropf, M. M., 1977, Location of new clear plaque genes on the P1 map, *Virology* **82**:362–368.

Scott, J. R., and Rownd, R. H., 1980, Workshop summary: Regulation of plasmid replication, in: *ICN/UCLA Symposium on Mechanistic Studies of DNA Replication and Genetic Recombination* (B. Alberts, ed.), pp. 1–8, Academic Press, New York.

Scott, J. R., and Vapnek, D., 1980, Regulation of replication of the P1 plasmid prophage, in: *ICN/UCLA Symposium on Mechanistic Studies of DNA Replication and Genetic Recombination* (B. Alberts, ed.), pp. 335–345, Academic Press, New York.

Scott, J. R., Kropf, M., and Mendelson, L., 1977a, Clear plaque mutants of phage P7, *Virology* **76:**39–46.

Scott, J. R., Laping, J. L., and Chesney, R. H., 1977b, A phage P1 virulent mutation at a new map location, *Virology* **78:**346–348.

Scott, J. R., Chesney, R. H., and Novick, R. P., 1978a, Mutant in P1 plasmid maintenance, in: *Microbiology—1978* (D. Schlessinger, ed.), pp. 74–77, American Society for Microbiology, Washington.

Scott, J. R., West, B. W., and Laping, J. L., 1978b, Superinfection immunity and prophage repression in phage P1. IV. The c1 repressor bypass function and the role of c4 repressor in immunity, *Virology* **85:**587–600.

Scott, J. R., Kropf, M. M., Padolsky, L., Goodspeed, J. K., Davis, R., and Vapnek, D., 1982, Mutants of the plasmid prophage P1 with elevated copy number: Isolation and characterization, *J. Bacteriol.* **150:**1329–1339.

Segev, N., and Cohen, G., 1981, Control of circularization of bacteriophage P1 DNA in *Escherichia coli, Virology* **114:**333–342.

Segev, N., Laub, A., and Cohen, G., 1980, A circular form of bacteriophage P1 DNA made in lytically infected cells of *Escherichia coli.* I. Characterization and kinetics of formation, *Virology* **101:**261–271.

Selvaraj, G., and Iyer, V. N., 1980, A *dnaB* analog function specified by bacteriophage P7 and its comparison to the similar function specified by bacteriophage P1, *Mol. Gen. Genet.* **178:**561–566.

Sengstag, C., and Arber, W., 1983, IS2 insertion is a major cause of spontaneous mutagenesis of the bacteriophage P1: Non-random distribution of target sites, *EMBO J.* **2:**67–71.

Sengstag, C., and Arber, W., 1987, A cloned DNA fragment from bacteriophage P1 enhances IS2 insertion, *Mol. Gen. Genet.* **206:**344–351.

Sengstag, C., Shepherd, J. C. W., and Arber, W., 1983, The sequence of the bacteriophage P1 genome region serving as hot target for IS2 insertion, *EMBO J.* **2:**1777–1781.

Shafferman, A., Geller, T., and Hertman, I., 1978, Genetic and physical characterization of P1*dlw* prophage and its derivatives, *Virology* **86:**115–126.

Shafferman, A., Geller, T., and Hertman, I., 1979, Identification of the P1 compatibility and plasmid maintenance locus by a mini P1 *lac⁺* plasmid, *Virology* **96:**32–37.

Shapiro, J. A., 1977, Appendix B, in: *Bacterial Plasmids in DNA Insertion Elements, Plasmids and Episomes* (A. L. Bukhari, J. A. Shapiro, and S. L. Adhya, eds.), p. 634, Cold Spring Harbor Laboratory, Cold Spring Harbor, NY.

Sherratt, D., Dyson, P., Boocock, M., Brown, L., Summers, D., Stewart, G., and Chan, P., 1984, Site-specific recombination in transposition and plasmid stability, *Cold Spring Harbor Symp. Quant. Biol.* **49:**227–233.

Shields, M. S., Kline, B. C., and Tam, J. E., 1986, A rapid method for the quantitative measurement of gene dosage: Mini-F plasmid concentration as a function of cell growth rate, *J. Microbiol. Methods* **6:**33–46.

Simon, M., and Silverman, M., 1983, Recombinational regulation of gene expression in bacteria, in: *Gene Function in Prokaryotes* (J. Beckwith, J. Davies, and J. A. Gallant, eds.), pp. 211–227, Cold Spring Harbor Laboratory, Cold Spring Harbor, New York.

Simon, M., Zieg, J., Silverman, M., Mandel, G., and Doolittle, R., 1980, Phase variation: Evolution of a controlling element, *Science* **209:**1370–1374.

Smith, D. W., Garland, A. M., Herman, G., Enns, R. E., Baker, T. A., and Zyskind, J., 1985, The importance of state of methylation of *oriC* GATC sites in initiation of DNA replication in *Escherichia coli, EMBO J.* **4:**1319–1327.

Smith, G. R., 1983, General recombination, in: *Lambda II.* (R. W. Hendrix, J. W. Roberts, F. W. Stahl, and R. A. Weisberg, eds.), pp. 175–209, Cold Spring Harbor Laboratory, Cold Spring Harbor, New York.

Smith, H. W., 1972, Ampicillin resistance in *Escherichia coli* by phage infection, *Nature New Biol.* **238**:205–206.

Smith, T. A. G., and Hays, J. B., 1985, Repair and recombination of non-replicating UV-irradiated phage DNA in *E. coli*. II. Stimulation of RecF-dependent recombination by excision repair of cyclobutane pyrimidine dimers and of other photoproducts, *Mol. Gen. Genet.* **201**:393–401.

Søgaard-Anderson, L., Rokeach, L. A., and Molin, S., 1984, Regulated expression of a gene important for replication of plasmid F in *E. coli*, *EMBO J.* **3**:257–262.

Som, T., Sternberg, N., and Austin, S., 1981, A nonsense mutation in bacteriophage P1 eliminates the synthesis of a protein required for normal plasmid maintenance, *Plasmid* **5**:150–160.

Sommer, S., Bailone, A., and Devoret, R., 1985, SOS induction by thermosensitive replication mutants of miniF plasmid, *Mol. Gen. Genet.* **198**:456–464.

Stahl, F. W., 1979, Specialized sites in generalized recombination, *Annu. Rev. Genet.* **13**:7–24.

Stalker, D. M., and Helinski, D. R., 1985, DNa segments of the IncX plasmid R485 determining replication incompatibility with plasmid R6K, *Plasmid* **14**:245–254.

Sternberg, N., 1978, Demonstration and analysis of P1 site-specific recombination using λ-P1 hybrid phages constructed *in vitro*, *Cold Spring Harbor Symp. Quant. Biol.* **43**:1143–1146.

Sternberg, N., 1979, A characterization of bacteriophage P1 DNA fragments cloned in a lambda vector, *Virology* **96**:129–142.

Sternberg, N., 1981, Bacteriophage P1 site-specific recombination. III. Strand exchange during recombination at *lox* sites, *J. Mol. Biol.*, **150**:603–608.

Sternberg, N., 1985, Evidence that adenine methylation influences DNA-protein interactions in *Escherichia coli*, *J. Bacteriol.* **164**:490–493.

Sternberg, N., and Austin, S., 1981, The maintenance of the P1 plasmid prophage, *Plasmid* **5**:20–31.

Sternberg, N., and Austin, S., 1983, Isolation and characterization of P1 minireplicons, λ-P1:5R and λ-P1:5L, *J. Bacteriol.* **153**:800–812.

Sternberg, N., and Coulby, J., 1987a, Recognition and cleavage of the bacteriophage P1 packaging site (*pac*). I. Differential processing of the cleaved ends *in vivo*, *J. Mol. Biol.* **194**:453–468.

Sternberg, N., and Coulby, J., 1987b, Recognition and cleavage of the bacteriophage P1 packaging site (*pac*). II. Functional limits of *pac* and location of *pac* cleavage termini, *J. Mol. Biol.* **194**:469–480.

Sternberg, N., and Hamilton, D., 1981, Bacteriophage P1 site-specific recombination. I. Recombination between *loxP* sites, *J. Mol. Biol.* **150**:467–486.

Sternberg, N., and Hoess, R., 1983, The molecular genetics of bacteriophage P1, *Annu. Rev. Genet.* **17**:123–154.

Sternberg, N., and Weisberg, R., 1975, Packaging of prophage and host DNA by coliphage lambda, *Nature* **256**:97–103.

Sternberg, N., Austin, S., Hamilton, D., and Yarmolinsky, M., 1978, Analysis of bacteriophage P1 immunity by using lambda P1 recombinants constructed *in vitro*, *Proc. Natl. Acad. Sci. USA* **75**:5594–5598.

Sternberg, N., Austin, S., and Yarmolinsky, M., 1979, Regulatory circuits in bacteriophage P1 as analyzed by physical dissection and reconstruction., *Contrib. Microbiol. Immunol.* **6**:89–99.

Sternberg, N., Hamilton, D., Austin, S., Yarmolinsky, M., and Hoess, R., 1980, Site-specific recombination and its role in the life cycle of bacteriophage P1, *Cold Spring Harbor Symp. Quant. Biol.* **45**:297–309.

Sternberg, N., Hamilton, D., and Hoess, R., 1981a, P1 site-specific recombination. II. Recombination between *loxP* and the bacterial chromosome, *J. Mol. Biol.* **150**:487–507.

Sternberg, N., Powers, M., Yarmolinsky, M., and Austin, S., 1981b, Group Y incompatibility and copy control of P1 prophage, *Plasmid* **5**:138–149.

Sternberg, N., Hoess, R., and Abremski, K., 1983, The P1 *lox*-Cre site-specific recombination system: Properties of *lox* sites and biochemistry of *lox*-Cre interactions, in: *Mechanisms of DNA Replication and Recombination, ICN-UCLA Symposia on Molecular and Cellular Biology*, New Series (N. Cozzarelli, ed.), Vol. 10, pp. 671–684, Alan R. Liss, New York.

Sternberg, N., Sauer, B., Hoess, R., and Abremski, K., 1986, An initial characterization of the bacteriophage P1 *cre* structural gene and its regulatory region, *J. Mol. Biol.* **187**:197–212.

Stodolsky, M., 1973, Bacteriophage P1 derivatives with bacterial genes: A heterozygote enrichment method for the selection of P1*dpro* lysogens, *Virology* **53**:471–475.

Stoleru, G. H., Gerbaud, G. R., Bouanchaud, D. H., and LeMinor, L., 1972, Etude d'un plasmide transférable déterminant la production d'H$_2$S et la résistance à la tétracycline chez "*Escherichia coli*," *Ann. Inst. Pasteur.* **123**:743–754.

Streiff, M. B., Iida, S., and Bickle, T. A., 1987, Expression and proteolytic processing of the *darA* anti-restriction gene product of bacteriophage P1, *Virology* **157**:167–171.

Streisinger, G., Emrich, J., and Stahl, M. M., 1967, Chromosome structure in phage T4. III. Terminal redundancy and length determination, *Proc. Natl. Acad. Sci. USA* **57**:292–295.

Summers, D. K., and Sherratt, D. J., 1984, Multimerization of high copy number plasmids causes instability: ColE1 encodes a determinant essential for plasmid monomerization and stability, *Cell* **36**:1097–1103.

Susskind, M. M., and Botstein, D., 1978, Molecular genetics of bacteriophage P22, *Microbiol. Rev.* **42**:385–413.

Swack, J. A., Pal, S. K., Mason, R. J., Abeles, A. L., and Chattoraj, D. K., 1987, P1 plasmid replication: measurement of initiator protein concentration *in vivo*, *J. Bacteriol.* **169**:3737–3742.

Szostak, J. W., Orr-Weaver, T. L., Rothstein, R. J., and Stahl, F. W., 1983, The double-stranded-break repair model for recombination, *Cell* **33**:25–35.

Tabuchi, A., 1985, Nucleotide sequence of the replication region of plasmid R401 and its incompatibility function, *Microbiol. Immunol.* **29**:383–393.

Takano, T., 1971, Bacterial mutant defective in plasmid formation. Requirement for the *lon* plus allele, *Proc. Natl. Acad. Sci. USA* **68**:1469–1473.

Takano, T., 1977, Mechanism of defective lysogenization by phage P1 in a *lon*-mutant of *Escherichia coli* K12, *Microbiol. Immunol.* **21**:573–581.

Takano, T., and Ikeda, S., 1976, Phage P1 carrying kanamycin resistance gene of R factor, *Virology* **70**:198–200.

Tanimoto, K., and Iino, T., 1984, An essential gene for replication of the mini-F plasmid from origin I, *Mol. Gen. Genet.* **196**:59–63.

Teather, R. M., 1974, The localization and timing of cell division in *Escherichia coli*, Ph.D. dissertation, University of Edinburgh.

Terawaki, Y., and Rownd, R., 1972, Replication of R factor *Rts1* in *Proteus mirabilis*, *J. Bacteriol.* **109**:492–498.

Terawaki, Y., Takayasu, H., and Akiba, T., 1967, Thermosensitive replication of a kanamycin resistance factor, *J. Bacteriol.* **94**:687–690.

Terawaki, Y., Kakizawa, Y., Takayasu, H., and Yoshikawa, M., 1968, Temperature sensitivity of cell growth in *Escherichia coli* associated with the temperature sensitive R(KM) factor, *Nature* **219**:284–285.

Ting, R. C., 1962, The specific gravity of transducing particles of bacteriophage P1, *Virology* **16**:115–121.

Tokino, T., Murotsu, T., and Matsubara, K., 1986, Purification and properties of the mini-F plasmid-encoded E protein needed for autonomous replication control of the plasmid, *Proc. Natl. Acad. Sci. USA* **83**:4109–4113.

Tomás, J. M., and Kay, W. W., 1984, Effect of bacteriophage P1 lysogeny on lipopolysaccharide composition and the lambda receptor of *Escherichia coli*, *J. Bacteriol.* **159**:1047–1052.

Tomás, J., Regué, M., Parés, R., Jofre, J., and Kay, W. W., 1984, P1 bacteriophage and tellurite sensitivity in *Klebsiella pneumoniae* and *Escherichia coli, Can. J. Microbiol.* **30**:830–836.

Tominaga, A., and Enomoto, M., 1986, Magnesium-dependent plaque formation by bacteriophage P1*cinC*(−) on *Escherichia coli* C and *Shigella sonnei, Virology* **155**:284–288.

Touati-Schwartz, D., 1978, Two replication functions in phage P1: *ban*, an analog of *dnaB*, and *bof*, involved in the control of replication, in: *DNA Synthesis—Present and Future*, (I. Molineux, and M. Kohiyama, eds.), pp. 683–692, Plenum Press, New York.

Touati-Schwartz, D., 1979a, A *dnaB* analog, *ban*, specified by bacteriophage P1: Genetic and physiological evidence for functional analogy and interactions between the two products, *Mol. Gen. Genet.* **174**:173–188.

Touati-Schwartz, D., 1979b, A new pleiotropic bacteriophage P1 mutation, *bof*, affecting *c*l repression activity, the expression of plasmid incompatibility and the expression of certain constitutive prophage genes, *Mol. Gen. Genet.* **174**:189–202.

Toussaint, A., Lefebvre, N., Scott, J. R., Cowan, J. A., DeBruijn, F., and Bukhari, A. I., 1978, Relationships between temperate phages Mu and P1, *Virology* **89**:146–161.

Trawick, J. D., and Kline, B. C., 1985, A two-stage molecular model for control of mini-F replication. *Plasmid* **13**:59–69.

Tsutsui, H., Fujiyama, A., Murotsu, T., and Matsubara, K., 1983, Role of nine repeating sequences of the mini-F genome for expression of F-specific incompatibility phenotype and copy number control, *J. Bacteriol.* **155**:337–344.

Tucker, W. T., Miller, C. A., and Cohen, S. N., 1984, Structural and functional analysis of the *par* region of the pSC101 plasmid, *Cell* **38**:191–201.

Tye, B. K., Chan, R. K., and Botstein, D., 1974a, Packaging of an oversize transducing genome by *Salmonella* phage P22, *J. Mol. Biol.* **85**:485–500.

Tye, B. K., Huberman, J. A., and Botstein, D., 1974b, Non random circular permutation of phage P22 DNA, *J. Mol. Biol.* **85**:501–532.

Tyler, B. M., and Goldberg, R. B., 1976, Transduction of chromosomal genes between enteric bacteria by bacteriophage P1, *J. Bacteriol.* **125**:1105–1111.

Uhlin, B. E., and Nordström, K., 1985, Preferential inhibition of plasmid replication *in vivo* by altered DNA gyrase activity in *Escherichia coli, J. Bacteriol.* **162**:855–857.

Velleman, M., Dreisekelmann, B., and Schuster, H., 1987, Multiple repressor binding sites in the genome of bacteriophage P1, *Proc. Natl. Acad. Sci. USA* **84**:5570–5574.

Vetter, D., Andrews, B. J., Roberts-Beatty, L., and Sadowski, P. D., 1983, Site-specific recombination of yeast 2-μm DNA *in vitro, Proc. Natl. Acad. Sci. USA* **80**:7284–7288.

Wada, C., Akiyama, Y., Ito, K., and Yura, T., 1986, Inhibition of F replication in *htpR* mutants of *Escherichia coli* deficient in sigma 32 protein, *Mol. Gen. Genet.* **203**:208–213.

Walker, D. H. Jr., and Anderson, T. F., 1970, Morphological variants of coliphage P1, *J. Virol.* **5**:765–782.

Walker, D. H. Jr., and Walker, J. T., 1975, Genetic studies of coliphage P1. I. Mapping by use of prophage deletions, *J. Virol.* **16**:525–534.

Walker, D. H. Jr., and Walker, J. T., 1976a, Genetic studies of coliphage P1. II. Relatedness to phage P7, *J. Virol.* **19**:271–274.

Walker, D. H. Jr., and Walker, J. T., 1976b, Genetic studies of coliphage P1. III. Extended genetic map, *J. Virol.* **20**:177–187.

Walker, J. T., and Walker, D. H. Jr., 1980, Mutations in coliphage P1 affecting host cell lysis, *J. Virol.* **35**:519–530.

Walker, J. T., and Walker, D. H. Jr., 1981, Structural proteins of coliphage P1, in: *Progress in Clinical and Biological Research* (M. S. DuBow, ed.), Vol. 64, pp. 69–77, Alan R. Liss, New York.

Walker, J. T., and Walker, D. H. Jr., 1983, Coliphage P1 morphogenesis. Analysis of mutants by electron microscopy, *J. Virol.* **45**:1118–1139.

Walker, J. T., Iida, S., and Walker, D. H. Jr., 1979, Permutation of the DNA in small-headed virions of coliphage P1, *Mol. Gen. Genet.* **167**:341–344.

Wall, J. D., and Harriman, P. D., 1974, Phage P1 mutants with altered transducing abilities for *Escherichia coli*, *Virology* **59**:532–544.

Wandersman, C., and Yarmolinsky, M., 1977, Bipartite control of immunity conferred by the related heteroimmune plasmid prophages, P1 and P7 (formerly φAmp), *Virology* **77**:386–400.

Watanabe, T., Nishida, H., Ogata, C., Arai, T., and Sato, S., 1964, Episome-mediated transfer of drug-resistance in *Enterobacteriaceae*. III. Transduction of resistance factors, *J. Bacteriol.* **82**:202–209.

Watkins, C. A., and Scott, J. R., 1981, Characterization of bacteriophage D6, *Virology* **110**:302–317.

West, B. W., and Scott, J. R., 1977, Superinfection immunity and prophage repression in phage P1 and P7. III. Induction by virulent mutants, *Virology* **78**:267–276.

Wickner, S. H., and Chattoraj, D. K., 1987, Replication of miniP1 plasmid DNA *in vitro* requires two initiation proteins encoded by the *repA* gene of phage P1 and the *dnaA* gene of *Escherichia coli*, *Proc. Natl. Acad. Sci., USA* **84**:3668–3672.

Wierzbicki, A., Kendall, M., Abremski, K., and Hoess, R., 1987, A mutational analysis of the bacteriophage P1 recombinase Cre, *J. Mol. Biol.* **195**:785–794.

Windle, B. E., 1986, Characterization of a P1 bacteriophage encoded function, *ref*, that stimulates homologous recombination in *E. coli*, Ph.D. dissertation, University of Maryland, Catonsville.

Windle, B. E., and Hays, J. B., 1986, A phage P1 function that stimulates homologous recombination of the *Escherichia coli* chromosome, *Proc. Natl. Acad. Sci. USA* **83**:3885–3889.

Wolfson, J. S., Hooper, D. C., Swartz, M. N., Swartz, M. D., and McHugh, G. L., 1983, Novobiocin-induced elimination of F'*lac* and mini-F plasmids from *Escherichia coli*, *J. Bacteriol.* **156**:1165–1170.

Womble, D. D., and Rownd, R. H., 1986, Regulation of λdv plasmid replication. A quantitative model for control of plasmid λdv replication in the bacterial cell division cycle, *J. Mol. Biol.* **191**:367–382.

Womble, D. D., and Rownd, R. H., 1987, Regulation of mini-F plasmid DNA replication. A quantitative model for control of plasmid mini-F replication in the bacterial cell division cycle, *J. Mol. Biol.* **195**:99–114.

Woods, W., and Egan, B., 1976, Prophage induction of non-inducible coliphage 186, *J. Virol.* **14**:1349–1366.

Yamaguchi, K., and Masamune, Y., 1985, Autogenous regulation of synthesis of the replication protein in plasmid pSC101, *Mol. Gen. Genet.* **200**:362–367.

Yamaguchi, K., and Yamaguchi, M., 1984, The replication origin of pSC101: Replication properties of a segment capable of autonomous replication, *J. Gen. Appl. Microbiol.* **30**:347–358.

Yamamoto, Y., 1982, Phage P1 mutant with decreased abortive transduction, *Virology* **118**:329–344.

Yarmolinsky, M., 1977, Genetic and physical structure of bacteriophage P1 DNA, in: *DNA Insertion Sequences, Episomes and Plasmids* (A. I. Bukhari, J. A. Shapiro, and S. L. Adhya, eds.), pp. 721–732, Cold Spring Harbor Laboratory, Cold Spring Harbor, NY.

Yarmolinsky, M., 1984, Bacteriophage P1, in: *Genetic Maps. 1984* (S. J. O'Brien, ed.), Vol. 3, pp. 42–54, Cold Spring Harbor Laboratory, Cold Spring Harbor, NY.

Yarmolinsky, M., 1987, Bacteriophage P1, in: *Genetic Maps, 1987* (S. J. O'Brien, ed.), Vol. 4, pp. 38–47, Cold Spring Harbor Laboratory, Cold Spring Harbor, NY.

Yarmolinsky, M. B., and Stevens, E., 1983, Replication-control functions block the induction of an SOS response by a damaged P1 bacteriophage, *Mol. Gen. Genet.* **192**:140–148.

York, M., and Stodolsky, M., 1981, Characterization of P1*argF* derivatives from *Escherichia coli* K12 transduction. I. IS*1* elements flank the *argF* gene segment, *Mol. Gen. Genet.* **181**:230–240.

York, M. K., and Stodolsky, M., 1982a, Characterization of P1*argF* derivatives from *Esche-*

richia coli K12 transduction. II. Role of P1 in specialized transduction of *argF*, *Virology*
120:130–145.

York, M. K., and Stodolsky, M., 1982b, Characterization of P1*argF* derivatives from *Escherichia coli* K12 transduction. III. P1*Cm*13*argF* derivatives, *Virology* **123**:336–343.

Yoshida, Y., and Mise, K., 1984, Characterization of generalized transducing phage φW39 heteroimmune to phage P1 in *Escherichia coli* W39, *Microbiol. Immunol.* **28**:415–426.

Yuan, R., 1981, Structure and mechanism of multifunctional restriction endonucleases, *Annu. Rev. Biochem.* **50**:285–315.

Yun, T., and Vapnek, D., 1977, Electron microscopic analysis of bacteriophages P1, P1Cm and P7. Determination of genome sizes, sequence homology and location of antibiotic resistance determinants, *Virology* **77**:376–385.

Zabeau, M., and Roberts, R. J., 1979, The role of restriction endonuclease in molecular genetics, *Mol. Genet.* **3**:1–63.

Zabrovitz, S., Segev, N., and Cohen, G., 1977, Growth of bacteriophage P1 in recombination-deficient hosts of *Escherichia coli*, *Virology* **80**:233–248.

Zieg, J., Silverman, H., Hilmen, M., and Simon, M., 1977, Recombinational switch for gene expression, *Science* **196**:170–172.

Zyskind, J. W., Cleary, J. M., Brusilow, W. S. A., Harding, N. E., and Smith, D. W., 1983, Chromosomal replication origin from the marine bacterium *Vibrio harveyi* functions in *Escherichia coli: oriC* consensus sequence, *Proc. Natl. Acad. Sci. USA* **80**:1164–1168.

CHAPTER 10

Bacteriophage T5 and Related Phages

D. James McCorquodale and Huber R. Warner

I. INTRODUCTION

Bacteriophages T5 and BF23 have been the two most extensively investigated of the T5 group, which also includes bacteriophages PB, BG3, and 29 alpha. Unusual features of T5, BF23, and PB include the existence of interrupted phosphodiester bonds ("nicks") at specific locations in only one of the strands of their duplex DNA, a two-step mechanism for transfer of their DNA into host cells, and large terminal repetitions in their DNA.

Our understanding of the molecular events occurring during infection by T5 group bacteriophages has progressed in several areas and has been aided by the newer techniques of cloning and sequencing, which were barely emerging at the time of the last review (McCorquodale, 1975*). Because most work has utilized either T5 itself or the closely related BF23, two phages that differ most demonstrably in the host cell receptor that each utilizes for irreversible adsorption, this review will switch back and forth between T5 and BF23 depending on which one has generated the most information for a given event in their life cycle.

---

*This review, published in 1975, will be used to refer to work well established by 1975, which will not be reviewed in detail here.

D. JAMES McCORQUODALE • Department of Biochemistry, Medical College of Ohio, Toledo, Ohio 43699. HUBER R. WARNER • National Institute on Aging, National Institutes of Health, Bethesda, Maryland 20892.

A. Phage Structure

Bacteriophages T5 and BF23 have one of the most common morphologies among bacteriophages, having an icosahedral head and a long, flexible, noncontractile tail. Both phages belong to morphological group B of Bradley (1967), to which bacteriophage lambda also belongs. The tail of both is attached to one of the head apices, and it has three angled and one straight tail fiber at its distal end. The head of T5 has an average diameter of 90 nm, and the tail has a diameter of 12 nm with a length of 190 nm from where it attaches to the head to the position of a ringlike structure near the end of the tail. A transition from tubular to conical form occurs at this ringlike structure such that the cone adds another 12 nm to the length of the tail. At the tip of the cone is attached a single straight tail fiber about 50 nm in length (Saigo, 1978), bringing the total length of the tail to about 250 nm. Three angled tail fibers are attached to the end of the tubular form of the tail (Williams and Fisher, 1974), and their attachment generates the ringlike structure at this location (Saigo, 1978). A mature T5 or BF23 particle contains at least 15 different structural polypeptides (Zweig and Cummings, 1973a,b); some of these are summarized in Table I.

TABLE I. Structural Polypeptides of Bacteriophage T5 and BF23[a]

Band in SDS–PAGE[b]	Structural gene	MW ($\times 10^{-3}$)		Structural polypeptide	Copies/ particle
		T5	BF23		
1	ltf	125	110	L-shaped tail fibers	6–9
2	D18–19	108	107	Straight tail fiber	5
3	D16	103	105	Tail	5
4	D17	75	75–80	Tail	5
5	oad	67	60	Receptor-binding protein	3 or less
6	N4	58	58	Major tail	120
7		44	46	Minor head	12
8	D20–21	32	33	Major head	730
8a		30	ND[c]	Head	< 12
8b		28	ND	Head	26
9		22	22	Tail	< 16
10	N5	19	20	Head	114
11		18	18	Head	20
12		17	17	?	< 20
13		15.5	15.5	?	< 20

[a]Data largely from Zweig and Cummings (1973a) and Heller (1984).
[b]Polyacrylamide gel electrophoresis in the presence of sodium dodecyl sulfate.
[c]Not determined.

B. DNA Structure

Bacteriophage T5, the best-studied member of this group, has a genome length of about 121,300 bp, with terminal repetitions that are each 8.4% of the total genome length, or about 10,160 bp each (Rhoades, 1982). Although little of the sequence of the DNA has been determined, all evidence obtained so far suggests that the sequences of the left and right terminal repetitions are identical.

The noncontinuous nature of one strand of T5 DNA was first recognized by Abelson and Thomas (1966). The interruptions in the DNA are nicks characterized by 3'-hydroxyl and 5'-phosphoryl termini and can be eliminated by treatment of the DNA with DNA ligase (Jacquemin-Sablon and Richardson, 1970). The origin and role of these nicks remain to be elucidated, although at least four T5-induced endonucleases capable of nicking double-stranded DNA have been identified (Rogers and Rhoades, 1976). However, the nicks produced in ligase-repaired T5 DNA by these endonucleases do not correspond to those found in native parental T5 DNA. Furthermore, the nicks do not appear to be essential for phage replication, because mutants producing interruption-free progeny have been isolated (Rogers et al., 1979a).

The interruptions observed can be divided into two classes. The major nicks occur specifically at 7.9%, 18.5%, 32.6%, 64.8%, and 99.5% from the left end of the genome in 80–90% of the population (Rhoades, 1977a), whereas the minor nicks occur in a variable, but fairly nonrandom manner. The nonrandom location of the interruptions implies that the base sequence surrounding the nicks may be constant, and that has been confirmed by Nichols and Donelson (1977a,b). They have determined that the sequence surrounding most, if not all, of the interruptions is –NPu pGCGCN–. However, at least two of the major interruptions have the 5' sequence pGCGCGGTG. If the nicking enzyme cleaved at every GCGC sequence in T5 DNA, the probability of nicking would be once every 512 bp or roughly 237 nicks per molecule. Furthermore, both strands would be nicked. That the DNA is nicked only once every 24,000 bp indicates that the recognition site is longer than 4 bp, probably in the range of 6–8 bp. However, occasional nicking may occur at "partial" recognition sites, leading to variable but nonrandom nicking.

Rogers et al. (1979a) noted that their interruption-deficient mutants fell into two classes. Mutants either lacked one of the major interruptions, and mutants affecting each of the five major sites were obtained, or the mutants lacked all of the interruptions. The most reasonable hypothesis is that the former class results from alteration of the recognition sequence for the endonuclease, and these mutations are neither recessive nor dominant to wild type as expected. Sequence analysis of these mutants and sensitivity of these mutants to restriction enzymes recognizing the pGCGC sequence would confirm this hypothesis. The latter class probably represents mutations in the structural gene for the specific endonuclease and are codominant with wild type, indicating a cis-acting protein (M. Rhoades, personal communication). Furthermore, Rogers et al. (1979a) found that this class included two complementation groups, which they designated sciA and sciB. Preliminary studies on these mutants have revealed that sciA mutants do not lack any of the four known T5-induced endonucleases, that amber sciA mutants can be isolated, and that the sciA gene lies toward the extreme right end of the unique sequence DNA, near the right end of the D region. Recombination between sciA and sciB is infrequent, suggesting the two genes may lie close together.

C. The Replication Cycle

The major events occurring following infection of *Escherichia coli* by phage T5 have been discussed earlier (McCorquodale, 1975) and will be only briefly summarized here. Details will be elaborated later in the chapter. Following attachment of the phage to the cell envelope, the first 7.9% of the DNA is injected into the cell and expressed. This DNA segment is called the first-step transfer (FST) DNA, and the products of its genes are referred to as pre-early proteins. Two pre-early proteins are necessary for completion of DNA injection, and others effect the turnoff of host macromolecular synthesis and inactivate presumably hostile host functions. Soon after completion of DNA injection (second-step transfer), at about 4 min after infection at 37°C, the early proteins first appear, and their synthesis continues for about 20 min or longer. About 8 min after synthesis of the early proteins begins, the synthesis of the late proteins begins and continues until lysis at about 35–50 min after infection. Typical burst sizes are 200–500 progeny per infected cell.

II. THE GENETIC AND PHYSICAL MAP

After an initial burst of activity, the genetic mapping of the genome has languished in recent years. This can be blamed mostly on the large size of the genome compared to the number of investigators studying this phage. Furthermore, even though mutants continue to be isolated, many of the mutations have not been accurately mapped because of the dearth of nearby markers and the absence of easily identifiable phenotypes. Other complications arise both from heterozygotes produced during crosses between phages with mutations in the terminally repetitive, or A region of the phage, and from the high recombination rate between markers on different genetic regions of the phage.

Figure 1 indicates the current state of knowledge about the arrangement of genes in the T5 genome. The genome is divided into five major genetic regions: A, C, B, D and A', the locations of which can be discerned by the letter in the letter-number designation for most genes. Maximum recombination occurs between any gene in one region and any gene in another region. A and A' are the terminally redundant regions of the DNA and include all of the pre-early genes. Early genes are located on all three remaining genetic units—C, B, and D—whereas late genes are located only in the D region. Genes located only by genetic crosses are identified by vertical lines extending across the linear map, whereas genes localized more exactly by correspondence between the genetic and physical maps are "boxed in." Three regions of the DNA can be deleted without significantly affecting phage replication, and these are indicated on the map with shaded boxes and identified as *del-1*, *del-2*, and *del-3*. The major advance in mapping T5 genes has been the establishment of correspondence between parts of the genetic map and the physical map, defined largely by restriction endonuclease sites. The known restriction sites in the T5 genome for 14 different restriction endonucleases and in the BF23 genome for five different restriction endonucleases are included in Fig. 1.

A. Pre-Early Genes

Brunel and Davison (1979) have isolated mutants of T5 that cannot grow in hosts that have the *Eco*RI restriction endonuclease. These *ris* mutants have *Eco RI* restriction sites within their FST DNA, where no such sites appear in wild-type T5. This restriction system is normally inactivated by some function coded by the FST DNA. However, if the phage DNA carries a restriction site within its FST DNA, cleavage at this site can occur before the restriction endonuclease is inactivated, rendering the phage sensitive to the host's *Eco*RI restriction system. *Eco*RI sites within the rest of the T5 DNA are not cleaved because by the time the DNA carrying them enters the cell, the host's restriction endonuclease has been inactivated.

The availability of these *ris* mutants allows precise localization of the *ris* mutation by cleavage with *Eco*RI and measurement of the sizes of the resulting DNA fragments by gel electrophoresis. The *ris* mutants can also be genetically mapped in relation to known *amber* mutations in the terminal repetition. By such mapping, Brunel and Davison (1979) placed T5 genes A1, A2, and A3 into the left half of the terminal repetition with genes A2 and A3 located between *ris*1 and *ris*3, which defines a 1.8-kb *Eco*RI fragment. Gene A1 was placed to the left of genes A2 and A3 by virtue of an *amber* marker in gene A1 (Fig. 1). Because their map gives the size of the terminal repetition of T5 as 8.9 kbp, which is 1.26 kbp shorter than the value of 10.16 kbp reported by Rhoades (1982), there is an uncertainty of the exact placement of genes A1, A2, and A3. However, if the T5 that they used had a small deletion (of 1.26 kbp) in its terminal repetition

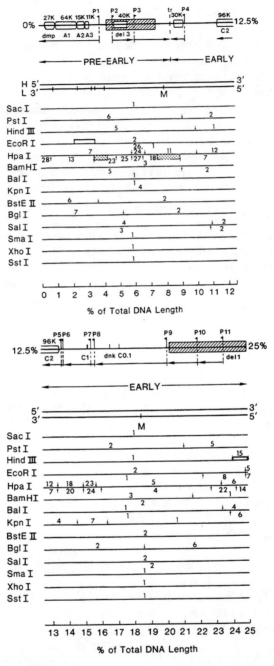

FIGURE 1. Genetic, physical, and transcriptional maps of T5 and BF23. The upper heavy line represents the genetic map with genes localized by open blocks when their sizes are known or by a vertical line when only a marker has been mapped. Three nonessential regions, which are physically deletable, are shown by shaded boxes and designated *del*-1, *del*-2, and *del*-3. Promoters are indicated by "flags" with their tips pointed in the direction of transcription. Directions and extents of transcription are further indicated by horizontal arrows below the genetic map. The vertical dashed lines represent the boundaries of the terminal repetitions (tr). The map is divided into pre-early, early, and late regions, as indicated. The heavy double lines represent the double-stranded DNA of T5 with major (M) and

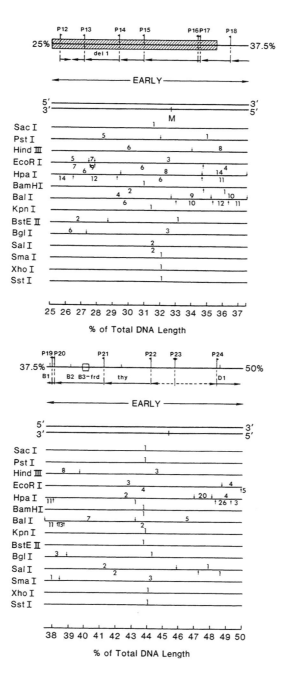

minor nicks indicated where they occur in the L-strand by hatch marks. Restriction sites are shown for T5 DNA (downward-pointing arrows) and for BF23 DNA (upward-pointing arrows). Shaded boxes above the restriction maps define T5 DNA fragments that have been cloned, whereas stippled boxes below the restriction maps define BF23 DNA fragments that have been cloned. Sections of DNA defined by restriction sites are numbered according to size (Rhoades, 1982). Most of the data for restriction endonuclease sites in T5 DNA are from Rhoades (1982). The *Sac*I data are from Davison (1981). The data for restriction endonuclease sites in BF23 DNA are from Lange-Gustafson and Rhoades (1979).

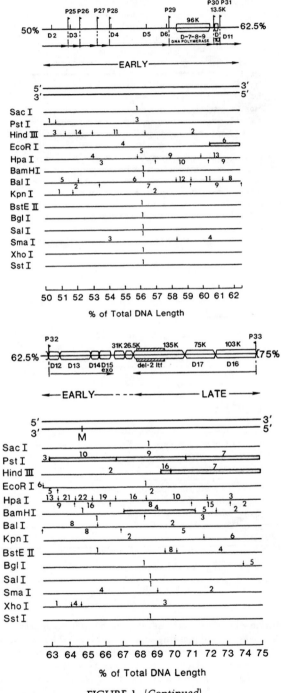

FIGURE 1. (Continued)

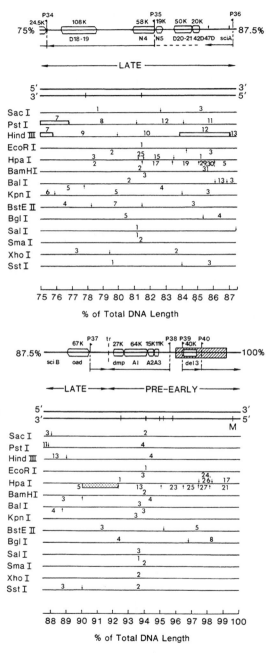

FIGURE 1. (*Continued*)

in a region to the right of *ris3*, where an analogous deletion has been verified in BF23, their placements of the *ris* mutations and genes A1, A2, and A3 in relation to the left end of the T5 DNA molecule are accurate. Their order for genes A1, A2, and A3 also confirms the order previously reported by Beckman *et al.* (1973).

Blaisdell and Warner (1986) have used restriction fragments from the ends of T5 DNA as a template with a coupled transcription-translation system to place not only genes A1, A2, and A3, but also the T5 deoxynucleoside-5'-monophosphatase (*dmp*) gene within 6.3 kbp from the left end of T5 DNA. In addition, they show that the pre-early gene coding for a 40K protein probably is located at a position 6.3 kbp from the left end of T5 DNA, and an early gene coding for a 30K protein lies somewhere between the right end of the left terminal repetition and the *Sal*I site 13.2 kbp from the left end of T5 DNA. The placement of the pre-early gene coding for the 40K protein is in accord with the absence of this gene in the deletion mutant BF23st(4). This deletion mutant has about 3.8 kbp deleted from each terminal repetition, and the deletion spans most of the right half of its terminal repetitions (from 4.4% to 7.6% in the case of the left terminal repetition) (Shaw *et al.*, 1979). This deletion mutant does not synthesize the 40K pre-early polypeptide. These pre-early gene placements are in excellent agreement with Brunel and Davison (1979) and require that *dmp* be located to the left of gene A1, because there is insufficient room for it to the right of gene A3.

B. Early Genes

The only early T5 genes coding for proteins that have been cloned are located in *Pst*I fragment 10, *Eco*RI fragment 6, and BamH1 fragment 4 (Kaliman *et al.*, 1986; Brunel *et al.*, 1979; Davison *et al.*, 1981; Heller and Krauel, 1986). PstI fragment 10 encodes early genes D12, D13, D14, and D15 as well as a small portion of D11. The map for this region in Fig. 1 is based on Kaliman *et al.* (1986), which is congruent with the genetic map (McCorquodale, 1975). Davison *et al.* (1981) also cloned *Pst*I fragment 10 but arrived at a different size for the D14 gene and considered that only part, rather than all, of the D12 gene is in this fragment. They also did not demonstrate the presence of gene D13 in *Pst*I fragment 10, perhaps owing to the lack of an authentic D13 mutant.

*Eco*RI fragment 6 encodes early gene D10 (Brunel *et al.*, 1979), which had previously been omitted from the genetic map (McCorquodale, 1975) because genetic crosses did not confidently locate it. The placement of gene D10 in *Eco*RI fragment 6 agrees with the indications from genetic crosses that it maps between D9 and D11 (hence its numerical designation). In Fig. 1, gene D10 is placed between the left end and the center of *Eco*RI fragment 6, because the gene for T5 DNA polymerase (D7-8-9) extends into the left end of *Eco*RI fragment 6 (see below), and gene D11 extends just into the left end of *Pst*I fragment 10.

*Bam*H1 fragment 4 codes for the L-shaped tail fibers (Heller and Krauel, 1986), and the corresponding structural gene (*ltf*) is mapped accordingly in Fig. 1.

The inability to clone most early T5 genes has prompted the use of marker rescue techniques to localize T5 genes on defined restriction fragments. Cells can be transformed with various restriction fragments and subsequently infected with a particular *amber* mutant of T5. Those restriction fragments containing the corresponding wild-type T5 gene can rescue the *amber* mutant by allowing production of wild-type phages. In this way, it was shown that early gene C2 was localized to *Eco*RI fragment 2 and that the T5 dihydrofolate reductase gene (B3) was localized to position 40 in *Hpa*I fragment 2 (Peter *et al.*, 1979). (Positions are given as percent of the total DNA length starting from the left end.) Thus, in Fig. 1, gene B3 is precisely located, whereas gene C2 lies somewhere between position 8.4 and 21.5. Fujimura *et al.* (1985) have shown by marker rescue that the T5 DNA polymerase gene is near the left end of *Sma*I fragment 4 and is delineated by the *Sma*I/*Bal*I site at position 58.3 and the *Bal*I site at position 61.3. This places the DNA polymerase gene immediately to the left of D10, which is located in *Eco*RI fragment 6 (Brunel *et al.*, 1979). They also showed that "gene" D7 maps in the DNA polymerase gene, which now includes the three markers D7, D8, and D9. From the size of the polypeptide fragments generated by *amber* mutants D8, *am*1, D9, and *am*6, the direction of transcription was shown to be from D7, near the left end of the gene, toward D9, near the right end of the gene (Schneider *et al.*, 1985). Early genes that have not been cloned or localized by marker rescue are placed on the map of Fig. 1 according to their frequency of genetic recombination with physically localized genes.

The early gene region contains a large section with nonessential deletable genes, which extends from about position 20 to position 36 and is designated the *del*-1 region (Okada, 1980; Von Gabain *et al.*, 1976). This region codes for at least 24 different tRNAs, which include at least one tRNA for each of the 20 amino acids incorporated into proteins (Hunt *et al.*, 1976, 1980; Desai *et al.*, 1986; Chen *et al.*, 1976; Ikemura *et al.*, 1978; Ksenzenko *et al.*, 1982; Ikemura and Ozeki, 1975). The locations in the DNA for tRNA genes of T5 and BF23 are shown in Fig. 2. There is good agreement, with respect to position in respective maps, between the regions that code for T5 tRNAs and those that code for BF23 tRNAs. The fact that these tRNA genes are not essential for viability raises the question why they are so well conserved in T5 and BF23. The most likely role they play is to ensure that the process of translation in T5-infected cells proceeds with high efficiency, with the codon usage dictated by the T5 or BF23 genome. They also serve a protective role in that some of the tRNA genes can be mutated to yield tRNAs that suppress nonsense mutations

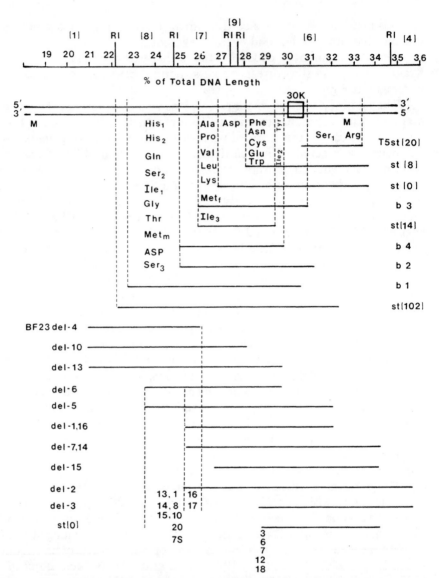

FIGURE 2. Map of the *del*-1 region of T5 and BF23. The top scale is the percent of total DNA length for both T5 and BF23 and includes the *Eco*R1 restriction sites in BF23 DNA in the *del*-1 region. The numbers in parentheses between the *Eco*R1 sites denote the same DNA fragments as in the *Eco*R1 restriction map of BF23 in Fig. 1. The heavy double lines represent the double-stranded phage DNA in the *del*-1 region and include the two major nicks (M) present in T5 DNA in this region. The horizontal lines below the heavy double lines represent the regions of DNA deleted in various deletion mutants of T5 (designated on the right) and of BF23 (designated on the left). Locations of individual tRNA genes are indicated by the amino acid that they accept, in the case of T5, or by a number, in the case of BF23. The box between 30% and 31% designated 30K represents the required coding length for a 30K protein coded for in this region as reported by Okada (1980). The mapping data for T5 have been taken largely from Hunt *et al.* (1980), Desai *et al.* (1986), and Ksenzenko *et al.* (1982b); the data for BF23 are from Okada (1980).

within the phage genome (Macaluso *et al.*, 1982; Okada *et al.*, 1978). Such suppressor mutations within the phage genome itself would overcome nonsense mutations that arise in any gene of T5 or BF23 except in those pre-early genes that are necessary for completion of phage DNA transfer (Okada *et al.*, 1978).

Not all of the *del*-1 region codes for tRNAs, since Ksenzenko *et al.* (1982) have shown that stable low-molecular-weight RNAs of as yet unknown function are coded for by genes in this region. Also, Okada (1980) has reported that a 30K protein is coded for in the *del*-1 region of BF23 (Fig. 2).

The tRNA genes have been mapped approximately by deletion mapping (Fig. 2). During such mapping, Okada (1980) has reported that deletions in the *del*-1 region of BF23 larger than about 11% yield phage DNA molecules too short to be packaged. Thus the *del*-1 region cannot be deleted in its entirety in a single phage DNA molecule and still yield viable phages.

Some portions of the *del*-1 region have been cloned and sequenced (Ksenzenko *et al.*, 1982b; Kryukov *et al.*, 1983; Desai *et al.*, 1986). Nucleotide sequencing has localized some tRNA genes precisely and has revealed that tRNA genes overlap, which seems a wasteful process, because a given RNA transcript can be processed into one or the other mature tRNA but not into both tRNAs. Nevertheless, overlapping genes now appear quite common in T5 with occurrences revealed in the late gene region (Brunel *et al.*, 1983), in the early gene region surrounding gene D15 (Kaliman *et al.*, 1986), and in the tRNA gene region (Desai *et al.*, 1986).

C. Late Genes

Late genes of T5 start to the right of early gene D15 with the first known late gene being *ltf* (Heller and Krauel, 1986). A viable deletion mutant (*del*-2), reported by Rhoades *et al.* (1980), has a portion of the *ltf* gene deleted (Heller, 1984), indicating that the L-shaped tail fibers for which it codes are unessential, in confirmation with an earlier report (Saigo, 1978). The placement of the *del*-2 deletion within the *ltf* gene is also compatible with earlier reports that this region is not silent, because phage polypeptides are made from this region in plasmid-containing minicells (Brunel *et al.*, 1981), and plasmids into which *Pst*I fragment 9 and *Pst*I fragment 7 are inserted make the cells "unhealthy" (Davison *et al.*, 1981).

Brunel *et al.* (1981) reported that *Pst*I fragment 9 codes for four polypeptides with M_r = 35,000, 31,000, 26,500, and 14,500, and that *Hind*III fragment 7 also codes for four polypeptides with M_r = 92,500 79,000, 24,500, and 14,500. Because the polypeptide with M_r = 14,500 is common to both fragments, it is most likely coded by the overlap region between

PstI fragment 9 and HindIII fragment 7. Such mapping places the coding region for the 35,000, 31,000, and 26,500 proteins to the left of that for the 14,500 protein, which places them in the del-2 region. The coding region for the 92,500, 79,000, and 24,500 proteins would then map to the right of that for the 14,500 protein.

However, Heller and Krauel (1986) have cloned T5 BamH1 fragment 4 and shown that it codes for a polypeptide with an M_r = 125,000 which may arise from a precursor with M_r = 150,000. In addition, they showed that cells containing a recombinant pBR322 plasmid with a BamH1 fragment 4 insert would complement Saigo's (1978) mutant that lacked L-shaped tail fibers. We have therefore mapped the ltf gene according to Heller and Krauel (1986). Such mapping indicates that the polypeptide with M_r = 14,500 and at least one of the other three polypeptides arising from clones of PstI fragment 9, represent out-of-frame starts and stops within the ltf gene. The other two polypeptides arising from clones of PstI fragment 9 could be accommodated to the left of the ltf gene and are so indicated in the map.

Heller (1984) has mapped genes D17, D16, D18–19, N4, D21, and oad by constructing hybrid phages of T5 and BF23. We have placed them in Fig. 1 according to Heller's map. Such mapping places gene D17 to the right of the ltf gene, and gene D16 to the right of gene D17. Both genes D17 and D16 therefore reside within HindIII fragment 7. Plasmids containing HindIII fragment 7 give rise to polypeptides with M_r = 92,500, 79,000, and 24,500 (Brunel et al., 1981). Two of these molecular weights approximate those given for the products of genes D17 (75,000) and D16 (103,000) (Table I) (Heller, 1984). The third molecular weight (24,500) could correspond to a polypeptide whose gene resides between genes D16 and D18–19, and is indicated in the map of Fig. 1. Why recombinant plasmids containing HindIII fragment 7 will not complement T5 mutated in gene D17 remains unknown.

HindIII fragment 12 has been cloned into lambda Charon-21A and into pBR322, and it directs the synthesis of polypeptides with M_r = 33,000, 31,000, and 20,000 in the latter vector (Brunel et al., 1981). The polypeptide with M_r = 31,000 is considered to be a breakdown product of the polypeptide with M_r = 33,000. However, in lambda Charon-21A, HindIII fragment 12 directs the synthesis of polypeptides with M_r = 34,000 or 36,000, depending on its orientation, and with M_r = 20,000. The polypeptide with M_r = 20,000 is therefore independent of vector and orientation and most likely is an intact gene with its own promoter and terminator. The portion of HindIII fragment 12 that codes for a polypeptide with variable M_r and that is dependent upon vector (33,000 in pBR322) and orientation (34,000 or 36,000 in lambda Charon-21A) complements amber mutations in gene D20-21, but it is considered not to be the entire D20-D21 gene (Brunel et al., 1981). Indeed, the product of gene D20–21 has an M_r = 50,000, so this gene must be on one end of HindIII fragment 12. By deletion mapping, Brunel *et al.* (1981) place it on the left end of this fragment, and the gene coding for the polypeptide with M_r = 20,000 is assigned to the center of *Hind*III fragment 12, near the *Bgl*I site. These genes are placed accordingly in the map of Fig. 1.

Heller and Bryniok (1984) have placed another gene into the region spanned by genes D17 to D20–21. Their order in this region is D17, D16, D18–19, N4, N5, D20–21, and *oad*, which is the same as the order previously published (McCorquodale, 1975), but with the addition of genes N5 and *oad*. In addition, they detect a tail polypeptide that has an M_r = 108,000, which is the product of gene D18–19. In the map of Fig. 1, these genes are placed between *del-2* and D20–21 in accordance with established order and size. However, it should be noted that, although we place *oad* in accordance with Heller (1984), the reported recombinational distance between genes *oad* and N5 (Heller and Bryniok, 1984) would seem to place it to the immediate right of gene N5.

SciA and *sciB*, the genes whose products determine the single-strand interruptions in T5, map at positions 86–87 and 88–89, respectively (M. Rhoades, personal communication). They are so placed on the map in Fig. 1, which leaves only the space between the gene coding for the protein with M_r = 20,000 and the *sciA* gene for genes 42D–28D and 47D. This space is very limited so we suggest that the protein with M_r = 20,000 might be the product of gene 42D–28D, which would then place gene 47D immediately to the left of the *sciA* gene.

The T5 DNA between the left end of the *ltf* gene and the beginning of the right terminal repetition could code for 1,117,500 daltons of protein. Of this coding capacity, the structural proteins of T5 account for about 800,000 daltons, of which only about 490,000 have been located in the map of Fig. 1. Gaps in the map of this region and mapped genes coding for late proteins of unknown function (Fig. 1), however, can easily accommodate the remaining 310,000-dalton coding capacity necessary for the remainder of the structural proteins.

III. PHAGE ATTACHMENT AND DNA INJECTION

A. Adsorption to Host Cells

Heller and Schwarz (1985) have proposed that after collision between a host cell and T5, the angled tail fibers bind reversibly to polymannose O-antigens of the *Escherichia coli* host cell, thereby accelerating the rate of adsorption by a factor of 15 (Heller and Braun, 1979, 1982). The single straight tail fiber may then probe the surface of the outer membrane for adhesion sites (Bayer, 1968), and when one is located, the single straight tail fiber may insert itself into the membrane to form a pore. The receptor-binding protein, located distal to the conical part of the tail at the straight tail fiber attachment site, then binds irreversibly to the cellular receptor *fhu*A in the case of T5 and *btu*B in the case of BF23 (Heller and Schwarz, 1985).

B. Injection of Parental DNA

After attachment of the phage to the cell surface, only 7.9% of the parental DNA is initially injected into the cell (Lanni, 1968). This DNA is called the first-step transfer (FST) DNA. Two genes in this DNA segment must be expressed before the remaining DNA can be brought into the cell. The control of this two-step injection of the parental DNA remains a mystery, although several hypotheses have been developed, and some details of the injection process have been uncovered. The most plausible hypothesis is the idea that injection of the DNA is blocked by some discontinuity in the DNA near base pair 9500 (Shaw *et al.*, 1979). Speculations about this discontinuity have included (1) a single-strand interruption, (2) some unusual DNA structure such as a cruciform, and (3) the existence of a specific DNA-binding protein for this discontinuity. Even though a major interruption does occur at about 7.9% from the left end of the genome, the first option seems least likely because of the isolation of a phage mutant that specifically lacks this interruption (Rogers *et al.*, 1979b). This phage transfers its DNA into the cells in two steps, as usual, and is viable.

Determination of the DNA sequence around position 8.1, which is a recent estimate of the location of the first-step stop signal in T5 (Heusterspreute *et al.*, 1987) could provide evidence for the second option. In fact, Heusterspreute *et al.* (1987) have found that a sequence of 937 bp's around position 8.1 contains several palindromes and direct repeats, and note that some of the 10-bp repeats have a sequence very similar to dnaA protein-binding sites. J. S. Wiest and D. J. McCorquodale (unpublished results) have extended this sequence in BF23 to the right end of the terminal repetition (only 71 more bases), where additional, but shorter, direct repeats and palindromes exist.

The findings of Heusterspreute *et al.* (1987) point to the possibility of a host protein binding specifically to T5 DNA at the first-step stop signal. Phage-specific proteins should also interact with the first-step stop signal to facilitate the completion of phage DNA transfer. So far, only two DNA-binding proteins induced by T5 have been identified, the A2 and D5 proteins (Snyder and Benzinger, 1981; McCorquodale *et al.*, 1979). The A2 protein binds to T5 DNA and other DNAs, but there is no evidence that the A2 protein binds to a specific sequence in T5 DNA. The D5 protein, which binds nonspecifically to T5 DNA (Rice *et al.*, 1979), is not involved in DNA transfer, because it is an early protein.

Complete injection of the DNA requires both the A1 and A2 proteins (McCorquodale, 1975). Whereas the A1 gene product is a pleiotropic protein with many roles, the A2 protein is known to be involved only in DNA injection. The purification and characterization of this protein has been undertaken by Benzinger and his colleagues (Snyder and Benzinger, 1981). The protein has a molecular weight of about 15,000, which agrees with gel patterns obtained using A2 mutants (McCorquodale et al., 1977). The A2 protein also binds to the A1 protein, which in turn binds to membranes. Therefore, the potential for forming at least quaternary complexes of A1 and A2 proteins, T5 parental DNA, and host membranes suggests that membranes can play a role in both stopping and completing transfer of parental T5 DNA into the cell. Furthermore, Snyder (1984) has shown that when the A2 protein is produced after infection, an outer membrane protein, *ompF*, is lost from the *E. coli* envelope. The amino acid sequence of the A2 protein is being determined (Fox et al., 1982) to seek clues about how the protein accomplishes these various functions.

Whereas the actual role of the A1 and A2 proteins in control of the two-step transfer process has not been elucidated, it is clear that once the A1 and A2 proteins have been synthesized energy is no longer required for second-step transfer (Maltouf and Labedan, 1983). Thus, metabolic poisons such as azide, dinitrophenol, arsenate, cyanide, and carbonyl cyanide-m-chlorophenyl hydrazone do not prevent second-step transfer, suggesting this is a passive process which does not require ATP, a membrane potential, or an electrochemical gradient of protons. This would also rule out an active role for RNA polymerase, using transcription to drive the transfer. Thus, Maltouf and Labedan (1983) suggested that the DNA enters by diffusion, using histonelike proteins (possibly A1 and A2?) to "pack" the DNA and bind it to the membrane.

Finally, the state of the membrane during injection of T5 DNA is critical. When T5 phage and *E. coli* are mixed at 0°C, the DNA does become weakly attached to the cell envelope, but actual injection of the FST DNA does not occur until the temperature is raised to 15°C or above (Labedan, 1976). This injection failure is due, at least partially, to rigidity of the membrane. When the state of membrane is manipulated by changing the fatty-acid composition of the membrane, or by adding local anesthetics, DNA injection is correlated directly with membrane fluidity (Labedan, 1984).

Because the phage genome contains large terminal redundancies, the question arises: Which end enters first? Saigo (1976) has separated phage progeny into three major bands by isopycnic centrifugation and found that the lightest band lacks about 7–8% of its DNA. Analysis of fragments produced by various restriction enzymes indicates that the right-hand terminal redundancy is missing from these phage. Although these phage adsorb and inject their DNA normally (as assayed by host DNA degradation), they fail to form progeny or to replicate their DNA. These results suggest that the left end does enter first and that the right end is not essential until DNA replication begins.

The above conclusion is supported by three other lines of evidence.

Labedan *et al.* (1973) examined naked DNA extending out from infected cells from the FST-DNA that had been transferred into the cells. The sizes of the DNA fragments generated by mechanical shear from this extracellular phage DNA were compatible with exclusive left-end transfer, if a nick occurred in T5st(0) DNA at about 35% from the left end. The conclusion was questioned on the basis of the existence of a nick at this position, but we now know that a minor nick does occur at about 44% in wild-type T5 DNA, which would be at about 36% in T5st(0) DNA. Thus, their conclusion was correct. Herman and Moyer (1975) concluded that the left end of T5 DNA always enters the host cell first on the basis of the sequence by which the nicks in T5 DNA were ligated. Shaw and Davison (1979) came to the same conclusion on the basis of the size of *Eco*R1 restriction fragments generated from the FST DNA of *ris* mutants. Thus, the polarity of T5 DNA transfer, by which the left terminal repetition always enters host cells first, is well established.

IV. TRANSCRIPTION

A. Temporal Pattern of Transcription

Expression of T5 and BF23 genes follows a well-defined pattern (McCorquodale, 1975). The first portion of T5 and BF23 DNA to enter host cells is the left terminal repetition, which contains only pre-early genes. These pre-early genes are therefore the first phage genes to be expressed after infection. Their expression is mostly shutoff about 8 min after infection by a mechanism that requires the product of pre-early gene A1. The product of gene A1 is also necessary, together with the product of pre-early gene A2, for completion of phage DNA transfer to host cells, which is complete by 2–3 min after infection. Thus, the 3- 4-min delay in the turn-on of early gene expression after infection is due to this two-step DNA transfer mechanism. Even though late genes enter host cells at essentially the same time as early genes, they are not expressed until several minutes later, because their expression depends on the presence of the protein products of early genes C2, D5, and D15. Expression of late genes commences about 12 min after infection and continues until lysis. Shutoff of expression of many, but not all, early genes occurs about 20 min after infection.

Pre-early genes are transcribed by unmodified host RNA polymerase, early genes by host RNA polymerase modified by the protein products of pre-early gene A3 and possibly A1, and late genes by an RNA polymerase further modified by the protein products of early gene C2 and an early protein with $M_r = 15K$. Host sigma factor is present with the RNA polymerase at all times during the infective cycle. Host RNA polymerase is required for all essential T5 transcription, since rifampin inhibits all essential classes of T5 transcription (Beckman *et al.*, 1972). Efficient transcription of most late genes requires a conformationally strained state of the DNA, because inactivation of the host's DNA gyrase virtually eliminates late, but not early or pre-early transcription (Constantinou et al., 1986).

B. Transcription Maps

Transcription maps of T5 and BF23 have been refined by the detection of phage promoters in cloned fragments, by nucleotide sequencing, and by direct observation of transcription with the electron microscope. A summary of transcriptional units is incorporated into the genetic map of Fig. 1.

Using electron microscopy, Stueber et al. (1978) observed 41 promoters in T5 DNA, of which 30 serve early genes, five serve late genes, and six serve pre-early genes with three promoters repeated in each pre-early region. This distribution gives a total of 38 different T5 promoters. In the map of Fig. 1 promoter 28 consists of two promoters (28a and 28b) close together, but this is not indicated in the figure. The positions of T5 promoters, taken from the data of Stueber et al. (1978), appear to be appropriately located at the beginnings of genes. For instance, promoter P1 serves pre-early genes A3, A2, A1, and dmp in that order in what seems to be a single operon. The sizes of these genes dictate a mRNA with $M_r = 1,000,000$ if they are transcribed as a unit, and Sirbasku and Buchanan (1970) have reported a mRNA with such an M_r. Interestingly, this putative operon uses the intact strand for transcription, perhaps because if the interrupted strand were used, transcription would be interrupted by the nicks if they had not yet been repaired, which would lead to delay of the infective cycle. The positioning of promoter P1 is in general agreement with Blaisdell and Warner (1986), who place P1 close to the beginning of gene A3.

The data of Stueber et al. (1978) indicate that transcription of the terminal repetition containing pre-early genes proceeds from right to left for genes A3, A2, A1, and dmp, but from left to right for other pre-early genes (Fig. 1). Transcription of early genes proceeds mainly from right to left in the region from 8.3% to 46%, but from left to right in the region from 46% to 67%. Transcription of late genes proceeds mainly from right to left, except from late promoter P37. However, not all promoters needed for transcription of T5 genes appear to have been detected by Stueber et al. (1978), because none of their promoters would serve the sciB gene or the region in del-1 between 23.7% and 25.5% (Fig. 1). Also, by nucleotide sequencing, Brunel et al. (1983) detected three left-to-right late promoters, of which two are in PstI fragment 7 and one is in HindIII fragment 12. The data of Stueber et al. (1978) also place two promoters in PstI fragment 7, but as right-to-left promoters, and the left-to-right late promoter in HindIII fragment 12 was not detected by Stueber et al. (1978).

These studies illustrate that unambiguous location and polarity of T5 promoters requires the determination of nucleotide sequences and also footprinting to confirm that a putative T5 promoter revealed by a nucleotide sequence is indeed a specific binding site for RNA polymerase. In this last regard, it should be pointed out that the identification of T5 promoters by visualization of transcripts with the electron microscope and the expression of cloned T5 genes from T5 promoters have all been done with unmodified host RNA polymerase. It follows that most early and late T5 promoters are recognized by unmodified RNA polymerase. However, Melin and Dehlin (1987) have reported that one late T5 promoter is not utilized in uninfected *E. coli* but is utilized in T5-infected *E. coli* and in uninfected *Bacillus subtilis*. These findings indicate that some late T5 promoters are not recognized by unmodified RNA polymerase in *E. coli* and may require a modified RNA polymerase for recognition.

From the foregoing, the role of modified forms of RNA polymerase, which are clearly generated in T5-infected cells, would seem to be one of regulation. Clearly, promoters for pre-early genes must be recognized by unmodified host RNA polymerase, because it is the only polymerase available at the time of infection. Modification of the host RNA polymerase by the products of genes A1 and A3 occurs very soon after infection and before the completion of phage DNA transfer into host cells. The product of gene A1 is necessary for the shutoff of pre-early gene expression and can be found loosely complexed to host RNA polymerase (McCorquodale *et al.*, 1981). The product of gene A3 (gpA3) is more tightly bound to host RNA polymerase and is necessary for the abortive response to T5 and BF23 infection by hosts that carry a *Col*Ib plasmid (for a review of this abortive infection, see Duckworth *et al.*, 1981).

However, in a normal infection, the role of gpA3 is not at all clear. It does not seem necessary for the recognition of early T5 promoters, because they are readily recognized by unmodified RNA polymerase both *in vitro* and *in vivo* (see above). A possible role for gpA3 is to modify host RNA polymerase such that it cannot recognize late T5 promoters. Such a role would explain why late genes are not expressed as soon as they enter host cells. It also explains why further modifications by products of early genes, not only of RNA polymerase (by gpC2 and an early 15K protein) but also of the phage DNA (by gpD5, gpD15, and host DNA gyrase), are necessary for recognition of late T5 promoters. Only one or two steps in the above model for regulated expression of T5 genes have been tested by *in vitro* studies. Szabo and Moyer (1975) have compared the transcriptional specificities of unmodified RNA polymerase from uninfected cells with that of RNA polymerases from T5-infected cells that have been modified with only the early 15K protein in both *Col*Ib⁺ and *Col*Ib⁻ cells. No differences in specificities were found, and it was concluded that all three RNA polymerases could transcribe pre-early, early, and late genes, and with equal efficiency. These experiments seem to rule out possible covalent alterations to host subunits of RNA polymerase as a means to change the specificity of RNA polymerase in an infected cell, but they do not clearly address the question of transcriptional specificities of host RNA polymerase modified with only the pre-early 11K protein (now known to be gpA3) or with gpA3, C2, and the 15K early protein (Szabo et al., 1975). Thus, the crucial experiment with RNA polymerase modified only with the pre-early 11K protein has yet to be done. McCorquodale et al. (1981) have shown that host RNA polymerase modified with gpA1 transcribes pre-early genes very poorly but transcribes early and late genes well. These latter findings support the model of transcriptional regulation by modifications to host RNA polymerase.

Brunel et al. (1981) have summarized the behavior of the T5 DNA fragments they have cloned in terms of expression of T5 genes. They conclude that EcoR1 fragment 6 contains an early promoter serving gene D10, and that HindIII fragment 12 contains a late promoter serving gene D21. These conclusions are in agreement with the transcriptional map of Fig. 1. More recently, Brunel et al. (1983) have shown, by nucleotide sequencing, that late T5 promoters and terminators overlap or are very close together. Overlapping has also been detected by Kaliman et al. (1986) in the case of early gene D15. Nucleotide sequencing revealed that its initiation codon overlaps the termination codon of the preceding gene. These findings indicate that coding sequences are tightly arranged in the late region and in at least some portions of the early region.

The transcriptional map of T5 shown in Fig. 1 is in general agreement with previously published maps (Von Gabain and Bujard, 1977; Davison and LaFontaine, 1984). Some differences, however, should be pointed out. Von Gabain and Bujard (1977) indicate that major pre-early transcription of genes A1, A2, and A3 occurs in the right half rather than the left half of the terminal repetition. Chen and Bremer (1976) and Chen (1976) also conclude that major pre-early transcription occurs in the right half of the terminal repetition. Von Gabain and Bujard (1977) indicate that late T5 genes occupy about 35% of the total length of T5 DNA, whereas estimates by Chen (1976) yield a maximum of 22% of the total length. The data used to construct Fig. 1 indicate that between 24% and 25% of the total T5 DNA contains late genes, assuming that no late genes are to the left of gene D15.

C. Promoters

An interesting feature of T5 promoters is their strength. By measuring both relative rates of formation and stabilities of RNA polymerase-promoter complexes, Von Gabain and Bujard (1977) demonstrated that the level of transcriptional activity both in vivo and in vitro correlates with the rate of formation of the RNA polymerase-promoter complex and not with the stability of the complex. A comparison of T5 promoters showed that their strengths varied over a six- to sevenfold range with early promoters being the strongest, pre-early promoters being intermediate in strength, and late promoters being the weakest. Furthermore, early T5 promoters are stronger than other promoters tested, including those of phages lambda, T7, and fd and of plasmids pML21 and pSC101 (Von Gabain and Bujard, 1979).

Gentz et al. (1981) have shown that the high strength of T5 promoters apparently does not allow them to be cloned but that they can be cloned if a strong transcriptional stop signal is placed downstream from the T5 promoter. In this manner, Gentz and Bujard (1985) have cloned several strong T5 promoters, assessed their efficiency in vivo, and sequenced 11 of them. Strong T5 promoters were found among those that serve pre-early, early, and late genes, and their nucleotide sequences showed several regions of homology. The region around the promoter that contacts the RNA polymerase is from nucleotide +20 to −55 and averages 75% AT (nucleotide +1 is the nucleotide that initiates the RNA transcript). Higher AT-rich regions occur between −56 and −36 as well as between −12 and +8. Highly conserved regions occur around +7, −11, −10, −33, and −43, and the distance between the −10 region, which corresponds to the consensus sequence, TATAAT, and the −33 region, which corresponds to the consensus sequence, TTGACA, is almost always 17 nucleotides. Perfect homologies occur at positions −7, −11, −12, −34, and −43. Pre-early promoters have homologies around +7 where the sequence ATGAGAG occurs, whereas early promoters conserve the sequence TTGA around this position. Not all T5 promoters in these three classes have the foregoing properties. Brunel et al. (1983) have identified late T5 promoters, but although they are AT-rich, their sequences around −10 and −33 often vary considerably from the consensus sequences for these regions. Ksenzenko et al. (1982a) also have identified a late T5 promoter that is GC-rich.

Some early T5 promoters are so strong that during in vitro transcription they outcompete all other promoters present in a plasmid. The result is that 90% or more of all transcripts are initiated at the T5 promoter. In addition, most strong T5 promoters will accept G^{Me}-ppp-A as an initiator of transcription to yield capped RNA transcripts (Gentz and Bujard, 1985). Because of these properties, strong T5 promoters appear to be useful for producing large amounts of specific transcripts that could be used for both prokaryotic and eukaryotic translation (Stueber et al., 1984).

D. Coupled Transcription and Replication

Late genes are not transcribed until early genes C2, D5, and D15 are expressed. The protein product of gene C2 binds to RNA polymerase

(Szabo *et al.*, 1975) and presumably allows late promoters to be recognized. However, late transcription *in vivo* requires gpD5, a DNA-binding protein (McCorquodale *et al.*, 1979; Rice *et al.*, 1979), and gpD15, a nuclease that presumably modifies the DNA template (Moyer and Rothe, 1977). In addition to these requirements for the products of early genes, late T5 transcription requires host DNA gyrase (Constantinou *et al.*, 1986), presumably to facilitate melting of late phage promoters and/or genes. Late T5 transcription appears to be very complex in the infected cell and may be difficult to reproduce *in vitro*. Along these lines, Ficht and Moyer (1980) isolated a transcription-replication enzyme complex that included RNA polymerase, gpC2, gpD9 (the T5 DNA polymerase), gpD5 (a T5 DNA-binding protein), and gpD15 (a T5 nuclease), together with T5 DNA. The existence of such a complex suggests that the processes of late transcription and DNA replication are integrated in T5-infected cells.

V. EFFECTS ON HOST METABOLISM

A. Host DNA Degradation

Degradation of host DNA in T5-infected cells is unusually rapid and complete, and the thymine present is not reused for phage DNA synthesis (Crawford, 1959; Zweig *et al.*, 1972). Only about 10% of the host DNA remains in acid-insoluble form about 10 min after infection, and the remaining 90% of the DNA bases are actually excreted into the medium (Pfefferkorn and Amos, 1958; Warner *et al.*, 1975), requiring that all dNMPs used for phage DNA synthesis be synthesized *de novo*. In spite of this very active process, no host or T5-induced protein has been directly implicated in the degradation of host DNA to deoxyribonucleotides. A brief report of an endonuclease appearing soon after infection (Fielding and Lunt, 1969) has not been substantiated. Furthermore, this induced activity lacks the needed substrate specificity. A1 mutants fail to degrade host DNA (Lanni, 1969; McCorquodale and Lanni, 1970), but no evidence has been reported confirming that the A1 gene product either has DNase activity or combines with any other protein having DNase activity. Furthermore, intermediates in this degradative process have not been characterized. Only one of the final steps in this process, dephosphorylation of 5'-dNMPs, is known to be performed by a phage-induced enzyme (Warner *et al.*, 1975; Mozer *et al.*, 1977).

The induction of such an effective degradation system by T5 suggests that the phage must subsequently also inhibit this process before phage DNA synthesis can begin. However, the induction of this inhibitor has not been directly demonstrated. If it does exist, one would predict that the gene is not located on the FST DNA and is expressed only after transfer of the remaining DNA into the cell. An alternative to induction of this putative inhibitor is to control degradation by the absence or presence of specific sequences in the DNA, as discussed below.

B. Synthesis of Host Macromolecules

As described in detail in earlier reviews, T5 has profound effects on synthesis of host-specific macromolecules (McCorquodale, 1975). Host DNA, RNA, and protein synthesis are all halted soon after infection with phage T5, and the host DNA is rapidly degraded to acid-soluble fragments. It is tempting to conclude that all three effects may be due to this rapid and rather complete degradation of host DNA. Experiments related to this point center around use of the A1 mutant, a pleiotropic mutant unable to degrade host DNA, to complete transfer of phage DNA into the cell, to shut off synthesis of host RNA and protein, and to shut off expression of T5 pre-early genes (Lanni, 1969; McCorquodale and Lanni, 1970). Because A1 mutants, which fail to degrade host DNA, still do not permit the continued synthesis of host DNA, it was argued that host DNA degradation is not the cause of the shutoff of host DNA synthesis. The discovery that T5 induces a 5′-deoxyribonucleotidase within 2 min after infection suggested a possible explanation for the shutoff (Warner et al., 1975). This enzyme could deplete the pools of dUMP and dTMP present in the infected cell so that dTTP and DNA synthesis would be effectively blocked until the 5′-deoxyribonucleotidase is eventually inhibited upon transfer of the remaining 92% of the DNA into the cell. However, when this hypothesis was tested using deoxyribonucleotidase (dmp) mutants by following the incorporation of thymidine into DNA following infection of E. coli with wild-type T5, T5-dmp, or T5-A1-dmp, no evidence was obtained indicating that 5′-deoxyribonucleotidase is responsible for shutoff of host DNA synthesis (Mozer et al., 1977).

The A1 protein may play either a direct or an indirect role in turning off host RNA and protein synthesis. Because the only mutants isolated so far that fail to degrade the host DNA are A1 mutants, it has not been possible to distinguish between the failure to degrade host DNA and the failure to shut off host RNA and protein synthesis. Chloramphenicol prevents the shutoff of host RNA synthesis (Sirbasku and Buchanan, 1970), indicating that protein synthesis is required for shutoff; presumably the A1 protein is responsible, but the actual mechanism remains to be established.

C. Inhibition of Host Enzymes

The inhibition of several E. coli enzymes and processes after infection with T5 has been reported, but none of these inhibitors have been purified, and most remain poorly characterized. These enzymes and processes include restriction endonuclease *EcoRI* (Davison and Brunel, 1979), uracil-DNA glycosylase (Warner *et al.*, 1980), DNA methylases (Hausmann and Gold, 1966), host-cell reactivation (Chiang and Harm, 1976a,b), *rec BC* exonuclease (Sakaki, 1974), and poly-A-polymerase (Ortiz *et al.*, 1965). The rationale for the inhibition of the last five phenomena has not been established, because little follow-up work has been reported. It can be assumed that some intermediate in DNA replication or packaging may be sensitive to attack by the *rec BC* exonuclease, necessitating its inhibition. The inhibition of poly-A-polymerase is probably related to modification of the host DNA-dependent RNA polymerase by T5-induced proteins as discussed elsewhere.

1. Uracil-DNA Glycosylase

The inhibition of the uracil-DNA glycosylase is another curious property of phage T5, as there is no obvious reason why this occurs. This enzyme removes uracil from DNA, and the DNA is subsequently nicked at the baseless sites created by this depyrimidination. An argument can be made that the inhibition of uracil-DNA glycosylase results from a need to rigorously control the number and location of nicks in the T5, but the existence of "minor" nicks (Rhoades, 1977a) renders this suggestion rather unlikely. Some purification and characterization of this inhibitor have been accomplished (H. Warner, unpublished results). The inhibitor is presumed to be a protein, and it migrates on Sephadex gel columns as if it has a molecular weight of about 15,000 daltons. The partially purified inhibitor inhibits uracil-DNA glycosylase purified from *E. coli* but not from wheat germ (Blaisdell and Warner, 1983). This was interpreted to mean that the inhibitor binds to the enzyme rather than to the DNA. No further insights into the mechanism of action of the inhibitor or the reason for its induction have been obtained.

2. Restriction Endonuclease *EcoRI*

The clearest rationale can be developed for the *EcoRI* inhibitor, even though the inhibition of the *EcoRI* restriction enzyme following T5 infection has not been actually shown. What has been shown is that a "restriction protection system" is developed after infection, but before injection of the last 92% of the parental DNA (Davison and Brunel, 1979). This conclusion is drawn from the observation that none of the six naturally occurring *EcoRI* sites lie in the FST DNA, but when *ris* mutants sensitive to *EcoRI* restriction are isolated, all of the *ris* mutations occur in the FST DNA. The simplest explanation for this result is that T5 FST DNA contains a gene coding for a protein that can inhibit *EcoRI*. To our knowledge, the existence of this putative inhibitor has not been confirmed directly. It is interesting that all *ris* mutants isolated so far contain a new restriction site in the FST DNA, suggesting that the restriction protection system may have some other essential role during T5 replication. Further experimentation is clearly necessary to elucidate both the nature of the restriction protection system and its role, if any, in T5 replication.

3. DNA Methylation

The observation that phage T5 inhibits DNA methylation soon after infection is consistent with the complete absence of methylated bases in parental T5 DNA (Gefter et al., 1966). Whether the genetic information for this process resides on the FST DNA is difficult to ascertain from the published data (Hausmann and Gold, 1966), because the host used for those experiments does not adsorb T5 readily. Hence, it is likely that the decrease in DNA methylation occurs much earlier than 10 min after infection. The mechanism of this inhibition has not been elucidated but is apparently not due to the induction of an S-adenosylmethionine-cleaving enzyme, as is the case with T3 (Gefter et al., 1966). Because both 6N-methyladenine and 5-methylcytosine are absent from T5 DNA, both DNA methylases must be inhibited, presumably by a common mechanism.

The observations of Hausmann and Gold (1966) have not been substantiated by Chernov et al. (1985). These latter workers found only a slow and partial reduction of the dam, dcm, and EcoRII DNA methylase activities after infection with phage T5, and they could find no evidence for the presence of a methylase inhibitor in the extracts from infected cells. However, they did confirm that although T5 DNA synthesized in vivo does not contain methylated bases, it can be methylated by these enzymes in vitro.

The reason for this absence of DNA methylation is unknown, but it could be responsible for the stability of phage DNA in a cell where near-total degradation of the host DNA is occurring. If the phage-induced nuclease is specific for methylated DNA, then unmethylated phage DNA would be spared. Such a mechanism would also explain the induction of a 5'-deoxyribonucleotidase activity by T5 (Warner et al., 1975). This enzyme is active on all 5'-deoxyribonucleotides tested (Mozer and Warner, 1977) and is presumed to be active on methylated nucleotides as well. The presence of this enzyme results in the excretion of free bases derived from the host DNA before phage DNA synthesis begins; hence, these methylated bases would not be reincorporated into the progeny phage DNA. This phage-induced 5'-deoxyribonucleotidase is discussed in more detail below.

4. Host-Cell Reactivation

Chiang and Harm (1976a,b) have shown that UV-irradiated T5, in contrast to T1, T3, and T7, are not reactivated in repair-proficient E. coli.

Furthermore, superinfecting T5 prevent reactivation of other UV-irradiated phage—e.g., T1. Expression of only FST genes other than the A1 and
A2 genes is required for the inactivation of this repair system. Chiang and
Harm concluded that the T5-induced protein inferred from their experiments must have some essential role in phage replication other than the
inhibition of the host UV repair system.

The induction of these various inhibitors suggests possible roles for
the various low-molecular-weight polypeptides induced during expression of the FST genes. McCorquodale *et al.* (1977) demonstrated the
induction of at least five small polypeptides with molecular weights of
13K, 8.3K, 7.1K, 5.5K, and 3.3K. Because host function inhibitors would
most likely be synthesized very early, it is reasonable to assume that
most, if not all, of these are involved in the inhibitory processes outlined
above. In fact, either some of these proteins may be involved in more than
one of these six inhibitory functions, or some of the bands observed by
McCorquodale *et al.* (1977) may contain more than one polypeptide.

VI. NUCLEOTIDE METABOLISM

A. Induction of New Enzymes

Bacteriophages whose DNA contains unusual bases must alter the
pathways for biosynthesis of deoxyribonucleotides. A well-documented
example of this is *E. coli* phage T4, which must synthesize 5-hy-
droxymethyl deoxycytidine 5'-triphosphate at the expense of deoxycytidine 5'-triphosphate. This new pathway requires the induction of
three new enzyme activities not found in the uninfected cell; the enzymatic activities required for this task have been summarized by Mathews and Allen (1983). Because these activities are not found in the
uninfected cell, these enzymes are essential for phage replication under
most circumstances.

A more common observation is that phages often induce the synthesis of enzyme activities similar to activities already existing in the
uninfected cell. In general, these enzymes are not essential for phage
replication, although a failure to induce a particular enzyme may reduce
the eventual burst size of progeny from the infected cell.

Phage T5 DNA contains only the four normal bases (adenine, cyto-
sine, guanine, and thymine) and therefore does not induce the synthesis
of enzymes with new substrate specificities. However, it does induce a
battery of enzyme activities which mimic the activities of enzymes
found in uninfected cells. These activities are summarized in Table II. Of
the six known phage-induced enzymes directly affecting the synthesis
and metabolism of deoxyribonucleotides in T5-infected cells, three have
low base specificity and act on deoxyribonucleotides containing any of
the four bases found in T5 DNA. In this way, the phage-induced
ribonucleotide reductase and 5'-deoxyribonucleotidase activities mimic

TABLE II. T5 Phage-Induced Enzymes Involved in Deoxyribonucleotide Metabolism

Enzyme	Reaction	Alternate substrates	Mutation	Effect of enzyme deficiency on phage replication
5′-Deoxyribonucleotidase	dTMP → dT	dAMP, dCMP, dGMP	dmp (A)[a]	DNA synthesis and phage production are delayed 10 min; normal burst size
Ribonucleotide reductase	UDP → dUDP	ADP, CDP, GDP	?	
dUTPase	dUTP → dUMP	—	dut (D?)	Burst size reduced twofold
Thymidylate synthetase	dUMP → dTMP	—	thy (B)	Burst size reduced twofold
dTMP kinase	dTMP → dTDP	dAMP, dCMP, dGMP	dnk (C)	DNA synthesis and burst size are reduced threefold
Dihydrofolate reductase	DHF → THF	—	B3 (B)	Burst size reduced greater than tenfold

[a]Letter in parentheses indicates genetic segment on which gene is located.

the known *E. coli* enzymes. In contrast, *E. coli* has four distinct kinases for the phosphorylation of dAMP, dCMP, dGMP, and dTMP. It would appear that a way to conserve coding capacity on phage genomes of limited size is to code for multifunctional enzymes whenever possible.

B. 5'-Deoxyribonucleotidase

The most curious enzyme induced by phage T5 may be the enzyme that degrades 5'-deoxyribonucleotides to their corresponding deoxyribonucleosides plus inorganic phosphate. This enzyme is induced immediately after infection (Warner *et al.*, 1975), peaks about 4 min after infection, and then essentially disappears by about 15 min after infection. The kinetics of induction and disappearance are consistent with the idea that the information for induction is carried by the FST DNA and that the enzyme is later inhibited by a protein induced when the remaining DNA enters the cell (Berget *et al.*, 1976). It is obvious why this enzyme must be inhibited before phage DNA synthesis can begin, but it is not apparent why the enzyme is induced at all.

We have hypothesized above that it could be induced to rid the cell of 5'-deoxyribonucleotides containing methylated bases. The only obvious result of enzyme induction is, in fact, the loss of the host DNA degradation products from the cell in the form of deoxyribonucleosides and free bases, most of which cannot be taken up later, when DNA synthesis begins, because there are no kinases for deoxyadenosine, deoxycytidine, and deoxyguanosine in these cells (Warner *et al.*, 1975). Thus, all of these deoxyribonucleotides are wasted by the phage rather than reused for phage DNA synthesis. A delay of about 10 min for phage DNA synthesis is actually observed when the cells are infected with a T5 mutant unable to induce this enzyme (Mozer *et al.*, 1977), although the eventual burst size is normal. This 10-min delay could result from an inability to synthesize stable progeny DNA until the pool of methylated 5'-dNMPs has been sufficiently diluted by the synthesis of dADP and dCDP *de novo*.

C. Reduction of Ribonucleotides

The reduction of ribonucleotides is increased markedly by infection of *E. coli* by phage T5. This stimulation appears to be due to the induction of at least two proteins: thioredoxin and an activity analogous to the *E. coli* ribonucleotide reductase (Eriksson and Berglund, 1974). The thioredoxin induced by T5 has been purified and is clearly distinct from the *E. coli* thioredoxin, as it cannot serve as hydrogen donor for the *E. coli* ribonucleotide reductase. Although a 20-fold increase in ribonucleotide reductase activity after infection has been reported, the exact nature of the phage-induced activity remains obscure. The ribonucleotide reductase induced by T5 either is a totally new enzyme with a subunit immunologically similar to the B2 subunit of the *E. coli* enzyme or is composed of a phage-induced subunit complexed with *E. coli* B2 subunit. The enzyme can accept hydrogen from *E. coli* thioredoxin.

D. Synthesis of dTTP

At least three enzymes directly affecting the rate of synthesis of dTTP are induced by phage T5 as indicated below, although none are essential for phage replication.

$$\text{UDP} \xrightarrow{?} \text{dUDP} \xrightarrow{\text{Host}} \text{dUTP} \xrightarrow{dut} \text{dUMP} \xrightarrow{thy} \text{dTMP} \xrightarrow{dnk} \text{dTDP} \xrightarrow{\text{Host}} \text{dTTP}$$

If these enzymes are not essential, why are they induced? The most obvious answer is that in most cases, burst size is increased substantially. This is true for dUTPase, thymidylate synthetase, and dNMP kinase (Table II). This indicates that the supply of doexyribonucleoside 5'-triphosphates, and in particular dTTP, may be rate-limiting for DNA synthesis in the T5-infected cell.

However, the absence of dUTPase has other effects. The failure to induce this enzyme leads to a small but significant increase in the amount of uracil incorporated into the progeny phage (Warner *et al.*, 1979), and this incorporation is greatly increased when the phage is also unable to induce thymidylate synthetase (Swart and Warner, 1985). When T5 *thy dut* phages are grown in wild-type *E. coli* hosts, almost 10% of the thymine normally present in the phage DNA is replaced by uracil. These double mutants have a burst size only about 20% of normal, and they are very difficult to grow in these hosts. These observations suggest that the rate of production of dTTP in uninfected cells is not sufficient to maintain an adequate pool of dTTP for normal phage DNA synthesis in T5-infected cells.

The T5 phage-induced dihydrofolate reductase provides a completely different picture. The product of this reaction, tetrahydrofolic acid, is required stoichiometrically during the synthesis of dTMP from dUMP. Mutants unable to induce dihydrofolate reductase produce very few phage progeny and can only be grown under conditionally permissive conditions (Hendrickson and McCorquodale, 1972). It is not clear why dihydrofolate reductase deficiency in T5-infected cells is near-lethal, whereas T4 phages unable to induce this enzyme are able to replicate reasonably well and produce normal-looking plaques (Hall, 1967). Perhaps the enzymes involved in deoxyribonucleotide biosynthesis in T5-infected cells are more crucial for normal phage production than they are in T4-infected cells, because the host DNA degradation products cannot be recycled into phage DNA. Thus, T5 is completely dependent on deoxyribonucleotides synthesized *de novo*.

VII. DNA REPLICATION

Phage T5 induces a new DNA polymerase (Orr *et al.*, 1965), and this polymerase is essential for DNA replication in the phage-infected cell (DeWaard *et al.*, 1965). The enzyme is a monomer with a molecular weight of about 96,000 (Fujimura and Roop, 1976a) and contains an associated 3' → 5' exonuclease activity (Fujimura and Roop, 1976b). The gene for this protein actually includes three separate loci identified earlier—D7, D8, and D9 (Fujimura *et al.*, 1985), as determined by marker rescue experiments, as described above.

The enzyme is enzymatically similar to many other prokaryotic DNA polymerases, with several exceptions (Fujimura *et al.*, 1981). The first is that the enzyme catalyzes strand displacement when using a nick as a primer terminus, even though it has no 5' → 3' exonuclease activity. Also, the polymerase activity is unusually processive in the absence of auxiliary proteins (Das and Fujimura, 1979). Finally, the active sites for the polymerase and 3' → 5' exonuclease activities may be distinct, as suggested by experiments with heat-sensitive mutants with a defective DNA polymerase gene (Fujimura and Roop, 1976b). The 3' → 5' exonuclease has been characterized as quasiprocessive on both single- and double-stranded substrates (Das and Fujimura, 1980). Studies with accurately mapped DNA polymerase mutants have indicated that at least part of the essential domains lie within the carboxyl terminal 15% of the enzyme (Schneider *et al.*, 1985).

Studies to elucidate the mechanism of DNA replication by this enzyme have been carried out both *in vitro* and *in vivo*. *In vitro*, the enzyme can replicate a nicked circle readily by the rolling-circle mechanism of synthesis, but use of a nicked linear duplex for the primer-template results in a highly branched product (Fujimura *et al.*, 1981). Somewhat surprisingly, the DNA-binding protein induced by gene D5, which is essential for DNA replication *in vivo*, inhibits the polymerase *in vitro* regardless of the primer-template used (Fujimura and Roop, 1983). The relationship of these observations to *in vivo* replication remains unclear. Everett and Lunt (1980a) have observed the existence of both slow- and fast-sedimenting forms of T5 DNA in infected cells. The slow form moves at a rate similar to that of DNA extracted from mature T5 phages; while linear, this form is actually 12% shorter than mature T5 DNA, as analyzed by electron microscopy (Everett and Lunt, 1980b).

The appearance of a fast-sedimenting form *in vivo* suggests that concatameric DNA is an intermediate produced during DNA replication. Bourguignon *et al.* (1976), using electron microscopy to examine early intermediates, concluded that replication involved multiple origins resulting in the formation of large branched molecules. The results have been extended by Everett (1981), who showed that replicating DNA contains single-stranded regions closer together than mature-length T5 DNA. Everett also suggested that the exonuclease coded by the D15 gene may be responsible for the formation of these gaps from nicks rather than for the generation of the nicks, as suggested by Moyer and Rothe (1977). These studies have not elucidated the mechanism of the formation and processing of concatemeric DNA, as the results support roles for both bidirectional and rolling-circle mechanisms.

Several phage-induced proteins are essential for normal DNA replication in T5-infected cells, but the roles of few of these proteins have been elucidated. The only proteins known to be absolutely essential for DNA replication include DNA polymerase and the D5 DNA-binding protein, as indicated above (Hendrickson and McCorquodale, 1972). However, DNA synthesis is markedly reduced by the failure to induce several other proteins, including the products of genes B1, B2, B3 (dihydrofolate reductase), C2, D1, D2, D13, D14, D15 (exonuclease), and D19. Herman and Moyer (1975) have inferred that T5 may induce a new DNA ligase, but this suggestion remains unconfirmed. Evidence has also been obtained for the involvement of at least the B subunit of host DNA gyrase in T5 DNA replication (Constantinou et al., 1986).

ACKNOWLEDGMENTS. This review was started while one of the authors (D. J. M.) held a National Science Foundation grant (No. PCM-8208370), and some of the work referred to in this review was supported by that grant.

REFERENCES

Abelson, J., and Thomas, C. A. Jr., 1966, The anatomy of the T5 bacteriophage DNA molecule, J. Mol. Biol. 18:262.

Bayer, M. E., 1968, Adsorption of bacteriophages to adhesions between wall and membrane of Escherichia coli, J. Virol. 2:346.

Beckman, L. D., Witonsky, P., and McCorquodale, D. J., 1972, Effect of rifampin on the growth of bacteriophage T5, J. Virol. 10:179.

Beckman, L. D., Anderson, G. C., and McCorquodale, D. J., 1973, Arrangement on the chromosome of the known pre-early genes of bacteriophages T5 and BF23, J. Virol. 12:1191.

Berget, S. M., Mozer, T. J., and Warner, H. R., 1976, Early events after infection of Escherichia coli by bacteriophage T5. II. Control of bacteriophage-induced 5'-nucleotidase activity, J. Virol. 18:71.

Blaisdell, P., and Warner, H. R., 1983, Partial purification and characterization of a uracil-DNA glycosylase from wheat germ, J. Biol. Chem. 258:1603.

Blaisdell, P., and Warner, H. R., 1986, Cell free transcription and translation of isolated restriction fragments localize bacteriophage T5 pre-early genes, J. Virol. 57:759.

Bourguignon, G. J., Sweeney, T. K., and Delius, H., 1976, Multiple origins and circular structures in replicating T5 bacteriophage DNA, J. Virol. 18:245.

Bradley, D. E., 1967, Ultrastructure of bacteriophages and bacteriocins, *Bacteriol. Rev.* **31**:230.

Brunel, F., and Davison, J., 1979, Restriction insensitivity in bacteriophage T5. III. Characterization of *EcoRI*-sensitive mutants by restriction analysis, *J. Mol. Biol.* **128**:527.

Brunel, F., Davison, J., and Merchez, M., 1979, Cloning of bacteriophage T5 DNA fragments in plasmid pBR322 and bacteriophage lambda gtWES, *Gene* **8**:53.

Brunel, F., Davison, J., Ha-Thi, V., and Reeve, J., 1981, Cloning of bacteriophage T5 DNA fragments. III. Expression in *Escherichia coli* mini-cells, *Gene* **16**:107.

Brunel, F., Thi, V. H., Pilaete, M. F., and Davison, J., 1983, Transcription regulatory elements in the late region of bacteriophage T5 DNA, *Nucleic Acids Res.* **11**:7649.

Chen, C., 1976, Transcription map of bacteriophage T5, *Virology* **74**:116.

Chen, C., and Bremer, H., 1976, Identification of single-stranded DNA fragments of bacteriophage T5, *Virology* **74**:104.

Chen, M. J., Locker, J., and Weiss, S. B., 1976, The physical mapping of bacteriophage T5 transfer tRNAs, *J. Biol. Chem.* **251**:536.

Chernov, A. P., Venozhinskis, M. T., and Kanopkaite, S. I., 1985, Activity of DNA methylases in *E. coli* cells infected with bacteriophage T5, *Biokhimiya* **50**:74.

Chiang, T., and Harm, W., 1976a, On the lack of host-cell reactivation of UV-irradiated phage T5. I. Interference of T5 infection with the host cell reactivator of phage T1, *Mutat. Res.* **36**:121.

Chiang, T., and Harm, W., 1976b, On the lack of host-cell reactivation of UV-irradiated phage T5. II. Further characterization of the repair inhibition exerted by T5 infection. *Mutat. Res.* **36**:135.

Constantinou, A., Voelkel-Meiman, K., Sternglanz, R., McCorquodale, M. M., and McCorquodale, D. J., 1986, Involvement of host DNA gyrase in growth of bacteriophage T5, *J. Virol.* **57**:875.

Crawford, L. V., 1959, Nucleic acid metabolism in *Escherichia coli* infected with phage T5, *Virology* **7**:359.

Das, S. K., and Fujimura, R. K., 1979, Processiveness of DNA polymerases. A comparative study using a simple procedure, *J. Biol. Chem.* **254**:1227.

Das, S. K., and Fujimura, R. K., 1980, Mechanism of 3' → 5' exonuclease associated with phage T5-induced DNA polymerase: Processiveness and template specificity, *Nucleic Acids Res.* **8**:657.

Davison, J., 1981, Mapping of the *XhoI* and *SacI* restriction sites on the bacteriophage T5 DNA, *Gene* **16**:97.

Davison, J., and Brunel, F., 1979, Restriction insensitivity in bacteriophage T5. I. Genetic characterization of mutants sensitive to *EcoRI* restriction, *J. Virol.* **29**:11.

Davison, J., and LaFontaine, D., 1984, Direction of transcription in bacteriophage T5 first-step transfer DNA, *J. Virol.* **50**:629.

Davison, J., Brunel, F., Merchez, M., and Ha-Thi, V., 1981, Cloning of bacteriophage T5 DNA fragments. II. Isolation of recombinants carrying T5 *PstI* fragments, *Gene* **16**:99.

Desai, S. M., Vaughan, J., and Weiss, S. B., 1986, Identification and location of nine T5 bacteriophage tRNA genes by DNA sequence analysis, *Nucleic Acids Res.* **14**:4197.

DeWaard, A., Paul, A. V., and Lehman, I. R., 1965, The structural gene for deoxyribonucleic acid polymerase in bacteriophages T4 and T5, *Proc. Natl. Acad. Sci. USA* **54**:1241.

Duckworth, D. H., Glenn, J., and McCorquodale, D. J., 1981, Inhibition of bacteriophage replication by extrachromosomal genetic elements, *Microbiol. Rev.* **45**:52.

Eriksson, S., and Berglund, O., 1974, Bacteriophage-induced ribonucleotide reductase systems. T5- and T6-specific ribonucleotide reductase and thioredoxin, *Eur. J. Biochem.* **46**:271.

Everett, R. D., 1981, DNA replication of bacteriophage T5. 3. Studies on structure of concatameric T5 DNA, *J. Gen. Virol.* **52**:25.

Everett, R. D., and Lunt, M. R., 1980a, DNA replication of bacteriophage T5. 1. Fractionation of intracellular T5 DNA by agarose gel electrophoresis, *J. Gen. Virol.* **47**:123.

Everett, R. D., and Lunt, M. R., 1980b, DNA replication of bacteriophage T5. 2. Structure and properties of the slow sedimenting form of intracellular T5 DNA, *J. Gen. Virol.* **47**:133.

Ficht, T. A., and Moyer, R. W., 1980, Isolation and characterization of a putative bacterio- phage T5 transcription-replication enzyme complex from infected *Escherichia coli*, *J. Biol. Chem.* **255**:7040.

Fielding, P. E., and Lunt, M. R., 1969, A new deoxyribonuclease activity from bacteria infected with T5 bacteriophage, *FEBS Lett.* **5**:214.

Fox, J. W., Barish, A., Snyder, C. E., and Benzinger, R., 1982, Amino acid sequence of the bacteriophage T5-coded A2 protein, *Biochem. Biophys. Res. Commun.* **106**:265.

Fujimura, R. K., and Roop, B. C., 1976a, Characterization of DNA polymerase induced by bacteriophage T5 with DNA containing single strand breaks, *J. Biol. Chem.* **251**:2168.

Fujimura, R. K., and Roop, B. C., 1976b, Temperature-sensitive DNA polymerase induced by a bacteriophage T5 mutant. Relationship between polymerase and exonuclease ac- tivities, *Biochemistry* **15**:4403.

Fujimura, R. K., and Roop, B. C., 1983, Interaction of a DNA-binding protein, the product of gene D5 of bacteriophage T5, with double-stranded DNA: Effects on T5 DNA poly- merase functions *in vitro*, *J. Virol.* **46**:778.

Fujimura, R. K., Das, S. K., Allison, D. P., and Roop, B. C., 1981, Replication of linear duplex DNA *in vitro* with bacteriophage T5 DNA polymerase, *Prog. Nucl. Acid Res. Mol. Biol.* **26**:49.

Fujmura, R. K., Tavtigian, S. V., Choy, T. L., and Roop, B. C., 1985, Physical locus of the DNA polymerase gene and genetic maps of bacteriophage T5 maps, *J. Virol.* **53**:495.

Gefter, M., Hausmann, R., Gold, M., and Hurwitz, J., 1966, The enzymatic methylation of ribonucleic acid and deoxyribonucleic acid. X. Deoxyribonucleic acid methylase in bacteriophage-infected *Escherichia coli*, *J. Biol. Chem.* **241**:1995.

Gentz, R., and Bujard, H., 1985, Promoters recognized by *Escherichia coli* RNA polymerase selected by function: Highly efficient promoters from bacteriophage T5, *J. Bacteriol.* **164**:70.

Gentz, R., Langner, A., Chang, A. C. Y., Cohen, S. N., and Bujard, H., 1981, Cloning and analysis of strong promoters is made possible by the downstream placement of a RNA termination signal, *Proc. Natl. Acad. Sci. USA* **78**:4936.

Hall, D. H., 1967, Mutants of bacteriophage T4 unable to induce dihydrofolate reductase activity, *Proc. Natl. Acad. Sci. USA* **58**:584.

Hausmann, R., and Gold, M., 1966, The enzymatic methylation of ribonucleic acid and deoxyribonucleic acid. IX. Deoxyribonucleic acid methylase in bacteriophage-infected *Escherichia coli*, *J. Biol. Chem.* **241**:1985.

Heller, K. J., 1984, Identification of the phage gene for host receptor specificity by analyzing hybrid phages of T5 and BF23, *Virology* **139**:11.

Heller, K. J., and Braun, V., 1979, Accelerated adsorption of bacteriophage T5 to *Escherichia coli* F, resulting from reversible tail fiber-lipopolysaccharide binding, *J. Bacteriol.* **139**:32.

Heller, K. J., and Braun, V., 1982, Polymannose O antigens of *Escherichia coli* F, the binding sites for the reversible adsorption of the bacteriophage T5+ via the L-shaped tail fibers, *J. Virol.* **41**:222.

Heller, K. J., and Bryniok, D., 1984, O-antigen dependent mutant of bacteriophage T5, *J. Virol.* **49**:20.

Heller, K. J., and Krauel, V., 1986, Cloning and expression of the *ltf* gene of bacteriophage T5, *J. Bacteriol.* **167**:1071.

Heller, K. J., and Schwarz, H., 1985, Irreversible binding to the receptor of bacteriophages T5 and BF23 does not occur with the tip of the tail. *J. Bacteriol.* **162**:621–625.

Hendrickson, H. E., and McCorquodale, D. J., 1972, Genetic and physiological studies of bacteriophage T5. III. Patterns of deoxyribonucleic acid synthesis induced by mutants of T5 and identification of genes influencing the appearance of phage-induced di- hydrofolate reductase and deoxyribonuclease, *J. Virol.* **9**:981.

Herman, R. C., and Moyer, R. W., 1975, In vivo repair of bacteriophage T5 DNA: An assay for viral growth control, Virology 66:393.

Heusterspreute, M., Ha-Thi, V., Tournis-Gamble, S., and Davison, J., 1987, The first-step transfer-DNA infection-stop signal of bacteriophage T5, Gene 52:155.

Hunt, C., Hwang, L. T., and Weiss, S. B., 1976, Mapping of two isoleucine tRNA isoacceptor genes in bacteriophage T5 DNA, J. Virol. 20:63.

Hunt, C., Desai, S. M., Vaughan, J., and Weiss, S. B., 1980, Bacteriophage T5 transfer RNA. Isolation and characterization of tRNA species and refinement of the tRNA gene map, J. Biol. Chem. 255:3164.

Ikemura, T., and Ozeki, H., 1975, Two-dimensional polyacrylamide-gel electrophoresis for purification of small RNAs specified by virulent coliphages T4, T5, T7, and BF23, Eur. J. Biochem. 51:117.

Ikemura, T., Okada, K., and Ozeki, H., 1978, Clustering of transfer RNA genes in bacterio-phage BF23, Virology 90:142.

Jacquemin-Sablon, A., and Richardson, C. C., 1970, Analysis of the interruptions in bacte-riophage T5 DNA, J. Mol. Biol. 47:477.

Kaliman, A. V., Krutilina, A. I., Kryukov, V. M., and Bayev, A. A., 1986, Cloning and DNA sequence of the 5'-exonuclease gene of bacteriophage T5, FEBS Lett. 195:61.

Kryukov, V. M., Ksenzenko, V. N., Kaliman, A. V., and Bayev, A. A., 1983, Cloning and DNA sequence of the genes for two bacteriophage T5 tRNAs, FEBS Lett. 158:123.

Ksenzenko, V. N., Kamynina, T. P., Pustoshilova, N. M., Kryukov, V. M., and Bayev, A. S., 1982a, Cloning of bacteriophage T5 promoters, Mol. Gen. Genet. 185:520.

Ksenzenko, V. N., Kamynina, T. P., Kazantsev, S. I., Shlyapnikov, M. G., Kryukov, V. M., and Bayev, A. A., 1982b, Cloning of genes for bacteriophage T5 stable RNAs, Biochim. Biophys. Acta 697:235.

Labedan, B., 1976, A very early step in the T5 DNA injection process, Virology 75:368.

Labedan, B., 1984, Requirement for a fluid host cell membrane in injection of coliphage T5 DNA, J. Virol. 49:273.

Labedan, B., Crochet, M., Legault-Demare, J., and Stevens, B. J., 1973, Location of the first step transfer fragment and single-strand interruptions in T5st(0) bacteriophage DNA, J. Mol. Biol. 75:213.

Lange-Gustafson, B., and Rhoades, M., 1979, Physical map of bacteriophage BF23 DNA: Restriction enzyme analysis, J. Virol. 30:923.

Lanni, Y. T., 1968, First-step-transfer deoxyribonucleic acid of bacteriophage T5, Bacteriol. Rev. 32:227.

Lanni, Y. T., 1969, Functions of two genes in the first-step-transfer DNA of bacteriophage T5, J. Mol. Biol. 44:173.

Macaluso, A., Midthun, K., Bender, R. A., and Botstein, D., 1982, Nonsense suppressor mutants of bacteriophage T5, Virology 117:275.

Maltouf, A. F., and Labedan, B., 1983, Host cell metabolic energy is not required for injection of bacteriophage T5 DNA, J. Bacteriol. 153:124.

Mathews, C. K., and Allen, J. R., 1983, DNA precursor biosynthesis, in: Bacteriophage T4 (C. K. Mathews, E. M. Kutter, G. Mosig, and P. B. Berget, eds.), pp. 59–70, American Society for Microbiology, Washington.

McCorquodale, D. J., 1975, The T-odd bacteriophages, CRC Crit. Rev. Microbiol. 4:101.

McCorquodale, D. J., and Lanni, Y. T., 1970, Patterns of protein synthesis in Escherichia coli infected by amber mutants in the first-step-transfer DNA of T5, J. Mol. Biol. 48:133.

McCorquodale, D. J., Shaw, A. R., Shaw, P. K., and Chinnadurai, G., 1977, Pre-early polypep-tides of bacteriophage T5 and BF23, J. Virol. 22:480.

McCorquodale, D. J., Gossling, J., Benzinger, R., Chesney, R., Lawhorne, L., and Moyer, R., 1979, Gene D5 product of bacteriophage T5: DNA-binding protein affecting DNA rep-lication and late gene expression, J. Virol. 29:322.

McCorquodale, D. J., Chen, C. W., Joseph, M. K., and Woychik, R., 1981, Modification of

RNA Polymerase from *Escherichia coli* by pre-early gene products of bacteriophage T5, *J. Virol.* **40**:958.

Melin, L., and Dehlin, E., 1987, Functional comparison of an "early" and a "late" promoter active DNA-segment from coliphage T5 in *Bacillus subtilis* and in *Escherichia coli,* *FEMS Microbiol. Lett.* **41**:141.

Moyer, R. W., and Rothe, C. T., 1977, Role of T5 gene D15 in the generation of nicked bacteriophage T5 DNA, *J. Virol.* **24**:177.

Mozer, T. J., and Warner, H. R., 1977, Properties of deoxynucleoside 5′-monophosphatase induced by bacteriophage T5 after infection of *Escherichia coli, J. Virol.* **24**:635.

Mozer, T. J., Thompson, R. B., Berget, S. M., and Warner, H. R., 1977, Isolation and characterization of a bacteriophage T5 mutant deficient in deoxynucleoside 5′-monophosphatase activity, *J. Virol.* **24**:642.

Nichols, B. P., and Donelson, J. E., 1977a, The nucleotide sequence at the 3′-termini of three major T5 DNA fragments, *Virology* **83**:396.

Nichols, B. P., and Donelson, J. E., 1977b, Sequence analysis of the nicks and termini of bacteriophage T5 DNA, *J. Virol.* **22**:520.

Okada, K., 1980, Physical map of the dispensable region of the genome of *E. coli* bacteriophage BF23, *Gene* **8**:369.

Okada, K., Ohira, M., and Ozeki, H., 1978, Nonsense suppressor mutants of bacteriophage BF23, *Virology* **90**:133.

Orr, C. W. M., Herriott, S. T., and Bessman, M. J., 1965, The enzymology of virus-infected bacteria. VII. A new deoxyribonucleic acid polymerase induced by bacteriophage T5, *J. Biol. Chem.* **240**:4652.

Ortiz, P. J., August, J. T., Watanabe, M., Kaye, A. M., and Hurwitz, J., 1965, Ribonucleic acid–dependent ribonucleotide incorporation. II. Inhibition of polyriboadenylate polymerase activity following bacteriophage infection, *J. Biol. Chem.* **240**:423.

Peter, G., Hanggi, U. J., and Zachau, H. G., 1979, Localization of the dihydrofolate reductase gene on the physical map of bacteriophage T5, *Mol. Gen. Genet.* **175**:333.

Pfefferkorn, E., and Amos, H., 1958, Deoxyribonucleic acid breakdown and resynthesis in bacteriophage T5 infection, *Virology* **6**:299.

Rhoades, M., 1977a, Localization of single-chain interruptions in bacteriophage T5 DNA. I. Electron microscopic studies, *J. Virol.* **23**:725.

Rhoades, M., 1977b, Localization of single-chain interruptions in bacteriophage T5 DNA. II. Electrophoretic studies, *J. Virol.* **23**:737.

Rhoades, M., 1982, New physical map of bacteriophage T5 DNA, *J. Virol.* **43**:566.

Rhoades, M., Schwartz, J., and Wahl, J. M., 1980, New deletion mutant of bacteriophage T5, *J. Virol.* **36**:622.

Rice, A. C., Ficht, T. A., Holladay, L. A., and Moyer, R. W., 1979, The purification and properties of a double-stranded DNA-binding protein encoded by the gene D5 of bacteriophage T5, *J. Biol. Chem.* **254**:8042.

Rogers, S. G., and Rhoades, M., 1976, Bacteriophage T5-induced endonucleases that introduce site-specific single-chain interruptions in duplex DNA, *Proc. Natl. Acad. Sci. USA* **73**:1577.

Rogers, S. G., Godwin, E. A., Shinosky, E. S., and Rhoades, M., 1979a, Interruption-deficient mutants of bacteriophage T5. I. Isolation and general properties, *J. Virol.* **29**:716.

Rogers, S. G., Hamlett, N. V., and Rhoades, M., 1979b, Interruption-deficient mutants of bacteriophage T5. II. Properties of a mutant lacking a specific interruption, *J. Virol.* **29**:726.

Saigo, K., 1976, A new bacteriophage T5 particle with a DNA lacking the terminal repetition at the "right hand end," *J. Mol. Biol.* **107**:369.

Saigo, K., 1978, Isolation of high density mutants and identification of nonessential structural proteins in bacteriophage T5: Dispensability of L-shaped tail fibers and a secondary head protein, *Virology* **85**:422.

Sakaki, Y., 1974, Inactivation of the ATP-dependent DNase of *Escherichia coli* after infection with double-stranded DNA phages, *J. Virol.* **14**:1611.

Schneider, S. S., Roop, B. C., and Fujimura, R. K., 1985, Identification by immunobinding assay of the polypeptide coded by the DNA polymerase gene of bacteriophage T5 and its *amber* mutants and the direction of transcription of the gene, *J. Virol.* **56**:245.

Shaw, A. R., and Davison, J., 1979, Polarized injection of the bacteriophage T5 chromosome, *J. Virol.* **30**:933.

Shaw, A. R., Lang, D., and McCorquodale, D. J., 1979, Terminally redundant deletion mutants of bacteriophage BF23, *J. Virol.* **29**:220.

Sirbasku, D. A., and Buchanan, J. M., 1970, Patterns of ribonucleic acid synthesis in T5-infected *Escherichia coli*. II. Separation of high molecular weight ribonucleic acid species by disc electrophoresis on acrylamide gel columns, *J. Biol. Chem.* **245**:2679.

Snyder, C. E. Jr., 1984, Bacteriophage T5 gene A2 protein alters the outer membrane of *Escherichia coli*, *J. Bacteriol.* **160**:1191.

Snyder, C. E. Jr., and Benzinger, R. H., 1981, Second-step transfer of bacteriophage T5 DNA: Purification and characterization of T5 gene A2 protein, *J. Virol.* **40**:248.

Stueber, D., Delius, H., and Bujard, H., 1978, Electron microscopic analysis of *in vitro* transcriptional complexes: Mapping of promoters of the coliphage T5 genome, *Mol. Gen. Genet.* **166**:141.

Stueber, D., Ibrahimi, I., Cutler, D., Dobberstein, B., and Bujard, H., 1984, A novel *in vitro* transcription-translation system: Accurate and efficient synthesis of single proteins from cloned DNA sequences. *EMBO J.* **3**:3143.

Swart, W. J., and Warner, H. R., 1985, Isolation and partial characterization of a bacteriophage T5 mutant unable to induce thymidylate synthetase and its use in studying the effect of uracil incorporation into DNA on early gene expression, *J. Virol.* **54**:86.

Szabo, C., and Moyer, R. W., 1975, Purification and properties of a bacteriophage T5–modified form of *Escherichia coli* RNA polymerase, *J. Virol.* **15**:1042.

Szabo, C., Dharmgrongartama, B., and Moyer, R. W., 1975, The regulation of transcription in bacteriophage T5–infected *Escherichia coli*, *Biochemistry* **14**:989.

Von Gabain, A., and Bujard, H., 1977, Interaction of *E. coli* RNA polymerase with promoters of coliphage T5: The rates of complex formation and decay and their correlation with *in vitro* and *in vivo* transcriptional activity, *Mol. Gen. Genet.* **157**:301.

Von Gabain, A., and Bujard, H., 1979, Interaction of *Escherichia coli* RNA polymerase with promoters of several coliphage and plasmid DNAs, *Proc. Natl. Acad. Sci. USA* **76**:189.

Von Gabain, A., Hayward, G. S., and Bujard, H., 1976, Physical mapping of the *Hind*III, *Eco*RI, *Sal* and *Sma* restriction endonuclease cleavage fragments from bacteriophage T5 DNA, *Mol. Gen. Genet.* **143**:279.

Warner, H. R., Drong, R. F., and Berget, S. M., 1975, Early events after infection of *Escherichia coli* by bacteriophage T5. I. Induction of a 5'-nucleotidase activity and excretion of free bases, *J. Virol.* **15**:273.

Warner, H. R., Thompson, R. B., Mozer, T. J., and Duncan, B. K., 1979, The properties of a bacteriophage T5 mutant unable to induce deoxyuridine 5'-triphosphate nucleotidohydrolase. Synthesis of uracil-containing T5 deoxyribonucleic acid, *J. Biol. Chem.* **254**:7534.

Warner, H. R., Johnson, L. K., and Snustad, D. P., 1980, Early events after infection of *Escherichia coli* by bacteriophage T5. III. Inhibition of uracil-DNA glycosylase activity, *J. Virol.* **33**:535.

Williams, R. C., and Fisher, H. W., 1974, *An Electron Micrographic Atlas of Viruses*, Charles C. Thomas, Springfield, IL.

Zweig, M., and Cummings, D. J., 1973a, Structural proteins of bacteriophage T5, *Virology* **51**:443.

Zweig, M., and Cummings, D. J., 1973b, Cleavage of head and tail proteins during bacteriophage T5 assembly: Selective host involvement in the cleavage of a tail protein, *J. Mol. Biol.* **80**:505.

Zweig, M., Rosenkranz, H. S., and Morgan, C., 1972, Development of coliphage T5: Ultrastructural and biochemical studies, *J. Virol.* **9**:526–543.

CHAPTER 11

Bacteriophage SPO1

CHARLES STEWART

I. INTRODUCTION

SPO1 is a large virulent bacteriophage of *Bacillus subtilis*, first isolated from soil in Osaka, Japan (Okubo *et al.*, 1964). Because of its complex sequence of gene action and the isolation of mutations altering that sequence, SPO1 has been an object of intensive study, which has provided fundamental insights into the mechanisms of sequential gene action. The study of other aspects of SPO1 biology has not progressed as far but has identified several interesting phenomena that require further investigation. This review will summarize current knowledge of SPO1.

II. THE VIRION

An SPO1 particle consists of an icosahedral head; a tail consisting of sheath, tube, and baseplate; and a neck that attaches the head to the tail. The head diameter is about 87 nm, and the dimensions of sheath and baseplate are about 19×140 nm and 25×60 nm, respectively. When contraction of the tail is induced *in vitro*, the configuration of the base plate is altered, and the sheath contracts to about 27×63 nm, exposing the 9×142 nm tube (Okubo *et al.*, 1964; Truffaut *et al.*, 1970; Parker and Eiserling, 1983a; Parker *et al.*, 1983).

Abbreviations used in this chapter: gp, gene product (thus gp28 is the product of gene 28); *sus*, suppressor-sensitive; *ts*, temperature-sensitive (thus *sus*28 is a suppressor-sensitive mutant, affected in gene 28); hmUra, hydroxymethyluracil; HMdUMP, hydroxylmethyl deoxyuridylic acid; kb, kilobases; HPUra, 6-(p-hydroxyphenylazo)-uracil; PAGE, polyacrylamide gel electrophoresis.

CHARLES STEWART • Department of Biology, Rice University, Houston, Texas 77251.

Fifty-three polypeptides were identified by PAGE of the subassemblies of the bacteriophage particle (Parker and Eiserling, 1983b). Twenty-eight of these are in the baseplate, six in the sheath and tube, three in the neck, and 16 in the head.

The head contains a double-stranded DNA molecule of 140–145 kbp including a 12.4-kbp terminal redundancy (Davison et al., 1964; Truffaut et al., 1970; Cregg and Stewart, 1978; Lawrie et al., 1978; Pero et al., 1979; Parker et al., 1983; Curran and Stewart, 1985a). The DNA contains hmUra in place of thymine, causing an anomalously high CsCl buoyant density. The single strands separate readily in CsCl (Okubo et al., 1964; Truffaut et al., 1970).

III. OTHER hmUra-CONTAINING B. SUBTILIS PHAGES

Several other B. subtilis phages, including SP8, SP5C, SP82, 2C, φe, and Hl, also contain hmUra in place of thymine. To the extent that they have been measured, all of these phages are similar to SPO1 with respect to size and structure of virions, length of DNA, GC content, CsCl buoyant density of both native and denatured DNA, antigenic specificity, DNA:DNA hybridization, genetic and restriction maps, time course of polypeptide synthesis, species of RNA transcripts and polypeptides synthesized, and regulation of the sequence of transcription (Kallen et al., 1962; Davison, 1963; Marmur and Greenspan, 1963; Marmur et al., 1963; Green, 1964; Davison et al., 1964; Kahan, 1966; Aposhian and Tremblay, 1966; Truffaut et al., 1970; Liljemark and Anderson, 1970; Green and Laman, 1972; Arwert and Veneman, 1974; Spiegelman and Whiteley, 1974; Lawrie and Whiteley, 1977; Lawrie et al., 1978; Reeve et al., 1978; Panganiban and Whiteley, 1981; Hoet et al., 1983; Coene et al., 1983). However, there are significant differences between each of these phages, so we cannot assume that any particular result, concerning any other phage, is applicable to SPO1. This review will focus on SPO1, about which the most is known, but will also discuss data regarding the other phages, which can illuminate areas not studied for SPO1. More detailed discussions of some of the other phages may be found in earlier reviews (Hemphill and Whiteley, 1975; Rabussay and Geiduschek, 1977; Geiduschek and Ito, 1982).

IV. MAPS OF THE SPO1 GENOME

Figure 1 shows the genetic and restriction maps of SPO1, and Table I lists the known gene functions. Genes with related functions tend to be clustered, with genes 35 and 1–3 affecting virion assembly, 4–6 affecting head formation, 7–20 affecting tail formation, 21–23 and 27–32 affecting

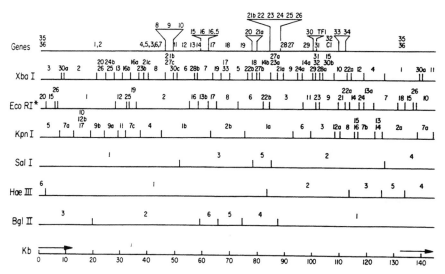

FIGURE 1. Restriction and genetic maps of the SPO1 genome. The restriction and genetic maps are taken from Pero *et al.* (1979), Okubo *et al.* (1972), and Curran and Stewart (1985a). The arrows show the position of the terminal redundancy. The EcoRI* site between EcoRI* 3 and EcoRI* 22b is not there in some strains of SPO1 (e.g., Pero *et al.*, 1979). Where the order of two or more adjacent genes or fragments has not been determined with respect to each other, their numbers are placed one above the other. The positions of genes 6, 7, 10, 15, 22, 23, 24, 25, and 26 have been determined only by genetic mapping; those of all other genes have been determined from the positions of the restriction fragments on which they are located (Curran and Stewart, 1985a). In most cases, the latter positions are the same as those determined from genetic mapping, but when those determinations differ, the positions given are those determined from the restriction map. For several pairs of genes (1 and 2, 8 and 9, and 20 and 21a), their position with respect to the rest of the genome was determined from the restriction map, but their position with respect to each other was determined only from the genetic map. Mutation C1 was mapped adjacent to gene 32 but was not shown to be in a separate gene (Okubo *et al.*, 1972). TF1 has not been defined by mutation, but its position was determined precisely by nucleotide sequencing (Greene *et al.*, 1984). Eleven additional genes have been located to the right of genes 35 and 36 in the terminal redundancy, by *in vitro* expression from specific restriction fragments, although none of them have been defined by mutation (Perkus and Shub, 1985). Their positions are shown in an expanded map of the terminal redundancy in Fig. 3.

DNA synthesis, and 33 and 34 affecting regulation of gene action. About 50 genes have been mapped of nearly 200 that may be expected. More detailed maps of certain sections of the genome have also been constructed and will be discussed below. About 65% of the genomic sequences have been cloned in *B. subtilis*, primarily as EcoRI*, XbaI, and KpnI fragments (Curran and Stewart, 1985a).

TABLE I. Summary of Genes of SPO1

Gene[p]	Phenotype of mutant[a]	Activity of gene product	Gene product purified or identified by PAGE band	Cloned restriction fragment on which gene located[dd]
A. Mapped genes				
1	D+. Deficient virion assembly[b]		x	KpnI 9b
2	D+ or DD. Deficient virion assembly[c]		x	KpnI 9b
4	D+ or DA. Deficient head formation[c]			XbaI 23b, KpnI 4
5	D+. Polyheads formed[d]			XbaI 21c, KpnI 4
3	D-int or DA. Deficient virion assembly[c]			XbaI 8, KpnI 4
6	D+. Deficient head formation[b,e]	Part of head structure[o]	o	[ee], [ee]
7	D+. Deficient tail formation[b]			XbaI 21b
8	D+. Deficient tail formation[f]			XbaI 21b
9	D+. Deficient tail formation[b]			[ee]
10	D+. Deficient tail formation[c]			XbaI 30c, EcoRI 2Δ[f]
11	D+ or D-int. Deficient tail formation[b]			EcoRI 2Δ[f]
12	D+ or D-int. Deficient tail formation[b]			EcoRI 16
13	D+ or D-int. Deficient tail formation[c]			EcoRI 16, XbaI 7
14	D+ or D-int. Deficient tail formation[c]			[ee]
15	D+. Deficient tail formation[b]			BglII 6, XbaI 7, EcoRI 13b
16	D+. Deficient tail formation[b]			BglII 6, XbaI 7, EcoRI 13b
16.5	D+[g]			BglII 6, XbaI 19, EcoRI 17
17	D+. Deficient tail formation[b]			EcoRI 17

Gene	Phenotype	Function		Restriction fragment
18	D+. Deficient tail formation[c]			ee
19	D+. Deficient tail formation.[c] Deficient lytic enzyme.			ee
20	D+. Deficient tail formation[b]			XbaI 18
21a,21b	DO. Deficient in initiation of replication[h,i]	Necessary for initiation of replication but not for elongation[h,i]		XbaI 18 (Gene 21a only)
22	DO	Necessary for elongation stage of replication[h]		
23	DO			
24	D+	HMdUMP kinase[p]		
25	D+			
26	D+			
28	DO. Deficient in transcription of middle and late genes[k]	σ factor that confers middle gene specificity on host RNA polymerase[q,r,s]	q,x,z,aa	XbaI 9; XbaI 21a[gg]
27	DO.[g] Deficient in transcription of late genes[j]		y,z	XbaI 9[gg]
29	DO. Hmase⁻	dUMP hydroxymethylase[t]	bb	XbaI 29
30a,b,c	DO[l]	Necessary for elongation stage of replication[h,l]	l,x	
TF1	m	Specifically inhibits in vitro transcription of SPO1 DNA[u]	cc	EcoRI 11[hh]
31	DO	DNA polymerase[v]	v,x,aa	EcoRI 23; XbaI 15
32	DO. Deficient in initiation of replication[h]	Necessary for initiation of replication but not for elongation[h]	x	
C1	Clear plaques with distinct edges			
33	D+. Deficient in transcription of late genes[k]	Part of σ factor that confers late gene specificity on RNA polymerase[s,w]	w	EcoRI 21; XbaI 10[ii]
34	D+. Deficient in transcription of late genes[k]	Part of σ factor that confers late gene specificity on RNA polymerase[s,w]	w	EcoRI 21; XbaI 10[ii]

(Continued)

TABLE 1. (Continued)

Gene[pp]	Phenotype of mutant[a]	Activity of gene product	Gene product purified or identified by PAGE band	Cloned restriction fragment on which gene located[dd]
35	D⁺. Deficient in virion assembly[c]			EcoRI 18Δ[ff]
36	D⁺			EcoRI 18Δ[ff]
e9	m		ii	
e6	m		ii	
e18	m		ii	
e4	m		ii	
e15	m		ii	
e3	m		ii	
e20	m		ii	
e16	m		ii	
e7	m		ii	
e21	m		ii	
e12	m		ii	

B. Unmapped mutations

Mutation	Phenotype
bh	Plaques with black halo[n]
hd	Forms plaques on host mutant on which wild-type does not plaque[n]
Nal[R]	Resistant to nalidixic acid[kk]

C. Phage-specific gene products whose activities are known but for which the SPO1 gene has not been identified

1. Inhibitor of thymidylate synthetase[ll]
2. dTTPase-UPTase[mm]
3. dCMP deaminase[mm]
4. Deoxythymidylate-5'-nucleotidase[oo]

[a] All data in this column were taken from Okubo et al., 1972, except where indicated by another reference. Abbreviations for replication phenotype: D⁺, replication positive; DD, replication delayed; DA, replication arrested prematurely; D-int, partial, but not normal, replication; DO, replication negative.
[b] Fujita, 1971
[c] These data have never been published, and the original data are no longer accessible [F. Eiserling, personal communication]. The data were used by E. P. Geiduschek in writing an earlier review (Geiduschek and Ito, 1982), and it is from that review that this information is taken.
[d] Parker and Eiserling, 1983b.
[e] Levner, 1972a.
[f] Parker et al., 1983.

[g] Stewart, 1984.

[h] Glassberg et al., 1977b.

[i] Mutants N6 and F2 are affected in two different cistrons, 21a and 21b, since the two mutations are located on different restriction fragments, at least 2.8 kb apart, and since the gene 21a product is expressed from its restriction fragment (Curran and Stewart, 1985a,b). However, the two mutants frequently do not complement each other or the mutant cs 21-1, which was used to identify the replication initiation phenotype. Because of the inconsistency of complementation patterns, it is not clear which of these two cistrons is responsible for the replication initiation phenotype. Each has a DO phenotype.

[j] Greene et al., 1982.

[k] Fujita et al., 1971.

[l] Cistron 30 was originally divided into subcistrons 30a, 30b, and 30c, represented by mutations O52, F26, and O81, respectively (Okubo et al., 1972). Each of these mutants is replication-negative. They sometimes show substantial complementation with each other (Okubo et al., 1972; our unpublished results) and also show different patterns of complementation with a variety of ts mutants. For instance, O52 complements all of the SPO1 ts mutants that fail to complement F26 and/or O81, but all three fail to complement one SP82 ts mutant. Thus, it is unclear from complementation tests whether these mutations represent one, two, or three separate genes. Each of O52 and O81 is missing a different PAGE band (Reeve et al., 1978). However, the band missing in O52 infection appears at late times and is dependent upon gp34 activity. Thus, it seems likely that polarity or multiple mutations are involved, so no firm conclusion can be drawn.

[m] No mutation isolated. Gene identified physically.

[n] Okubo et al., 1964.

[o] Shub, 1975.

[p] Gene 23 mutants fail to complement SP82 mutant H20, which is stated to be deficient in HMdUMP kinase (Kahan, 1971; unpublished results, this laboratory).

[q] Fox et al., 1976.

[r] Duffy and Geiduschek, 1977.

[s] Talkington and Pero, 1977.

[t] Okubo et al., 1972.

[u] Wilson and Geiduschek, 1969; Wilhelm et al., 1972.

[v] De Antoni et al., 1985.

[w] Fox, 1976.

[x] Reeve et al., 1978.

[y] Costanzo et al., 1983.

[z] Chelm et al., 1981a.

[aa] Heintz and Shub, 1982.

[bb] Alegria et al., 1968; Kunitani and Santi, 1980.

[cc] Johnson and Geiduschek, 1972.

[dd] All data in this column were taken from Curran and Stewart (1985a) and refer to cloning in B. subtilis, except where indicated by another reference.

[ee] These genes were not tested for marker rescue by cloned fragments. However, assuming the correctness of the genetic map as shown in Fig. 1, at least four, and possibly all six, of these genes must be on cloned fragments.

[ff] EcoRI* fragment that comprises the rightmost 5.3 kb of EcoRI fragment 2. EcoRI* fragment 18Δ consists of a portion of EcoRI fragment 18.

[gg] Gene 28 is split between these two fragments. It and gene 27 have also been cloned in E. coli on subfragments of EcoRI 3 (Costanzo and Pero, 1983; Costanzo et al., 1983].

[hh] Greene et al., 1984; Cloned in E. coli on EcoRI 11 and on certain subfragments thereof.

[ii] Also cloned in E. coli and B. subtilis on subfragments of EcoRI 21 (Costanzo et al., 1984].

[jj] Perkus and Shub, 1985.

[kk] Alonso et al., 1981.

[ll] Haslam et al., 1967.

[mm] Roscoe, 1969a; Price et al., 1972; Dunham and Price, 1974a,b.

[nn] Roscoe and Tucker, 1964; Okubo et al., 1972; Nishihara et al., 1967; Swanton et al., 1975.

[oo] Aposhian and Tremblay, 1966.

[pp] The genes in this column are listed in map order, which differs from numerical order in two places.

V. REGULATION OF SPO1 GENE ACTION

A. Sequential Modifications of RNA Polymerase

1. Description of the System

During SPO1 infection, there is a complex sequence of gene actions. By hybridization competition, Gage and Geiduschek (1971a) distinguished six classes of transcripts on the basis of their times of synthesis, and these can be subdivided into additional classes, as discussed below. We do not have a complete understanding of the regulation of this sequence, but a major portion can be explained by sequential modifications of RNA polymerase.

Three different polymerases, A, B, and C, are active at early, middle, and late times, respectively, each transcribing a different set of genes. This can be seen from hybridization of labeled transcripts to Southern blots of SPO1 EcoRI* digests (Talkington and Pero, 1977). RNA labeled at early (1–4 min), middle (7–12 min), or late (later than 18 min) times during infection produces three substantially different patterns, and the same three patterns are produced by RNA synthesized by the three polymerases *in vitro*.

Mutations in SPO1 genes 28, 33, and 34 prevent portions of the normal sequence of gene action (Fujita *et al.*, 1971) and have permitted elucidation of the mechanisms by which these changes in RNA polymerase take place. Polymerase A is the host polymerase with sigma factor σ^{43}. [This sigma factor has traditionally been called σ^{55}, based on its apparent molecular weight form gel electrophoresis. However, nucleotide sequencing has now shown its correct molecular weight to be about 43,000 (Gitt *et al.*, 1985).] SPO1 gene 28 is an early gene (Reeve *et al.*, 1978; Heintz and Shub, 1982) that specifies a sigma factor found in RNA polymerase B in place of σ^{43} (Duffy *et al.*, 1973, 1975a,b, 1977; Fox *et al.*, 1976). SPO1 genes 33 and 34 are middle genes (Fox and Pero, 1974; Pero *et al.*, 1975) that specify a sigma factor found in RNA polymerase C (Fox, 1976). Thus, host RNA polymerase transcribes early genes, which include gene 28. Gp28 replaces the host sigma factor, changing the specificity of the RNA polymerase so it now transcribes middle genes, which include genes 33 and 34. The products of these genes replace gp28, changing the specificity of the RNA polymerase so it now transcribes late genes. An illuminating discussion of such cascades of sigma factors has been provided by Losick and Pero (1981).

No changes in any of the core proteins of the RNA polymerase have been detected during SPO1 infection. Infection of a rifamycin-resistant host mutant remains resistant to rifamycin throughout the infection (Geiduschek and Sklar, 1969), showing that the host β protein continues to be used, and no change in the electrophoretic properties of any of the core proteins has been reported.

SPO1 proteins, identified as bands on polyacrylamide gel electrophoresis, can be categorized as early, middle, or late on the basis of dependence on gp28, 33, and 34. By these criteria, about 20 proteins have been assigned to each class (Reeve *et al.*, 1978; Heintz and Shub, 1982), and these are clearly underestimates, since they include only proteins clearly visible and resolvable on one-dimensional gels. For instance, the late proteins presumably include at least the 53 structural proteins mentioned above.

2. Sequences and Characterization of Promoters and Terminators

Each of the three RNA polymerases recognizes a different promoter sequence. Promoters specific for each polymerase were found by assaying for binding of each polymerase to restriction fragments. They were located within specific fragments by electron microscopy of polymerase: DNA complexes, by subdivision of specific fragments with other restriction enzymes, by size analysis of *in vitro* transcripts, and by S1 mapping. Sequences were determined for three early, eight middle, and five late promoters (Talkington and Pero, 1978, 1979; Lee *et al.*, 1980a; Chelm *et al.*, 1981a; Lee and Pero, 1981; Romeo *et al.*, 1981; Costanzo, 1983; Costanzo and Pero, 1983; Pero, 1983; Greene *et al.*, 1984, 1986a). Each of the three types shows a different consensus sequence in both the -35 and -10 regions, as indicated in Table II. The three early promoters (P_E4, 5, and 6 in Fig. 3) are identical in the regions shown. The eight middle promoters (including P^{MII} 3, 7, and 8 in Fig. 4) are identical for the -10 region, but there is some variation in the -35 region, with ($P_{MII}8$) diverging strikingly from the others. The late promoters show some variation in the -10 region and more in the -35 region. As expected, the early promoter sequence is closely similar to that of host promoters recognized by σ^{43} (Johnson *et al.*, 1983; Stewart *et al.*, 1983). It is also similar to the consensus sequence of *E. coli* promoters (Rosenberg and Court, 1979), differing by one nucleotide in each of the -10 and -35 regions, and, indeed, the *E. coli* polymerase is able to transcribe an SPO1 early promoter (Lee *et al.*, 1980).

Cloning permits the isolation of fragments of SPO1 DNA that contain thymine instead of hmUra and thus has made it possible to test the

TABLE II. Consensus Sequences of SPO1 Promoters[a]

Type of promoter	-35 Region	-10 Region
Early	TTGACT	ATAAT
Middle	AGGAGA–A–TT	TTT–TTT
Late	CGTTAGA	GATATT

[a]Dashes indicate nucleotide positions for which there is no consensus. T represents hydroxymethyluracil.

importance in promoter function of the hmUra for T substitution. Lee *et al.* (1980b) found that early promoters function as well in thymine-containing DNA as with hmUra, but that transcription from middle promoters could be detected only when they contained hmUra. Romeo *et al.* (as reported by E. P. Geiduschek, 1986, personal communication), using an S1 protection assay, were able to detect substantial middle promoter function from T-DNA, although they too found substantially greater function with hmUra-DNA. Both middle and late genes can be expressed from T-DNA *in vivo*, since fragments cloned in *B. subtilis* were able to complement mutants in those genes independently of detectable vector promoters (Curran and Stewart, 1985b). Thus, hmUra is not required for the activity of any of the SPO1 promoters, although substitution of T for hmUra substantially decreases the activity of middle promoters. T for hmUra substitution in either strand of the middle promoter decreases the effectiveness of that promoter, showing that hmUra residues in each strand play an important role in promoter function (Choy *et al.*, 1986).

Several transcription terminators, all but one from the terminal redundancy, have also been sequenced (Brennan and Geiduschek, 1983; Greene *et al.*, 1984). They show sequences similar to those of *E. coli* terminators, with a region of dyad symmetry followed by a stretch of hmUra residues. There are only slight differences between the sequences of efficient terminators and those of the two less efficient ones that have been sequenced. One of the latter has a greater than usual distance between the two regions mentioned above. The other has a C within its stretch of hmUras. The inefficiency of the latter, as well as of two others that have not been sequenced, can be overcome by decreasing the ribonucleoside triphosphate concentration in the transcription mixture (Brennan, 1984).

3. Mechanism of Function of gp28, 33, and 34

Genes 28, 33, and 34 have been cloned and sequenced (Costanzo and Pero, 1983; Costanzo *et al.*, 1984). They contain, respectively, 220, 101, and 197 amino acids, and each has an unusually high proportion of charged amino acids. Gp28 and gp34 show significant homology to σ^{43} and to *E. coli* σ^{70} (Gribskov and Burgess, 1986).

Gp28 confers upon either *B. subtilis* or *E. coli* core RNA polymerase the ability specifically to recognize middle promoters (Duffy and Geiduschek, 1977; Costanzo and Pero, 1984), although the combination with *E. coli* polymerase is not as efficient. *In vitro*, at ionic strengths comparable to those *in vivo*, gp28 is able to displace σ^{43} from the RNA polymerase core. However, this happens fairly slowly, and under certain conditions the reverse reaction is favored, so this mechanism may not be sufficient to account for the *in vivo* transition from early to middle transcription (Chelm *et al.*, 1982a). Once transcription has begun, gp28 can dissociate from the RNA polymerase, although it retains greater affinity for the

RNA polymerase than does σ^{43} under similar conditions (Chelm et al., 1981b).

There is a 4-bp overlap between the terminus of gene 33 and the origin of gene 34, suggesting a translational coupling which may act to assure the presence of equimolar quantities of the two proteins.

4. Transcription Mapping

The approximate sites at which early, middle, and late transcriptions take place have been located on the restriction map by hybridizing RNA, synthesized at various times, to Southern blots of EcoRI* digests of SPO1 DNA (Talkington and Pero, 1977; Pero et al., 1979). These data are summarized in Fig. 2. There is a tendency for the different categories of transcripts to be clustered, with the major early cluster in the terminal redundancy, two middle clusters in the left and right halves of the unique region, and the major late cluster between the middle clusters. The clustering is not absolute, with smaller quantities of each category scattered about the genome. These results, from analyses of in vivo transcription, are generally consistent with those from in vitro transcription of, or RNA polymerase binding to, separated restriction fragments (Talkington and Pero, 1978; Brennan et al., 1981; Chelm et al., 1981a; Romeo et al., 1981).

In several regions of the genome, individual transcription units have been mapped precisely. Most of this mapping has employed in vitro as-

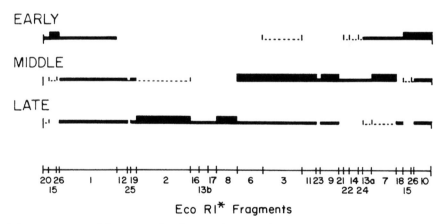

FIGURE 2. General locations of early, middle, and late sequences on the SPO1 genome. The thick, thin, and dotted lines represent the relative amounts of early, middle, and late labeled RNA that hybridize to each of the indicated EcoRI* restriction fragments. EcoRI* fragments 13a and 13b and fragments 14 and 15 were not resolved on these Southern blots, so the hybridized RNA for these bands was arbitrarily assigned to one (or both) member(s) of each pair of DNAs on the basis of the amount of DNA hybridizing to restriction fragments neighboring each member of the pair. The data are taken from Talkington and Pero (1977) and Pero et al. (1979) and have been reproduced with permission from the American Society for Microbiology.

says, although, whenever available, *in vivo* assays have confirmed the *in vitro* results (e.g., Romeo *et al.*, 1981; Brennan and Geiduschek, 1983). I will discuss each of these regions in turn.

a. Transcription Units in the Terminal Redundancy

All major early promoters are located in the terminal redundancy, as determined either by binding RNA polymerase to specific restriction fragments (Talkington and Pero, 1978) or by the formation of ternary complexes of restriction fragments, RNA polymerase, and nascent RNA (Romeo *et al.*, 1981). Several weaker early promoters were detected outside the terminal redundancy by R loop mapping and *in vitro* transcription (Chelm *et al.*, 1981a; Romeo *et al.*, 1981) and will be discussed below. No early promoters were detected in the left half of the unique region of the genome.

By electron microscopy of complexes of RNA polymerase and restriction fragments, 12 early promoters were mapped at specific sites within the terminal redundancy (Romeo *et al.*, 1981), and a 13th was detected by *in vitro* transcription (Brennan *et al.*, 1981). Transcription from the nine leftmost of these promoters proceeds rightward, and that from the four rightmost proceeds leftward. All transcripts terminate in a central termination region, which prevents transcription in one direction from overlapping that in the other. In addition, before reaching the major terminator, several of the transcripts reach other terminators, at which they may be terminated under certain conditions or with lesser efficiency (Brennan *et al.*, 1981; Brennan and Geiduschek, 1983; Brennan, 1984).

Figure 3 shows the positions of these promoters and terminators on the restriction map of the terminal redundancy. The genes specifying 11 of the early proteins have been mapped within the terminal redundancy by *in vitro* transcription and translation of restriction fragments (Perkus and Shub, 1985), and the positions of these and the two previously known genes are also shown in Fig. 3. The transcripts necessary for synthesis of most of these proteins can be provided by one or more of the 13 promoters shown. However, the expression of two of the proteins, from two restriction fragments, cannot be accounted for by promoters known to be in those fragments, suggesting either that anomalous expression is occurring in the *in vitro* transcription-translation system or that there are additional promoters not identified by the other techniques.

Many of these early promoters are among the strongest promoters known (Romeo *et al.*, 1981). It has been suggested (Geiduschek and Ito, 1982) that the purpose of this concentration of strong promoters is to permit the SPO1 early genes to compete effectively with host genes for RNA polymerase. We do not know the role of any of these early gene products, but they may include proteins whose activity is to shut off host functions, as discussed below.

Evidence for the presence of two middle promoters within the terminal redundancy has also been presented. The gene specifying protein e16 is expressed *in vitro* by either RNA polymerase A or B, and its expression *in vivo* is enhanced by gp28 (Perkus and Shub, 1985). The promoter from which this gp28-dependent transcription takes place has not been identified by other means, although a tentative location is indicated in Fig. 3. A middle promoter sequence has been found in the major termination region (Brennan and Geiduschek, 1983), but no gene product has been observed, and its position is not appropriate for e16.

Thirteen genes have been identified within this region, and there remains space, not occupied by known genes, for about five more. If this space is occupied by genes, and if similar densities of genes are characteristic of the rest of the genome, SPO1 would have a total of about 190 genes.

b. Transcription Units within the Right-Hand Cluster of Middle Genes

Middle promoters have been searched for systematically only within EcoRI fragments 19 and 25 [about 4 kb in the left-hand cluster of middle genes (Talkington and Pero, 1979; Lee and Pero, 1981)] and within HaeIII fragment 2 (about 28.5 kb within the right hand cluster of middle genes (Chelm *et al.*, 1981a). A transcription map of the latter region is shown in Fig. 4.

Three early and 13 middle promoters have been identified within this region (Chelm *et al.*, 1981a). Each of the early promoters causes transcription in the same direction as that caused by a nearby middle promoter, suggesting that these may be responsible for genes transcribed at both early and middle times. In fact, hybridization-competition shows that RNA transcribed *in vitro* from P_E15, one of the early promoters, is effectively competed by either early or middle *in vivo* RNAs (Chelm *et al.*, 1981a).

Several of these promoters have been identified with specific genes, also shown in Fig. 4. P_E13 is just upstream from genes 27 and 28, and $P_{MII}3$ is between 27 and 28, suggesting that gene 28 is transcribed early and that gene 27 is transcribed at both early and middle times (Chelm *et al.*, 1981a; Costanzo and Pero, 1983). $P_{MII}7$ is just upstream from the TF1 gene, which is followed by a normal termination sequence (Greene *et al.*, 1984), defining a monocistronic transcription unit whose transcript has been observed by S1 mapping (unpublished work cited in Greene *et al.*, 1984). Genes 33 and 34 are downstream from $P_{MII}12$, and in the correct orientation, but the transcript has not been examined.

c. Transcription within the Left-Hand Cluster of Middle Genes

Five middle promoters were located within the adjacent EcoRI* fragments 12, 25, and 19 in the left-hand cluster (Talkington and Pero, 1979;

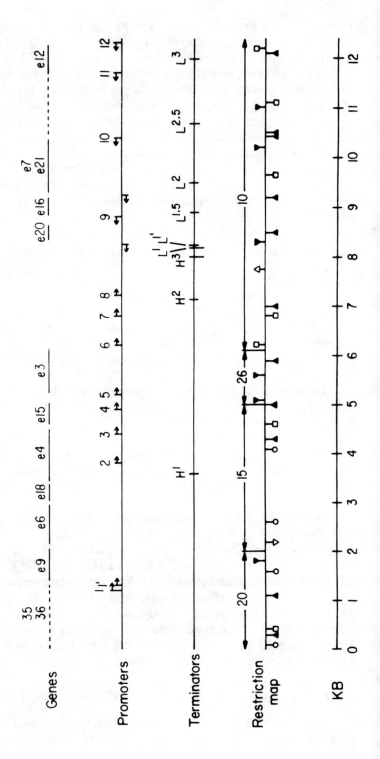

FIGURE 3. Transcription map of the terminal redundancy. The restriction map is taken from Pero et al. (1979), Romeo et al. (1981), Brennan et al. (1981), Perkus and Shub (1985), and E. P. Geiduschek (personal communication). Small adjustments have been made to reconcile differences in positions reported by different groups. The numbers of the EcoRI* fragments are indicated. HaeII, ▼; KpnI, □; BstN1, △; EcoRI, ⏐; TaqI, ▲; HpaII, ○; BstEII, ○; HaeIII, ▽. Positions of early promoters, taken from Romeo et al. (1981), are indicated by the numbered sites above the promoters' line. Positions of two possible middle promoters are indicated below the promoters line. The leftmost of these is a middle promoter sequence, observed by Brennan and Geiduschek (1983), but not known to be functional. The rightmost of these is inferred from Perkus and Shub's (1985) conclusion that gene e16 is expressed from a middle promoter. Its position is shown just to the right of e16, but it could also be farther to the right. (There is evidence suggesting the presence of more than one middle promoter in this region; E. P. Geiduschek, personal communication.) The direction of transcription driven by each promoter is indicated by an arrow. The positions of transcription terminators are taken from Brennan et al. (1981) and Brennan and Geiduschek (1983). Those indicated as H are heavy-strand terminators, which terminate rightward transcription; the light-strand terminators, indicated as L, terminate leftward transcription. The efficient terminators in the central termination region are emphasized with longer lines. Transcription from each rightward promoter is terminated at the efficient terminator H3, and that from each leftward promoter is terminated at L', with intermediate terminators functioning less efficiently or under particular conditions. The gene positions are from Perkus and Shub (1985) and Curran and Stewart (1985a). Placements are based on the best arguments available, but in some cases those arguments are not conclusive, so the placements should be considered approximations. Where there is no basis for determining the order of two genes, they are placed one above the other. The length of the line for each gene represents the size of the gene, calculated from the estimated molecular weight of the gene product. In the e9 to e3 and e20 to e16 regions, the indicated genes occupy nearly all the space available, assuming no overlapping. In the e7, e21, e12 region, the dashed line represents space unaccounted for by the known genes. Any of those three genes could be any place within that dashed region, instead of their indicated position, with the only constraints being that e12 be to the right of e7 and e21, and that e21 not overlap a HaeII or BstEII site. Markers in genes 35 and 36 are located in EcoRI* fragment 18 (Curran and Stewart, 1985a). That in gene 35 is in the terminal redundancy (Cregg and Stewart, 1978) and that in gene 36 behaves as if it is too (Glassberg et al., 1977b), so they are probably both in EcoRI* 20, which is the portion of fragment 18 that is in the redundant region. It remains possible that part of one of these genes extends into the unique region. It is unlikely that either of these genes is e9, since gene 36 apparently is expressed from EcoRI* 18 (Curran and Stewart, 1985b), whereas e9 is not expressed from EcoRI* 20 (Perkus and Shub, 1985), and since e9 is an early gene, whereas gene 35 is required for viral assembly (Fujita, 1971) and is therefore probably a late gene. Therefore, genes 35 and 36 are placed to the left of e9. There is no basis for estimating the size of genes 35 and 36, so the dashed line merely indicates the region within which they are found.

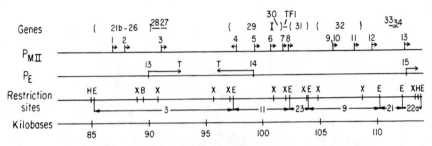

FIGURE 4. Transcription map of HaeIII fragment 2. This region includes the majority of the right-hand cluster of middle genes. The restriction map is an expansion of that in Fig. 1, with H, E, X, and B representing sites for HaeIII, EcoRI*, xbaI, and BglII, respectively. The EcoRI* fragment numbers have been provided for orientation. The vertical lines on lines P_E and P_{MII} show the positions of early and middle promoters, respectively, as determined by Chelm *et al.* (1981a). The arrows on the P_{MII} line show direction of transcription only. Positions of termination have not been determined. The arrows on the P_E line show the lengths of *in vitro* transcripts. T indicates the position of termination sites as calculated from the lengths of these transcripts. Where no T is shown, transcription ran off the end of the restriction fragment used, so the termination site was not determined. Each gene whose precise position has been determined by nucleotide sequencing is indicated with a horizontal line covering the position of the gene. Genes within parentheses are located somewhere within the segment circumscribed by those parentheses. References for the precise gene locations are Costanzo and Pero (1983), and Costanzo *et al.* (1983) for 27 and 28, Greene *et al.* (1984) for TFl, and Costanzo *et al.* (1984) for 33 and 34. The location of other genes was described in the legend to Fig. 1.

Lee and Pero, 1981), and there may be more promoters in this region, since parts of fragment 12 were not tested. All five of these promoters generate rightward transcription, but the position of termination was determined for only one of the transcripts. No known genes are located on any of these three fragments.

d. Transcription Units in the Late Transcription Cluster

Five late promoters, each driving rightward transcription, were identified in the adjacent EcoRI* fragments 16, 13b, and 17 (Costanzo, 1983). Genes 13 through 17 are located on these fragments but have not been identified with specific transcripts.

B. Additional Complexities in the Regulation of Gene Action

The story of regulation of the sequence of gene action, as told to this point, is elegant and straightforward: three different classes of genes are transcribed, in sequence, because of sequential substitutions of sigma factors in the RNA polymerase. However, the actual situation is more complicated, and there is little understanding of the additional complexities. The SPO1 genes can be subdivided into many more than three groups on the basis of time and regulation of activity. Moreover, several other mechanisms have been demonstrated during SPO1 infection which potentially could regulate SPO1 gene action, but which are not needed for the simple sequence described above. These additional complexities will be discussed in the following paragraphs.

1. Six Categories of Transcripts That Can Be Explained by the Three-Polymerase Sequence

Gage and Geiduschek (1971a) identified six categories of transcripts, based on time of synthesis, as indicated in Fig. 5. These categories can be explained in terms of the above sequence of RNA polymerase specificities. According to this explanation, "e" are genes transcribed only by polymerase A (the host polymerase). "em" are genes transcribed either by polymerase A or by polymerase B (the gp28-modified polymerase). An example of these might be gene 27, as indicated above. "m" are genes transcribed only by polymerase B; "m_1l" are genes transcribed by polymerase B or polymerase C (the gp33- and 34-modified polymerase). "m_2l" and "l" are transcribed only by polymerase C, with "l" being distinguished from "m_2l" by a requirement for DNA replication, as discussed in the following paragraph. These assignments are consistent with the effects of mutants, since e and em are made in $sus28$ infection, and e, em, m, and m_1l are made in $sus33$ or 34 infection (Fujita et $al.$, 1971).

2. Effect of DNA Replication on Late Gene Function

When SPO1 DNA replication is prevented, the total amount of late transcription is quite low in comparison with the amount of early and middle transcription or with the amount of late transcription when replication is permitted (Fujita et $al.$, 1971; Sarachu et $al.$, 1978). This could be due in part to gene dosage, but replication may also alter the specificity of transcription.

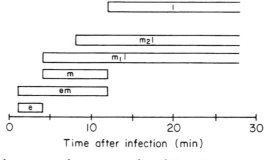

FIGURE 5. Time course of synthesis of six classes of SPO1 RNA. These data were determined by Gage and Geiduschek (1971a) and have been reproduced by permission of Academic Press, London. Under these conditions, the eclipse and latent periods are 25 and 33 min, respectively (Wilson and Gage, 1971). However, these values vary with experimental conditions. For instance, values of 40 min (Shub, 1966) and 50 min (Okubo et $al.$, 1964) have also been reported for the latent period.

By hybridization competition, it was shown that F30, a replication-negative mutant affected in cistron 22, prevents any detectable appearance of "1" RNA (Gage and Geiduschek, 1971b). This suggested that replication may be a prerequisite for "1" gene function. However, the effect of preventing replication by other means was not tested, leaving open the possibility that F30 affects transcription independently of its effect on replication. Several other replication-negative mutants, including another mutant affected in cistron 22, permitted synthesis of many late-specific transcripts, as measured by hybridization to Southern blots (Talkington and Pero, 1977; Greene et al., 1982), and of nearly all late-specific proteins, as measured by polyacrylamide gel electrophoresis (Heintz and Shub, 1982). The latter three papers do not contradict the first, since (1) in the Southern blot experiments, "1" transcripts may be obscured by "$m_2$1" transcripts hybridizing to the same fragment, and (2) only a small number of late proteins are observable in these gel electrophoresis experiments, and the synthesis of one of them, "1" 14, is largely prevented by replication-deficient mutants. Thus replication may be necessary to provide DNA competent for synthesis of certain late transcripts.

3. Additional Regulatory Mechanisms

Thus, in conjunction with an effect of DNA replication on late transcription, the known sequence of sigma substitutions can explain the six categories of transcripts identified by Gage and Geiduschek (1971a). However, the sequence of gene action is more complicated than this, so additional regulatory mechanisms, not yet understood, must be operative. The following paragraphs describe some of the additional complications and explain why each of several additional mechanisms is necessary.

a. Additional Mechanisms Regulating the Onset of Early Transcription

Under the simplest model of sigma substitution, all early genes should begin transcription immediately after infection, and new classes of transcripts should not appear until gp28 is available to begin middle transcription. However, the onset of early gene activity actually shows a more complex pattern, which requires at least one, and possibly several, additional regulatory events. Observations suggesting such heterogeneity in early gene activity include the following.

1. Each of the categories e and em may be subdivided into at least two categories on the basis of time of first appearance (Gage and Geiduschek, 1971b). It was possible that these differences in time

of appearance were caused only by differences in distance from transcription start sites. However, Perkus and Shub (1985) measured the time of *initiation* of the transcripts for five early genes by determining the time at which those transcripts become resistant to inhibition by rifampin and found a different time of initiation for each transcript. Assuming that the map in Fig. 3 shows the correct relationships between genes and promoters, these variations in time of initiation cannot be explained by variations in the strength of promoters, as measured *in vitro* (Romeo *et al.*, 1981). Thus, the time of initiation is apparently determined by some factor beyond the intrinsic strength of the promoter. Similarly, additional factors must play a role in determining the *amount* of expression of early genes, since the relative amount of expression of various early genes *in vivo* (Heintz and Shub, 1982) differs from that of those same genes *in vitro* (Perkus and Shub, 1985). The additional factors regulating time of initiation and amount of expression may be related, since there is a direct correlation between the time of initiation and the amount of *in vivo* expression of the five genes tested (Heintz and Shub, 1982).

2. In *sus*28 infection, several sequences do not begin transcription until middle times (Chelm *et al.*, 1981a, 1982b). Chloramphenicol prevents some of this additional transcription, so an additional SPO1 gene product(s) may be required. However, the chloramphenicol effect might be due to an alternative mechanism such as the exposure of rho-dependent termination sites. This effect of choramphenicol can also be seen from analysis of early *protein* synthesis. Several proteins are synthesized during *sus*28 infection, the synthesis of whose mRNAs seems to be inhibited by chloramphenicol. This is shown by the fact that they are not synthesized when *in vitro* translation is programmed by RNA made in the presence of chloramphenicol, nor are they synthesized *in vivo* when chloramphenicol is withdrawn and new transcription is prevented by rifampicin (Reeve *et al.*, 1978). On the other hand, when RNAs were observed by polyacrylamide gel electrophoresis, no difference was seen between those synthesized in the presence of chloramphenicol and those made in *sus*28 infection (Reeve *et al.*, 1978; Downard and Whiteley, 1981).

3. The shutoff of host DNA replication (see below) takes place in *sus*28 infection (unpublished results, this laboratory). Yet, it depends on transcription that is not initiated until the beginning of the middle time period (Gage and Geiduschek, 1971b).

4. One EcoRI* fragment (EcoRI*-12) specifies transcripts that are not synthesized until middle time but that are synthesized both in *sus*28 infection and in the presence of chloramphenicol (Talkington and Pero, 1977). This might be explained by an additional

regulatory mechanism or by transcription which begins at an early promoter in the adjacent fragment, and which reads through into fragment 12 after several minutes.

b. Additional Mechanism Regulating Onset of Late Gene Activity

Sus33 and sus34 do not always have identical effects. Some middle transcripts are turned off in sus33 infection but not in sus34 (Talkington and Pero, 1977; Greene et al., 1982). In minicells, some late proteins are not made, and the synthesis of those that are takes place in sus33 but not in sus34 infection (Reeve et al., 1978). Thus, there appear to be regulatory events that can be mediated by gp34 in the absence of gp33.

The following scenario is the simplest, although far from the only, interpretation of these data. Gp34 binds to RNA polymerase independently of gp33 and causes certain promoters to be recognized. These promoters are distinguished from other late promoters in that the others require both gp33 and 34 and in that the others cannot be expressed in minicells. Among the gene products produced from these promoters is one whose activity is to repress certain middle genes.

c. Additional Mechanism for Shutoff of Early Transcription

If substitution of gp28 for σ^{43} were complete, it would, in itself, provide a sufficient mechanism for the shutoff of early transcription. However, several observations suggest that that substitution is neither necessary nor sufficient for that shutoff.

1. In vitro measurement suggests that the efficiency with which gp28 substitutes for σ^{43} is not great enough to account for the essentially complete shutoff observed (Chelm et al., 1982a).
2. Sus28 infection permits shutoff of early transcription (Fujita et al., 1971), although chloramphenicol does not (Gage and Geiduschek, 1971b). Synthesis of several proteins is also shut off during sus28 infection (Perkus and Shub, 1985). Thus, shutoff of early genes can be accomplished without the transition from RNA polymerase A to B and appears to require the synthesis of some protein(s) other than gp28.
3. Yansura and Henner (1984) have constructed a B. subtilis plasmid in which the penicillinase gene from B. licheniformis is controlled by the E. coli lac repressor:operator system, so expression can be induced by isopropyl-β-D-thiogalactoside. The promoter for this gene is similar to those recognized by σ^{43} and is different from the middle and late promoters. We find (unpublished observations) that this gene can be induced at early, middle, or late times during SPO1 infection, arguing that some σ^{43} RNA polymerase continues to be active throughout infection.

4. Host rRNA synthesis continues throughout infection (Shub, 1966; Gage and Geiduschek, 1971a). Since the known promoters for the rRNA genes have sequences characteristic of σ^{43}-specific promoters (Stewart et al., 1983) and since the other known host polymerases recognize other promoter sequences (Johnson et al., 1983), this also suggests the persistence of σ^{43} RNA polymerase. However, it remains possible that there is another polymerase, unknown at present, that is specific to the rRNA genes.

5. sus34 mutants continue to transcribe, at middle times, several fragments that are ordinarily transcribed only at early times (Talkington and Pero, 1977). This suggests that normal shutoff of at least some early genes may require not only gp28 but also gp34 or something that is dependent upon it or upon the normal expression of gene 34.

6. Novobiocin prevents the shutoff of early transcription, while permitting the turn-on of middle transcription (Sarachu et al., 1980; Chelm et al., 1982b). This effect of novobiocin appears to be independent of its effect on replication, since prevention of replication by sus22 did not prevent this shutoff (Gage and Geiduschek, 1971b).

7. In cells irradiated with UV before infection, middle and late transcription are turned on, but little if any shutoff of early or middle transcription takes place (Heintz and Shub, 1982).

There are ad hoc arguments by which several of these observations could be explained in the context of the simple sigma substitution model. For instance, novobiocin or UV might permit partial but incomplete transition from one sigma to the next. sus28 might cause shutoff to occur by a mechanism that does not occur in wild-type infection. However, none of these ad hoc arguments appear strong, and the probability that each is valid seems very small.

d. Additional Mechanism for Shutoff of Middle Transcription

If shutoff of middle transcription occurs simply as a result of substitution of gp33, 34 for gp28, the prevention of that substitution should prevent that shutoff. However, shutoff of at least some middle transcription occurs in sus33 or sus34 infection (Fujita et al., 1971; Talkington and Pero, 1977). Moreover, a distinction is made between two categories of middle transcripts in that em transcripts are shut off, but at least some m transcripts are not (Fujita et al., 1971). Thus, it appears that shutoff of em, at least, can be accomplished by some mechanism beyond sigma substitution.

It is conceivable that complete removal of gp28 could be accomplished by replacement by gp34 alone. In paragraph b above, I have discussed such a replacement as a possible explanation for certain data. If so, that would vitiate the portion of the argument in the preceding paragraph that is based on shutoff occurring in *sus*33 mutants. However, the argument based on shutoff of em in *sus*34 mutants would remain.

e. Translational Control Mechanisms

Two observations suggest that expression of certain genes is regulated at the translational level:

1. In minicells, *sus*28 infection permits synthesis of several proteins that are not synthesized if chloramphenicol is present during the first 30 min of infection (and is then replaced by rifampicin, to permit protein synthesis but prevent further RNA synthesis). Several of these proteins *are* synthesized if RNA made in the presence of chloramphenicol is used to program *in vitro* translation. Thus, certain RNA molecules that are present are not translated (Reeve *et al.*, 1978). The simplest explanation is translational control, overcome by a process requiring protein synthesis, although the differences between the two experimental conditions prevent a definitive conclusion.
2. During late times, one of the RNA classes $(m_1 1)$ decreases in abundance, although its relative rate of synthesis remains constant, suggesting a differential rate of degradation (Gage and Geiduschek, 1971b).

f. Additional Shutoff Mechanisms

The six transcription categories defined by Gage and Geiduschek (1971a) require at least two shutoff mechanisms, for shutting off early and middle transcription. Observations of the time course of protein synthesis suggest that more than two shutoff mechanisms are operative. In minicells, there are at least three different times at which synthesis of specific proteins stops (Reeve *et al.*, 1978). In normal infection, there are at least five such times (Parker, 1979).

g. Extensive Heterogeneity in Time of Protein Synthesis

Many categories of proteins can be distinguished on the basis of times of appearance and disappearance of specific bands on polyacrylamide gel electrophoresis. As many as 13 categories can be identified in normal infection of normal cells (Parker, 1979; Heintz and Shub, 1982), and the number of categories observed can be increased by infecting under stressful conditions. Parker (1979) found at least 30 categories during infection of concentrated cells. In minicells (Reeve *et al.*, 1978), different proteins whose synthesis does not require gp28 first appear at 1–4, 7–10, 13–16, or 25–28 min after infection. Proteins that require gp28, but not gp34, first appear at 7–10, 13–16, or 19–22 min. Proteins that require both gp28 and gp34 first appear at 7–10, 13–16, 19–22, 25–28, or 37–40 min. Moreover, these do not include the full range of proteins, since some are not synthesized at all in minicells. Regulation under these abnormal conditions need not be the same as in normal infection. However, the fact that these differences occur shows that there are mechanisms available by which such distinctions can be made.

Some of these differences may reflect different distances from promoters or different translational efficiencies. However, many of the differences are too great for such explanations to be sufficient. Moreover, such explanations predict correlations between the times of turn-on and turn-off, and such correlations are usually not observed. Thus, some of these differences probably reflect additional regulatory mechanisms. However, it is not clear which of the differences are accounted for by mechanisms already discussed or how many of the differences can be accounted for by trivial explanations such as different distances from promoters, so no reasonable estimate can be made as to the number of additional regulatory mechanisms implied by these data.

4. Additional Proteins That Are, or May Be, Involved in Regulation of Gene Action

a. Gene 27 Product

A mutant defective in gene 27 shows a deficiency in late gene action (Greene et al., 1982; Heintz and Shub, 1982). The pattern of late transcription, as shown by hybridization to Southern blots, is significantly different from that shown by sus33 or 34, suggesting that the role of gp27 is different from that of gp33 and 34.

The gene 27 mutant is also deficient in DNA replication (Okubo et al., 1972; Stewart, 1984), and neither deficiency appears to be an indirect result of the other. Thus, gp27 may play a direct role in both replication and late transcription, although it remains possible that either deficiency is the result of a polar effect on a downstream gene.

We do not know the mechanism by which gp27 affects either replication or transcription. The gene has been sequenced (Costanzo et al., 1983), showing the product to be a basic protein with a high density of proline residues, but not revealing its function. If it does affect both processes directly, it would be similar to T4 gp45 (Wu and Geiduschek, 1975). Gp45 is an integral part of the T4 replication complex (Sinha et al., 1980) and also interacts directly with RNA polymerase (Ratner, 1974), but no comparable role for gp27 has been shown. The nucleotide sequences of the two genes show no extensive homologies.

b. TF1

TF1 is an SPO1-specified basic protein (Greene et al., 1984) that binds selectively to many sites on the SPO1 genome (Johnson and

Geiduschek, 1972, 1977) and specifically inhibits *in vitro* transcription of SPO1 DNA (Wilson and Geiduschek, 1969; Shub and Johnson, 1975). It affects the initiation rather than the elongation of RNA molecules and can interfere both with the formation of complexes between DNA and RNA polymerase and with their stability (Geiduschek *et al.*, 1977). Specific sites to which TF1 binds have been identified by DNA footprinting, and several have been sequenced (Greene and Geiduschek, 1985a,b; Greene *et al.*, 1986). No precise homologies are seen between these sites, but they share common features such as sequences of alternating purines and pyrimidines and AT-rich sequences.

The *in vivo* role of TF1 is unknown. Since it is expressed from a middle promoter and since it represses early genes *in vitro*, an appealing speculation is that it is responsible for the shutoff of early transcription. Four early promoters have been tested, and each overlaps a preferred site for TF1 binding (Greene and Geiduschek, 1985a,b). One middle promoter shows no such selectivity (Greene *et al.*, 1986). However, at least two preferred sites for TF1 binding were found in fragments without early promoters, so specificity is not complete, and a corresponding specificity of effect on transcription has not been demonstrated. The TF1 gene has been cloned and sequenced (Greene *et al.*, 1984), and the inferred amino acid sequence shows extensive similarity to those of the prokaryotic type II DNA-binding proteins, which may play a role in forming folded conformations of DNA (Tanaka *et al.*, 1984). Thus, it is possible that TF1 plays a role in DNA packaging, although it is not found in the phage particle (E. P. Geiduschek, personal communication). Alternatively, it may play a role in folding SPO1 DNA into a chromatinlike structure for proper function during infection (Greene *et al.*, 1987). The determination of the role of TF1 awaits isolation of mutations altering its gene.

c. mRNA Processing Enzyme

Posttranscriptional processing of early RNAs occurs for both SPO1 and SP82. Early RNAs produced during infection are substantially smaller than those produced by *in vitro* transcription of phage DNA, and the *in vitro* transcripts can be converted into RNAs of the same size as several of the *in vivo* RNAs by an enzyme extracted from uninfected *B. subtilis* (Downard and Whitely, 1981). By S1 mapping, the RNAs formed by *in vitro* processing were shown to be the same as those formed *in vivo* (Panganiban and Whiteley, 1983b). The enzyme has been purified and shown to be similar to *E. coli* RNAase III (Panganiban and Whiteley, 1983a,b). Three of the cleavage sites have similar nucleotide sequences, each capable of forming a stem and a loop, and the cleavage site in each is at the 5′ side of an adenylate residue within the loop (Panganiban and Whiteley, 1983b).

Thus, there is a processing activity that is potentially capable of regulating translation. However, no evidence that it does so has been obtained. *In vitro* cleavage of one of the SP82 transcripts into three sub-fragments causes no detectable changes in the proteins translated from that RNA *in vitro* (Panganiban and Whiteley, 1983a).

5. Summary of Regulation of SPO1 Gene Action

SPO1 employs a complex variety of mechanisms to regulate a complex sequence of gene actions. One of these mechanisms, the transition from early to middle to late transcription by sequential modification of RNA polymerase, is well understood. All of the others remain to be elucidated.

VI. SHUTOFF OF HOST ACTIVITIES

A. Introduction

Host DNA, RNA, and protein synthesis are shut off during SPO1 infection. Since SPO1 macromolecular synthesis proceeds, these shutoffs must be specific to the host systems and not merely the result of general disruption of the infected cells.

B. Host DNA Synthesis

Host DNA synthesis is shut off virtually completely by 5–7 min after infection (Shub, 1966; Wilson and Gage, 1971). This shutoff takes place in *sus*28 infection (unpublished results, this laboratory), so it does not require any of the known modifications of RNA polymerase. However, it remains completely sensitive to inhibition by rifamycin until 4 min after infection (Gage and Geiduschek, 1971b), suggesting that at least one of the necessary transcriptions is delayed beyond the normal time for early transcription.

Strong arguments have been provided against several of the obvious hypotheses as to the mechanism by which this shutoff occurs. There is no detectable degradation of the host genome (Yehle and Ganesan, 1972). The TTPase induced during infection might inhibit replication by limiting substrate. However, the shutoff of host replication is maintained in extracts of infected cells and is not relieved by excess TTP (Yehle and Ganesan, 1972). Moreover, in φe, a mutant deficient in TTPase was able to shut off host replication (Roscoe, 1969a). Also in φe, the possibility that shutoff is caused by the phage-induced inhibitor of thymidylate synthetase was eliminated by showing that shutoff still occurs in a strain in which the requirement for thymidylate synthetase is circumvented (Roscoe, 1969a).

C. Host RNA and Protein Synthesis

The synthesis of most host proteins is reduced by SPO1 infection to about 10% of their normal level (Shub, 1966). A substantial decrease in the intensity of host protein bands on polyacrylamide gel electrophoresis may be seen by 1–4 min after infection (Shub, 1966; Levinthal et al., 1967; Reeve et al., 1978; Heintz and Shub, 1982), and most bands have reached their minima by 6–8 min. Host RNA synthesis also decreases during infection (Gage and Geiduschek, 1971a), and the rate of decrease in protein synthesis after infection is comparable to that accomplished by treatment of uninfected cells with actinomycin (Shub, 1966). Thus, the shutoff of protein synthesis may be entirely explained by the shutoff of most mRNA synthesis within a few minutes after infection, although the possibility that there is also a direct effect at the translational level has not been ruled out. Presumably, these shutoffs are caused by the products of an early gene or genes, since substantial effects occur at times before middle genes become active.

The shutoff mechanism discriminates between different host transcripts, permitting continued synthesis of ribosomal RNA and of the mRNA for ribosomal proteins (Shub, 1966; Gage and Geiduschek, 1971a). However, even rRNA synthesis is not totally unaffected. The rate of rRNA synthesis decreases late in infection (Gage and Geiduschek, 1971a), and the normal increase in its rate of synthesis, in response to a shift to enriched media, does not occur during SPO1 infection (Webb and Spiegelman, 1984).

D. Shutoff Genes

None of the SPO1 genes responsible for shutoff of host functions has been identified, although a φe mutant deficient in shutoff of host replication has been isolated (Marcus and Newlon, 1971). Curran and Stewart (1985a) identified several SPO1 restriction fragments that may carry host shutoff genes. These fragments are refractory to cloning, as would be expected of fragments carrying host shutoff genes, and alternative reasons for unclonability, such as the disruption of vector function by overactive promoters, were eliminated. Several of these fragments are located in the terminal redundancy, which, as discussed above, includes at least 11 early genes with unknown functions.

VII. SPO1 DNA REPLICATION

A. Gene Products Involved in Replication

The products of at least 10 SPO1 genes (21a, 21b, 22, 23, 27, 28, 29, 30, 31, 32) are essential for replication. There are not likely to be many more replication genes, since selection of 150 additional replication-negative mutants did not reveal any additional genes (Glassberg et al., 1977a). The specific functions of some of these gene products have been identified and are summarized in Table I. Genes 27 and 28 have been discussed above. Genes 29 and 23 are involved in synthesis of the hydroxymethylated nucleotide, specifying, respectively, dUMP hydroxymethylase and HMdUMP kinase (Kahan, 1971; Okubo et al., 1972; unpublished results, this laboratory). Gene 31 specifies the DNA polymerase (DeAntoni et al., 1985), which appears at about the time that SPO1 DNA replication begins and shows template specificity different from that of host polymerase (Yehle and Ganesan, 1972, 1973). Genes 21 and 32 are required specifically for initiation of replication, as shown by the fact that temperature-sensitive mutations in these genes permit completion of an ongoing round of replication after shift to the restrictive temperature but do not permit initiation of a new round (Glassberg et al., 1977b). Nothing is known of the activities of gp22 and 30 except that they are probably directly involved in polymerization, since replication is stopped abruptly by a shift to restrictive temperature (Glassberg et al., 1977a,b). In contrast, temperature shift of mutants affected in gene 23 does not cause replication to stop rapidly, presumably because the gene 23 product does not participate directly in polymerization.

Different mutants affected in gene 29 show different effects of temperature shift (Glassberg et al., 1977a), suggesting that, in addition to precursor synthesis, gp29 may participate directly in polymerization, possibly as part of a replication complex such as that suggested for the analogous enzyme in T4 infection (Chiu et al., 1976, Tomich et al., 1974). Mutants affected in several other genes (including 2, 3, 11, 12, 13, and 14) are partially deficient in replication (Okubo et al., 1972). The nature and extent of the deficiency differ between mutants affected in the same gene, so the significance of these deficiencies is uncertain, and nothing further is known about the function of any of these genes.

DNA gyrase appears to play an essential role, since gyrase inhibitors nalidixic acid (Gage and Fujita, 1969) and novobiocin (Sarachu et al., 1980) both inhibit SPO1 replication. Host mutants resistant to the two antibiotics make SPO1 infection resistant to novobiocin but not to nalidixic acid, while an SPO1 mutation provides resistance to nalidixic acid (Alonso et al., 1981). The host gyrase includes the products of two different genes, gyrA and gyrB, which specify resistance to nalidixic acid and novobiocin, respectively (Sugino and Bott, 1980; Piggott and Hoch, 1985), so these data suggest that the SPO1 genome specifies a product that modifies or substitutes for the bacterial gyr A product but that SPO1 uses the host gyrB product. To my knowledge, none of the mapped SPO1 mutants has been tested for gyrase activity, nor has the nalidixic acid–resistant mutation been mapped.

Inhibition of SPO1 replication by nalidixic acid requires higher concentrations than does inhibition of host replication, offering a means of inhibiting host replication while permitting SPO1 replication to proceed.

The same facility is offered by HPUra (Brown, 1970; Marcus and Newlon, 1971; Lavi *et al.*, 1974).

Several additional enzymes are known to function in replication of one or another of the hmUra phages, primarily in the nucleotide metabolism necessitated by the substitution of hmUra for T (summarized in Kornberg, 1980). These include (1) an inhibitor of thymidylate synthetase (Haslam *et al.*, 1967), (2) a dTTPase-dUTPase (Roscoe, 1969b; Price *et al.*, 1972; Dunham and Price, 1974a,b), (3) a dCMP deaminase (Roscoe and Tucker, 1964; Okubo *et al.*, 1972; Nishihara *et al.* 1967), and (4) a dTMPase (Aposhian and Tremblay, 1966). No SPO1 genes have been shown to specify any of these activities. φe mutants that had 1–5% of the normal activity for the dUTPase-dTTPase had normal burst sizes (Dunham and Price, 1974a).

Host mutants temperature-sensitive for DNA polymerase III, the polymerase essential for *B. subtilis* replication, were also temperature-sensitive for replication of φe or SPO1 DNA (Lavi *et al.*, 1974). This observation is confusing, since HPUra inhibits DNA polymerase III but does not inhibit replication of the hmUra phages. At present, the resolution of this apparent conflict requires introduction of an *ad hoc* hypothesis, such as that pol III may have two activities, both of which are inactivated in the *ts* mutant, and only one of which is inhibited by HPUra. A host mutation affecting DNA polymerase I has no effect on SPO1 infection (Yehle and Ganesan, 1972).

B. Structure of Replicating DNA

Infecting DNA is modified soon after infection so that, if extracted, it is inactive in transfection (Levner, 1972b). This modification is correlated in time with the introduction of a small number of single-strand breaks. Recovery from this modification takes place about the time that replication begins, and there is no evidence as to which, if either, of these events causes the other. However, initiation of replication of an infecting genome is hastened if the culture has been preinfected with another SPO1 genome (Cregg and Stewart, 1977), so it is not the time of residence of the infecting DNA within the infected cell that determines the time at which replication begins.

The origins and directions of replication were determined by infecting with light phage in heavy medium, and following the shift in density of genetic markers in different portions of the genome. For the first round of replication, SPO1 DNA replicates from at least two origins, using at least three growing points. Most of the genome is replicated right to left, from an origin near gene 32 to a terminus between genes 2 and 10. Replication of the region to the left of gene 3 requires at least one other origin. The region to the right of gene 32 may either use a third origin or be replicated rightward from the origin near gene 32 (Glassberg *et al.*, 1977c; Curran and Stewart, 1985a).

Sedimentation-velocity experiments show that much of the replicating SPO1 DNA is eventually found in concatemers of up to 20 genomes in length (Levner and Cozzarelli, 1972). There are two forms of concatemeric DNA, F and VF, which sediment, respectively, about three and five times as fast as mature DNA. Newly replicated DNA is found first in VF, then in F, and finally in mature DNA. VF can be converted into more slowly sedimenting material by pronase, or, less effectively, by RNase. Since much of this more slowly sedimenting material approximates the velocity of F, it appears that VF consists of concatemeric DNA in complex with protein and RNA, and that VF is converted to F by releasing the DNA from the complex (Levner and Cozzarelli, 1972).

Restriction digests of the concatemers show that they are formed by head-to-tail joining of individual genomes, with overlapping of the terminal redundancies, so there is only one redundant region present at each join (Cregg and Stewart, 1978; this laboratory, unpublished observations).

Because all known DNA polymerases require primers and polymerize only in the 5' to 3' direction, an additional mechanism must be available for synthesis of the 5' ends of linear DNA molecules. Watson (1972) suggested that it is for the purpose of providing such a mechanism that concatemers are formed, by overlap of the complementary protruding 3' ends. This hypothesis predicts that, when the first dimer has formed, the entire genome will have replicated except for the terminal redundancy involved in the join. Thus, genetic markers in the rest of the genome will be completely replicated, whereas some of those in the terminal redundancy will remain unreplicated. This is precisely the result observed when replication stops after ts32 has been shifted to restrictive temperature (Glassberg et al., 1977b; Cregg and Stewart, 1978), supporting the Watson hypothesis. The SPO1 terminal redundancy is much longer than necessary for this purpose, and its length makes the process more complicated. However, alternative explanations of the ts32 result seem even more improbable.

Apparently gp32, in addition to preventing initiation, prevents resolution of concatemers. There must be a way of resolving concatemers without initiating another round of replication, since the gene 21 mutant also prevents initiation of the second round but leaves all markers equally replicated (Glassberg et al., 1977b). Resolution of concatemers apparently requires a late function, since concatemers accumulate during sus34 infection, but does not require packaging, since it takes place in a mutant that fails to make heads (Levner, 1972b).

C. Regulation of Replication

Incorporation of labeled precursors into SPO1 DNA begins about 10 min after infection (Wilson and Gage, 1971). The capacity to replicate becomes insensitive to rifamycin at about 8 min (Gage and Geiduschek, 1971b), which is consistent with the time of synthesis of those DNA synthesis enzymes that have been tested, including dCMP deaminase (Wilson and Gage, 1971); dUMP hydroxymethxlase (Alegria *et al.*, 1968); DNA polymerase (Yehle and Ganesan, 1972); and gp30c, gp31, and gp32 (Reeve *et al.*, 1978; Heintz and Shub, 1982). Gp30a is anomalous; see footnote (*l*) to Table I.

In mixed infection of SPO1 and SP82, SP82 predominates in the burst, and the site on the genome that determines this difference is close to a replication origin (Stewart and Franck, 1981). Nothing is known about the mechanism of this predominance, but it is conceivable that it reflects a negative regulation of replication.

D. Two Clusters of Replication Genes

There are two clusters of replication genes, 21–23 and 29–32, which behave differently in a variety of seemingly unrelated ways: (1) Cold-sensitive mutations tend to fall into genes 21–23, heat–sensitive mutations into 29–32 (Glassberg *et al.*, 1977a). (2) Of 34 genes tested, seven could not be cloned on any available restriction fragments. Six of these are in a cluster that includes 21–23 (Curran and Stewart, 1985a). (3) Proteins specified by genes in the second cluster can readily be detected on polyacrylamide gels; those specified by genes in the first cluster cannot (Reeve *et al.*, 1978). (4) The SP82 dominance effect could be seen most strikingly with respect to mutations in the second cluster (Stewart and Franck, 1981). There is a ready explanation for this (see above), but its validity has not been demonstrated. (5) Mutations in the two clusters have been reported to have different effects on late transcription (Añon, quoted in Rabussay and Geiduschek, 1977). (6) In sequential complementation experiments, mutants affected in the second cluster could be rescued more effectively by infection with a second mutant than could mutants affected in the first cluster (this laboratory, unpublished observations).

There are a few exceptions to these general distinctions, and not all of the genes were tested in each case. However, there is a clear impression that there are general differences between these two clusters that would be interesting if they could be understood. The two clusters are separated by several genes not necessary for replication, and by genes 27 and 28, whose behavior does not place them in either group.

E. 2C Replication

Several aspects of replication have been studied more thoroughly in 2C than in SPO1, finding, for instance, the apparent discontinuous replication of both strands, and extensive recombination resulting in dispersion of parental DNA among progeny molecules (Hoet *et al.*, 1976, 1979).

VIII. LATE FUNCTIONS

A. Lysis

Little is known about mechanisms of lysis by SPO1. A lytic activity can be extracted from SPO1-infected cells (Okubo *et al.*, 1972), but its significance is uncertain since (1) the appearance of lytic activity in SP82-infected cells is not sufficient for lysis (Stewart *et al.*, 1971, 1972), and (2) an increase in extractable lytic activity can be evoked by introduction of any exogenous DNA, including bacterial DNA from the same strain of *B. subtilis* (Stewart and Marmur, 1970). The time of appearance of this activity has not been measured for SPO1, but it appears at late time in 2C or SP82 infection (Pene and Marmur, 1967; Stewart *et al.*, 1971). An SPO1 mutant, affected in gene 19, is deficient in lytic activity (Okubo *et al.*, 1972), but this could be due to either a direct or indirect effect on the lytic enzyme. Gene 19 has also been reported to be essential for tail formation (Table I).

B. Morphogenesis

A number of mutations affect head or tail structure (Table I), and at least 53 proteins have been identified as components of the subassemblies of the phage particle (Parker and Eiserling, 1983b). Yet there has been little integration of the genetic and physical analyses. Only one gene product has been identified as a structural protein (gp6 is part of the head: Shub, 1975), and only one step in morphogenesis has been analyzed (gp5 activity is necessary for proteolytic processing to generate the mature form of the major head protein: Parker and Eiserling, 1983b).

Presumably, the structural and morphogenetic proteins are specified by late genes, since the genes affecting head or tail structure are located primarily in a late transcription region of the genome, since the appearance of heads and tails is dependent on gp33 and 34 (Fujita *et al.*, 1971), and since the major antigenic activity of the virion appears at late times (Wilson and Gage, 1971), but no direct observation of the time of appearance, or the gp33, gp34 dependence, of individual SPO1 structural proteins has been published. Several of the SP82 structural proteins have been shown to be late proteins (Hiatt and Whiteley, 1978).

IX. MISCELLANEOUS FUNCTIONS

A. Adsorption and Penetration

Little is known about the early stages of SPO1 infection. The contractile tail (Parker and Eiserling, 1983a) presumably plays a role in injection of DNA. By marker rescue analysis of markers injected before interruption by blending, McAllister (1970) showed that the SP82 genome enters in linear fashion, with the left end (comparable to the right end of SPO1) entering first.

B. Transfection

B. subtilis cells made competent for transformation can be transfected by SPO1 DNA, yielding bursts that are substantial, although smaller than those produced by infection (Okubo *et al.*, 1964). Successful transfection requires that the same cell receive several SPO1 genomes, which undergo recombination with each other.

This phenomenon has been investigated more extensively with SP82. Transfecting DNA is apparently susceptible to a cellular nuclease which causes sufficient damage that recombination between several genomes is necessary to form a single intact genome (Green, 1966). Infection results in inhibition of this nuclease, which thus causes no problem for normal infection (McAllister and Green, 1972; King and Green, 1977). The effect of the nuclease on transfection can be diminished by the presence of UV-damaged DNA (either exogenously added or generated by irradiation of recipient cells) (Epstein and Mahler, 1968).

C. Recombination

Little is known about mechanisms of recombination or DNA repair in SPO1 infection. Mutations in the host uvr-1 or pol A genes are mildly deficient in reactivation of UV-treated SPO1 (Ferrari *et al.*, 1977), but I am not aware of any mutation of either *B. subtilis* or SPO1 having been shown to affect SPO1 recombination. The recombination frequency can be estimated to be roughly 0.001% per base pair, although the applicability of such an estimate is limited by such distorting factors as interference, nonproportionality, site-to-site variation, and variation in experimental conditions. The estimate is based on measurements of recombination frequency at two sites where the exact distance between mutations has been determined by nucleotide sequencing. There are 359 base pairs between mutations F14 and F4, in genes 33 and 34 (Costanzo *et al.*, 1984), and the measured frequency of recombination is 0.5% (Okubo *et al.*, 1972). The distance from HA20 to O33, in genes 27 and 28, respectively, is between 117 and 314 bp (Costanzo and Pero, 1983; Costanzo *et al.*, 1983; Curran and Stewart, 1985a), and the recombination frequency is 0.1%. (The distance from HA20 to F21, also in gene 28, is known precisely, but the recombination frequency has not been reported.)

X. SUMMARY

SPO1 is a fertile field for further investigation. Its fruitfulness as an experimental system has been shown by the elegant and detailed elucidation of the regulation of sequential gene action by sequential sigma substitutions. A firm groundwork for future investigation has been laid in the extensive genetic, restriction, and transcription maps that have been prepared and in the analysis of several of the functional systems that has already been done. This analysis has defined a multitude of important problems for further research, including the mechanisms of (1) the many unexplained regulatory events in the sequence of SPO1 gene action, (2) the events involved in shutoff of host activities, (3) initiation and propagation of DNA replication, (4) concatemer formation and resolution, (5) cell lysis, and (6) particle assembly.

ACKNOWLEDGMENTS. I am grateful to Richard Burgess, Peter Geiduschek, and Jan Pero for providing information prior to publication, and to Peter Geiduschek, Jan Pero, and David Shub for critical reading of the manuscript. Unpublished work from this laboratory, discussed herein, was supported by grants PCM77-24086 and PCM-79001163 from the National Science Foundation and by grant GM-28488 from the National Institutes of Health.

REFERENCES

Alegria, A. H., Kahan, F. M., and Marmur, J., 1968, A new assay for phage hydroxymethylases and its use in *Bacillus subtilis* transfection, *Biochemistry* **7**:3179.

Alonso, J. C., Sarachu, A. N., and Grau, O., 1981, DNA gyrase inhibitors block development of *Bacillus subtilis* bacteriophage SPO1, *J. Virol.* **39**:855.

Aposhian, H. V., and Tremblay, G. Y., 1966, Deoxythymidylate 5'-nucleotidase. Purification and properties of an enzyme found after infection of *B. subtilis* with phage SP5C*, *J. Biol. Chem.* **241**:5095.

Arwert, F., and Veneman, G., 1974, Transfection of *Bacillus subtilis* with bacteriophage H1 DNA: Fate of transfecting DNA and transfection enhancement in *B. subtilis* uvr$^+$ and uvr$^-$ strains, *Mol. Gen. Genet.* **128**:55.

Brennan, S. M., 1984, Ribonucleoside triphosphate concentration-dependent termination of bacteriophage SPO1 transcription *in vitro* by *B. subtilis* RNA polymerase, *Virology* **135**:555.

Brennan, S. M., and Geiduschek, E. P., 1983, Regions specifying transcriptional termination and pausing in the bacteriophage SPO1 terminal repeat, *Nucleic Acids Res.* **11**:4157.

Brennan, S. M., Chelm, B. K., Romeo, J. M., and Geiduschek, E. P., 1981, A transcriptional map of the bacteriophage SPO1 genome. II. The major early transcription units, *Virology* **111**:604.

Brown, N. C., 1970, 6-(p-Hydroxyphenylazo)-uracil: A selective inhibitor of host DNA replication in phage-infected *Bacillus subtilis*, *Proc. Natl. Acad. Sci. USA* **67**:1454.

Chelm, B. K., Romeo, J. M., Brennan, S. M., and Geiduschek, E. P., 1981a, A transcriptional map of the bacteriophage SPO1 genome. III. A region of early and middle promoters (the gene 28 region), *Virology* **112**:572.

Chelm, B. K., Beard, C., and Geiduschek, E. P., 1981b, Changes in the association between

Bacillus subtilis RNA polymerase core and two specificity-determining subunits during transcription, *Biochemistry* **20**:6564.

Chelm, B. K., Duffy, J. J., and Geiduschek, E. P., 1982a, Interaction of *Bacillus subtilis* RNA polymerase core with two specificity-determining subunits, *J. Biol. Chem.* **257**:6501.

Chelm, B. K., Greene, J. R., Beard, C., and Geiduschek, E. P., 1982b, The transition between early and middle gene expression in the development of phage SPO1: Physiological and biochemical aspects, in: *Molecular Cloning and Gene Regulation in Bacilli* (A. T. Ganesan, S. Chang, and J. A. Hoch, eds.), pp. 345–395, Academic Press, New York.

Chiu, C.-S., Tomich, P. K., and Greenberg, G. R., 1976, Simultaneous initiation of synthesis of bacteriophage T4 DNA and of deoxyribonucleotides, *Proc. Natl. Acad. Sci. USA* **73**:757.

Choy, H. A., Romeo, J. M., and Geiduschek, E. P., 1986, Activity of a phage-modified RNA polymerase at hybrid promoters: Effects of substituting thymine for hydroxymethyluracil in a phage SPO1 middle promoter, *J. Mol. Biol.* **191**:59.

Coene, M., Hoet, P., and Cocito, C., 1983, Physical map of phage 2C DNA: Evidence for the existence of large redundant ends, *Eur. J. Biochem.* **132**:69.

Costanzo, M. C., 1983, Bacteriophage SPO1 regulatory genes, Ph.D. Thesis, Harvard University, Cambridge, MA.

Costanzo, M., and Pero, J., 1983, Structure of a *Bacillus subtilis* bacteriophage SPO1 gene encoding a RNA polymerase φ factor, *Proc. Natl. Acad. Sci. USA* **80**:1236.

Costanzo, M., and Pero, J., 1984, Overproduction and purification of a bacteriophage SPO1-encoded RNA polymerase sigma factor, *J. Biol. Chem.* **259**:6681.

Costanzo, M., Hannett, N., Brzustowicz, L., and Pero, J., 1983, Bacteriophage SPO1 gene 27: Location and nucleotide sequence, *J. Virol.* **48**:555.

Costanzo, M., Brzustowicz, L., Hannett, N., and Pero, J., 1984, Bacteriophage SPO1 genes 33 and 34. Location and primary structure of genes encoding regulatory subunits of *Bacillus subtilis* RNA polymerase, *J. Mol. Biol.* **180**:533.

Cregg, J. M., and Stewart, C. R., 1977, Timing of initiation of DNA replication in SPO1 infection of *Bacillus subtilis*, *Virology* **80**:289.

Cregg, J. M., and Stewart, C. R., 1978, Terminal redundancy of "high frequency of recombination" markers of *Bacillus subtilis* phage SPO1, *Virology* **86**:530.

Curran, J. F., and Stewart, C. R., 1985a, Cloning and mapping of the SPO1 genome, *Virology* **142**:78.

Curran, J. F., and Stewart, C. R., 1985b, Transcription of *Bacillus subtilis* plasmid pBD64 and expression of bacteriophage SPO1 genes cloned therein, *Virology* **142**:98.

Davison, P. F., 1963, The structure of bacteriophage SP8, *Virology* **21**:146.

Davison, P. F., Freifelder, D., and Holloway, B. W., 1964, Interruptions in the polynucleotide strands in bacteriophage DNA, *J. Mol. Biol.* **8**:1.

DeAntoni, G. L., Besso, N. E., Zanassi, G. E., Sarachu, A. N., and Grau, O., 1985, Bacteriophage SPO1 DNA polymerase and the activity of viral gene 31, *Virology* **143**:16.

Downard, J. S., and Whiteley, H. R., 1981, Early RNAs in SP82- and SPO1-infected *Bacillus subtilis* may be processed, *J. Virol.* **37**:1075.

Duffy, J. J., and Geiduschek, E. P., 1973, Transcription specificity of an RNA polymerase fraction from bacteriophage SPO1-infected *B. subtilis*, *FEBS Lett.* **34**:172.

Duffy, J. J., and Geiduschek, E. P., 1975, RNA polymerase from phage SPO1-infected and uninfected *Bacillus subtilis*, *J. Biol. Chem.* **250**:4530.

Duffy, J. J., and Geiduschek, E. P., 1977, Purification of a positive regulatory subunit from phage SPO1-modified RNA polymerase, *Nature* **270**:28.

Duffy, J. J., Petrusek, R. L., and Geiduschek, E. P., 1975, Conversion of *Bacillus subtilis* RNA polymerase activity *in vitro* by a protein induced by phage SPO1, *Proc. Natl. Acad. Sci. USA* **72**:2366.

Dunham, L. T., and Price, A. R., 1974a, Deoxythymidine triphosphate-deoxyuridine triphosphate nucleotidohydrolase induced by *Bacillus subtilis* bacteriophage φe, *Biochemistry* **13**:2667.

Dunham, L. T., and Price, A. R., 1974b, Mutants of *Bacillus subtilis* bacteriophage φe defective in dTTP-dUTP nucleotidohydrolase, *J. Virol.* **14**:709.

Epstein, H. T., and Mahler, I., 1968, Mechanisms of enhancement of SP82 transfection. *J. Virol.* **2**:710.

Ferrari, E., Siccardi, A. G., Galizzi, A., Canosi, U., and Mazza, G., 1977, Host cell reactivation of *Bacillus subtilis* bacteriophages, *J. Bacteriol.* **131**:382.

Fox, T. D., 1976, Identification of phage SPO1 proteins coded by regulatory genes 33 and 34, *Nature* **262**:748.

Fox, T. D., and Pero, J., 1974, New phage-SPO1-induced polypeptides associated with *Bacillus subtilis* RNA polymerase, *Proc. Natl. Acad. Sci. USA* **71**:2761.

Fox, T. D., Losick, R., and Pero, J., 1976, Regulatory gene 28 of bacteriophage SPO1 codes for a phage-induced subunit of RNA polymerase, *J. Mol. Biol.* **101**:427.

Fujita, D. J., 1971, Studies on conditional lethal mutants of bacteriophage SPO1, Ph.D. Thesis, University of Chicago, Chicago.

Fujita, D. J., Ohlsson-Wilhelm, B. M., and Geiduschek, E. P., 1971, Transcription during bacteriophage SPO1 development: Mutations affecting the program of viral transcription, *J. Mol. Biol.* **57**:301.

Gage, L. P., and Fujita, D. J., 1969, Effect of nalidixic acid on deoxyribonucleic acid synthesis in bacteriophage SPO1-infected *Bacillus subtilis*, *J. Bacteriol.* **98**:96.

Gage, L. P., and Geiduschek, E. P., 1971a, RNA synthesis during bacteriophage SPO1 development: Six classes of SPO1 RNA, *J. Mol. Biol.* **57**:279.

Gage, L. P., and Geiduschek, E. P., 1971b, RNA synthesis during bacteriophage SPO1 development. II. Some modulations and prerequisites of the transcription program, *Virology* **44**:200.

Geiduschek, E. P., and Ito, J., 1982, Regulatory mechanisms in the development of lytic bacteriophages in *Bacillus subtilis*, in: *The Molecular Biology of Bacilli* I (D. A. Dubnau, ed.), pp. 203–245, Academic Press, New York.

Geiduschek, E. P., and Sklar, J., 1969, Continual requirement for a host RNA polymerase component in a bacteriophage development, *Nature (Lond.)* **221**:833.

Geiduschek, E. P., Armelin, M. C. S., Petrusek, R., Beard, C., Duffy, J. J., and Johnson, G. G., 1977, Effects of the transcription inhibitory protein, TF1, on phage SPO1 promoter complex formation and stability, *J. Mol. Biol.* **117**:825.

Gitt, M. A., Wang, L.-F., and Doi, R. H., 1985, A strong sequence homology exists between the major RNA polymerase σ factors of *B. subtilis* and *E. coli*, *J. Biol. Chem.* **260**:7178.

Glassberg, J., Slomiany, R. A., and Stewart, C. R., 1977a, Selective screening procedure for the isolation of heat- and cold-sensitive, DNA replication-deficient mutants of bacteriophage SPO1 and preliminary characterization of the mutants isolated, *J. Virol.* **21**:54.

Glassberg, J., Franck, M., and Stewart, C. R., 1977b, Initiation and termination mutants of *Bacillus subtilis* bacteriophage SPO1, *J. Virol.* **21**:147.

Glassberg, J., Franck, M., and Stewart, C. R., 1977c, Multiple origins of replication for *Bacillus subtilis* phage SPO1, *Virology* **78**:433.

Green, D. M., 1964, Infectivity of DNA isolated from *Bacillus subtilis* bacteriophage SP82, *J. Mol. Biol.* **10**:438.

Green, D. M., 1966, Intracellular inactivation of infective SP82 bacteriophage DNA, *J. Mol. Biol.* **22**:1.

Green, D. M., and Laman, D., 1972, Organization of gene function in *Bacillus subtilis* bacteriophage SP82G, *J. Virol.* **9**:1033.

Greene, J. R., and Geiduschek, E. P., 1985a, Site-specific DNA binding by the bacteriophage SPO1-encoded type II DNA binding protein, *EMBO J.* **4**:1345.

Greene, J. R., and Geiduschek, E. P., 1985b, Interaction of a virus-coded type II DNA binding protein, in: *Sequence Specificity in Transcription and Translation* (R. Calendar, ed.), pp. 255–269, Alan R. Liss, New York.

Greene, J. R., Chelm, B. K., and Geiduschek, E. P., 1982, SPO1 gene 27 is required for viral late transcription, *J. Virol.* **41**:715.

Greene, J. R., Brennan, S. M., Andrew, D. J., Thompson, C. C., Richards, S. H., Heinrikson, R. L., and Geiduschek, E. P., 1984, Sequence of bacteriophage SPO1 gene coding for transcription factor 1, a viral homologue of the bacterial type II DNA-binding proteins, *Proc. Natl. Acad. Sci. USA* **81**:7031.

Greene, J. R., Morrissey, L. M., Foster, L. M., and Geiduschek, E. P., 1986, DNA binding by the bacteriophage SPO1-encoded type II DNA-binding protein, TF1: Formation of nested complexes at a selective binding site, *J. Biol. Chem.* **261**:12820.

Greene, J. R., Appelt, K., and Geiduschek, E. P., 1987, Prokaryotic chromatin: Site-selective and genome-specific DNA-binding by a virus-coded type II DNA-binding protein, in: *DNA: Protein Interactions and Gene Regulation* (E. B. Thompson and J. Papaconstantinou, eds.), pp. 57–65, University of Texas Press, Austin.

Gribskov, M., and Burgess, R. R., 1986, Sigma proteins from *E. coli*, *B. subtilis*, phage SPO1 and T4 are homologous proteins, *Nucleic Acids Res.* **14**:6745.

Haslam, E. A., Roscoe, D. H., and Tucker, R. G., 1967, Inhibition of thymidylate synthetase in bacteriophage-infected *Bacillus subtilis*, *Biochim. Biophys. Acta* **134**:312.

Heintz, N., and Shub, D. A., 1982, Transcriptional regulation of bacteriophage SPO1 protein synthesis *in vivo* and *in vitro*, *J. Virol.* **42**:951.

Hemphill, H. E., and Whiteley, H. R., 1975, Bacteriophages of *Bacillus subtilis*, *Bacteriol. Rev.* **39**:257.

Hiatt, W. R., and Whiteley, H. R., 1978, Translation of RNAs synthesized *in vivo* and *in vitro* from bacteriophage SP82 DNA, *J. Virol.* **25**:616.

Hoet, P., Fraselle, G., and Cocito, C., 1976, Recombinational-type transfer of viral DNA during bacteriophage 2C replication in *Bacillus subtilis*, *J. Virol.* **17**:718.

Hoet, P. P., Fraselle, G., and Cocito, C., 1979, Discontinuous duplication of both strands of virus 2C DNA, *Mol. Gen. Genet.* **171**:43.

Hoet, P., Coene, M., and Cocito, C., 1983, Comparison of the physical maps and redundant ends of the chromosomes of phages 2C, SPO1, SP82 and φe, *Eur. J. Biochem.* **132**:63.

Johnson, G. G., and Geiduschek, E. P., 1972, Purification of the bacteriophage SPO1 transcription factor 1, *J. Biol. Chem.* **247**:3571.

Johnson, G. G., and Geiduschek, E. P., 1977, Specificity of the weak binding between the phage SPO1 transcription-inhibitory protein, TF1, and SPO1 DNA, *Biochemistry* **16**:1473.

Johnson, W. C., Moran, C. P., and Losick, R., 1983, Two RNA polymerase sigma factors from *Bacillus subtilis* discriminate between overlapping promoters for a developmentally related gene. *Nature* **302**:800.

Kahan, E., 1966, A genetic study of temperature-sensitive mutants of the *subtilis* phage SP82, *Virology* **30**:650.

Kahan, E., 1971, Early and late gene function in bacteriophage SP82, *Virology* **46**:634.

Kallen, R. G., Simon, M., and Marmur, J., 1962, The occurrence of a new pyrimidine base replacing thymine in a bacteriophage DNA: 5-hydroxymethyl uracil, *J. Mol. Biol.* **5**:248.

King, J. J., and Green, D. M., 1977, Inhibition of nuclease activity in *Bacillus subtilis* following infection with bacteriophage SP82G, *Biochem. Biophys. Res. Commun.* **74**:492.

Kornberg, A., 1980, *DNA Replication*, W. H. Freeman, San Francisco.

Kunitani, M. G., and Santi, D. V., 1980, On the mechanism of 2'-deoxyuridylate hydroxymethylase, *Biochemistry* **19**:1271.

Lavi, U., Nattenberg, A., Ronen, A., and Marcus, M., 1974, *Bacillus subtilis* DNA polymerase III is required for the replication of virulent bacteriophage φe, *J. Virol.* **14**:1337.

Lawrie, J. M., and Whiteley, H. R., 1977, A physical map of bacteriophage SP82 DNA, *Gene* **2**:233.

Lawrie, J. M., Downard, J. S., and Whiteley, H. R., 1978, *Bacillus subtilis* bacteriophages SP82, SPO1 and φe: A comparison of DNAs and of peptides synthesized during infection, *J. Virol.* **27**:725.

Lee, G., and Pero, J., 1981, Conserved nucleotide sequences in temporally controlled bacteriophage promoters, *J. Mol. Biol.* **152**:247.

Lee, G., Talkington, C., and Pero, J., 1980a, Nucleotide sequence of a promoter recognized by *Bacillus subtilis* RNA polymerase, *Mol. Gen. Genet.* **180**:57.

Lee, G., Hannett, N. M., Korman, A., and Pero, J., 1980b, Transcription of cloned DNA from

Bacillus subtilis phage SPO1. Requirement for hydroxymethyluracil-containing DNA by phage-modified RNA polymerase, *J. Mol. Biol.* **139**:407.

Levinthal, C., Hosoda, J., and Shub, D., 1967, The control of protein synthesis after phage infection, in: *The Molecular Biology of Viruses* (J. S. Colter and W. Paranchych, eds.), pp. 71–87, Academic Press, New York.

Levner, M. H., 1972a, Replication of viral DNA in SPO1-infected *Bacillus subtilis*. II. DNA maturation during aborative infection, *Virology* **48**:417.

Levner, M. H., 1972b, Eclipse of viral DNA infectivity in SPO1-infected *Bacillus subtilis*, *Virology* **5n**:267.

Levner, M. H., and Cozzarelli, N. R., 1972, Replication of viral DNA in SPO1-infected *Bacillus subtilis*. I. Replicative intermediates, *Virology* **48**:402.

Liljemark, W. F., and Anderson, D. L., 1970, Structure of *Bacillus subtilis* bacteriophage φ25 and φ25 deoxyribonucleic acid, *J. Virol.* **6**:107.

Losick, R., and Pero, J., 1981, Cascades of sigma factors. *Cell* **25**:582.

Marcus, M., and Newlon, M. C., 1971, Control of DNA synthesis in *Bacillus subtilis* by φe, *Virology* **44**:83.

Marmur, J., and Greenspan, C. M., 1963, Transcription *in vivo* of DNA from bacteriophage SP8, *Science* **142**:387.

Marmur, J., Greenspan, C. M., Palecek, E., Kahan, F. M., Levine, J., and Mandel, M., 1963, Specificity of the complementary RNA formed by *Bacillus subtilis* infected with bacteriophage SP8, *Cold Spring Harbor Symp. Quant. Biol.* **28**:191.

McAllister, W. T., 1970, Bacteriophage infection: Which end of the SP82G genome goes in first? *J. Virol.* **5**:194.

McAllister, W. T., and Green, D. M., 1972, Bacteriophage SP82G inhibition of intracellular deoxyribonucleic acid inactivation process in *Bacillus subtilis*, *J. Virol.* **10**:51.

Nishihara, M., Chrambach, A., and Aposhian, H. V., 1967, The deoxycytidylate deaminase found in *Bacillus subtilis* infected with phage SP8, *Biochemistry* **6**:1877.

Okubo, S., Strauss, B., and Stodolsky, M., 1964, The possible role of recombination in the infection of competent *Bacillus subtilis* by bacteriophage deoxyribonucleic acid, *Virology* **24**:552.

Okubo, S., Yanagida, T., Fujita, D. J., and Ohlsson-Wilhelm, B. M., 1972, The genetics of bacteriophage SPO1, *Biken J.* **15**:81.

Panganiban, A. T., and Whiteley, H. R., 1981, Analysis of bacteriophage SP82 major "early" *in vitro* transcripts, *J. Virol.* **37**:372.

Panganiban, A. T., and Whiteley, H. R., 1983a, Purification and properties of a new *Bacillus subtilis* RNA processing enzyme, *J. Biol. Chem.* **258**:12487.

Panganiban, A. T., and Whiteley, H. R., 1983b, *Bacillus subtilis* RNAase III cleavage sites in phage SP82 early mRNA, *Cell* **33**:907.

Parker, M. L., 1979, Structure and composition of *Bacillus subtilis* bacteriophage SPO1, Ph.D. Thesis, University of California, Los Angeles.

Parker, M. L., and Eiserling, F. A., 1983a, Bacteriophage SPO1 structure and morphogenesis. I. Tail structure and length regulation, *J. Virol.* **46**:239.

Parker, M. L., and Eiserling, F. A., 1983b, Bacteriophage SPO1 structure and morphogenesis. III. SPO1 proteins and synthesis, *J. Virol.* **46**:260.

Parker, M. L., Ralston, E. J., and Eiserling, F. A., 1983, Bacteriophage SPO1 structure and morphogenesis. II. Head structure and DNA size, *J. Virol.* **46**:250.

Pene, J. J., and Marmur, J., 1967, Deoxyribonucleic acid replication and expression of early and late bacteriophage functions in *Bacillus subtilis*, *J. Virol.* **1**:86.

Perkus, M. E., and Shub, D. A., 1985, Mapping the genes in the terminal redundancy of bacteriophage SPO1 with restriction endonucleases, *J. Virol.* **56**:40.

Pero, J., 1983, A procaryotic model for the developmental control of gene expression, in: *Gene Structure and Regulation in Development* (S. Subtelny and F. C. Kafatos, eds.), pp. 227–233, Alan R. Liss, New York.

Pero, J., Tjian, R., Nelson, J., and Losick, R., 1975, *In vitro* transcription of a late class of phage SPO1 genes, *Nature* **257**:248.

Pero, J., Hannett, N. M., and Talkington, C., 1979, Restriction cleavage map of SPO1 DNA: General location of early, middle, and late genes, *J. Virol.* **31**:156.

Piggot, P. J., and Hoch, J. A., 1985, Revised genetic linkage map of *Bacillus subtilis, Microbiol. Rev.* **49**:158.

Price, A. R., Dunham, L. F., and Walker, R. L., 1972, Thymidine triphosphate nucleotidohydrolase and deoxyuridylate hydroxymethylase induced by mutants of *Bacillus subtilis* bacteriophage SP82G, *J. Virol.* **10**:1240.

Rabussay, D., and Geiduschek, E. P., 1977, Regulation of gene action in the development of lytic bacteriophages, in: *Comprehensive Virology 8* (H. Fraenkel-Conrat and R. R. Wagner, eds.), pp. 1–196, Plenum Press, New York.

Ratner, D., 1974, The interaction of bacterial and phage proteins with immobilized *Escherichia coli* RNA polymerase, *J. Mol. Biol.* **88**:373.

Reeve, J. N., Mertens, G., and Amann, E., 1978, Early development of bacteriophages SPO1 and SP82G in minicells of *Bacillus subtilis, J. Mol. Biol.* **120**:183.

Romeo, J. M., Brennan, S. M., Chelm, B. K., and Geiduschek, E. P., 1981, A transcriptional map of the bacteriophage SPO1 genome. I. The major early promoters, *Virology* **111**:588.

Roscoe, D. H., 1969a, Synthesis of DNA in phage-infected *B. subtilis, Virology* **38**:527.

Roscoe, D. H., 1969b, Thymidine triphosphate nucleotidohydrolase: A phage-induced enzyme in *Bacillus subtilis, Virology* **38**:520.

Roscoe, D. H., and Tucker, R. G., 1964, The biosynthesis of a pyrimidine replacing thymine in bacteriophage DNA, *Biochem. Biophys. Res. Commun.* **16**:106.

Rosenberg, M., and Court, D., 1979, Regulatory sequences involved in the promotion and termination of RNA transcription. *Annu. Rev. Gen.* **13**:319.

Sarachu, A. N., Añon, M. C., and Grau, O., 1978, Bacteriophage SPO1 development: Defects in a gene 31 mutant, *J. Virol.* **27**:483.

Sarachu, A. N., Alonso, J. C., and Grau, O., 1980, Novobiocin blocks the shutoff of SPO1 early transcription, *Virology* **105**:13.

Shub, D. A., 1966, Functional stability of messenger RNA during bacteriophage development, Ph.D. Thesis, Massachusetts Institute of Technology, Cambridge, MA.

Shub, D. A., 1975, Nature of the suppressor of *Bacillus subtilis* HA101B, *J. Bacteriol.* **122**:788.

Shub, D. A., and Johnson, G. G., 1975, Bacteriophage SPO1 DNA- and RNA-directed protein synthesis *in vitro:* The effect of TF-1, a template-selective transcription inhibitor, *Mol. Gen. Genet.* **137**:161.

Shub, D. A., Swanton, M., and Smith, D. H., 1979, The nature of transcription selectivity of bacteriophage SPO1-modified RNA polymerase, *Mol. Gen. Genet.* **172**:193.

Sinha, N. K., Morris, C. F., and Alberts, B. M., 1980, Efficient *in vitro* replication of double-stranded DNA templates by purified T4 bacteriophage replication system, *J. Biol. Chem.* **255**:4290.

Spiegelman, G. B., and Whiteley, H. R., 1974, *In vivo* and *in vitro* transcription by ribonucleic acid polymerase from SP82-infected *Bacillus subtilis, J. Biol. Chem.* **249**:1483.

Stetler, G. L., King, G. J., and Huang, W. M., 1979, T4 DNA-delay proteins, required for specific DNA replication, form a complex that has ATP-dependent DNA topoisomerase activity, *Proc. Natl. Acad. Sci. USA* **76**:3737.

Stewart, C. R., 1984, Dissection of HA20, a double mutant of bacteriophage SPO1, *J. Virol.* **49**:300.

Stewart, C. R., and Franck, M., 1981, Predominance of bacteriophage SP82 over bacteriophage SPO1 in mixed infections of *Bacillus subtilis, J. Virol.* **38**:1081.

Stewart, C. R., and Marmur, J., 1970, Increase in lytic activity in competent cells of *Bacillus subtilis* after uptake of deoxyribonucleic acid, *J. Bacteriol.* **101**:449.

Stewart, C. R., Cater, M., and Click, B., 1971, Lysis of *Bacillus subtilis* by bacteriophage SP82 in the absence of DNA synthesis, *Virology* **46**:327.

Stewart, C. R., Click, B., and Tole, M. F., 1972, DNA replication and late protein synthesis during SP82 infection of *Bacillus subtilis*, *Virology* **50**:653.

Stewart, G. C., and Bott, K. F., 1983, DNA sequence of the tandem ribosomal RNA promoter for *B. subtilis* operon rrnB, *Nucleic Acids Res.* **11**:6289.

Sugino, A., and Bott, K. F., 1980, *Bacillus subtilis* deoxyribonucleic acid gyrase, *J. Bacteriol.* **141**:1331.

Swanton, M., Smith, D. H., and Shub, D. A., 1975, Synthesis of specific functional messenger RNA *in vitro* by phage-SPO1 modified RNA polymerase of *Bacillus subtilis*, *Proc. Natl. Acad. Sci. USA* **72**:4886.

Talkington, C., and Pero, J., 1977, Restriction fragment analysis of temporal program of bacteriophage SPO1 transcription and its control by phage-modified RNA polymerases, *Virology* **83**:365.

Talkington, C., and Pero, J., 1978, Promoter recognition by phage SPO1-modified RNA polymerase, *Proc. Natl. Acad. Sci. USA* **75**:1185.

Talkington, C., and Pero, J., 1979, Distinctive nucleotide sequences of promoters recognized by RNA polymerase containing a phage-coded "σ-like" protein, *Proc. Natl. Acad. Sci. USA* **76**:5465.

Tanaka, I., Appelt, K., Dijk, J., White, S. W., and Wilson, K. S., 1984, 3-A° resolution of a protein with histone-like properties in prokaryotes, *Nature* **310**:376.

Tomich, P. K., Chiu, C.-S., Wovcha, M. G., and Greenberg, G. R., 1974, Evidence for a complex regulating the *in vivo* activation of early enzymes induced by bacteriophage T4, *J. Biol. Chem.* **249**:7613.

Truffaut, N., Revet, B., and Soulie, M. O., 1970, Etude comparative des DNA de phages 2C, SP8*, SP82, φe, SPO1 et SP50, *J. Biochem.* **15**:391.

Watson, J. D., 1972, Origin of concatemeric T7 DNA, *Nature New Biol.* **239**:197.

Webb, V. B., and Spiegelman, G. B., 1984, Ribosomal RNA synthesis in uninfected and SPO1am 34 infected *B. subtilis*, *Mol. Gen. Genet.* **194**:98.

Wilhelm, J. M., Johnson, G., Haselkorn, R., and Geiduschek, E. P., 1972, Specific inhibition of bacteriophage SPO1 DNA-directed protein synthesis by the SPO1 transcription factor, TF1, *Biochem. Biophys. Res. Commun.* **46**:1970.

Wilson, D. L., and Gage, L. P., 1971, Certain aspects of SPO1 development, *J. Mol. Biol.* **57**:297.

Wilson, D. L., and Geiduschek, E. P., 1969, A template-selective inhibitor of *in vitro* transcription, *Proc. Natl. Acad. Sci. USA* **62**:514.

Wu, R., and Geiduschek, E. P., 1975, The role of replication proteins in the regulation of bacteriophage T4 transcription, *J. Mol. Biol.* **96**:513.

Yansura, D. G., and Henner, D. J., 1984, Use of the *Escherichia coli lac* repressor and operator to control gene expression in *Bacillus subtilis*, *Proc. Natl. Acad. Sci. USA* **81**:439.

Yehle, C. O., and Ganesan, A. T., 1972, Deoxyribonucleic acid synthesis in bacteriophage SPO1-infected *Bacillus subtilis*. I. Bacteriophage deoxyribonucleic acid synthesis and fate of host deoxyribonucleic acid in normal and polymerase-deficient strains, *J. Virol.* **9**:263.

Yehle, C. O., and Ganesan, A. T., 1973, Deoxyribonucleic synthesis in bacteriophage SPO1-infected *Bacillus subtilis*. II. Purification and catalytic properties of a deoxyribonucleic acid polymerase induced after infection, *J. Biol. Chem.* **248**:7456.

CHAPTER 12

Viruses of Archaebacteria

WOLFRAM ZILLIG, WOLF-DIETER REITER,
PETER PALM, FELIX GROPP, HORST NEUMANN,
AND MICHAEL RETTENBERGER

I. INTRODUCTION

The archaebacteria constitute the third distinct urkingdom of life, beside eubacteria and eucytes (eukaryotic nucleus and cytoplasm) (Woese and Fox, 1977; Woese et al., 1978; Fox et al., 1980). They exhibit a characteristic mosaic of features, some of them—e.g., their lipids—unique to the group (for review see Langworthy, 1985); others—e.g., the organization of genes in operons (Konheiser et al., 1984; Hamilton and Reeve, 1985; Reeve et al., 1986; Reiter et al., 1987a) and the existence of ribosome-binding sites in mRNAs (Reiter et al., 1987a, and literature cited therein)—of eubacterial quality; and a third type—e.g., the ADP ribosylatability of their EFIIs by diphtheria toxin (Kessel and Klink, 1982)—of eukaryotic quality. Most interestingly, features of a fourth group—e.g., the structures of 5S rRNAs, initiator tRNAs, and DNA-dependent RNA polymerases and the occurrence of introns in tRNA genes—are highly divergent in different archaebacteria (Zillig et al., 1985a). Phylogenetically, the archaebacterial kingdom is deeply divided into three major branches (Woese and Olsen, 1986; Klenk et al., 1986): (1) the methanogens (Methanococcales, Methanobacteriales, and Methanomicrobiales) (for review see Whitman, 1985) plus extreme halophiles (Halobacteriales and Thermoplasmales) (for review see Kushner, 1985); (2) the sulfur-dependent extremely thermophilic Thermococcales (Woese and Olsen, 1986; Zillig et al., 1987); and (3) the sulfur-dependent, extremely thermophilic Thermoproteales plus Sulfolobales (for review see Stetter and Zillig, 1985) (Fig. 1).

WOLFRAM ZILLIG, WOLF-DIETER REITER, PETER PALM, FELIX GROPP, HORST NEUMANN, AND MICHAEL RETTENBERGER • Max-Planck-Institut für Biochemie, 8033 Martinsried, Federal Republic of Germany.

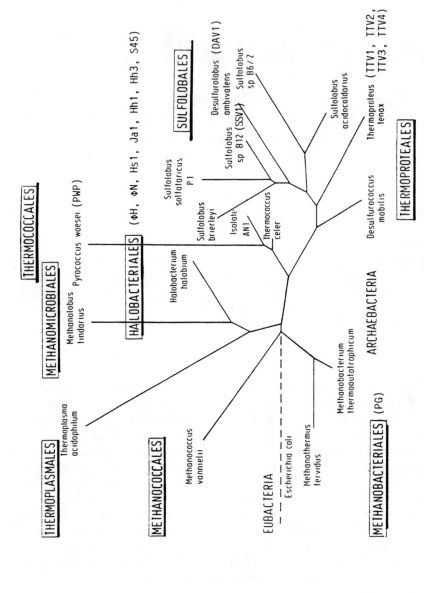

FIGURE 1. Phylogenetic tree of archaebacteria on the basis of DNA-rRNA cross-hybridization data (Klenk *et al.*, 1987). The known archaebacterial viruses are listed at the phylogenetic positions of their hosts.

Through their pattern of features, archaebacteria increase our perspective for viewing the early branching in evolution. Furthermore, some of them (e.g., *Thermococcus*) appear extremely primitive and thus testify to the nature of early forms of life (Woese and Olsen, 1986).

Investigation of virus-host systems in the other kingdoms has been of paramount importance in our understanding of genome and gene organization, signal structures, and control mechanisms in gene expression. The nucleic acids of viruses have served as templates in developing and studying *in vitro* gene expression. Therefore, the availability of archaebacterial viruses for the comparative molecular biological analysis of genetic machineries of archaebacteria appears highly desirable.

Two viruses of extremely halophilic archaebacteria, phage Hs 1 of *Halobacterium salinarium* (Torsvik, 1982; Torsvik and Dundas, 1974, 1978, 1980) and *Halobacterium* phage Ja 1 (Wais *et al.*, 1975), had been isolated before their hosts were recognized as archaebacteria. In this branch of the archaebacteria including the methanogens and halophiles, however, the virus studied in most detail on the molecular level has been phage ϕH of *Halobacterium halobium* (Schnabel *et al.*, 1982a). Other viruses of this branch include the *Halobacterium* phages Hh 1 and Hh 3 (Pauling, 1982), S 45 (Daniels and Wais, 1984), ϕN (Vogelsang-Wenke, 1984; Vogelsang-Wenke and Oesterhelt, 1986), and phage PG of an isolate of *Methanobrevibacter smithii* (Bertani and Baresi, 1986). A viruslike particle of *Methanococcus* sp. has recently been discovered (A. Wood and J. Konisky, personal communication).

For the Thermococcales, only a prospective viruslike particle produced by *Pyrococcus woesei* (Zillig *et al.*, 1987) has been discovered, and this requires further analysis. For the Thermoproteales plus Sulfolobales, the viruses TTV 1, 2, 3, and 4 of *Thermoproteus tenax* (Janekovic *et al.*, 1983) and the viruslike particles SSV 1 of *Sulfolobus* sp. B12 (Martin *et al.*, 1984) and DAV 1 of *Desulfurolobus ambivalens* (Zillig *et al.*, 1986a) have been discovered in novel isolates.

A short overview (Zillig *et al.*, 1986b) and a detailed review (Reiter *et al.*, 1987b) of archaebacterial viruses have already been published.

II. VIRUSES OF EXTREMELY HALOPHILIC AND METHANOGENIC ARCHAEBACTERIA

A. Origin

Halobacteriophages have been isolated from natural high salinity habitats (Ja 1, S 45), but also from fermented anchovy sauce (Hh 1 and Hh 3). Halophage Hs 1 was found in a culture of *H. salinarium* strain 1 18 years after the isolation of this strain from salted codfish. The phages ϕH and ϕN appeared "spontaneously" in laboratory cultures of *H. halobium* strains. The *Methanobrevibacter smithii* phage PG and its host were

TABLE I. Bacteriophages of Extremely Halophilic and Methanogenic Archaebacteria

	Authors	Head diameter (nm)	Tail length (nm)	Contractile tail?	DNA kbp	Per-cent G+C	Proteins Major	Proteins Minor	Eclipse	Latency period (h)	Burst size
ΦH	Schnabel et al., 1982a	64	170	+	59	64	3 (80, 46, 27 kd)	10	5.5	7	170
ΦN	Vogelsang-Wenke and Oesterhelt, 1986	55	80	−	56	70	1 (53 kd)	12	10	14	400
Hs1	Torsvik and Dundas, 1974, 1978	50	120	+	?[b]	?	?		12	17	324 ± 134
Ja1	Wais et al., 1975	90[a]	150[a]	−	230	?	?		2	6	140
Hh1	Pauling, 1982	60	100	−	37.2	67	1 (35 kd)	Several (5)	6	12	1100 (in 2 cycles)[c]
Hh3	Pauling, 1982	75[a]	50	?	29.4	62	1 (47 kd)	Several (4)	5	8	425
S45	Daniels and Wais, 1984	40[a]	70[a]	?	?	?	?		?	?	1300[c]
P6	Bertani and Baresi, 1986	?	?	?	50		?		?	7–9	26

[a] Measured by reviewer from figure in description paper.
[b] Data not available.
[c] Persistent infection (cells survive phage liberation at least partially).

isolated from rumen fluid. For a listing of all phages of this group including references, see Table I.

B. Morphology

All viruses of extremely halophilic archaebacteria known to date and phage PG of a rumen isolate classified as *Methanobrevibacter smithii* have isometric, icosahedric heads, the sizes of which differ considerably, and tails, which only for φH and Hs 1 have been described as contractile (Table I; Fig. 2). Tail fibers have been seen in φH and φN; a collar has been observed in Hh 3.

C. DNA

All these viruses resemble certain *E. coli* phages—e.g., λ or the T series in containing double-stranded linear DNAs, ranging in size from about 30 to 230 kbp (Table I). The G+C contents of DNAs of *Halobacterium* phages were found to vary between 62% and 70% (Table I).

D. Proteins

Similar to the protein patterns of eubacteriophages of comparable morphology, the SDS–PAGE patterns of the proteins of halobacteriophages are complex, exhibiting 1–3 major and in some cases more than 12 minor bands (Table I).

E. Ion Requirements

With two remarkable exceptions, the ion requirements of halobacteriophages parallel those of their hosts that live in habitats of high salinity. Phage Ja 1, for example, requires 3.5 M NaCl and 0.05 M Mg^{2+}; phage φH requires 3.5 M NaCl or KCl in the absence of 0.1 M Mg^{2+} and (similar to phage Hh 3) at least 1 M NaCl or KCl in its presence. The exceptions are phage Hh 1, which is stable in a low-salt medium (Pauling, 1982), and phage φN, which has a half-life of 14 hr in distilled water (Vogelsang-Wenke, 1984; Vogelsang-Wenke and Oesterhelt, 1986).

F. Phage-Host Relationships

The host ranges of different halobacteriophages differ considerably. In view of the probable identity of the species *H. halobium, H. salinarium,* and *H. cutirubrum* (Pfeifer, 1987), and since the possibility of

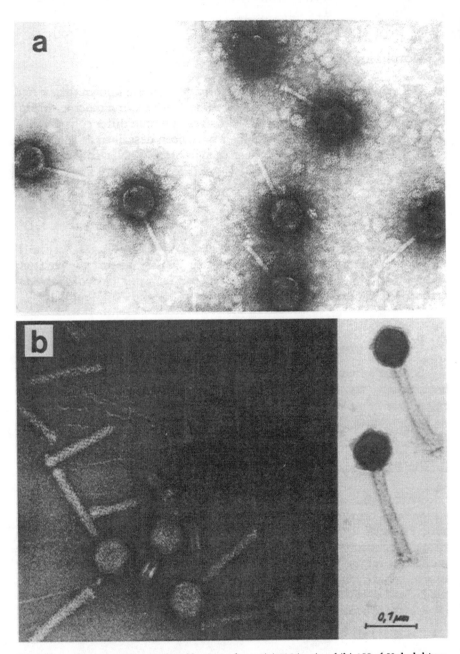

FIGURE 2. Electron micrographs of bacteriophages (a) φN (top) and (b) φH of *H. halobium*. Negative staining. Bar (0.1 μm) is for both figures. (a) Electron micrograph I. Scholz, courtesy of H. Vogelsang-Wenke. (b) Some contracted tails (left) and tail end fibers (right).

TABLE II. Host Ranges of Halobacteriophages and *Methanobrevibacter* Phage PG of *M. smithii*

ΦH	*H. halobium* R₁ (no restriction), WT, NRL, O7, L33, not SY1
ΦN	*H. halobium* NRL/JW, R₁ (plating efficiency 0.025 after 1 passage)
Hs1	*H. salinarium* strain 1.7
Ja1	*H. cutirubrum* lab. strain, *H. halobium* "Delft," "Utah 18"
Hh1	*H. halobium* ATCC 29341, *H. curirubrum* NRC 34001 (efficiency 1.38×10^{-5}), not *H. salinarium* 1 and 1.7
Hh3	*H. halobium* ATTC 29341, *H. cutirubrum* NRC 34001 (efficiency 8×10^{-5}), not *H. salinarium* 1 and 1.7
S45	*H. cutirubrum* NRC 34001, *H. halobium* R₁, *H. halobium* NTHC 5, *H.* sp. NTHC 54
PG	*Methanobrevivacter smithii* strain G, not 5 other *M. smithii* strains and several unclassified rumen isolates

host-controlled modification-restriction has not been considered in some cases, however, the data collected in Table II should be viewed with caution. Three different restriction-modification systems have been demonstrated with halophage S 45, two of them in the same host strain and at least one of them involving a restriction endonuclease (Daniels and Wais, 1984). An unclassified restriction endonuclease (neither type I nor type II) has been purified from *H. halobium* NRC 817 (Schinzel, 1985).

All halophages described so far are capable of lytic multiplication and are released from infected hosts after a latency period of between 6 and 17 h with burst sizes between 140 and 1300 (Table I). At least in one case, lysis is not an immediate consequence of phage production. The liberation of Hh 1 proceeds in two steps, the first of which is not followed by a significant decrease in host viability (Pauling, 1982).

True lysogeny—i.e., the existence of persistently infected hosts carrying a latent prophage genome—has been shown for phage ΦH (Schnabel and Zillig, 1984). The prophage of ΦH exists in the cell in the closed circular state, similar to that of *E. coli* phage P 1 (Ikeda and Tomizawa, 1968). Cultures of the lysogenic *H. halobium* R₁-4 contain 10⁴ to 10⁵ pfu/ml, probably owing to occasional spontaneous lysis induction. Lysis could not be induced by UV irradiation (H. Schnabel, personal communication).

So-called carrier states have been reported for the interaction of the halobacteriophages Hs 1 (Torsvik and Dundas, 1980), Hh 1, and Hh 3 (Pauling, 1982) with their hosts. They could also exist for the phages ΦH and S 45. A carrier state has been defined as a situation in which the host persistently carrying and continually producing virus survives and multiples. The carrier state thus appears as an equilibrium between virus and cell multiplication, in contrast to lysogeny, in which the proviral genome harbored by the host remains latent until induction. A shift of the equilibrium can lead to the segregation of cured host cells or, in the other direction, to lytic virus production. Because of different ionic strength dependence of phage adsorption and bacterial growth rate, the carrier relationship of phage Hs 1 with its host is confined to high ionic strength (Torsvik and Dundas, 1980). Immunity to superinfection has been described for bacteria surviving infection with the phages Hh 1 and Hh 3 and carrying these phages (Pauling, 1982), but it appears questionable whether the survivors are really in carrier states or rather represent true lysogens.

G. Molecular Biology of Halobacteriophage φH

On the molecular level, halobacteriophage φH is by far the most extensively studied virus of the branch of archaebacteria comprising the methanogens and extreme halophiles.

1. Packaging of φH DNA

An unsharp and a sharp substoichiometric band (3 and 5 in Fig. 3) in the Pst 1 restriction fragment pattern of the DNA of φH 1, the major variant of bacteriophage φH (see below), and the formation of circular duplexes with single-stranded tails upon reannealing of denatured φH 1

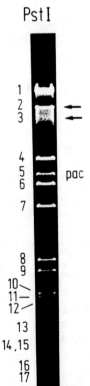

FIGURE 3. Pst 1 restriction endonuclease fragment pattern of DNA of bacteriophage φH 1. Arrows indicate terminal fragments. The substoichiometric band designated pac represents the fragment containing the pac site from which packaging started.

FIGURE 4. Circular duplex obtained by renaturation of denatured φH 1 DNA. Single-strand tails at lower right are the result of terminal redundancy.

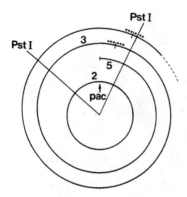

FIGURE 5. Model for the packaging of φH DNA from the concatemeric precursor from Schnabel *et al.* (1982a). The inner circle represents the unit genome; the spiral surrounding it, the concatemeric precursor. Packaging starts at pac. Headful cleavage regions are dotted. Numbers correspond to Pst fragments (see Fig. 13).

DNA (Fig. 4) prove partial, circular permutation (about 10% of the genome length). This suggests a headful packaging mechanism starting from a "pac" site of the concatemeric replication product and proceeding to a headful (about 59 kbp—i.e., 3% more than the unit genome length) with some inaccuracy as shown in Fig. 5 (Schnabel *et al.*, 1982a). This situation resembles the packaging of the DNA of *Salmonella* phage P22 (Jackson *et al.*, 1978). Since about 50% of the phage DNA molecules start from the pac site, the other 50% being packaged processively, the permutation is limited to about 10% of the genome length.

2. Genetic Variability of Phage φH

The virus propagated from the original spontaneous lysate of a laboratory culture of *H. halobium* R_1 turned out to be a mixture. Five variants distinguished by insertions, deletions, and inversions in their DNAs (Fig. 6) were isolated from single plaques of this mixture (Schnabel *et al.*, 1982b). Thirty-four of 38 plaques formed by plating φH on *H. halobium* R_1 yielded a major variant, φH1; the remaining four plaques yielded φH 2, φH 3, and φH 6. Phage φH 4 was obtained by plating the original mixture on a wild-type strain with low plating efficiency. Three variants—φH 5 (Schnabel *et al.*, 1982b), φH 7, and φ 8 (Schnabel *et al.*, 1984)—were isolated from single plaques obtained by plating the highly unstable φH 2. A further variant, φHL 1, was isolated from a single plaque obtained by plating the original φH mixture with high m.o.i. (multiplicity of infection) on a lawn of *H. halobium* R_1 (L), which is immune to infection by φH 1–8 (Schnabel, 1984).

The genetic instability involves an insertion element of 1.895 kbp, ISH 1.8 (Schnabel *et al.*, 1984), in φH variants 1–6 adjacent to the right end (site B) of the so-called L region, which is about 12 kbp long and contains genes for both immunity and early functions in the lytic cycle (Fig. 6). In φH 2 and φH 5, a second copy of ISH 1.8 is present in opposite orientation immediately to the left of the L region (site A; see Fig. 6). In φH 7 and φH 8, it is present only at site A, whereas the B site copy is

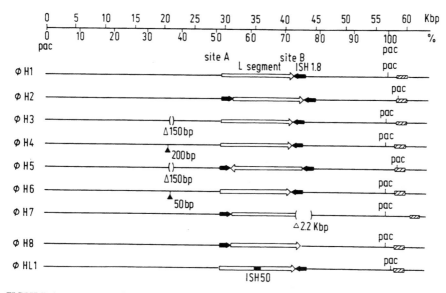

FIGURE 6. Genomes of variants of bacteriophage φH as they exist in phage particles. Hatched bars indicate headful cleavage regions for first packaging round. Open arrows represent the L region. Solid arrows indicate ISH 1.8 sequences. Solid bar within L region of ΦHL 1 is ISH 50/23.

excised, precisely in φH 8 and with about 2.2 kbp flanking sequences, in φH 7 (Fig. 6).

ISH 1.8 has been sequenced (Schnabel et al., 1984). Unlike most eubacterial and other archaebacterial IS elements, the copy of ISH 1.8 integrated at the A site of φH 2 is flanked by neither inverted repeats nor adjacent duplications but by two tetranucleotide sequences which in φH 1, where ISH 1.8 is absent from this site, are joined into an octanucleotide sequence (Fig. 7). In contrast, in the B site in φH 2 DNA, the same two flanking tetranucleotide sequences are double-framed by two adjacent direct repeats of five nucleotide pairs each. In φH 8, where ISH 1.8 is absent from this site, the flanking tetranucleotide sequences are again joined into the corresponding octanucleotide, which remains framed by the direct repeats (Fig. 7). Since no integration event has been observed directly, it is unclear whether the octanucleotide sequence is the integration site or part of it or belongs to ISH 1.8 and is left behind after excision. Whether this element, which also occurs in the host's genome in two copies (Schnabel et al., 1982b; Pfeifer, 1987), is required for the phage remains an open question, though no variant has so far been found in which it is absent from both sites A and B.

A frequent change of genome structure is the inversion of the L region (Schnabel et al., 1982b). Plating of φH 2 or φH 5, in which the L regions have opposite orientations, leads to 10% single plaques with inverted L regions. This can be understood as a consequence of legitimate recombination between the two inverted copies of ISH 1.8. In H. halobium R_1L, derived with high frequency (1 out of 35) from the defec-

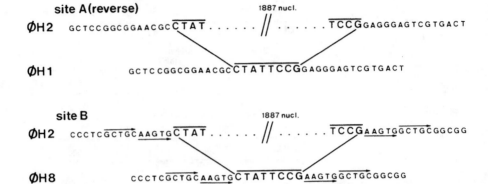

FIGURE 7. Sequences flanking ISH 1.8 in A and B site of φH 2 and corresponding sequences from φH 1 site A and φH 8 site B, not containing ISH 1.8 (for details see text).

tive lysogen R_1-3 essentially containing a circularized φH 2 genome, the circularized L region, or L plasmid, is the only relic of the phage genome. Its circularization point is within a specific inverted 9-bp repeat close to one end of the ISH 1.8 sequence. Circularization of the L region in φH 2 prophage genome therefore appears to be site-specific recombination between short, direct repeats resulting from the opposite orientation of the two copies of ISH 1.8. About 20 copies of the L plasmid have been found in strain R_1L (U. Blaseio, personal communication), indicating that the origin of replication of the viral genome is situated within the L region.

A frequent event not involving ISH 1.8 is the deletion of a 150-bp sequence outside of the L region, in the left arm of the phage DNA, observed both in φH 3 and in φH 5, which is frequently formed from φH 2. Short insertions have been observed in φH 4 and φH 6 in the same DNA region. The large deletion in and adjacent to the right end of the L region of φH 7 could have resulted from imprecise deletion of the B site copy of ISH 1.8 in φH 2.

The insertion of another IS element of known sequence, ISH 50/23, previously found in the host genome (Xu and Doolittle, 1983; Pfeifer *et al.*, 1984) into the L region of φH 1 has been observed in phage φHL 1 selected for its ability to form plaques on the immune *H. halobium* strain R_1L (Schnabel, 1984a).

About 1% of the phage particles produced upon infection of R_1L with φHL 1 contain a duplicated L region (Schnabel, 1984b). In about 50% of these genomes, the ISH 50/23 insert is present in the left copy; in the other 50%, it is present in the right copy of the L tandem. This has been explained as the result of legitimate recombination occurring between phage genome and L plasmid, with equally high frequency to the left and to the right of the ISH 50/23 insert situated close to the center of the L region of the phage (Fig. 8). Beside and in concert with transposition,

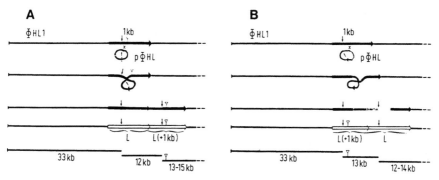

FIGURE 8. Model for the duplication of the L region in φHL 1 by legitimate recombination with the plasmid pφHL. The site of the crossover is in A to the left, in B to the right of the 1 Kb insert of ISH 50/23.

legitimate recombination might thus be a major cause of genome rearrangement in *H. halobium*.

The variability of the φH DNA sequence appears to be a consequence of the genetic mobility of the host, which belongs to the small group of halobacteria characterized by high mutation rates in a number of genes, by the existence of a minor fraction of DNA of only 58% G+C content beside a major fraction of DNA of 68% G+C content, and by the occurrence of many families of middle repetitive sequences of which several have been characterized as IS elements (Pfeifer, 1987). The limitation of observable changes by the smaller size of the phage genome and by the viability requirements of the phage facilitates the investigation of some of the mechanisms involved.

Frequent recombination events, possibly of different types, have also been observed in other archaebacteria—e.g., in the *H. halobium-salinarium–cutirubrum* group (see Pfeifer, 1987), in *Desulfurococcus* (J. Kjems and R. Garrett, personal communication), and in the genome of the virus TTV 1 of *Thermoproteus tenax* (see Section III.B).

3. Lysogeny and Immunity Caused by Phage φH

Many of the host cells surviving infection by bacteriophage φH do not carry the phage genome and are sensitive to infection. Few are sensitive when they carry the phage genome. Twenty-one out of 25 single colonies derived from survivors present in PEG precipitates of φH lysates carried φH DNA and proved to be immune against φH infection (Schnabel and Zillig, 1984). Nine of these were able to produce phages with low frequency, and 12 did not yield phages. One of the true lysogens, R_1-4, contained the complete circularized genome of phage φH6, minus the terminal redundancy, resulting in a genome length of 57 kbp and produced 104–105 pfu/ml, corresponding to spontaneous virus production with normal burst size by one out of 10^7 cells. Two of the non-virus-producing strains were characterized as defective lysogens—one, R_1-3, carrying the circularized genome of ϕH 2 with some uncharacterized DNA rearrangements, and the other, R_1-24, carrying the circularized genome of ϕH 1, with no recognizable change in the restriction pattern.

Strain R_1L containing only the circularized L region of the phage genome is nevertheless immune against ϕH infection (Schnabel, 1984a). Phage ϕHL 1, which contains an insert of ISH 50/23 about 2 kb upstream of the start of the major early transcript (see below) and which is suppressed in R_1L and in true and defective lysogens, is able to overcome the immunity (i.e., it is able to produce the major early transcript and to multiply in this host), though with only 10% of the efficiency of wild-type phage in a nonimmune host. This is most probably due to an enhancer function of the IS element (F. Gropp, unpublished). The production of the major early transcript in the host R_1 is much stronger after infection with ϕHL 1 than after infection with ϕH 1. ϕHL 1 is, however, unable to overcome the immunity conferred by the entire prophage genome. Since the same "immunity transcript" from the left end of the L region is produced as well in R_1L as in lysogens, this indicates a second immunity function to be encoded outside the L region (F. Gropp, unpublished).

4. Expression of ϕH Genes

The expression of the ϕH genome has been investigated by one of us (F. Gropp, unpublished), mainly by Southern and Northern hybridization techniques (Figs. 9, 10). In the (defective) lysogen R_1-24, containing the entire circularized ϕH1 genome as prophage, the left and the central part of the L region, small pieces of the "left" and the "right" arm outside of the L region and ISH 1.8 are transcribed. The right part of the L region is not transcribed. In R_1L, the transcribed fraction of the L region coincides with that expressed in the complete lysogen.

In the lytic cycle, transcription of the ϕH genome follows a time course (Fig. 9). ISH 1.8, which is present in two copies in the uninfected cell (see above), is transcribed before infection and with increased efficiency after infection. Early regions transcribed within the first hour after infection include the right (Fig. 10) and the central but not the extreme left part of the L region. Middle genes begin to be transcribed within the second hour after infection from regions situated at the extreme left and right ends of the linear phage map. Late genes start to be transcribed within the third hour after infection from the rest of the left arm of the linear phage map, including one of the two stretches outside of the L region transcribed in the (defective) lysogen R_1-24, whereas the other (right) of these is silent during the lytic cycle.

Since the translation inhibitor puromycin given upon infection pro-

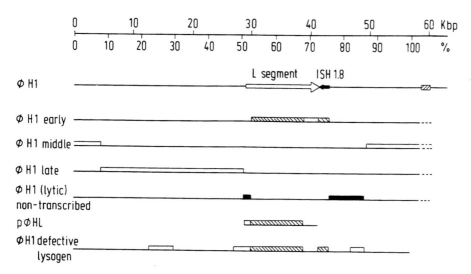

FIGURE 9. Transcription of the genome of bacteriophage φH 1 as analyzed by Southern and Northern hybridization. The first line shows the physical map; the three lines below show different periods in the lytic transcription program; the two lines at the bottom show different immune states. The lowest line shows the gene expression in a complete prophage (pφH 1). The second-lowest is the defective prophage pφHL. Transcribed regions are indicated by white bars. Those parts that are transcribed both in lysogens and during the lytic cycle are indicated by hatching. Black bars represent DNA sequences not transcribed during lytic growth of φH.

hibits middle and late transcription, early translation must be required for a modification of the transcription system. The mechanism of this positive control remains to be elucidated.

A remarkable characteristic of φH immunity compared to that operating in eubacterial lysogens (e.g., of coliphage λ) is the constitutive transcription of several RNAs from the central part of the L region. Two major early transcripts, one of 1.7 kbp and the other of 3.7 kbp, are transcribed from the right of the L region of φH 1 only during the lytic cycle. They appear to start from the same promoter (Fig. 10). The larger

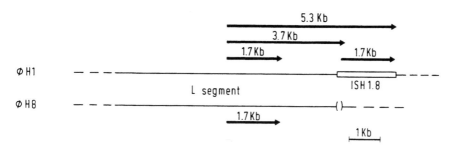

FIGURE 10. Early major transcripts in lytic cycle of bacteriophages φH 1 and φH 8, the latter lacking ISH 1.8 in the B site.

transcript terminates within the extreme left region of the ISH 1.8 sequence; the shorter terminates to the left of the deletion present in φH 7. Neither is transcribed from the L plasmid or from the complete prophage genome or from the φH 1 genome in infected immune hosts carrying the L plasmid or the entire prophage. Since the infecting phage clearly entered the cell, as shown by the increase in ISH 1.8 expression and by the transcription of a piece to the left of the L region in φH 1 infected R_1L, this suppression is a *trans* effect, most probably involving two transcripts (500 and 950 bases) exclusively produced in the immune condition. The ISH 50/23 insert about 2 kbp upstream of the start of the major early promoter in φHL 1 appears to overcome immunity by strongly enhancing transcription, clearly by a *cis* effect, since the presence of this element in the host genome has no effect, and a transcript of ISH 50/23 was barely detectable in φHL 1–infected R_1L cells.

The second level of immunity operates in lysogens containing the entire prophage and should be encoded in a sequence outside of the L region.

Further elucidation of the lytic and the lysogenic transcription programs of φH and the controls operating therein should not only uncover the signal structures involved but possibly yield new insights into regulation mechanisms used in halophilic archaebacteria.

H. Evolution of Phages of Halobacteria and Methanogens

Like many eubacteriophages, all viruses of extremely halophilic and methanogenic archaebacteria known to date are of the types A and B defined by Bradley (1967), with one exception (A. Wood and J. Konisky, see below). Most of the known features fall into the familiar patterns for such eubacteriophages, except perhaps the extreme genetic variability of the genome of halobacteriophage φH, which, however, appears to be a special feature of its host rather than a general property of halobacterial viruses (see below).

It has been suggested that the distribution of a certain virus type over a wide host range indicates the existence of this type before the splitting of lineages leading to the present hosts (see Reanney and Ackermann, 1982). If this is true, head and tail viruses of the types A and B (Bradley, 1967) should already have existed before the separation of eubacteria and methanogenic + halophilic archaebacteria more than 3, possibly 4 billion years ago (Pflug, 1982).

I. Viruslike Particle from *Methanococcus* sp.

A viruslike particle (VLP) has been isolated by A. Wood and J. Konisky (unpublished) from the culture supernatant of a *Methanococcus*

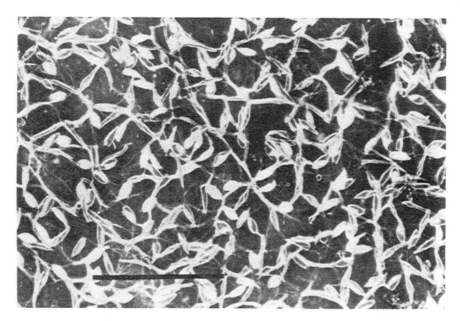

FIGURE 11. Viruslike particles from culture supernatant of a *Methanococcus voltae* isolate. Filaments are added *E. coli* phage M 13. Bar represents 1 μm. Courtesy of A. Wood.

isolate closely related to *M. voltae,* by PEG precipitation and CsCl gradient centrifugation. The particle, resembling SSV 1 of *Sulfolobus* sp. B 12 (see Section III.C.1) but apparently of higher plasticity and lower stability (see Fig. 11), consists of one major protein of 13 kD and contains a cccDNA of 23 kb which occurs also linearly integrated in the host genome. Homologous sequences have been demonstrated in *M. voltae,* which does not produce the VLP. Infectivity has not been shown.

Thirty percent of the subclones after plating have lost the capacity to produce the VLP and have suffered a deletion of 80% of the integrated VLP DNA. This indicates a frequent recombination event. The features of this system resemble those of the *Sulfolobus* sp. B 12-SSV 1 system (see below), suggesting that this type of VLP could have arisen before the separation of the major branches of the archaebacterial kingdom.

III. VIRUSES OF EXTREMELY THERMOPHILIC, SULFUR-DEPENDENT ARCHAEBACTERIA

A. Viruslike Particle of *Pyrococcus woesei*

A culture of the ultrathermophile *P. woesei,* which represents a third branch of the archaebacteria (Zillig *et al.,* 1987; Woese and Olsen, 1986; Klenk *et al.,* 1986) and is able to grow at temperatures up to 104°C, lysed completely after it reached the stationary growth phase (Zillig et al., 1987). Concomitantly, about 100 polyhedric, probably icosahedric particles of about 32 nm diameter (Fig. 12) and simple, defined protein composition (1 major, 2 minor proteins) were liberated from each cell. Thin sections of lysing cells exhibited clusters, sometimes distorted crystals, of the particles (Fig. 12). Occasionally, the particles were seen in linear arrangement inside tubules with apparently closed ends (Fig. 12). These structures were frequently found in exponentially growing cells. Similar tubules enclosing virus particles have been demonstrated in RNA virus-infected plant cells (Hitchborn and Hills, 1968; Conti and Lovisolo, 1971; Martelli and Russo, 1977).

No nucleic acid was found in the particles, nor was there any second particle fraction of higher buoyant density, which could contain nucleic acid. No plasmid or unusual RNA fraction was seen in lysing cells. The nature and location of the genes coding for the production of these particles are unknown. In view of the defined composition and structure and the large "burst size" (50–100 per cell), it appears probable that the particles are empty or defective virions.

Similar lysis phenomena, and, occasionally, polyhedric particles have been observed in cultures of Thermococcus celer (Fig. 13).

B. The Thermoproteus tenax Viruses TTV 1, 2, 3, and 4

1. Occurrence and Isolation

Three types of rod-shaped to filamentous particles were observed after transferring the isolate Kra 1 of Thermoproteus tenax from heterotrophic to autotrophic growth conditions (Janekovic et al., 1983). Sometimes, mass production of only one or the other of them, occasionally accompanied by lysis, was observed, but often mixed populations were found (Fig. 14). The viruses TTV 1, 2, and 3 were isolated from lysates. Only TTV 1 could be propagated at will, either from virus-carrying cells or by lytic multiplication.

TTV 2 was found associated with all inspected clones of T. tenax Kra 1, probably in a lysogenic state. A small number of TTV 2 particles (107/ml) were continually and spontaneously produced in autotrophic T. tenax cultures. Production of larger amounts was occasionally observed. TTV 3 was produced only a few times in autotrophic cultures of the original isolate of T. tenax Kra 1. TTV 4 was observed as an agent lysing both heterotrophic and autotrophic mass cultures of T. tenax during enrichment of organisms from fresh samples of solfataric sources at the Krafla in Iceland, the origin of T. tenax Kra 1 (R. Hensel and M. Rettenberger, unpublished).

Rod-shaped particles of various length and diameter, differing from the known TTVs, but possibly representing related viruses, were fre-

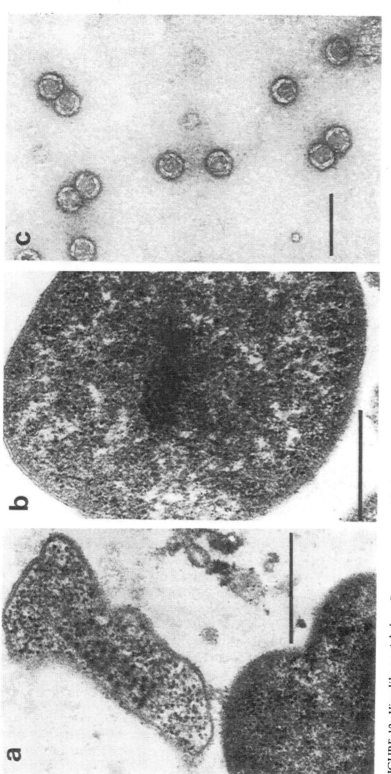

FIGURE 12. Viruslike particle from *Pyrococcus woesei*. (a) Thin section of cell exhibiting tubular structures containing particles. Bar 0.5 μm. (b) Cluster (distorted crystal) of particles in thin section of lysing cell. Bar 0.5 μm. (c) Free viruslike particles, negative staining. Bar 100 nm.

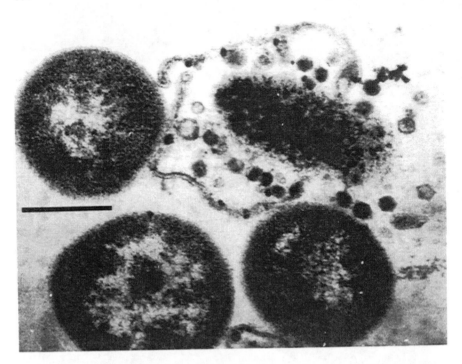

FIGURE 13. Viruslike particles from lysing culture of *Thermococcus celer*. Electron micrograph negatively stained. Bar 0.5 μm.

quently seen in electron micrographs of field samples, especially from solfataric sources of the Krafla in Iceland (D. Janekovic and W. Zillig, unpublished observation).

2. Structure

The TTVs are all rod-shaped or filamentous, with different lengths and diameters (Fig. 14; Table III) (Janekovic *et al.*, 1983; M. Rettenberger, unpublished). The largest of them is TTV 3 with 2.5 μm length and 0.03 μm diameter. They all consist of several proteins and a linear, double-stranded DNA. Since only one of them, TTV 1 (length 0.4 μm, diameter 0.04 μm), has been studied in some detail on the molecular level, it is unclear whether their morphological resemblance is justified for considering them members of one virus family.

Thin sections of TTV 1 show an outer membrane surrounding a core. Lipids can be extracted by solvents. One polar lipid fraction, however, forms water-soluble micelles and can be purified by sucrose-gradient centrifugation and ethanol precipitation. In electron micrographs, it appears to consist of fragments of the virus envelope. Upon hydrolysis with 1 M HCl for 1 hr at 100°C, glucose is liberated, and the residual lipid becomes chloroform-soluble (W. Zillig and W. Schäfer, unpublished).

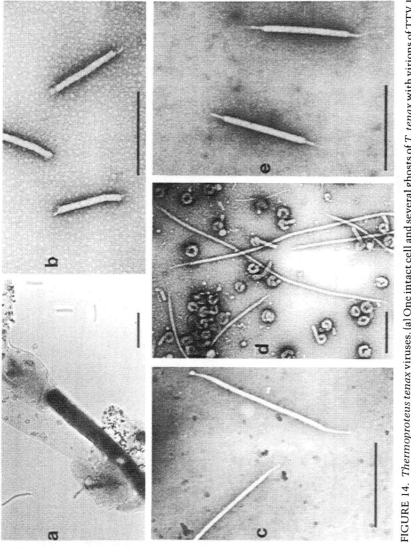

FIGURE 14. *Thermoproteus tenax* viruses. (a) One intact cell and several ghosts of *T. tenax* with virions of TTV 1 and TTV 2. Pt round shadowed, bar 1 μm; (b) TTV 1; (c) TTV 2; (d) TTV 3; (e) TTV 4; all negatively stained. Bars 0.5 μm.

TABLE III. Viruses of Sulfur-Dependent Archaebacteria

Virus	Host	Morphology of virion	Approximate size of virion	DNA	Approximate size of DNA (kb)	Proteins (numbers are mol. wt. in kD)	Lysogeny	Remarks
TTV 1	*Thermoproteus tenax* Kra 1	Rod	40 × 400 nm	ds, lin.	16	3 major (14, 15, 27[a]) 1 minor (45[a])	± ("carrier state")	Variants, 2 DNA-binding proteins, lipid membrane
TTV 2	*Thermoproteus tenax* Kra 1	Flexible rod	20 × 1250 nm	ds, lin.	16	2 major (both 15.5, diff. charges)	+	
TTV 3	*Thermoproteus tenax* Kra 1	Flexible rod	30 × 2500 nm	ds, lin.	27	1 major (13)	?	
TTV 4	*Thermoproteus tenax* Kra 1	Rod	30 × 500 nm	ds, lin.	17	2 major (13, 18)	?	
SSV 1	*Sulfolobus* sp. B 12	Short-tailed lemon	60 × 100 nm	ds, ccc and site sp. integrated	15.463	1 major (VP1, 7.7) 2 minor (VP2, 8.8) (VP3, 9.7)	+ (induction by uv)	Infectivity not shown, DNA-binding proteins (1 virus encoded), hydrophobic envelope
DAV 1	*Desulfurolobus ambivalens*	Short-tailed lemon	60 × 100 nm	ds, ccc	7	?	+	Correspondence of DNA not proven
CWP	*Pyrococcus woesei*	Icosahedron	30 nm ⌀	?	?	1 major (41[a]) 2 minor (26, 25)	+ (induction by energy depletion)	No corresponding nucleic acid identified

[a]These proteins are subject to transition of apparent molecular weight upon phenol treatment. The molecular weights given are of the phenol-transformed proteins.

In thin-layer chromatography, the extracted lipids yield spots coinciding with constituents of the host lipids, though they occur in different proportion (S. Thurl and M. Rettenberger, unpublished). The membrane thus appears to be independently assembled from the pool of host lipids rather than directly derived from the host membrane.

Sucrose-gradient centrifugation of fragments obtained by partial disruption of the virus with EDTA, Triton X-100, and/or organic solvents (e.g., ethyl ether) (Fig. 15a,b) yields, besides the above-mentioned membrane fragments (Fig. 15c), a band consisting of short, hollow filaments of apparently helical structure closed at the ends and sometimes branched or with sharp bends (Fig. 15f), containing two proteins, P 3 and P 4. In addition, an opalescent band with higher sedimentation rate was observed containing very long, hollow filaments of the same diameter as

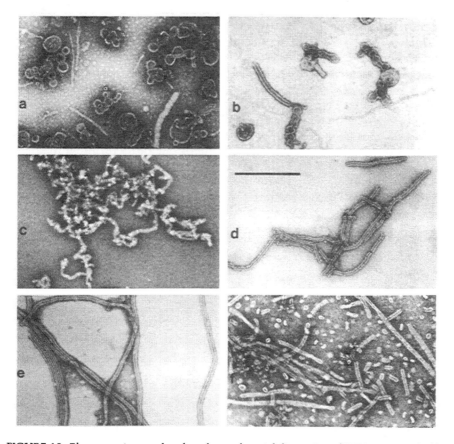

FIGURE 15. Electron micrographs of products of partial disruption of TTV 1 virions. (a, b) TTV 1 after treatment with Triton X-100, exhibiting details of internal structure; (c) isolated lipid envelopes; (d) core consisting of linear, double-stranded DNA and two basic proteins; (e) long filaments consisting of hydrophobic major protein; (f) short, closed filaments consisting of major and minor hydrophobic protein. Negative staining. Bar 0.5 μm.

those in the slower-sedimenting band but with open ends (Fig. 15e), yielding mainly protein P 3 with very little P 4. These filaments can be dissociated by sodium dodecyl sulfate. They reassociate upon removal of the detergent by hydrophobic interaction. The fastest-sedimenting band from the sucrose gradient contains the DNA and two strongly basic proteins, P 1 and P 2, in about equimolar ratio. The particles in this fraction have roughly virus length and, again, helical structure (Fig. 15d). They are not dissociated by phenol or upon CsCl gradient centrifugation.

The model derived from these results (Fig. 16) depicts TTV 1 consisting of (1) a core of linear, double-stranded DNA densely covered by equimolar amounts of the two DNA-binding proteins P 1 and P 2 wound up into a helix such that the DNA, which in its stretched state would have a length of about 5 μm, fits into the 0.4 μm length of the virus particle; (2) an inner envelope consisting of protein P 3, which, by hydrophobic interaction in a helical array forms a hollow tube, the length of which is determined by that of the core, and which is closed by caps of protein P 4, which yields bent rather than straight aggregates; and (3) an asymmetric unit membrane, consisting of an outer glucolipid layer and an inner layer assembled from the pool of host lipid components.

Only a small fraction of the isolated DNA of TTV 1 can be terminally phosphorylated by polynucleotide kinase even after removal of prospective terminal phosphate residues by alkaline phosphatase. This indicates a covalent terminal masking. Further hints in this direction are the failure to clone the whole viral genome or one of its terminal fragments

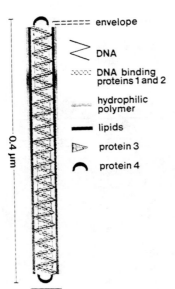

FIGURE 16. Model of the structure of TTV 1 derived from fragmentation experiments.

and the hydrophobic behavior of the DNA in phenol extraction and ethanol precipitation.

Although the other three viruses also have a rod shape and contain linear, double-stranded DNA, their similarity to TTV 1 appears limited. TTV 2 contains only two basic major proteins of equal molecular weight (M. Rettenberger and W. Zillig, unpublished). In electron micrographs, a core slipping out of the envelope is sometimes apparent, and tail fibers have been seen. TTV 3 can be fragmented to yield a helical core and an inner envelope resembling those obtained from TTV 1. Like TTV 2, the straight TTV 4 rod contains only two major proteins, one of them rather hydrophobic. No membrane lipids have been found in this virus (M. Rettenberger, unpublished). The DNAs of TTV 1, TTV 2, and TTV 3 have unrelated restriction patterns and no homologies recognizable by Southern hybridization (H. Neumann, unpublished).

3. Interaction with the Host

TTV 1 has sometimes been found attached with its end to the ends of filaments that most probably represented pili of the host (Janekovic et al., 1983). TTV 4 can adsorb to subcellular particles, probably surrounded by membranes and possibly with distorted S layers, via its distinct long tail fibers (D. Janekovic and W. Zillig, unpublished).

Plating of a TTV 1 carrying culture of T. tenax yields virus-carrying and cured, virus-sensitive clones. The "wild-type" virus can be lytically multiplied on sensitive host cells not carrying the viral genome. Lysis occurs between 24 and 48 hr after infection and yields between 10^{10} and 2×10^{10} virus particles per milliliter of lysate.

TTV 1–carrying cells lyse and produce virus in heterotrophic and autotrophic culture—i.e., with CO_2 as sole carbon source and S and H_2 as energy source. Lysis begins when the sulfur is consumed, indicating that it is, in some unknown way, a consequence of energy depletion. The undefined relationship between host and virus, which ranges from curing to lysis, might be called a carrier state as opposed to lysogeny, where the provirus is a latent but stable component of the genome of the lysogen.

A clone of T. tenax, in which a piece of the TTV 1 genome, contained in the Cla 1 fragment 5, is integrated into the host's genome, is resistant against infection by the wild-type virus. A variant virus, distinguished from the wild type by an inversion in a distant region (Cla 1 fragments 4 and 6), which appeared 2 weeks after infection with wild-type TTV 1 in the resistant T. tenax strain, is, however, able to multiply lytically in this strain (H. Neumann and W. Zillig, unpublished). It thus appears possible that the resistance of this clone is caused by an immunity function encoded in the integrated fragment, which acts on a gene situated in the varied region of the TTV 1 genome.

Since no lawns of T. tenax can so far be grown on plates (only single colonies can be obtained), a plaque test is not available, a situation hindering genetic work.

In contrast to TTV 1, TTV 2 is truly temperate. All subclones of *T. tenax* Kra 1 inspected so far are TTV 2 lysogens (I. Holz and W. Zillig, unpublished). They usually produce the virus in small quantities, but sometimes between 10^9 and 10^{10} virus particles per milliliter are liberated during a sudden lysis. It is not known, however, how virus production can be induced optionally. The relationship of TTV 3, which has only been observed by chance, and *T. tenax* is unknown. TTV 4, which infects and lyses *T. tenax* both in autotrophic and in heterotrophic culture, is highly virulent and, similar to *E. coli* phage T 1, easily spread (R. Hensel and M. Rettenberger, unpublished). No lysogeny or virus carrier state has so far been observed.

4. Genetic Variability of TTV 1

Six variants of the TTV 1 genome, including the primary isolate, have been observed (Fig. 17) (H. Neumann, unpublished). One of these caused lysis 2 weeks after infection of a host strain which initially appeared to be immune to the "wild type" and carried a piece of the TTV 1 genome (see above). This variant, M3, proved able to multiply by a lytic cycle in this host and therefore seems to have lost its response to immunity suppression. The others appeared spontaneously, either in uninfected culture or late in infected cultures. Such lysates usually contained one variant. In one case two different variants turned up in about equal amounts. The variants are distinguished from the original isolate and from each other by deletions and inversions (Fig. 17). Insertions have not been found. The changes are restricted to a region comprising the neigh-

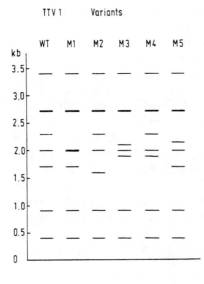

FIGURE 17. Cla 1 restriction endonuclease fragment patterns of variants of TTV 1, schematic.

boring Cla 1 fragments 4 and 6, distant from fragment 5, which contains a sequence apparently conferring immunity to *T. tenax* when integrated into its genome (see Section III.B.3).

This observation indicates a certain genetic mobility in *T. tenax*. Though the genesis of the changes could in no case be followed, the absence of insertion indicates that the mechanisms involved might differ from those known to operate in *Halobacterium halobium–salinarium–cutirubrum* (Pfeifer, 1987).

5. Genes and Reading Frames

The genes for the DNA-binding proteins P 1 and P 2 of the virus have been cloned and mapped in adjacent regions of the Cla 1 fragments 1 and 2 (H. Neumann, unpublished).

Sequencing of the Cla 1 fragment 6, which is one of the two pieces of the TTV 1 genome involved in variant formation, revealed a peculiar reading frame with a triplicate repeat of a 22 amino acid sequence, starting with *val thr* followed by ten *pro thr* pairs with some, usually functionally equivalent, amino acid substitutions (Zillig *et al.*, 1986b). These sequences are separated by a repeat of a specific sequence of nine amino acids. The C-terminal copy differs from the upstream one only by the substitution of a *val* by an *ile*. The presence of typical third-base exchanges in the corresponding nucleotide sequence suggests a function of this reading frame on the protein level. Since the recombination processes causing variant formation do not occur within this sequence, it is probably not involved in the genetic variation of the TTV 1 genome.

C. Viruses of Sulfolobales

1. Viruslike Particle SSV 1 from *Sulfolobus* sp. B 12

a. The System

Sulfolobus sp. B 12, an obligately heterotrophic isolate originating from a hot spring at Beppu, Japan, contains a 15.5-kbp plasmid, pSB12. The whole plasmid sequence also occurs site-specifically integrated into the host chromosome (Yeats *et al.*, 1982). The plasmid is amplified upon UV irradiation and between 4 and 16 hr after UV induction some 10^{10} lemon-shaped particles containing this plasmid DNA are released (Fig. 18). This occurs essentially without lysis of the host, which then resumes growth and returns to the normal "lysogenic"—i.e., nonproductive—state (Martin *et al.*, 1984). The final titer of these particles after induction is some 10^{10} per milliliter. After dilution of the culture into fresh medium, the process can be repeated.

Sulfolobus sp. B 12 has a DNA-dependent RNA polymerase with an

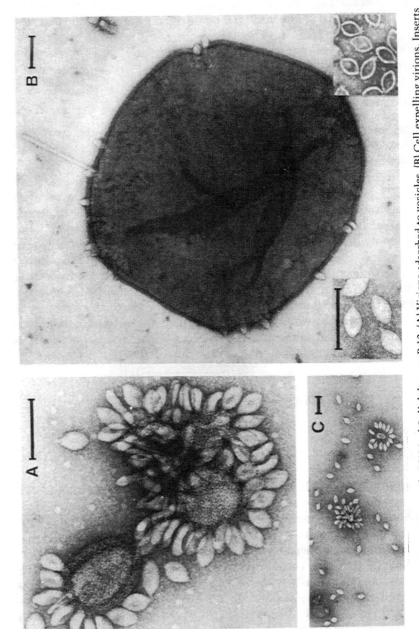

FIGURE 18. Viruslike particle SSV 1 of *Sulfolobus* sp. B 12. (A) Virions adsorbed to vesicles. (B) Cell expelling virions. Inserts magnified viruslike particles (left) and empty envelopes (right). (C) Overview of free and adsorbed virions. Bars 0.2 μm except in right insert in B (0.1 μm).

SDS-PAGE pattern very similar to that of *S. solfataricus* (Zillig *et al.*, 1985b) and is similar to the latter *Sulfolobus* species also in rRNA-DNA cross-hybridization (Klenk *et al.*, 1986). The original designation SAV 1 for the viruslike particle was based on an erroneous classification of the host and was therefore changed to SSV 1. A formal description of strain B 12 has not been published.

The UV dose yielding optimal induction is low compared to the doses influencing the growth of other *Sulfolobus* species (S. Yeats, unpublished). Cells approaching the stationary state yield a higher virus titer than exponentially growing cells (Reiter, 1985). A low number of viruslike particles are produced without UV induction, especially in the approach to stationary state.

b. Structure

The viruslike particle has the shape of a lemon, appears rather flexible, and carries short tail fibers on one end (Fig. 18). It is able to adsorb with this end to a class of vesicles derived from cells, which appear to differ from a class of nonadsorbing similar vesicles by a blurred appearance of their envelopes (Fig. 18) (Martin *et al.*, 1984). This suggests the ability of SSV 1 to interact with a receptor possibly situated in the membrane. The SSV 1 coat contains no lipid (A. Gambacorta and M. De Rosa, personal communication) and is assembled from a major and a minor protein (VP 1 and VP 3) with 73 and 92 amino acid residues, respectively. Both coat proteins are highly hydrophobic, the major protein to such an extent it is soluble in ethanol/chloroform 2:3 (Reiter *et al.*, 1987a). VP 1 and VP 3 are highly homologous to each other. They share a hydrophobic sequence of 20 amino acids close to the carboxyl termini. Less conserved hydrophobic regions are found close to the amino termini. Within the envelope SSV 1 contains the circular DNA as its genome. Polyamines and a strongly basic protein, VP 2 (29 basic residues in a total of 74 amino acids), are bound to the packaged SSV 1 DNA (Reiter *et al.*, 1987a). Isolated virions contain only a small amount of undamaged circular DNA. The major fraction of the DNA is nicked or even shows double strand breaks. This fraction increases in proportion with the time after the onset of virus production and with decreasing pH of the medium (Reiter, 1985). The cccDNA isolated from the virus particle is fully positively supercoiled. In contrast, the plasmid isolated from cells exhibits a broad spectrum of supertwist isomers (Nadal *et al.*, 1986), possibly because maximal supertwist does not allow DNA function—e.g., in replication and transcription.

The virus coat is extremely stable under extreme conditions such as high temperature (97°C), acidity (0.5 M perchloric acid), and agents that denature proteins (6 M urea and 7 M guanidinium chloride). SSV 1 is disintegrated by chloroform but not by diethyl ether. Exposure to high pH (> 11.5) or 0.01% N-lauroylsarcosine leads to the release of the DNA and the DNA-binding protein and to the formation of empty envelopes (Reiter, 1985).

The isolated DNA-binding protein VP 2 can form a complex with DNA which resembles that formed by certain cellular DNA-binding proteins of *Sulfolobus* (R. Reinhardt, personal communication). In addition to this virus-encoded protein, a small amount of a host-encoded 6-kD DNA-binding protein has been identified in purified virions (Reiter *et al.*, 1987a).

Induced cultures produce a small number of oversize virions besides the normal size class, suggesting errors in assembly (Martin *et al.*, 1984) (upper left corner of Fig. 18A).

c. Forms of Existence of the Viral Genome within the Cell

As mentioned above, the viral genome is present in the cell both in ccc form and site-specifically integrated. The whole virus DNA comprising 15.5 kbp has been sequenced (P. Palm and B. Grampp, unpublished). The circular DNA contains a specific 44-bp sequence that is present as a direct repeat at the ends of the copy integrated in the chromosome (S. Yeats and P. Palm, unpublished; reviewed in Zillig *et al.*, 1986b). Besides this copy, fragments of the viral genome, at least in one case containing many point mutations, have been found at different locations within the host genome and might thus represent residues of earlier integration or recombination events (W.-D. Reiter, unpublished).

In a line of *Sulfolobus* sp. B 12, the plasmid is present in substoichiometric quantity compared to the integrated complete SSV 1 genome. Induction of virion production is, however, unimpaired. It therefore appears possible that the integrated SSV 1 genome rather than the plasmid functions as provirus. The plasmid would then represent unpackaged viral DNA.

d. Is SSV 1 a Virus?

SSV 1 must be termed a viruslike particle rather than a virus, because it can so far only be propagated by induction of a lysogen and not by multiplication in a sensitive host. Several different *Sulfolobus* species neither adsorbed nor multiplied SSV 1 (S. Yeats, unpublished). Clearly, a cured host strain would be the best candidate for infection studies. Attempts to cure *Sulfolobus* sp. B 12 of SSV 1 failed, however, possibly because of the existence of two forms of the SSV 1 genome in the cell. An integration site free of the viral genome could not be detected (S. Yeats, unpublished), showing that excision does not occur frequently or that it is highly reversible.

The virus does not appear to adsorb to intact cells, even of its host *Sulfolobus* sp. B 12 (S. Yeats, unpublished). Cell-cell contacts observed

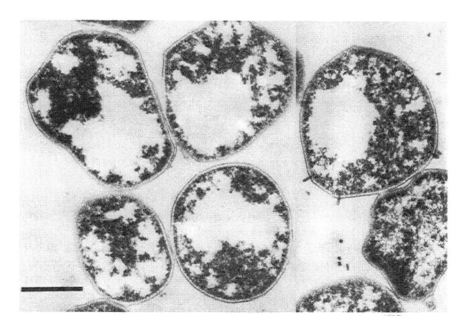

FIGURE 19. Thin sections of virion producing cells of *Sulfolobus* sp. B 12 showing cell–cell contacts and (on the right) some virions being expelled. Bar 0.5 μm.

after the onset of virus production in thin sections of induced cells (Fig. 19) and the existence of cell-derived vesicles competent for SSV 1 adsorption, besides incompetent, similar vesicles, suggest that infection might require changes in the cell envelope.

In spite of the failures in proving infectivity, the ability of SSV 1 to adsorb to cell-derived structures, to be induced and liberated, and the packaging of intact cccDNA into the SSV 1 particle speak strongly for its viral nature. Its value as a lucid genome for studying replication and gene expression in *Sulfolobus* remains unaffected by the nonavailability of a sensitive host.

e. Gene Expression

Nine different transcripts of the whole SSV 1 genome have been identified by Northern hybridization and approximately mapped by probing with single-stranded DNA clones (W.-D. Reiter, *et al.*, 1987c) (Fig. 20). The 5′ ends of these transcripts (T1–T6) and of an additional transcript downstream of the VP operon (T9; see below) have been determined by S1 endonuclease mapping. Eight of these transcripts are coordinately amplified after UV induction, but they are also found in uninduced cells, especially in the exponential growth phase. Three of these RNAs (T1, T2,

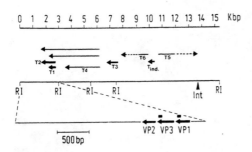

FIGURE 20. Map of SSV 1 transcripts. T1, T2, T3, T4, and RNA ind were mapped with S 1 endonuclease; the other transcripts were mapped with M 13 clones from sequencing. The primary transcript character of T1 + T2 and T3 was confirmed by *in vitro* capping. On the bottom is a blow-up of the region containing the VP operon, showing the arrangement of the genes for the structural proteins of the virion.

T3) can be considered major transcripts, and two of these (T1, T2) start at the same site and code for structural proteins of the virus. The third major transcript (T3) maps at a distant location. Its function is unknown. Only one transcript of about 0.3 kb, T_{ind}, which is almost undetectable in uninduced cells, appears to be specifically induced by UV irradiation and is maximally produced 3–4 hr after induction, just before the onset of virion liberation. In contrast to all other transcripts, which remain at a high level after UV irradiation, the amount of T_{ind} strongly decreases after about 4 hr. Since this transcript does not contain a long open reading frame, a regulatory role involving RNA-RNA or RNA-DNA interaction appears possible.

Surprisingly, there is no straightforward connection between transcription of the SSV 1 genome and virus release. With the exception of the inducible transcript T_{ind}, all RNAs are also detectable in high levels in exponentially growing uninduced cells (comparable to the level after UV irradiation), so that even translational control is not excluded. Alternatively, virion production might not result from enhanced gene expression but from the availability of high amounts of packageable DNA after UV irradiation.

f. The VP Operon

The genes of the three structural proteins, VP 1, VP 2, and VP 3, are found closely linked in a cluster (Fig. 21). Ten nucleotides separate the termination codon of VP 1 and the initiation codon of VP 3, and only one nucleotide separates the stop codon of VP 3 from the start of VP 2 (Reiter et al., 1987a) (Fig. 21). The mature VP 1 has a glutamic acid residue at its amino terminus and thus appears to be the processing product of a precursor starting 13 amino acid residues upstream, as indicated by the presence of two putative Shine+Dalgarno sequences before the corresponding GUG codon. Duplicate Shine-Dalgarno sequences are also found in front of the other two genes—adjacent to each other immediately in front of the AUG of the VP 3 gene and overlapping directly upstream of the initiation codon of the VP 2 gene.

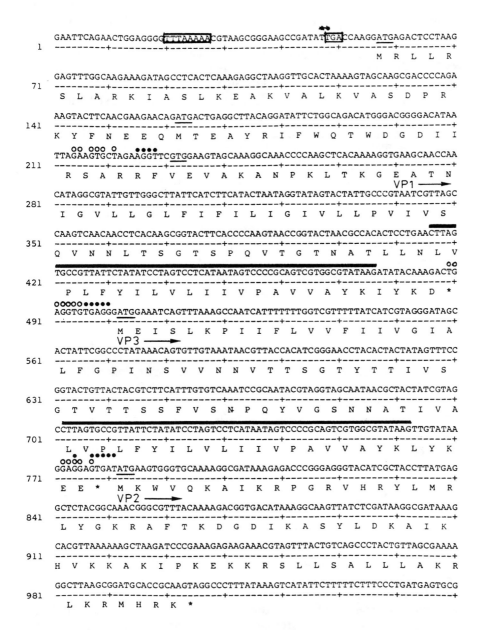

FIGURE 21. Section of the total sequence of SSV 1 containing the VP operon. Boxes in uppermost line enclose characteristic promoter sequences. Shine-Dalgarno sequences are dotted, initiation codons are underlined, and the transcript start mapped with the S1 nuclease technique is labeled by stars. The exact nucleotide sequence repeats in the genes for VP 1 and VP 3 are indicated by bars above the DNA sequence.

A striking feature of the genes encoding VP 1 and VP 3 is the exact repetition of a 61-bp nucleotide sequence within the coding region. One copy of this repeat encodes a highly hydrophobic amino acid sequence of VP 1, and the other one encodes an identical amino acid sequence in VP 3. The absence of third-base exchanges and conservative amino acid substitutions clearly indicates an additional function other than on the protein level, possibly in replication, recombination, or transcription, perhaps as a transcriptional enhancer.

The whole region encoding VP 1, VP 2, and VP 3 is covered by a major transcript (T2), starting about 220 nucleotides upstream of the amino-terminal glutamic acid codon of VP 1 and terminating about 60 bp downstream of VP 2. The three linked genes thus form an operon. Several other operons are known in methanogens (Konheiser et al., 1984; Hamilton and Reeve, 1985; Reeve et al., 1986). This one, however, is the first operon (and set of protein genes) known for a sulfur-dependent archaebacterium and an archaebacterial virus. A second, overlapping transcript (T1) shares the 5' end with the first one but terminates immediately downstream of the VP 1 gene. Thus, as in several similar situations in eubacteria, the relative expression of sequential genes within a transcriptional unit is modulated by partial read-through past a terminator.

g. Signal Structures for the Initiation of SSV 1 Transcription

S1 nuclease and primer extension transcript mapping of the 5' ends of three coordinately regulated SSV 1 RNAs (T1 + T2, T3, T4) showed that a conserved AT-rich sequence ("box A"; see Fig. 22) is found about 20 nucleotides upstream of these ends. In addition, the sequence TGA or AGA (T5 and T6), defining a box B, was found at these 5' ends. In the cases of T1 + T2 and T3, it was shown by in vitro capping that the 5' ends of the transcripts are indeed transcription starts (Reiter et al., 1988). Very similar sequence elements constitute the putative promoters of the 16S/23S rRNA operon (Reiter et al., 1987d) and the 5S rRNA gene in Sulfolobus sp. B 12. It therefore appears that the recognition signals for transcription initiation on the SSV 1 genome are "standard" Sulfolobus promoters. A different situation is encountered for the UV-inducible 0.3-kb RNA of SSV 1. Though a "box B" motif is present (TGG), the −20 bp region deviates significantly from the "box A" sequence. This is not unexpected, since an inducible RNA should be subject to a different control from transcripts that are more or less constitutively expressed. The comparison of promoter sequences of SSV 1 with promoter sequences of stable RNA genes in Methanococcus (Wich et al., 1986a,b) revealed the existence of a consensus promoter sequence in these archaebacteria (Fig. 22) which belong to different branches of the kingdom. On the other hand, subclasses of promoters were recognizable. Stable rRNA promoters contain the sequence TGC (or CGC and GGC in T. tenax) rather than TGA in box B. The promoters of T3, T4 + T7 + T8, and T9 are also highly homologous to each other in the region between

16S rRNA (1)	AGAAGTTAGATTTATATGGGATTTCAGAACAATATGTATAATGCGGATGCCCCCGCGGGA
5S rRNA (2)	TATTTAATTTTTTTATATGTGTTATGAGTACTTAATTTTGCCCACCCGGCCACAGTGAGCG
SSV1, T1+T2 (3)	ACTGGAGGGGTTTAAAAACGTAAGCGGGAAGCCGATATTGACCAAGGATGAGACTCCTAA
T9	AAGTAGGCCCTTTATAAAGTCATATTCTTTTTCTTTCCCTGATGAGTGCGTTAGGGGATG
T3	AGTTAGGCTCTTTTTTAAAGTCTACCTTCTTTTTCGCTTACAATGAGGAAGTCCCTTCTAG
T4+T7+T8	AAGATAGCCCTTTTTTAAAGCCATAAATTTTTTATCGCTTAATGAAGTGGGGACTATTATT
T6	TAGAGTAAAGTTTAAATACTTAT..ATAGATAGAGTATAGATAGAGGGTTCAAAAAATGG
T5	TAGAGTAAGACTTAAATACTAATTTATACATAGAGTATAGATAGAGTGGGATGGGAATAT
T_ind	TTAGTCGACTCTGTGTATCTTATGTATCTTATACAAAAAATATGGGATGTGCAAAATCTG
S.acidocaldarius RNAP-C gene (4)	CTGTGAGAAGTTTATATGGAGATATTGTTCAAGTAGTATATGGTGATGATGCAGTGCATC
T.tenax 16S + 23S rRNA (5)	AAAAATTTTTAATTTAGGGTGTTTTTAGGATGGTCGCGCCTTAATT

T.tenax cons. tRNA genes (5)	TTT AAT	GGC GG
M.vannielii cons. t and tRNA genes (6)	TTTATATA	TGCAAGT
H.halobium rRNA P2-P5 cons. (7)	TTAAGTAA	TGCGAACG
H.halobium ISH 1.8 (8)	GTCACAAGAGTTATCTCAAATTGGGTGTCTCGTATCTGCTAAGGCCAAATGGAGTATCATC	

T5	GATAGATAGAGTATAGATAGAGTAAGACTTAAATACTAATTTATACATAGAGTATAGATAGAGTGG
T6	TATAGATAGAGTATAGATAGAGTAAAGTTTAAATACTTAT..ATAGATAGAGTATAGATAGAGGGT

FIGURE 22. Sequences upstream of starts of SSV 1 transcripts in comparison with transcripts of rRNA and tRNA genes of various archaebacteria. Homologies between all sequences are in boxes; homologies within groups are underlined. Dots indicate primer extension mapped 5' ends. Shine-Dalgarno sequences are overlined. References: (1) W.-D. Reiter and P. Palm, unpublished; (2) W.-D. Reiter and W. Voos, unpublished; (3) W.-D. Reiter and P. Palm, unpublished; (4) G. Pühler and W. Zillig, unpublished; (5) Wich et al., 1986b; (6) Wich et al., 1986a; (7) Chant et al., 1986; (8) F. Gropp, unpublished.

boxes A and B. The promoters for T5 and T6 are embedded into an almost perfect direct repeat of a direct repeat (see Fig. 22).

h. Termination of SSV 1 Transcription

S$_1$ endonuclease mapping of the termination regions of the two overlapping major SSV 1 transcripts from the VP 1/VP 2/VP 3 operon each yielded a series of bands corresponding to frayed 3' ends close to oligo T clusters (W.-D. Reiter, unpublished). Stem and loop structures resembling eubacterial terminators could not be recognized (P. Palm, unpublished). The termination region of the longer transcript, which includes the whole operon, overlaps the prospective start site of a transcript (T9) following immediately downstream (Reiter et al., 1987d).

i. SSV 1: Perspectives

The entire nucleotide sequence of SSV 1, two classes of transcripts more or less covering the whole genome, and the major structural components of the virion are known. Understanding the function of the inducible transcript should lead to the elucidation of an unusual mode of control of replication and virus production via UV induction. One of the goals of the investigation of archaebacterial virus host systems defined in the introduction, the understanding of signal structures, mechanisms, and controls of genes expression in archaebacteria, thus appears in close reach with this system.

2. Plasmid pSL 10 and the Viruslike Particle DAV 1

Desulfurolobus ambivalens, strain Lei 10, is an archaebacterium that can grow as an obligate chemolithoautotroph deriving energy either by the oxidation of sulfur with O_2 to sulfuric acid, like many *Sulfolobus* isolates, or by the reduction of sulfur with H_2 to H_2S, like *Thermoproteus*, in both cases with CO_2 as sole carbon source (Segerer et al., 1985; Zillig et al., 1985b, 1986a). It thus appears as a link between sulfur-oxidizing and sulfur-reducing extremely thermophilic sulfur-dependent archaebacteria, though by its features it clearly belongs to the *Sulfolobaceae*.

Upon growth by sulfur reduction under conditions close to the tolerable limits of pH and temperature (pH 3.5 and 87°C) and in the absence of yeast extract as growth stimulant, a plasmid pSL 10 present in low copy number and under oxidative growth conditions is amplified about 20-fold (Zillig et al., 1985b). A subclone of *D. ambivalens* cured of this plasmid has not lost the ability to grow by sulfur reduction. Thus, the plasmid does not harbor genes essential for this capacity, as was suggested previously (Zillig et al., 1985b).

A viruslike particle morphologically closely resembling SSV 1 has occasionally been found under conditions leading to the amplification of pSL 10 (Fig. 23). Infection of *D. ambivalens* by SSV 1 can be excluded, however, since no SSV 1 DNA was found within the cells. The two large Bam H 1 fragments of pSL 10 weakly hybridized with nick-translated SSV 1 DNA, indicating same homology (S. Yeats, unpublished). It thus appears possible that pSL 10 is the "provirus" of an SSV 1-like particle. This system, in which the proviral genome exists solely in the plasmid state, allowing the easy isolation of cured recipient strains, with strong genetic markers offered by the metabolic properties of *S. ambivalens*, and the availability of plating procedures make the plasmid a candidate for the development of a vector for sulfur-dependent archaebacteria.

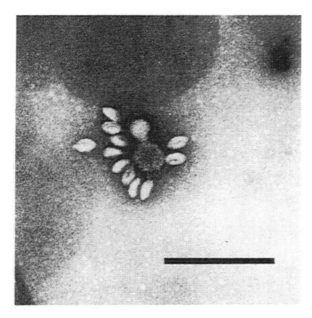

FIGURE 23. Viruslike particles DAV 1 from *Desulfurolobus ambivalens*, adsorbed to vesicle. Electron micrograph, negatively stained. Bar 0.5 μm.

3. Viruslike Particles in *Sulfolobus* sp. B 6

Sulfolobus sp. B 6, another isolate from Beppu, Japan, harbors a plasmid of 35 kb which shows some genetic variability (P. McWilliam and S. Yeats, unpublished; reviewed in Stetter and Zillig, 1985). Viruslike particles forming crystalline arrays resembling the crystals of adenovirus particles found in HeLa cells (I. Scholz, unpublished) were seen in thin sections of cells of this strain (Fig. 24) (I. Scholz, unpublished), and a labile particle yielding DNA could be shown to exist in cell extracts (W. Zillig, unpublished). However, it was not possible to induce the production of these particles by defined procedures (e.g., UV irradiation) such that this system so far evaded further investigation.

4. Conclusions

Whereas the viruses of extremely halophilic and methanogenic archaebacteria known to date, with one exception, represent bacteriophages of the types A and B (Bradley, 1967), the viruses of the sulfur-dependent archaebacteria isolated so far define new virus families, with but slight resemblances to known viruses of both eukaryotes and eubacteria. The failure to find A- and B-type bacteriophages in sulfur-dependent archaebacteria and eukaryotes and their frequent demonstration in extreme halophiles (and occurrence in methanogens) appear as one of several "di-

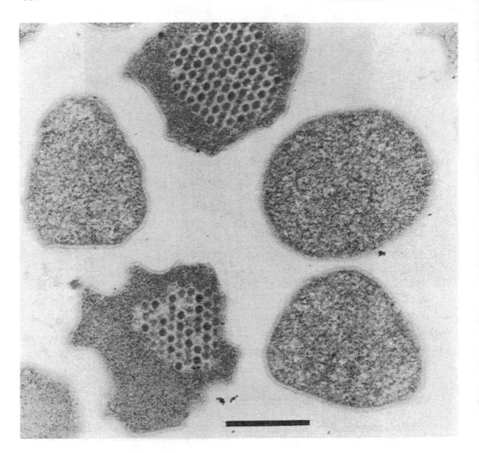

FIGURE 24. Crystals of viruslike particles in thin sections of *Sulfolobus* sp. B 6. Bar 1 μm.

vergent" features distinguishing the major branches of the archaebacterial kingdom. This is in line with putative relations between these
branches and the eukaryotes and eubacteria (Zillig *et al.*, 1985a). Firm
conclusions regarding the phylogeny of the viruses as well as of their
hosts are not possible at our present level of understanding, but unique
structures and controls have come into focus.

ACKNOWLEDGMENTS. Except where stated otherwise, all electron micrographs were made by Davorin Janekovic. Thanks are due J. Trent and D.
Grogan for critically reading the manuscript, and H. Vogelsang-Wenke
and A. Wood for the communication of unpublished results.

REFERENCES

Bertani, G., and Baresi, L., 1986, Looking for gene transfer mechanism in methanogenic bacteria, in: *Archaebacteria '85* (O. Kandler and W. Zillig, eds.), Gustav Fischer Verlag, Stuttgart.

Bradley, D. E., 1967, Ultrastructure of bacteriophages and bacteriocins, *Bacteriol. Rev.* **31**:230–314.

Chant, J., Hui, I., de Jong-Wong, D., Shimmin, L., and Dennis, P. P., 1986, The protein synthesizing machinery of the archaebacterium *Halobacterium cutirubrum:* Molecular characterization, *Syst. Appl. Microbiol.* **7**:106–114.

Conti, M., and Lovisolo, O., 1971, Tubular structures associated with maize rough dwarf virus particles in crude extracts: Electron microscopy study, *J. Gen. Virol.* **13**:173–176.

Daniels, L. L., and Wais, A. C., 1984, Restriction and modification of halophage S45 in *Halobacterium, Curr. Microbiol.* **10**:133–136.

Fiala, G., and Stetter, K. O., 1986, *Pyrococcus furiosus* sp. nov. represents a novel genus of marine heterotrophic archaebacteria growing optimally at 100°C, *Arch. Microbiol.* **145**:56–61.

Fiala, G., Stetter, K. O., Jannasch, H. W., Langworthy, T. A., and Madon, J., 1986, *Staphylothermus marinus* sp. nov. represents a novel genus of extremely thermophilic submarine heterotrophic archaebacteria growing up to 98°C, *Syst. Appl. Microbiol.* **8**:106–113.

Forterre, P., Elie, C., and Kohiyama, M., 1984, Aphidicolin inhibits growth and DNA synthesis in halophilic archaebacteria, *J. Bacteriol.* **159**:800–802.

Forterre, P., Nadal, M., Elie, C., Mirambeau, G., Jaxel, C., and Duguet, M., 1986, Mechanisms of DNA synthesis and topoisomerisation in archaebacteria—reverse gyration *in vitro* and *in vivo, Syst. Appl. Microbiol.* **7**:67–71.

Fox, G. E., Stackebrandt, E., Hespell, R. B., Gibson, J., Maniloff, J., Dyer, T. A., Wolfe, R. S., Balch, W. E., Tanner, R. S., Magrum, L. J., Zablen, L. B., Blakemore, R., Gupta, R., Bonen, L., Lewis, B. J., Stahl, D. A., Luehrsen, K. R., Chen, K. N., and Woese, C. R., 1980, The phylogeny of prokaryotes, *Science* **209**:457–463.

Hamilton, P. T., and Reeve, J. N., 1985, Structure of genes and an insertion element in the methane producing archaebacterium *Methanobrevibacter smithii, Mol. Gen. Genet.* **200**:47–59.

Hitchborn, J. H., and Hills, G. J., 1968, A study of tubes produced in plants infected with a strain of turnip yellow mosaic virus, *Virology* **35**:50–70.

Ikeda, H., and Tomizawa, J., 1968, Prophage P1, an extrachromosomal replication unit, *Cold Spring Harbor Symp. Quant. Biol.* **33**:791–798.

Jackson, E. N., Jackson, D. A., and Deans, R. J., 1978, *Eco*RI analysis of bacteriophage P22 DNA packaging, *J. Mol. Biol.* **118**:365–388.

Janekovic, D., Wunderl, S., Holz, I., Zillig, W., Gierl, A., and Neumann, H., 1983, TTV1, TTV2 and TTV3, a family of viruses of the extremely thermophilic, anaerobic, sulfur reducing archaebacterium *Thermoproteus tenax, Mol. Gen. Genet.* **192**:39–45.

Kessel, M., and Klink, F., 1982, Identification and comparison of eighteen archaebacteria by means of the diptheria toxin reaction, *Zbl. Bakteriol. Hyg. I. Abt. Orig.* **C3**:140–148.

Klenk, H.-P., Haas, B., Schwass, V., and Zillig, W., 1986, Hybridization homology—a new parameter for the analysis of phylogenetic relations, demonstrated with the urkingdom of the archaebacteria, *J. Mol. Evol.* **24**:167–173.

Kohiyama, M., Nakayama, M., and Mahrez, K. B., 1986, DNA polymerase and primase-reverse transcriptase from *Halobacterium halobium, Syst. Appl. Microbiol.* **7**:79–82.

Konheiser, U., Pasti, G., Bollschweiler, C., and Klein, A., 1984, Physical mapping of genes coding for two subunits of methyl CoM reductase component C of *Methanococcus voltae, Mol. Gen. Genet.* **198**:146–152.

Kushner, D. J., 1985, The halobacteriaceae, in: *The Bacteria* (I. C. Gunsalus, J. R. Sokatch, L.

N. Ornston, C. R. Woese, and R. S. Wolfe, eds.), Vol. VIII, pp. 171–214, Academic Press, Orlando, FL.

Langworthy, T. A., 1985, Lipids of archaebacteria, in: *The Bacteria* (I. C. Gunsalus, J. R. Sokatch, L. N. Ornston, C. R. Woese, and R. S. Wolfe, eds.), Vol. VIII, pp. 459–497, Academic Press, Orlando, FL.

Martelli, G. P., and Russo, M., 1977, Plant virus inclusion bodies, in: *Advances in Virus Research* (M. A. Lauffer, F. B. Bang, K. Maramorosch, and K. M. Smith, eds.), Vol. 21, pp. 175–266, Academic Press, New York.

Martin, A., Yeats, S., Janekovic, D., Reiter, W.-D., Aicher, W., and Zillig, W., 1984, SAV 1, a temperate u.v.-inducible virus-like particle from the archaebacterium *Sulfolobus acid-ocaldarius* isolate B12, *EMBO J.* **3:**2165–2168.

Morgan, H. W., and Daniel, R. M., 1982, Isolation of a new species of sulphur reducing extreme thermophile, in: *Proceedings of the XIII International Congress of Microbiology*, Boston, August 1982.

Nadal, M., Mirambeau, G., Forterre, P., Reiter, W.-D., and Duguet, M., 1986, Positively supercoiled DNA in a virus-like particle of an archaebacterium, *Nature* **321:**256–258.

Pauling, C., 1982, Bacteriophages of *Halobacterium halobium:* Isolation from fermented fish sauce and primary characterization, *Can. J. Microbiol.* **28:**916–921.

Pfeifer, F., 1987, Genetics of halobacteria, in: *Halophilic Bacteria* (F. Rodriguez-Valera, ed.), pp. 105–133, CRC Press, Boca Raton, FL.

Pfeifer, F., Friedman, J., Boyer, H. W., and Betlach, M., 1984, Characterization of insertions affecting the expression of the bacterio-opsin gene in *Halobacterium halobium*, *Nucleic Acids Res.*, **12:**2489–2497.

Pflug, H. D., 1982, Early diversification of life in the archaean, *Zbl. Bakteriol. Hyg.* I. Abt. Orig. **C3:**53–64.

Reanney, D. C., and Ackermann, H.-W., 1982, Comparative biology and evolution of bacteriophages, in: *Advances in Virus Research* (M. A. Lauffer, F. B. Bang, K. Maramorosch, and K. M. Smith, eds.), Vol. 27, pp. 205–280, Academic Press, New York.

Reeve, J. N., Hamilton, P. T., Beckler, G. S., Morris, C. J., and Clarke, C. H., 1986, Structure of methanogen genes, *Syst. Appl. Microbiol.* **7:**5–12.

Reiter, W.-D., 1985, Das virusartige Partikel SSV 1 von *Sulfolobus solfataricus* Isolat B12: UV-Induktion, Reinigung und Charakterisierung, Diploma thesis, Eberhard-Karls-Universität, Tübingen, FRG.

Reiter, W.-D., Palm, P., Henschen, A., Lottspeich, F., Zillig, W., and Grampp, B., 1987a, Identification and characterization of the genes encoding three structural proteins of the *Sulfolobus* virus-like particle SSV1, *Mol. Gen. Genet.* (in press).

Reiter, W.-D., Zillig, W., and Palm, P., 1987b, Archaebacterial viruses, in: *Advances in Virus Research* (K. Maramorosch, F. Murphy, and A. Shatkin, eds.), Academic Press, New York (in press).

Reiter, W.-D., Palm, P., Yeats, S., and Zillig, W., 1987c, Gene expression in archaebacteria: Physical mapping of constitutive and UV-inducible transcripts from the *Sulfolobus* virus-like particle SSV1, *Mol. Gen. Genet.* **209:**270–275.

Reiter, W.-D., Palm, P., Voos, W., Kaniecki, J., Grampp, B., Schulz, W., and Zillig, W., 1987d, Putative promoter elements for the ribosomal RNA genes of the thermoacidophilic archaebacterium *Sulfolobus* sp. strain B12, *Nucl. Acids Res.* **15:**5581–5595.

Reiter, W.-D., Palm, P., and Zillig, W., 1988, Analysis of transcription in the archaebacterium *Sulfolobus* indicates that archaebacterial promoters are homologous to eukaryotic pol II promoters, *Nucl. Acids Res.* **16:**1–19.

Rohrmann, G. F., Cheney, R., and Pauling, C., 1983, Bacteriophages of *Halobacterium halobium:* Virion DNAs and proteins, *Can. J. Microbiol.* **29:**627–629.

Schinzel, R., 1985, DNA-polymerisierende, DNA-restringierende und DNA-modifizierende Aktivitäten in *Halobacterium halobium*, Ph.D. Thesis, Bayerische Julius-Maximilians-Universität, Würzburg, FRG.

Schnabel, H., 1984a, An immune strain of *Halobacterium halobium* carries the invertible L segment of phage φH as a plasmid, *Proc. Natl. Acad. Sci. USA* **81:**1017–1020.

Schnabel, H., 1984b, Integration of plasmid pϕHL into phage genomes during infection of *Halobacterium halobium* R₁-L with phage ϕHL1, *Mol. Gen. Genet.* **197**:19–23.

Schnabel, H., and Zillig, W., 1984, Circular structure of the genome of phage ϕH in a lysogenic *Halobacterium halobium*, *Mol. Gen. Genet.* **193**:422–426.

Schnabel, H., Zillig, W., Pfäffle, M., Schnabel, R., Michel, H., and Delius, H., 1982a, *Halobacterium halobium* phage ϕH, *EMBO J.* **1**:87–92.

Schnabel, H., Schramm, E., Schnabel, R., and Zillig, W., 1982b, Structural variability in the genome of phage ϕH of *Halobacterium halobium*, *Mol. Gen. Genet.* **188**:370–377.

Schnabel, H., Palm, P., Dick, K., and Grampp, B., 1984, Sequence analysis of the insertion element ISH1.8 and of associated structural changes in the genome of phage ϕH of the archaebacterium *Halobacterium halobium*, *EMBO J.* **3**:1717–1722.

Segerer, A., Stetter, K. O., and Klink, F., 1985, Two contrary modes of chemolithotrophy in the same archaebacterium, *Nature* **313**:787–789.

Stetter, K. O., and Zillig, W., 1985, *Thermoplasma* and the thermophilic sulfur-dependent archaebacteria, in: *The Bacteria* (I. C. Gunsalus, J. R. Sokatch, L. N. Ornston, C. R. Woese, and R. S. Wolfe, eds.), Vol. VIII, pp. 85–170, Academic Press, Orlando, FL.

Torsvik, T., 1982, Characterization of four bacteriophages for *Halobacterium*, with special emphasis on phage Hs1, in: *Archaebacteria* (O. Kandler, ed.), p. 351, Gustav Fischer Verlag, Stuttgart.

Torsvik, T., and Dundas, I. D., 1974, Bacteriophage of *Halobacterium salinarium*, *Nature* **248**:680–681.

Torsvik, T., and Dundas, I. D., 1978, Halophilic phage specific for *Halobacterium salinarium* str. 1, in: *Energetics and Structure of Halophilic Microorganisms* (S. R. Caplan and M. Ginzburg, eds.), pp. 609–614, Elsevier North-Holland, Amsterdam.

Torsvik, T., and Dundas, I. D., 1980, Persisting phage infection in *Halobacterium salinarium* str. 1, *J. Gen. Virol.* **47**:29–36.

Vogelsang-Wenke, H., 1984, Charakterisierung des Bakteriophagen ϕN aus *Halobacterium halobium* NRL/JW, Diploma Thesis, Bayerische Julius-Maximilians-Universität, Würzburg, FRG.

Vogelsang-Wenke, H., and Oesterhelt, D., 1986, Halophage ϕN, in: *Archaebacteria '85* (O. Kandler and W. Zillig, eds.), pp. 403–405, Gustav Fischer Verlag, Stuttgart.

Wais, A. C., Kon, M., MacDonald, R. E., and Stollar, B. D., 1975, Salt-dependent bacteriophage infecting *Halobacterium cutirubrum* and *H. halobium*, *Nature* **256**:314–315.

Wich, G., Sibold, L., and Böck, A., 1986a, Genes for tRNA and their putative expression signals in *Methanococcus*, *Syst. Appl. Microbiol.* **7**:18–25.

Wich, K. G., Leinfelder, W., and Böck, A., 1986b, Genes for stable RNA in the extreme thermophile *Thermoproteus tenax*: Introns and transcription signals, *EMBO J.* **6**:523–528.

Whitman, W. B., 1985, Methanogenic bacteria, in: *The Bacteria* (I. C. Gunsalus, J. R. Sokatch, L. N. Ornston, C. R. Woese, and R. S. Wolfe, eds.), Vol. VIII, pp. 4–84, Academic Press, Orlando, FL.

Woese, C. R., and Fox, G. E., 1977, Phylogenetic structure of the prokaryotic domain: The primary kingdoms, *Proc. Natl. Acad. Sci. USA* **74**:5088–5090.

Woese, C. R., and Olsen, G. J., 1986, Archaebacterial phylogeny: Perspectives on the urkingdoms, *Syst. Appl. Microbiol.* **7**:161–177.

Woese, C. R., Magrum, L. J., and Fox, G. E., 1978, Archaebacteria, *J. Mol. Evol.* **11**:245–252.

Xu, W. L., and Doolittle, W. F., 1983, Structure of the archaebacterial transposable element ISH50, *Nucleic Acids Res.* **11**:4195–4199.

Yeats, S., Mc William, P., and Zillig, W., 1982, A plasmid in the archaebacterium *Sulfolobus acidocaldarius*, *EMBO J.* **1**:1035–1038.

Zabel, H.-P., Fischer, H., Holler, E., and Winter, J., 1985, *In vivo* and *in vitro* evidence of eukaryotic α-type DNA-polymerases in methanogens. Purification of the DNA-polymerase of *Methanococcus vannielii*, *Syst. Appl. Microbiol.* **6**:111–118.

Zillig, W., Stetter, K. O., Schäfer, W., Janekovic, D., Wunderl, S., Holz, I., and Palm, P., 1981, *Thermoproteales:* A novel type of extremely thermoacidophilic anaerobic archaebacteria isolated from Icelandic solfataras, Zbl. Bakteriol. Hyg. I. Abt. Orig. C2:205–227.

Zillig, W., Holz, I., Janekovic, D., Schäfer, W., and Reiter, W.-D., 1983, The archaebacterium Thermococcus celer represents a novel genus within the thermophilic branch of the archaebacteria, Syst. Appl. Microbiol. 4:88–94.

Zillig, W., Schnabel, R., and Stetter, K. O., 1985a, Archaebacteria and the origin of the eukaryotic cytoplasm, Curr. Top. Microbiol. Immunol. 114:1–18.

Zillig, W., Yeats, S., Holz, I., Böck, A., Gropp, F., Rettenberger, M., and Lutz, S., 1985b, Plasmid-related anaerobic autotrophy of the novel archaebacterium Sulfolobus ambivalens, Nature 313:789–791.

Zillig, W., Yeats, S., Holz, I., Böck, A., Rettenberger, M., Gropp, F., and Simon, G., 1986a, Desulfurolobus ambivalens, gen. nov., sp. nov., an autotrophic archaebacterium facultatively oxidizing or reducing sulfur, Syst. Appl. Microbiol. 8:197–203.

Zillig, W., Gropp, F., Henschen, A., Neumann, H., Palm, P., Reiter, W.-D., Rettenberger, M., Schnabel, H., and Yeats, S., 1986b, Archaebacterial virus host systems, Syst. Appl. Microbiol. 7:58–66.

Zillig, W., Holz, I., Klenk, H.-P., Trent, J., Wunderl, S., Janekovic, D., Imsel, E., and Haas, B., 1987, Pyrococcus woesei sp. nov., an ultra-thermophilic marine archaebacterium, representing a novel order, Thermococcales, Syst. Appl. Microbiol. 9:62–70.

CHAPTER 13

Temperate Bacteriophages of *Bacillus subtilis*

Stanley A. Zahler

I. INTRODUCTION

Bacillus subtilis is a gram-positive sporulating soil eubacterium. Like its gram-negative counterpart in molecular biology, *Escherichia coli*, it is prototrophic, easy to cultivate, and relatively harmless. Genetic exchange mechanisms are known for *B. subtilis*: DNA transformation, either "natural" (in strain 168 or its derivatives) or using protoplasts; generalized transduction, usually carried out by the very large bacteriophage PBS1; and specialized transduction. A number of useful plasmids have been described or constructed, although most of them originated in other gram-positive bacteria; shuttle vectors useful both in *B. subtilis* and in *E. coli* have been produced. A transposon, Tn917, can be used for shotgun mutagenesis; it has been engineered in ways that make it useful for producing transcriptional fusion insertions (with the *E. coli* β-galactosidase gene, for example; see Youngman *et al.*, 1985). A useful system for studying phage infections in minicells has been developed (Reeve, 1977). Almost one-half as many genes have been identified in *B. subtilis* as in *E. coli*. The sporulation process is considered a useful analogue of developmental and cellular differentiation in higher organisms and has been subjected to extensive study. *B. subtilis* has been known for some time to make use of what was originally thought to be a unique mechanism of metabolic control: it produces multiple RNA polymerases, which differ from each other in the polypeptides they use as σ factor.

STANLEY A. ZAHLER • Section of Genetics and Development, Division of Biological Sciences, Cornell University, Ithaca, New York 14853.

Useful reviews of these and other aspects of *B. subtilis* molecular biology can be found in Dubnau (1982, 1985).

Temperate bacteriophages of *Bacillus subtilis* have been known since 1965, when Okubo and Romig first described phage SPO2. In 1969 the temperate phage φ105 was described by Birdsell *et al.*, and also the large thymine-converting phage φ3T was isolated and described by Tucker. Phage SP16 was described by Mele in 1972. In 1977 Warner *et al.* discovered that *B. subtilis* strain 168 and almost all of its derivatives are lysogenic for a phage, related to φ3T, called SPβ.

The structures of the cell walls of *B. subtilis* strains 168 and W23 differ strikingly (Chin *et al.*, 1966; Glaser *et al.*, 1966). Almost all studies of temperate bacteriophages of *Bacillus subtilis* have been carried out with hosts that are derivatives of *B. subtilis* strain 168. It is not clear to what extent this has limited the variety of phages isolated. Certainly, most of the phages that have been examined cannot infect *B. subtilis* strain W23, probably because they cannot adsorb to that strain. [Phage SP16 is an exception to this generalization; it was originally isolated on strain W23 (see Parker and Dean, 1986).] Thus this review will consider what is likely to be only a fraction of the kinds of temperate *B. subtilis* phages existing in nature.

An earlier review of *B. subtilis* phages by Hemphill and Whiteley (1975) can be consulted. The review by Rutberg (1982) on temperate phages of *B. subtilis* is particularly useful, especially on the genetics and replication of phages φ105 and SPO2.

All temperate phages of *B. subtilis* that have been studied (which does not include phage SP16) produce a lysogenic state in which the phage genome is inserted linearly into the bacterial chromosome. In this situation the phage genome is called a prophage. No case of a plasmidlike phage genome during lysogeny (as occurs with the *Escherichia coli* phage P1; Yarmolinsky, 1982) has been reported in *B. subtilis*. In fact, *B. subtilis* phages in general seem to be less exotic than those of *E. coli*; no RNA phages, single-stranded DNA phages, or transposonlike phages (like coliphage Mu; Bukhari *et al.*, 1977) have been reported in *B. subtilis*.

Dean *et al.* (1978a) have divided the temperate phages of *B. subtilis* into four groups based upon serology, immunity, host range, and adsorption site. This useful division corroborated and expanded an earlier report by Wilson *et al.* (1974). Group I consists of phage φ105 and its close relatives. Group II consists of the similar, and slightly related, phage SPO2. Group III consists of the large phages φ3T, ρ11, SPβ, and their relatives; I will include with group III the distantly related phage H2. Group IV consists of phage SP16. An additional group, the defective temperate phage PBSX and its relatives, may be added to those four; I will call them group V phages. Phage ρ18 (Dean *et al.*, 1978a) differs from all of these, and more extensive characterization may show that it should be

TABLE I. Characteristics of Typical Temperate *B. subtilis* Phages

	Group				
	I	II	III	IV	V
Phage	ϕ105	SPO2	SPβ	SP16	PBSX
Head (nm)	52 × 52[a]	50 × 50[b]	72 × 82[c]	61 × 61[d]	45 × 45[e]
Tail (nm)	10 × 220[a]	10 × 180[b]	12 × 358[c]	12 × 192[d]	20 × 200[e]
DNA in head (kb)	40[f]	40[f]	126[c]	53–60[g]	13[h]
G + C (mole%)	43.5[a]	43[i]	31[c]	37.8[d]	43
Prophage chromosome site (degrees)	244[i,k]	15[l,k]	188[m,k]	?	112[n,k]

[a] Birdsell *et al.* (1969).
[b] Boice *et al.* (1969).
[c] Warner *et al.* (1977).
[d] Mele, J., cited by Parker and Dean (1986).
[e] Steensma *et al.* (1978).
[f] Chow *et al.* (1972).
[g] Parker and Dean (1986).
[h] Anderson and Bott (1985).
[i] Romig (1968).
[j] Rutberg (1969).
[k] Piggot and Hoch (1985).
[l] Inselburg *et al.* (1969); Smith and Smith (1973).
[m] Zahler *et al.* (1977).
[n] Anderson and Bott (1985); Anderson (1984).

placed in a separate group. A brief summary of some of the characteristics of the members of the five groups is given in Table I.

Yasbin *et al.* (1980) have constructed useful strains of *B. subtilis* that have lost the SPβ prophage and that carry a mutation that prevents the expression of the defective group V PBSX prophage for which strain 168 is lysogenic.

II. GROUP I PHAGES: ϕ105, ρ6, ρ10, ρ14

A. General Description

The group I phages that have been described include ϕ105 (Birdsell *et al.*, 1969), ρ6, ρ10, and ρ14 (Dean *et al.*, 1978a). They are relatively small phages, with genomes consisting of linear double-stranded DNA of approximately 38.5–40.1 kilobase pairs (kb) (Rudinski and Dean, 1978; Anaguchi *et al.*, 1984a). Heteroduplex analysis (Rudinski and Dean, 1978), antigenic properties, adsorption site (as indicated by the finding that bacterial mutants resistant to a clear-plaque mutant of ϕ105 are resistant to all members of the group), homoimmunity, host range (Dean *et al.*, 1978a), and restriction endonuclease analysis (Perkins *et al.*, 1978) all indicated that the four phages are closely related, with 80–90% DNA homology.

The phage particles of members of this group have icosahedral heads of approximate diameter 50–52 nm and flexible, noncontractile tails 177–220 nm long (Birdsell *et al.*, 1969; Boice *et al.*, 1969). The DNA of φ105 has a base composition of 43.5 mol% G + C, about the same as that of the host's DNA (Birdsell *et al.*, 1969). φ105 has a burst size of 100–200 and a latent period of 40 min at 37°C (Birdsell *et al.*, 1969).

The prophage of φ105 lies between the *leu* and *phe* genes on the chromosome of *B. subtilis* (Rutberg, 1969; Dean *et al.*, 1976a). [Note that the "*ilvA1*" mutation mentioned by Rutberg (1969), Peterson and Rutberg (1969), and Armentrout and Rutberg (1970) is actually in *ilvB* or *ilvC*, which are closely linked to *leu*.] Piggot and Hoch (1985) have estimated that the φ105 chromosomal attachment site may more accurately lie between the *hem* and *rod* genes at about 245° on the 360° genetic map of *B. subtilis*. Dean *et al.* (1976a) showed that the linkage between *pheA* and *leuB* was decreased from about 53% in nonlysogens to about 20% in φ105 lysogens, when measured by cotransduction with the large generalized transducing phage PBS1. The prophages of the other members of the group have not been located on the chromosome of *B. subtilis*.

Rutberg (1969) and Armentrout and Rutberg (1970) isolated more than 40 temperature-sensitive or suppressible mutations in genes essential for the replication or maturation of φ105 and classified them into 11 cistrons by genetic mapping and complementation tests. Other mutants available include several that have suffered deletions in a region extending from about 55% to about 70% from the left end of the standard φ105 genetic map (Flock, 1977). Deletions that mapped in the 55–65% region (by heteroduplex electron microscopy) did not eliminate any genes needed for growth or lysogenization of the phage. Some deletions that extended into the 65–70% region prevented lysogenization. Clear-plaque mutations, both spontaneous and mutagen-induced, were identified in the 65–70% region (Scher *et al.*, 1978), and a temperature-sensitive "repressor" mutation (*cts23*) described by Armentrout and Rutberg (1971) was also located there.

There are two sources of confusion in interpreting the literature on φ105. First, the early publications on restriction endonuclease fragmentation of the phage DNA contained two minor errors that were perpetuated for several years. In the *EcoR* I–generated genome map, a 1.1-kbp fragment (*EcoR* I–G) was erroneously positioned, and a smaller fragment (*EcoR* I–J) was reported that apparently does not exist (Perkins *et al.*, 1978; Scher *et al.*, 1978.) Corrected restriction maps were published by Bugaichuk *et al.* (1984), Lampel *et al.* (1984), Anaguchi *et al.* (1984a), Marrero and Yasbin (1986), and Ellis and Dean (1986).

A much more serious source of confusion arose early in the study of φ105. A genetic map of the phage was generated, primarily by determining the frequency of wild-type recombinants when pairs of temperature-sensitive mutants infected bacteria simultaneously (Rutberg, 1969; Armentrout and Rutberg, 1970). This map was compared to genetic maps of the prophage generated by marker-rescue experiments and by transformation and transduction experiments using the linked *phe* and *ilv* (or *leu*) bacterial markers (Peterson and Rutberg, 1969; Armentrout and Rutberg, 1970; Scher *et al.*, 1978). The genetic map of the phage seemed to be colinear with the genetic map of the prophage, and the conclusion was drawn that φ105 integrates into the *B. subtilis* chromosome at, or very close to, the ends of the vegetative phage DNA. This conclusion corroborated, and was corroborated by, the failure of Chow and Davidson (1973) to find evidence for heteroduplex structures between φ105 phage DNA and φ105 prophage DNA that would be expected if the prophage were a circular permutation of the phage DNA. The arguments leading to the conclusion are given in Rutberg's review (1982). This conclusion was almost certainly incorrect. Marrero and Yasbin (1986) and Ellis and Dean (1986) showed by Southern blot analysis of digests of φ105 phage and of prophage DNA that there was a site located about 64% from the left end of the phage DNA that acted as an attachment site for permuted insertion of φ105 DNA into the *B. subtilis* chromosome.

Examination of the papers listed in the paragraph above suggests that the best genetic and restriction maps one can construct at this time for the φ105 phage and prophage DNAs are those shown in Fig. 1. It is slightly more likely that the *B. subtilis pheA* gene lies to the left of the prophage shown in Fig. 1 than to its right.

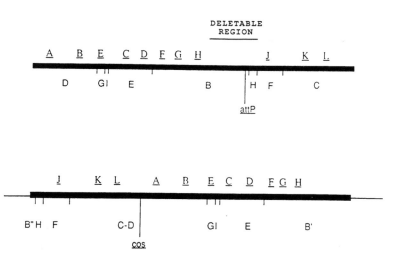

FIGURE 1. Approximate maps of φ105 phage (above) and prophage (below) DNA. Genes are shown above the lines; *Eco*R I fragments are below the lines. The prophage *Eco*R I C-D fragment (sometimes called *Eco* I-A) consists of fragments C and D of the phage map, joined at the *cos* site. The prophage *Eco*R I-B" and B' fragments result from splitting the phage *Eco*R I-B fragment at *attP*.

Phage ϕ105 DNA has complementary 3' single-stranded cohesive ends (cos) 7 bases long (Ellis and Dean, 1985). Their sequences are

$$5'-GCGCTCC-3'$$
$$3'/CGCGAGG-5'$$

Coliphage λ has complementary 5' single-stranded cohesive ends 12 bases long. The sequences of the two phage termini do not have any obvious similarity except that they are both rich in G–C pairs.

It seems clear that ϕ105 must have a life cycle similar in many respects to that of coliphage λ. Its DNA spontaneously cyclizes *in vitro* (Chow *et al.*, 1972; Scher *et al.*, 1977) and presumably *in vivo* as well. Insertion of the cyclized phage DNA at its *att* site probably occurs at a single host site between the *leuD* and *pheA* genes, resulting in a linear prophage molecule.

Lysogens of ϕ105 (and of phage SPO2) are only poorly transformable by the "natural" transformation system of *B. subtilis* (Peterson and Rutberg, 1969; Yasbin and Young, 1972; Yasbin *et al.*, 1973, 1975a,b; Garro and Law, 1974.) This is due, in large part at least, to the partial induction of the phage by the procedures that make the cells competent for transformation (Yasbin *et al.*, 1975a).

Marrero and Lovett (1980) stated that neither ϕ105 nor SPO2 phages could transduce the small plasmids pUB110 and pC194. However, either phage could transduce plasmids into which small fragments of the same phage had been cloned. Flock (1983) showed that a plasmid carrying the *cos* site of ϕ105 could be transduced by ϕ105 very efficiently (1 transductant per 10^2 plaque-forming particles). Plasmids into which foreign DNA had been cloned could also be transduced efficiently by ϕ105 if they contained the ϕ105 *cos* site. The process was described by Flock as the *in vivo* equivalent of *in vitro* cosmid packaging with coliphage λ (Collins and Hohn, 1978). Flock (1983) also found that plasmid pC194 could be transduced at low efficiency by ϕ105 and at higher efficiency by deletion mutants of ϕ105. This discrepancy between the Marrero and Lovett (1980) and Flock (1983) papers has not been explained.

B. The Lytic Cycle

The functions of most of the essential genes of ϕ105 are not known. The K gene, and possibly the J gene, are probably required for DNA replication (Rutberg, 1982).

Induction of ϕ105cts23 prophage by heat inactivation of the defective "repressor" protein results in the initiation of several rounds of replication of prophage DNA before excision of the prophage from the chromosome is completed. This replication does not stop at the prophage boundaries; the nearby chromosomal genes *phe* and *leu* are replicated as well, and they remain linked genetically to prophage markers (Armentrout and Rutberg, 1971; Rutberg, 1973). A similar phenomenon occurs with coliphage λ (Imae and Fukasana, 1970).

During phage replication in induced lysogens, concatenated phage DNA is found intracellularly. There may be a "rolling-circle" stage in the replication of φ105 (Romig, 1968; Rutberg and Rutberg, 1970).

Osburne and Sonenshein (1980) showed that φ105 growth is inhibited by the RNA polymerase inhibitors rifampin and lipiarmycin in *B. subtilis* strains that are sensitive to these drugs, and resistant when grown in mutants that are resistant to the drugs. They concluded that phage φ105 does not produce its own RNA polymerase but, rather, uses that of its host.

C. φ105 "Repressor" Activity

The expression of some of the φ105 genes is prevented by a phage-encoded protein that I will call the φ105 "repressor." There has been no formal demonstration that this repressor protein binds to φ105 DNA to prevent the transcription of phage genes, although analyses by Cully and Garro (1985) and Dhaese *et al.* (1985b) suggest that the amino acid sequence of the cloned repressor has similarities to those of known DNA-binding repressors.

Cully and Garro (1980, 1985) have described three complementation classes of φ105 mutants that are defective in the establishment or the maintenance of lysogeny. They call the genes in which the different classes occur *cI*, *cII*, and *cIII*, in conscious imitation of the coliphage λ immunity system (see review, Wulff and Rosenberg, 1983). The gene that they cloned and sequenced, which is probably their gene *cI*, lies in the *Eco*R I–F fragment, close to its border with the *Eco*R I–H fragment. (Note that the order of the genes in Fig. 1 of Cully and Garro (1985) is reversed from the usual presentation.) When the cloned gene is carried on a multicopy plasmid, it protects nonlysogenic bacteria from lysis by φ105*csi6*, a clear-plaque mutant of φ105. It is probable that *csi6* is in the same gene as the temperature-sensitive mutation *cts23*. The open reading frame believed to encode the repressor protein is sufficient to code for a protein 144 amino acids long with a molecular weight of 16,521. It is preceded by a strong ribosome-binding site and a transcription start signal suitable for RNA polymerase containing the σ^{43} polypeptide; this is the major RNA polymerase species found during vegetative growth. (The σ^{43} polypeptide was earlier called σ^{55}.)

Transcription and translation of different plasmids containing the putative repressor gene in a *B. subtilis* minicell system resulted in the biosynthesis of a protein of approximate mass 18 kD that the authors believe is the active repressor protein. Other, slightly larger proteins (19.3, 19.5, or 20 kD) are synthesized by minicells carrying three different cloned segments that also encode the 18-kD protein. It is not clear whether these larger proteins are products resulting from the fusion of vector and insert sequences or are the result of the translation of an open reading frame that overlaps (in the +1 frame) 17 bp at the 3' end of the *cl* coding sequence and encodes a 158–amino acid protein of predicted molecular weight 18,408. This second open reading frame is not preceded by a strong ribosome-binding site.

The same cloned gene, studied and sequenced by Dhaese *et al.* (1984, 1985b), protected cells that carried it from lysis by wild-type φ105. These authors pointed out an ambiguity in determining the starting codon of the repressor gene and suggested that it may in fact start at a GTG triplet 9 bp upstream from the ATG that Cully and Garro (1985) believed to be the start. This would encode a polypeptide chain of 147 amino acids, with a calculated molecular weight of 16,745. [The M_r calculations of Dhaese *et al.* (1985b) ignored the starting formylmethionine residues.] Dhaese *et al.* (1985b) pointed out a weak ribosome-binding site upstream of the second, out-of-phase open reading frame and suggested that it might be translated in a reaction coupled to the translation of the active gene. By deleting most of the second gene, they showed that protection against lysis by φ105 is indeed encoded by the upstream gene; the function of the second gene is unknown. Dhaese *et al.* (1985a) have also cloned a different gene, on the *PstI*–E fragment of φ105 DNA, that protects *B. subtilis* cells from lysis by φ105 when the gene is present on a multicopy plasmid. (This gene is located on the *EcoRI*–B fragment of φ105 DNA.) There is no evidence at present that the *PstI*–E gene is transcribed and translated; it may titrate out a *trans*-acting substance.

Dhaese *et al.* (1984) have cloned a promoter from a region very close to the repressor gene that is negatively controlled by the product of the repressor gene and have cloned several other promoters that are also under negative control of the repressor. Dhaese *et al.* (1985a) have also shown that the cloned *PstI*–E fragment gene does not exert control over transcription from this promoter.

Osburne *et al.* (1985) cloned the *EcoR* I–F fragment from φ105cts23. On a multicopy plasmid, it protected *B. subtilis* from lysis by φ105 at 30°C but not at 45°C. They also cloned a promoter from the same fragment (probably the same one studied by Dhaese *et al.* (1984)) that was under the negative control of the repressor; when a chloramphenicol acetyltransferase gene was transcribed from the promoter, the repressor prevented the expression of the enzyme at low temperature but not at high temperature. More than 50 times more mRNA was produced from the promoter at high temperature than at low temperature, suggesting strongly that the control was at the level of transcription.

D. *In Vivo* Specialized Transduction by φ105

Shapiro *et al.* (1974) showed that mitomycin C-induced lysates of φ105 lysogens contained phage particles that could transduce bacterial genes close to the prophage insertion site; in particular, the *ilvBC*, *leuC*, *pheA*, and *nic* genes were transduced. Specialized transduction was much more efficient if the recipient auxotroph was lysogenic for φ105, at least in part because nonlysogens were often lysed by the free phage present in such experiments. About one transductant per 10^6 plaque-forming particles was found for the nearest of the tested genes, *leuC*. The authors tried unsuccessfully to find transductants that could release high-frequency transducing (HFT) lysates, as can be done with coliphage λ and with some of the group III *B. subtilis* phages. The explanation for this failure is not clear, although a reexamination of the situation looking for the transduction of markers closer to the phage attachment site (*hem*, for example) might permit the isolation of HFT-producing strains.

E. φ105 and ρ14 as Cloning Vehicles

Phage φ105 has been used as a cloning vehicle, mostly for *B. subtilis* genes, in situations where the investigators want to examine the expression of a single copy of the gene *in vivo*. Most studies involving this use of φ105 have dealt with cloned sporulation (*spo*) genes. (The phenotypes of Spo mutants permit the classification of *spo* genes on the basis of the stage of sporulation that is blocked, and they are labeled *spo0*, *spoI*, *spoII*, *spoIII*, *spoIV*, or *spoV*, followed by a letter designating the particular gene giving that phenotype. There are probably more than 100 *spo* cistrons.)

The most popular cloning technique was pioneered by Iijima *et al.* (1980) and is called the "prophage transformation method." φ105 DNA was digested with a restriction endonuclease such as *Eco*R I and ligated with a similarly digested chromosomal DNA preparation. This mixture was used to transform a φ105 lysogen that was defective in the gene that was to be cloned, with selection for the wild-type allele of the desired gene. Some of the transformants were due to the replacement of the defective gene by its wild-type allele, but others were caused by the insertion of a chimera of the wild-type gene plus φ105 DNA into the φ105 prophage. Mitomycin C induction of a mixture of transformants caused an enrichment for φ105 particles that carried the wild-type allele of the desired gene. The "specialized transducing phages" (or "converting phages") were defective and required a helper phage (often, a clear-plaque mutant) to maturate. Since lysogens for specialized transducing phages can be made heterogenotic for the gene carried by the phage, it is possible to study dominance and *cis-trans* relationships between alleles. The genes could also be subcloned onto plasmid vectors.

Iijima *et al.* (1980) used the method to clone the *metB* gene of *B. subtilis.* Other papers using essentially these methods include the following. Kawamura *et al.* (1981b) and Hirochika *et al.* (1982) cloned the *spoOB* and *spoOF* genes into φ105 using DNA that had been enriched for those genes by prior cloning into phage ρ11, a member of group III of *B. subtilis* temperate phages. Yamada *et al.* (1983) cloned the *spoVE* gene in φ105. Kudoh *et al.* (1984) cloned the *spoOA* gene, using DNA enriched for the gene. Detailed studies of the cloned *spoOA* and *"spoOC"* genes have been presented by Ikeuchi *et al.* (1985). Ayaki and Kobayashi (1984) cloned the *spoIIG* gene. Anaguchi *et al.* (1984b) cloned the *spoIIC* gene into φ105 from enriched DNA originating in a specialized transducing phage of ρ11. Fujita and Kobayashi (1985) cloned the *spoIVC* gene into φ105.

Jenkinson and Mandelstam (1983) cloned the *lys* and *spoIIIB* genes into φ105. They found that lysogens of the defective lysine-transducing phage, φ105d*lys*, could be induced with mitomycin C; they produced lysates that contained φ105 particles that lacked tails. The addition *in vitro* of a φ105 lysate, or of φ105 tails, to a lysate of φ105d*lys* particles gave rise to active specialized transducing particles for the *lys* gene. They used DNA extracted from φ105d*lys*, which is about 2 kb smaller than φ105 DNA, as starting material for the isolation of a specialized transducing phage carrying the *spoIIIB* gene. The φ105d*spoIIIB* phage that they isolated had lost the *lys* gene; it also could be supplied with tails *in vitro* to become capable of transduction.

Jenkinson and Deadman (1984) cloned a selectable chloramphenicol resistance gene into the genome of φ105d*lys*, using the prophage transformation method. The insert contained a *Bgl* II target; neither φ105 nor φ105d*lys* has any *Bgl* II targets. This construct, φ105dCm^r, had lost the *lys* gene and some phage DNA. Its use as a prophage permitted the authors to clone the *metC* gene of *B. subtilis*, present on a 6-kb *Bgl* II fragment, by the same method. The resulting phage, φ105d(Cm^{r}met), contained less DNA than wild-type φ105. *Bgl* II fragments of DNA could be ligated into its DNA (causing the loss of the *metC* gene), and transformants resistant to chloramphenicol could then be selected from the ligation mixture. The authors suggested several possible uses of this phage for cloning purposes.

A deletion mutant of φ105, φ105DI:1t (Flock, 1977), was used by Savva and Mandelstam (1984) to clone the *spoIIA* and *spoVA* genes. To improve the probability of getting useful insertions into the phage DNA, they cloned the *Eco*R I–E fragment of φ105 into the *E. coli* plasmid pUC9 to generate a plasmid they called pDSMU7. Then they cloned *Hind* III fragments of the *B. subtilis* chromosome into the *Hind* III site of the cloned *Eco*R I–E fragment. This DNA preparation was then used to transform a *B. subtilis* strain lysogenic for the deletion mutant of φ105 and carrying a mutation in the appropriate *spo* gene. Again, the specialized transducing phages isolated were defective and produced no tails; but transducing particles could be produced by the addition of phage tails *in vitro*. The *spoIIA* gene was present on a 6.95-kb fragment of DNA, considerably larger than any previously reported to be cloned in ϕ105 (generally under 4 kb). Presumably the use of a deletion mutant permitted the cloning of larger pieces without making packaging the DNA in the phage head impossible.

Errington (1984) produced a useful cloning vector, ϕ105J9, by starting with the deletion mutant ϕ105DI:It. DNA fragments of at least 4 kb in length could be cloned into ϕ105J9's *Bam*H I or *Xba* I sites, or into one of its two *Sal* I sites, to give specialized transducing phages that are not defective; the phages can form plaques and lysogenize normally. The method Errington used for cloning was to concatenize purified ϕ105J9 DNA by ligating the cohesive ends together. The concatemers were then cut into single-genome-length fragments with *Bam*H I and ligated to bacterial DNA fragments cut with *Bam*H I (or with a different enzyme that produced overlaps complementary to *Bam*H I single-strand ends— e.g., *Bcl* I or *Bgl* II). The ligated mixture, consisting largely of genome-size circles of DNA containing inserts, was then used to transform protoplasts of *B. subtilis*. The cloning of *lys, spoIIA, spoIIC* (= *spoIID*; Lopez-Diaz *et al.*, 1986), and *spoVA* in ϕ105J9 were described. At present this seems to be the most useful vector for cloning in ϕ105. However, Errington was successful in cloning only four of the 15 genes he tried to clone.

Lopez-Diaz *et al.* (1986) first used the pDSMU7 method of Savva and Mandelstam (1984) to clone the *spoIID* gene and then subcloned it into the ϕ105J9 vector of Errington (1984). It is possible that this combination of methods will be more efficient than any single method, particularly if subcloning the gene onto plasmids is desired.

Dean *et al.* (1978a,b) and Kroyer *et al.* (1980) have suggested the use of phage ρ14 for cloning purposes. Its advantage over ϕ105 is said to be the existence of a *Sal* I target, potentially useful for cloning, in a nonessential region of its immunity region. Phage ρ14 also has a single target for *Bgl* II.

III. GROUP II PHAGE: SPO2

A. Relationship to ϕ105

Phage SPO2 was first described by Okuba and Romig (1965). Physically, it is very similar to ϕ105, and its DNA was partly homologous to that of ϕ105 (Boice, 1969; Boice *et al.*, 1969; Chow *et al.*, 1972). Dean *et al.* (1978a) showed that SPO2 was serologically related to ϕ105. Unlike ϕ105, it was able to form plaques on *B. globigii*. Its adsorption site on *B. subtilis* was apparently different from that of ϕ105, but the SPO2 adsorption site may be identical to that of the group III phages ρ11 and ϕ3T, as determined by tests with mutant bacteria resistant to various phages

(Dean et al., 1978a). Boice (1969), Boice et al. (1969), and Dean et al. (1978a) showed that SPO2 and φ105 were heteroimmune.

The DNA homology between SPO2 and φ105 was restricted to a region making up about 14% of their genomes (5.7 kb), lying in the central region of the phage DNAs; even within that region, homology was only partial (Chow et al., 1972). These authors have suggested that this region (detected by the electron microscopy of DNA heteroduplexes) may encode the tail proteins that, presumably, are inactivated by antiserum. The other regions of the two phages' DNA did not have detectable homology. A restriction endonuclease map of SPO2 has been prepared by Yoneda et al. (1979b).

Rutberg et al. (1972) showed that φ105 used the host's DNA polymerase III for the synthesis of phage DNA during the lytic cycle; SPO2 did not, but rather it encoded its own DNA polymerase, which has been cloned (Rutberg et al., 1981) and sequenced (Raden and Rutberg, 1984). (The sensitivity of the B. subtilis DNA polymerase III to the specific inhibitor 6(p-hydroxyphenylazo)uracil and the resistance of the phage-encoded enzyme to the drug were used to make the original determinations.)

Rutberg et al. (1972) found that a clear-plaque mutant of SPO2 was unable to complement 10 of 11 suppressible (sus) or temperature-sensitive (ts) mutants of φ105 in mixed infections. The 11th sus mutation, in gene J of φ105, was very poorly complemented. Furthermore, no wild-type recombinants with the immunity of φ105 were detected during crosses between SPO2 and φ105. The authors concluded that the two phages are essentially unrelated.

B. Characteristics of SPO2

SPO2 has a linear genome about 40 kb in length. Its ends are constant, not permuted. It inserts into the B. subtilis chromosome in a permuted form. The bacterial attachment site lies near the ribosomal protein gene clusters (Inselburg et al., 1969; Smith and Smith, 1973); Piggot and Hoch (1985) estimate that it lies between the nonA and divIVC genes. A linear genetic map with 17 cistrons was constructed by the complementation of sus mutants by Yasunaka et al. (1970). Yoneda et al. (1979b) produced a circular physical map of the SPO2 genome, indicating where the complementary ends (the cos site) lay. Graham et al. (1982) correlated the restriction endonuclease map and the genetic map, locating 11 of the 17 complementation groups on specific fragments of SPO2 DNA. Graham et al. (1979) isolated a variety of deletion mutants of SPO2. Most were clear-plaque mutants and were not examined in detail. Two of the deletion mutants that could still lysogenize B. subtilis and could be maintained as stable lysogens were studied further. Although the two deletions were not identical, they overlapped almost completely; each was about 3.5 kb long.

As expected, all of the 17 essential complementation groups were complemented by these deletion mutants.

As is the case with lysogens of φ105, lysogens of SPO2 are only poorly transformable (Peterson and Rutberg, 1969; Yasbin and Young, 1972; Yasbin et al., 1973, 1975a,b). An interesting and apparently unrelated phenomenon was described by Marrero et al. (1981) and Marrero and Lovett (1982). The plasmid pC194 (formerly called pCM194), a *Staphylococcus aureus* plasmid frequently used in B. *subtilis* molecular biology, is incompatible with the SPO2 prophage. If a lysogen carrying SPO2 is transformed for pC194 (either naturally or via protoplast transformation), the SPO2 prophage is lost frequently; eventually only nonlysogens are left in the culture, if selection for the plasmid marker (chloramphenicol resistance) is maintained. The prophage loss was not seen with φ105, nor did other plasmids cause it with SPO2. The expression of the *recE* gene (homologous with *recA* of *E. coli*) was not needed for the loss. Chimeric plasmids made by fusing the gram-positive plasmid pUB110 to pC194 were compatible with the SPO2 prophage. Chimeric plasmids made by fusing the gram-negative plasmid pBR322 to pC194 were incompatible with the SPO2 prophage and caused its loss. The patterns observed suggested to the authors that replication of pC194, using its own replication machinery, was incompatible with the retention of the SPO2 prophage. They have considered several possible models for the incompatibility.

Marrero and Lovett (1980) showed that although neither φ105 nor SPO2 phages could transduce small plasmids, either phage can transduce plasmids into which small fragments of the same phage have been cloned. If the *cos* site of SPO2 was cloned on a plasmid, the frequency of transduction of the plasmid by SPO2 was very high (1 transductant per 100 plaque-forming units, compared with $1-30 \times 10^{-6}$ for other SPO2 fragments). The plasmid containing the SPO2 *cos* site, pPL1010, and some derivatives of it, were shown by Marrero et al. (1984) to be transducible into some strains of B. *subtilis* other than strain 168 and its derivatives, and into B. *amyloliquefaciens* strains as well. SPO2 cannot form plaques on these non-168 strains.

IV. GROUP III PHAGES: φ3T, SPβ, ρ11, SPR, Z, IG1, IG3, IG4, AND H2

A. General Description

The large temperate phages of group III that have been examined seem all to be similar in appearance. They have icosahedral heads and long, flexible tails with a complex structure at the tail terminus. They contain single linear DNA molecules with measured sizes of 114–129 kb. H. E. Hemphill and his co-workers (personal communication) have found that there are seven major proteins in the phage particles of SPβ, Z, and ρ11, and six in φ3T, distinguishable by their electrophoretic mobility under denaturing conditions. (One of the φ3T bands may comigrate with another one.) Five of the bands are of similar sizes in the four phages.

Cregg and Ito (1979) have published a detailed restriction endonuclease map of the genome of phage φ3T. Kawamura et al. (1979b) have constructed such a map for phage ρ11. Fink et al. (1981), Fink and Zahler (1982a), and Spancake and Hemphill (1985) have produced such maps for SPβ. R. G. Wake (personal communication) has kindly pointed out an error in the map in Fink and Zahler (1982a); the three smallest BamH I fragments of SPβ DNA are actually 2.0, 1.4, and 0.6 kb in size rather than the published 1.1, 0.4, and 0.03 kb. Kevin Mangan in this laboratory has reexamined the sizes of the fragments and agrees with Wake's correction.

There are three lines of evidence that suggest that the chromosome of B. subtilis strain 168 contains relics of group III phages, just as the chromosome of E. coli strain K12 seems to contain relics of lambdoid phages. (1) Rowe et al. (1986) have cloned a region of the B. subtilis chromosome very close to the terminus of replication that has strong homology with the DNA of phage SPβ. Stroynowski (1981a) has also shown homology between φ3T DNA and B. subtilis chromosomal DNA. (2) Spontaneous deletions (called "citD" deletions) that start within the SPβ prophage and extend to near the replication terminus, between the citK and gltAB regions, have been described (see Zahler, 1982). It is likely that they represent recombinations between the prophage DNA and the DNA of the homologous region described by Rowe et al. (1986). Other spontaneous deletions extend from the prophage past gltAB (Zahler, 1982; Weiss et al., 1983; Rowe et al., 1986) and may represent recombinations between the prophage of SPβ and another region of homology between gltAB and citB. (3) Insertions of an integration-deficient mutant (int5) of SPβ may frequently occur in regions of homology on the chromosome. Such insertions have been found near gltAB, between thyA and glnA, and near the dal regions of the chromosome (Zahler, 1982; Gardner et al., 1982). Schneider and Anagnostopoulos (1983) have proposed that recombinations between such homologous sites of SPβ-like DNA may be responsible for the occurrence of massive inversions and other unusual genetic events that give rise to the rearranged B. subtilis chromosomes that have been studied in Anagnostopoulos's laboratory.

Sargent et al. (1985) have shown that protoplasts of SPβ lysogens degrade their own DNA during osmotic stress at least 10 times as rapidly as do nonlysogens, and they suggested that a protein encoded by the phage——possibly induced by protoplasting conditions——may be responsible.

B. Isolation

The first member of group III that was characterized was φ3T (formerly called φ3; Tucker, 1969). It was isolated from soil on B. subtilis strain NCTC 3610, which is believed to be the parent from which strain 168 was derived. Tucker (1969) showed that φ3T converted Thy⁻ B. subtilis strains to Thy⁺ and caused the appearance of thymidylate synthetase activity in the cells. Phage ρ11, isolated from soil by J. A. Hoch and characterized by Dean et al. (1976b), is similar to φ3T in almost all respects. It was isolated on a derivative of strain 168.

Warner et al. (1977) showed that B. subtilis strain 168 is lysogenic for phage SPβ, a member of group III. This discovery was made possible by the discovery of a strain, "su3⁺" (= CU1050 = BGSC 1A459), derived from strain 168, that had spontaneously lost the SPβ prophage and could therefore be used as a plating host for SPβ. Although phage particles of SPβ had been seen during an earlier electron microscope study (Eiserling, 1964), no suitable plating bacterium had been available.

These facts have led to a problem first pointed out by H. E. Hemphill. Since φ3T and ρ11 were first isolated in lysogens of SPβ, and since they are still grown in SPβ lysogens in some laboratories, φ3T and ρ11 may now be contaminated with SPβ genetic material, picked up through recombination. Different batches of φ3T or ρ11 may have different contributions of SPβ in their genomes. Zahler (1982) suggested that B. subtilis strains 1L1 and 1L3 of the Bacillus Genetic Stock Center (BGSC) should become the standard sources of φ3T and ρ11, respectively; these strains do not carry SPβ.

Phage SPR was isolated and studied by Trautner and his co-workers (1980). It is present in the prophage form in B. subtilis strain R, a strain that produces the restriction endonuclease BsuR I. Some confusion arose over the isolation of SPR, since the Trautner laboratory at first believed that SPR was actually SPβ. The two phages differ in several respects.

Phages Z (Hemphill et al., 1980), IG1, IG3, and IG4 (Fernandes et al., 1983, 1986) were isolated from soil using nonlysogenic hosts derived from strain 168. Since they are all coimmune with SPβ, they cannot form plaques on strain 168.

Phage H2 is present in the prophage state in B. amyloliquefaciens strain H (Zahler et al., 1987a). It can cause lysis but not visible plaques on ordinary strains of nonlysogenic derivatives of B. subtilis strain 168, but produces plaques only on Thy⁻ nonlysogenic derivatives (Zahler et al., 1987b). It is less closely related to other members of group III than are any of the others (Weiner, 1986).

C. Phage Immunity; Chromosome Attachment Sites

Phages SPβ, Z, SPR, IG1, IG3, IG4, and H2 form one immunity group, in that strains lysogenic for one phage cannot be lysed by another. Phages φ3T and ρ11 can make plaques on SPβ lysogens and form a second immunity group. However, the relationships are not reciprocal; SPβ cannot form plaques on φ3T or ρ11 lysogens (Dean et al., 1978a; Fernandes et al., 1986).

The chromosomal attachment site(s) of SPβ (Zahler *et al.*, 1977) and of φ3T (Williams and Young, 1977; Odebralski and Zahler, 1982) lie between the *ilvA* and *kauA* genes on the *B. subtilis* chromosome map (Piggot and Hoch, 1985) and about 132 kb from the chromosome's replication terminus (Rowe *et al.*, 1986). The relationship between *attSPβ* and *attφ3T* is not clear; they may be identical. The attachment site for the H2 prophage is a few kilobases farther from the chromosome terminus, between the *metB* and *tyrA* genes (Zahler *et al.*, 1987b). Locations of the attachment sites of other members of group III are not known.

D. Host Range and Serology

The host ranges of most of the group III phages are generally restricted to *B. subtilis* strains. None of them can infect *B. subtilis* strain W23. There are apparently strains of *B. globigii* on which φ3T and SPβ can form plaques (Dean *et al.*, 1978a). IG4 can infect a strain of *B. pumilis* (Fernandes *et al.*, 1986).

Youngman (see Zahler *et al.*, 1982) and Weiner (1986) isolated plaque-forming group III phages onto which the *Streptococcus faecalis* transposon Tn917 had transposed. This allowed Weiner to test for the host range of SPβ, H2, and SPR without demanding that phage replication and plaque formation occur. He found that H2 could infect (and make resistant to erythromycin and lincomycin, the MLS^r characteristic of Tn917) the seven strains of *B. amyloliquefaciens* that were tested: N, K, F, T, I, P, and H. Neither SPβ nor SPR could infect any of the *B. amyloliquefaciens* strains. All three could infect the strain of *B. pumilis* that was tested. The ability to convert *B. subtilis* strains to antibiotic resistance was not strongly inhibited by the presence of group III prophages, even in Rec$^-$ bacteria.

It is possible to isolate mutants of nonlysogenic derivatives of strain 168 that are resistant to clear-plaque mutants of group III phages. Estrela *et al.* (1986) have shown that a mutant isolated as resistant to φ3Tc became resistant simultaneously to ρ11, SPβ, Z, Ig1, IG3, and IG4 but not to SpR. The bacterial mutation responsible for the resistance, called *pha-3*, was mapped to a position close to the *gta* genes (glycosylation of teichoic acid) studied by Yasbin *et al.* (1976), in which mutations also gave rise to resistance to various phages. The "*gtaB*" mutation described by Rosenthal *et al.* (1979) and Zahler (1982) was apparently actually in the *pha* gene. Estrela *et al.* (1986) showed that mutations in *pha-3*, *gtaB*, and *gtaC* caused resistance to φ3Tc. Weiner (1986) found that SPβ::Tn917 could still infect the *pha-3* mutant and make it antibiotic-resistant, although at lowered efficiency.

Antiserum against SPβ inactivates SPβ, φ3T, ρ11, IG1, IG3, and IG4 but not SPR (Fernandes *et al.*, 1986) or H2 (Weiner, 1986). Antiserum against SPR inactivates only SPR (Noyer-Weidner *et al.*, 1983). Antiserum against H2 inactivates H2 but not SPβ or SPR (Weiner, 1986).

E. Betacin Production

Two of the group III phages, SPβ and Z, cause bacteria in which they are present as prophage to excrete a bacteriocinlike proteinaceous substance that kills nearby nonlysogenic bacteria on agar plates (Hemphill *et al.*, 1980). A mutant of SPβ lacking this ability, SPβ*bet-1*, has been isolated. Lysogens of SPβ and of Z are not killed by betacin; this implies the existence of a tolerance gene ("*tol*") as well. These genes must be expressed in the prophage state.

F. Thymidylate Synthetase

Each of the group III phages except SPβ carries a gene for thymidylate synthetase, which is expressed by the prophage and which converts thymine auxotrophs of *B. subtilis* to prototrophy. The genes for thymidylate synthetase have been cloned from φ3T (gene *thyP3*: Ehrlich *et al.*, 1976; Graham *et al.*, 1977; Duncan *et al.*, 1977, 1978; Stroynowski, 1981a,b; Spancake *et al.*, 1984) and from ρ11 (gene *thyP11*: Graham *et al.*, 1977). They are expressed from their own promoters in *E. coli* and in *B. subtilis*, and the cloned genes make Thy⁻ auxotrophs of either species prototrophic. There is strong homology between the *thyP3* gene and the *B. subtilis thyA* gene, which encodes the major bacterial thymidylate synthetase (Stroynowski, 1981a). There is also homology between a DNA fragment adjacent to *thyP3* and one adjacent to *thyA*. It should be noted that *B. subtilis*, unlike all other known organisms, has two thymidylate synthetases, encoded by the unlinked genes *thyA* and *thyB* (Neuhard *et al.*, 1978). A φ3T lysogen, then, must have three *thy* genes; and presumably a φ3T/Z double lysogen has four.

The homology between *thyP3* and *thyA* is sufficient to permit *thyA⁻* mutants of *B. subtilis* to be transformed to Thy⁺ by DNA from *thyP3*. In lysogens that carry the SPβ prophage, two kinds of transformants have been observed (Stroynowski, 1981b). If the *thyP3* gene was contained on a small DNA fragment, the bacterial *thyA* gene was replaced by *thyP3*. If the *thyP3* gene was contained on a larger DNA fragment, the *thyP3* gene was more likely to enter the region of the SPβ prophage that is homologous to φ3T regions surrounding the *thyP3* gene. The result of the latter event is an SPβ prophage carrying *thyP3*. Spancake *et al.* (1984) have studied the *thyP3*-carrying SPβ phages (which they called SPβT) released from such transformants. The *thyP3* gene resides at the right end of the SPβ prophage and is roughly central in the whole

DNA of SPβT phage particles. Extensive phage DNA deletions accompany the production of viable SPβT particles.

G. Other Mutations of Group III Phages

Temperature-sensitive mutants of phage SPβ have been isolated by Hemphill (cited in Zahler, 1982) and in my laboratory (unpublished experiments). None have been characterized physiologically. It is not possible to isolate nonsense mutations of SPβ using the usual plating bacteria, strain CU1050, because that strain carries a nonsense suppressor (sup-3). We have isolated a derivative of CU1050, called strain CU3069, that has lost the suppressor mutation and several other markers carried by CU1050 (Zahler et al., 1987b). Using this strain as host, we have shown that it is possible to isolate suppressible (sus) mutations of SPβ in essential genes (unpublished results).

Clear-plaque mutants of group III phages have been observed frequently. Warner et al. (1977) described one, SPβc1, that is unable to lysogenize B. subtilis.

Rettenmeier et al. (1979) found that certain mutants (φ1m) of the virulent phage φ1 were unable to grow in lysogens of SPβ, although they grew well in nonlysogens. Lysogens carrying a mutant of SPβ they called SPβmpi no longer prevented productive infection by φ1m. This system is strongly reminiscent of the relationship between the rex gene of coliphage λ and the rII genes of phage T4 (review, Court and Oppenheim, 1983).

Rosenthal et al. (1979) described an SPβ mutant, SPβc2, which seems to have a temperature-sensitive "repressor." Lysogens carrying SPβc2 could be induced to lyse by a brief treatment at elevated temperature. The c^+ and c2 alleles could complement the c1 clear-plaque mutation; presumably the three are alleles. Bacteria lysogenic for SPβc2 are cured of the prophage by growth at 50°C. (SPβ cannot replicate above about 45°C.)

Zahler (1982) described a mutation, int5, that greatly decreased the ability of SPβ to lysogenize B. subtilis at the usual bacterial attachment site (attSPβ) between ilvA and kauA. The int5 mutation could be complemented by SPβc1, permitting the integration of SPβc2int5 into attSPβ. Such a lysogen could not be induced efficiently by mitomycin C or by heat. Apparently the defect responsible for impaired integration also prevented normal excision of the prophage. Bacteria lysogenic for SPβc2int5 are killed by growth at 50°C; many of the rare survivors carry deletions that include much of the SPβ prophage and bacterial DNA extending through the kauA and citK genes (Zahler, 1982) and occasionally through the gltAB genes as well (Rowe et al., 1986). This suggests that the SPβ prophage has a gene ("kil") that is lethal to the host if it is expressed, as is true of coliphage λ (Greer, 1975).

Many deletion mutants of group III phages have been isolated. Kawamura *et al.* (1979) described a number of deletion mutants of ρ11c3, a clear-plaque mutant of ρ11, some of which had lost as much as 9 kb of DNA. Fink *et al.* (1981) and especially Spancake and Hemphill (1985) described common deletion mutants of SPβ. Most of the mutants fell into one or another of three classes with deletions of 11.8, 14.0, or 14.2 kb. These deletions lay within a 27-kb contiguous region that lies near the center of the SPβ genome. Although no essential genes lay in the region, it was transcribed both in the SPβ prophage and during phage replication. The phage attachment site (*attP*) at which the vegetative phage DNA separates during integration as prophage lies within this region and is deleted in some of the mutants.

H. DNA Methyltransferases

Several of the group III phages encode DNA methyltransferase enzymes that methylate the 5-carbons of cytidylic acids in DNA. These enzymes are expressed only during the vegetative growth of the phages. The particular cytidylic acids affected differ from phage to phage and lie within the sequences corresponding to one or another restriction endonuclease target. The phage DNAs are thus protected from the endonucleases. The phages are not known to encode restriction endonucleases. The finding that bacteria lysogenic for SPβ or for SPR do not methylate the cytidylates of certain DNA sequences unless they are first induced by various treatments that cause phage production (e.g., mitomycin C, UV light) served as an indication that there might be phage-encoded methylases. Hints that such enzymes might be encoded by the phages can be found in papers by Arwert and Rutberg (1974), Bron and Murray (1975), Bron *et al.* (1975), Gunthert *et al.* (1976), and Cregg and Ito (1979).

Cregg *et al.* (1980) showed that phage φ3T DNA contains *Hae* III targets that are protected from the endonuclease activity of *Hae* III (or its *Bacillus* isoschizomer *Bsu*R I) in mature φ3T DNA and that this protection is due to the methylation of *Hae* III targets during phage replication. It should be noted that the host strains used by Cregg *et al.* were lysogenic for phage SPβ. Trautner *et al.* (1980) showed a similar activity in cells in which SPR was replicating.

Trautner and his colleagues (Trautner *et al.*, 1980; Noyer-Weidner *et al.*, 1981, 1985; Behrens *et al.*, 1983; Montenegro *et al.*, 1983; Gunthert and Trautner, 1984 (review); Buhk *et al.*, 1984; Tran-Betcke *et al.*, 1986) showed that φ3T, ρ11, and SPβ all carry methyltransferase genes. In each of those three phages, the gene encodes a single 50-kD enzyme that recognizes two different restriction enzyme targets: the *Hae* III (= *Bsu*R I) sequence GGCC, and the *Fnu*4H I sequence GCNGC where N indicates any nucleotide. (Cytidylates that are methylated by the phage methylase enzymes are underlined.) Phage SPR, on the other hand, encodes a single

enzyme that recognizes the *Hae* III target (GGCC) and the CCGG target
that is recognized by the isoschizomeric restriction endonucleases *Hpa* II
and *Msp* I (= *Bsu*F). It should be noted that phage SPR was misidentified
as SPβ in Trautner *et al.* (1980) and in several other publications, listed in
Noyer-Weidner *et al.* (1983). Each of the four genes has been cloned and is
expressed both in *E. coli* and in *B. subtilis.* The cloned fragments carry
their own promoters. The sequences of the enzymes and of the DNA that
encodes them are markedly homologous, and the difference between the
activity of the SPR enzyme and the activities of the other three enzymes
seems to be localized in small, nonhomologous 33–amino acid sequences
that lie centrally in the polypeptides (Buhk *et al.*, 1984; Tran-Betcke *et
al.*, 1986). Mutants deficient in one of the two methylation activities can
be isolated, as can mutants lacking both activities. Tran-Betcke *et al.*
(1986) also reported that SPR DNA cannot be digested by the restriction
endonuclease *Bst*N I; target: CC(AT)GG. The parentheses indicate that
both CCAGG and CCTGG are suitable targets. If this is due to SPR-
encoded methyltransferase activity, it is not clear if the known methyltransferase of SPR is responsible or if there is a second enzyme produced. Kiss and Baldauf (1983) also cloned the gene for the methyltransferase of phage SPR (which they called SPβB in their paper.)

Phages Z (Gunthert and Trautner, 1984; Noyer-Weidner *et al.*, 1985),
IG1, IG3, and IG4 (Fernandes *et al.*, 1986) have no known DNA methyltransferases.

Phage H2 (Weiner, 1986) encodes two DNA methyltransferases. Both
have been cloned and expressed in *E. coli* and in *B. subtilis.* One enzyme is

TABLE II. DNA Methyltransferases of Group III Phages

Phage	Gene name	Endonuclease protection against	Target sequence[a]
SPR	*psm*SPR[b]	*Hae* III	GGCC
		Hpa II, *Msp* I	CCGG
	?[c]	*Eco*R II	CC(AT)GG
SPβ	*psm*SPβ[c]	*Hae* III	GGCC
φ3T	*psm*φ3T	*Fnu*4H I	GCNGC
ρ11	*psm*ρ11[c,d]		
H2	*psm*H2[e]	*Hae* III	GGCC
		*Fnu*4H I	GCNGC
		*Bsp*1286 I	G(GAT)GC(CAT)C
	*bam*M2[e]	*Bam*H I	GGATCC

[a] Cytidylates known to be methylated by the phage enzyme are underlined. Any nucleotide within a
pair of parentheses makes a suitable target for the restriction enzyme.
[b] Buhk *et al.* (1984).
[c] Tran-Betcke *et al.* (1986).
[d] H. E. Hemphill (personal communication) has reported that ρ11 DNA cannot be cut by *Sst* I (target:
GAGCTC).
[e] Weiner (1986).

much like the *Hae* III–*Fnu*4H I enzymes of SPβ and its relatives, having a molecular weight of about 50 kD, but it also carries a third activity. In addition to methylating *Hae* III and *Fnu*4H I targets, this enzyme methylates the target of *Bsp*1286 I: G(GAT)GC(CAT)C. The DNA encoding this enzyme hybridizes to that of the *Hae* III–*Fnu*4H I enzymes of φ3T and SPβ. The second methyltransferase of phage H2 recognizes the target of *Bam*H I, an enzyme made by the strain of *Bacillus amyloliquefaciens* in which H2 was lysogenic. Its target sequence is GGAT̲CC. The *Bam*H I methylase gene is physically distinct from the H2 *Hae* III-*Fnu*4H I-*Bsp*1286 I methylase gene, and no homology could be detected between the two cloned genes. Unlike the *Hae* III–*Fnu*4H I methylase gene, the *Bam*H I methylase gene did not have its own promoter for expression in *E. coli* or in *B. subtilis*; no promoter activity could be detected in clones carrying at least 2 kb of DNA upstream of the structural gene. Promoters for *E. coli* or *B. subtilis* had to be supplied for expression of the *Bam*H I DNA methyltransferase. Table II summarizes the known DNA methyltransferases of the group III phages.

I. Specialized Transduction

Phage SPβ gave specialized transduction of bacterial genes close to its normal attachment site (*attSPβ*) on the *B. subtilis* chromosome (Zahler *et al.*, 1977; review, Zahler, 1982). These included the *kauA* and *citK* genes that lie counterclockwise to the attachment site (Rosenthal *et al.*, 1979) and the *ilvA*, *thyB*, and *ilvD* genes that lie clockwise to the attachment site (Fink and Zahler, 1982b). The specialized transduction process is similar to that found with coliphage λ, except that the ratio of transducing particles to plaque-forming particles in high-frequency transducing lysates is usually about 1:1000 instead of 1:1. All of the specialized transducing phages described except one have been defective; the one exception carried only the *ilvA* gene (Fink and Zahler, 1983).

By using SPβ mutants carrying the integration-defective mutation *int5* (Zahler, 1982), it was possible to find SPβ prophages inserted in abnormal sites on the *B. subtilis* chromosome. Lipsky *et al.* (1981) described specialized transducing phages from one such insertion that carried the *dal* and *sup-3* or the *dal*, *sup-44*, and *ddl* markers. Gardner *et al.* (1982) studied the control of glutamine synthetase production in a heterogenote by using a specialized transducing phage derived from an SPβc2int5 insertion close to *glnA*.

By infecting bacteria that lacked the normal *attSPβ*, Mackey and Zahler (1982) found an SPβc2 prophage that had inserted into the *sdh* (= *citF*) genes. From that insertion they were able to find defective specialized transducing phages that carried the 13 known genes lying between *sdh* and *leuD*. They used these phages to study the *ilvBNC-leu* operon and developed methods for studying *cis-trans* and dominance relationships in that operon (Mackey and Zahler, 1982; Vandeyar *et al.*, 1986). The availability of an SPβ phage with the transposon Tn*917* in its genome made it relatively simple to isolate new insertions in unusual locations; new specialized transducing phages could be isolated from them (Zahler *et al.*, 1982).

Defective specialized transducing phages from phage φ3T inserted in its normal attachment site were able to transduce the *kauA* and *citK* markers of *B. subtilis* (Odebralski and Zahler, 1982).

Phage H2 was able to give specialized transduction of the nearby chromosomal markers *metB*, *ilvD*, and *ilvA* (Zahler *et al.*, 1987b).

J. Cloning in Group III Phages

Phages ρ11, φ3T, and SPβ have been used as cloning vectors. Sporulation genes (*spo*) have frequently been cloned in them. Like the smaller phage, φ105, they have the advantage of being present in a single copy in lysogenic bacteria, and cloning into ρ11 is more efficient than into φ105. However, because of the large genome size of the group III phages, it is difficult to reisolate the cloned gene from them. It is sometimes desirable to subclone a gene from ρ11 into φ105 before trying to isolate it for sequencing, for example.

The prophage transformation method, described in section II.E, was first used with phage ρ11 (Kawamura *et al.*, 1979a). Both defective and plaque-forming transducing phages were produced, carrying the *hisA* or *lys* gene of *B. subtilis*. The *hisA*-containing phage was unstable (Kawamura *et al.*, 1981a). Nomura *et al.* (1979) cloned the α-amylase gene of *B. subtilis* into ρ11. Yoneda *et al.* (1979a) cloned the α-amylase gene of *B. amyloliquefaciens* into φ3T. Kawamura *et al.* (1980) cloned the *spoOF* gene of *B. subtilis* into ρ11. Hirochika *et al.* (1981) cloned the sporulation gene *spoOB* into ρ11. Shinomiya *et al.* (1981) moved the gene for a thermostable α-amylase from a soil thermophile into *B. subtilis* and then cloned the gene into ρ11. Two genes for gluconate utilization, coding for an uptake protein and a kinase, were cloned in ρ11 by Fujita *et al.* (1983, 1986). Ikeuchi *et al.* (1983) cloned the *spoOA* gene (and the "*spoOC*" gene, which later was shown to be within the *spoOA* cistron) into ρ11. Yamazaki *et al.* (1983) cloned α-amylase genes from a hyperproducing strain of *B. subtilis* into ρ11. Anaguchi *et al.* (1984b) cloned *spoIIC* into ρ11. Ayaki and Kobayashi (1984) cloned the *spoIIG* gene into ρ11.

Ferrari and Hoch (1982, 1983) took a different approach. They constructed a shuttle plasmid called pFH7 that contains the *E. coli* plasmid pBR322, the *Staphylococcus aureus* plasmid pHV14 (which can replicate in *B. subtilis*), and a fragment of SPβ DNA. This plasmid is unstable in *B. subtilis*, but when it integrates into the SPβ prophage, it becomes very stable. The integration is apparently via a Campbell-type reaction. Bacteria that are lysogenic for SPβ::pFH7 can be induced; they release SPβ particles that convert new cells to chloramphenicol resistance. Such lysogens can be transformed with pBR322-related plasmids that contain cloned genes. The pFH7 lying within the SPβ prophage serves as a site of homology with the incoming plasmid DNA, which can then insert (possibly via a Campbell-type reaction) into pFH7. From such transformants, SPβ particles containing both pFH7 and the cloned gene can be induced. Ferrari and Hoch (1983) reported the cloning of the wild-type allele of the temperature-sensitive mutation *tms-26* in this manner.

Ordal *et al.* (1983) used a variant of this method to clone several chemotaxis (*che*) genes from *B. subtilis*. They used SPβ*c2*, a heat-inducible mutant of SPβ, and isolated deletion mutants of it to permit the cloning of larger pieces of DNA. The DNA was subcloned from coliphage λ clones containing *che* genes, directly into pFH7, and then moved by transformation into bacteria lysogenic for deletion mutants of SPβ*c2*. Induction of the lysogens permitted the isolation of transducing SPβ phages carrying *che* genes (and the chloramphenicol resistance marker of pFH7). The largest chromosomal fragment cloned in this report was 11.7 kb in length; pFH7 is an additional 10 kb. Thus the transducing particles carry more than 20 kb of inserted DNA. The method was particularly valuable in this case, because many Che⁻ mutants cannot be transformed.

V. GROUP IV PHAGE: SP16

The medium-size temperate phage SP16 has been the focus of one thesis (Mele, 1972) and three short articles (Thorne and Mele, 1974; Dean *et al.*, 1978a; Parker and Dean, 1986). SP16 was isolated with *B. subtilis* strain W23 as host and is the only temperate phage known that can infect that strain. It infects strain 168 as well as some strains of *B. amyloliquefaciens* and *B. licheniformis*. It is unique in its antigenic properties, immunity, and host range (Thorne and Mele, 1974; Dean *et al.*, 1978a). Attempts to demonstrate generalized transduction with SP16 failed (Mele, 1972).

The phage has a head diameter of 61 nm and a tail of length 192 nm. The G + C molar content of its genome is given as 37.8%, based on its T_m value (Mele, 1972). Its DNA content is given by Parker and Dean (1986) as 52.8 kb (measured by the electrophoretic mobility of its fragments after digestion with *Kpn* I) or 60.0 kb (measured as the length of spread molecules by electron microscopy). After partial digestion of the DNA by exonuclease III and annealing, circular molecules of length 48.9 kb were detected, suggesting terminal redundancy of 10–15%. Parker and Dean (1986) also deduced that the termini of the phage particles were circularly permuted.

Thorne and Mele (1974) found that the efficiency of plating by SPO2 on lysogens of SP16 was reduced to 10^{-4} of the efficiency on nonlysogens. This was not due to a restriction-modification system induced by the SP16 prophage or to phage immunity; the rare plaques found on 168 (SP16) were due to mutants of SPO2. The mutants were not altered in their ability to adsorb to the host.

VI. GROUP V DEFECTIVE PHAGES: PBSX, PBSZ, *et al.*

If the cells of *B. subtilis* strain 168 are exposed to DNA-damaging treatments (mitomycin C, UV light, etc.), the cells lyse after an incubation period of an hour or so and release particles of a defective bacteriophage called PBSX (Seaman *et al.*, 1964) [= SPα (Eiserling, 1964) = μ (Ionesco *et al.*, 1964) = ϕ3610 (Stickler *et al.*, 1965) = PBSH (Haas and Yoshikawa, 1969)]. The defective particles can be detected by electron microscopy. The burst size is several hundred. The particles adsorb to and kill certain strains of related bacilli. In particular, PBSX kills the cells of *B. subtilis* strain W23, acting like a bacteriocin. PBSX is defective in a number of interesting ways: (1) Although it looks like a small-headed phage with a complex tail structure, its head does not contain phage DNA; rather, each phage head contains a randomly selected 13-kb fragment of bacterial chromosomal DNA (Okamoto *et al.*, 1968a,b; Siegel and Marmur, 1969; Anderson and Bott, 1985). DNA extracted from PBSX can be used as a source of *B. subtilis* chromosomal DNA (Kawamura *et al.*, 1979a). (2) Under certain conditions—e.g., low Mg^{2+} concentration (Okamoto *et al.*, 1968a; Steensma *et al.*, 1978)—the contractile tail sheaths contract in a manner quite different from that of, for example, the T-even coliphages: the distal end of the sheath remains attached to the tail core, and the proximal portion of the sheath detaches from the core and increases in diameter while decreasing in length (Stickler *et al.*, 1965; Steensma, 1981a). (3) Although the PBSX particles adsorb efficiently to the walls of certain strains of *B. subtilis*, the DNA in the phage heads is not injected (Okamoto *et al.*, 1968a). Thus, PBSX cannot transduce markers to new strains of bacteria. PBSX does not adsorb to the cells of strain 168.

Strain W23, on the other hand, produces a slightly different defective phage, PBSZ, that similarly kills cells of strain 168. Steensma *et al.* (1978) have described similar phages found in lysates from strains of *B. subtilis*, *B. amyloliquefaciens*, *B. licheniformis*, *B. pumilis*, *B. funicularius*, and *B. laterosporus*. They were able to distinguish at least five different morphological types, differing according to the lengths of their tails and by the range of strains their defective particles could kill. The phage heads were about 45 nm in diameter, and the tails were about 20 nm in diameter, ranging in length from 185 nm to 265 nm for different phages. The tail length for PBSX is 200 nm; for PBSZ, 255 nm. Steensma (1981a) has studied the adsorption of PBSZ to the cells of strain 168 and has shown that tail contraction is induced by low Mg^{2+} concentration.

The prophage region of PBSX lies on the *B. subtilis* 168 chromosome in a region where the nearby markers are *metA*-PBSX-*sapA/phoS-metC* (Anderson, 1984). The *sapA/phoS* markers control alkaline phosphatase activity. Several mutations affecting PBSX have been isolated within the *metA-metC* region: *xin* (Thurm and Garro, 1975b) prevents the induction of PBSX; *xhi* (Buxton, 1976) is heat-inducible for PBSX; *xtl* (Thurm and Garro, 1975b) causes the loss of tail structures; *xhd* (Thurm and Garro, 1975b) causes the production of particles with defective heads; and *xki* (Buxton, 1976; Steensma, 1981b) causes the production of normal-looking particles that cannot kill cells of strain W23 (they lack an 85-kD protein that may be responsible for the killing effect of PBSX). Steensma (1981b) and Thurm and Garro (1975a) have studied the synthesis of the structural proteins of PBSX. Buxton (1980) has isolated *B. subtilis* 168 chromosomal deletions that remove much or all of the PBSX prophage.

Anderson *et al.* (1982) cloned a region of the *B. subtilis* chromosome from the region between *metA* and *metC* that permits the autonomous replication of plasmids in *B. subtilis*. This region is presumably the origin of replication of PBSX phage DNA. They also cloned and characterized DNA that lies adjacent to this fragment (Anderson, 1984) and includes PBSX DNA. None of the cloned fragments complemented any of the presumed prophage mutations described above. The cloned DNA is partly homologous to regions in *B. subtilis* strain W23 that presumably encode PBSZ.

Anderson and Bott (1985) have studied the nature of the DNA packaged into PBSX particles and have reviewed other papers that examined similar questions. The DNA is a reasonably random sample of the bacterial chromosome, with some excess of a marker (*purA*) close to the bacterial origin of replication. There is little or no enrichment for markers close to the PBSX prophage.

The evolutionary history of the group V defective prophages is rather mysterious. Presumably there was once a nondefective precursor of PBSX and its family. We can imagine that the precursor prophage mutated to become defective, and perhaps that its host mutated so that it could no longer adsorb the lethal defective phage particles. It is more difficult to imagine the pressures that led to the development of a family of defective prophages with different lengths of tails and different patterns of adsorption and killing.

ACKNOWLEDGMENTS. I thank my colleagues who have given freely of their time and knowledge. In particular, my thanks go to Linda Anderson, H. Ernest Hemphill, Donald Dean, Thomas Trautner, and Michael Weiner for sharing unpublished information. This work was supported in part by grant GM33152 from the National Institute of General Medical Services.

REFERENCES

Anaguchi, H., Fukui, S., and Kobayashi, Y., 1984a, Revised restriction maps of *Bacillus subtilis* bacteriophage φ105 DNA, *J. Bacteriol.* **159:**1080–1082.

Anaguchi, H., Fukui, S., Shimotsu, H., Kawamura, F., Saito, H., and Kobayashi, Y., 1984b, Cloning of sporulation gene *spoIIC* in *Bacillus subtilis*, *J. Gen. Microbiol.* **130:**757–760.

Anderson, L. M., 1984, Molecular studies of the Bacillus subtilis defective bacteriophage, PBSX, Ph.D. Thesis, University of North Carolina, Chapel Hill.

Anderson, L. M., and Bott, K. F., 1985, DNA packaging by the *Bacillus subtilis* defective bacteriophage PBSX, *J. Virol.* **54:**773–780.

Anderson, L. M., Ruley, H. E., and Bott, K. F., 1982, Isolation of an autonomously replicating DNA fragment from the region of defective bacteriophage PBSX of *Bacillus subtilis*, *J. Bacteriol.* **150:**1280–1286.

Armentrout, R. W., and Rutberg, L., 1970, Mapping of prophage and mature deoxyribonucleic acid from temperate *Bacillus* bacteriophage φ105 by marker rescue, *J. Virol.* **6:**760–767.

Armentrout, R. W., and Rutberg, L., 1971, Heat induction of φ105 in *Bacillus subtilis:* Replication of the bacterial and bacteriophage genomes, *J. Virol.* **8:**455–468.

Arwert, F., and Rutberg, L., 1974, Restriction and modification in *Bacillus subtilis*. Induction of a modifying activity in *Bacillus subtilis* 168, *Mol. Gen. Genet.* **133:**175–177.

Ayaki, H., and Kobayashi, Y., 1984, Cloning of sporulation gene *spoIIG* in *Bacillus subtilis*, *J. Bacteriol.* **158:**507–512.

Behrens, B., Pawlek, B., Morelli, G., and Trautner, T. A., 1983, Restriction and modification in *Bacillus subtilis:* Construction of hybrid λ and SPP1 phages containing a DNA methyltransferase gene from *B. subtilis* phage SPR, *Mol. Gen. Genet.* **189:**10–16.

Birdsell, D. C., Hathaway, G. M., and Rutberg, L., 1969, Characterization of temperate *Bacillus* bacteriophage φ105, *J. Virol.* **4:**264–270.

Boice, L. B., 1969, Evidence that *Bacillus subtilis* bacteriophage SPO2 is temperate and heteroimmune to bacteriophage φ105, *J. Virol.* **4:**47–49.

Boice, L., Eiserling, F. A., and Romig, W. R., 1969, Structure of *Bacillus subtilis* phage SPO2 and its DNA: Similarity of *Bacillus subtilis* phages SPO2, φ105 and SPP1, *Biochem. Biophys. Res. Commun.* **34:**398–403.

Bron, S., and Murray, K., 1975, Restriction and modification in *B. subtilis*. Nucleotide sequence recognized by restriction endonuclease R.*BsuR* from strain R, *Mol. Gen. Genet.* **143:**25–33.

Bron, S., Murray, K., and Trautner, T. A., 1975, Restriction and modification in *B. subtilis*. Purification and general properties of a restriction endonuclease from strain R, *Mol. Gen. Genet.* **143:**13–23.

Bugaichuk, U. D., Deadman, M., Errington, J., and Savva, D., 1984, Restriction enzyme analysis of *Bacillus subtilis* bacteriophage φ105 DNA, *J. Gen. Microbiol.* **130:**2165–2167.

Buhk, H.-J., Behrens, B., Tailor, R., Wilke, K., Prada, J. J., Gunthert, U., Noyer-Weidner, M., Jentsch, S., and Trautner, T. A., 1984, Restriction and modification in *Bacillus subtilis:* Nucleotide sequence, functional organization and product of the DNA methyltransferase gene of bacteriophage SPR, *Gene* **29:**51–61.

Bukhari, A. I., Ljungquist, E., De Bruijn, F., and Khatoon, H., 1977, The mechanism of bacteriophage Mu integration, in: *DNA Insertion Elements, Plasmids, and Episomes* (A. I. Bukhari, J. A. Shapiro, and S. L. Adhya, eds.), pp. 249–261, Cold Spring Harbor Laboratory, Cold Spring Harbor, NY.

Buxton, R. S., 1976, Prophage mutation causing heat inducibility of defective *Bacillus subtilis* bacteriophage PBSX, *J. Virol.* **20:**22–28.

Buxton, R. S., 1980, Selection of *Bacillus subtilis* 168 mutants with deletions of the PBSX prophage, *J. Gen. Virol.* **46:**427–437.

Chin, T., Burger, M. M., and Glaser, L., 1966, Synthesis of teichoic acids. VI. The formation of multiple wall polymers in *Bacillus subtilis* W-23, *Arch. Biochem. Biophys.* **116**:358–367.

Chow, L. T., Boice, L., and Davidson, N., 1972, Map of the partial sequence homology between DNA molecules of *Bacillus subtilis* bacteriophages SPO2 and φ105, *J. Mol. Biol.* **68**:391–400.

Chow, L. T., and Davidson, N., 1973, Electron microscope study of the structures of the *Bacillus subtilis* prophages SPO2 and φ105, *J. Mol. Biol.* **75**:257–264.

Collins, J., and Hohn, B., 1978, Cosmids: A type of plasmid gene-cloning vector that is packageable in vitro in bacteriophage λ heads, *Proc. Natl. Acad. Sci. USA* **75**:4242–4246.

Court, D., and Oppenheim, A. B., 1983, Phage lambda's accessory genes, in: *Lambda II* (R. W. Hendrix, J. W. Roberts, F. W. Stahl, and R. A. Weisberg, eds.), pp. 251–277, Cold Spring Harbor Laboratory, Cold Spring Harbor, NY.

Cregg, J. M., and Ito, J., 1979, A physical map of the genome of temperate phage φ3T, *Gene* **6**:199–219.

Cregg, J. M., Nguyen, A. H., and Ito, J., 1980, DNA modification induced during infection of *Bacillus subtilis* by phage φ3T, *Gene* **12**:17–24.

Cully, D. F., and Garro, A. J., 1980, Expression of superinfection immunity to bacteriophage φ105 by *Bacillus subtilis* cells carrying a plasmid chimera of pUB110 and *Eco*RI fragment F of φ105 DNA, *J. Virol.* **34**:789–791.

Cully, D. F., and Garro, A. J., 1985, Nucleotide sequence of the immunity region of *Bacillus subtilis* bacteriophage φ105: Identification of the repressor gene and its mRNA and protein products, *Gene* **38**:153–164.

Dean, D. H., Arnaud, M., and Halvorson, H. O., 1976a, Genetic evidence that *Bacillus* bacteriophage φ105 integrates by insertion. *J. Virol.* **20**:339–341.

Dean, D. H., Orrego, J. C., Hutchison, K. W., and Halvorson, H. O., 1976b, New temperate bacteriophage for *Bacillus subtilis*, ρ11, *J. Virol.* **20**:509–519.

Dean, D. H., Fort, C. L., and Hoch, J. A., 1978a, Characterization of temperate phages of *Bacillus subtilis*, *Curr. Microbiol.* **1**:213–217.

Dean, D. H., Perkins, J. B., and Zarley, C. D., 1978b, Potential temperate bacteriophage molecular vehicle for *Bacillus subtilis*, in: *Spores VII* (G. Chambliss and J. C. Vary, eds.), pp. 144–149, American Society for Microbiology, Washington.

Dhaese, P., Hussey, C., and Van Montagu, M., 1984, Thermo-inducible gene expression in *Bacillus subtilis* using transcriptional regulatory elements from temperate phage φ105, *Gene* **32**:181–194.

Dhaese, P., Dobbelaere, M.-R., and Van Montagu, M., 1985a, The temperate *B. subtilis* phage φ105 genome contains at least two distinct regions encoding superinfection immunity, *Mol. Gen. Genet.* **200**:490–492.

Dhaese, P., Seurinck, J., De Smet, B., and Van Montagu, M., 1985b, Nucleotide sequence and mutational analysis of an immunity repressor gene from *Bacillus subtilis* temperate phage φ105, *Nucleic Acids Res.* **13**:5441–5455.

Dubnau, D. A. (ed.), 1982, *Molecular Biology of the Bacilli*, Vol. 1, Academic Press, New York.

Dubnau, D. A. (ed.), 1985, *Molecular Biology of the Bacilli*, Vol. 2, Academic Press, New York.

Duncan, C. H., Wilson, G. A., and Young, F. E., 1977, Transformation of *Bacillus subtilis* and *Escherichia coli* by a hybrid plasmid pCD1, *Gene* **1**:153–167.

Duncan, C. H., Wilson, G. A., and Young, F. E., 1978, Mechanism of integrating foreign DNA during transformation of *Bacillus subtilis*, *Proc. Natl. Acad. Sci. USA* **75**:3664–3668.

Ehrlich, S. D., Bursztyn-Pettegrew, I., Stroynowski, I., and Lederberg, J., 1976, Expression of the thymidylate synthetase gene of the *Bacillus subtilis* bacteriophage φ3T in *Escherichia coli*, *Proc. Natl. Acad. Sci. USA* **73**:4145–4149.

Eiserling, F. A., 1964, Ph.D. Thesis, University of California, Los Angeles.

Ellis, D. M., and Dean, D. H., 1985, Nucleotide sequence of the cohesive single-stranded ends of *Bacillus subtilis* temperate bacteriophage φ105, *J. Virol.* **55**:513–515.

Ellis, D. M., and Dean, D. H., 1986, Location of the *Bacillus subtilis* temperate bacteriophage φ105 *attP* attachment site, *J. Virol.* **58**:223–224.

Errington, J., 1984, Efficient *Bacillus subtilis* cloning system using bacteriophage vector φ105J9, *J. Gen. Microbiol.* **130**:2615–2628.

Estrela, A. I., De Lencastre, H., and Archer, L. J., 1986, Resistance of a *Bacillus subtilis* mutant to a group of temperate bacteriophages, *J. Gen. Microbiol.* **132**:411–415.

Fernandes, R. M., De Lencastre, H., and Archer, L. J., 1983, Two newly isolated temperate phages of *Bacillus subtilis*, *Broteria-Genetica* **4(79)**:27–33.

Fernandes, R. M., De Lencastre, H., and Archer, L. J., 1986, Three new temperate phages of *Bacillus subtilis*, *J. Gen. Microbiol.* **132**:661–668.

Ferrari, E., and Hoch, J. A., 1982, System for complementation and dominance analyses in *Bacillus*, in: *Molecular Cloning and Gene Regulation in Bacilli* (A. T. Ganesan, S. Chang, and J. A. Hoch, eds.), pp. 53–61, Academic Press, New York.

Ferrari, E., and Hoch, J. A., 1983, A single copy, transducible system for complementation and dominance analyses in *Bacillus subtilis*, *Mol. Gen. Genet.* **189**:321–325.

Fink, P. S., Korman, R. Z., Odebralski, J. M., and Zahler, S. A., 1981, *Bacillus subtilis* bacteriophage SPβc1 is a deletion mutant of SPβ, *Mol. Gen. Genet.* **182**:514–515.

Fink, P. S., and Zahler, S. A., 1982a, Restriction fragment maps of the genome of *Bacillus subtilis* bacteriophage SPβ, *Gene* **19**:235–238.

Fink, P. S., and Zahler, S. A., 1982b, Specialized transduction of the *ilvD-thyB-ilvA* region mediated by *Bacillus subtilis* bacteriophage SPβ, *J. Bacteriol.* **150**:1274–1279.

Fink, P. S., and Zahler, S. A., 1983, SPβc2pilvA: Plaque-forming bacteriophages that transduce the *Bacillus subtilis* ilvA gene, *Abstr. Am. Soc. Microbiol.* **1983**:111.

Flock, J. I., 1977, Deletion mutants of temperate *Bacillus subtilis* bacteriophage φ105, *Mol. Gen. Genet.* **155**:241–247.

Flock, J.-I., 1978, Transfection with replicating DNA from the temperate *Bacillus* bacteriophage φ105 and with T4-ligase treated φ105 DNA: The importance in transfection of being longer than one genome-length, *Mol. Gen. Genet.* **163**:7–15.

Flock, J.-I., 1983, Cosduction: Transduction of *Bacillus subtilis* with phage φ105 using a φ105 cosplasmid, *Mol. Gen. Genet.* **189**:304–308.

Fujita, M., and Kobayashi, Y., 1985, Cloning of sporulation gene *spoIVC* in *Bacillus subtilis*, *Mol. Gen. Genet.* **199**:471–475.

Fujita, Y., Fujita, T., Kawamura, F., and Saito, H., 1983, Efficient cloning of genes for utilization of D-gluconate of *Bacillus subtilis* in phage ρ11, *Agric. Biol. Chem.* **47**:1679–1682.

Fujita, Y., Nihashi, J.-I., and Fujita, T., 1986, The characterization and cloning of a gluconate (*gnt*) operon of *Bacillus subtilis*, *J. Gen. Microbiol.* **132**:161–169.

Gardner, A., Odebralski, J., Zahler, S., Korman, R. Z., and Aronson, A. I., 1982, Glutamine synthetase subunit mixing and regulation in *Bacillus subtilis* partial diploids, *J. Bacteriol.* **149**:378–380.

Garro, A. J., and Law, M. F., 1974, Relationship between lysogeny, spontaneous induction, and transformation efficiency in *Bacillus subtilis*, *J. Bacteriol.* **120**:1256–1259.

Glaser, L., Ionesco, H., and Schaeffer, P., 1966, Teichoic acids as components of a specific phage receptor in *Bacillus subtilis*, *Biochim. Biophys. Acta* **124**:415–417.

Graham, S., Sutton, S., Yoneda, Y., and Young, F. E., 1982, Correlation of the genetic map and the endonuclease site map of *Bacillus subtilis* bacteriophage SPO2, *J. Virol.* **42**:131–134.

Graham, S., Yoneda, Y., and Young, F. E., 1979, Isolation and characterization of viable deletion mutants of *Bacillus subtilis* bacteriophage SPO2, *Gene* **7**:69–77.

Graham, R. S., Young, F. E., and Wilson, G. A., 1977, Effect of site-specific endonuclease digestion on the *thyP3* gene of bacteriophage φ3T and the *thyP11* gene of bacteriophage ρ11, *Gene* **1**:169–180.

Greer, H., 1975, The *kil* gene of bacteriophage lambda, *Virology* **66**:589–604.

Gunthert, U., and Trautner, T. A., 1984, DNA methyltransferases of *Bacillus subtilis* and its bacteriophages, *Curr. Top. Microbiol. Immunol.* **108**:11–22.

Gunthert, U., Pawlek, B., Stutz, J., and Trautner, T. A., 1976, Restriction and modification in *Bacillus subtilis:* Inducibility of a DNA methylating activity in non-modifying cells, *J. Virol.* **20**:188–195.

Haas, M., and Yoshikawa, H., 1969, Defective bacteriophage PBS H in *Bacillus subtilis.* I. Induction, purification, and physical properties of the bacteriophage and its deoxyribonucleic acid, *J. Virol.* **3**:248–260.

Hemphill, H. E., and Whiteley, H. R., 1975, Bacteriophages of *Bacillus subtilis, Bacteriol. Rev.* **39**:257–315.

Hemphill, H. E., Gage, I., Zahler, S. A., and Korman, R. Z., 1980, Prophage-mediated production of a bacteriocin-like substance by SPβ lysogens of *Bacillus subtilis, Can. J. Microbiol.* **26**:1328–1333.

Hirochika, H., Kobayashi, Y., Kawamura, F., and Saito, H., 1981, Cloning of sporulation gene *spoOB* of *Bacillus subtilis* and its genetic and biochemical analysis, *J. Bacteriol.* **146**:494–505.

Hirochika, H., Kobayashi, Y., Kawamura, F., and Saito, H., 1982, Construction and characterization of φ105 specialized transducing phages carrying sporulation genes *spoOB* and *spoOF* of *Bacillus subtilis, J. Gen. Appl. Microbiol.* **28**:225–229.

Iijima, T., Kawamura, F., Saito, H., and Ikeda, Y., 1980, A specialized transducing phage constructed from *Bacillus subtilis* phage φ105, *Gene* **9**:115–126.

Ikeucki, T., Kudoh, J., and Kurahashi, K., 1983, Cloning of sporulation genes *spoOA* and *spoOC* of *Bacillus subtilis* onto ρ11 temperate bacteriophage, *J. Bacteriol.* **154**:988–991.

Ikeuchi, T., Kudoh, J., and Kurahashi, K., 1985, Genetic analysis of *spoOA* and *spoOC* mutants of *Bacillus subtilis* with a φ105 prophage merodiploid system, *J. Bacteriol.* **163**:411–416.

Imae, Y., and Fukasawa, T., 1970, Regional replication of the bacterial chromosome by derepression of prophage lambda, *J. Mol. Biol.* **54**:585–597.

Inselburg, J. W., Eremenko-Volpe, T., Greenwald, L., Meadow, W. L., and Marmur, J., 1969, Physical and genetic mapping of the SPO2 prophage on the chromosome of *Bacillus subtilis, J. Virol.* **3**:627–628.

Ionesco, H., Ryter, A., and Schaeffer, P., 1964, Sur un bacteriophage heberge par ia souche Marburg de *Bacillus subtilis, Ann. Inst. Pasteur* **107**:764–776.

Jenkinson, H. F., and Deadman, M., 1984, Construction and characterization of recombinant phage φ105 d(Cmʳmet) for cloning in *Bacillus subtilis, J. Gen. Microbiol.* **130**:2155–2164.

Jenkinson, H. F., and Mandelstam, J., 1983, Cloning of the *Bacillus subtilis lys* and *spoIIIB* genes in phage φ105, *J. Gen. Microbiol.* **129**:2229–2240.

Kawamura, F., Saito, H., and Ikeda, Y., 1979a, A method for construction of specialized transducing phage ρ11 of *Bacillus subtilis, Gene* **5**:87–91.

Kawamura, F., Saito, H., Ikeda, Y., and Ito, J., 1979b, Viable deletion mutants of *Bacillus subtilis* phage ρ11, *J. Gen. Appl. Microbiol.* **25**:223–226.

Kawamura, F., Saito, H., Hirochika, H., and Kobayashi, Y., 1980, Cloning of sporulation gene, *spoOF*, in *Bacillus subtilis* with ρ11 phage vector, *J. Gen. Appl. Microbiol.* **26**:345–355.

Kawamura, F., Mizukami, T., Anzai, H., and Saito, H., 1981a, Frequent deletion of *Bacillus subtilis* chromosomal fragment in artificially constructed phage ρ11*phisA* ⁺, *FEBS Lett.* **136**:244–246.

Kawamura, F., Shimotsu, H., Saito, H., Hirochika, H., and Kobayashi, Y., 1981b, Cloning of *spoO* genes with bacteriophage and plasmid vectors in *Bacillus subtilis,* in: *Sporulation and Germination* (H. S. Levinson, A. L. Sonenshein, and D. J. Tipper, eds.), American Society for Microbiology, Washington.

Kiss, A., and Baldauf, F., 1983, Molecular cloning and expression in *Escherichia coli* of two modification methylase genes of *Bacillus subtilis, Gene* **21**:111–119.

Kroyer, J. M., Perkins, J. B., Rudinski, M. S., and Dean, D. H., 1980, Physical mapping of *Bacillus subtilis* phage ρ14 cloning vehicles: Heteroduplex and restriction enzyme analyses, *Mol. Gen. Genet.* **177**:511–517.

Kudoh, J., Ikeuchi, T., and Kurahashi, K., 1984, Identification of the sporulation gene *spoOA* product of *Bacillus subtilis. Biochem. Biophys. Res. Commun.* **122**:1104–1109.

Lampel, J. S., Ellis, D. M., and Dean, D. H., 1984, Reorienting and expanding the physical map of temperate *Bacillus subtilis* bacteriophage φ105, *J. Bacteriol.* **160**:1178–1180.

Lipsky, R. H., Rosenthal, R., and Zahler, S. A., 1981, Defective specialized SPβ transducing bacteriophages of *Bacillus subtilis* that carry the *sup-3* or *sup-44* gene, *J. Bacteriol.* **148**:1012–1015.

Lopez-Diaz, I., Clarke, S., and Mandelstam, J., 1986, *spoIID* operon of *Bacillus subtilis:* Cloning and sequence, *J. Gen. Microbiol.* **132**:341–354.

Mackey, C. J., and Zahler, S. A., 1982, Insertion of bacteriophage SPβ into the *citF* gene of *Bacillus subtilis* and specialized transduction of the *ilvBC-leu* genes, *J. Bacteriol.* **151**:1222–1229.

Marrero, R., Chiafari, F. A., and Lovett, P. S., 1981, High-frequency elimination of SPO2 prophage from *Bacillus subtilis* by plasmid transformation, *J. Virol.* **39**:318–320.

Marrero, R., and Lovett, P. S., 1980, Transductional selection of cloned bacteriophage φ105 and SPO2 deoxyribonucleic acids in *Bacillus subtilis, J. Bacteriol.* **143**:879–886.

Marrero, R., and Lovett, P. S., 1982, Interference of plasmid pCM194 with lysogeny of bacteriophage SPO2 in *Bacillus subtilis, J. Bacteriol.* **152**:284–290.

Marrero, R., and Yasbin, R. E., 1986, Evidence for circular permutation of the prophage genome of *Bacillus subtilis* bacteriophage φ105, *J. Virol.* **57**:1145–1148.

Marrero, R., Young, F. E., and Yasbin, R. E., 1984, Characterization of interspecific plasmid transfer mediated by *Bacillus subtilis* temperate bacteriophage SPO2, *J. Bacteriol.* **160**:458–461.

Mele, J., 1972, Biological characterization and prophage mapping of a lysogenizing bacteriophage for *Bacillus subtilis*, Ph.D. Thesis, University of Massachusetts, Amherst.

Montenegro, M. A., Pawlek, B., Behrens, B., and Trautner, T. A., 1983, Restriction and modification in *Bacillus subtilis:* Expression of the cloned methyltransferase gene from *B. subtilis* phage SPR in *E. coli* and *B. subtilis, Mol. Gen. Genet.* **189**:17–20.

Neuhard, J., Price, A. R., Schack, L., and Thomassen, E., 1978, Two thymidylate synthetase in *Bacillus subtilis, Proc. Natl. Acad. Sci. USA* **75**:1194–1198.

Nomura, S., Yamane, K., Masuda, T., Kawamura, F., Mizukami, T., Saito, H., Takatsuki, A., Yamasaki, M., Tamura, G., and Maruo, B., 1979, Construction of transducing phage ρ11 containing α-amylase structural gene of *Bacillus subtilis, Agric. Biol. Chem.* **43**:2637–2638.

Noyer-Weidner, M., Pawlek, B., Jentsch, S., Gunthert, U., and Trautner, T. A., 1981, Restriction and modification in *Bacillus subtilis:* Gene coding for a *Bsu*R-specific modification methyltransferase in the temperate bacteriophage φ3T, *J. Virol.* **38**:1077–1080.

Noyer-Weidner, M., Jentsch, S., Pawlek, B., Gunthert, U., and Trautner, T. A., 1983, Restriction and modification in *Bacillus subtilis:* DNA methylation potential of the related bacteriophages Z, SPR, SPβ, φ3T, and ρ11, *J. Virol.* **46**:446–453.

Noyer-Weidner, M., Jentsch, S., Kupsch, J., Berbauer, M., and Trautner, T. A., 1985, DNA methyltransferase genes of *Bacillus subtilis* phages: Structural relatedness and gene expression, *Gene* **35**:143–150.

Odebralski, J. M., and Zahler, S. A., 1982, Specialized transduction of the *kauA* and *citK* genes of *Bacillus subtilis* by bacteriophage φ3T, *Abstr. Am. Soc. Microbiol.* **1982**:130.

Okamoto, K., Mudd, J. A., Mangan, J., Huang, W. M., Subbaiah, T. V., and Marmur, J., 1968a, Properties of the defective phage of *Bacillus subtilis, J. Mol. Biol.* **34**:413–428.

Okamoto, K., Mudd, J. A., and Marmur, J., 1968b, Conversion of *Bacillus subtilis* DNA to phage DNA following mitomycin C induction, *J. Mol. Biol.* **34**:429–437.

Okubo, S., and Romig, W. R., 1965, Comparison of ultraviolet sensitivity of *Bacillus subtilis* bacteriophage SPO2 and its infectious DNA, *J. Mol. Biol.* **14**:130–142.

Ordal, G. W., Nettleton, D. O., and Hoch, J. A., 1983, Genetics of *Bacillus subtilis* chemotaxis: Isolation and mapping of mutations and cloning of chemotaxis genes, *J. Bacteriol.* **154**:1088–1097.

Osburne, M. S., and Sonenshein, A. L., 1980, Inhibition by lipiarmycin of bacteriophage growth in *Bacillus subtilis*, *J. Virol.* **33**:945–953.

Osburne, M. S., Craig, R. J., and Rothstein, D. M., 1985, Thermoinducible transcription system for *Bacillus subtilis* that uses control elements from temperate phage φ105, *J. Bacteriol.* **163**:1101–1108.

Parker, A. P., and Dean, D. H., 1986, Temperate *Bacillus* bacteriophage SP16 genome is circularly permuted and terminally redundant, *J. Bacteriol.* **167**:719–721.

Perkins, J. B., Zarley, C. D., and Dean, D. H., 1978, Restriction endonuclease mapping of φ105 and closely related temperate *Bacillus subtilis* phages ρ10 and ρ14, *J. Virol.* **28**:403–407.

Peterson, A. M., and Rutberg, L., 1969, Linked transformation of bacterial and prophage markers in *Bacillus subtilis* 168 lysogenic for bacteriophage φ105, *J. Bacteriol.* **98**:874–877.

Piggot, P. J., and Hoch, J. A., 1985, Revised genetic linkage map of *Bacillus subtilis*, *Microbiol. Rev.* **49**:158–179.

Raden, B., and Rutberg, L., 1984, Nucleotide sequence of the temperate *Bacillus subtilis* bacteriophage SPO2 DNA polymerase gene L, *J. Virol.* **52**:9–15.

Reeve, J. N., 1977, Bacteriophage infection of minicells. A general method for identification of "*in vivo*" bacteriophage directed polypeptide biosynthesis, *Mol. Gen. Genet.* **158**:73–79.

Rettenmeier, C. W., Gingell, B., and Hemphill, H. E., 1979, The role of temperate bacteriophage SPβ in prophage-mediated interference in *Bacillus subtilis*, *Can. J. Microbiol.* **25**:1345–1351.

Romig, W. R., 1968, Infectivity of *Bacillus subtilis* bacteriophage deoxyribonucleic acids extracted from mature particles and from lysogenic hosts, *Bacteriol. Rev.* **32**:349–357.

Rosenthal, R., Toye, P. A., Korman, R. Z., and Zahler, S. A., 1979, The prophage of SPβc2dcitK₁, a defective specialized transducing phage of *Bacillus subtilis*, *Genetics* **92**:721–739.

Rowe, D. B., Iismaa, T. P., and Wake, R. G., 1986, Nonrandom cosmid cloning and prophage SPβ homology near the replication terminus of the *Bacillus subtilis* chromosome, *J. Bacteriol.* **167**:379–382.

Rudinski, M. S., and Dean, D. H., 1978, Evolutionary considerations of related *B. subtilis* temperate phages φ105, ρ14, ρ10 and ρ6 as revealed by heteroduplex analysis, *Virology* **99**:57–65.

Rutberg, L., 1969, Mapping of a temperate bacteriophage active on *Bacillus subtilis*, *J. Virol.* **3**:38–44.

Rutberg, L., 1973, Heat induction of prophage φ105 in *Bacillus subtilis*: Bacteriophage-induced bidirectional replication of the bacterial chromosome, *J. Virol.* **12**:9–12.

Rutberg, L., 1982, Temperate bacteriophages of *Bacillus subtilis*, in: *Molecular Biology of the Bacilli* (D. A. Dubnau, ed.), pp. 247–268, Academic Press, New York.

Rutberg, L., Armentrout, R. W., and Jonasson, J., 1972, Unrelatedness of temperate *Bacillus subtilis* bacteriophages SPO2 and φ105, *J. Virol.* **9**:732–737.

Rutberg, L., Raden, B., and Flock, J.-I., 1981, Cloning and expression of bacteriophage SPO2 DNA polymerase gene L in *Bacillus subtilis*, using the *Staphylococcus aureus* plasmid pC194, *J. Virol.* **39**:407–412.

Rutberg, L., and Rutberg, B., 1970, Characterization of infectious deoxyribonucleic acid from temperate *Bacillus subtilis* bacteriophage φ105, *J. Virol.* **5**:604–608.

Sargent, M. G., Davies, S., and Bennett, M. F., 1985, Potentiation of a nucleolytic activity in *Bacillus subtilis*, *J. Gen. Microbiol.* **131**:2795–2804.

Savva, D., and Mandelstam, J., 1984, Cloning of the *Bacillus subtilis* spoIIA and spoVA loci in phage φ105DI:It, *J. Gen. Microbiol.* **130**:2137–2145.

Scher, B. M., Dean, D. H., and Garro, A. J., 1977, Fragmentation of *Bacillus* bacteriophage φ105 DNA by restriction endonuclease *Eco*RI: Evidence for complementary single-stranded DNA in the cohesive ends of the molecule, *J. Virol.* **23**:377–383.

Scher, B. M., Law, M. F., and Garro, A. J., 1978, Correlated genetic and *Eco*RI cleavage map of *Bacillus subtilis* bacteriophage φ105 DNA, *J. Virol.* **28**:395–402.

Schneider, A.-M., and Anagnostopoulos, C., 1983, *Bacillus subtilis* strains carrying two non-tandem duplications of the *trpE-ilvA* and the *purB-tre* regions of the chromosome, *J. Gen. Microbiol.* **129**:687–701.

Seaman, E., Tarmy, E., and Marmur, J., 1964, Inducible phages of *Bacillus subtilis*, *Biochemistry* **3**:607–613.

Shapiro, J. M., Dean, D. H., and Halvorson, H. O., 1974, Low-frequency specialized transduction with *Bacillus subtilis* bacteriophage φ105, *Virology* **62**:393–403.

Shinomiya, S., Yamane, K., Mizukami, T., Kawamura, F., and Saito, H., 1981, Cloning of thermostable α-amylase gene using *Bacillus subtilis* phage ρ11 as a vector, *Agric. Biol. Chem.* **45**:1733–1735.

Siegel, E. C., and Marmur, J., 1969, Temperature-sensitive induction of bacteriophage in *Bacillus subtilis* 168, *J. Virol.* **4**:610–618.

Smith, I., and Smith, H., 1973, Location of the SPO2 attachment site and the bryamycin resistance marker on the *Bacillus subtilis* chromosome, *J. Bacteriol.* **114**:1138–1142.

Spancake, G. A., and Hemphill, H. E., 1985, Deletion mutants of *Bacillus subtilis* bacteriophage SPβ, *J. Virol.* **55**:39–44.

Spancake, G. A., Hemphill, H. E., and Fink, P. S., 1984, Genome organization of SPβ *c2* bacteriophage carrying the *thyP3* gene, *J. Bacteriol.* **157**:428–434.

Steensma, H. Y., 1981a, Adsorption of defective phage PBSZ1 to *Bacillus subtilis* 168 Wt, *J. Gen. Virol.* **52**:93–101.

Steensma, H. Y., 1981b, Effect of defective phages on the cell membrane of *Bacillus subtilis* and partial characterization of a phage protein involved in killing, *J. Gen. Virol.* **56**:275–286.

Steensma, H. Y., Robertson, L. A., and Van Elsas, J. D., 1978, The occurrence and taxonomic value of PBS X-like defective phages in the genus *Bacillus*, *Antonie van Leeuwenhock* **44**:353–366.

Stickler, D. J., Tucker, R. G., and Day, D., 1965, Bacteriophage-like particles released from *Bacillus subtilis* after induction with hydrogen peroxide, *Virology* **26**:142–145.

Stroynowski, I. T., 1981a, Distribution of bacteriophage φ3T homologous deoxyribonucleic acid sequences in *Bacillus subtilis* 168, related bacteriophages, and other *Bacillus* species, *J. Bacteriol.* **148**:91–100.

Stroynowski, I. T., 1981b, Integration of the bacteriophage φ3T-coded thymidylate synthetase gene into the *Bacillus subtilis* chromosome, *J. Bacteriol.* **148**:101–108.

Thorne, C. B., and Mele, J., 1974, Prophage interference in *Bacillus subtilis* 168, *Microbial Genet. Bull.* **36**:27–29.

Thurm, P., and Garro, A. J., 1975a, Bacteriophage-specific protein synthesis during induction of the defective *Bacillus subtilis* bacteriophage PBSX, *J. Virol.* **16**:179–183.

Thurm, P., and Garro, A. J., 1975b, Isolation and characterization of prophage mutants of the defective *Bacillus subtilis* bacteriophage PBSX, *J. Virol.* **16**:184–191.

Tran-Betcke, A., Behrens, B., Noyer-Weidner, M., and Trautner, T. A., 1986, DNA methyltransferase genes of *Bacillus subtilis* phages: Comparison of their nucleotide sequences, *Gene* **42**:89–96.

Trautner, T. A., Pawlek, B., Gunthert, U., Canosi, U., Jentsch, S., and Freund, M., 1980, Restriction and modification in *Bacillus subtilis:* Identification of a gene in the temperate phage SPβ coding for a *Bsu*R specific modification methyltransferase, *Mol. Gen. Genet.* **180**:361–367.

Tucker, R. G., 1969, Acquisition of thymydylate synthetase activity by a thymine-requiring mutant of *Bacillus subtilis* following infection by the temperate phage φ3, *J. Gen. Virol.* **4**:489–504.

Vandeyar, M. A., Mackey, C. J., Lipsky, R. H., and Zahler, S. A., 1986, The *ilvBC-leu* operon of

Bacillus subtilis, in: Bacillus *Molecular Genetics and Biotechnology Applications* (A. T. Ganesan and J. A. Hoch, eds.), pp. 295–306, Academic Press, New York.

Warner, F. D., Kitos, G. A., Romano, M. P., and Hemphill, H. E., 1977, Characterization of SPβ: A temperate bacteriophage from *Bacillus subtilis* 168M, *Can. J. Microbiol.* **23**:45–51.

Weiner, M., 1986, Characterization of bacteriophage H2, Ph.D. Thesis, Cornell University, Ithaca, NY.

Weiss, A. S., Smith, M. T., Iismaa, T. P., and Wake, R. G., 1983, Cloning DNA from the replication terminus region of the *Bacillus subtilis* chromosome, *Gene* **24**:83–91.

Williams, M. T., and Young, F. E., 1977, Temperate *Bacillus subtilis* bacteriophage φ3T: Chromosomal attachment site and comparison with temperate bacteriophages φ105 and SPO2, *J. Virol.* **21**:522–529.

Wilson, G. A., Williams, M. T., Baney, H. W., and Young, F. E., 1974, Characterization of temperate bacteriophages of *Bacillus subtilis* by the restriction endonuclease *Eco*RI: Evidence for three different temperate bacteriophages, *J. Virol.* **14**:1013–1016.

Wulff, D. L., and Rosenberg, M., 1983, Establishment of repressor synthesis, in: *Lambda II* (R. W. Hendrix, J. W. Roberts, F. W. Stahl, and R. A. Weisberg, eds.), pp. 53–73, Cold Spring Harbor Laboratory, Cold Spring Harbor, NY.

Yamada, H., Anaguchi, H., and Kobayashi, Y., 1983, Cloning of the sporulation gene *spoVE* in *Bacillus subtilis*, *J. Gen. Appl. Microbiol.* **29**:477–486.

Yamazaki, H., Ohmura, K., Nakayama, A., Takeichi, Y., Otozai, K., Yamasaki, M., Tamura, G., and Yamane, K., 1983, α-Amylase genes (*amyR2* and *amyE*⁺) from an α-amylase-hyperproducing *Bacillus subtilis* strain: Molecular cloning and nucleotide sequences, *J. Bacteriol.* **156**:327–337.

Yarmolinsky, M., 1982, Bacteriophage P1, in: *Genetic Maps*, Vol. 2 (S. J. O'Brien, ed.), pp. 34–43, National Cancer Institute, Frederick, MD.

Yasbin, R. E., Wilson, G. A., and Young, F. E., 1973, Transformation and transfection in lysogenic strains of *Bacillus subtilis* 168, *J. Bacteriol.* **113**:540–548.

Yasbin, R. E., Wilson, G. A., and Young, F. E., 1975a, Transformation and transfection in lysogenic strains of *Bacillus subtilis:* Evidence for selective induction of prophage in competent cells, *J. Bacteriol.* **121**:296–304.

Yasbin, R. E., Wilson, G. A., and Young, F. E., 1975b, Effect of lysogeny on transfection and transfection enhancement in *Bacillus subtilis*, *J. Bacteriol.* **121**:305–312.

Yasbin, R. W., Maino, V. C., and Young, F. E., 1976, Bacteriophage resistance in *Bacillus subtilis* 168, W23 and interstrain transformants, *J. Bacteriol.* **125**:1120–1126.

Yasbin, R. E., Fields, P. I., and Andersen, B. J., 1980, Properties of *Bacillus subtilis* 168 derivatives freed of their natural prophages, *Gene* **12**:155–159.

Yasbin, R. E., and Young, F. E., 1972, The influence of temperate bacteriophage φ105 on transformation and transfection in *Bacillus subtilis*, *Biochem. Biophys. Res. Commun.* **47**:365–371.

Yasunaka, A., Tsukamato, H., Okubo, S., and Horiuchi, T., 1970, Isolation and properties of suppressor-sensitive mutants of *Bacillus subtilis* bacteriophage SPO2, *J. Virol.* **5**:819–821.

Yoneda, Y., Graham, S., and Young, F. E., 1979a, Cloning of a foreign gene coding for α-amylase in *Bacillus subtilis*, *Biochem. Biophys. Res. Commun.* **91**:1556–1564.

Yoneda, Y., Graham, S., and Young, F. E., 1979b, Restriction-fragment map of the temperate *Bacillus subtilis* bacteriophage SPO2, *Gene* **7**:51–68.

Youngman, P. J., Suber, P., Perkins, J. B., Sandman, K., Igo, M., and Losick, R., 1985, New ways to study developmental genes in spore-forming bacteria, *Science* **228**:285–291.

Zahler, S. A., 1982, Specialized transduction in *Bacillus subtilis*, in: *Molecular Biology of the Bacilli* (D. A. Dubnau, ed.), Vol. 1, pp. 269–305, Academic Press, New York.

Zahler, S. A., Korman, R. Z., Rosenthal, R., and Hemphill, H. E., 1977, *Bacillus subtilis* bacteriophage SPβ: Localization of the prophage attachment site, and specialized transduction, *J. Bacteriol.* **129**:556–558.

Zahler, S. A., Korman, R. Z., Odebralski, J. M., Fink, P. S., Mackey, C. J., Poutre, C. G., Lipsky, R. H., and Youngman, P. J., 1982, Genetic manipulations with phage SPβ, in: *Molecular Cloning and Gene Regulation in* Bacillus (J. A. Hoch, S. Chang, and A. T. Ganesan, eds.), pp. 41–50, Academic Press, New York.

Zahler, S. A., Korman, R. Z., Thomas, C., and Odebralski, J. M., 1987a, Temperate bacteriophages of *Bacillus amyloliquefaciens, J. Gen. Microbiol.* **133**:2933–2935.

Zahler, S. A., Korman, R. Z., Thomas, C., Fink, P. S., Weiner, M. P., and Odebralski, J. M., 1987b, H2, a temperate bacteriophage isolated from *Bacillus amyloliquefaciens* strain H, *J. Gen. Microbiol.* **133**:2937–2944.

Index